GaAs FET PRINCIPLES and TECHNOLOGY

The Artech House Microwave Library

Introduction to Microwaves by Fred E. Gardiol
Microwaves Made Simple: Principles and Applications by W. Stephen Cheung and Frederic H. Levien
Microwave Tubes by A. S. Gilmour, Jr.
Electric Filters by Martin Hasler and Jacques Neirynck
Nonlinear Circuits by Martin Hasler and Jacques Neirynck
Microwave Technology by Erich Pehl
Receiving Systems Design by Stephen J. Erst
Microwave Mixers by Stephen A. Maas
Feedback Maximization by B.J. Lurie
Applications of GaAs MESFETs by Robert Soares, et al.
GaAs Processing Techniques by Ralph E. Williams
GaAs FET Principles and Technology by James V. DiLorenzo and Deen D. Khandelwal
Dielectric Resonators, Darko Kajfez and Pierre Guillon, eds.
Modern Spectrum Analyzer Theory and Applications by Morris Engelson
Design Tables for Discrete Time Normalized Lowpass Filters by Arild Lacroix and Karl-Heinz Witte
Microwave Materials and Fabrication Techniques by Thomas S. Laverghetta
Handbook of Microwave Testing by Thomas S. Laverghetta
Microwave Measurements and Techniques by Thomas S. Laverghetta
Principles of Electromagnetic Compatibility by Bernhard E. Keiser
Linear Active Circuits: Design and Analysis by William Rynone, Jr.
The Design of Impedance-Matching Networks for Radio-Frequency and Microwave Amplifiers by Pieter L.D. Abrie
Microwave Filters, Impedance Matching Networks, and Coupling Structures by G.L. Matthaei, Leo Young, and E.M.T. Jones
Analysis, Design, and Applications of Fin Lines by Bharathi Bhat and Shiban Koul
Microwave Engineer's Handbook, 2 vol., Theodore Saad, ed.
Microwave Integrated Circuits, Jeffrey Frey and Kul Bhasin, eds.
Computer-Aided Design of Microwave Circuits by K.C. Gupta, Ramesh Garg, and Rakesh Chadha
Microstrip Lines and Slotlines by K.C. Gupta, R. Garg, and I.J. Bahl
Advanced Mathematics for Practicing Engineers by Kurt Arbenz and Alfred Wohlhauser
Microstrip Antennas by I.J. Bahl and P. Bhartia
Antenna Design Using Personal Computers by David M. Pozar
Microwave Circuit Design Using Programmable Calculators by J. Lamar Allen and Max Medley, Jr.
Stripline Circuit Design by Harlan Howe, Jr.
Microwave Transmission Line Filters by J.A.G. Malherbe
Electrical Characteristics of Transmission Lines by W. Hilberg
Microwave Diode Control Devices by Robert V. Garver
Tables for Active Filter Design by Mario Biey and Amedeo Premoli
Ferrite Control Components, 2 vol., Lawrence Whicker, ed.
Microwave Remote Sensing, 3 vol., by F.T. Ulaby, R.K. Moore, and A.K. Fung

GaAs FET PRINCIPLES and TECHNOLOGY

James V. DiLorenzo
EDITOR IN CHIEF
Deen D. Khandelwal
ASSOCIATE EDITOR

ARTECH HOUSE, INC.
610 Washington Street
Dedham, Massachusetts 02026

Printed and bound in the United States of America by Bookcrafters, Inc., Chelsea, MI.

Library of Congress Catalog No.: 81-69399
Standard Book Number: 0-89006-118-11

Third printing, August, 1985.

Contents

Preface

This book is intended to provide a comprehensive look at the materials and device technology necessary for high performance GaAs FET devices. The editors hoped to provide the reader with a complete look at the science and art of the design, fabrication and application of GaAs FETs. Therefore, we have included chapters on material technology, device technology, microwave circuit technology, digital integrated circuit technology, and finally, a look into the future on new physics concepts which may provide the basis of a similar book 5 to 10 years hence.

Throughout this book, the term GaAs FET is used to mean GaAs MESFET, i.e., a metal semiconductor field effect transistor in GaAs with Schottky gate and operating in the depletion mode. This decision was made since GaAs FET has become the common vernacular in the technical community.

The editors are grateful to the contributors to this book who obviously put much time and effort into producing the material that this book contains. The editors are indebted to Ms. Claire Kosakowski for her tireless efforts in the administrative details required to complete this book.

One of us (D.K.) would like to acknowledge the encouragements of his wife, Savita, and Dr. Leo Young during the course of putting this book together.

April 2, 1982

Contributors

P. Asbeck
J.R. Barker
R. Bennett
R.S. Butlin
H.M. Cox
L.A. D'Asaro
J.V. DiLorenzo
I. Drukier
R.C. Eden
F. Eisen
W.M. Ford
H. Fukui
J. Hallais
F. Hasegawa
L. Hollan
H. Huang
A. Hughes
J.C.M. Hwang
J.C. Irvin
C. Kirkpatrick
K. Lehovec
P.F. Lindquist
H.M. Macksey
W.C. Niehaus
D. Parker
A. Peake
D.H. Phillips
W.O. Schlosser
V. Sokolov
J.A. Turner
G.D. Vendelin
B.M. Welch
S.H. Wemple
D. Wilson
C.E. Wood
R. Zuleeg

Biographies

Peter M. Asbeck received the B.S. and Ph.D. degrees in 1969 and 1975, respectively, in electrical engineering from the Massachusetts Institute of Technology. Prior to joining Rockwell International, where he is presently employed, he worked at Phillips Laboratories, Briarcliff Manor, New York, where he was involved with the characterization, physics, and application of double hetero-structure lasers. He is presently active in the area of ion implantation and semiconductors.

John R. Barker received the B.Sc. in theoretical physics from the University of Edinburgh, United Kingdom in 1966, the M.Sc. from the University of Durham in 1967 and the Ph.D. from the University of Warwick in 1969. His research has included work in theory of semiconductors, electroluminescence, hot electrons, quantum transport and submicron semiconductor devices and systems. Since 1970 he has been a Lecturer for the Physics Department at Warwick University and is an Affiliate Professor of Physics at Colorado University (1979-present).

Herbert M. Cox received the B.A., M.S., and Ph.D. degrees in mineralogy from the Ohio State University, Columbus, in 1969, 1971, and 1973, respectively, after serving four years in the U.S. Air Force.

He joined the Technical Staff of Bell Laboratories, Murray Hill, New Jersey, in 1973. Since that time, he has been responsible for the development of materials growth techniques by chemical vapor deposition in support of GaAs FET programs.

He is a member of the Electrochemical Society.

L.A. D'Asaro was born in Buffalo, New York on January 20, 1927. He received the B.S. and M.S. degrees in physics from Northwestern University in Evanston, Illinois in 1949 and 1950. He then attended Cornell University, Ithaca, New York where he was awarded the Ph.D. degree in 1955. Since that time he has been at Bell Laboratories, Murray Hill, New Jersey, where he has been concerned with development of semiconductor devices. At the present time he is fabricating GaAs microwave Schottky-barrier field-effect transistors.

Dr. D'Asaro is a member of the Institute of Electrical and Electronics Engineers, the American Physical Society, Sigma Xi, Phi Beta Kappa and Gamma Alpha.

James V. DiLorenzo received the M.S. and Ph.D. degrees from the State University of New York at Stonybrook and pursued post-doctoral work at Yale University all in chemical physics in 1965, 1967 and 1968, respectively. In 1969, he joined Bell Laboratories to study chemical vapor deposition of compound semiconductor materials. He was particularly interested in how this related to gallium arsenide (GaAs) crystal growth. This work resulted in substantial improvements in the purity, control and understanding of the epitaxial growth of GaAs materials.

Following this work, he began studies on parameters important to the design and fabrication of microwave devices. From 1970 to 1973, he worked on Gunn and IMPATT devices.

In 1973, he became supervisor of a group working in the area of microwave materials and processes. This work centered on the development of GaAs field effect transistors, which resulted in obtaining the first transistors with noise figures less than 1 decibel and power greater than 10 watts at 4 GHz.

In 1980, he assumed the responsibility of Head of the Special Materials and Devices Department concerned with research in high speed materials, devices and circuits for electronic and optical applications.

I. Drukier received his B.S. in 1966 and M. Eng. (E) in 1967 in electrical engineering from Cornell University and was awarded the Ph.D. in the same field from the Polytechnic Institute in Brooklyn in 1973.

In July of 1973, he joined RCA at the David Sarnoff Research Center in Princeton, New Jersey. In March 1976, he joined the Microwave Semiconductor Corporation in Somerset, New Jersey and is now Vice President for GaAs Products.

In 1980, he was co-recipient of the Beatrice Winner Award for Editorial Excellence at the International Solid State Circuits Conference.

Richard C. Eden received the B.S. degree from Iowa State University, Ames, in 1961, the M.S. degree from the California Institute of Technology, Pasadena, in 1962 and the Ph.D. degree from Stanford University, Stanford, California in 1967, all in electrical engineering.

He was employed at the U.S. Naval Ordnance Laboratory, White Oak, Maryland, in the design of specialized ordnance testing instrumentation (1961). After receiving the M.S. degree, he worked at the Advanced Systems Development Laboratory of IBM, San Jose, California. In 1968, after completing the Ph.D. program and a one year research associateship at Stanford, he joined the Science Center, Rockwell International, Thousand Oaks, California. In 1980, Dr. Eden left Rockwell to become Vice President of Engineering of Vedette Energy Research, later becoming a founder of a split-off from this organization, Enhanced Energy Systems (now a subsidiary of Brunswick Corp.). In 1981 Dr. Eden became a founder and Vice President of Research and Development of GigaBit Logic, Inc., a new venture dedicated to the commercial manufacture of GaAs integrated circuits.

Dr. Eden is a member of the American Physical Society, the Society of Photo-Optical Instrumentation Engineers, the Electrochemical Society, Eta Kappa Nu, Tau Beta Pi, Phi Kappa Phi, Phi Eta Sigma, Pi Mu Epsilon and Sigma Xi. He received the Rockwell Engineer of the Year award in 1977.

Fred Eisen received a B.S. degree in physics from Caltech in 1951 and a Ph.D. in physics from Princeton University in 1956. His research has included work in crystals, ion beam analysis and ion implantation in GaAs. He is presently Director of GaAs Electronic Device Research at the Rockwell International Microelectronics Research and Development Center, Thousand Oaks, California.

W.M. Ford began his career in Semiconductor Materials at Shockley Semiconductor Laboratories in 1957. In 1959 he left Shockley Labs to establish a company whose business was the manufacture of silicon crystals called Knapic Electrophysics, of which Mr. Ford was Engineering Manager. From 1960 through 1964 Mr. Ford worked as a consultant for Rheem Semiconductor and for Allegheny Electronic Chemicals Company in the development of machines and processes for silicon crystal growth and evaluation. From 1964 through 1966 at Raytheon Semiconductor and at I.T.T. Semiconductor he developed processes for deposition of silicon oxides and nitrides. At Hewlett-Packard from 1967 to mid-1980 Mr. Ford, as a technical staff member, developed processes for the bulk and epi growth of GaAs and bulk growth of GAP. Mr. Ford is now GaAs Materials Manager at Harris Microwave Semiconductor, Inc. where he is responsible for development of semi-insulating GaAs substrates for FET and IC applications.

Hatsuaki Fukui was born in Yokohama, Japan. He received the Doctor of Engineering degree in electrical engineering from Osaka University, Osaka, Japan. He has worked at Osaka City University, Osaka (1949-1954), Shimada Physical and Chemical Industrial Company, Tokyo (1955), Tokyo Tushin Kogyo (the former name of Sony Corporation), Tokyo (1955-1962), and in 1962, he joined Bell Laboratories, Murray Hill, New Jersey. He has worked in the fields of microwave semiconductor devices, GaAs bulk-effect devices, Si avalanche diodes and their circuit applications, research and development of electrooptical devices and subsystems for PICTURE-PHONE® use. He supervised the development of new techniques for the vacuum-deposition of III-V compounds and is now involved in the GaAs FET development project.

Dr. Fukui is a member of the IEEE.

J.P. Hallais, Master of Science and Chemical Engineer (Ecole Nationale Supérieure de Chimie de Lille) in 1964 received his Doctoral Engineering degree at the University d'Orsay in 1970. The same year he joined les Laboratoires d'Electronique et de Physique Appliquée (LEP) where he was involved in research on III-V compounds and materials growth.

Fumio Hasegawa received the B.S., M.S., and Dr. of Eng. degrees in electronics engineering from the Tohoku University, Sendai, Japan, in 1963, 1965 and 1973, respectively. In 1965 he joined Central Research Laboratories, Nippon Electric Company, Ltd., Kawasaki, Japan, where he worked on the development of GaAs vapor phase epitaxy, Gunn diodes, GaAs IMPATT diodes, and GaAs MESFETs. From 1974 to 1975, he was with the University of Sheffield, England, as a Senior Research Fellow. From 1977 he was the Research Manager of GaAs FETs and Vapor Phase Epitaxy Group, NEC Central Research Laboratories. Since 1981 he has been an Associate Professor at the Institute of Material Science, University of Tsukuba.

L. Hollan attended the Ecole Nationale Supérieure de Chimie in Paris (ENSCP) and became chemical engineer in 1962 and simultaneously Master of Science at the Sorbonne. Working in the Centre National de la Recherche Scientifique (CNRS), he obtained his Ph.D. at the Faculté des Sciences in Orsay in 1966. Since then, he has been working at the Philips Research Organization at RTC in Suresnes and in the Laboratoires d'Electronique et de Physique Appliquée (LEP), engaged in the field of semiconductor materials and devices.

Ho Huang is Head, Microwave Processing Technology of RCA Laboratories. He has the responsibility of monolithic GaAs FETs, GaAs TE diodes, Si monolithic microwave IMPATT diode-amplifiers, Si PIN diodes, flight-qualified components for space applications, and engineering production of special components.

Dr. Huang received his Ph.D. degree from Cornell University in 1967. He has since worked on microwave solid state device physics, device design, device fabrication technology, amplifiers and oscillators.

He is a senior member of the IEEE.

James C.M. Hwang graduated in 1970 with the B.S. degree in physics from the National Taiwan University, Taipei, Taiwan. Between 1971 and 1976 he did his graduate studies at Cornell University, Ithaca, New York, and received the M.S. and Ph.D. degrees in materials science and engineering.

He spent two years at IBM T.J. Watson Research Center, Yorktown Heights, New York, as a post-doctoral fellow. Since 1978, he has been with Bell Laboratories, Murray Hill, New Jersey. His main activities have been in the preparation and properties of thin-film materials including metals, semiconductors, and dielectrics. He is currently engaged in the molecular beam epitaxy of compound semiconductors for novel device applications.

Dr. Hwang is a member of the Electrochemical Society and the Metallurgical Society.

John C. Irvin (born 1929) obtained his B.A. in 1949 at Miami University (Ohio), his M.A. in 1953, and Ph.D. (physics) in 1957 at the University of Colorado, and has been with Bell Laboratories since 1957.

Mr. Irvin has engaged in research on the properties of bulk Si and diffused layers in Si and on Si interfaces. He has been involved in the development of microwave semiconductor devices including GaAs varactors, mixer diodes, Gunn diodes, IMPATT diodes, and, most recently, the reliability aspects of the GaAs FET. Mr. Irvin is a senior member of the IEEE, and a member of the American Physical Society, Sigma Xi, and Phi Beta Kappa.

Deen D. Khandelwal is a Member of the Professional Staff at Martin Marietta Orlando Aerospace where he is responsible for the millimeter wave semiconductor device technology development.

Previously, Dr. Khandelwal was with General Electric Space Division where as a Technical Staff Engineer, he provided a base for semiconductor devices for communication satellite applications in general and GaAs FETs in particular. His specific contributions in the area of GaAs FETs included device technology evaluation for reliability for both low noise and power GaAs FETs. Other activities at General Electric included IMPATT diode power combiners, circular waveguide filters, Ku-band receiver as a low cost ground station for TV signals, liquid crystal study for millimeter wave applications, etc.

Prior to joining General Electric, Dr. Khandelwal was with TRW where he was Manager, Microwave Communication Devices, developing power bipolar transistors, amplifiers and IMPATT diodes. There he also initiated and established a capability for millimeter wave IMPATT diodes.

Conilee G. Kirkpatrick received the B.S. degree in physics in 1969 and the A.M. degree in physics in 1970 from Washington University, St. Louis, Missouri, and the Ph.D. degree in electrical engineering from the University of Illinois-Urbana in 1974. In 1974 she joined the Corporate Research and Development Center of the General Electric Corporation, where her research centered on beam effects in semiconductors studied through a variety of analytical techniques, especially electron microscopy. In 1969 she joined Rockwell International Science Center where her work involved ion implantation and characterization of GaAs and indium phosphide. She subsequently became Project Leader, Ion Implantation, for the Microelectronics Research and Development Center. She is presently Manager of the Materials Preparation Section where efforts are centered on GaAs bulk growth, ion implantation, encapsulation, and characterization. She is a member of the American Physics Society, the Electrochemical Society, and the Society of Women Engineers.

Kurt Lehovec received the Ph.D. degree in Physics from the University of Prague, Czechoslovakia, in 1941.

From 1947-1952, he was Consultant to the U.S. Army Signal Corps. at Fort Monmouth, New Jersey. In 1952, he joined the Sprague Electric Company in North Adams, Massachusetts, to establish their semiconductor activities and held various positions including Director of Semiconductor Research and Development and Director of Research. He is now a Professor of Electrical Engineering and Materials Science at the University of Southern California, Los Angeles.

Dr. Lehovec is a fellow of the American Physical Society and of the IEEE.

Peter F. Lindquist received the S.B. in metallurgy from M.I.T. in 1961 and the Ph.D. in materials science from Stanford University in 1970. He has worked in the areas of metallurgical failure analysis, X-ray diffraction, transmission electron microscopy, crystal growth, and solid state device physics and technology. Since 1970, he has been with Hewlett-Packard Company and is presently an R&D project manager in the Optoelectronics Division, where he is responsible for process technology development of visible light-emitting diodes.

H. Michael Macksey received the B.S. degree in physics from the University of Michigan, Ann Arbor, in 1968 and the M.S. and Ph.D. degree in physics from the University of Illinois, Urbana-Champaign, in 1970 and 1972, respectively. Since 1973 he has been with the Central Research Laboratories, Texas Instruments, Incorporated, Dallas, Texas, where he has been engaged in the development of GaAs field-effect transistors. This has included investigations of device design, channel parameter optimization, metallization systems, device fabrication technology, and device reliability. These studies have resulted in the development of both low noise and high-power GaAs FETs operating from 3 to 30 GHz. Recently, this work has been extended to GaAs monolithic microwave integrated circuits.

William C. Niehaus received his Ph.D. from the University of Notre Dame in 1968. In the same year, he joined Bell Laboratories, Murray Hill, New Jersey, where he had worked in the microwave device area developing silicon IMPATT, GaAs IMPATT, and GaAs Gunn diodes. In the fall of 1981 he became Western Electric Department Chief in charge of microwave GaAs material and devices.

Howard Phillips is the Director of The Aerospace Corporation's Electronics Research Laboratory (since 1980). The laboratory is engaged in gallium arsenide and silicon VLSI device research. Prior to joining Aerospace in 1980, he provided technical leadership and direction for the Lockheed Microelectronics Center and programs to develop GaAs and silicon microelectronics technologies.

Earlier experience includes work with KSPI radio (communications consultant, 1962); Pan American Guided Missiles Range Division (communication engineer, 1963-1964); The University of Oklahoma (faculty instructor, 1965); Rockwell International Electronics Research Center (MTS, Program Manager and Group Leader, 1966-1969 and 1972-1977), and the Lockheed Microelectronics Center (Manager, 1977-1980).

He received a Ph.D. degree in electrical engineering and computer science, conferred with honors from the University of New Mexico in 1972, a masters degree in nuclear engineering from the University of Oklahoma, 1966 and a B.S. in electrical engineering from Oklahoma State University in 1963.

W.O. Schlosser is supervisor of the Microwave Device Group at Bell Laboratories in Murray Hill, New Jersey.

Since joining Bell Laboratories in 1966, Mr. Schlosser has helped develop circuits and combining techniques for IMPATT diodes and has worked on path-length modulators for the millimeter waveguide system and high speed modulation of semiconductor lasers. He has worked on GaAs FETs since 1973.

Mr. Schlosser received the Diplom Ingenieur and the Dr. Ing. degrees from the Technical University of Darmstadt, West Germany. He is a member of the IEEE.

Vladimir Sokolov received the Ph.D. degree in electrical engineering in 1973, from the University of Wisconsin, Madison, Wisconsin.

From 1974 to 1975 he was a Senior Engineer in the Electronics Division, Defense Systems Department, of Northrop Corporation where he was engaged in the analysis, design and evaluation of microstrip circuits (up to 18 GHz) for ECM applications.

From 1975 to 1981 he was a Member of the Technical Staff at the Central Research Laboratories of Texas Instruments, where his work included the development of microwave FET and IMPATT Hybrid amplifiers in the 8-20 GHz range. He also was engaged in the development of GaAs FET monolithic circuits at X-band frequencies.

Currently, Dr. Sokolov is with the Honeywell Corporate Technology Center, and is a Staff Scientist in the Gallium Arsenide Electronics Section of the Electro-Optics Department. His responsibilities include the development of GaAs monolithic integrated circuits for microwave and mm-wave systems applications.

George Vendelin received the B.S., M.S., and Degree of Engineer in electrical engineering from Stanford University in 1959, 1961 and 1963 respectively. He has worked at Texas Instruments (1963-1968), Signetics (1968-1972), Fairchild (1972-1974), Varian (1974-1975), as a private consultant (1975-1977), Dexcel (1977-1979), and presently with Eaton Corporation as Technical Director in the Sunnyvale R&D Center. He has worked on numerous microwave integrated circuits and has been associated with GaAs MESFETs since 1972. He has published over 30 technical articles and has written a book entitled *Design of Amplifiers and Oscillators by the S-Parameter Method* which is scheduled for publication in early 1982.

Bryant M. Welch is Manager, GaAs Integrated Circuit Processing Section, Rockwell International/Microelectronics Research and Development Center, Thousand Oaks, California.

He received the A.A. degree in engineering from Miami Date Junior College, Miami, Florida in 1963 and a B.S. degree in mathematics from the California State University, Northridge, California in 1968.

From 1963 to 1965 he was associated with Westinghouse Electric Corporation, where he was involved in the early development of Si bipolar integrated circuits. Since 1965 he has been employed with Rockwell International and is presently the manager of the GaAs Integrated Circuit Processing Section. He is currently responsible for both the process sustaining and development activities on all of the GaAs digital IC programs, with special emphasis on the development of GaAs LSI/VLSI.

Mr. Welch is a member of the IEEE and the Electrochemical Society.

Stuart Wemple was born in Rockford, Illinois. He received a B.S. in mechanical engineering from Northwestern University, an M.S. in mechanical engineering from Caltech and a Ph.D. in electrical engineering from M.I.T. Since graduation in 1963 he has engaged in a variety of solid state research and development efforts at Bell Laboratories, including ferroelectricity, electro-optic materials, magnetic bubble materials and most recently, power GaAs FET design and analysis.

D. Wilson is the GaAs IC manager at Harris Microwave Semiconductors and has the responsibility for defining the market potential of monolithic GaAs ICs, selection of the best candidates, circuit design of these ICs and their subsequent testing.

Previously, Mr. Wilson was the project leader and principal investigator on company sponsored and external GaAs IC product development programs at Lockheed Microelectronics Center. He has a M.S. in physics and is a member of the IEEE and American Physical Society.

Colin E.C. Wood attended Harris College of Advanced Technology, Preston, Lancashire (1963-1967). In 1966 he took a course in industrial chemistry leading to a higher national diploma in chemistry (HMB) and a graduate of the royal institute of chemistry (GRIC). From 1967-1970 he attended Nottingham University, Department of Physical Chemistry. He performed research in mass-spectrometry and vapor pressure determination of organic explosive compounds leading to the Ph.D. in Physical Chemistry.

From 1970-1978 Dr. Wood was a research scientist at Mullar Research Laboratories, Cross Oak Lane, Salfords, Surrey. Since 1978, when he was appointed a permanent research associate for the Electrical Engineering Department and National Research and Resource Facility for Submicron Structures at Cornell, Dr. Wood has worked on MBE and related processes.

R. Zuleeg is Principal Staff Scientist of Solid State Electronics at McDonnell Douglas Astronautics Company, California, and in this capacity directs and supervises the semiconductor device and integrated circuit research and development program of the company. He is actively engaged in semiconductor research and development for the past 26 years and published numerous papers in leading professional journals.

He is also a Lecturer of Semiconductor Device Physics and Electronics at the University of California, Irvine, since 1972. In 1975 he was a Visiting Professor at RMIT (Royal Melbourne Institute of Technology, Australia).

Chapter 1

Semi-Insulating GaAs Substrates

Introduction

In this chapter, P.F. Lindquist and W.M. Ford review the status of semi-insulating GaAs technology. The editors decided to include a rather complete and comprehensive review of this technology for two reasons: firstly, it is obviously a critical element in the technology of GaAs FETs; secondly, although it has been studied extensively for nearly twenty years, progress in understanding the electrical properties of semi-insulating GaAs has been slow. Therefore, in many cases, the technology remains an "art" since it lacks fundamental understanding as a base. Semi-insulating substrates are used either as starting material for epitaxial GaAs or as material for direct ion implantation. The authors discuss first the various models for the conduction mechanism and the nature of deep levels in GaAs. The levels associated with oxygen, chromium and other transition metals are reviewed as well as two, three and four level conduction models of semi-insulating GaAs. Various crystal growth techniques such as directional solidification, horizontal growth techniques (Bridgman, gradient freeze), liquid encapsulated Czochralski and the Gremmelaer techniques are discussed. Material characterization techniques are reviewed, with particular emphasis on thermal conversion, which is a major problem in ion implantation. Substrate - active layer interface effects, "back-gating" and light sensitivity are discussed in terms of their influence on FET devices.

Semi-Insulating GaAs Substrates

P. F. Lindquist

Hewlett-Packard Co., Optoelectronics Division,
Palo Alto, California.

W. M. Ford

Harris Microwave Semicontuctor,
Sunnyvale, California.

INTRODUCTION

High resistivity GaAs has been the subject of a great amount of study for nearly twenty years. Initially this material (termed "semi-insulating" by Allen[1]) was produced accidentally during attempts to obtain extremely pure GaAs.[2-5] Later it was intentionally produced by additions of either oxygen[6,7] or chromium[8] to the melt. More recently, the development of GaAs FETs has provided the practical motivation to understand this material more completely so that substrates with controlled and stable characteristics can be produced.

Progress in understanding the electrical properties of semi-insulating GaAs has been hindered by the difficulty in making and interpreting measurements on high resistivity materials. Even today some controversy remains regarding the nature of the conduction mechanism in this material and the role of native defects or complexes relative to impurities. A related problem has been the difficulty in obtaining reliable impurity analyses, since several of the most common electrically active impurities are detectable only by mass spectrometry.

Semi-insulating GaAs for use as substrates for FETs must meet the following requirements:

1. Thermal stability (high resistivity maintained during epitaxial growth or annealing of ion implanted active layer).
2. Absence of undesirable substrate-active layer interface effects ("looping" of characteristics, "back-gating", light sensitivity).
3. No degradation of active layer properties by outdiffusion of impurities from substrate during thermal processing.
4. Lowest possible total impurity concentration consistent with (1).
5. Lowest possible density of crystalline defects (dislocations, stacking faults, precipitates).

The difficulty in achieving these requirements in bulk material has led to the development of buffer layer technology [9,10] , in which a relatively thick, high resistivity epitaxial layer is grown on the semi-insulating substrate. The active layer is then produced by either a second epitaxial growth or by ion implantation into the buffer layer. This procedure has been successful in masking undesirable substrate properties, but at the cost of an additional, rather difficult process. The practical goal of present activity in semi-insulating GaAs development is to be able to form the active layer directly on (or in) the substrate, with the fulfillment of the above five requirements. The progress made in understanding this material during the past few years, together with improved crystal growth technology, suggests that this goal may be achieved in the near future.

In this chapter, the present status of semi-insulating GaAs technology is summarized. First, the various models for the conduction mechanism and the nature of deep levels in GaAs are discussed. The section on crystal growth reviews the various growth techniques with special emphasis on the liquid-encapsulated Czochralski (LEC) process. Finally, the various electrical, optical and analytical techniques used to characterize semi-insulating GaAs are described and typical results are indicated.

1. CONDUCTION MECHANISM OF SEMI-INSULATING GaAs

The nature of the conduction mechanism of semi-insulating GaAs has been the subject of controversy up to the present day. Various energy level models for this material have been proposed, based on experimental measurements of either temperature-dependent transport phenomena (e.g. Hall effect) or optically stimulated effects (absorption, photoconductivity, photoluminescence, etc.). These models are described in this section in the order of their development, after a brief review is given of the present state of understanding of the nature of the important deep level impurities in GaAs.

1.1 Deep Levels in GaAs

Oxygen

Oxygen is an unavoidable impurity in both bulk and epitaxial GaAs. As a substitutional impurity on the As sublattice, it is expected to behave as a donor. Theoretical calculations have indicated that it should form a deep level.[11] Experimental work on the role of oxygen in GaAs has been hindered by the difficulty in obtaining reliable chemical determinations of its concentration.

A characteristic deep level in GaAs was first encountered in studies of high-purity GaAs which was occasionally found to be semi-insulating. Early studies interpreted the semi-insulating behavior in terms of an unidentified deep acceptor automatically compensating the background shallow donors.[1,3] Oxygen was suggested as a likely candidate for this compensating deep level.[1] The activation energy was found to be 0.76 eV below the conduction band. Blanc and Weisberg[12] showed that the same data could be interpreted more naturally in terms of a deep donor compensating a net excess of shallow acceptors. The deep donor concentration was estimated to be of the order of 10^{16} /cm^3, but no attempt was made to identify the level with oxygen.

Haisty, Mehal and Stratton[4] prepared semi-insulating GaAs by intentional doping with both Zn and O. The results were interpreted by an elaboration of the Blanc-Weisberg model, and the deep donor was attributed to oxygen since the same deep level was observed as in the earlier studies. Other studies of GaAs grown under various pressures of O_2 also provided indirect evidence of a deep level associated with the presence of oxygen.[6,13]

More recent investigations of GaAs containing oxygen by photoconductivity techniques[14-16] have further confirmed an important donor level at $E_c - 0.75$ eV, although proof of its identification with oxygen was not presented. Photocapacitance measurements have indicated the dominant level in GaAs:O at $E_c - 0.79$ eV.[17] An investigation of GaAs containing both O and transition metal impurities Fe and Ni showed an activation energy of 0.78 eV which was attributed to oxygen. In this work, the oxygen concentration was measured by both radioisotope and mass spectrographic techniques and found to be in the 3-5 x 10^{16} cm^{-3} range.[18]

Mircea et al. [19] have termed the ubiquitous 0.75 eV level in GaAs the "A-center", to emphasize the fact that its identification with substitutional oxygen has not been proven. They studied the outdiffusion of this center from epitaxial GaAs and determined its diffusion coefficient as a function of temperature. In addition, the donor nature of the center was established. These results were viewed as favoring the interpretation of the A-centers as oxygen occupying arsenic sites.

Very recent work by Huber et al. [20] has provided evidence to the contrary: careful mass spectrographic determinations of oxygen showed that the concentrations obtained were significantly lower than the A-center concentrations deduced from transient capacitance measurements.

At this time, it is clear that GaAs commonly exhibits a deep donor level at about $E_c - 0.75$ eV. While theoretical and some experimental studies support the identification of this level as O_{As}, other investigations suggest that it may be a native defect. Future experiments utilizing new techniques will undoubtedly help to clarify the role of oxygen in GaAs.

Chromium

Cr is the most common intentionally added deep level impurity in semi-insulating GaAs. Cr has been demonstrated to behave as an acceptor, as postulated in the early work of Cronin and Haisty.[8] However, the various techniques used to study GaAs:Cr have produced somewhat conflicting results concerning the level of Cr in the GaAs band gap. Much of the uncertainty stems from the complexity of transition metal impurities in a zinc-blende crystal environment, as described later in this section.

Both optical measurements and thermally activated transport measurements have been used to study the Cr level in GaAs. For semi-insulating material, the measurement of the dark Hall coefficient versus temperature is confined to a fairly narrow temperature range in the vicinity of 300°K because of the extremely high resistivity at lower temperatures.

Since the Fermi level is near the middle of the band gap, conduction by both holes and electrons must be considered. The two-carrier Hall coefficient is given by

$$R_H = -\frac{r_H}{q} \cdot \frac{n\mu_n^2 - p\mu_p^2}{(n\mu_n + p\mu_p)^2} \tag{1}$$

where r_H is the effective scattering factor (ratio of Hall to drift mobility), q is the electronic charge, n and p the carrier concentrations and μ_n and μ_p the respective carrier mobilities. In general, it has been found that R_H is negative for GaAs:Cr, indicative of n-type material, while thermal probe measurements indicate p-type material ($p > n$). [8] This can be readily interpreted from equation (1) when it is realized that the mobility ratio $\mu_n/\mu_p \cong 10$; all that is required to have $R_H < 0$ is that $n(\mu_n/\mu_p)^2 > p$.

The apparent ambiguity of the conductivity type of GaAs:Cr has led to some confusion in the literature regarding the placement of the Cr level in the band gap. Measurement of the dark Hall coefficient versus temperature gives an activation energy E(0) corresponding to T = 0°K when $R_H T^{3/2}$ is plotted against 1/T. (No information is obtained regarding the position of the level at 300°K.) Depending on whether the material is considered n or p type, the energy E(0) will be interpreted either as $E_C - E_A$ or $E_A - E_V$, where E_C and E_V are the conduction and valence band edges. For example, Haisty and Cronin [21] assigned the experimental activation energy of 0.79 eV to $E_A - E_V$, whereas Allen [22], citing the same work, interpreted the transition as $E_C - E_A$ on the basis of the n-type Hall effect. Allen's experiments led to the relationship

$$E_C - E_A = 0.81 \pm 0.02 - 8 \times 10^{-4}\,T\,(^\circ K).$$

The temperature coefficient was determined by the variation of the photoconductivity threshold with temperature.

Balagurov et al. [23] found $E_C - E_A = 0.81$ eV and $E_A - E_V = 0.79$ eV by analyzing dark Hall data using the two carrier expression (1). A temperature-independent mobility ratio $b = \mu_n/\mu_p = 15$ was assumed. The two values summed to give the "effective" band gap at $T = 0^\circ K$ of 1.6 eV, greater than the accepted value of 1.522 eV.[24]

Look[25] performed an extensive investigation of Cr-doped GaAs by measuring both the Hall coefficient and the magnetoresistance as a function of magnetic field strength. The mixed conductivity expressions were used to derive separate values for μ_n, μ_p, n, and p. This study showed that the material was indeed p-type ($p > n$), with the Cr acceptor lying 0.69 eV above the valence band ($T = 0^\circ K$), compared to 0.79 eV obtained earlier by Haisty and Cronin. A refined version of Look's study was recently published[26], which presented more accurate determinations of the intrinsic carrier concentration and the mobilities without changing any basic conclusions of the earlier work.

The representative Hall effect studies just discussed show that the Cr acceptor level is situated about 0.7 - 0.8 eV above the valence band at $T = 0^\circ K$, with a possible slight shift at $300^\circ K$. These studies have not required the consideration of the nature of the Cr atom with its unfilled d shell in the GaAs crystal lattice. This will be described briefly in the following summary of optical measurements on GaAs:Cr.

Cr atoms occupy substitutional sites on the Ga sublattice of GaAs. For neutrality with respect to the lattice, the Cr atom, with an outer electron configuration $3d^5 4s^1$, must give up three electrons to form Cr^{+3} ($3d^3$). This corresponds to the neutral (empty) acceptor state. If the host crystal is n-type with $N_D > N_{Cr}$, each Cr atom will acquire an additional electron to form the ionized acceptor Cr^{+2} ($3d^4$). The ratio of Cr^{+2} to Cr^{+3} can be increased by co-doping with a shallow donor.[27] Crystal field theory shows that the Cr^{+2} level is split into a ground state, denoted by $^5T_2$ in group theory notation, and a first excited state denoted by ^5E.[28,29] Each of these levels is in turn split into two closely spaced levels by a Jahn-Teller distortion.[30] Using the photo-ESR technique, Kaufmann and Schneider [31] deduced the band diagram for the Cr impurity shown in Figure 1. The ESR data clearly indicated the existence of the doubly ionized acceptor Cr^+ ($3d^5$) lying about 0.6 eV below the conduction band. This level is also inferred from photoconductivity data.[32] In Figure 1, the separation of 0.75 eV between the lower $^5T_2$ level of Cr^{+2} and the valence band is in good agreement with the Hall data. This value corresponds to the onset of the photo-ESR signal of Cr^+, as well as optical absorption.

It is now believed that the excited state(s) ^5E lie close to the conduction band[27,31,33] and that the frequently observed photoluminescence[34] and photoconductivity[16,28] peaks at $\sim$ 0.85 eV correspond to the $^5T_2 \rightarrow {}^5E$ transition within the Cr^{+2} center.

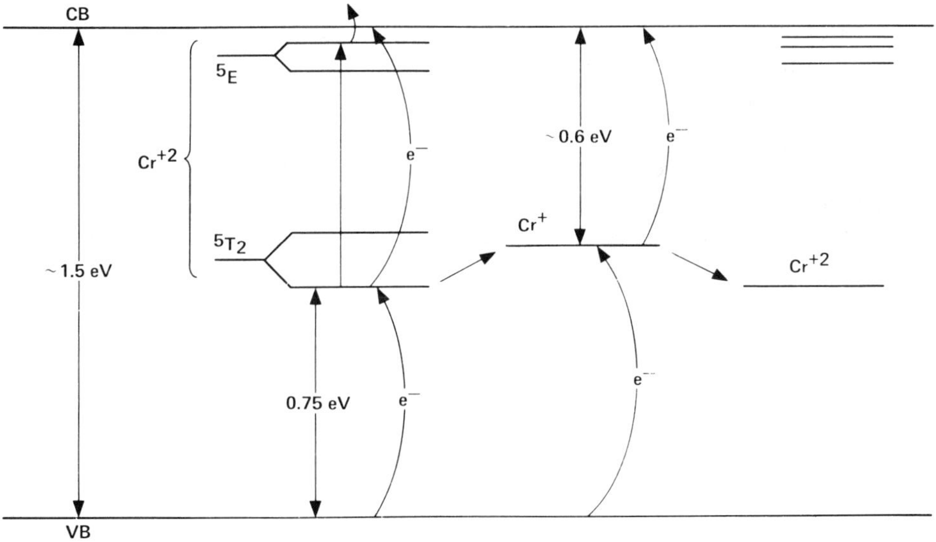

Figure 1/Cr levels in GaAs according to ESR data. After Kaufmann and Schneider [31].

Thus the peak energy roughly corresponds to the separation between the Cr^{+2} ground state and the conduction band. At T = 0°K, this gives the energy separation from the valence band of 1.52 − 0.85 ∿ 0.67 eV, very close to the activation energy measured by Look. [25] The optical measurements are therefore consistent with the Hall data within ∿ 0.1 eV for the placement of the Cr^{+2} level. There is considerable uncertainty regarding the location of the Cr^{+3} level, however. Koschel et al. [33] reported a luminescence band centered at 0.56 eV in bulk semi-insulating GaAs:Cr which was not observed in n-type GaAs epi lightly doped with Cr. Thus they postulated that the 0.56 eV band was due to the Cr^{+3} level which would be absent in n-type material (all Cr acceptors ionized). Plesiewicz[35] studied "p-type" semi-insulating GaAs:Cr in order to characterize the Cr^{+3} state. A level E_v + 0.77 eV, attributed to Cr^{+3} was deduced from photoconductivity threshold measurements at 95°K. Since this level is very near the one found by Haisty and Cronin[21] and by Kaufmann and Schneider[31], it is more likely that the Cr^{+2} level was in fact being studied. A more conclusive experiment would involve a study of strongly p-type GaAs co-doped with a shallow acceptor and with Cr, which would insure that no Cr^{+2} would be involved. Kaufmann and Schneider [31] pointed out that Cr^{+3} has never been observed in tetrahedral coordination in any semiconductor, and that the Cr^{+3} ground state would probably lie below the Fermi level in semi-insulating or n-type GaAs, favoring its conversion to Cr^{+2}. They concluded that Cr^{+2} must be the most stable state of Cr in semi-insulating GaAs. Recent photo-EPR measurements by Stauss and Krebs[36] support the view that the Cr^{+3} level lies below the Cr^{+2} level, and thus is present in negligible concentration except in heavily p-type GaAs:Cr.

The above discussion can be summarized as follows: (a) the thermal activation and optical data are in general agreement in locating the dominant Cr acceptor level at 0.7 - 0.8 eV above the valence band; (b) the dominant Cr acceptor is singly ionized Cr^{+2} ($3d^4$) and thus can be treated by ordinary semiconductor statistics; (c) the thermal and optical measurements generally apply to very low temperatures; therefore, the exact location of the acceptor level at 300°K is not well known. In the later discussion it will be assumed that the level E_v + 0.79 eV for Cr is applicable to 300°K. Table I summarizes the results of a number of investigations of the position of the Cr level(s) in GaAs.

Other Transition Metals

The properties of other transition metal impurities in GaAs have been studied in numerous investigations, particularly in the Russian literature. These elements are of less practical importance because for GaAs the highest resistivities are obtained with Cr.[43] There is general agreement that the elements Fe, Ni, Co, and Mn form increasingly shallow acceptor levels in the order listed. Sc, Ti, and V have been studied less extensively. Of these, Sc and Ti also appear to be acceptors, while some evidence indicates that V may behave as a relatively shallow donor.[43] ESR studies have shown that in all but the narrow gap semiconductors InSb and InAs, the transition metals occupy cation sites and behave electrically as singly-ionized acceptors. [44] Table II summarizes the transition metal energy level positions in GaAs according to a number of studies. A recent publication has described a theoretical model for transition metal impurities in GaAs which predicts energy levels very close to those determined experimentally.[53]

1.2 Models for Electrical Conductivity of Semi-Insulating GaAs

Two and Three-Level Models

Allen[1] and Gooch, Hilsum and Holeman[3] first described a two-level model to explain the semi-insulating property of certain "undoped" GaAs crystals. The model involved a single shallow donor level and a compensating deep acceptor level. This model was able to account for the experimental observation that the semi-insulating material was always n-type, but it required extremely close compensation to obtain a low value of electron concentration (high resistivity). The possibility of mixed conduction was not considered. An "auto-compensation" mechanism was postulated to provide the required near-equality of ionized shallow donors and deep acceptors. [1] The electron concentration in the two-level model is given by

$$n \approx \frac{1}{x - 1} N_c \, e^{(E_A - E_C)/kT} \tag{2}$$

where $x = N_A / N_D$. This equation assumes the Fermi level to be far from the conduction band, and spin degeneracy is neglected ($\beta = 1$).

Blanc and Weisberg[12] showed that the auto-compensation mechanism was not required to explain the existence of n-type semi-insulating material. They proposed instead a three-level model consisting of a shallow donor, a deep donor and a shallow acceptor. The requirement for semi-insulating material is then

$$N_A > N_D, N_{DD} > (N_A - N_D). \tag{3}$$

The material cannot be p-type as long as $N_{DD} > N_A$. This model was later confirmed by studies which demonstrated the donor type behavior of the $E_c - 0.75$ eV level[2,4,6] and is still used to interpret studies of high-purity undoped GaAs.[54]

According to the Blanc-Weisberg model, the electron concentration n is given by

$$n = \frac{N_c N_{DD} e^{-E_D/kT}}{2n + N_c e^{-E_D/kT}} + \frac{n_i^2}{n} - (N_A - N_D) \tag{4}$$

which approximates to

$$n = \frac{N_C N_{DD}}{2(N_A - N_D)} e^{-E_D/kT} \tag{5}$$

in the high resistivity regime (cf. equation (2)). Here E_D is the energy separation between the conduction band and the deep donor level.

For Cr-doped material, the "inverse" of the Blanc-Weisberg model has been used to establish the deep acceptor behavior of Cr.[8] In this case the requirement for semi-insulating material becomes

$$N_D > N_A, N_{AA} > (N_D - N_A). \tag{6}$$

A similar point of view was taken by Balagurov et al. [23] and Inoue and Ohyama [39]. Zucca [40] has challenged the adequacy of this model for Cr-doped GaAs on the grounds that the presence of oxygen (or other species responsible for the $E_C - 0.75$ eV level) is neglected. This is a valid criticism since numerous studies have demonstrated the simultaneous presence of the deep donor and deep acceptor levels[16, 17,28]. Zucca proposed another three-level model consisting of a shallow donor, a deep donor (O) and a deep acceptor (Cr). The main objections to this model are: (a) there is no smooth transition to the Blanc-Weisberg model as the deep acceptor concentration goes to zero, i.e., no shallow acceptor level is considered, and (b) application of the model to crystals with various concentrations of Cr resulted in different apparent values of the Cr energy level in each case. Nevertheless it is clear that an adequate model for GaAs:Cr must include both the deep acceptor level and the characteristic deep donor level at $E_C - 0.75$ eV.

Four-Level Model

Lindquist[55] described the electrical properties of Cr-doped GaAs in terms of the four-level model illustrated in Figure 2. A deep donor and a deep acceptor are involved, as well as shallow donors and acceptors. The model clearly reduces to those described earlier as one or another of the impurity level concentrations is allowed to approach zero. A disadvantage of such a general model is that no tractable analytical expression exists except in certain limiting cases, so that computer solutions must be obtained. Since in semi-insulating GaAs the Fermi level is approximately at mid-gap, the shallow levels are completely ionized, whereas the deep levels are only partially ionized.

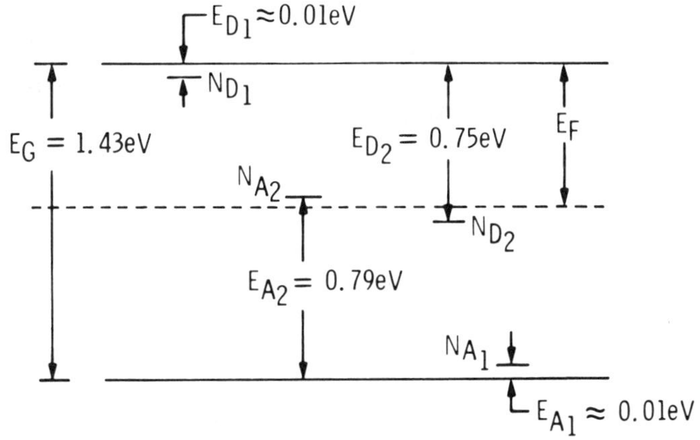

Figure 2*/Four-level band diagram for semi-insulating GaAs, after Lindquist [55]. A_2 denotes Cr level and D_2 denotes deep donor (oxygen?) level.*

The Fermi level position is determined by the neutrality condition

$$n + \sum_i N_{Ai}^{(-)} = p + \sum_j N_{Dj}^{(+)}. \tag{7}$$

The fractional ionization of the deep donor and acceptor are given respectively by

$$\frac{N_{Dj}^{(+)}}{N_{Dj}} = \left[1 + \frac{1}{\beta_D} \exp\left(\frac{E_{Dj} - E_f}{kT}\right)\right]^{-1} \tag{8a}$$

$$\frac{N_{Ai}^{(-)}}{N_{Ai}} = \left[1 + \beta_A \exp\left(\frac{E_f - (E_G - E_{Ai})}{kT}\right)\right]^{-1} \tag{8b}$$

For deep impurity levels the degeneracy factors β_D and β_A are not known, but for purposes of computation they can be assumed to be 1/2 and 4 respectively, as in the case of shallow levels [45, 56]. (Balagurov et al. [23] have calculated an effective degeneracy factor of 9.6 for Cr in GaAs based on a comparison of electrical data and chemical analyses. This value assumed that all the Cr was electrically active, however.)

For typical deep level impurity concentrations in the 10^{16}-10^{17} cm^{-3} range [55], equations (8) show that many centers are in the neutral state in semi-insulating GaAs at 300°K. Therefore neutral impurity scattering must be considered in the calculation of the electron mobility, as discussed by Look [25] and Lindquist [55].

Once the Fermi level has been calculated by equation (7) and the electron and hole drift mobilities determined, the resistivity and Hall mobility are given by

$$\rho = \left[q(n\mu_n + p\mu_p)\right]^{-1} \tag{9}$$

$$\mu_H = r_H \left| \frac{n\mu_n^2 - p\mu_p^2}{n\mu_n + p\mu_p} \right| \tag{10}$$

where r_H is the effective Hall factor. Typically, Cr-doped semi-insulating bulk GaAs has a resistivity close to 10^9 ohm-cm, particularly when co-doped with a shallow donor such as Te.[54] The Hall mobility of this material generally lies in the range 100 – 1000 cm^2/volt-sec, which is indicative of p $\gg$ n since $\mu_n/\mu_p \approx 10$ (eq. (10)). High-purity "undoped" GaAs generally has lower resistivity (10^7 ohm-cm range) and higher Hall mobility, indicative of n-type material.[54] Figure 3 gives some insight into the variation of the resistivity with concentrations of impurities in the four-level model.[55] In this plot the acceptor concentrations are fixed, and resistivity is given as a function of shallow donor concentration with the deep donor concentration as a parameter. The resistivity peaks at $\sim 2 \times 10^9$ ohm-cm for $N_{D1} \approx N_{A1}$. For $N_{D1} < N_{A1}$, the resistivity depends strongly on the deep donor concentration. The deep donor thus assists in maintaining the resistivity at a high level when there is a net excess of shallow acceptors, as expected from the Blanc-Weisberg model.

The background shallow donors in bulk GaAs are mainly S and Si, while the dominant shallow acceptor is either C or a vacancy complex. [54,57,58] The evidence for and against the identification of the characteristic deep donor with oxygen has already been discussed. Thus, it appears that the four-level description is the most realistic and flexible in describing the dark electrical properties of semi-insulating GaAs in terms of the concentrations of electrically active impurities. A more complex model is required to describe the photoelectronic behavior.[28,36]

Other Models

The models just discussed assume that (a) the deep impurity levels in GaAs can be validly treated using ordinary semiconductor statistics; (b) the material can be viewed as reasonably homogenous; and (c) the conventional interpretation of an activation energy in terms of a single level is justified.

Regarding (a) it must be pointed out that the properties of deep levels in semiconductors are still not well understood.[43] In particular, the correct degeneracy factors are not well known.[23,40] Also, it is well known that bulk grown crystals contain many defects, i.e., dislocations and precipitates, as well as fluctuations in impurity concentrations[59] superimposed on the gradual variations caused by directional solidification.

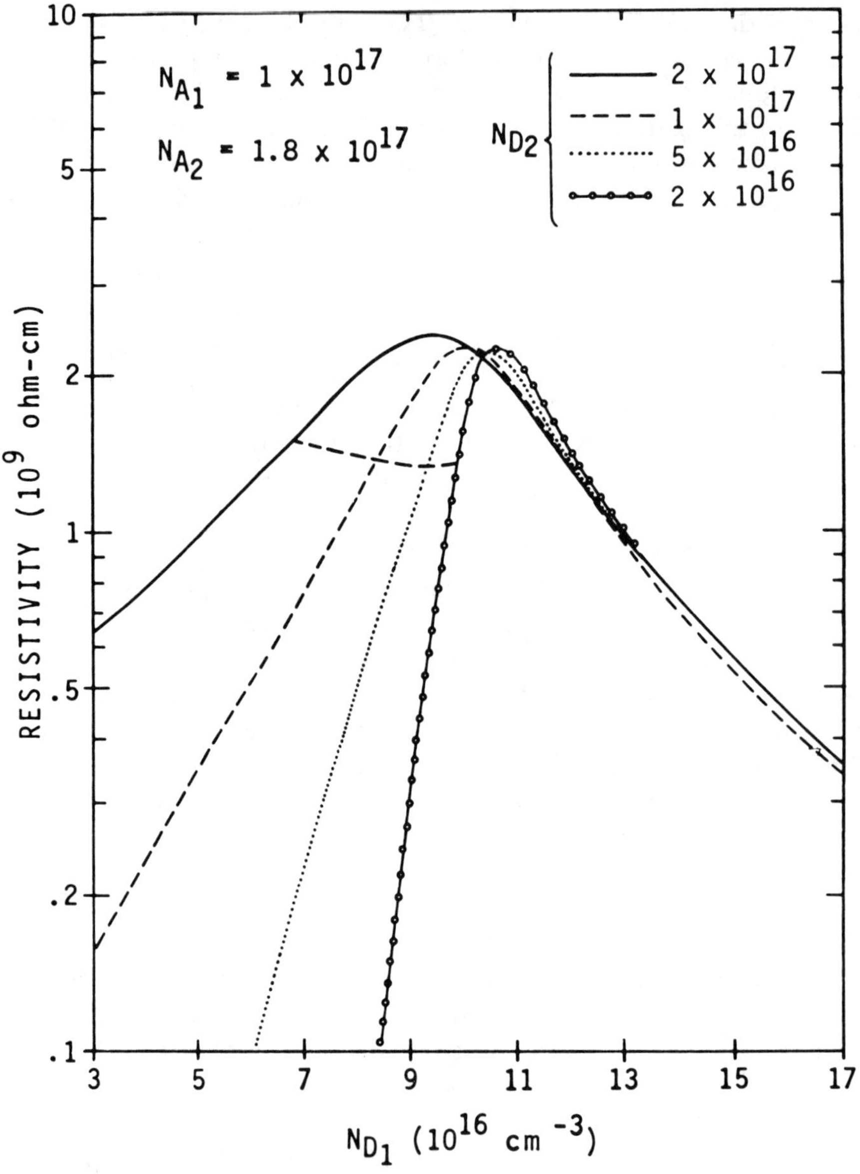

Figure 3/Variation of resistivity with doping according to four-level model. Horizontal dashed line is locus of zero Hall mobility.

Recently, Robert et al. [60] have proposed a model for the conduction mechanism in semi-insulating GaAs based on the spatial fluctuation of impurity concentrations. These fluctuations were shown to result in potential fluctuations of the order of 100 meV within the material. These affect the free electron concentration and the mobility, and may cause the interpretation of transport data in terms of a "homogeneous" model to be in error.

Recent publications by Mullin, Ashby et al. [61, 62] have challenged the straight-forward interpretation of a measured activation energy as indicating a single level located in the gap at the value of the activation energy from either the conduction or valence band. From a study of the thermal activation energies of the ohmic and space-charge limited current, these authors have found some evidence that conduction in semi-insulating GaAs is "non-extrinsic" in the sense discussed by Roberts and Schmidlin. [63] In this view, the conductivity is controlled by a "dominant hole level" N_q and a "dominant electron level" N_m which lie on either side of the "level" conventionally associated with a measured activation energy. For ohmic conduction, the activation energy is given by

$$E = (E_c - E_m) + \frac{1}{2}(E_m - E_q) \tag{11}$$

whereas for SCL conduction the activation energy is just $E_c - E_m$. In applying these ideas to GaAs:Cr, Mullin et al. [61, 62] find $E_m = E_c - 0.4$ eV, $E_q = E_c - 0.98$ eV, with concentrations of 5×10^{12} cm^{-3} and 1.5×10^{15} cm^{-3}, respectively. The density of N_q was two orders of magnitude lower than the measured concentration of Cr, however, which led the authors to speculate that N_q is due only indirectly to Cr, perhaps through a Cr-O complex.

This model can be criticized on the grounds that (a) the energy levels are not consistent with the majority of studies on GaAs; (b) the necessary conditions for the "non-extrinsic" situation were found only in certain samples; and (c) the deep level concentrations are lower than required to over-compensate the typical concentrations of shallow donors and acceptors, so that the common occurrence of semi-insulating material would not be expected.

TABLE I. EXPERIMENTAL DETERMINATIONS of Cr LEVELS IN GaAs

REF	TECHNIQUE	MATERIAL	LEVEL	T(°K)	300°K LEVEL	COMMENTS
35	Photoconductivity	Bulk GaAs:Cr, Cu	$E_v + 0.77$ (Cr^{+3} ?)	95	$E_v + 0.69$	$E_c - E_t = \text{Const.}$
22	Hall (Dark) ($R_H T^{3/2}$ vs. 1/T)	Bulk GaAs:Cr	$E_c - 0.81$	0	$E_c - 0.57$ ($E_v + 0.86$)	$E(T) = E_o - \alpha T$ $\alpha = 8 \times 10^{-4}$ ev/°K
21	Hall (Dark) ($pT^{-3/2}$ vs. 1/T)	Bulk GaAs:Cr	$E_v + 0.79$	0	---	
25	Hall (Dark) ($pT^{-3/2}$ vs. 1/T)	Bulk GaAs:Cr (p-type)	$E_v + 0.69$	0	---	
16, 34	PL, Photo-conductivity	Bulk & Epi GaAs:Cr	$E_c - .838$ $E_c - .86$	4.2 20	---	
15, 28	Dark Hall, σ; Photo-Hall, Photoconductivity, TSC	Bulk	$E_c - 0.86$ $E_v + 0.90$	0	---	
33	PL	Bulk GaAs:Cr	$E_v + 0.67$ (Cr^{+2}) $E_c - 0.56$ (Cr^{+3} ?)	6	---	PL threshold at at $E_c - 0.837$ (Cf. Ref. 16)

31	Photo-ESR	Bulk GaAs:Cr	$E_v + 0.75$ (Cr^{+2}) $E - 0.6$ (Cr^{+})	10	---	
42	DLTS	Epi GaAs:Cr (n-type)	$E_v + 0.78$	∼ 300	---	
37, 38	Photo-Capacitance (DSDP)	VPE, LPE GaAs:Cr	$E_c - 0.65$	77	---	
23	Hall (Dark)	Bulk GaAs:Cr	$E_c - 0.81$ ($E_v + 0.79$)	0	---	
39	Hall (Dark) ($R_H T^{3/2}$ vs. 1/T)	Zone Melted GaAs:Cr	$E_v + 0.8$		---	
40	Dark Conductivity ($\rho T^{3/2}$ vs. 1/T)	Bulk GaAs:Cr	$E_c - 0.72$ $E_c - 0.83$ $E_c - 0.90$	0	---	Fits to 3-level model. Different values depending on Cr content of sample
41	Photo-thermo-electric effect	Bulk GaAs:Cr	$E_c - 0.56$	300	$E_c - 0.56$	

2. CRYSTAL GROWTH TECHNIQUES

In this section, several bulk crystal growth techniques which have been applied to the growth of semi-insulating GaAs will be briefly reviewed. The emphasis in the discussion will be on those aspects of crystal growth which relate to the incorporation of impurities. First, we will discuss the principles of directional solidification, which pertain to all bulk crystal growth methods.

2.1 Directional Solidification

Impurity incorporation in a directionally solidified crystal is non-uniform along the length of the crystal because of segregation. The segregation coefficient κ is defined as the ratio of the concentration of a particular impurity in the solid to the concentration of the same impurity in the liquid. The variation of the impurity concentration along the length of the crystal is described by the normal freezing equation[64]

$$C_s(g) = \kappa C_L (1 - g)^{\kappa - 1} \tag{12}$$

where κ is the equilibrium segregation coefficient, g is the fraction solidified, and C_L is the initial impurity concentration in the liquid. This equation assumes infinitely rapid diffusion in the liquid and negligible diffusion in the solid. Most important impurities in GaAs have values of κ less than unity, which means that C_s increases as solidification proceeds. Table III quotes segregation coefficients for a number of impurity elements in GaAs. Figure 4 illustrates the form of equation (12) for various values of κ. Note that for $\kappa < .01$, the impurity concentration varies by more than an order of magnitude over the length of the crystal. In the particular case of Cr, with $\kappa < 10^{-3}$, the rapidly increasing concentration near the bottom of the crystal can exceed the solubility limit. For this reason, Cr doped ingots are generally "cropped" at a g value of ~ 0.8, and the tip portion discarded. Radial variations of impurity concentrations can occur because of segregation as well as insufficient stirring of the melt. With this brief discussion of impurity incorporation, we turn now to descriptions of several important crystal growth techniques for semi-insulating GaAs.

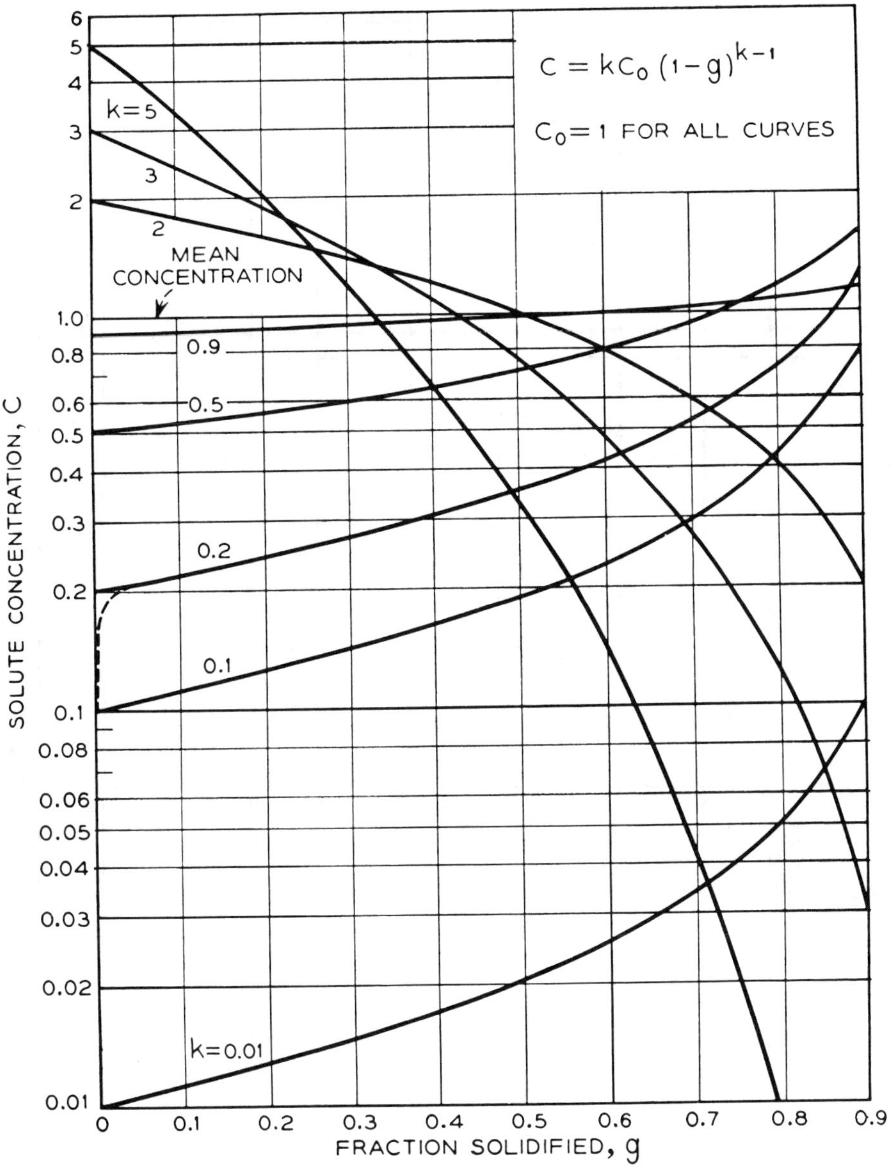

Figure 4*/Variation of solute concentration with fraction solidified for various values of segregation coefficient k. After Pfann [64].*

TABLE II. TRANSITION METAL IMPURITY LEVELS* IN GaAs

REF	Fe	Ni	Co	Mn	Sc	Ti	V	COMMENTS
21	$E_v + 0.52$	$E_v + 0.21$	$E_v + 0.16$	$E_v + 0.094$				
45		$E_v + 0.20$	$E_v + 0.16$	$E_v + 0.09$				
22	$E_v + 0.52$	$E_v + 0.21$ $E_v + 0.38$						0.21 Epi 0.38 – Bulk
46		$E_v + 0.35$						
47	$E_v + 0.50$							
48	$E_v + 0.5$	$E_v + 0.20 - 0.22$						
49	$E_v + 0.5$ $E_v + 0.35$							0.5 – T = 0°K 0.35 – 300°K
43							$E_c - 0.22$	Donor
50					$E_v + 0.57$ $E_v + 0.87$			
51							$E_c - 0.77$	
52						$E_c - 0.45$		

*ALL ACCEPTORS UNLESS OTHERWISE INDICATED

2.2 Horizontal Growth Techniques (Bridgman, Gradient Freeze)

The horizontal Bridgman technique has been employed for many years for the production of semi-insulating GaAs crystals.[7,65] Crystals as large as 3 Kg have been grown by this technique. The method involves the transport of As from one end of an evacuated tube to a boat containing Ga at the other end. A temperature difference is established along the tube such that the Ga boat is at the melting point of GaAs (1235°C), whereas the arsenic at the other end is held at about 614°C. At this temperature, the As vapor pressure is about one atmosphere. The geometry is illustrated in Figure 5. Stoichiometry is attained either by precise control of the temperature at the coldest point in the tube or by careful weighing of the Ga and As, taking account of the As in the vapor phase. After GaAs has been synthesized in the boat, the melt is cooled from one end such that the freezing interface moves along the boat at a rate in the range of 0.1 - 2 cm/hr. Two variations are used to accomplish this directional solidification. In the Bridgman method, cooling is accomplished by motion of the boat with respect to the furnace. In the "gradient freeze" method, both boat and furnace are stationary but the entire temperature profile along the boat is lowered, while maintaining a constant temperature gradient of about 1°C/cm. This is illustrated in Figure 6. (A recently developed growth technique uses a combination of both the above methods to assure that impurities segregate at a predictable position in the ingot, and to form a solid-liquid interface shape conducive to single crystal growth with a minimum dislocation density.)[66] The reactor vessel and the gallium boat are usually made of quartz for semi-insulating crystal growth.

Sources of impurities in the horizontal Bridgman/gradient freeze process are as follows:

(1) Starting materials (Ga, As)

(2) Quartz (reacted with melt and evaporated during sealing of the tube ("Bloom")) (Si)

(3) Backstreaming vacuum oil introduced during evacuation of the reactor tube (C).

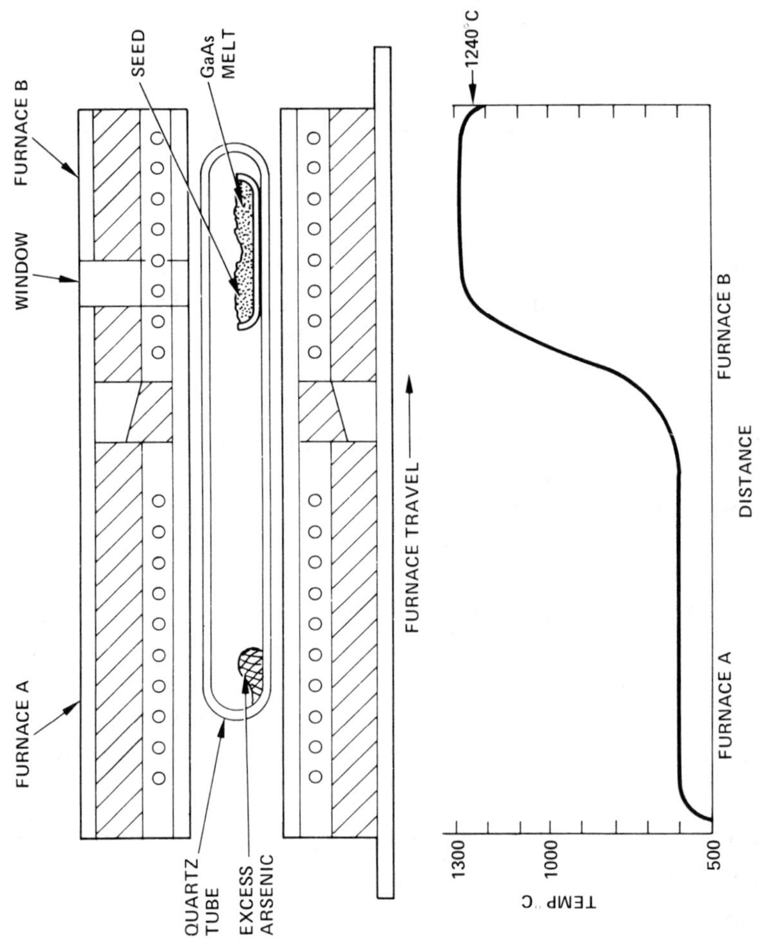

Figure 5*/Horizontal Bridgman configuration for GaAs crystal growth.*

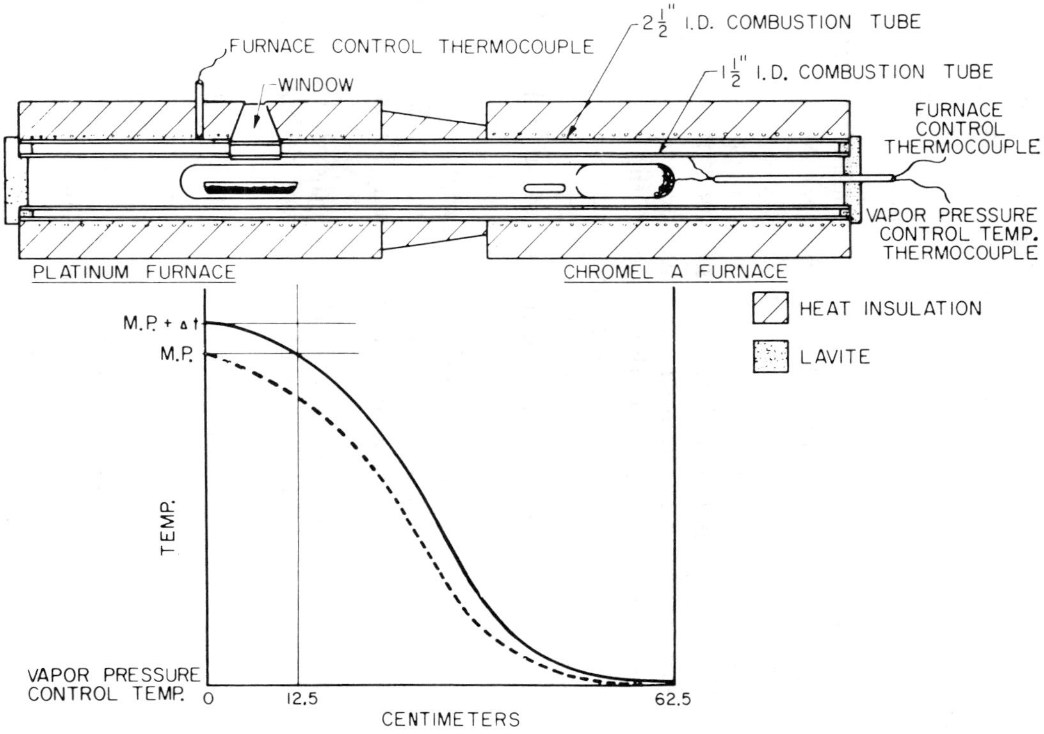

***Figure 6**/Gradient freeze furnace and temperature profile. After S.E. Miller, "Gradient Freeze Single Crystal Growth", Chapter 31 in Compound Semiconductors, Vol. 1: Preparation of III-V Compounds, R.K. Willardson and H.L. Goering, Eds., New York: Reinhold, 1962, p. 275.*

The major impurity incorporated in the Bridgman technique is Si from the reactions with quartz. The Si concentration in the crystal can be reduced significantly by controlled additions of oxygen.[6] The best reported electrical properties for GaAs grown with added O_2 are room temperature Hall mobilities of ~ 8000 cm^2/volt-sec (20,000-30,000 cm^2/volt-sec at 77°K) with carrier concentration in the 2-4 x 10^{15} cm^{-3} range.[6]

2.3 Liquid-Encapsulated Czochralski (LEC) Technique

The LEC technique is widely used for growth of compound crystals with one volatile constituent.[67-69] The essential features are illustrated in Figure 7. The crucible is charged with polycrystalline GaAs and with sufficient B_2O_3 to cover the melt to a depth of about 2 cm. The GaAs charge is melted and the seed is lowered through the molten B_2O_3 into the melt. The temperature must be held at a value which neither melts the seed nor allows the surface of the melt to freeze. The seed and the crucible are rotated in opposite directions at about 3 and 12 RPM, respectively. The growth rate is typically .1 to .5 in/hr., and is limited by the relatively large heat of fusion and small thermal conductivity of GaAs.

Seeds are usually cut to allow growth in the (111) direction with the "A" face (Ga face) exposed to the melt. Growth in the (100) direction is also possible, although the growth rate is considerably lower.

The polycrystalline GaAs for the charge is usually synthesized in a separate operation (similar to the gradient freeze growth technique), but can also be synthesized directly in the grower by a high pressure technique with a suitable chamber,[70] or by gas phase transport in a low pressure system.[71]

The crystal growth furnace consists of a "picket fence" type resistance heater[72], a cylindrical heat shield and susceptor, all made of UHP graphite, together with mechanisms for raising, lowering and rotating the seed and crucible. Temperature is sensed with a thermocouple on the heater, and is controlled to ± 0.5°C. The furnace atmosphere is N_2 at about two atmospheres pressure to suppress the escape of arsenic through the B_2O_3.

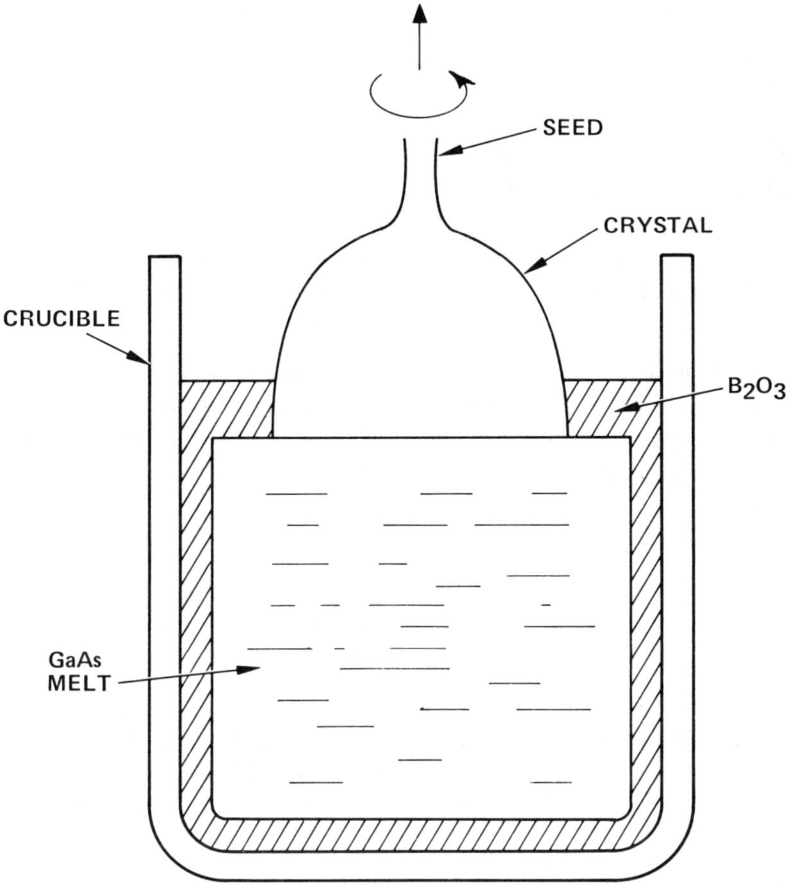

Figure 7*/Schematic illustration of GaAs crystal growth by the liquid-encapsulated Czochralski (LEC) technique. Crucible is enclosed in high pressure chamber (2 atm. N_2).*

Crucible materials used for LEC growth of GaAs include pyrolytic boron nitride, alumina, graphite coated with pyrolytic graphite, and quartz. The purity of the resulting crystals decreases in the order given.[70,71]

The B_2O_3 has a great influence on the purity of LEC crystals. Before crystal growth, the B_2O_3 must be dried (degassed) by heating under a vacuum of $< 10^{-3}$ Torr. This is usually done in a separate furnace, although the degassing can be done in the puller immediately before adding the GaAs charge. If the B_2O_3 contains water, silicon contamination can occur when quartz crucibles are used. Also, the moisture can cause loss of visibility through the B_2O_3 by promoting the formation of Ga and As compounds in the molten glass.

The B_2O_3 provides a purification of the GaAs because of its "gettering" action for many impurities. Table IV presents experimental data [73] for the reduction in concentration of several impurities below the values expected on the basis of the segregation coefficients using equation (12), for the bottom end of a crystal ($g = 0.99$). The actual concentrations in the crystal were determined by emission spectrographic analysis.

The LEC technique can produce very large crystals. Ingots weighing 6 Kg have been grown, although with high dislocation density ($10^4 - 10^6$/cm^2).[73]

In terms of electrical properties, the best results achieved to date for LEC growth of GaAs are: 300°K electron mobility of 7000 cm^2/V-sec for a semi-insulating ingot grown in quartz (derived using mixed conductivity analysis)[25], and 6300 cm^2/V-sec for a conducting GaAs crystal with carrier concentration in the 10^{14} cm^{-3} range.

2.4 Gremmelmaier Technique

The Gremmelmaier technique[74] is a variant of the Czochralski method, in which the melt and seed are sealed in an ampoule. A one atmosphere overpressure of As is maintained by introducing the required excess As before sealing. Rotation and translation are provided by magnetic coupling (Figure 8). This method has been used by Liebmann and Kampschulte[75] to grow GaAs crystals with room temperature Hall mobilities of 7000 - 8500 cm^2/volt-sec at carrier concentrations in the mid-10^{15} cm^{-3} range. In this work, aluminum nitride

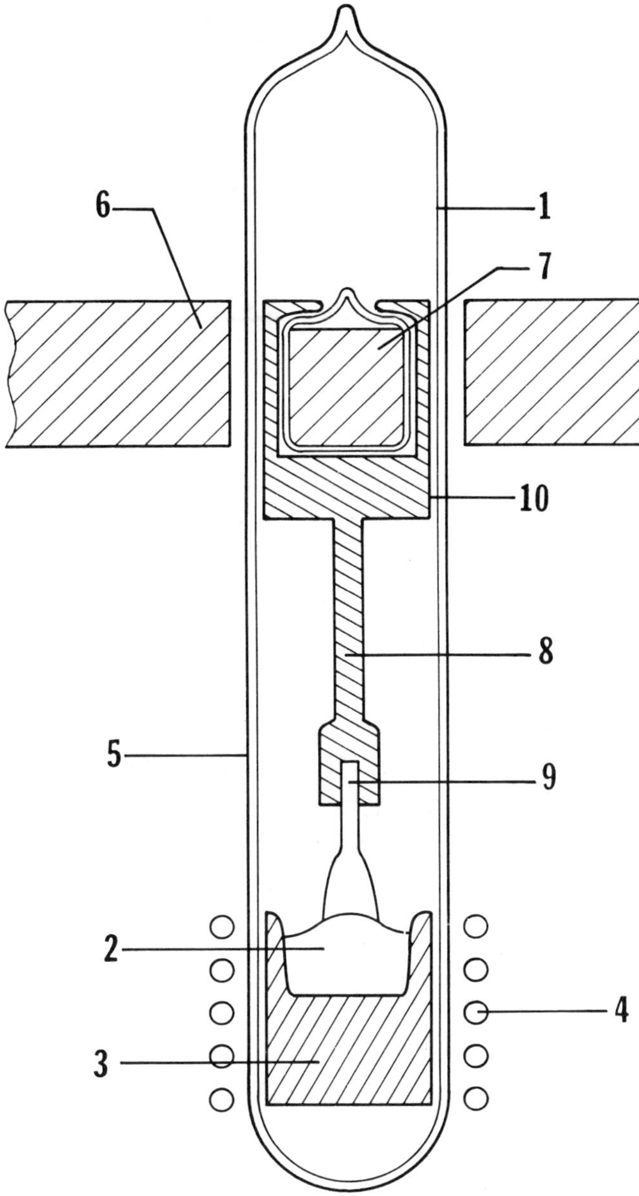

Figure 8/*Geometry of Gremmelmaier technique. After R. Gremmelmaier, Z. Naturforsch, 11A:511 (1956).*

TABLE III. SEGREGATION COEFFICIENTS† OF VARIOUS ELEMENTS IN GaAs

ELEMENT	k	ELEMENT	k
Al	.2	Mg	.1
C	.8	Mn	.02
Ca	.002	Ni	4E-5
Co	.0004	Pb	1E-5
Cr	.0006	S	.3
Cu	.002	Se	.3
Fe	.002	Si	.14
Ge	.01	Sn	.08
In	.007	Te	.06
		Zn	.4

†R.K. Willardson and W.P. Allred, *Int. Sym. on GaAs,* Reading (1966), p. 35.

TABLE IV. PURIFICATION OF GaAs BY B_2O_3 ("GETTERING")†

IMPURITY	C_L (ppmw)	(ppmw) C_s (CALCULATED)*	(ppmw) C_s (MEASURED)
Al	114	907	.05
Ca	2059	408	<.1
Cu	2100	416	8
Fe	572	113	10
Mg	41	258	.4
Si	30	220	5

*Calculated from $C_S = kC_L(1 - g)^{k-1}$ using g = 0.99 and k from Table III

†R.K. Willardson and W.P. Allred, *Int. Sym. on GaAs,* Reading (1966), p. 35.

crucibles were used. A special procedure was used to clean the quartz ampoule, and the charge was given a vacuum bake-out for 80 hours at 620°C before sealing the ampoule.

The results quoted in this section indicate that high purity, relatively uncompensated "undoped" GaAs can be produced by all the techniques discussed. The "undoped" crystals occasionally are semi-insulating, presumably due to the overcompensation of excess shallow acceptors by deep donors, as discussed earlier. Using the highest purity starting materials, the LEC technique has produced semi-insulating undoped crystals on a regular basis.

3. MATERIAL CHARACTERIZATION TECHNIQUES

This section gives a brief review of various techniques which have been used to characterize semi-insulating GaAs, or conductive GaAs containing deep-level impurities. For details the reader may consult the references cited.

3.1 Thermal Conversion Studies

Conversion of a thin surface layer of GaAs from a semi-insulating to a conducting state as a result of thermal treatment is a well-documented phenomenon, although the interpretation of the effect is still controversial. It is possible that several factors may be involved.

There is general agreement that the converted surface layer is p-type. At present there are two main interpretations of the conversion. Lum et al. [76-78], using low temperature photoluminescence, have found peaks after heat treatment which were identified as carbon shallow acceptors and carbon acceptor-arsenic vacancy complexes. The carbon appeared to originate from the graphite parts in the reactor.[78] Hydrocarbon solvents used in the substrate preparation process have been shown to affect the surface conversion[79], which further supports the role of carbon. Another interpretation, also based on photoluminescence measurements, is that the acceptor responsible for the conversion is Mn.[80,81] Whether the Mn originates from the bulk material or the reactor environment is still not clear. However, conversion is frequently observed in samples in which no Mn can be detected by emission spectroscopy.

It is of practical importance to be able to pre-screen semi-insulating GaAs to eliminate material which is susceptible to conversion by a particular heat treatment. Barrera[82] has described a useful technique to rapidly determine the magnitude of the conversion effect. After subjecting a chemically polished substrate wafer to a standard heat treatment (usually performed in an epi reactor at a typical growth temperature for about one hour in flowing, purified hydrogen), a simple metal test pattern is formed by a lifting technique. Values of the "conversion resistance" can then be readily measured on a curve tracer. The conversion resistance measured in this test was shown to correlate with RF performance parameters of FETs produced using substrates from the same ingot.

The net shallow acceptor concentration in a converted surface layer is typically in the 10^{16} cm^{-3} range.[77,80] The four-level model discussed in an earlier section indicates that if shallow donors are intentionally incorporated during growth, and just overcompensated by the Cr acceptors, the material should be more resistant to conversion. This has been verified experimentally with ingots doped with Cr and Te. Such material effectively resists conversion in a 750°C, one hour anneal, particularly for wafers near the bottom of the ingot where the impurity concentrations are higher.[83] Unfortunately, the high impurity concentrations result in other undesirable effects, to be described in a later section.

Thermal conversion has also been observed in capped annealing experiments simulating the activation anneal after ion implantation. In this case, the conversion in n-type [84] and material which successfully resists conversion in a free surface anneal appears susceptible to conversion in a nitride-capped anneal. There is evidence that with the capped anneal, conversion takes place by loss of Cr from the surface[85], which uncompensates background shallow donors.

3.2 Hall Effect and Related Techniques

Hall effect measurements are basic to the electrical characterization of semiconductors. Simultaneous measurements of the Hall coefficient R_H and the resistivity ρ permit one to determine the majority carrier concentration and Hall mobility μ_H at the temperature of measurement. These quantities can in turn be used to derive an estimate of the total concentration of electrically active impurities and the depth of the dominant impurity level. A review of the Hall technique applied specifically to conducting GaAs has been given by Blood.[86]

Special experimental difficulties are encountered when the Hall technique is applied to semi-insulating GaAs. The extremely high sample resistance can result in leakage of current around the sample, surface conduction effects, and very long time constants when the voltage, current or field polarities are switched. These problems can be minimized by use of a guarded system using unity-gain amplifiers.[87,88] This technique has been used successfully by Look.[25]

Mixed conductivity expressions for the resistivity and Hall coefficient must be used in interpreting Hall data on semi-insulating GaAs. Inclusion of the magnetoresistance effect due to mixed conduction allows individual values of n, p, μ_n, and μ_p to be derived.[25,26,89]

An extension of the Hall technique involving photoexcitation ("photo-Hall" effect) is widely used to study the properties of deep levels in semiconductors. The identity of the photoexcited carriers (electrons or holes) is determined by the Hall effect. This permits one to distinguish between hole emission and electron emission from a given deep level. This method has been applied to semi-insulating GaAs by Look [90], Lin and Bube[15,28], and others.[91,92]

3.3 Photoluminescence

Photoluminescence (PL) is a widely used technique for studying both shallow and deep levels in semiconductors. The luminescence is excited using a laser source with a wavelength shorter than that corresponding to the material band gap. A related technique, cathdoluminescence (CL) uses electron beam excitation. For high resolution, the sample is usually cooled to 4.2°K.

Turner et al. [13] studied the variation intensity of 0.65 eV luminescence from GaAs as a function of the O_2 partial pressure used during the growth of the crystals. The Cr level has been extensively studied by PL, both in bulk crystals[32,33,93] and in epi lightly doped with Cr.[34,94,95]

Shallow donor and acceptor levels can be identified using high-resolution PL: Bois and Beaudet[96] identified five shallow acceptor impurity levels in GaAs. Annealing studies involving PL measurements have characterized native defects in GaAs.[76,97] Recently, PL measurements have been used to determine the degree of compensation in GaAs epitaxial[98,99] and bulk[100] crystals.

3.4 Optical Absorption and Photoconductivity

These complementary techniques are basic to the study of impurity levels in semiconductors. They have been applied extensively to the study of Cr and other transition metal impurities in GaAs and to the deep donor frequently associated with oxygen. Basically, light of lower than band gap energy is absorbed at wavelengths corresponding to the energy separation between a localized level and a band, or between ground and excited states of a localized level. In the former case, the conductivity is enhanced due to the additional carriers in the band. The photo-Hall effect can be used to identify which band is involved.

Typical 300°K photoconductivity spectra for samples of semi-insulating GaAs are shown in Figure 9(a) - (c). Figure 9(a) corresponds to a Cr-doped sample which is on the n-type side of intrinsic, i.e., all the Cr is in the +2 state.[90,91] The peak at 0.87 eV is indicative of the intra-center transition in Cr^{+2} discussed earlier. The same peak is observed in optical absorption spectra [27]. For more "p-type" GaAs:Cr, this "resonance peak" disappears, but there is a rather abrupt threshold of photoconductivity at about 0.8 eV [35, 91] (Figure 9(b)). Figure 9(c) is a typical photoconductivity spectrum of "undoped" or "oxygen-doped" semi-insulating GaAs. In this case, there is no abrupt photoconductivity threshold, but, rather, a slowly increasing effect with increasing photon energy. Photoconductivity spectra of Fe [49], Sc [50], and Co and Ti [52] in GaAs have also been measured.

The effect of the simultaneous presence of both oxygen and transition metal impurities on photoconductivity has been studied by Andrianov et al. [13], and by Stocker [14, 16] using an AC photoconductivity method. The oxygen center was found to produce an oscillatory spectrum at low temperature indicative of a local vibrational mode.

Estimates of the Cr concentration of GaAs:Cr have been made using optical absorption in conjunction with radiotracer measurements[101] and neutron activation analysis or mass spectrometry.[102]

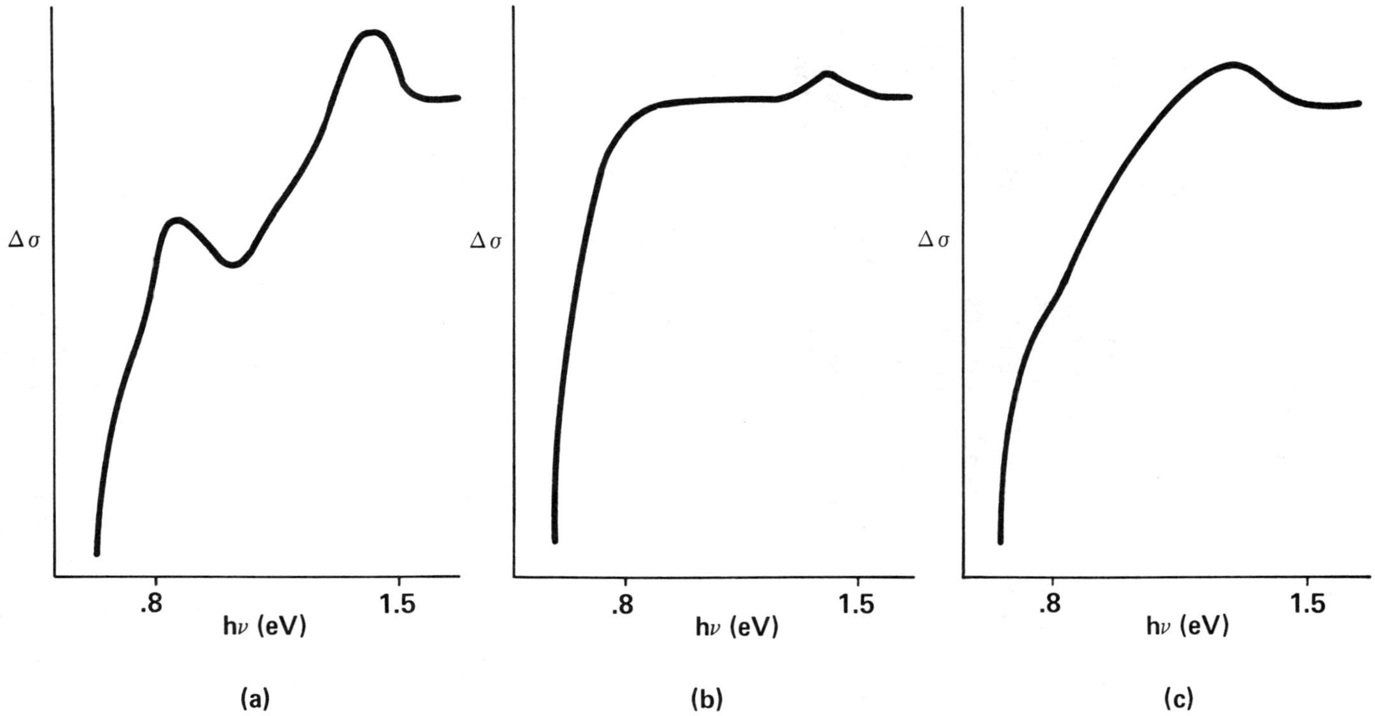

Figure 9/*Schematic photoconductivity curves for semi-insulating GaAs.*
(a) Cr-doped GaAs, all Cr in +2 state [28, 90]
(b) Cr-doped GaAs, more "p-type" [90].
(c) "Undoped" GaAs (no Cr) [15].

3.5 Deep Level Spectroscopy

Changes in the capacitance of a Schottky barrier in response to changes in illumination and/or temperature have been exploited in several techniques for studying deep levels in low-resistivity semiconductors. The best known of these is the DLTS technique developed by Lang.[103] In this method, the bias on a Schottky barrier is repetitively switched from forward to reverse at a particular frequency. The forward bias portion of the pulse fills majority carrier traps in the depletion region. During the longer reverse bias portion of the pulse, the traps lying above the Fermi level (for n-type material) are emptied at a rate which depends on the temperature of the measurement. As the majority carriers are liberated, the capacitance changes exponentially with a particular time constant. At a particular temperature, this time constant corresponds to a particular $\Delta C = C(t_1) - C(t_2)$, where t_1 and t_2 are fixed times measured after the onset of the reverse-bias portion of the pulse. In a DLTS run, the temperature is raised while keeping t_1 and t_2 fixed.

For a single deep level, ΔC passes through a maximum as the temperature is increased. The DLTS data are analyzed to give information on the emission rate of carriers from the deep levels, as well as estimates of the deep level depth and concentration. Details are given in reference 103.

This method has been applied to the study of deep levels in GaAs in numerous investigations.[42,104-106] A variation of the DLTS technique has recently been reported[107], in which the traps are filled optically rather than electrically ("optical DLTS"). This allows minority as well as majority carrier traps to be investigated using a Schottky barrier. Another technique, "minority carrier trap spectroscopy," makes use of optical generation of minority carriers in the bulk rather than in the depletion region.[108] This has been shown to overcome certain limitations of the optical DLTS method.

A different type of capacitance spectroscopy for studying deep levels in conductive material is the "double source differentiated photocapacitance" (DSDP) technique developed by White, Dean and Porteous. [37,38] This measurement is carried out at constant temperature, usually at 77°K. Traps are filled by a steady source of low intensity near-bandgap "priming" radiation. Traps are then selectively emptied by sweeping the wavelength of a narrow band "probing" radiation from low to high energy. The capacitance signal is differentiated to improve sensitivity. The DSDP spectrum consists of a plot of dC/dt versus photon energy. A positive peak corresponds to an electron trap and a negative peak to a hole trap in n-type material. Advantages of this method over the DLTS technique are its rapidity, better resolution due to optical rather than thermal trap emptying, the detectability of both majority and minority carrier traps with a Schottky barrier, and considerably simpler experimental apparatus. This technique is suitable for the rapid assessment of material quality, and has been used to select "good" VPE GaAs wafers for FET fabrication.[109]

The capacitance techniques described above cannot be applied to semi-insulating material. However, a recently developed technique termed "optical transient current spectroscopy" (OTCS) can determine deep level parameters in insulators.[110] Essentially a transient photoconductivity technique, the OTCS spectrum is a plot of transient current increment ΔI at two fixed measurement times after the light pulse, versus temperature (analogous to the ΔC vs T spectrum obtained in the DLTS technique). Good agreement has been demonstrated between the levels derived by OTCS on semi-insulating GaAs and by DLTS on conducting GaAs.

3.6 Impurity Analysis

In the study of semiconductor materials, it is of great importance to relate the concentrations of donors and acceptors derived from electrical measurements to impurity concentrations determined by chemical analytical techniques. Certain impurities are electrically inactive, such as isoelectronic elements B and N in GaAs. Also, only a fraction of a particular donor or acceptor impurity may be electrically active if its distribution in a crystal is non-uniform or if the solubility limit is exceeded. On the other hand, the electrically active species may outnumber the chemical impurities if native defects play a significant role in the electrical properties.

The most widely used analytical techniques for the determination of impurity concentrations in semiconductors are emission spectroscopy and spark source mass spectroscopy. Both are destructive techniques for analyzing impurities in bulk samples. Emission spectroscopy is primarily useful for the detection of metallic impurities. Reviews of this technique have been published by Wang.[111,112]

Mass spectrometry can in principle see all the impurity elements, although there are practical limitations due to interferences. A comprehensive review of spark source mass spectrometry (SSMS) applied to semiconductors has been given by Brown et al. [113]. A major motivation for using the SSMS technique is its ability to detect the important electrically active impurities carbon and oxygen. Early attempts to determine the concentrations of these elements in GaAs led to unreasonably high values in the 10^{18} to 10^{19} cm^{-3} range [114, 115]. More recent investigations using the conventional SSMS technique have determined C and O concentrations in the 10^{16} to 10^{17} cm^{-3} range. [18,116] Recently a liquid He cryopumped sample chamber has been used to eliminate contributions to the C and O intensities due to adsorbed gases on the chamber walls [117, 118]. Absolute calibrations of the C and O concentrations were obtained using the gamma photon activation method described by Baker and Williams. [119] The results of these studies showed that the true C and O concentrations in GaAs are much lower than previously thought, i.e., in the range 4×10^{15} to 1×10^{16} cm^{-3}.

Comparisons of SSMS results and Hall data for semi-insulating GaAs have been made. [55,116] Look[116] obtained reasonable agreement between the total ionized impurity concentrations inferred from the SSMS data, and the same quantity derived from the electron mobility obtained by Hall measurements. Lindquist[55] attempted to reconcile the SSMS concentrations and Hall data in terms of the four-level model described earlier. For good agreement between calculated and experimental values of ρ and μ_H, the required electrically active concentrations were usually significantly lower than the SSMS concentrations.

3.7 Ion Implantation

There is great interest in the fabrication of GaAs FETs and integrated circuits by direct ion implantation of substrate material. The implanted layers must have characteristics similar to implanted buffer epi or epitaxial active layers in order to produce high performance devices. It has been found that ultra-high purity "undoped" semi-insulating GaAs fulfills this requirement. Devices have been fabricated by direct implantation of this material which are comparable to those fabricated by implanting buffer epi. [120] Furthermore, measurements of the carrier concentration and Hall mobility of the implanted layer correlate well with device performance parameters, thus permitting substrate ingots to be screened. Implanted layers in the highest quality undoped semi-insulating GaAs have Hall mobilities near 5000 cm^2/volt-sec at 300°K, for $N_D - N_A = 1 \times 10^{17}$ cm^{-3}, and fractional activities of $\geqslant$ 70% (ratio of concentration of electrically active implanted element, after annealing, to total concentration implanted). If the production of this high-quality bulk material becomes routine, we may anticipate the eventual elimination of epitaxial growth of active layers.

4. SUBSTRATE-ACTIVE LAYER INTERFACE EFFECTS

The substrate-active layer or buffer-active layer interface is of vital importance to the performance of GaAs FETs because of its proximity to the conducting channel under the gate depletion region. It is essential that the material just beneath the active layer have negligible conductivity. For an active layer grown directly on a semi-insulating substrate, this means that no thermal conversion must take place during the epitaxial growth. Although a p-type converted layer would be isolated from the n-type active layer, the p-n junction would result in a depletion layer in the n-type epi which would change the pinch-off characteristic of the device. Also, the existence of a p-type layer under the active epi has been associated with anomalous features ("kinks") in the drain current characteristics.[121] For ion-implanted active layers, the substrate must remain semi-insulating after the activation anneal. For active layers grown on or implanted in n^- buffer epi, the buffer must either be completely depleted by the negative charge build-up at the substrate-buffer interface (to be discussed later), or be high resistivity itself by incorporation of deep levels (e.g., Fe or Cr) during growth. [122-124]

It has been found that Cr out-diffuses rapidly from a semi-insulating substrate during epitaxial growth.[125,126] The diffusivity of Cr in GaAs is about 7×10^{-14} cm^2/sec at a typical epi growth temperature of 750°C [127]. Thus, for a typical active layer growth time of about one minute to produce a layer 0.25 μm thick, the Cr penetrates a distance

$$x \approx \sqrt{Dt} = 0.02\ \mu m$$

or nearly 10 percent of the total layer thickness. This incorporation of Cr acceptors in the active layer near the interface reduces the electron concentration and the drift mobility, to the detriment of device performance. The following section is a brief discussion of several techniques used to characterize the variations of net doping concentration and mobility across an epitaxial layer grown on a semi-insulating substrate. The remaining sections describe the problems of "backgating" and light sensitivity.

4.1 Doping and Mobility Profiling

The capacitance-voltage (C-V) profiling method using a Schottky barrier on the epi surface is used to determine the variation of net doping concentration with distance into the epi, as well as to define the effective thickness of the epitaxial layer. Figure 10 illustrates a typical doping versus distance (n vs. x) plot for a uniformly doped epi layer, derived from the Schottky barrier C-V relation, and the corresponding geometry of the epitaxial layer. The actual epi layer thickness is T, whereas the C-V method always determines a thickness $T' < T$. T′ is usually defined to be the thickness corresponding to the onset of the drop in the indicated doping concentration. Beyond this point, the n vs. x plot is not a true representation of the actual doping profile because of the breakdown of the assumptions used in deriving the expressions for n and x from the C-V data.[128] The distance $T - T'$ depends on the doping level of the epitaxial layer and the properties of the substrate, and is a minimum for an intrinsic substrate. In this case, W_n (the depletion layer on the n-side of the substrate epi junction) is given by [129]

$$W_n = L/\sqrt{e} \approx .6L, \tag{13}$$

and the quantity δ is approximately 1.5L, where L is the extrinsic Debye length given by[128]

$$L = \sqrt{\frac{2k\,\epsilon_o}{qN_D}\left(\frac{kT}{q}\right)} \ . \tag{14}$$

For a typical epi doping of $1 \times 10^{17}\ cm^{-3}$, L is about 0.019 μm, so that the distance $T - T'$ is about 0.04 μm. The C-V method cannot determine the doping level at distances closer to the surface than the zero bias depletion layer with W(0), given by

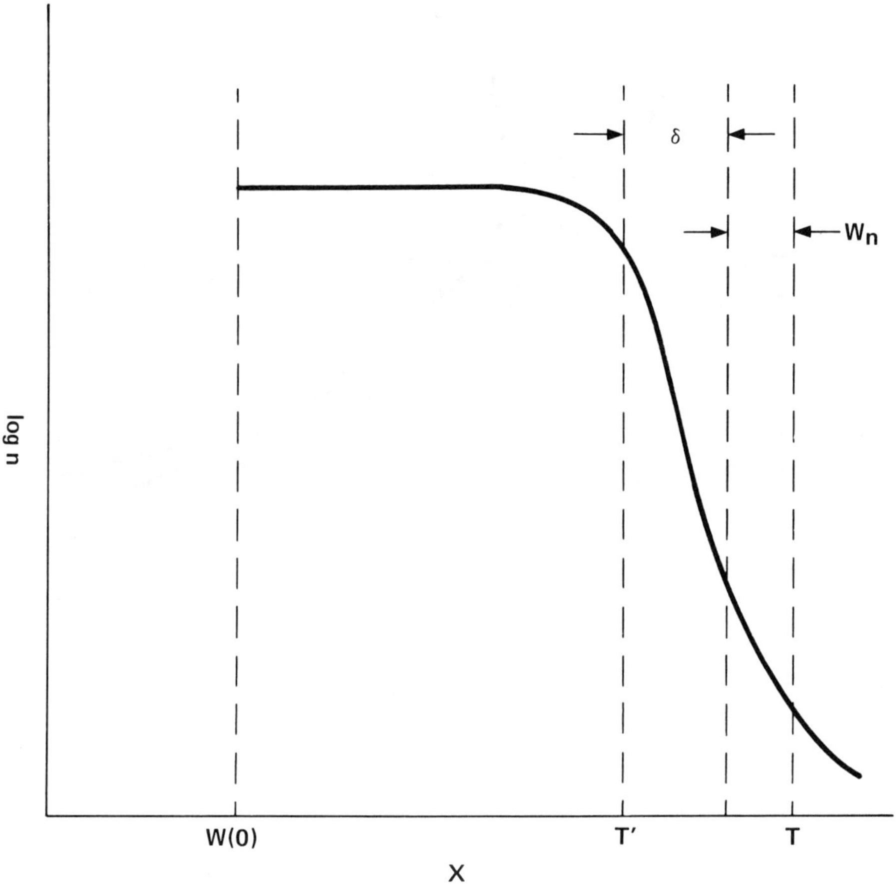

Figure 10*/Schematic doping vs. distance curve for n-type epitaxial layer on SI or p-type substrate. w(o) is zero-bias depletion layer width of Schottky barrier. T is the true epi layer thickness; T′ is the "apparent" thickness determined by the C-V measurement. w_n is depletion layer width in epitaxial layer caused by substrate-epi charge dipole.*

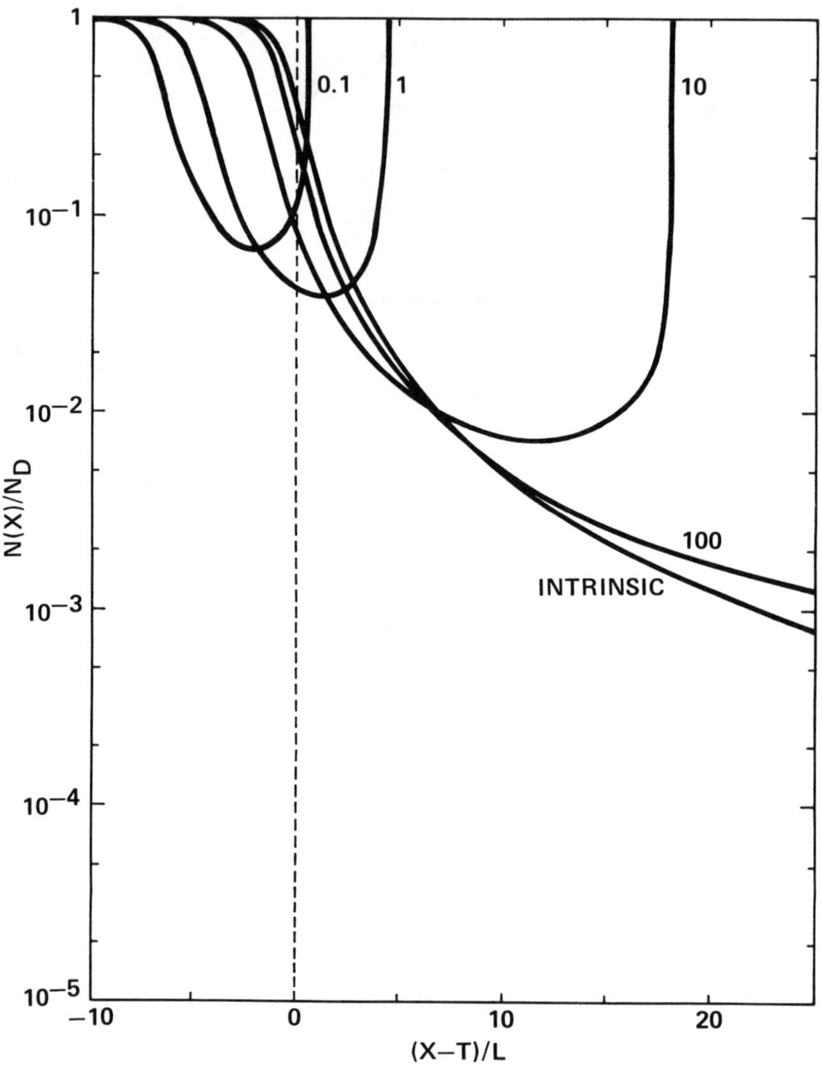

Figure 11*/Calculated doping vs. distance curves for various values of N_D/N_A. $N_D = 10^{17}$ cm^{-3}, $\Phi_B = 0.8$ volt, epi layer thickness $T =$ 0.3 μm.*

$$W(0) = L \sqrt{\frac{q\Phi_\beta}{kT} + \ln \frac{N_D}{N_c}} \tag{15}$$

where Φ_β is the Schottky barrier height and N_c is the conduction band effective density of states (for GaAs, $N_c = 4.5 \times 10^{17}$ cm^{-3} at 300° K). For $N_D = 10^{17}$ cm^{-3} and $\Phi_\beta = 0.8$ volt, W(0) = 5.43L = 0.1 μm.

The inability of the C-V techniques to measure the "true" epi layer thickness T is not a serious problem, since the layer is effectively completely pinched off for $x > T'$. The value of T' for a given T and N_D is reduced if a p-type converted layer exists just beneath the epi. Figure 11 illustrates n vs. x curves obtained for p-type substrates of various net acceptor concentrations N_A.[128,129] As N_A increases for fixed N_D, the initial drop in the n vs. x plot occurs closer to the epi surface, and the minimum in the n – x plot becomes shallower and is shifted toward the epi. The occurrence of such a U-shaped curve is a conclusive indication of a p-type converted layer.

It is important to determine the variation of mobility with distance into the active layer as well as the variation of doping concentration. The most direct approach to mobility profiling is to successively etch down a van der Pauw[130] mobility sample. Figure 12 illustrates the results for epi grown directly on a substrate compared to epi grown on a buffer layer.[131] The sharp drop in the mobility near the interface for the non-buffered sample indicates the diffusion of acceptors into the epi during growth, resulting in increased ionized impurity scattering. Tansley[132] developed a completely electrical method to replace the successive etching technique; the epi is increasingly pinched off by voltage applied to a Schottky barrier gate while the mobility under the gate depletion layer is measured by an AC van der Pauw technique. Pucel and Krumm[133] devised a "long gate" FET structure which permits the profile of drift mobility (rather than Hall mobility) to be determined by measuring the gate capacitance and transconductance as the gate is gradually pinched off. Their results also indicated increased compensation near the interface. These mobility profiling techniques based on a biased Schottky barrier gate are restricted to $x \leqslant T'$ as discussed previously.

4.2 "Back-Gating"

The term "back-gating" refers to pinch-off of the conducting channel of a FET caused by a voltage-variable depletion layer extending into the epi from the substrate-epi interface. This effect is sometimes seen when the substrate is biased negatively with respect to the active layer. Back-gating is most often observed in highly Cr-doped substrates. The phenomenon can be understood qualitatively by considering Figure 13, which shows the energy band, fixed charge and electric field profiles of a Cr-doped semi-insulating/n^+ junction. In the neutral semi-insulating material, the Fermi level lies very close to the deep donor and acceptor (Cr) levels, so that both levels are partially ionized and partially neutral. [55] The band bending at the interface results in deep acceptors becoming more completely ionized and the deep donors more neutral. This is equivalent to a build-up of negative charge on the substrate side of the interface, which is counteracted by the depletion of a portion of the n^+ epi. Thus, a dipole of fixed charge exists as in the case of a normal p-n junction. The energy separation y(0) between the conduction band edge and the Fermi level at the interface can be used to compute the "effective p-doping" $N_A{}^*$ of the substrate-epi junction. [129] $N_A{}^*$ is that value of acceptor doping of a p-type substrate which will cause exactly the same band profile in the n^+ epi as the semi-insulating substrate (see Figure 14). If y(0) is known, $N_A{}^*$ can be computed according to the asymmetric p-n junction formula

$$y(0) + \ln \frac{N_D}{N_C} = \frac{1}{(1 + N_D/N_A{}^*)} \left[\ln \frac{N_D N_A{}^*}{n_i^2} + \frac{N_D}{N_A{}^*} - 1 \right] \tag{16}$$

The SI-n^+ junction can then be treated formally as a p-n junction with p-side doping $N_A{}^*$. The voltage dependence of the n-side depletion layer W_n can then be shown to be [129]

$$W_n(V) = L \sqrt{\frac{1}{1 + x} \left\{ \frac{qV_D{}^{pn}}{kT} - (1 - x) \right\} + \frac{qV}{kT} \cdot \frac{1}{1 + x} \left(\frac{V_D{}^{pn}}{V_D{}^{SI}} \right) \left(\frac{V_D{}^{SI} + V}{V_D{}^{pn} + V} \right)} \tag{17}$$

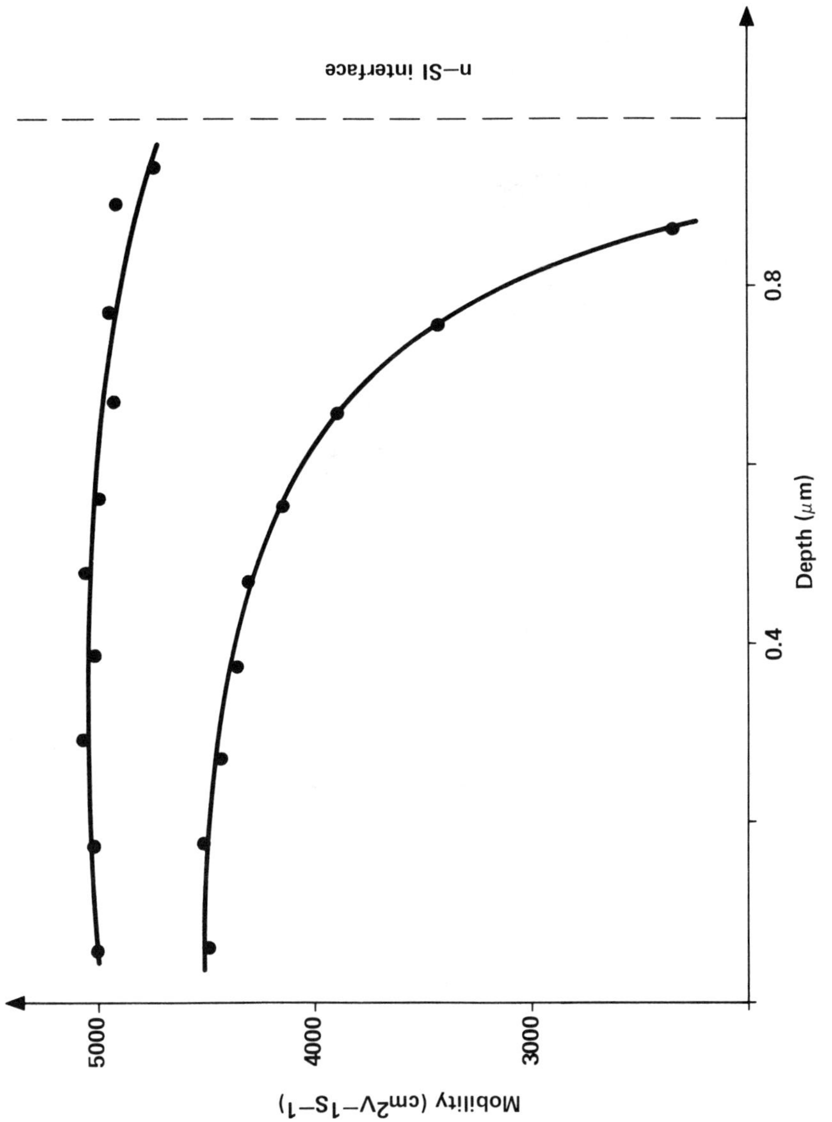

Figure 12/*Mobility profiles in buffered (top curve) and non-buffered (bottom curve) epitaxial layers. After Butlin et al. [131].*

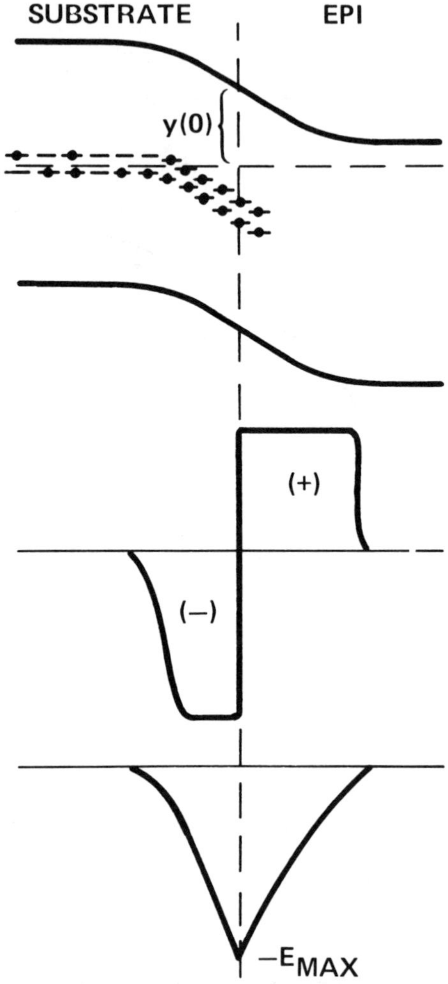

Figure 13/Schematic representation of the band, charge density, and built-in field profiles for a semi-insulating/n^+ junction. Charge dipole results from band bending in substrate, which alters the occupancy of the deep donor and acceptor levels.

where L is the Debye length, V is the total voltage drop across the SI $- n^+$ interface, $x = N_D/N_A^*$, and V_D^{SI} and V_D^{pn} are the diffusion potentials indicated in Figure 14. V_D^{pn} is given by

$$V_D^{pn} = \frac{kT}{q} \ln \left(\frac{N_D N_A^*}{n_i^2} \right) . \tag{18}$$

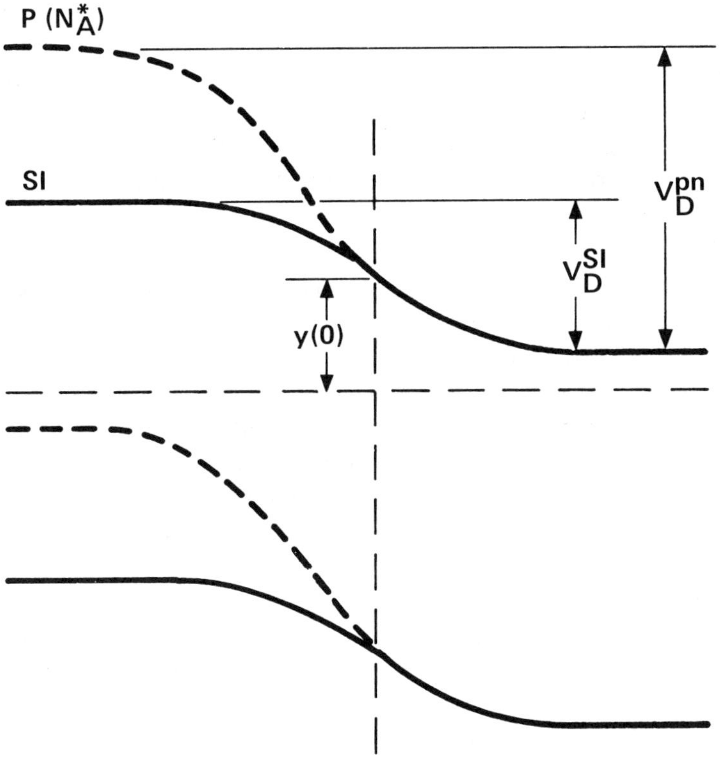

*Figure 14/Band diagrams for SI-n^+ and "equivalent" p-n^+ junctions. N_A * is "effective-p-doping" of substrate which results in same depletion of n^+ layer. V_D^{SI} and V_D^{pn} are the diffusion potentials of the two junctions.*

Figure 15 is an experimentally determined curve of W_n versus V for a particular substrate-epi combination. The solid curve is a plot of equation (17) using the values $V_D^{SI} = 0.7$ volt, $N_A^* = 2 \times 10^{16}$ cm^{-3}, and assuming that the entire bias is dropped across the interface. The good agreement indicates that the effective p-n junction model for back-gating is essentially correct. Values of V_D^{SI} and y(0) can be obtained by solving Poisson's equation for the SI-n^+ junction, using the four-level model to describe the substrate and taking account of the spatial variation of the ionization of the deep levels.

The back-gating phenomenon is a troublesome effect in FETs fabricated from epi grown directly on Cr-doped substrates. The effect can be caused also by a p-type converted layer beneath the epi, in which case the usual p-n junction formalism is applicable directly. Avoidance of back-gating requires substrate material which not only has a minimal concentration of deep impurities but which also is thermally stable. As indicated earlier, this combination is difficult to achieve.

4.3. Light Sensitivity

Light sensitivity of FET I-V characteristics is another troublesome phenomenon which is related to deep levels at the substrate-epi interface. Crossley et al. [109] showed that FETs fabricated on an n-layer grown directly on a semi-insulating substrate have light-sensitive characteristics, whereas no light effect was found for FETs fabricated with a buffer layer interposed between substrate and active layer. The occurrence of strong light sensitivity was correlated with a high concentration of Cr acceptors at the interface. [131] It has also been observed that a substrate-epi combination which shows a large back-gating effect in the dark does not back-gate under illumination. [134] Evidently the optical excitation empties the Cr centers, thereby eliminating space charge dipole at the interface. As expected, the light sensitivity of undoped or lightly doped semi-insulating material is significantly less than for highly Cr-doped material.

SUMMARY

In this chapter we have attempted to present an overview of the development of semi-insulating GaAs technology over the past decade. The many investigations carried out during this time have led to a reasonably clear understanding of the properties of this material and

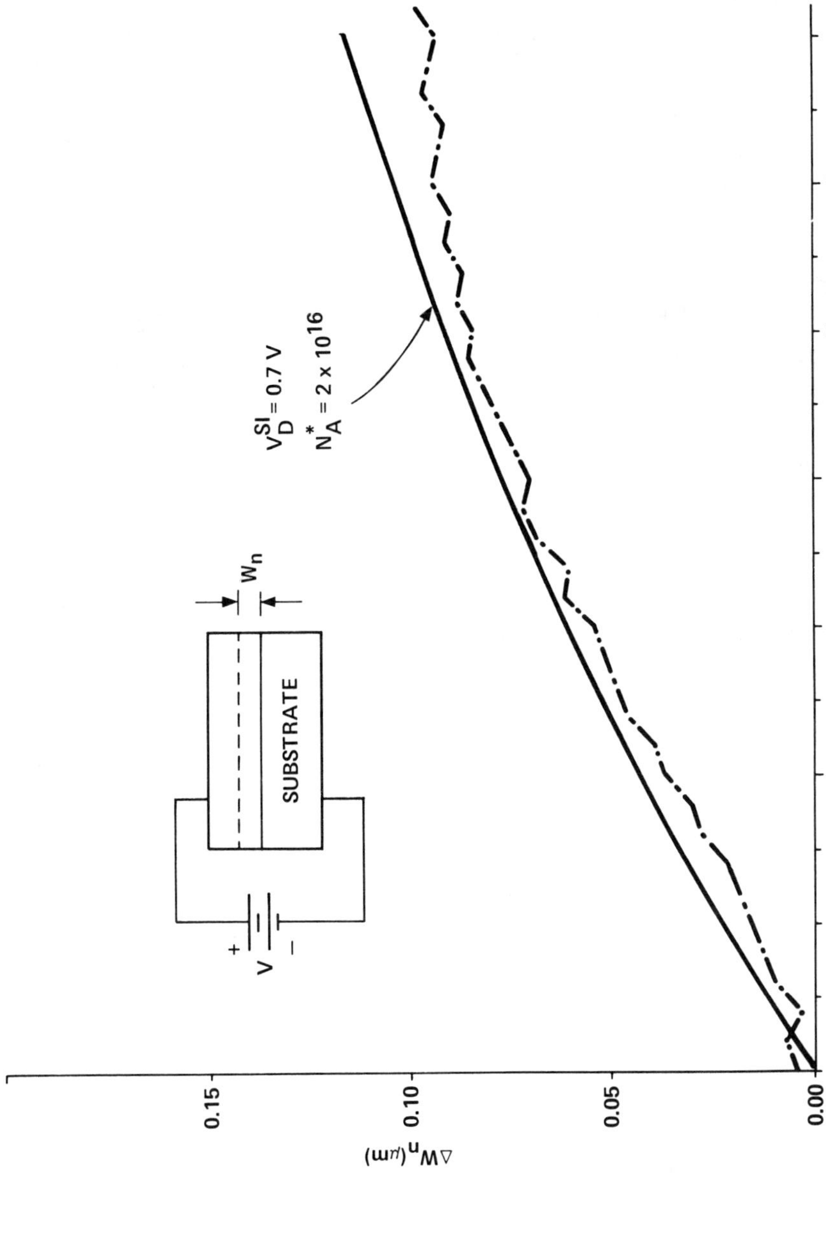

Figure 15/*Comparison of experimental back-gating curve with calculation using equation (17) in text.*

of the behavior of particular deep impurities. Of necessity, only a portion of the extensive literature on semi-insulating GaAs has been cited here.

As the properties of Cr-doped semi-insulating material have become better understood, it has become increasingly clear that the Cr levels near the interface create a number of undesirable effects which degrade the properties of microwave devices. Fortunately, advances in crystal growth technology, in starting material purity and in analytical methods have made possible the growth of high purity semi-insulating GaAs without Cr. This material, used as a substrate for either epi growth or direct implantation of an active layer, results in devices with performance comparable to the best buffer-layer devices. Future effort in semi-insulating GaAs will be focused on developing the capability of producing this material on a routine basis.

REFERENCES

[1] Allen, J.W., *Nature,* 187:403, 1960.
[2] Bube, R.H., *J. Appl. Phys.,* 31:315, 1960.
[3] Gooch. C.H., C. Hilsum, and B.R. Holeman, *J. Appl. Phys.,* 32:2069, 1961.
[4] Haisty, R.W., E.W. Mehal, and R. Stratton, *J. Phys. Chem. Solids,* 23:829, 1962.
[5] Kul'sreshta, A.P., and A.E. Yunovich, *Sov. Phys. -Solid State,* 7:2058, 1966.
[6] Woods, J.F., and N.G. Ainslie, *J. Appl. Phys.,* 34:1469, 1963.
[7] Woodall, J.M., *Electrochem. Technol.,* 2:167, 1964.
[8] Cronin, G.R., and R.W. Haisty, *J. Electrochem. Soc.,* 111: 874, 1964.
[9] Slaymaker, N.A., and J.A. Turner, Proc. 1st European Microwave Conf., A.5.1, 1973.
[10] Nozaki, T., M. Ogawa, H. Terao, and H. Watanabe in *Gallium Arsenide and Related Compounds (Deauville) 1974* (Inst. Phys. Conf. Ser. no. 24) p. 46. London: Adam Helger/The Institute of Physics, 1975.
[11] Ross, S.F., and M. Jaros, *Phys. Lett.,* 45A:355, 1973.
[12] Blanc, J., and L.R. Weisberg, *Nature,* 192:155, 1961.
[13] Turner, W.J., G.D. Pettit, and N.G. Ainslie, *J. Appl. Phys.,* 34:3274, 1963.
[14] Stocker, H.J., *Solid State Comm.,* 16:525, 1975.
[15] Lin, A.L., E. Omelianovski, and R.H. Bube, *J. Appl. Phys.,* 47:1852, 1976.
[16] Stocker, H.J., *J. Appl. Phys.,* 48:4583, 1977.
[17] Vasudev, P.K., and R.H. Bube, *Solid-State Electron.,* 21: 1095, 1978.
[18] Andrianov, D.G., E.M. Omel'yanovskii, E.P. Rashevskaya, and N.I. Suchkova, *Sov. Phys. Semicond.,* 10:637, 1976.
[19] Mircea, A., A. Mitonneau, L. Hollan, and A. Briere, *Appl. Phys.,* 11:153, 1976.
[20] Huber, A.M., et al., *J. Appl. Phys.,* 50:4022, 1979.
[21] Haisty, R.W. and G.R. Cronin, Proc. 7th Int'l. Conf. on Physics of Semiconductors, M-4, Paris, 1964.
[22] Allen, G.A., *Brit. J. Appl. Phys.,* 1:593, 1968.
[23] Balagurov, L.A., E.M. Omel'yanovskii, and L. Ya. Pervova, *Sov. Phys. Semicond.,* 8:1051, 1975.

[24] Panish, M.B. and H.C. Casey, Jr., *J. Appl. Phys.*, 40:163, 1969.
[25] Look, D.C., *J. Phys. Chem. Solids*, 36:1311, 1975.
[26] Betko, J., and K. Merinsky, *J. Appl. Phys.*, 50:4212, 1979.
[27] Ippolitova, G.K., E.M. Omel'yanovskii, and L. Ya. Pervova, *Sov. Phys. Semicond.*, 9:864, 1976.
[28] Lin, A.L., and R.H. Bube, *J. Appl. Phys.*, 47:1859, 1976.
[29] Ganapol'skii, E.M., *Sov. Phys. Solid State*, 16:1868, 1975.
[30] Vallin, J.T., and G.D. Watkins, *Phys. Rev.*, B9:2051, 1974.
[31] Kaufmann, U., and J. Schneider, *Solid State Comm.*, 20:143, 1976.
[32] Peka, G.P., and Yu. I. Karkhanin, *Sov. Phys. Semicond.*, 6:261, 1972.
[33] Koschel, W.H., S.G. Bishop, and B.D. McCombe, *Solid State Comm.*, 19:521, 1976.
[34] Stocker, H.J., and M. Schmidt, *J. Appl. Phys.*, 47:2450, 1976.
[35] Plesiewicz, W., *J. Phys. Chem. Solids*, 38:1079, 1977.
[36] Stauss, G.H., and J.J. Krebs in *Gallium Arsenide and Related Compounds (Edinburgh) 1976* (Inst. Phys. Conf. Ser. no. 33a) p. 84. London: Adam Helger/The Institute of Physics, 1977.
[37] White, A.M., P. Porteous, and P.J. Dean, *J. Electron. Mater*, 5:91, 1976.
[38] White, A.M., P.J. Dean, and P. Porteous, *J. Appl. Phys.*, 47: 3230, 1976.
[39] Inoue, T., and M. Ohyama, *Solid State Comm.*, 8:1309, 1970.
[40] Zucca, R., *J. Appl. Phys.*, 48:1987, 1977.
[41] Harper, J.G., H.E. Matthews, and R.H. Bube, *J. Appl. Phys.*, 41:3182, 1970.
[42] Lang, D.V., and R.A. Logan, *J. Electron. Mater.*, 4:1053, 1975.
[43] Milnes, A.G., *Deep Impurities in Semiconductors.* New York: John Wiley & Sons, 1973.
[44] Masterov, V.F., and B.E. Samorukov, *Sov. Phys. Semicond.*, 12:363, 1978.
[45] Brown, W.J., Jr., and J.S. Blakemore, *J. Appl. Phys.*, 43: 2242, 1972.

[46] Suchkova, N.I., D.G. Andrianov, E.M. Omel'yanovskii, E.P. Rashevskaya, and N.N. Solov'ev, *Sov. Phys. Semicond.*, 9:469, 1975.
[47] Lukicheva, N.I., O.V. Pelevin, and L. Ya. Pervova, *Sov. Phys. Semicond.*, 5:169, 1971.
[48] Kolchanova, N.M., D.N. Nasledov, and G.N. Talalakin, *Sov. Phys. Semicond.*, 4:106, 1970.
[49] Omel'yanovskii, E.M., L. Ya. Pervova, E.P. Rashevskaya, N.N. Solov'ev, and V.I. Fistul', *Sov. Phys. Semicond.*, 4:316, 1970.
[50] Nakagawa, H., and S. Zukotynski, *J. Appl. Phys.*, 46:4809, 1975.
[51] Vasil'ev, A.V., G.K. Ippolitova, E.M. Omel'yanovskii, and A.I. Ryskin, *Sov. Phys. Semicond.*, 10:341, 1976.
[52] Bekmuratov, M.F., and V.I. Murygin, *Sov. Phys. Semicond.*, 7:55, 1973.
[53] Partin, D.L., J.W. Chen, A.G. Milnes, and L.F. Vassamillet, *Solid State Electron.*, 22:455, 1979.
[54] Swiggard, E.M., S.H. Lee, and F.W. von Batchelder in *Gallium Arsenide and Related Compounds (St. Louis) 1978* (Inst. Phys. Conf. Ser. no. 45) p. 125. London: Adam Helger/ The Institute of Physics, 1979.
[55] Lindquist, P.F., *J. Appl. Phys.*, 48:1262, 1977.
[56] Blakemore, J.S., *Semiconductor Statistics.* New York: Pergamon Press, 1962.
[57] Logan, R.M., and D.T.J. Hurle, *J. Phys. Chem. Solids,* 32: 1739, 1971.
[58] Hurle, D.T.J., Proc. 6th Int'l. Symp. GaAs and Related Compounds, Edinburgh, 1976 (*Inst. Phys. Conf. Ser.*, 33A:113, 1977).
[59] Raymond, A., J.L. Robert, and B. Pistoulet in *Gallium Arsenide and Related Compounds (Edinburgh) 1976* (Inst. Phys. Conf. Ser. no. 33a) p. 105. London: Adam Helger/ The Institute of Physics, 1977.
[60] Robert, J.L., B. Pistoulet, A. Raymond, J.M. Dusseau, and G.M. Martin, *J. Appl. Phys.*, 50:349, 1979.
[61] Ashby, A., G.G. Roberts, D.J. Ashen, and J.B. Mullin, *Solid State Comm.*, 20:61, 1976.

[62] Mullin, J.B., D.J. Ashen, G.G. Roberts, and A. Ashby in *Gallium Arsenide and Related Compounds (Edinburgh) 1976* (Inst. Phys. Conf. Ser. no. 33a) p. 91. London: Adam Helger/The Institute of Physics, 1977.
[63] Roberts, G.G., and F.W. Schmidlin, *Phys. Rev.,* 180:785, 1969.
[64] Pfann, W.G., *Zone Melting.* New York: John Wiley & Sons, 1958.
[65] Richards, J.L., in *Compound Semiconductors,* Vol. 1, Preparation of III-V Compounds, ed. R.K. Willardson and H.L. Goering, p. 279. New York: Reinhold, 1962.
[66] Allred, W.P., Crystal Specialties, Inc., private communication.
[67] Metz, E.P.A., R.C. Miller, and R. Mazelsky, *J. Appl. Phys.,* 33:2016, 1962.
[68] Mullin, J.B., B.W. Straughan, and W.S. Brickell, *J. Phys. Chem. Solids,* 26:782, 1965.
[69] Bass, S.J., and P.E. Oliver in *Gallium Arsenide and Related Compounds (Reading) 1966* (Inst. Phys. Conf. Ser. no. 3) p. 41. London: Adam Helger/The Institute of Physics, 1967.
[70] AuCoin, T.R., R.L. Ross, M.J. Wade, and R.O. Savage, *Solid State Technol.,* 22:59, 1979.
[71] Pekarek, L., Czech. *J. Phys.,* 20:857, 1970.
[72] RF heating has also been used in LEC crystal growth of GaAs, although problems are encountered with arsenic deposition on the coils.
[73] Ford, W.M., AACG West 4th Conf. on Crystal Growth, Fallen Leaf Lake, California, 1979.
[74] Gremmelmaier, R., in *Compound Semiconductors,* Vol. 1, Preparation of III-V Compounds, ed. R.K. Willardson and H.L. Goering, p. 254. New York: Reinhold, 1962.
[75] Liebmann, W.K., and G. Kampschulte, *Solid-State Electron.,* 9:828, 1966.
[76] Lum, W.Y. H.H. Wieder, W.H. Koschel, S.G. Bishop, and B.D. McCombe, *Appl. Phys. Lett.,* 30:1, 1977.
[77] Lum, W.Y., and H.H. Wieder, *Appl. Phys. Lett.,* 31:213, 1977.
[78] Koschel, W.H., S.G. Bishop, B.D. McCombe, W.Y. Lum, and H.H. Wieder in *Gallium Arsenide and Related Compounds (Edinburgh) 1976* (Inst. Phys. Conf. Ser. no. 33a) p. 98. London: Adam Helger/The Institute of Physics, 1977.

[79] DeFevere, D., Hewlett-Packard Co., Microwave Semiconductor Division, private communication.
[80] Hallais, J., A. Mircea-Roussel, J.P. Farges, and G. Poiblaud in *Gallium Arsenide and Related Compounds (St. Louis) 1976* (Inst. Phys. Conf. Ser. no. 33b) p. 220. London: Adam Helger/The Institute of Physics, 1977.
[81] Zucca, R., in *Gallium Arsenide and Related Compounds (St. Louis) 1976* (Inst. Phys. Conf. Ser. no. 33b) p. 228. London: Adam Helger/The Institute of Physics, 1977.
[82] Barrera, J., Proc. 5th Biennial Electrical Eng. Conf., Cornell University, Ithaca, N.Y., 1975.
[83] Barrera, J., Hewlett-Packard Co., Microwave Semiconductor Division, private communication.
[84] Stolte, C., Hewlett-Packard Co., Corporate Solid State Laboratory, private communication.
[85] Huber, A.M., et al., *Appl. Phys. Lett.*, 34:858, 1979.
[86] Blood, P., *Acta Electron.*, 15:33, 1972.
[87] Colman, D., *Rev. Sci. Inst.*, 39:1946, 1968.
[88] Hemenger, P.M., *Rev. Sci. Inst.*, 44:698, 1973.
[89] Philadelpheus, A. Th., and P.C. Euthymiou, *J. Appl. Phys.*, 45:955, 1974.
[90] Look, D.C., *J. Electron. Mater.*, 7:147, 1978.
[91] Omel'yanovskii, E.M., A.N. Pantyukhov, L. Ya. Pervova, V.I. Fistul', and Yu. A. Vasil'ev, *Sov. Phys. Semicond.*, 9:1267, 1976.
[92] Erchak, D.P., V.F. Stel'makh, and V.D. Tkachev, *Sov. Phys. Semicond.*, 10:922, 1976.
[93] Domanevskii, D.S., and V.D. Tkachev, *Sov. Phys. Semicond.*, 4:1790, 1971.
[94] Gorelenok, A.T., B.V. Tsarenkov, and N.G. Chiabrishvili, *Sov. Phys. Semicond.*, 5:95, 1971.
[95] Kocot, K., and G.L. Pearson, *Solid State Comm.*, 25:113, 1978.
[96] Bois, D., and D. Beaudet, *J. Appl. Phys.*, 46:3882, 1975.
[97] Chatterjee, P.K., K.V. Vaidyanathan, M.S. Durschlag, and B.G. Streetman, *Solid State Comm.*, 17:1421, 1975.
[98] Nam, S.B., D.W. Langer, D.L. Kingston, and M.J. Luciano, *Appl. Phys. Lett.*, 31:652, 1977.
[99] Yu, P.W., *Solid State Comm.*, *27,* 1421, 1978.

[100] Guislain, H.J., L. DeWolf, and P. Clauws, *J. Electron. Mater.*, 6:541, 1977.
[101] Jones, C.E., Jr., and A.R. Hilton, *J. Electrochem. Soc.*, 113: 504, 1966.
[102] Martin, G.M., M.L. Verheijke, J.A.J. Jansen, and G. Poiblaud, *J. Appl. Phys.*, 50:467, 1979.
[103] Lang, D.V., *J. Appl. Phys.*, 45:3023, 1974.
[104] Henry, C.H., *J. Electron. Mater.*, 4:1037, 1975.
[105] Mitonneau, A., *Philips Res. Repts.*, 31:244, 1976.
[106] Martin, G.M., A. Mitonneau, and A. Mircea, *Electron. Lett.*, 13:191, 1977.
[107] Mitonneau, A., G.M. Martin, and A. Mircea in *Gallium Arsenide and Related Compounds (Edinburgh) 1976* (Inst. Phys. Conf. Ser. no. 33a) p. 73. London: Adam Helger/The Institute of Physics, 1977.
[108] Brunwin, R., B. Hamilton, P. Jordan, and A.R. Peaker, *Electron. Lett.*, 15:349, 1979.
[109] Crossley, I., I.H. Goodridge, M.J. Cardwell, and R.S. Butlin in *Gallium Arsenide and Related Compounds (St. Louis) 1976* (Inst. Phys. Conf. Ser. no. 33b) p. 289. London: Adam Helger/The Institute of Physics, 1977.
[110] Martin, G.M., and D. Bois, *J. Electrochem. Soc.* (to be published).
[111] M.S. Wang, *Appl. Spectroscopy,* 22:761, 1968.
[112] M.S. Wang, *Appl. Spectroscopy,* 24:60, 1970.
[113] Brown, R., R.D. Craig, and J.D. Waldron, in *Compound Semiconductors,* Vol. 1, Preparation of III-V Compounds, ed. R.K. Willardson and H.L. Goering, p. 106. New York: Reinhold, 1962.
[114] Rachmann, J., and R. Biermann, *Solid State Comm.*, 7:1771, 1969.
[115] Wolfe, C.M., G.E. Stillman, and E.B. Owens, *J. Electrochem. Soc.*, 117:129, 1970.
[116] Look, D.C., *J. Appl. Phys.*, 48:5141, 1977.
[117] Blackmore, G.W., J.B. Clegg, J.S. Hislop, and J.B. Mullin, *J. Electron. Mater.*, 5:401, 1976.
[118] Clegg, J.B., and E.J. Millet, *Philips Tech. Rev.*, 34:344, 1974.
[119] Baker, C.A., and D.R. Williams, *Talanta,* 15:1143, 1968.

[120] Stoneham, E.B., Hewlett-Packard Co., Santa Rosa Division, unpublished.
[121] Wang, P., Hewlett-Packard Co., Santa Rosa Division, private communication.
[122] Hoyt, P.L., and R.W. Haisty, *J. Electrochem. Soc.,* 113:296, 1966.
[123] Shibatomi, A., N. Yokoyama, H. Ishikawa, K. Dazai, and O. Ryuzan, *J. Crystal Growth,* 31:240, 1975.
[124] Otsubo, M., and H. Miki, *J. Electrochem. Soc.,* 124:441, 1977.
[125] Cox, H.M., and J.V. DiLorenzo in *Gallium Arsenide and Related Compounds (St. Louis) 1976* (Inst. Phys. Conf. Ser. no. 33b) p. 11. London: Adam Helger/The Institute of Physics, 1977.
[126] Tuck, B., G.A. Adegboyega, P.R. Jay, and M.J. Cardwell in *Gallium Arsenide and Related Compounds (St. Louis) 1978* (Inst. Phys. Conf. Ser. no. 45) p. 114. London: Adam Helger/ The Institute of Physics, 1979.
[127] Casey, H.C., Jr., in *Atomic Diffusion in Semiconductors,* ed. D. Shaw, p. 417. London: Plenum Press, 1973.
[128] Lehovec, K. and R. Zuleeg in *Gallium Arsenide and Related Compounds (Deauville) 1974* (Inst. Phys. Conf. Ser. no. 24) p. 292. London: Adam Helger/The Institute of Physics, 1975.
[129] Lindquist, P.F., Hewlett-Packard Co., Optoelectronics Division, unpublished.
[130] van der Pauw, L.J., *Philips Res. Repts.,* 13:1, 1958.
[131] Butlin, R.S., D. Parker, I. Crossley, and J. Turner in *Gallium Arsenide and Related Compounds (Edinburgh) 1976* (Inst. Phys. Conf. Ser. no. 33a) p. 237. London: Adam Helger/ The Institute of Physics, 1977.
[132] Tansley, T.L., *J. Phys. E: Sci. Inst.,* 8:52, 1975.
[133] Pucel, R.A., and C.F. Krumm, *Electron. Lett.,* 12:240, 1976.
[134] DeFevere, D., Hewlett-Packard Co., Microwave Semiconductor Division, private communication.

Chapter 2

Materials Technology for GaAs FETs

Introduction

In the previous chapter, the importance of substrates to a good materials technology for GaAs FETs was discussed. Perhaps even more important is the technology by which the active channel is produced. In this chapter molecular beam epitaxy (MBE), vapor phase epitaxy (VPE) and ion implantation (II) will be discussed as three materials approaches to the generation of active FET channels.

In the first section L. Hollan and J.P. Hallais discuss VPE for GaAs FETs. Their contribution includes a definition of the required material characteristics, the importance of the buffer layer, and a brief review of the main process of epitaxial growth for FET devices. The main body of their text discusses the growth of FET epi layers by the $AsCl_3$ transport method, the production method most commonly employed today. Hollan details the $AsCl_3$ process, covering the important growth parameters, the required equipment, the growth of the "buffer layer" – an essential step in obtaining high quality active films – and the growth of the active layer.

MBE is described by C.E.C. Wood. He focuses on the advantages of MBE over other growth techniques, presenting a comprehensive comparison of the major epi technologies and MBE. He reviews the impact MBE can make on device structures by its unique ability to control both the doping profile and thickness along the direction of growth, as well as the control of spatial uniformity of thickness and doping over large areas (2-3 inches).

An achievement unique to MBE is reviewed which involves the deposition of epitaxial Ge on GaAs for the formation on a nonalloyed – heterojunction ohmic contact.

Electron confinement by using heterojunction GaAlAs buffer layers is discussed. MBE allows selected area epitaxy which may lead to a whole new family of discrete and integrated devices with epitaxial areas selectively defined by suitable MBE techniques.

An alternative to epitaxial technology is the use of ion implantation. The motivation for the use of ion implantation in GaAs is the same as that which produced wide acceptance in silicon technology; namely, the ability to control the thickness and doping level of implanted layers by varying the dose and energy of the implanted ions.

As part of this chapter, F. Eisen discusses the state of ion implantation as it has been applied to the fabrication of GaAs FETs in various laboratories. He discusses substrate effects and effects during implantation and annealing which include the formation of surface conductivity layers and conversion of the substrates. Encapsulation and annealing of implanted GaAs are discussed in detail. Eisen states that silicon nitride of low oxygen content is the preferred encapsulant but discusses in addition the so-called "capless annealing" techniques.

The characteristics of the n-type dopants, silicon, sulfur, and selenium are reviewed and Eisen concludes that selenium and silicon are the preferred dopants for FET active layers. The applications of ion implantation technology to FETs are reviewed and ion implantation is stated to have a number of advantages such as control of doping profiles and the generation of selectively doped regions. In particular, Eisen's perception is that the uniformity and reproducibility which can be achieved by ion implantation are important advantages for the fabrication of GaAs integrated circuits.

Vapor Phase Epitaxy for GaAs FETs

L. Hollan
J. Hallais
Laboratoires d'Electronique et de Physique Appliquee,
Limeil Brevannes, France.

Currently, much attention is being devoted to the fabrication of GaAs low noise and power FET's. Low cost and high yield processes are needed for the preparation of the adequate GaAs material.

First, an analysis of the required material characteristics is made. Several growth techniques, all based on vapor phase epitaxy (VPE) have been investigated and are reviewed. This review mainly concentrates on the most used method, namely, the arsenic trichloride VPE. In this context, we discuss successively the different growth methods, the growth parameters, and the equipment. As the epitaxial material for FETs is composed of at least two layers, the buffer and the active layer, specific problems which arise during the growth of each are discussed separately.

1. DEFINITION OF THE REQUIRED MATERIAL CHARACTERISTICS

A field effect transistor of the Schottky barrier type (GaAs FET), shown schematically in Figure 1, consists of two ohmic contacts (the source and the drain) on either side of a Schottky barrier (the gate) realized on a thin, conductive active layer which has been previously formed on a semi-insulating substrate. The principle of the device implies complete depletion of the active layer by applying reverse bias to the gate (case number 2 in Figure 1). In terms of material, this means a coherence between the doping n of the layer and its thickness. Figure 2 gives the characteristic doping/thickness features for various values of the pinch off voltage; thus, for a - 2V bias (in the case of the low noise FET) the thickness of the active layer shall be 0.17 μm for a doping of 1.5 x 10^{17} cm^{-3} or 0.13 μm for a doping of 2.5 x 10^{17} cm^{-3}.

The gain of a transistor directly increases with the μn product, μ being the mobility of the majority carriers. Since the μn product is itself increasing with n, this implies to obtain an n doping as high as possible and, for any given n, the maximum of mobility. (For example, for n = 1 x 10^{17} cm^{-3}, μ = 4,600 cm^2/V.s. For other values, see Figure 3.) However, on account of the difficulty which exists, on the one hand, in mastering a very small active layer thickness, and on the other hand, in realizing a good barrier on a highly doped material, one is led to seek a compromise. This is illustrated in Figure 2, where the domains which may be utilized for the low noise or power FET are defined [1].

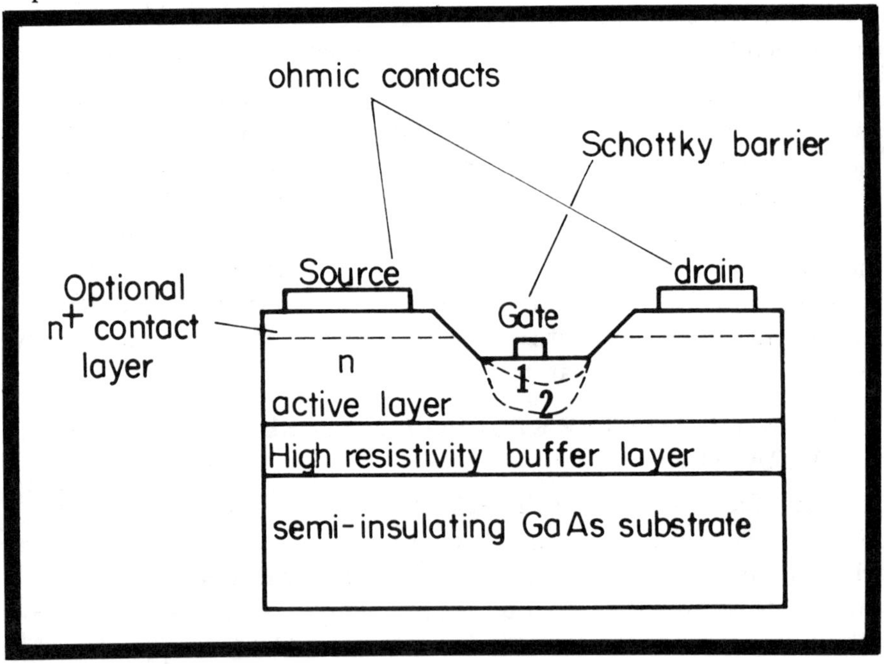

Figure 1/Schematic representation of a GaAs FET device with the recessed gate. The dotted line shows the limits of the depleted layer for gate bias value of 0 (1) and pinch off voltage (2), respectively.

The other essential characteristic of a transistor is its microwave noise. Such noise has several origins, one of which is linked to the material, and more especially to the interface of the active layer and the substrate. The quality of the interface is reflected in the mobility of the carriers; it is an accepted fact today that an increasing or at least a constant mobility near the interface is necessary for achieving

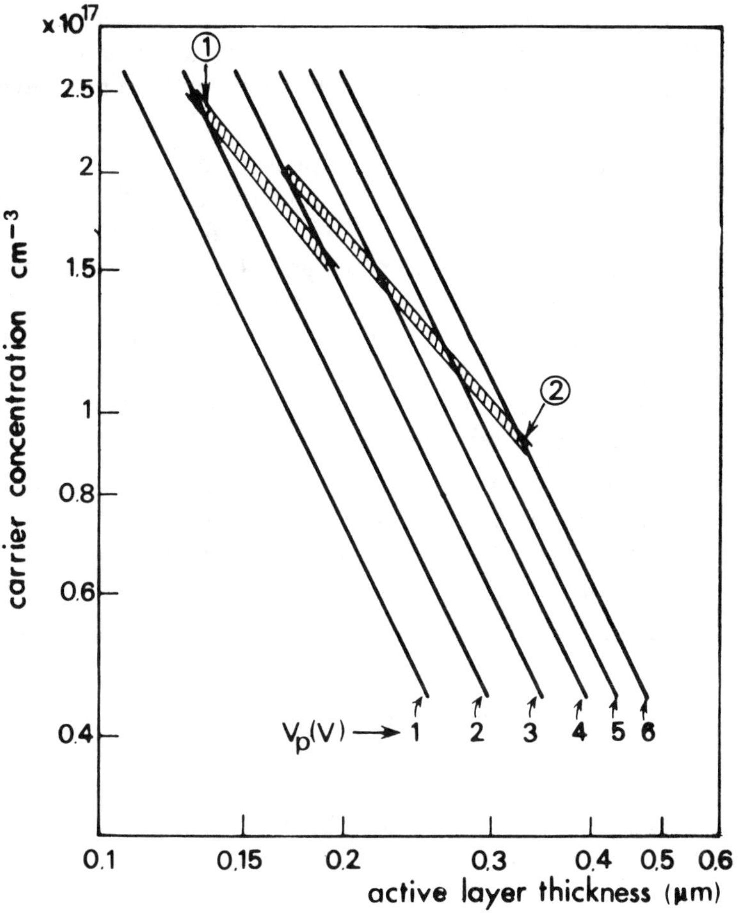

Figure 2/Relation between the thickness and the doping of an FET active layer under the gate. The curves are given for different pinch-off voltages (V_p). Striped areas show the respective domains of low noise (1) and power FETs (2).

a high performance transistor. It will be seen further along that the addition of an intermediate layer, the so-called buffer layer, constituting a screen of a few microns between the substrate and the noise generating interface, is necessary for the lowering of the noise factor. Such a buffer layer must have a high resistivity in order to avoid the leakage currents of the transistor.

The power FET requires the same quality of material as the low noise FET; only the range of the pinch off voltage is different (Figure 2). Uniformity in thickness and doping of the active layer is critical, especially for power devices, as the gate width can be several millimeters.

Using the recessed gate technology, it is possible either to grow a n^+ contact layer, about 0.5 μm thick, on top of the active layer or to increase the thickness of the n layer. The adjustment to the final thickness of the drain-source channel (Figure 1) is then performed on the as-grown material by means of a chemical etching (or plasma etching) controlled by a current meter. This geometry of the device improves the quality of the ohmic contact, decreases the surface contribution to the noise and also allows a better current distribution in the drain region of the power devices.

Lastly, the technology of the device, in particular the lithography used for gate and contacts definition, requires morphologically perfect sample surfaces, since any defect degrades the definition of the engraving or the mask itself.

To sum up, the field effect transistor requires a material with an active layer of the following characteristics:

- A coherence of the doping/thickness parameters within the limits which have been previously defined,
- A minimum dispersion of these two parameters,
- The highest possible mobility for the considered doping level and a constant or increasing mobility towards the interface;
- A morphologically perfect surface.

On the other hand, it requires a buffer layer having a high resistivity and, of course, a semi-insulating GaAs substrate.

Recent progress in material growth technology has made possible the fabrication of new FET structures which allow the experimental verification of some theoretical concepts. This deals with:

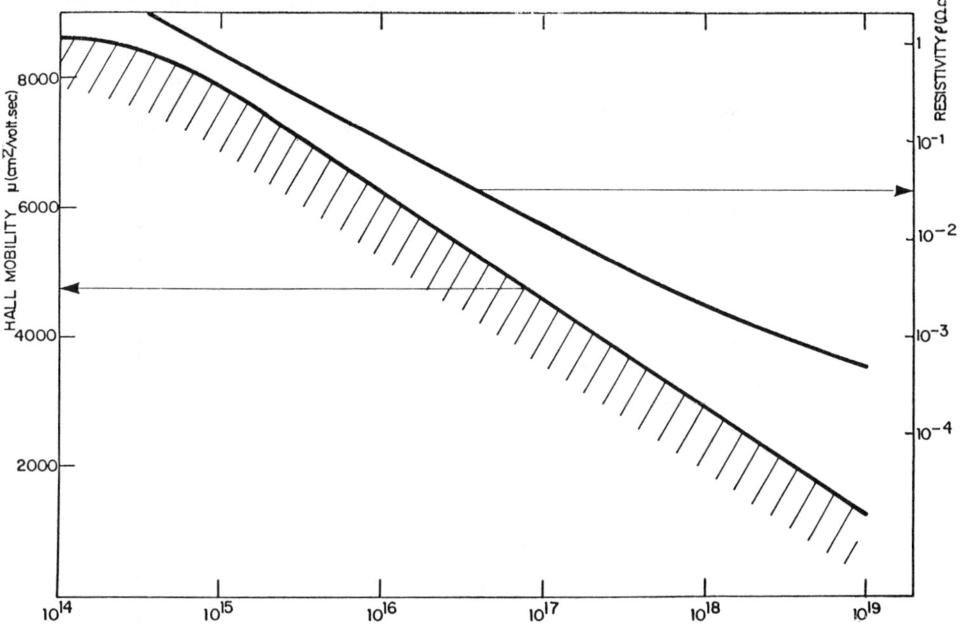

(3a) FREE CARRIER CONCENTRATION n (cm⁻³)

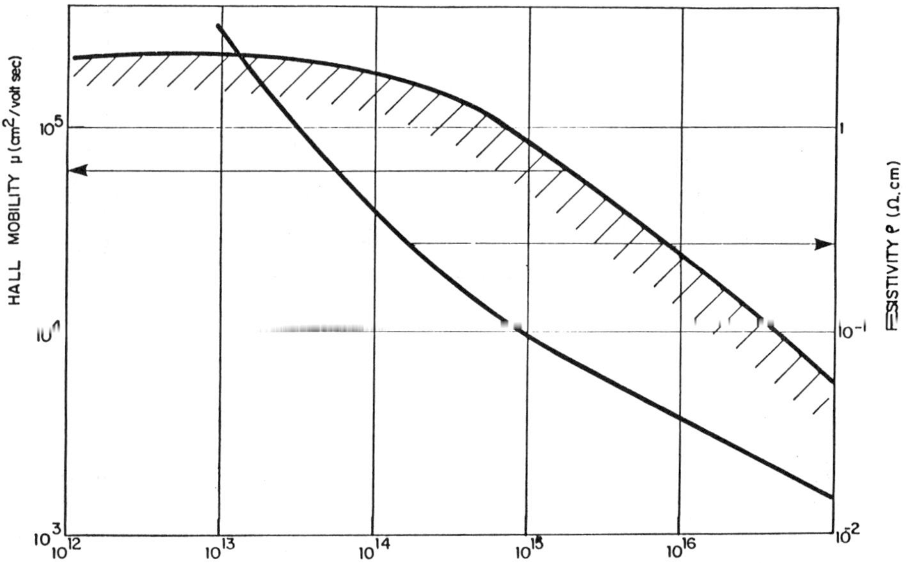

(3b) FREE CARRIER CONCENTRATION n (cm⁻³)

***Figure 3**/Compilation of experimental Hall mobilities measured at room temperature (3a) or liquid nitrogen temperature (3b) as a function of carrier concentration in n-type GaAs. The hatched region shows the data scatter. The resistivity values were calculated for the maximum mobility (solid line). Data were taken from more than 30 papers.*

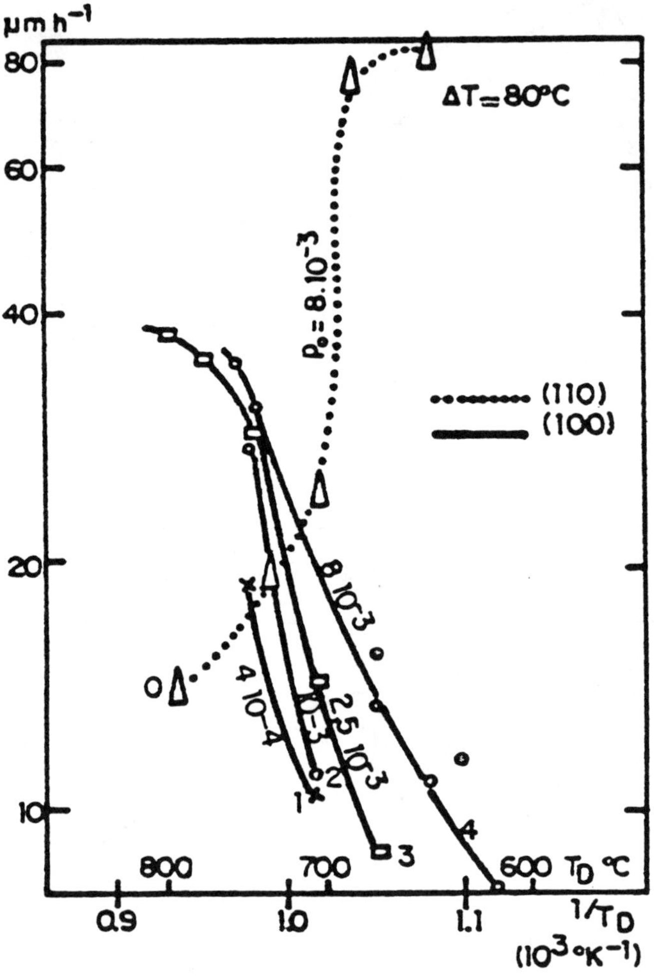

Figure 4/Effect of substrate temperature on GaAs growth rates for various $AsCl_3$ partial pressures (p_o) and for two crystallographic orientations of the substrate. T_D is the deposition temperature, ΔT is defined by the difference between T_S (source temperature) and T_D. The very different behaviour observed for (110) indicates a different growth kinetics (after Ref. 31).

i) Physical effects related to GaAs-(Al,Ga)As heterojunctions;
ii) Ballistic motion of the electrons in GaAs;
iii) Gain in majority carrier mobility obtained from (In,Ga)As active layers.

It has been shown [2] that growing a high resistivity (Al,Ga)As buffer layer, the injection of channel electrons into the substrate was significantly reduced. This may be of interest for power FET's where high currents and voltages are required. Recent work [3] on superlattice-like GaAs-(Al,Ga)As material demonstrates an increase of the electron mobility in the GaAs layer. Devices, so called HEMT (High Electron Mobility Transistors) [4] already fabricated with these materials have interesting characteristics at 77°K. It is also worth mentioning the potentialities of heterojunction FETs (H.J. FETs) [5] for normally-off devices and MISFET where the use of insulating (Al, Ga)As has been reported [6]. Furthermore, ballistic motion of the electrons in GaAs which would allow the fabrication of faster devices is currently studied [7]. All these new types of devices require sophisticated epitaxial structures involving very thin layers, sharp heterostructures and doping transitions.

The above mentioned structures are all at the exploratory stage and only GaAs low noise and power FETs are in development or production stages in many companies.

2. MAIN PROCESSES OF EPITAXIAL GROWTH

Several epitaxial processes are reviewed within the scope of the growth of material for FETs. More exhaustive descriptions of the growth methods will be found in the literature [8].

2.1 Liquid Phase Epitaxy (LPE)

The requirement of a high purity buffer layer might justify the choice of LPE. However, the necessity of a perfect control of the active layer thickness and of a perfect mirror-like surface morphology are two further requirements difficult to meet by this growth method. Nevertheless, some laboratories which have mastered this technique have succeeded in growing good quality material for high performance devices [9].

2.2 Molecular Beam Epitaxy (MBE)

In this same volume, a separate chapter is devoted to MBE growth. In principle, this technique has quite a number of advantages; in particular, the growth rate can be low, which allows a good monitoring of very thin layers, and also makes possible the suitable tailoring of the doping profiles. Furthermore, the possibility of using the low deposition temperatures (600°C or less) and of reaching very high doping levels (5×10^{19}) should also offer advantages.

Some devices equalling the best performances have already been realized by several laboratories [10, 11]. However, MBE has not yet entered the FET material production area, which is still dominated by "arsenic-trichloride VPE". On the other hand, it should be emphasized that probably MBE will find its widest scope in the field of new epitaxial structures based on heterojunctions.

2.3 Metallorganic Vapour Phase Epitaxy (MO-VPE)

In the case of GaAs deposition, this technique consists in the co-pyrolysis of a gallium metallorganic compound and arsine, following an overall reaction such as:

$$(CH_3)_3Ga + AsH_3 \xrightarrow{heat} GaAs + 3CH_4 \qquad (1)$$

Equation (1) is written for trimethyl-gallium, but other gallium compounds can be used as well. There is also a possibility of growing Al or In containing compounds using mixtures of metallorganic compounds and obtaining phosphides by flowing phosphine (PH_3) instead of arsine into the deposition chamber. Thus, the process becomes rather versatile and capable of producing complex heterostructure.

Although intensive study of the technique is still recent, and in spite of the difficulties met in reducing significantly the background doping level, the results obtained in GaAs FET materials are already encouraging [12]. The possibility of growing GaAs-(Al,Ga)As heterostructure for new FET materials is still at the exploratory stage [2] and it is very likely that the process will be developed in this new field.

3. VAPOR PHASE EPITAXY BY THE TRICHLORIDE TRANSPORT

3.1 The Methods

The principle and general conditions of vapor phase growth by means of the $AsCl_3/Ga/H_2$ method have already formed the subject of detailed descriptions [8, 13]. The volatile transported species are obtained by the reaction of $AsCl_3$ on Ga in the presence of H_2 carrier gas producing GaCl, As_4, HCl and, in smaller quantities, $GaCl_3$ and As_2. The main chemical equilibrium is then given by [14]:

$$4GaCl + As_4 + 2H_2 \underset{T_2}{\overset{T_1}{\rightleftharpoons}} 4GaAs + 4HCl \qquad T_2 > T_1 \tag{2}$$

This transport system has probably been chosen due to the availability of electronic grade $AsCl_3$, which is a liquid with vapour pressure behaviour like water, enabling easy control of the partial pressures in the 10^{-2}–10^{-3} mole fraction (M.F.) range. On the other hand, it is hampered by an unstable source kinetics characterized by "saturation" and "crust formation". As a matter of fact, the (p-T-x) diagram of GaAs shows that the As partial pressures ($P_{\Sigma As} \sim 10^{-4}$) at the usual temperatures ($T_s \sim 800°C$) are larger than those corresponding to an As saturated Ga solution. After a saturation period, there is a GaAs precipitation on the liquid Ga surface which takes the form of a thin crust [15, 16]. The extension of the crust over the whole surface takes place more or less rapidly depending on the temperature gradient, partial pressures, flow rate, and geometry. After complete crust formation, a new metastable equilibrium is established with two opposite processes: further saturation from the As excess diffusing through the crust and ablation of it by HCl attack, with a zero arsenic balance-sheet for the source ($P_{\Sigma As}in = P_{\Sigma As}out$). The practical consequences of these metastable source kinetics is that reproducibility is only obtained with careful optimization.

The same equilibrium (2) is also obtained in the so-called "hydride" method in which the starting materials are HCl, Ga, and AsH_3. Because of its higher background impurity level – mainly due to the starting products – this method is much less used (see, however, [17]). The same holds true for the method using a solid arsenic source [18].

The arsenic trichloride method has some variants. For example, solid GaAs sources avoid the above mentioned source kinetic problems, lower temperatures [19] or inert carrier gas [20, 21, 22]. The use of those inert gases has been introduced in order to avoid the silicon contamination which will be discussed later.

All these different transport methods are summarized in Table I.

References	Symbols	Currently used name
23, 24, 25	$AsCl_3/Ga/H_2$	Arsenic trichloride method
19	$AsCl_3/GaAs/H_2$	Arsenic trichloride with solid source
26, 27	$AsCl_3/Ga/H_2$ + addtl. $AsCl_3$	Nozaki's method
28, 17	$AsH_3/Ga/HCl/H_2$	Arsine method
	same with $AsCl_3$, HCl + As	Arsine with native HCl
18	$As/Ga/HCl/H_2$	Separated source method
20	$AsCl_3$/Ga/A or N_2 or He	Inert gas method
21, 22	$AsCl_3/Ar/Ga/H_2$	Mixed carrier gas method

Table 1/GaAs VPE Growth Methods

3.2 The Growth Parameters

The main parameters of the epitaxial layer are the thickness and doping level, both in absolute value and uniformity. It is therefore essential to control the growth parameters which have an influence on the growth rate and on the impurity incorporation.

Supersaturation: The driving force of growth is the temperature difference between the source and the deposition zone, following the thermodynamic functions associated with equilibrium (2). This parameter partly controls the growth rate on the substrate as well as the thickness uniformity. Too high values can lead to parasitic nucleation and consequently poor reproducibility.

Growth anisotropy: GaAs VPE by chloride transport is characterized by a strong anisotropy due to the predominant surface kinetic [18, 29] (see also Figure 4).

Consequently, there are very large variations in the growth rate as a function of the substrate orientation. The (100) orientation, which is exclusively used for FETs, represents a relative minimum of the growth rate. Perturbations of the substrate surface give rise to defects with prominent regular shape (i.e., pyramids). The use of a misorientation of some 3 to 6 degrees increases the growth rate [30] and gets rid of most of these defects.

Growth temperature: The growth temperature is one of the main parameters in any epitaxial process. The growth rate, τ, is usually represented in the $\log \tau = f(1/T_D)$ diagram [32] in order to point out a linear decrease roughly interpreted by a kinetic limitation. Figure 4 shows these types of curves at nearly constant supersaturation, which in the $AsCl_3$ method means that the source and substrate temperature difference $T_s - T_D = \Delta T$ is constant.

The partial pressures: The effect of the $AsCl_3$ partial pressure on both the growth rate (Figure 5) and impurity incorporation (Figure 6) is important and rather well known. The growth mechanisms which involve surface coverage depending on partial pressures explain the behaviour well [8].

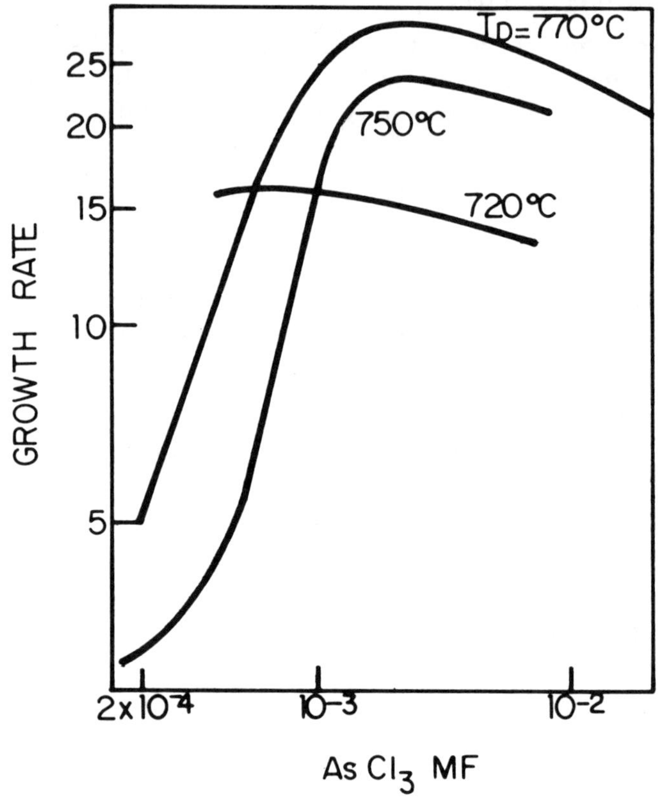

Figure 5*/Effect of the arsenic trichloride mole fraction on the growth rate at various deposition temperatures (after Ref. 33).*

3.3 Equipment

The success for reproducible high quality growth mainly depends upon the care taken in the build-up of the equipment used. The essential points are:

1. High purity carrier gas (i.e. if H_2 is used, it should be purified by palladium diffusion);
2. Non-reactive material for the circuits, PTFE, quartz or pyrex, possibly stainless steel, but, in this case, taking extreme care of humidity;

3. Leakage free circuitry (as few fittings as possible, high quality valves, flow controllers, etc.);
4. High purity (synthetic) quartz for the reactor, source, and substrate holder, with special care not to introduce any metallic impurities during the quartz blow;
5. Adequate design of the reactor exhaust allowing the elimination of the reaction by-products.

For GaAs growth, where relatively small surface areas are required, the type of reactor represented in Figure 7 has been used industrially after automation. It is capable of processing 20 cm^2 per run or more.

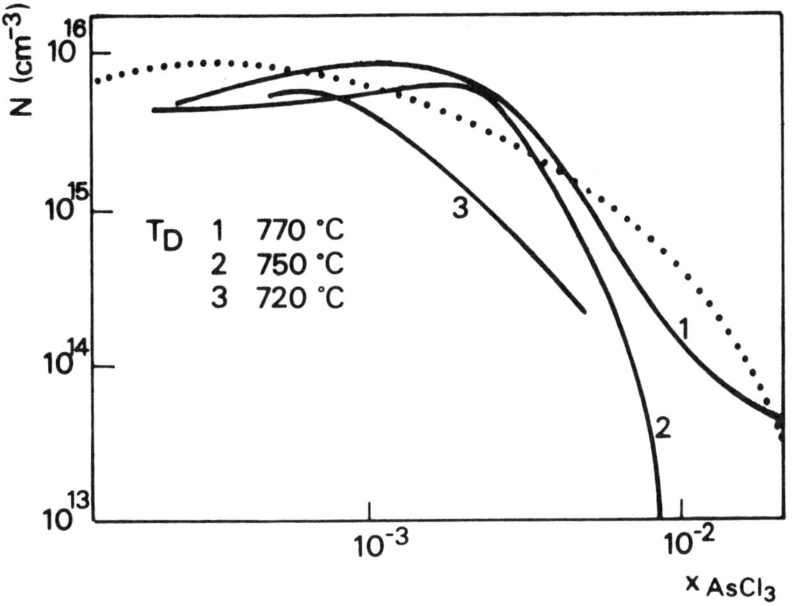

Figure 6/Effect of the arsenic trichloride mole fraction on the background doping level at various deposition temperatures. The dotted line gives the theoretical silicon activity curve (equilibrium (5) in the text) which has been arbitrarily translated (after Ref. 34).

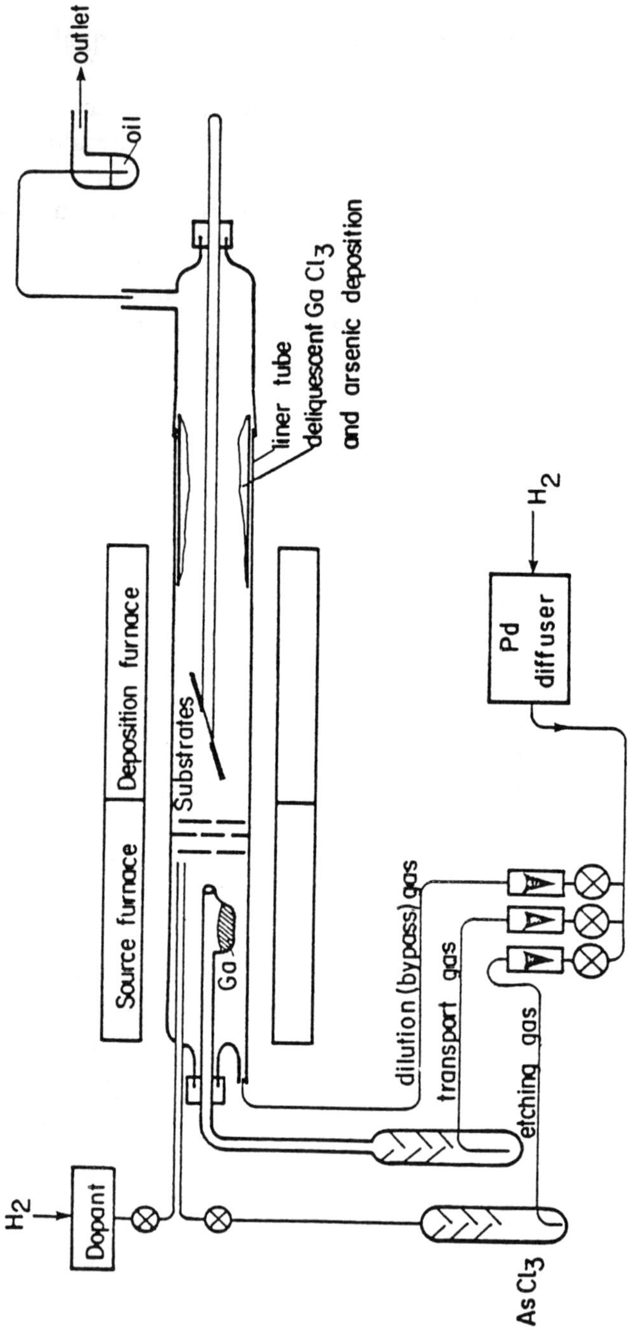

Figure 7/Schematic diagram of a GaAs epitaxial reactor based on the $AsCl_3$ method.

3.4 The Growth of the Buffer Layer

It is possible to realize FETs with the active epitaxial layer directly deposited on the semi-insulating substrate, but their performances remain modest. This is due to the poor quality of the epitaxial layer-substrate interface, which always remains more or less disturbed regardless of the precautions taken before and at the beginning of the growth. Crystallographic defects as well as doping anomalies may appear [35, 36, 37]. The growth of a buffer layer between the substrate and the active layer leads to an improvement of the performances of the transistor [27, 38] : for instance, according to T. Nozaki et al. [26], Figure 8 shows the evolution of the unilateral gain of an FET with or without a buffer layer. Such a buffer layer must have the resistive nature of the substrate in order to avoid any leakage current when the gate voltage reaches the pinch off.

Under usual conditions, the background doping is n type with a carrier concentration in the vicinity of 10^{14} cm^{-3}. At such carrier concentrations, the shallow acceptors (N_A) and the deep levels can no longer be neglected. A four level compensation model is used [39, 40] to describe this material:

$$n = N_D - N_A + N_{DD} - N_{DA} \tag{3}$$

where N_D, N_{DD}, and N_{DA} are shallow donors, deep donors, and deep acceptors respectively.

In the arsenic trichloride process, N_A is generally smaller than N_D. The concentration of the most important deep donor EL2 [41, 42], located fairly in the middle of the bandgap, does not depend very much on the growth conditions, and is in the range of $5.10^{13} - 2.10^{14}$.

In order to obtain high resistivity epitaxial layers, it is necessary to pin the Fermi level at the middle of the bandgap either by reducing N_D or by increasing N_{DA}.

The residual shallow donor is essentially silicon, for which the walls of the reactor constitute an inexhaustible source. The attack of silica by hydrochloric gas forms some chlorosilanes:

$$SiO_2 + nHCl + (4 - n)H_2 \rightleftharpoons 2H_2O + SiH_{(4-n)}Cl_n \tag{4}$$

They are reduced and incorporated at the surface of the sample according to equilibrium equation:

$$SiH_{(4-n)}Cl_n + (n - 2)H_2 \rightleftharpoons nHCl + Si_v \qquad (5)$$

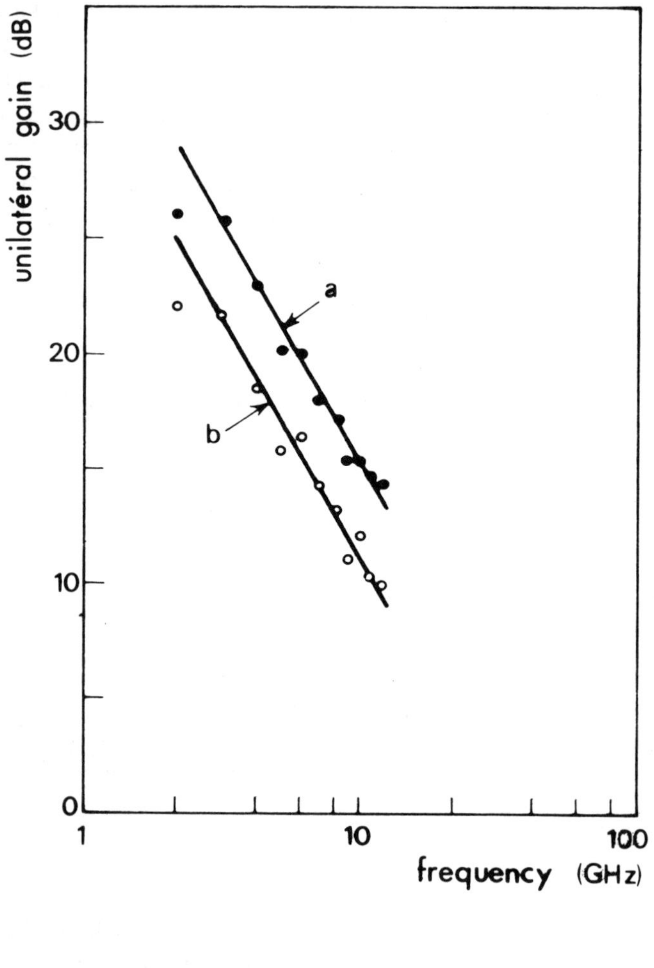

***Figure 8**/Power gain versus frequency curve of FETs obtained by the same technological process on material (a) with a buffer layer and (b) without a buffer layer (after Ref. 26).*

The thermodynamic study [34] shows that the total concentration of silicon in the vapour phase decreases simultaneously with an increase of the HCl partial pressure. So it is to be expected and it is also experimentally observed (see Figure 6) that an increasing $AsCl_3$ (and consequently HCl) partial pressure decreases the background doping. Attention should also be paid to other techniques leading to reduced background doping by silicon:

- The introduction of a small amount of oxygen [43], which inhibits the attack of the silica walls (equilibrium 4);
- The use of low growth temperatures (for example 630°C) which equally slows down this attack [19];
- The use of neutral carrier gases (nitrogen or argon) [20, 44] in which case equilibria (4) and (5) are suppressed.

As indicated in Equation (3), the other method to increase the resistivity of the buffer layer is by doping with deep acceptors such as iron or chromium. Iron is introduced in the gaseous phase in the form of $FeCl_2$ by means of an attack of metallic iron by HCl between 370°C and 550°C [45], and chromium in the form of CrO_2Cl_2 [46, 47].

In fact, this description of the growth of buffer layers is oversimplified, and, in practice, the terms of Equation (3) vary across the epitaxial layers. A relatively thick ($\geqslant 5\ \mu m$) buffer layer is schematically composed by (Figure 9):

1. An n type layer corresponding to the above described background silicon doping;
2. A high resistivity region near the interface where, for instance, $N_{DA} - N_{DD} \gtrapprox N_D - N_A$.

The existence of the high resistivity region, which is difficult to characterize because it is thin, resistive, and it is deserted by the N^-/S.I. barrier and the surface states as well, has given rise to quite a number of works [40, 48, 49]. Although present knowledge does not allow an unambiguous explanation of its formation, the following main origins can be assumed:

a) Impurity out-diffusion from the substrate;
b) Substrate surface pollution by impurity adsorption during preparation;
c) Evolution of the partial pressures (of the main products or the impurities) at the beginning of the growth.

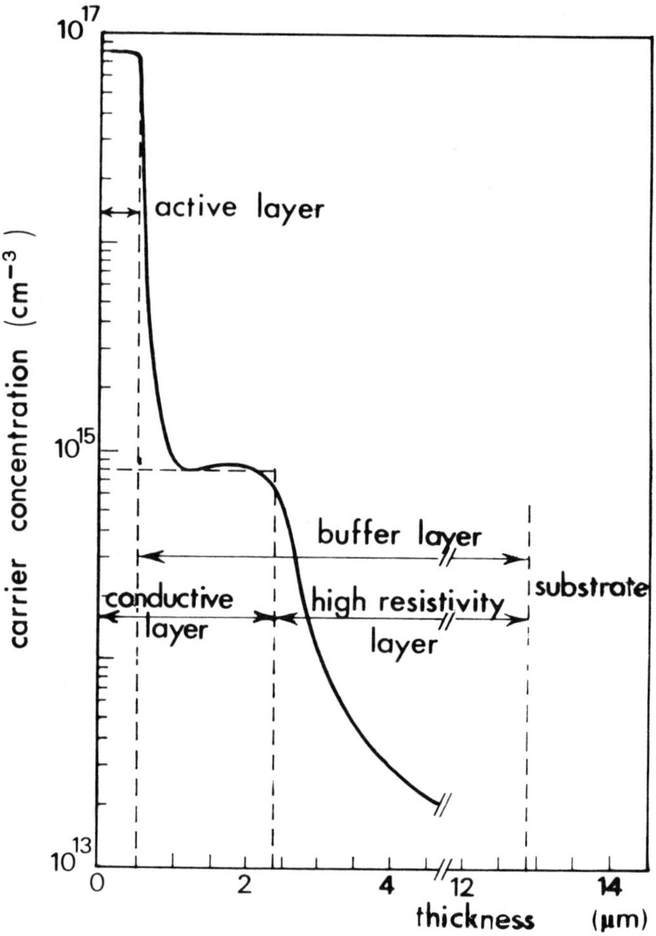

Figure 9/Free carrier concentration profile of an active layer-thick buffer layer structure. Note that in usual FET material, the buffer layer is only about 2-4 μm thick, the conductive layer (at 8 x 10^{14} level) is then avoided (after Ref. 27).

As a matter of fact, the substrate contains impurities either deliberately introduced (such as chromium), or not (such as silicium and carbon), which will be re-distributed during the heat treatments that are imposed by the epitaxial growth cycles [50]. Figure 10 gives an illustration of such a phenomenon, for a few impurities detected by means of secondary ion mass spectrometry (SIMS) in a chromium doped GaAs sample annealed at 750°C in hydrogen (for 24 hours in order to accentuate the effects). The diffusion of chromium has been the subject of numerous works [51, 52] which show the possibility of pollution of the epitaxial layer near the interface. However, considering the diffusion constant of chromium given by the following relation [51]:

$$D(cm^2 \cdot s^{-1}) = 6.3 \times 10^5 \exp(-3.4/kT) \quad (6)$$

(where kT is expressed in eV), and considering the relatively low growth temperatures ($T_D \leqslant 760°C$) and short diffusion times ($\leqslant 30$ min), this cannot allow a sufficient diffusion path to explain the high resistivity region. It should, however, be explicable by assuming anomalous Cr diffusion. There is also some evidence that too strong concentrations of chromium (about 10^{17} cm^{-3}) or of impurities with a high coefficient of diffusion (for example, Cu) are detrimental to the epitaxial layer quality. This has led a few authors [53] to set up a testing procedure for the preliminary selection of the crystals, which consists in an acceleration of the re-distribution phenomena by means of ion bombardment (krypton) and high temperature (850°C) annealing. Conversely, other authors [52], utilize the gettering on the crystallographic defects of the unpolished back surface of the substrate.

Surface treatment before growth is another parameter which plays a role in the occurrence of the high resistivity region. The effect can be the following: an adsorbed monolayer — which is always present when handled in air — could dope a 1 μm thick layer at 10^{18} cm^{-3}. An incomplete desorption during annealing and *in situ* etching can lead to compensation, for example, by carbon [40]. On the other hand, the *in situ* etched thickness may or may not correspond to the converted surface layer (Figure 10).

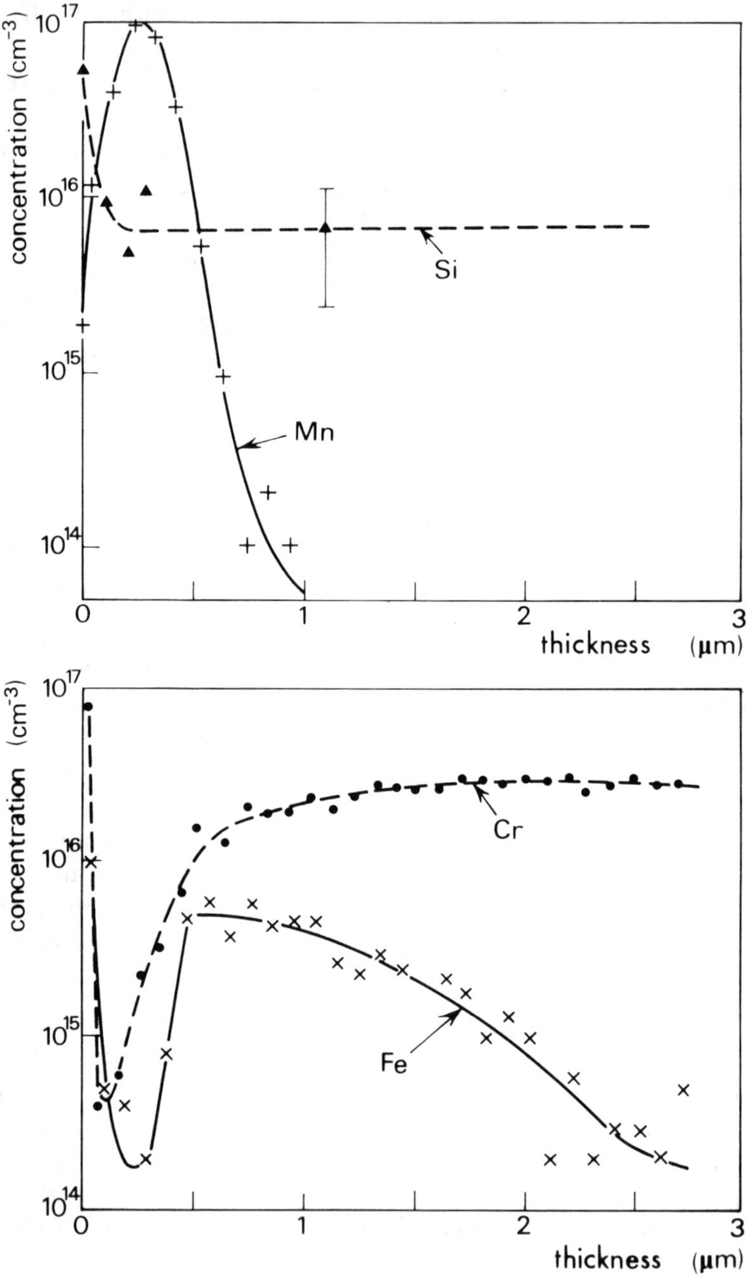

concentration (cm-3)
10^17
10^16
10^15
10^14
Si
Mn
0
1
2
3
thickness (μm)
concentration (cm-3)
10^17
10^16
10^15
10^14
Cr
Fe
0
1
2
3
thickness (μm)

To sum up, the origin of the high resistivity region is probably complex and depends in large part upon working conditions.

As far as the quality of the high resistivity layer is concerned, it appears that the interface region contains numerous deep levels [40, 53], among which a few impurities are identified according to the DLTS (Deep Level Transient Spectroscopy) or OTCS (Optical Transient Current Spectroscopy) spectra. Although it is difficult to measure, the electron mobility in such a resistive region may be rather high. Figure 11 shows a constant mobility region and a decreasing mobility region near the interface. The high resistivity layer close to the interface should be suitable as a buffer layer: however, the accumulation of deep levels carries a risk of making it photosensitive and it is reported that transistors which have been elaborated from such layers show a high noise factor [49]. To minimize this effect, as well as to have the high mobility layer (Figure 11) in contact with the active layer, it is suitable to increase the thickness of the buffer layer. On the other hand, if the thickness of the buffer layer is too important, the doped region previously mentioned appears in the vicinity of 10^{14} cm^{-3} (Figure 9). As a consequence, it is necessary to seek an optimum for the thickness of the buffer layer which is situated between a maximum value at which the doped region appears (leading to a leakage current), and a minimum value for which the buffer layer is dominated by the deep levels. At such an optimum thickness (between 2 and 4 μm), the quality of the top active layer and of its transition towards the buffer layer is secured; the mobility at such a point is roughly equal to that of the active layer deposited thereafter (4,500 to 5,000 $cm^2 \cdot V^{-1}s^{-1}$ for $n \cong 10^{17}$ at 300 K (Figures 3 and 14). Such a point is capital; it is admitted, indeed, that a mobility profile presenting a decrease right from the end of the active layer contributes to the limitation of the performances.

Figure 10/SIMS profile of manganese, silicon, chromium, and iron near the surface of a semi-insulating chromium doped GaAs substrate annealed under hydrogen at 750°C during 24 hours (after Ref. 50).

3.5 The Growth of the Active Layer

As previously shown, the active layer of a FET consists of a thin layer with precisely determined doping and thickness (Figure 2) which must also be very uniform over the whole wafer. High mobility of the carriers is needed and the surface of the layer must be without defect.

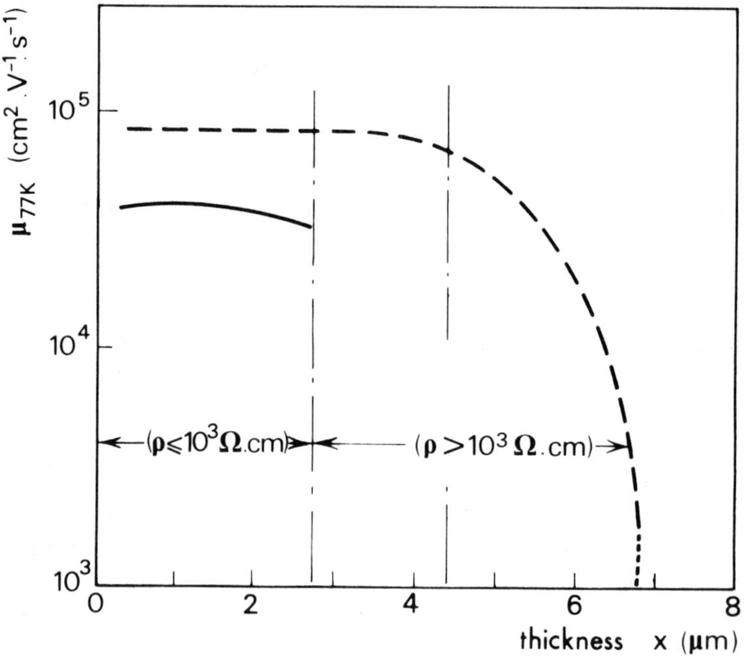

Figure 11/77 K mobility profile of a thick buffer layer as measured under illumination (dotted line) or in the dark (solid line). Note that in usual buffer layer (2-4 μm), the measurements are only possible under illumination (after Ref. 40).

Thickness and doping
It is not difficult to introduce the usual doping elements in the vapour phase. For n-type doping, sulphur can be chosen, in elemental form, or by using commercially available diluted H_2S. When elemental sulphur is used, the doping level is controlled by the temperature; the 20-100°C range covers nearly 3 orders of magnitudes (Figure 12)

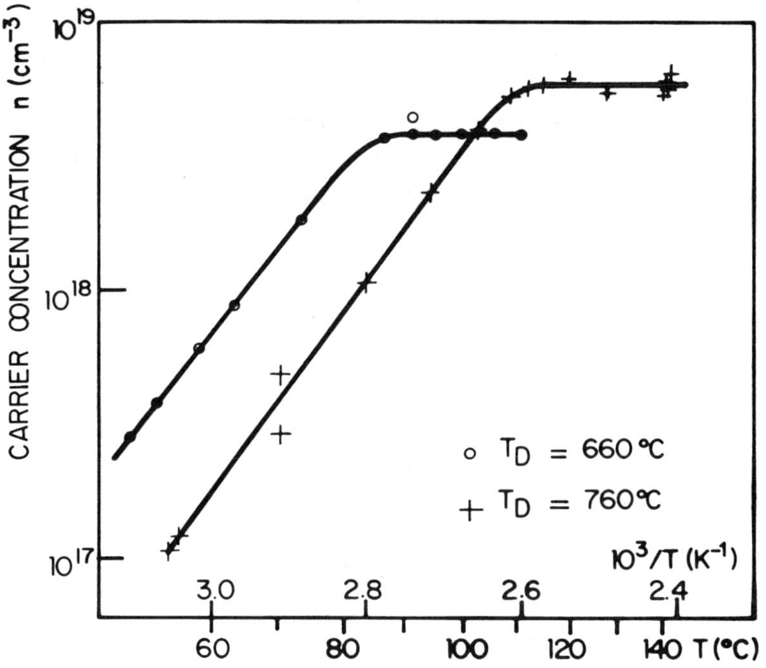

Figure 12/Carrier concentration as a function of the temperature of the sulphur doping source at two growth temperatures. The flow over the doping source represents 10% of the total flow (after Ref.56).

(from 10^{15} to 5×10^{18}), sulphur is transformed into H_2S in the reactor or by an additional furnace at 800°C [55]. Selenium, silicon, or tin can also be used, for example, by introducing $SiCl_4$ in the $AsCl_3$ or tin in the gallium source; the doping level in this last case can be monitored by the $AsCl_3$ M.F. [26, 59]. On the other hand, selenium allows the highest doping level ever published (10^{19}) with high mobility [56]; such a layer can be used as contact layer in the recessed gate technology (Figure 1).

The fact that the thickness and the doping of a layer only 2500 Å thick must be perfectly calibrated for the FET application makes its growth somewhat difficult. Let us point out that the average growth rate used in GaAs VPE is in the range of 10 to 25 μm/hour (30 to 50 Å/sec), so the growth time is about 1 minute only. This period of time starts after the growth of the buffer layer, generally by means of the introduction into the reactor of a supplementary flow carrying a dopant partial pressure. In this way, there will be some initial phenomena, linked to hydrodynamics and to diffusion through the "stagnant layer" [57], which are not easily controlled. The way to solve these problems by careful experimentation is typically in the field of "know-how." An example of a doping method [58] is given in Figure 13. In this case, the dopant is introduced in the downstream part of the long constant temperature deposition zone. After the growth of the buffer layer in the up-stream zone where the dopant retrodiffusion is negligible, the substrate is shifted into the downstream zone for the growth of the active layer. This provides the advantage of stabilized flows and partial pressures and avoids the above mentioned initial hydrodynamics problems. It allows the growth of FET material structures with well controlled and sharp transitions.

Morphology of the epitaxial layers

It is essential to achieve planar epitaxial growth free from hillocks, spikes, pyramids, terraces, and other microscopically visible defects. These defects are mainly due to a local pollution of the substrate surface by oxidation, dust, or other parasitic deposition, imperfect polishing, or defects in the substrate.

The parameters to be optimized are then: the substrate quality (absence of micro-precipitates and inclusions), substrate preparation, dust free introduction into the reactor, optimized *in situ* etching (flow, $AsCl_3$ M.F., time), and correct starting growth conditions (no Ga droplets, no anomalous supersaturation), as well as an adequate substrate misorientation (3 to 6°). As a matter of fact, on exactly (100) oriented surfaces, morphology features showing defect originated secondary nucleation often appear. On misoriented planes, it is possible to obtain nearly perfect, mirror-like morphologies.

Uniformity of doping and thickness

Without special care, the thin layers are not uniform in thickness from one edge of a wafer to the other; thickness may vary by 20 to 30%. The doping also varies in the same direction and 100% dispersions of the pinch off voltage can be observed because the two ef-

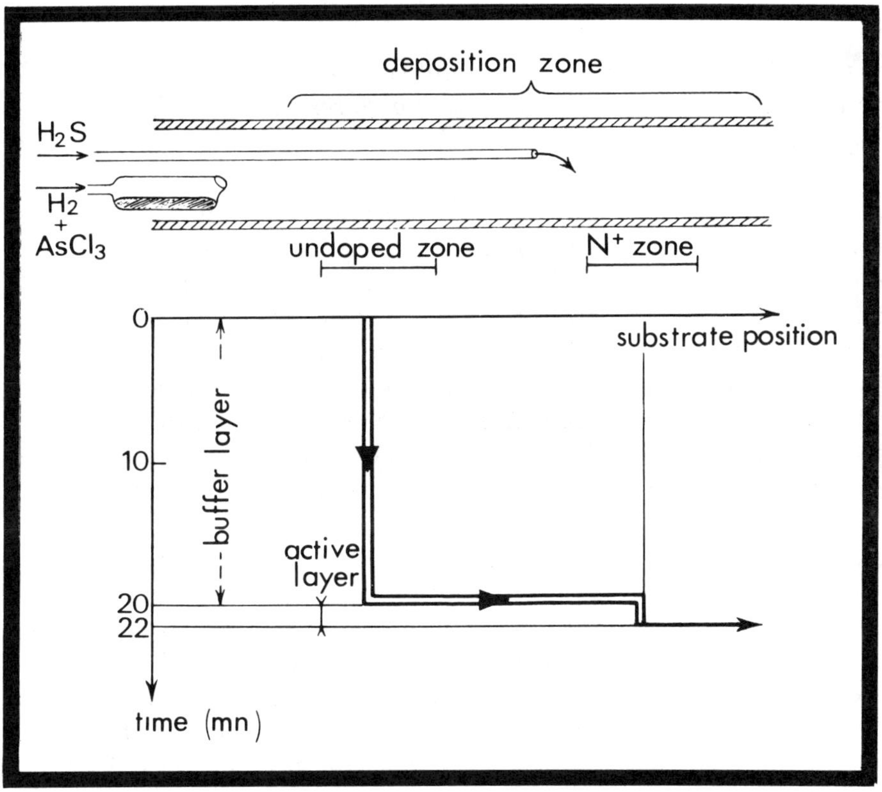

Figure 13/Principle of the method using a shift of the substrate during the deposition for the growth of the active layer-buffer layer structures (after Ref. 58).

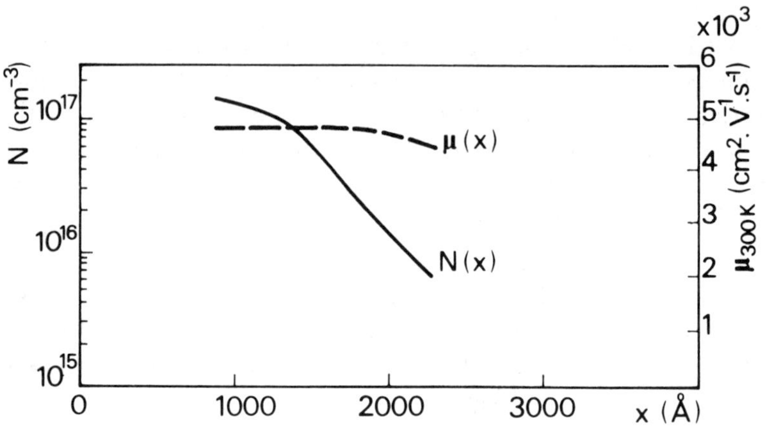

Figure 14/Room temperature mobility and carrier concentration profile of an active layer grown on a high resistivity buffer layer (after Ref. 54).

fects are cumulative. In the course of recent years, studies on uniformity have attempted to improve the results. In fact, in VPE the growth rate decreases in the direction of the flow as a consequence of the exhaustion of the gas phase along the wafer. In order to minimize such problems, one has to consider the hydrodynamics and the growth mechanisms. On the one hand, some studies carried out on hydrodynamics have attempted to reveal the existence of a stagnant gas layer and the role it played on the surface of the sample. Such a layer plays the part of a screen through which the molecules diffuse in order to reach the surface of the crystal. It is possible to modify the longitudinal distribution of the gas speed and, as a consequence of the foregoing, the thickness of the stagnant layer, by means of tilting up the sample with respect to the flow axis (Figure 15). In a direction perpendicular to the flow, the uniformity of the stagnant layer only depends upon an optimal value of the R ratio between the dimension of the sample-holder (or the substrate) and the internal diameter of the reactor. The speed profiles of the gases at the level of the sample for various values of R are represented in Figure 16 [57].

On the other hand, previous work on the growth mechanisms in the $AsCl_3$/Ga system has shown that the growth can be either diffusion or kinetically controlled depending upon the deposition temperature.

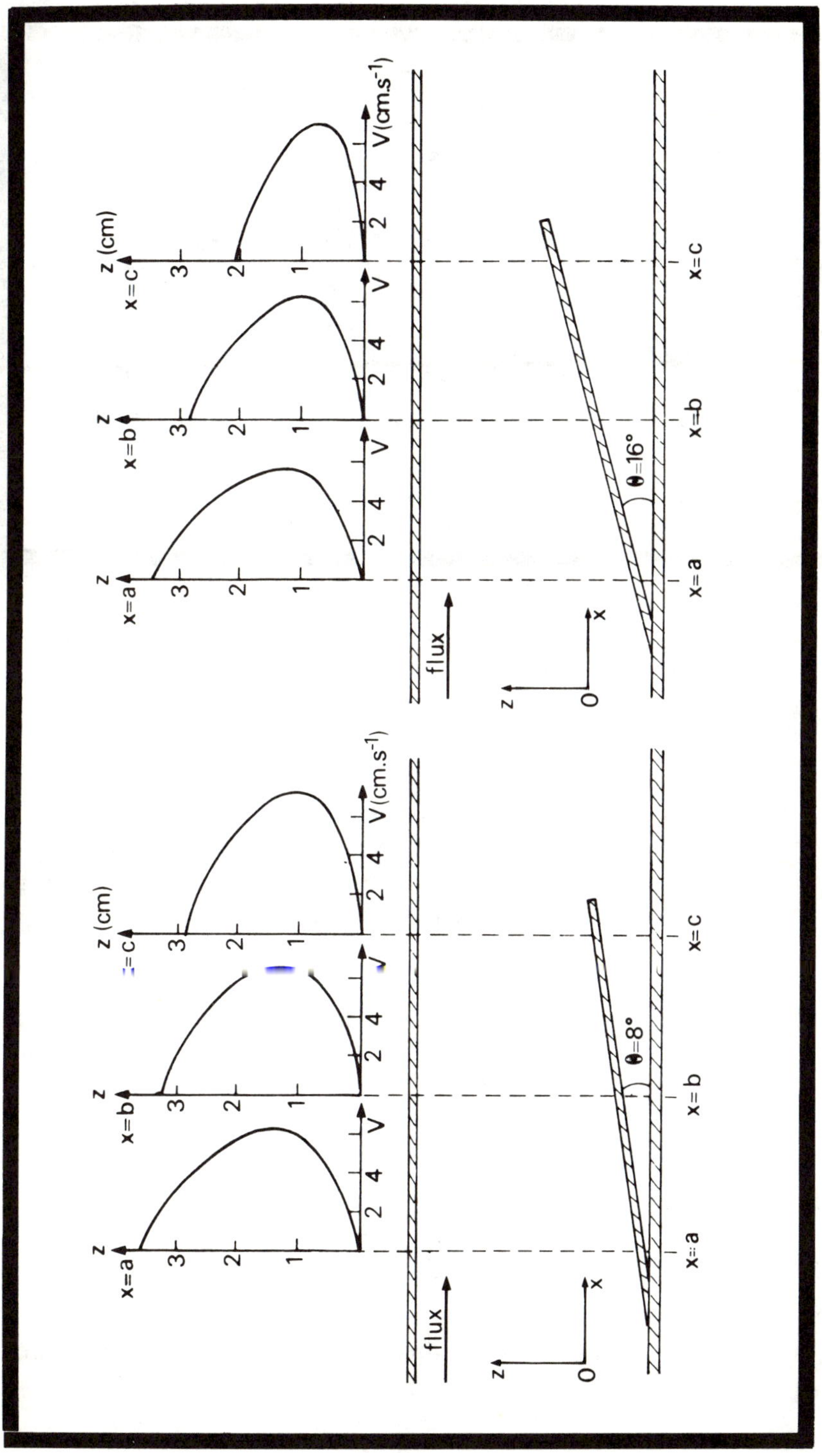

Figure 15*/Velocity profile of the flow along the reactor axis (after Ref. 57).*

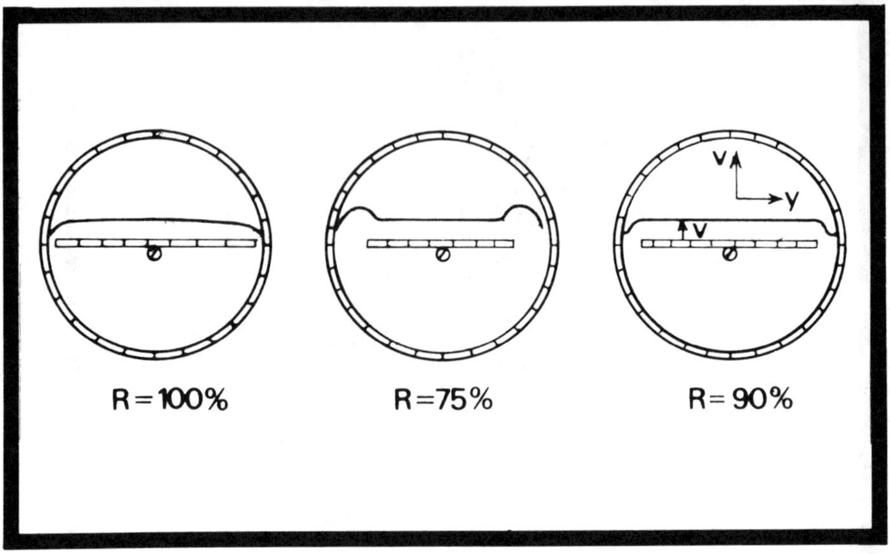

***Figure 16**/Velocity profile of the flow at 1 mm above the surface of the substrate perpendicular to the reactor. R is the ratio of the substrate width to the internal reactor diameter (after Ref. 57).*

In the low temperature region (Figure 4), the growth rate is kinetically controlled by some relatively slow surface process such as adsorption, desorption, or surface reaction [32]. In this region, the growth rate is less sensitive to the gas flow rate and is essentially determined by the substrate temperature. It is also clear from Figure 5 that the growth rate is a function of the $AsCl_3$ mole fraction. This is also related to the growth mechanism. It has been shown [60] that a careful optimization of both deposition temperature and $AsCl_3$ mole fraction could lead to extremely uniform thicknesses. This is illustrated in Figures 17 and 18.

Both effects, namely hydrodynamic and kinetic control, must be carefully considered in order to obtain uniformly doped layers. It is, indeed, likely that the incorporation of the doping impurity does not follow the same kinetic limitation as the growth rate.

Reproducibility is reasonably obtained using the $AsCl_3/Ga/H_2$ process [60] and it is used for development activities in many companies.

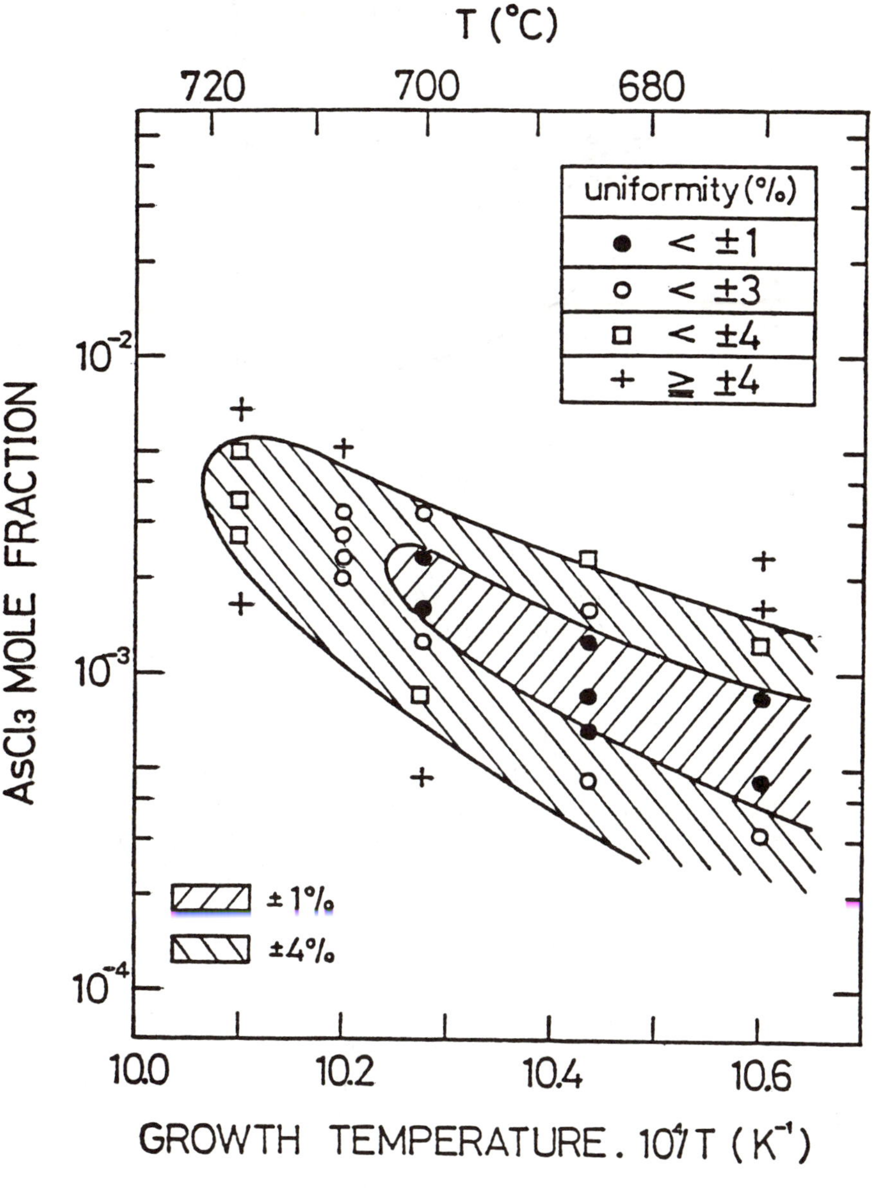

Figure 17/Thickness uniformity as a function of $AsCl_3$ mole fraction and growth temperature (after Ref. 60).

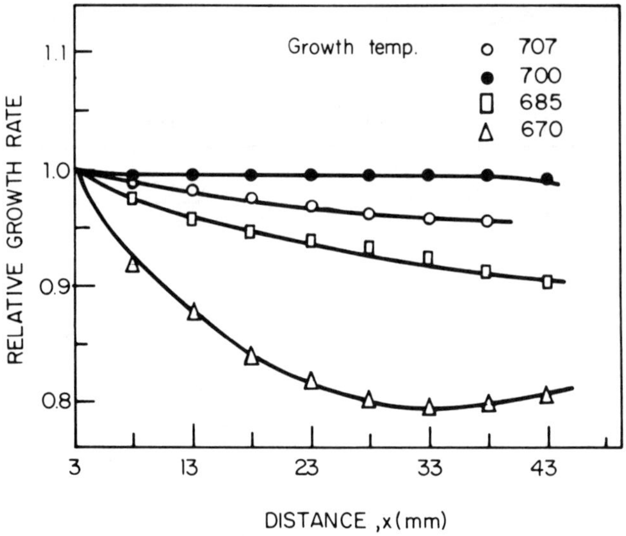

Figure 18/*Relative growth rate as a function of distance from the leading edge of substrate for different growth temperatures.* $AsCl_3$ *mole fraction:* 2.3×10^{-3} *(after Ref. 60).*

4. CONCLUSION

At present, the operational methods in research for the growth of the material for field effect transistors lead to peak results which are practically identical; the choice of method is therefore not vital at this level. This is, however, not the case at a development stage where the $AsCl_3/Ga/H_2$ method is still today the best technique for producing high quality material in a reproducible way. This constitutes an important scientific experience and has very likely formed the subject of the majority of work developed in the field of materials for microwave in the course of the last fifteen years. Such advancement is a result of the considerable experience acquired in the course of application. For these reasons the epitaxial process applied to the material for GaAs FETs is, for the present time, the main process being industrialized.

One should finally note the existence of ion implantation for realizing active layers for FETs in industrial applications. It has already led to remarkable results. However, only epitaxial growth processes allow the production of heterostructures, of thicker layers, and of some multiple layers with different dopings.

REFERENCES

[1] Boccon-Gibod, D., *Acta Electronica,* 23:2, 1980.

[2] Boccon-Gibod, D., J.P. André, P. Baudet, J.P. Hallais, *IEEE Trans. Electron Devices,* ED-27:1141, 1980.

[3] Dingle, R., H.L. Sormer, A.C. Gossard, W. Wiegmann, *Appl. Phys. Lett.,* 33:665, 1978.

[4] Mimura, T., S. Hiyamizu, T. Fujii, K. Nansu, *Jap. J. Appl. Phys.,* 19:L 225, 1980.

[5] Morcok, H., S. Bandy, R. Sankaran, G.A. Antypas, L. Bell, *IEEE Trans. Electron Devices,* ED-25:619, 1978.

[6] Casey, Jr., H.C., A.Y. Cho, D.V. Lang, E.H. Nicollian, P.W. Foy, *J. Appl. Phys.,* 50:3484, 1979.

[7] Eastman, L.F., et al. in *Gallium Arsenide and Related Compounds (Vienna) 1980* (Inst. Phys. Conf. Ser. no. 56) p. 185. London: Adam Helger/The Institute of Physics, 1980.

[8] Hollan, L., J.C. Brice, J.P. Hallais, "The Preparation of Gallium Arsenide", in *Current Topics in Materials Science,* vol. 5, E. Kaldis, ed., p. 1-217. Amsterdam: North-Holland Publ. Co., 1980.

[9] Kim, C.K., et al. in *Gallium Arsenide and Related Compounds (St. Louis) 1978* (Inst. Phys. Conf. Ser. no. 45) p. 305. Adam Helger/The Institute of Physics, 1979.

[10] Bandy, S.G., D.M. Collins, C.K. Nishimoto, *Electron. Lett.,* 15:218, 1979.

[11] Wataze, M., Y. Mitsui, T. Shimanoe, M. Nakatani, S. Mitsui, *Electron. Lett.,* 14:759, 1978.

[12] Nakanishi, T., Inst. Conf. on MOVPE, Ajaccio, 1981, to be published in the Proceedings.

[13] Hollan, L., *Acta Electronica,* 21:117, 1978.

[14] Boucher, A., L. Hollan, *J. Electrochem. Soc.,* 117:923, 1970.

[15] Shaw, D.W., *J. Cryst. Growth,* 8:117, 1971.

[16] Seki, H., A. Koukitu, M. Fujimoto, *Jap. J. Appl. Phys.,* 14:411, 1975.

[17] Stringfellow, G.B., G. Ham, *J. Electrochem. Soc.,* 124:18, 1977.

[18] Shaw, D.W., *Inst. Phys. Conf. Ser.,* 7:50, 1969.

[19] Hallais, J., D. Boccon-Gibod, J.P. Chané, J.M. Durand, L. Hollan, *J. Electrochem. Soc.,* 124:1290, 1977.

[20] Ozeki, M., K. Kithara, K. Nakai, A. Shibatomi, K. Dazai, S. Okawa, O. Ryuzan, *J. Appl. Phys.,* 16:1617, 1977.
[21] Seki, H., A. Koukitu, M. Fujimoto, *Jap. J. Appl. Phys.,* 16:847, 1977.
[22] Hallais, J.P., *Brevet Francais,* 76.06.832 du 10.03.76.
[23] Knight, J.R., D. Effer, P.R. Evans, *Sol. St. Electron,* 8:178, 1965.
[24] Effer, D., *J. Electrochem. Soc.,* 112:1020, 1965.
[25] Dilorenzo, J.V., *J. Cryst. Growth,* 17:189, 1972.
[26] Nozaki, T., et al. in *Gallium Arsenide and Related Compounds (Deauville) 1974* (Inst. Phys. Conf. Ser. no. 24) p. 46. London: Adam Helger/The Institute of Physics, 1975.
[27] Cox, H.M., and J.V. DiLorenzo in *Gallium Arsenide and Related Compounds (St. Louis) 1976* (Inst. Phys. Conf. Ser. no. 33b) p. 11. London: Adam Helger/The Institute of Physics, 1977.
[28] Tietjen, J.J., J.A. Amick, *J. Electrochem. Soc.,* 113:724, 1966.
[29] Hollan, L., C. Schiller, *J. Cryst. Growth,* 13-14:319, 1972.
[30] Hollan, L., C. Schiller, *J. Cryst. Growth,* 22:175, 1974.
[31] Hollan, L., J.M. Durand, R. Cadoret, *J. Electrochem. Soc.,* 124:135, 1977.
[32] Shaw, D.W., *J. Electrochem. Soc.,* 117:683, 1970.
[33] Hollan, L., *Inst. Phys. Conf. Ser.,* 24:22, 1975.
[34] Dilorenzo, J.V., G.E. Moore Jr., *J. Electrochem. Soc.,* 118: 1823, 1971.
[35] Hollan, L., J.P. Hallais, C. Schiller, *J. Cryst. Growth,* 9:165, 1971.
[36] Okamoto, H., S. Shakta, K. Sakai, *J. Appl. Phys.,* 44:1316, 1973.
[37] Hasegawa, F., *J. Electrochem. Soc.,* 119:930, 1972.
[38] Slaymaker, N.A., J.A. Turner, Proc. Eur. Microwave Conf., Bruxelles, 1973, A.5:1.
[39] Martin, G.M., J.P. Farges, G. Jacob, J.P. Hallais, G. Poiblaud, *J. Appl. Phys.,* 51:2840, 1980.
[40] Hurtes, C., L. Hollan, and M. Boulou in *Gallium Arsenide and Related Compounds (St. Louis) 1978* (Inst. Phys. Conf. Ser. no. 45) p. 342. London: Adam Helger/The Institute of Physics, 1979.
[41] Martin, G.M., A Mitonneau, A. Mircea, *Electron. Lett.,* 13:191, 1977.

[42] Mitonneau, A., G.M. Martin, A. Mircea, *Electron. Lett.*, 13: 666, 1977.
[43] According to unpublished data of the authors
[44] Komeno, J., K. Kitahara, S. Ohkawa, *J. Cryst. Growth,* 47:601, 1979.
[45] Nakai, K., K. Kitahara, A. Shibatome, S. Ohkawa, *J. Electrochem. Soc.,* 124:1635, 1977.
[46] Kato, Y., Y. Mori, K. Morizane, *J. Cryst. Growth,* 47:12, 1979.
[47] Cox, H.M., J.V. Dilorenzo, L.A. D'Arsaro, Proc. 1st GaAs IC Symp., Lake Tahoe, 1979, paper no. 12.
[48] Fairman, R.D., F.J. Morin, and J.R. Oliver in *Gallium Arsenide and Related Compounds (St. Louis) 1978* (Inst. Phys. Conf. Ser. no. 45) p. 134. London: Adam Helger/The Institute of Physics, 1979.
[49] Crossler, I., et al. in *Gallium Arsenide and Related Compounds (St. Louis) 1976* (Inst. Phys. Conf. Ser. no. 33b) p. 289. London: Adam Helger/The Institute of Physics, 1977.
[50] Clegg, J.B., G.B. Scott, J.P. Hallais, A. Mircea-Roussel, *J. Appl. Phys.,* 52(2):1110, 1981.
[51] Kasahara, J., N. Watanabe, *Jap. J. Appl. Phys.,* 19:L 151, 1980.
[52] Magee, T.J., J. Peng, J.D. Dong, C.A. Evans Jr., V.R. Deline Jr., R.M. Malbon, *Appl. Phys. Lett.,* 35:277, 1979.
[53] Fairman, R.D., F.J. Morin, and J.R. Oliver in *Gallium Arsenide and Related Compounds (St. Louis) 1978* (Inst. Phys. Conf. Ser. no. 45) p. 134. London: Adam Helger/The Institute of Physics, 1979,
[54] Chané, J.P., J.P. Hallais, *Acta Electronica,* 23:11, 1980.
[55] Hales, M.C., J.R. Knight, C.W. Wilkins, Inst. of Physics, Bristol, 1971, *Inst. Phys. Conf. Ser.,* 9:52.
[56] Hollan, L., M. Boulou, J.P. Chané, *J. Electr. Mat.,* 10:193, 1981.
[57] Chané, J.P., *J. Electrochem. Soc.,* 127:913, 1980.
[58] Hollan, L., and A. Mircea in *Gallium Arsenide and Related Compounds (Boulder) 1972* (Inst. Phys. Conf. Ser. no. 17) p. 217. London: Adam Helger/The Institute of Physics, 1973.
[59] Feng, M., V. Eu, T. Zielinski, H.B. Kim, *Inst. Phys. Conf. Ser.,* 56:1, 1980.
[60] Komeno, J., et al. in *Gallium Arsenide and Related Compounds (Vienna) 1980* (Inst. Phys. Conf. Ser. no. 56) p. 9. London: Adam Helger/The Institute of Physics, 1981.

Molecular Beam Epitaxy for Microwave Field Effect Transistors

C. E. C. Wood

Department of Electrical Engineering and National Research and Resource Facility for Submicron Structures, Cornell University, Ithaca, New York.

In recent years GaAs has increasingly fulfilled its promise as an exciting binary compound for many active microwave and optical devices. The family of field effect devices are among the most important for both communication and computation applications. Although molecular beam epitaxy has been used to improve the state of the art of microwave mixers [1, 2] and varactors [3, 4, 5], this chapter concentrates on the impact of molecular beam epitaxy as a technique for future preparation and improvement of field effect transistors.

1. ACTIVE LAYER PREPARATION – BULK DEVICES

The techniques available for preparation of GaAs layers suitable for FET fabrication are of two basic types. The first type relies on the electrical modification of surface regions of semi-insulating substrates by dopant diffusion or ion implantation, and the second on growth of doped epitaxial layers on semi-insulating substrates. Dopant diffusion in III-V compounds is notoriously uncontrollable due to the variability in stoichiometry, defect and compensating acceptor densities, etc., in commercially available semi-insulating substrates. If controllable diffusion were possible, the diffuse nature of the leading edge of resulting profiles would still lead to unpredictable gate pinch-off characteristics.

Ion implantation has made spectacular improvements in the last few years; however, substrate variability problems have necessitated costly in-house boule growth or inefficient substrate qualification methods. High activation efficiencies are now regularly reported [6] with respectable electron mobilities over large areas with fast throughput. However, diffuse dopant profiles and the need for further processing in the form of capping, annealing, etc., together with poor reproducibility of implantation, make molecular beam epitaxy an attractive alternative.

2. EPITAXIAL METHODS

Historically, epitaxial methods have proved most reliable for preparation of active layers with required doping and thickness characteristics.

Liquid phase epitaxy consistently produces high quality epitaxial layers but suffers in many cases from poor turn round, thickness control, and morphology and may ultimately be limited by its lack of versatility in terms of free carrier profile control and multilayer growth.

Chemical vapor phase epitaxy is currently the workhorse for GaAs FET devices [8]. Indeed, the ability to grow on large substrate areas and to generate relatively sophisticated free carrier profiles with relatively low apparatus cost will probably continue, at least for the near future, to be the main epitaxial technique for GaAs device layer preparation.

There are certain limitations (in terms of uniformity, profile control, heterojunction buffer layer growth, and ohmic contacts) that may eventually lead to its replacement by MBE or organo-metallic chemical vapor deposition. The latter technique overcomes the inability of the chloride transport process to grow AlGaAs buffers and, because of its lower growth temperatures (typically 600°C), reduces problems of substrate defect and impurity outdiffusion.

The advantages described above are unique to MBE and may lead it to displace other epitaxial techniques for many device applications. Table I gives a comparative analysis of the advantages of the various active layer preparation techniques.

3. FREE CARRIER PROFILE SYNTHESIS

To optimize FETs, layers with a fairly complicated free donor concentration profile with depth must be considered (see Figure 1).

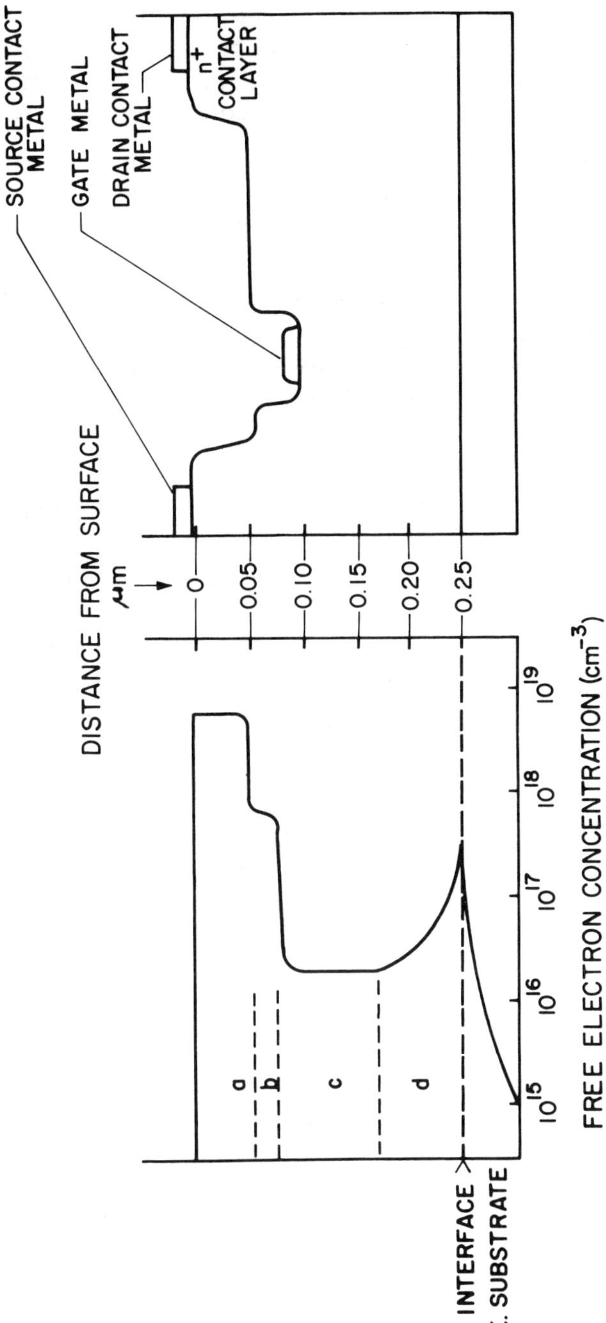

***Figure 1**/Schematic drawing of free electron concentration V_s depth (left), and cross-section through source-gate-drain region through linear power FET (right).*

In order to reduce ohmic resistances at source and drain contacts, high doping levels are required at the surface (sub-layer a in Figure 1). This and other approaches for improving ohmic contacts are covered later.

Secondly, source-contact-to-channel and channel-to-drain contact parasitic series resistances should be as low as possible. This can be achieved by increasing doping levels under the ohmic contact region [10] (sub-layer b in Figure 1) or by regrowth techniques [11] in which the channel region is defined by conventional lithography and etching techniques. Then, n^+ regions can be grown over the whole area, followed by conventional processing steps to open up gate notches. An alternative technique involves selective ion implantation into source and drain contact regions of epitaxially grown active layers for more planar structures.

Schottky breakdown voltages can be increased by a reduced doping level in the active layer [12] at and just below the final gate metal depth (sub-layer c in Figure 1). Buried regions of reduced doping level increase the source-to-channel and channel-to-drain resistances, but can also increase transconductance linearity if the doping level in the fourth region (sub-layer d in Figure 1) increases sufficiently rapidly toward the buffer layer interface [13]. This latter profile can be achieved by predeposition tin doping [14] or by lamella doping [15] during MBE.

Channel pinch-off can be positive only if the free electron density drops sufficiently rapidly at the bottom of the channel region (below sub-layer d in Figure 1).

Finally, although not strictly part of the carrier profile, the use of heterojunction electron confinement type buffer layers has been shown to decrease substrate space-charge shunt current at high drain biases near pinch-off [16].

The resulting compromise doping profile shown in Figure 1 is relatively easily achieved by MBE, especially if the apparatus is equipped with microprocessor controlled furnaces and shutters.

Control of spacial uniformity of thicknesses and doping concentration profiles becomes increasingly important as profiles become more complicated. Recent developments in MBE machine designs enable very close control of thickness and doping uniformity over large areas (2-3 inches) by continuous rotation of the substrate during epitaxy.

4. OHMIC CONTACTS

Apart from the internal device impedance limited by bulk layer saturation velocity (V_{sat}), doping level x thickness product, and source-drain separation, there can be significant series resistances associated with source and drain ohmic contacts which effectively reduce device gain, operation frequency, and increase noise.

In order to minimize thermal dissipation in power devices and obtain high operational frequencies, contact resistances should be reduced as much as possible.

Typically, ohmic contacts to GaAs FETs are made by alloying complex evaporated multilayer structures, e.g., Au:Ge/Au/Ni/Au at elevated temperatures (typically 400°C) for various times (~ 30 seconds). Such contacting procedures are extremely time consuming, expensive and, at best, unreproducible. Alloying also produces rough (pitted) metal/GaAs interfaces which can give rise to high field regions, current crowding with associated local joule heating, enhanced electron migration, filament formation and eventual device failure. The biggest single problem with alloyed contacts is the high values of specific contact resistances (R_c) they obtain. As devices become smaller and active layers become thinner, then R_c values become comparable with R_T (the total source to drain series resistances).

By modeling ohmic contacts as Schottky barriers, the depletion region (associated with metal-GaAs interface states) is sufficiently thin for tunneling to occur. It can be seen from the simplification [17] $R_c \propto \exp \ell_D \phi_b^{1/2}$ that the higher the doping level in the semiconductor under the metal and the lower the Schottky barrier height, then the lower will be the associated R_c.

$$\propto \exp \phi_b (N_D - N_A)^{-1/2}$$

where

ϕ_b = Schottky barrier height

$(N_D - N_A) \approx$ Free electron concentration under the contact metal

By molecular beam epitaxy it is possible to "superdope" the surfaces of GaAs up to $\geqslant$ 3 x 10^{19} cm^{-3} . Such surfaces have yielded non-alloyed contact resistances of the order 1-2 x 10^{-6} Ωcm^2 [18].

Further reduction in R_c is virtually impossible without introduction of deep level centers because of the fixed barrier height between most metals and GaAs (0.7-0.8 eV). Much lower reproducible values of R_c are easily obtained by molecular beam epitaxy using thin ($\sim$ 250 Å), heavily arsenic doped ($\geqslant 10^{20}$ cm^{-3}) germanium layers between the GaAs and the contact metal [19].

Germanium has two advantages over GaAs in the formation of non-alloyed contacts. Firstly, the metal-Ge barrier height is lower (e.g., Ni-Ge = 0.45 eV [20] and the higher solid solubility of As in Ge compared to donors in GaAs [12] enables much heavier doping levels ($> 10^{20}$ cm^{-3}). MBE deposition of 250 Å Ge on MBE grown GaAs surfaces in arsenic ambients ($\sim 10^{-9}$ Torr) at $\sim$ 280°C, and subsequent metallization without alloying reproducibly produce contact resistances well below 10^{-7} Ωcm^2.

Further advantages of this heterojunction contact are the good lattice constant match (better than 0.1% mismatch), low differential thermal expansion coefficient, good high temperature stability, and flatness of MBE grown epitaxial-germanium/GaAs interfaces. Further, by adding a heating stage, it is possible to prepare such Ge contacts in conventional evaporators.

5. ELECTRON CONFINEMENT

FET active layers are conventionally grown on undoped or chromium doped (semi-insulating) buffer layers to avoid structural and impurity-outdiffusion problems, which are found when buffer layers are not used. No matter how defect or trap-free, space charge associated currents are experienced in conventional homoepitaxial buffers used in epitaxially grown FETs [21]. The associated diffuse (semi-circular) current paths are problematical, leading to poor gate pinch-off characteristics, lower than predicted values of operating frequency (f_{max}), lower mutual conductance (g_m), and lower associated gain.

The presence of moderately shallow electron traps in buffer layers can also give rise to audio-frequency "DC" looping, noisy transfer characteristics, and, at high bias conditions, can result in avalanche induced device failure. These above problems can be overcome to a large extent by using lattice-matched undoped or high resistivity buffer layers which have a lower electron affinity than the active layer [16]. Alignment of the active and buffer layer Fermi levels causes an interfacial discontinuity in the conduction band which can act as a barrier to electron flow, providing a sharp interface can be prepared and the interface trap density is low. MBE is well suited to the preparation of abrupt interfaces in one of several such promising heterojunction systems. The conduction band discontinuity ΔE_c between GaAs and the closely latticed matched alloys of $Al_xGa_{1-x}As$ is given approximately by

$$\Delta E_c \approx 0.8 \, \Delta E_G$$

where ΔE_G is the band gap difference [22].

When $x = 1$, ΔE_G can be as high as 0.72 eV, leading to a discontinuity barrier of $\sim$ 0.58 eV. It is still possible for a reduced number of channel electrons to overcome such barriers and enter AlGaAs buffer layers, especially at higher operating biases, but the very low mobilities experienced in this alloy system keep the total buffer current down, resulting in a high output impedance ($\sim$ 2000 Ω mm) [16] at pinch off.

The controlled introduction of deep acceptor centers such as Cr is also possible in AlGaAs buffer layers, thereby creating a less mobile space charge situation, which again increases the RF output impedance, and significantly reduces buffer or substrate power dissipation and frequency lowering problems mentioned above. The ease with which very sharp interfaces may be prepared, and alloy compositions, dopant concentrations and types can be controlled give MBE significant advantages for device improvement by carrier confinement.

6. SELECTED AREA EPITAXY

The need for electrical isolation between discrete elements of integrated device circuits and the need for isolation of metal contact pads and beam leads from electrically active regions of semiconductors has traditionally been accomplished by selected area ion (H^+ or B^+) [23] bombardment or by removal of selected regions of active layers by chemical mesa etching. The former requires sophisticated implantation equipment and can lead to deep level centers which may be unstable to migration over extended time periods or under extreme operating conditions [24]. On the other hand, mesa etching does achieve excellent stable isolation, but requires closely controlled etching processes and, as the resulting surfaces are not flat, severe problems with focusing and resolution of lithographic mask exposures are experienced, together with mechanically and electrically weak regions of evaporated metal contacts over the mesa edges.

The technique of selected area epitaxy pioneered at Bell Labs [25] overcomes these problems by growth of regions of good electrical quality, single crystal layers in a matrix of submicron dimension polycrystalline material. The technique uses planar lithographically defined areas of epitaxy resists on substrates produced prior to epitaxy by plasma, thermal, or anodic oxidation or by deposition of SiO_2 or other high thermal stability material. MBE growth on such epitaxy resists is polycrystalline and highly resistive ($\geqslant 10^6\ \Omega$cm). The resolution between single and polycrystalline regions is better than a micron in certain crystallographic directions. The crystallite size does, however, increase with film thickness reducing resistivity much above the $\sim 3\ \mu$m total [26]. The whole grown area is essentially planar, allowing interconnects, beam leads, and simple address leads to be deposited without further processing or problems with lithography or electrical or structurally weak metallizations. Typically, the native or other oxide resist is of the order 100-300 Å thick and, if a suitable etchant can be found, the polycrystalline material may be selectively lifted off, leaving behind conventional mesa regions of electrically active layers for discrete devices [27].

Shadow masking techniques [28] allow selective area epitaxy without the polycrystalline matrix and will possibly find use in future IC technology. However, resolution is limited over large areas to $\sim 5\ \mu$m because of vibrational and differential thermal expansion runout problems between mask and substrate.

7. ALLOY SEMICONDUCTOR FETs

As power FET requirements become more demanding and optimization of technology allows the fundamental limits of GaAs to be approached, further improvement may require consideration of other binary or ternary semiconductors.

The low field mobility of InAs is very high, but the small band gap (0.3 eV) means that thermal noise will be high and conventional

Schottky barrier formation is impossible. The inability to make semi-insulating InAs also means that suitable lattice matching bulk substrates will not be available.

Semi-insulating InP can be obtained by either Cr or Fe doping and, as its lattice parameter is larger than GaAs, it is a useful material as a substrate for $In_{.53}Ga_{.47}As$. This particular alloy has approximately 50% higher room temperature drift mobility than GaAs and the 0.7 eV band gap overcomes thermal noise problems. Although Schottky barrier formation is still a problem ($\phi_b \sim 0.2$ eV) [29], shallow p-n junction formation [30] or thin lattice matched InAlAs between InGaAs and gate metals look promising [31, 32].

By 1980, good quality epitaxial latticed InGaAs had been grown on InP, matched by MBE and used in high gain (x10) OPFET detectors [33]. Encouraging LPE InGaAs/InP JFETs have been prepared. As with AlGaAs buffers for GaAs FETs, it is possible to use InAlAs as electron confining buffer layers for InGaAs FETs [31]. Quaternary materials in the InGaAsP system do not exhibit the high mobilities predicted theoretically [34], presumably because of increased alloy scattering. At the time of writing they do not appear to offer any advantages over InGaAs alloys.

8. RECENT DEVELOPMENTS

Limited ultimate purity (and hence compensated films), high concentrations of deep levels (noise-generating traps), limited growth rates (slow turn round times), and inability to grow on large areas with good spatial uniformity have all been shortcomings ascribed to MBE as a competitive growth technique. However, recent developments in machine design and understanding of growth techniques have overcome all of these problems.

Attention to effusion source design and the materials used in their construction have enabled more than 8 laboratories to reproducibly lower background acceptor densities to the low 10^{14} cm^{-3} range. As this concentration is less than 1% of currently required active layer doping levels, there is effectively no further problem with residual impurity levels.

Increasing substrate temperatures and reduced arsenic to gallium flux ratios during growth have noticeably reduced deep level electron trap densities to below 10^{12} cm^{-3} [24] and hole trap densities to below 10^{13} cm^{-3}. These figures are comparable – if not lower than – the best achieved by any other epitaxy techniques and certainly less than layers produced by direct implantation into chromium doped substrates. Growth rates greater than 5 μm/hour have been recently investigated with surprisingly increased electron mobilities and no obvious degradation of other material properties [35]. Thus turn round time can be increased five fold over those using the currently accepted 1 μm per hour growth rates. Increased growth rates coupled with the now generally accepted sample exchange interlock make turn round much faster than other techniques.

Uniformity of thickness and doping levels over large areas have been limited until recently by the intensity distribution of beams from Knudsen type effusion cells. The introduction of continuous off-axis substrate rotation now allows growth over 3 inch diameter substrates with uniformity better than 2% [36].

The only criticism that cannot be dismissed is that of basic machine costs, which are typically over $200,000. However, MBE apparatus manufacturing companies will no doubt forsee the large market for inexpensive rudimentary development machines and produce machines to satisfy this market.*

*For further reading on the practical development of FET devices using MBE grown films, the reader is referred to references 37-43, and on the techniques of MBE to references 44-46.

TABLE I

Comparative Evaluation of Competitively Viable Techniques for FET Layer Preparation

	Implantation	LPE	CVD	MOCVD	MBE
Investment	⩾ $200K	∿ $40K	∿ $40K	∿$100K	∿$300K
Run up time	∿ 2 months	∿ 6 Mos.	∿ 6 Mos.	∿ 6 Mos.	2 weeks
Thro'put time (per slice)	∿ 5 min	∿ 12 hrs.	∿ 2 hrs.	∿ 2 hrs.	∿ 1 hour
Uniformity doping	Excellent	Excellent	Fair	Fair	Good
thickness	"	Poor	Good	Good	Good
Reproducibility of doping and thickness	Good	Good	Excellent	Poor	Excellent
Profile control	Poor	Poor	Good	Good	Excellent
Low field mobility	Fair	Excellent	Excellent	Good	Good
Background impurity densities cm^{-3}	∿ 10^{16}	∿ 10^{14}	∿ 10^{14}	∿ 10^{15}	low 10^{14}
Trap density (Electron)	∿ 10^{16}	∿ 10^{12}	∿ 10^{13}	∿ 10^{14}	∿ 10^{12}
Trap density (hole)		∿ 10^{13}	∿ 10^{13}	∿ 10^{15}	∿ 10^{13}
Substrate impurity out diffusion	–	Fair	Poor	Good	Good
Thickness control	–	1000Å	1000Å	∿100Å	1% of thickness
Selective epitaxy		Yes	Yes	Yes	Yes
Heterojunction buffer	No	Yes	No	Yes	Yes
Heterojunction contacts	No	No	No	No	Yes
N^+ contacts	Yes	Yes	Yes	Yes	Yes
Metalization	No	No	No	No	Yes

REFERENCES

[1] Schneider, M.V., R.A. Linke, and A.Y. Cho, *Appl. Phys. Lett.*, 31:219, 1977.

[2] Meeks, E.L., G.N. Hill, and D.W. Covington, 1st Int. Symp. on Molecular Beam Epitaxy, Paris, April 1978.

[3] Cho, A.Y., and F.K. Reinhart, *J. Appl. Phys.*, 45:1812, 1974.

[4] Wood, C.E.C., J. Woodcock, and J.H. Harris in *Gallium Arsenide and Related Compounds (St. Louis) 1978* (Inst. Phys. Conf. Ser. no. 45) p. 28, London: Adam Helger/The Institute of Physics, 1979.

[5] Covington, D., and W.H. Hicklin, *Electronics Lett.*, 14:759, 1978.

[6] Eldridge, G.W., H.M. Hobogood, and D.L. Barret, 38th Ann. Device Research Conference, Cornell U., *Abstracts*, June 1980.

[7] Blom, G.M., S.L. Blank and J.M. Woodall, ed., *Liquid Phase Epitaxy.* Amsterdam: North Holland, 1979.

[8] Hewitt, V.S., et al., in *Gallium Arsenide and Related Compounds (St. Louis) 1976* (Inst. Phys. Conf. Ser. no. 33a) p. 246. London: Adam Helger/The Institute of Physics, 1977.

[9] Depuis, R.D., and P.D. Dapkus, *Appl. Phys. Lett.*, 31:466, 1977.

[10] Ohata, K., H. Itoh, and F. Hasegawa, *Proc. 25th Int. Electron Device Meeting,* Washington, Nov. 1979, p. 277.

[11] Judaprawira, S., G.M. Metze, C.E.C. Wood and L.F. Eastman, to be published.

[12] Sze, S.M., *Physics of Semiconductor Devices.* New York: Wiley, 1969.

[13] Williams, R.E., and D.W. Shaw, *IEEE Trans. Electron. Dev.*, ED-25:600, 1978.

[14] Wood, C.E.C., D. Desimone, and S. Judaprawira, *J. Appl. Phys.*, 51:2074, 1980.

[15] Wood, C.E.C., et al., *J. Appl. Phys.*, 51:383, 1980.

[16] Hallais, J., et al. in *Gallium Arsenide and Related Compounds (St. Louis) 1978* (Inst. Phys. Conf. Ser. no. 45) p. 361. Lon-

don: Adam Helger/The Institute of Physics, 1979. And W.I. Wang et al., *Appl. Phys. Lett.*, 38:708, 1981.

[17] Padovani, F.A., in *Semiconductors and Semimetals*, vol. 7A, ed. R.K. Willardon and A.C. Beer, p. 75. New York: Academic Press, 1971.

[18] DiLorenzo, J.V., W.C. Niehaus, and A.Y. Cho, *J. Appl. Phys.*, 50:50, 1979.

[19] Stall, R.A., et al., *Electronics Lett.*, 15:800, 1979.

[20] Chan, E.Y., and H.C. Card, *IEEE Trans. Electron Devices*, ED-27:78, 1980.

[21] Borden, P.G., *Electronics Lett.*, 15:307, 1979.

[22] Dingle, R.A., *Festkoerperprobleme*, 15:21, 1975.

[23] Donnelly, J.P., and F.J. Leonberger, *Solid State Electronics*, 20:183, 1977.

[24] Stall, R.A., et al., *Electronics Lett.*, 16:171, 1980.

[25] Cho, A.Y., and W.C. Bellamy, *J. Appl. Phys.*, 46:783, 1975.

[26] Hiyamizu, S., et al., *J. Electrochem. Soc.*, 127:1562, 1980.

[27] Cho, A.Y., J.V. DiLorenzo, and G.E. Mahoney, *IEEE Trans. Electron. Dev.*, ED-24:1186, 1977.

[28] Tsang, W.T., and M. Ilegems, *Appl. Phys. Lett.*, 31:301, 1977.

[29] Kajiyama, K., Y. Mizushima, and S. Sakata, *Appl. Phys. Lett.*, 23:458, 1973.

[30] Lehaney, R.F., et al., *IEEE Electron. Device Lett.*, EDL-1:110, 1980.

[31] Ohno, H., et al., *IEEE Electron Device Lett.*, EDL-1:154, 1980.

[32] Ohno, H., C.E.C. Wood, and L.F. Eastman, *Appl. Phys. Lett.*, 52(6):4033, 1981.

[33] Gammel, J., et al., *IEEE J. Quantum Electron.*, QE:17:269, 1981.

[34] Littlejohn, M.A., J.R. Hauser, and T.H. Glisson, *Appl. Phys. Lett.*, 30:242, 1977.

[35] Chai, Y.G., *Appl. Phys. Lett.*, 37:379, 1980.

[36] Cho, A.Y., and K.Y. Cheng, *Appl. Phys. Lett.*, 38:360, 1981.

[37] Naganuma, M., et al., *Jap. J. Appl. Phys.,* 14:581, 1975.

[38] Cho, A.Y., and D.R. Chen, *Appl. Phys. Lett.,* 28:30, 1976.

[39] Wood, C.E.C., *Appl. Phys. Lett.,* 29:746, 1976.

[40] Cho, A.Y., et al., *J. Appl. Phys.,* 48:346, 1977.

[41] Wood, C.E.C., S. Judaprawira and L.F. Eastman, Int. Electron. Device Meeting, Washington, Nov. 1979.

[42] Wataze, M., et al., *Electron. Lett.,* 14:759, 1978.

[43] Bandy, S., D.M. Collins, and C.K. Nishimoto, *Electronics Lett.,* 15:218, 1979.

[44] Cho, A.Y., and J.R. Arthur, in *Progress in Solid State Chemistry,* vol. 10. ed. J.O. McCaldin, p. 157. Elmsford, NJ: Pergamon Press, 1975.

[45] Chang, L.L., and R. Ludeke in *Epitaxial Growth,* pt. A., ed. J.W. Mathews, p. 37. New York: Academic Press, 1975.

[46] Wood, C.E.C., in *Physics of Thin Films,* vol. II, eds. G. Hass and M. Francombe. New York: Academic Press, 1980.

Implantation into GaAs

F. Eisen, C. Kirkpatrick, P. Asbeck

Rockwell International Science Center,
Thousand Oaks, California.

1. INTRODUCTION

Ion implantation has been applied to the fabrication of GaAs FETs in several laboratories [1-4] and is widely used in the fabrication of GaAs integrated circuits [5]. The motivation for these applications of ion implantation to GaAs lies in the ability to control the thickness and doping level of implanted layers by varying the energy and dose of the implanted ions and in the very good uniformity of doped layers which can be reproducibly fabricated. The reproducibility and uniformity of doping which can be achieved using ion implantation are helpful in attaining good yields of discrete FETs and are essential to the attainment of acceptable yields of integrated circuits. The strong requirement for uniformity in IC fabrication accounts for the widespread application of ion implantation in the development of GaAs ICs. In addition, it is possible to develop a planar IC process technology employing implantation of semi-insulating GaAs into localized regions. Use of such a planar process should result in higher yields than processes requiring mesa etching.

N-type layers with thicknesses of several thousand angstroms and a carrier concentration of about 10^{17} electrons/cm^3 are required for the active regions of GaAs FETs, the exact requirements contingent upon the intended frequency of operation and whether the device is a low noise or power FET. By suitable selection of dopant ion and implantation conditions, it is easily possible to produce layers of thicknesses ranging from about 1000-5000Å using an implanter with a 400 keV maximum energy. Implantation of n-type dopants has other applications to GaAs FETs and ICs besides the FET active layer. In the Rockwell Schottky diode FET logic, a second implantation step is used to form the active regions of Schottky barrier switching diodes [5]. This implantation results in an electron concentration of about $8 \times 10^{16}/cm^3$ extending to a depth of about 4000Å. Also, the source and drain contacts of FETs may be implanted in order to increase their conductivity and improve device performance. This is especially important for power FETs where the higher doping of the contacts results in increased breakdown voltage for the devices. The discussion in this paper will be confined to the use of n-type dopants to meet the kinds of needs mentioned above. Implantation of p-type dopants has been used to form junction FETs for IC applications [6]. However, since this application of ion implantation to FETs and ICs has been much more limited than the implantation of n-type dopants, the discussion in this paper will concentrate on n-type doping.

In most cases the application of ion implantation to GaAs FETs and ICs involves implantation into semi-insulating GaAs. This may be either a semi-insulating buffer layer grown upon semi-insulating substrates, which are usually chromium-doped, or the semi insulating substrate-material itself. The most economical process is implantation into semi-insulating substrates. In order to obtain reproducible implantation doping profiles, it is necessary to carefully select the substrate material. The methods which are used to carry out this selection process are discussed in Section 2 of this chapter. A discussion of the properties of these substrates and some of the phenomena which have been observed when they are subjected to heat treatment are also included in this section.

The post-implantation annealing temperature required to obtain good activation of implanted n-type dopants in GaAs must be sufficiently high so that dissociation of the GaAs occurs during annealing. A variety of techniques to prevent this dissociation have been investigated. The most common technique employed in device fabrication involves encapsulation of the implanted material with a dielectric layer such as silicon nitride or silicon dioxide. The use of other dielectric materials has been investigated and research involving various capless annealing techniques has been reported. Several laboratories have employed a proximity annealing technique for device fabrication in which the implanted surfaces of two wafers are in intimate contact [3, 7]. Section 3 contains a discussion of results obtained using various annealing techniques.

The n-type dopants for GaAs which have received the most attention are silicon, sulfur, selenium, and tellurium. The first three dopants can be utilized to meet most needs of n-type implantation doping of GaAs. Tellurium doping results are quite similar to those obtained with selenium; however, it is possible to obtain a given n-type layer thickness at a lower implantation energy with selenium than with tellurium. Silicon offers the advantage of having an appreciably higher projected range at a given energy than does selenium. In addition, it is possible to obtain retrograde doping profiles with silicon implantation which may have advantages for power FETs. The range of implanted sulfur atoms at a given energy is almost as great as that of silicon. Sulfur also exhibits extensive diffusion during annealing in some cases. The magnitude of this diffusion has been found to depend upon the encapsulant used during annealing. These diffusion effects are reproducible and can be used to advantage where doped layers of substantial thickness are required. A detailed review of both high and low dose implantation results utilizing silicon, sulfur, and selenium is presented in Section 4.

The final section of this paper contains a discussion of some of the applications of ion implantation to the fabrication of GaAs FETs and ICs. This discussion will cover the variety of implantation techniques that have been employed in device fabrication, illustrate the uniformity and reproducibility of implantation results, and present the performance characteristics of some of the discrete FETs and integrated circuits which have been fabricated utilizing ion implantation.

2. SUBSTRATE EFFECTS

The growth of GaAs is less developed and more complex than the production of Si and, as a result, currently available substrates are, in general, less pure and less reproducible than Si wafers. Unless adequate selection procedures are used, the lack of GaAs substrate reproducibility can cause large variations in implanted layer characteristics, which are most evident in the low dose ($1 - 3 \times 10^{12}$ cm^{-2}) donor implants typically used to obtain FET channel layers.

To date, most experience has been accumulated with commercially available Cr doped semi-insulating substrates. These are typically grown by the Horizontal Bridgman or Gradient Freeze techniques, which produce up to 1 kg in weight. In most instances, ingots produced without deliberate doping display n-type conductivity, with electron densities in the range 10^{15} to 5×10^{16} cm^{-3}. The residual doping in these ingots has been attributed principally to Si, which can be incorporated into the melt from decomposition of the (typically quartz) growth apparatus [8]. High resistivity ($\rho > 10^7\ \Omega$ cm) ingots are produced by adding Cr to the melt in concentrations between 2×10^{16} and 2×10^{17} cm^{-3}; Cr acts as a deep acceptor in GaAs and compensates the residual donors [9]. Additions of oxygen to the growth system are frequently made; the oxygen partially suppresses the quartz decomposition [10], and may itself become incorporated in the growing crystal as a deep donor or in complexes. Variations of stoichiometry in different ingots are also possible, based on the Ga/As ratio used in the melt. The compensation mechanism and charge balance in typical ingots is not simple and involves at least one set of deep donors and one set of deep acceptors [11, 12]. The detailed relationship between the various growth parameters and impurity densities and the performance of semi-insulating GaAs substrates is not properly understood and is the subject of considerable current effort.

One manifestation of the difficulties that arise with improper GaAs substrates is the formation of conducting surface layers when the (originally semi-insulating) substrates are subjected to a high temperature (700 - 900°C) anneal similar to the one used to activate implanted species and remove implant damage [13, 14]. Such "surface conversion" is particularly detrimental to the processing of integrated circuits, which can rely on the semi-insulating nature of the substrate for inter-device isolation. The characteristics of the conducting layer have been found to depend on the particular technique used for the anneal. When anneals are carried out in vacuum or in a H_2 ambient without the use of a cap, the converted layer is generally found to be p-type [14]. On the other hand, when a cap of Si_3N_4 is used, an n-type layer is observed if surface conversion occurs. In both cases, the peak carrier concentration is generally below $1 \times 10^{17}\ cm^{-3}$, and the depth is generally less than several micrometers. It has been found that GaAs ingots which are resistant to the formation of p-type layers in H_2 anneals are not necessarily free of surface conversion in capped anneals; the inverse of this is also true, so that the substrate requirements for processes involving uncapped heat treatments (for example, epitaxial growth) and capped heat treatments (ion implantation) are, in general, distinct.

When GaAs substrates of poor quality are used for ion-implantation, variations in carrier concentration and effective depth of the implanted layer are observed. In typical donor implants which involve the use of a Si_3N_4 cap during the post-implant anneal, the most prominent manifestation of unqualified substrates is the occurrence of greater than 100% apparent activation of the implanted species, together with the formation of an extended "tail" to the corresponding carrier density profile (which can penetrate up to 1 μm into the substrate). The distortion of the carrier density profile is particularly detrimental in FET channel applications in which control over the pinchoff voltage V_p is desired, because the contribution of spurious donors to V_p is weighted more heavily at greater depths. There are substantial indications that the "deep tail" phenomenon is strongly correlated with n-type surface conversion in unimplanted samples, and, in fact, the effects are roughly additive: the electron density profile in material which exhibits surface conversion and has been implanted with donors consists of independent contributions due to the implanted species and the spurious donors [13, 15, 16]. A correlation has been observed between the occurrence of n-type surface conversion and the position of the GaAs substrate within the original ingot; surface conversion is observed more frequently in the first-to-freeze (front) portion of the ingot than in the last-to-freeze (tail) of the crystal. This correlation may be related to the Cr concentration in the ingot, which typically increases from front to tail due to the very low ($\sim$6-8 x 10^{-4}) Cr distribution coefficient at the growth temperature [17]. In general, the uniformity of material behavior across a given wafer (or cross-section of the ingot) has been found to be very good.

Reports have also been published of GaAs substrates that consistently show very low (< 50%) activation of low dose donor implants. As in the case of surface conversion, the implant results with a given ingot appear to be a function of the annealing technique used. A particularly emphatic example is given by Donnelly, who observed that the activation of donor implants depended on the deposition rate of the pyrolytic Si_3N_4 cap which they used [13].

To fulfill the need for accurately controlled implanted layer characteristics, screening procedures for the GaAs substrate material are used to eliminate unqualified ingots from device processing. These screening procedures typically involve subjecting wafers from the front and tail ends of an unknown ingot to a variety of tests, which include (a) performing a low-dose donor implant, followed by appropriate (generally capped) annealing and measurement of the implant profile; (b) annealing the wafer without implantation and testing for surface conversion; and (c) performing an implant of ions of an inert gas (such as Kr), followed by annealing and testing for surface conversion. In this last test, lattice disorder is introduced to roughly the same extent as in a Se implant; any damage-enhanced contribution to surface conductivity changes can thus be sensitively detected [1]. Experience has shown that implantation results with good reproducibility can be obtained with the "qualified" substrates that pass these tests.

The mechanisms responsible for the variability of implant characteristics in different substrates are not fully understood. A variety of possible effects must be considered: (1) compensation of a portion of the implanted donors must be expected *a priori* due to the fact that the number of deep acceptors (Cr) introduced into the material typically exceeds the number of residual donors; (2) additional donors may be formed during the anneal, and since type conversion appears to be an exclusively surface rather than bulk phenomenon, these donors must be introduced from surface-related processes; (3) acceptors already present in the material may be removed, producing a net excess of donors; (4) the atomic profile or electrical activation of implanted donors may be a function of the substrate characteristics, because of mechanisms such as diffusion-enhancement due to non-stoichiometry, or complex formation.

Recent investigations of the behavior of Cr in GaAs appear to offer explanations for a variety of substrate phenomena [15, 16, 18-20]. Measurements by secondary ion mass spectrometry (SIMS) have shown that when Cr doped GaAs substrates are capped and annealed at temperatures typically used in implantation processes, a redistribution of the Cr occurs. Figure 1 illustrates the measured Cr profiles in a number of samples capped with reactively sputtered Si_3N_4 and annealed at 850°C. The Cr concentration is decreased from its bulk level over a distance of several microns; in addition, a strong accumulation of Cr is observed within 500-1000Å of the Si_3N_4 - GaAs interface (not apparent on the scale of Figure 1). The Cr profiles are in good agreement with an error function fit, characteristic of diffusion with a concentration independent diffusion coefficient D, with $D \approx 10^{-11}$ cm^2/s. The general characteristics of the profiles and the specific value of D were found to be identical in both qualified and unqualified substrates, and did not change when substrates were implanted with Se or Kr ions at low fluences [15].

As noted earlier, an expected consequence of the decrease in the density of Cr acceptors near the GaAs surface is the formation of a region with excess donor concentration. This interpretation is supported by the detailed comparison of Cr profiles and carrier density profiles in samples that display n-type surface conversion, and in Se implanted samples that display excess activation. According to this model, the excess donor density observed at the surface is directly related to the residual net donor density present in the substrate, while the thickness of the n-type region is a function of the bulk Cr concentration, and its diffusion length during the anneal. In order for substrates to maintain high surface resistivity after the anneal, the undercompensated region at the surface must be sufficiently thin that it is fully depleted, due to the Fermi level pinning at the sample surface and in the bulk. This criterion may be translated into the requirements that the residual donor density be low and the bulk Cr concentration be at least moderately high ($\gtrsim 10^{16}/cm^3$).

Another consequence of the Cr redistribution phenomenon is the relative insensitivity of donor implantation results to the bulk Cr content of qualified substrates. As observed earlier, partial compensation of implanted donors is expected in substrates that contain a substantial excess of Cr over residual donors. The experimentally observed reproducibility of implant characteristics is inconsistent with the probable variation of excess bulk Cr density which will occur among different qualified ingots or among wafers selected from different positions of the same qualified ingot. This

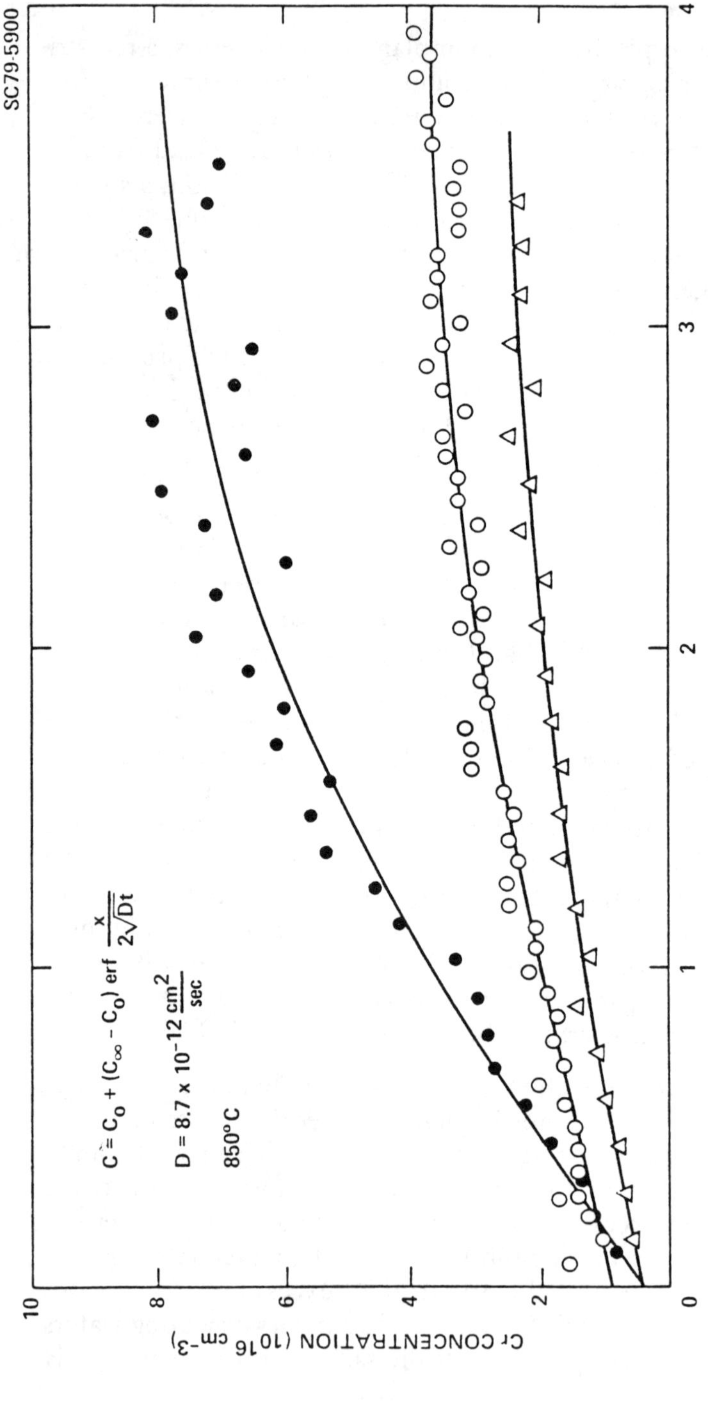

Figure 1/*Cr distribution measured by SIMS in substrates annealed with a Si_3N_4 cap. The experimental data have been fit with error function curves corresponding to a diffusion coefficient D = 8.7×10^{-12} cm^2/sec.*

dilemma can be resolved by noting that the Cr concentration in the region of the implant is reproducibly reduced to low levels during the anneal.

The driving force for the Cr diffusion is not currently known. It has been suggested that the essential cause is the low solubility of Cr in GaAs at the temperature of the anneal. Other possible effects include gettering by lattice disorder at the cap-GaAs interface or by chemical reactions in that region.

A number of other contributions to n-type surface conversion during capped anneals have been suggested, and it is likely that one or more of them contribute to the excess donor density, depending on the annealing technology used. These include: indiffusion of n-type impurities present on the substrate surface prior to capping (due to the difficulty of preparing pure GaAs surfaces); indiffusion of Si from caps of Si_3N_4 or SiO_2 [21] ; and indiffusion of vacancies of Ga or As [22]. Vacancies on either sub-lattice may have a direct doping effect, or they may affect doping by complexing with donors and acceptors already present in the material. Changes in vacancy concentrations can lead to a change of the lattice location of the amphoteric impurities Si and C, which changes their electrical activity from donor-like to acceptor-like (or vice versa) [22]. While a number of these mechanisms would appear to be equally applicable to qualified and unqualified substrates, the magnitude of the doping effect produced probably depends upon the extent of compensation in different ingots. Moreover, different defect reactions may occur according to the nature and density of complexes originally present in the material.

The nature of p-type layers formed on the surface of semi-insulating GaAs during (uncapped) anneals in H_2 atmospheres has been investigated by photoluminescence, transport measurement as a function of temperature and SIMS [14]. It has been found that accumulations of Mn occur at the surface. Also, the ionization energy of the Mn acceptor (~ 0.1 eV) was found to agree with the temperature dependence of resistivity in the p layer. The reason for the Mn accumulation is not currently known. Among the possible sources of the element are contamination from the H_2 gas stream, which is typically passed through Pd purifiers, and migration from the bulk of the substrate.

Considerable work has been done to produce substrates for ion-implanted layers by different methods, in an effort to overcome the difficulties found in the use of boat-grown Cr-doped substrates. Epitaxial growth of very high purity buffer layers (with net donor densities below 10^{14} cm^{-3}) has been possible for some years. More recent work has been successful in producing semi-insulating layers by incorporating Cr during growth in VPE and LPE. Preliminary results indicate that substrates free of surface conversion problems can be reproducibly obtained [23]. Another very promising area of research is the growth of ingots from the melt with the liquid-encapsulated Czochralski (LEC) technique [24]. Particularly when crucibles of pyrolitic boron-nitride are used, undoped ingots of high resistivity ($\rho > 10^6$ $\Omega/\square$) can be produced. The use of LEC undoped or Cr doped semi-insulating substrates in ion-implantation has been reported recently, with very encouraging results [25, 26].

3. ENCAPSULATION AND ANNEALING

Annealing of the implanted GaAs is necessary to remove the radiation damage and activate the dopants introduced by ion implantation. For the lower doses used in FET fabrication, a typical anneal cycle involves temperatures from 800-900°C to achieve maximum carrier concentrations. The compound GaAs, however, is subject to dissociation at temperatures greater than 600°C. Therefore, the semiconductor surface must be protected by an encapsulating layer or special ambient to prevent decomposition and inhibit outdiffusion of the implanted impurities.

A poor quality encapsulant or inappropriate ambient can result in a number of detrimental effects. Inappropriate annealing techniques can cause unusually low or high activation of the implanted dopant, buried high resistivity layers, and high defect levels. Since a deposited thin film cap material is under internal stress and has a different coefficient of expansion, wafer warpage can be an additional consideration. If the encapsulant is in place during implant, knock-ons into the substrate can be a factor influencing the number of defects.

A number of thin films have been explored as possible encapsulants for implanted GaAs. The question of encapsulant superiority depends not only upon the material, but also the method and parameters of deposition as well as the particular implanted dopant. For n-type dopants, pyrolytic Si_3N_4 [28], sputtered Si_3N_4 [29], SiO_2 [30], evaporated Al [31], sputtered AlN [32], and double layers of As-doped SiO_2 over Si_3N_4 [33] have all been applied with some degree of success.

The particular capping technique used for protection of the GaAs surface during annealing can have pronounced effects on the doping profile and activation of implanted dopants, as shown in Figure 2 [34]

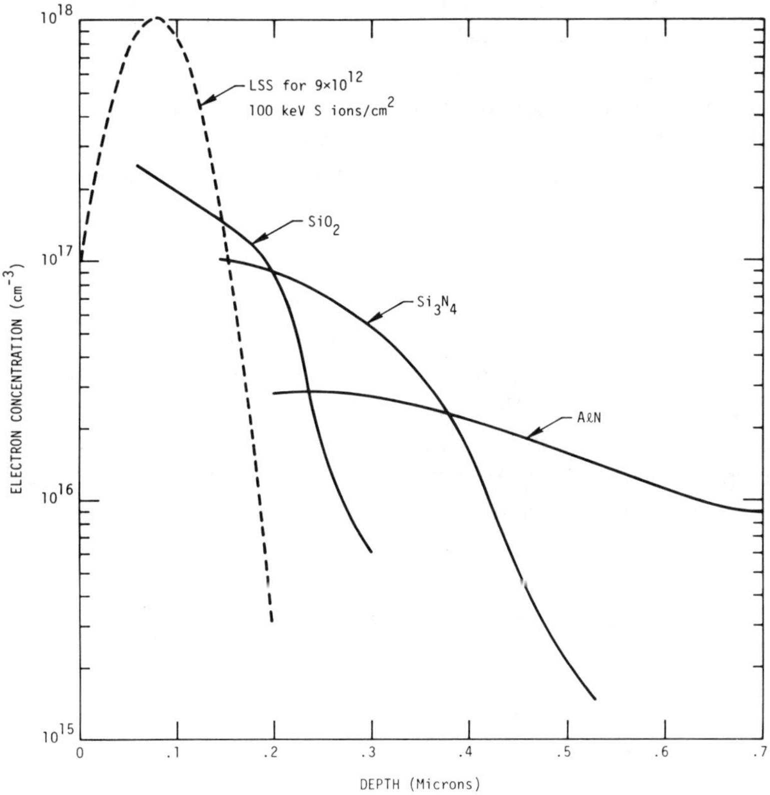

Figure 2*/Electron concentration profiles obtained from Schottky barrier C-V measurements for semi-insulating GaAs samples implanted at 350°C with 9 x 10^{12} 100 keV S ions/cm^2 and annealed at 850°C for 30 minutes with the indicated caps.*

for low dose sulfur implants. The SiO_2 cap results in a more shallow profile and higher electron concentrations than Si_3N_4 or AlN for this case. Differences in the outdiffusion of S^+ to the surface or into the cap are believed to be responsible for these variations.

Silicon dioxide was the earliest capping material employed [35], however, gallium can diffuse through this layer [36]. Provided it has a low oxygen content, silicon nitride can alleviate this problem, as well as retard As^+ outdiffusion and oxygen indiffusion. Lower trapping densities have also been observed for GaAs capped with Si_3N_4 [39]. These factors have resulted in the general acceptance of Si_3N_4 as a suitable encapsulant material. A layer of SiO_2 is often deposited over the thin Si_3N_4 film for protection against physical damage. In this laboratory, reactively sputtered Si_3N_4 layers which display very good adherence through 30 min, 850°C annealing cycles, have been used successfully to cap implanted n-type layers for reproducible profiles and activation levels.

The presence of the cap and the substrate temperature during the implantation process can have a strong impact on the resulting dopant profiles. Implantation through Si_3N_4 has been shown to increase the total deep level concentration in GaAs [40] compared to bare surface implants. For implants through thin Si_3N_4 layers, low beam currents together with elevated sample temperatures have resulted in reduced defect concentrations and more Gaussian profiles [41]. The substrate temperature is a particularly important parameter for high dose implants, where heating of the substrate can prevent formation of an amorphous layer and allow higher doping levels to be achieved. In Figure 3, carrier concentration profiles are shown for S^+ implanted at 350°C and room temperature. At the higher fluence, considerable differences in electron concentration and profile are observed for the two different implant temperatures. In the lower dose case, the effect is much less pronounced. For the doses typically employed in FET fabrication ($\leqslant 10^{13}$ ions/cm^2) room temperature implants are generally found to produce satisfactory doping levels together with reproducible profiles.

In addition to the encapsulation techniques involving the deposition of thin films, a number of techniques have been developed in which the sample is annealed without a protective thin film layer. GaAs

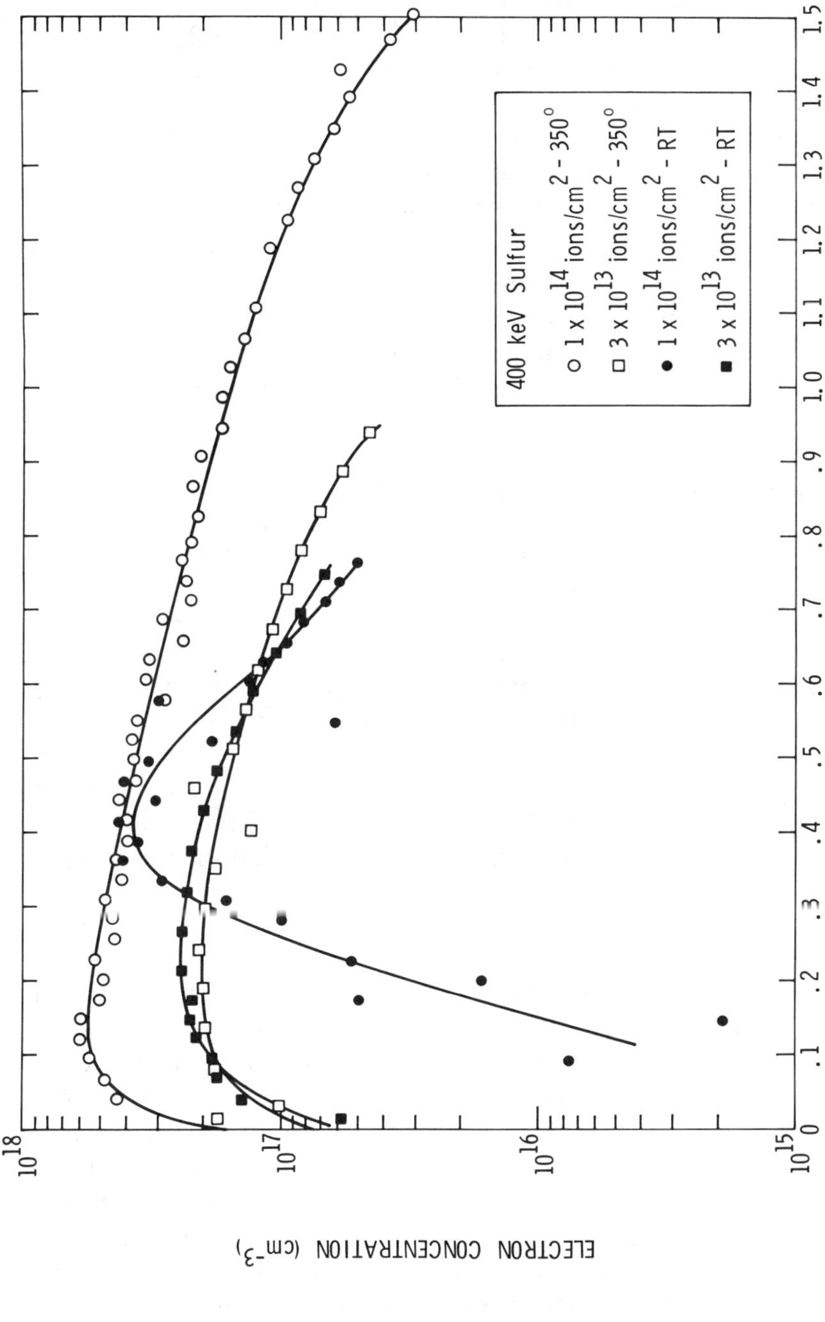

Figure 3*/Electron concentration profiles for GaAs samples implanted with the indicated doses of 400 keV sulfur ions at the indicated temperatures and annealed at 850°C for 30 minutes.*

samples embedded in powdered graphite and annealed in H_2 have resulted in good surfaces with little apparent As loss [42] and profiles similar to those from samples annealed with a reactively sputtered nitride cap. Another technique, CAT annealing, utilizes hydrogen flow with a melt of GaAs in Ga [43]. An overpressure of arsenic has also been used as a capless annealing ambient [44]. All these techniques attempt to keep As from escaping from the GaAs surface.

Some recent studies have employed laser or electron beam irradiation to anneal the lattice damage and activate implanted dopants in GaAs [45, 46]. Although high electron concentrations have been achieved for high dose implants, mobilities have generally remained poor, and for low implant doses, low levels of activation have been observed thus far.

4. CHARACTERISTICS OF n-TYPE DOPANTS

Typical electron concentration profiles for low dose silicon sulfur or selenium implants in GaAs are shown in Figure 4. These results were obtained from measurements of capacitance versus voltage using aluminum Schottky barriers on the surface of the implanted and annealed samples. Calculated profiles of the implanted species obtained using LSS range statistics [47-49] are also shown in Figure 4. The calculated and measured profiles are quite similar for silicon and selenium, indicating that a high percentage of the implanted atoms are effective in doping the GaAs and that there is only limited diffusion of these dopants atoms during implantation or post-implantation annealing. The calculated and measured profiles for sulfur are quite different from each other. Only about 20% of the implanted sulfur is effective in doping the GaAs and the depth of the doped layer is much greater than expected from the sulfur range statistics. Data obtained from samples implanted with a radioactive sulfur isotope indicate that these phenomena are due to diffusion of the sulfur which occurs during post-implantation annealing [34]. A significant fraction of the implanted sulfur is lost from the GaAs as a result of this diffusion. This loss of implanted sulfur accounts for the apparent low doping efficiency (sheet electron concentration/implanted dose) of the implanted sulfur. The doping efficiency of implanted sulfur has been observed to increase with increasing implantation energy [34]. This effect is probably also a consequence of the out diffusion of sulfur during annealing.

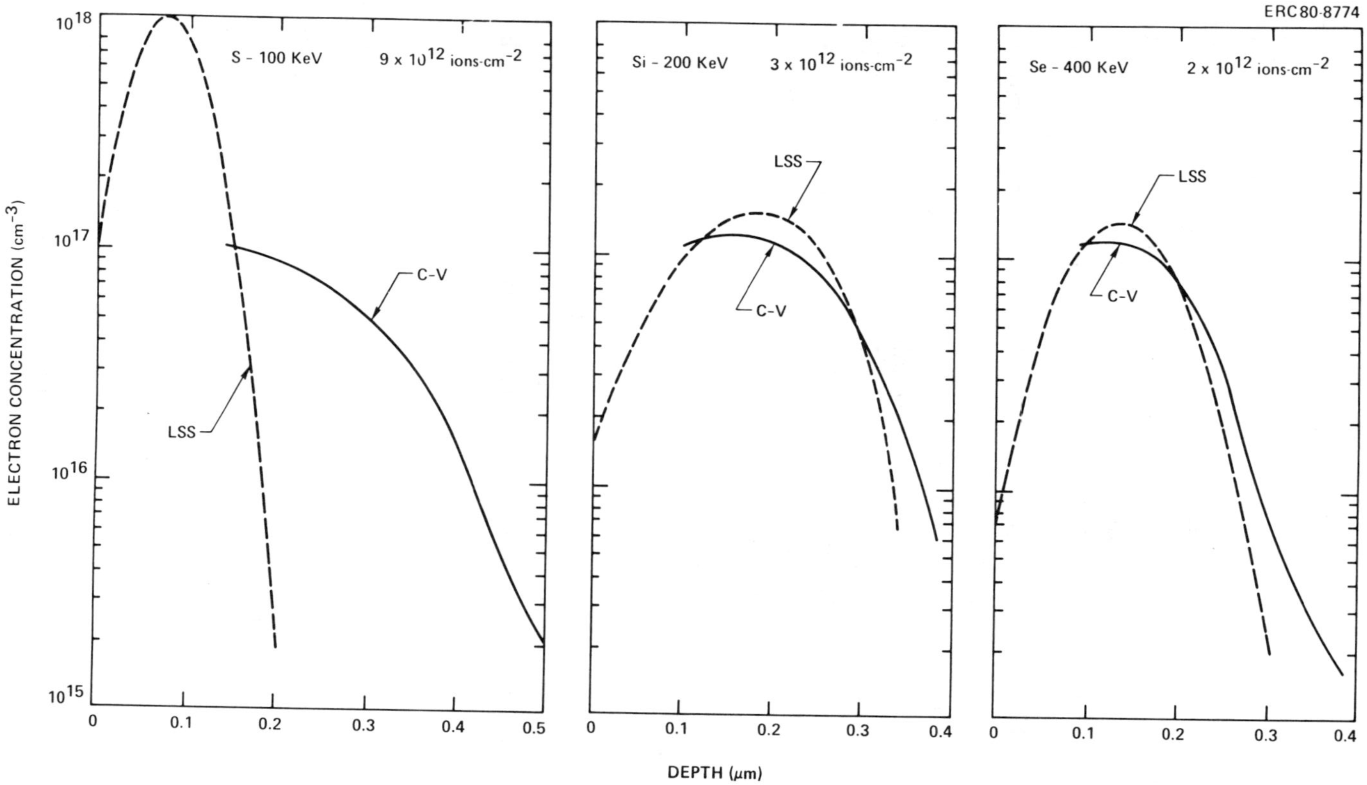

Figure 4*/Comparison of electron concentration profiles determined from C-V measurements and calculated [47-49] distribution of the implanted species for (a) 100 keV sulfur, (b) 200 keV silicon and (c) 400 keV selenium. The samples were annealed at 850°C for 30 minutes.*

The electron mobility in ion implanted layers, such as those for which data are shown in Figure 4, has been observed to be comparable to that measured for good epitaxial layers with similar electron concentrations. For example, electron mobilities of about 5000 cm^2/v-sec have been measured in selenium implanted layers exhibiting electron concentration profiles similar to that shown in Figure 4 for the selenium implanted sample. Recently, Jervis et al. [40] have shown that the deep trap density in ion implanted layers in GaAs can be quite low. The deep traps which they observed in their work appear to be related to properties of the substrate material rather than to implantation effects.

The preceding discussion involved low dose implants which produced a maximum electron concentration of about $10^{17}/cm^3$, and for which the doping efficiencies for silicon and selenium implantations were close to 100%. For substantially higher dose implants, which are intended to produce as high an electron concentration as possible, the doping efficiency is observed to be appreciably lower than 100% [27, 29, 32]. The carrier concentrations achieved are found to be functions of implantation temperature, unlike the lower dose cases discussed above, and of the annealing cap.

Carrier concentration profiles measured in samples implanted with 1×10^{14} 400 keV selenium ions at several different implant temperatures are shown in Figure 5 [50]. The activation of the selenium in the samples implanted at room temperature is much lower than in those implanted at elevated temperatures ranging from 250°C to 500°C. This kind of temperature dependence has been observed for all the group VI dopants in GaAs. The activation of high dose selenium implants in GaAs has been observed to depend upon the material used for the annealing cap. The data in Figure 5 were obtained utilizing a pyrolytically deposited silicon nitride cap. Similar results have been obtained using an aluminum oxi-nitride cap [32]. However, the use of sputtered silicon nitride has resulted in somewhat lower peak doping levels than those shown in Figure 5 [32].

Carrier concentration profiles measured in samples implanted with 1×10^{14} 400 keV sulfur ions/cm^2 were shown in Figure 3. The sample implanted at 350°C showed a much higher activation of the implanted sulfur than did the one implanted at room temperature.

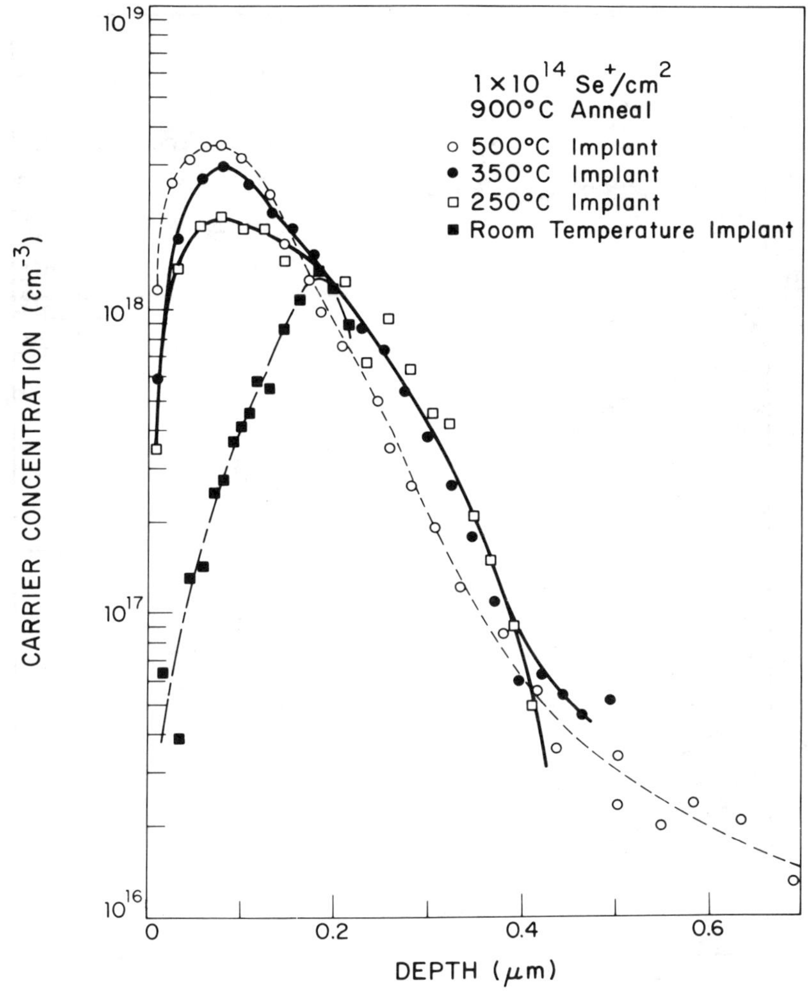

400 kV Se^+ IMPLANTED IN GaAs THROUGH 700Å Si_3N_4

Figure 5/Electron concentration profiles for GaAs samples implanted and annealed under the indicated conditions.

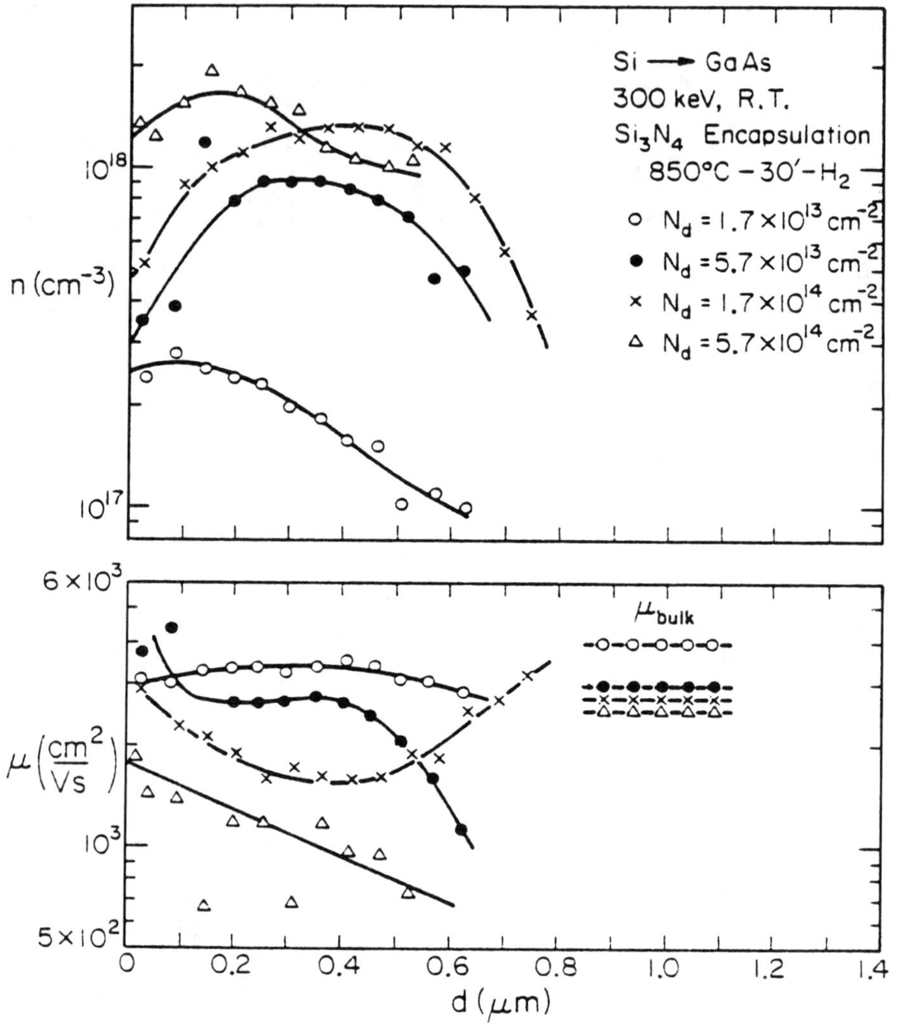

Figure 6*/Electron concentration n and mobility μ versus depth d. The profiles are for four samples implanted with four different doses, but annealed at the same temperature, 850°C. The* μ_{bulk} *values are estimated as in Figure 2.*

The n-type layer in the sample implanted at 350° extends to depths well beyond 1 μm, exhibiting the extensive diffusion effects which were mentioned previously for sulfur implants. As a result of this diffusion, the maximum electron concentration is much lower than was shown in Figure 5 for selenium implanted samples.

In contrast to the results for selenium and sulfur implants discussed above, good activation of high dose silicon implants performed at room temperature has been observed. Electron concentration and mobility profiles for several different implantation doses are shown in Figure 6 [51]. Maximum electron concentrations in the range of 1 to 2 x 10^{18}/cm^3 were obtained. This is of considerable practical interest since the processing of FETs and ICs is simpler if the implantation steps can be carried out at room temperature.

Higher electron concentrations than those shown above have been obtained by employing either double implantation or laser or e-beam annealing of the implanted samples. The implantation of a group VI dopant leads to a gallium deficiency in GaAs assuming that that dopant resides on arsenic site after annealing. It has been suggested that implantation of gallium would compensate for this deficiency and might make it possible to achieve higher electron concentrations than can be obtained utilizing selenium implantation alone [52]. Recently, Inada et al. have reported obtaining electron concentrations as high as 2 x 10^{19} /cm^3 in samples which were implanted with both selenium and gallium and which were annealed at 950°C [53]. Electron concentrations of about 2 x 10^{19} /cm^3 have also been observed in samples which were annealed utilizing a pulsed laser beam or a pulsed electron beam [54 - 56]. However, such annealing techniques have not been successfully applied to samples implanted with the lower doses (approximately 3 x 10^{12} ions/cm^2) required for FET active layers.

The properties of silicon, sulfur, and selenium as n-type implanted dopants in GaAs can be summarized as follows: For doses of a few x 10^{12} ions/cm^2, which are required for FET active layers, selenium and silicon show good activation and only limited diffusion during post implantation annealing. Sulfur exhibits low apparent activation and extensive diffusion at both low and high doses. Silicon may be the most useful dopant for achieving high concentrations in devices since implantation at elevated temperatures is not required in order to obtain good activation of the implanted silicon. Special techniques such as double implants or laser or e-beam annealing usually seem to be required in order to achieve carrier concentrations above 3 to 4 x 10^{18} /cm^2.

5. APPLICATIONS OF ION IMPLANTATION

The application of ion implantation technology to the fabrication of FETs offers a number of advantages over the epitaxial processes developed earlier. Ion implantation into semi-insulating GaAs substrates provides the opportunity to develop new fabrication approaches not previously feasible through epitaxial techniques. In addition to versatility, implantation offers proven uniformity, reproducibility, and lower cost. The technique is not bound by traditional fabrication methods which can limit the ultimate density and yield of small geometry devices. Using multiple implants into isolated regions of a semi-insulating substrate, optimization of individual circuit elements can be achieved in a manner not realizable through epitaxy.

For implementation in FET processing, the n-type dopants which have received the most attention are S, Si, Se, and Te. Of these, the first three have been most utilized in device fabrication. Although implanted Te exhibits characteristics similar to Se, its penetration depth is so shallow that high energies are needed to implant it to the depths generally of interest. This is shown by Table I, in which the projected range and straggling of the four major n-type dopants, as calculated by LSS theory [48], are shown for several energies. The actual depth profile of the implanted species can be deeper due to diffusion, which can be appreciable for S.

Doping by implantation is particularly appropriate for Schottky barrier GaAs FET applications. Typically, such devices have an n-type channel region approximately 2000Å thick, doped to a level $\sim 10^{17}$ electrons/cm^3, formed on a semi-insulating substrate. Implantations of S [2, 57-61], Si [3, 62, 63] and Se [1, 64-66] have been used effectively to fabricate these layers. In a substantial number of these efforts, the implant has been used to form an n-type layer over the entire surface of the semi-insulating substrate. Isolation is generally created by mesa etching. Of course, isolation can also be achieved

TABLE I

Projected Ranges (R_p) and Standard Deviations (ΔR_p) for Major n-Type Dopants Used in Implanted FETs [48]

		Energy (keV)		
Ion		**300**	**360**	**400**
Sulfur	$R_p(\mu m)$	0.2292	0.2758	0.3067
	$\Delta R_p(\mu m)$	0.0899	0.1021	0.1096
Silicon	$R_p(\mu m)$	0.2632	0.3160	0.3507
	$\Delta R_p(\mu m)$	0.1003	0.1131	0.1210
Selenium	$R_p(\mu m)$	0.1028	0.1233	0.1371
	$\Delta R_p(\mu m)$	0.0438	0.0510	0.0557
Tellurium	$R_p(\mu m)$	0.0707	0.0833	0.0917
	$\Delta R_p(\mu m)$	0.0287	0.0333	0.0364

by implanting localized active regions in a well-qualified semi-insulating substrate [66]. Lower specific contact resistance in the source and drain contact regions can also be produced by implantation in these areas [66, 67]. This can also result in improved performance for low noise and power FETs.

Since the characteristics of a GaAs FET device are strongly correlated with the n-type layer thickness and doping, the uniformity and reproducibility advantages of the implantation technique have important impact on the yield of devices which display appropriate behavior. In Figure 7, the degree of control which can be achieved is indicated by the close grouping of the S parameters for two randomly selected GaAs FETs from each of three different wafers. A further indication of this uniformity is shown in Figure 8, by histograms of pinchoff voltage V_p. FETs produced by ion implantation have displayed noise and gain parameters equivalent to those measured from epitaxial devices [60]. In addition, the drift in these parameters over time has been observed to be less for the implanted GaAs FETs [60]. These results indicate the potential advantages in yield and uniformity achievable through implantation.

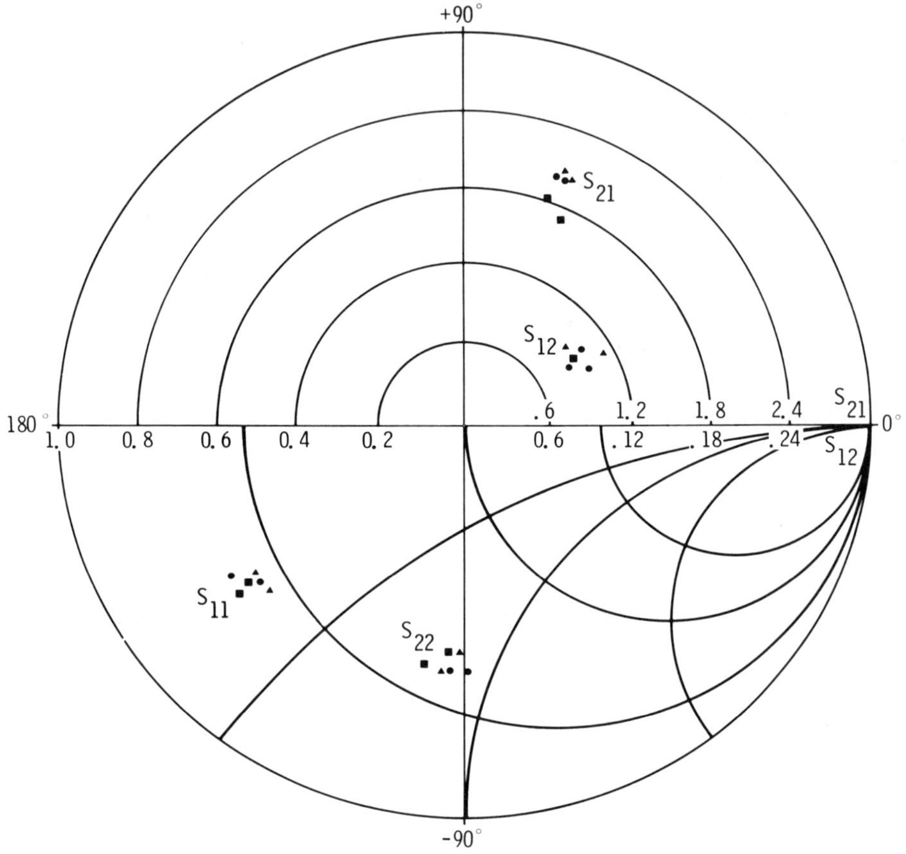

Figure 7*/S-parameters for two randomly selected Se implanted GaAs FETs from three Se implanted wafers, measured at 10 GHz [8].*

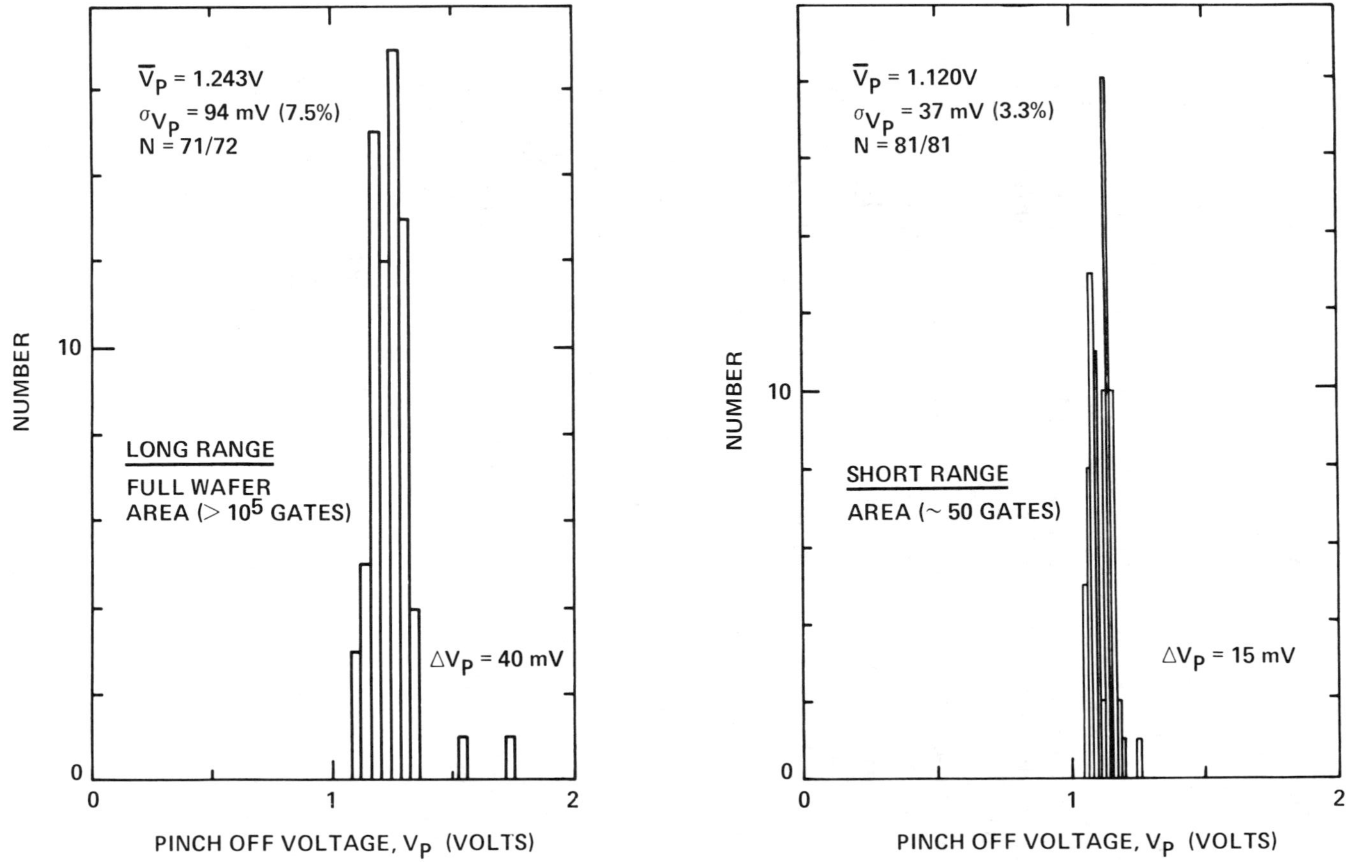

Figure 8*/Histograms of planar GaAs FET pinchoff voltages showing long and short range statistical distributions, indicating implanted layer uniformity.*

In the past several years, work involving the processing of GaAs integrated circuits has been undertaken in several laboratories [68-70] in which GaAs FETs provide the active elements. Because of their higher electron mobility compared to silicon, GaAs integrated circuits offer potential for higher speed operation. The greater availability of quality semi-insulating GaAs substrates, along with process development, have been factors in promoting this semiconductor for integrated circuit fabrication. In addition, the parasitic capacitances are considerably less for active device layers on semi-insulating substrates.

Implantation is a key process in GaAs integrated circuit for providing continuous active layers or localized doped regions in semi-insulating material [68]. An example of the latter approach has been developed at Rockwell International, where two implantation steps are utilized in the process. The first implant, which forms the thin (1000-1500Å) FET channels, uses Se, a dopant which displays little diffusion during post-implantation annealing. The second implant, which forms the high speed switching diodes for the circuit design being implemented, uses S to produce limited doping levels for minimal capacitance and low sheet resistances. The deep diffusion of the S is an asset in producing such layers. Both the S and Se implants are used for level shifting diodes and to enhance the doping under ohmic contact regions. By these techniques, implantation applied to the fabrication of integrated circuits allows the doping in different types of devices composing the integrated circuit to be individually optimized. As for the FETs, the uniformity and reproducibility which can be achieved through implantation is an important advantage for the fabrication of GaAs integrated circuits, since each device must be functional for a working circuit. Tests on FETs produced by this technique have indicated that this dual implantation technique produces the uniformity needed throughout the wafers for good yield of operational circuits.

REFERENCES

[1] Higgins, J.A., R.L. Kuvas, F.H. Eisen, and D.R. Ch'en, *IEEE Trans. Elec. Devices,* ED-25:587, 1978.

[2] Hunsperger, R.G., and N. Hirsch, *Electron. Lett.,* 9:1, 1973.

[3] Ohata, K., T. Nozaki, and N. Kawamura, *IEEE Trans. Elec. Devices,* ED-24:1129, 1977.

[4] Doerbeck, F.H., H.M. Macksey, G.Brehm, and W.R. Frensley, *Electron Lett.,* 15:577, 1979.

[5] Eden, R.C., B.M. Welch, R. Zucca, and S.I. Long, *IEEE J. Solid State Circuits,* SC-14:221, 1979.

[6] Troeger, G.L., et al., 1979, *IEDM Tech. Digest,* p. 497.

[7] Mandal, R.P. and W.R. Scoble, *Gallium Arsenide and Related Compounds 1978* (Inst. of Phys. Conf. Ser. 45), p. 462. Bristol: The Institute of Physics, 1979.

[8] Cochran, C.N., and L.M. Foster, *J. Electrochem. Soc.,* 109: 149, 1962.

[9] Cronin, G.R., and R.W. Haisty, *J. Electrochem. Soc.,* 111:874, 1964.

[10] Woods, J.F., and N.G. Ainslie, *J. Appl. Phys.,* 34:1469, 1963.

[11] Zucca, R., *J. Appl. Phys.,* 48:1987, 1977.

[12] Lindquist, P.F., *J. Appl. Phys.,* 48:1262, 1977.

[13] Donnelly, J.P., C.O. Bozler, and W.T. Lindley, *Sol. St. Electronics,* 20:273, 1977.

[14] R. Zucca, *Gallium Arsenide and Related Compounds 1976* (Inst. Phys. Conf. Ser. 33b), p. 228. Bristol: The Institute of Physics, 1977.

[15] Asbeck, P.M., et al., *IEEE Electro. Dev. Lett.,* March 1980.

[16] Favennec, P.N., and H.L. Haridon, *Appl. Phys. Lett.,* 35:699, 1979.

[17] Martin, G.M., M.L. Verheijke, J.A.J. Jansen, and G. Poibland, *J. Appl. Phys.,* 50:467, 1979.

[18] Huber, A.M., et al., *Appl. Phys. Lett.,* 34:858, 1979.

[19] Evans, C.A., Jr., V.R. Deline, T.W. Sigmon, and A. Lidow, *Appl. Phys. Lett.,* 35:291, 1979.

[20] Tuck, B., G.A. Adegboyega, P.R. Jay, and M.J. Cardwell, *Gallium Arsenide and Related Compounds 1978* (Inst. Phys. Conf. Ser. 45), p. 114. Bristol: The Institute of Physics, 1979.

[21] Kushiro, Y., T. Seimiya, O. Sinbori, and T. Kobayasi, *J. Appl. Phys.,* 48:1636, 1977.
[22] Lum, W.Y., et al., *Appl. Phys. Lett.,* 30:1, 1977.
[23] For a review of epitaxial growth of GaAs layers, see Chapter 2 of this volume.
[24] See Chapter 1 of this volume.
[25] Swiggard, E.M., S.H. Lee, and F.W. Van Batchelder, *Gallium Arsenide and Related Compounds 1978* (Inst. Phys. Conf. Ser. 45), p. 125. Bristol: The Institute of Physics, 1979.
[26] Fairman, R.D., and J.R. Oliver, *Semi-Insulating III-V Materials,* p. 83. Orpington, Kent: Shiva Publishing Ltd., 1980.
[27] Donnelly, P., W.T. Lindley, C.E. Hurwitz, *Appl. Phys. Lett.,* 27:41, 1975.
[28] Harris, J.S., et al., *Appl. Phys. Lett.,* 21:601, 1972.
[29] Eisen, F.H., et al., *Sol. St. Electron.,* 20:219, 1977.
[30] Harris, J.S., and F.H. Eisen, *Rad. Eff.,* 7:123, 1971.
[31] Sealy, B.J., and R.K. Surridge, *Thin Solid Films,* 26:119, 1975.
[32] Gamo, K., et al., *Sol. St. Electron.,* 20:213, 1977.
[33] Lidow, A., J.F. Gibbons, and T. Magee, *Appl. Phys. Lett.,* 31:158, 1977.
[34] Eisen, F.H., and B.M. Welch, in *Ion Implantation in Semiconductors,* ed. F. Chernow, J.A. Borders and D.K. Brice, p. 97. New York: Plenum Press, 1977.
[35] Foyt, A.G., J.P. Donnelly, and W.T. Lindley, *Appl. Phys. Lett.,* 14:73, 1969.
[36] Gyulai, J., J.W. Mayer, I.V. Mitchell, and V. Rodriguez, *Appl. Phys. Lett.,* 17:332, 1970.
[37] Yokota, K., et al., *Jpn. J. Appl. Phys.,* 17:1881, 1978.
[38] Beelk, E.C., A.E. Glaccum, P.L.F. Hemment, and B.J. Sealy, *Rad Eff.,* 22:253, 1974.
[39] Yokayama, N., S. Ohkawa, T. Sakurai, T. Furaya, and H. Ishikawa, *Appl. Phys. Lett.,* 28:405, 1976.
[40] Jervis, T.R., D.W. Woodard, and L.F. Eastman, *Elect. Lett.,* 15:619, 1979.
[41] Favennec, P.N., L. Henry, and H.L. Haridon, *Sol. St. Elect.,* 21:705, 1978.
[42] Immorlica, A.A., and F.H. Eisen, *Appl. Phys. Lett.,* 29:94, 1976.

[43] Malbon, R.M., D.H. Lee, and J.M. Whelan, *J. Electrochem. Soc.,* 123:1413, 1976.
[44] Kasahara, J., M. Arai, and N. Watanabe, *J. Appl. Phys.,* 50:541, 1979.
[45] Campisano, S.U., et al., *Sol. St. Elect.,* 21:485, 1978.
[46] Golovchenko, J.A., and T.N.C. Venkatesan, *Appl. Phys. Lett.,* 32:147, 1978.
[47] Lindhard, J., M. Scharff and H. Schiott, Kg. Danske Vid. Selskab., *Mat. Fys. Medd.,* 33(14):1, 1963.
[48] Gibbons, J.F., W.S. Johnson, and S.W. Myrolie, *Projected Range Statistics.* Stroudsburg: Dowden, Hutchinson and Ross, Inc., 1975.
[49] Winterbon, K.B., *Ion Implantation Range and Energy Deposition Distributions,* vol. 2. New York: Plenum, 1975.
[50] Donnelly, J.P., *Gallium Arsenide and Related Compounds, 1976,* (Inst. of Phys. Conf. Ser. 33b), p. 166. Bristol: The Institute of Physics, 1977.
[51] Tandon, J.L., N.A. Nicolet, and F.H. Eisen, *Appl. Phys. Lett.,* 34:165, 1979.
[52] Heckingbottom, R., and T. Ambridge, *Rad. Eff.,* 17:31, 1973.
[53] Inada, T., S. Kato, T. Hara, and N. Toyoda, *Appl. Phys. Lett.,* 50:4466, 1979.
[54] Sealy, B.J., S.S. Kular, K.G. Stephens, R. Croft, and A. Palmer, *Elec. Lett.,* 14-22:720, 1978.
[55] Inada, T., K. Tokunaga, and S. Taka, *Appl. Phys. Lett.,* 35: 546, 1979.
[56] Tandon, J.L. and F.H. Eisen, in *Laser-Solid Interactions and Laser Processing-1978,* AIP Conference Proceedings No. 50, ed. by S.D. Ferris, H.J. Leamy, J.M. Poate, p. 616. New York: American Institute of Physics, 1979.
[57] Hunsberger, R.G., and N. Hirsch, *Solid State Electronics,* 18: 349, 1975.
[58] Welch, B.M., F.H. Eisen, and J.A. Higgins, *J. Appl. Phys.,* 45: 3685, 1974.
[59] Higgins, J.A., B.M. Welch, F.H. Eisen, and G.D. Robinson, *Electron. Lett.,* 12:17, 1976.
[60] Kellner, W., H. Kniepkamp, D. Ristow, H. Heingle, and H. Boroffles, *Solid State Electron.,* 20:459, 1977.

[61] Miyutani, T., S. Ishida, and M. Fujimoto, *Electron. Lett.,* 12: 341, 1976.
[62] Stonebaum, E., T.S. Tan, and J. Gladstone, *Tech. Dig. 1977 Int. Electron Devices Mtg.,* Washington, D.C. p. 330.
[63] Kung, J.K., R.M. Malbon, and D.H. Lee, *Electron. Lett.,* 13: 187, 1977.
[64] Higgins, J.A., F. Eisen, B. Welch, G. Robinson, and W. Hill, *Gallium Arsenide and Related Compounds, 1976,* (Inst. Phys. Conf. Ser. 336) p. 236. Bristol: The Institute of Physics, 1977.
[65] Higgins, J.A., B.M. Welch, F.H. Eisen, and G.D. Robinson, *Electron. Lett.,* 12:462, 1976.
[66] Higgins, J.A., R.L. Kuvas, W.A. Hill, G.D. Robinson, and D.R. Ch'en, *Proc. 6th Bienn. Elec. Eng. Conf.,* p. 337. Ithica: Cornell University Press, 1977.
[67] Stoneham, E., T.S. Tan, and J. Gladstone, *Tech. Dig. 1977 Int. Electron. Devices Mtg.,* Washington, D.C., p. 330.
[68] Welch, B.M. and R.C. Eden, *Tech. Dig. 1977 Electron Devices Mtg.,* Washington, D.C., p. 205.
[69] Van Tuyl, R.L., C.A. Liechti, R.E. Lee, and E. Gawan, *IEEE J. Solid State Circuits,* SC-12:485, 1977.
[70] Zuleeg, R., J.K. Notthoff, and K. Lehovec, *IEEE Trans. Electron Devices,* ED-25:628, 1978.

Chapter 3

Low Noise GaAs FETs

Introduction

In this chapter, two papers are presented on low noise GaAs FETs from two leading laboratories in the research and development of the materials and device technology, as well as commercial supply of the devices.

The first paper, by J.A. Turner et al., presents design principles for achieving optimum devices. Using the theoretical work of H. Fukui et al., and R.A. Purcel et al., Turner., et al. show the theoretical dependence of noise figure (NF) on gate length, gate metal thickness, source series resistance, specific contact resistance, and source drain spacing. They describe an "etched channel" geometry which optimizes the epi structure for minimum resistance.

A technology is described for defining gates as small as .3 μm by electron beam lithography and "lift off," which has led to devices with an NF of 1.76 dB and an associated gain of 12.2 dB at 14 GHz.

The second paper describes a design approach to state-of-the-art low noise GaAs FETs. In addition, Hasegawa gives a brief review of design approaches used in the industry to reduce source and gate parasitic resistance, as well as technology trends to reduce gate length.

He discusses in detail the experimental optimization of the effect of a high resistance "buffer layer," the optimum doping density, the effect of the doping steepness at the interface of the active layer and the "buffer layer," and the effect of the gate "recess region." By fabricating a 0.5 μm gate FET with what is termed a deeply recessed gate, an NF of 1.68 dB and an associated gain of 10.7 dB at 12 GHz is obtained.

The Noise and Gain Performance of Submicron Gate Length GaAs FETs

J. A. Turner
R. S. Butlin
D. Parker
R. Bennett
A. Peake
A. Hughes

Plessey Research (Caswell) Ltd.,
Allen Clark Research Centre, England.

1. INTRODUCTION

Noise predictions of Pucel[1] and Fukui[2] show that, provided material quality and low contact resistances are maintained, then gate length is an important factor in determining the noise performance of a GaAs FET. Conventional lithographic processes using in-contact ultra-violet exposure techniques place limits on the ultimate dimension that can be achieved generally to values around 0.5 microns. It will be seen later in this paper that dimensions around 0.2 microns are desirable if low noise, high frequency performance is to be obtained. One technique that allows such dimensions to be realized is electron beam lithography. A suitable choice of resist technology and fabrication technique has enabled devices with 0.3 micron gate lengths to be obtained and this, as the theories predict, has led to improved gain and noise performance.

This paper will review the existing theories and relate them to practical device structures, describe the particular technologies used to manufacture the device, and present the results obtained on short gate length structures.

2. DEVICE DESIGN CONSIDERATIONS

For many years GaAs FET technologists relied mainly on instinct in their quest for improved noise and gain performance from the device. It was mainly by empirical techniques that improvements in performance were obtained, and such parameters as the carrier density of the material in which the FET was made, the source to gate separations, etc. were all varied to obtain the best performance. The papers by Pucel and Fukui have changed this approach, for in these publications the material parameters, and the geometry of the FET are directly related to the performance of the device. Specifically, the quality of the material at the substrate-layer interface, the gate length, the gate metal loss and source resistance all affect the noise performance of the transistor.

The schematic diagram of Figure 1 shows the major parasitic elements in a FET produced by what is now recognized to be the optimum technique – the etched channel approach.

The noise figure of such a design is given by the expression in Equation 1.

$$F_o = 1 + kf\,L^{5/6}\left(\frac{N}{a}\right)^{1/6}\Bigg[\underbrace{\frac{3.3W^2}{hL}}_{\text{term 1}} + \underbrace{0.6W^2\sqrt{\frac{\rho f}{hL}}}_{\text{term 2}} + \underbrace{\frac{1.8}{N}\left(\frac{L_{SG_1}}{a_1} + \frac{L_{SG_2}}{a_2}\right)}_{\text{term 3}} + \underbrace{\sqrt{\frac{0.18\,R_c}{Na_2}}}_{\text{term 4}}\Bigg]^{1/2} \qquad (1)$$

where:

F_o	=	noise factor
k	=	material quality factor (0.033)
f	=	frequency of operation in GHz
N	=	carrier density in 10^{16} cm^{-3}
W	=	unit gate width in mm
h	=	gate metallization thickness in microns
L	=	gate length in microns
ρ	=	gate metal resistivity in 10^{-6} Ωcm

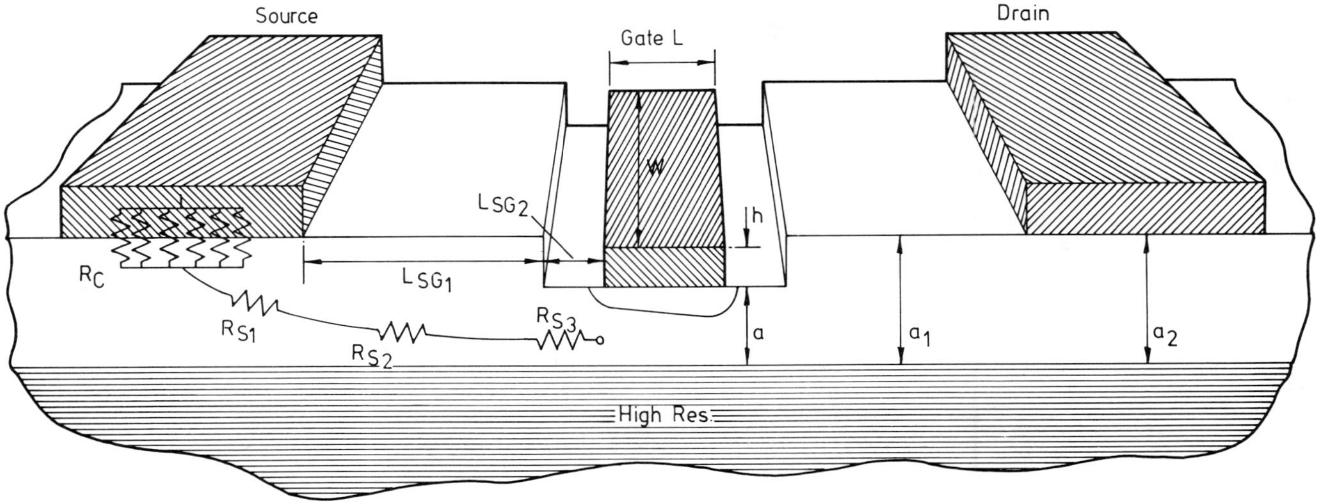

$$F_o = 1 + kfL^{5/6}\left(\frac{N}{a}\right)^{1/6}\left[\frac{3.3W^2\rho}{hL} + 0.6W^2\sqrt{\frac{\rho f}{hL}} + \frac{1.8}{N}\left(\frac{L_{SG1}}{a_1} + \frac{L_{SG2}}{a}\right) + \sqrt{\frac{0.18\,R_C}{Na_2}}\right]^{\frac{1}{2}}$$

***Figure 1**/Schematic Diagram of FET*

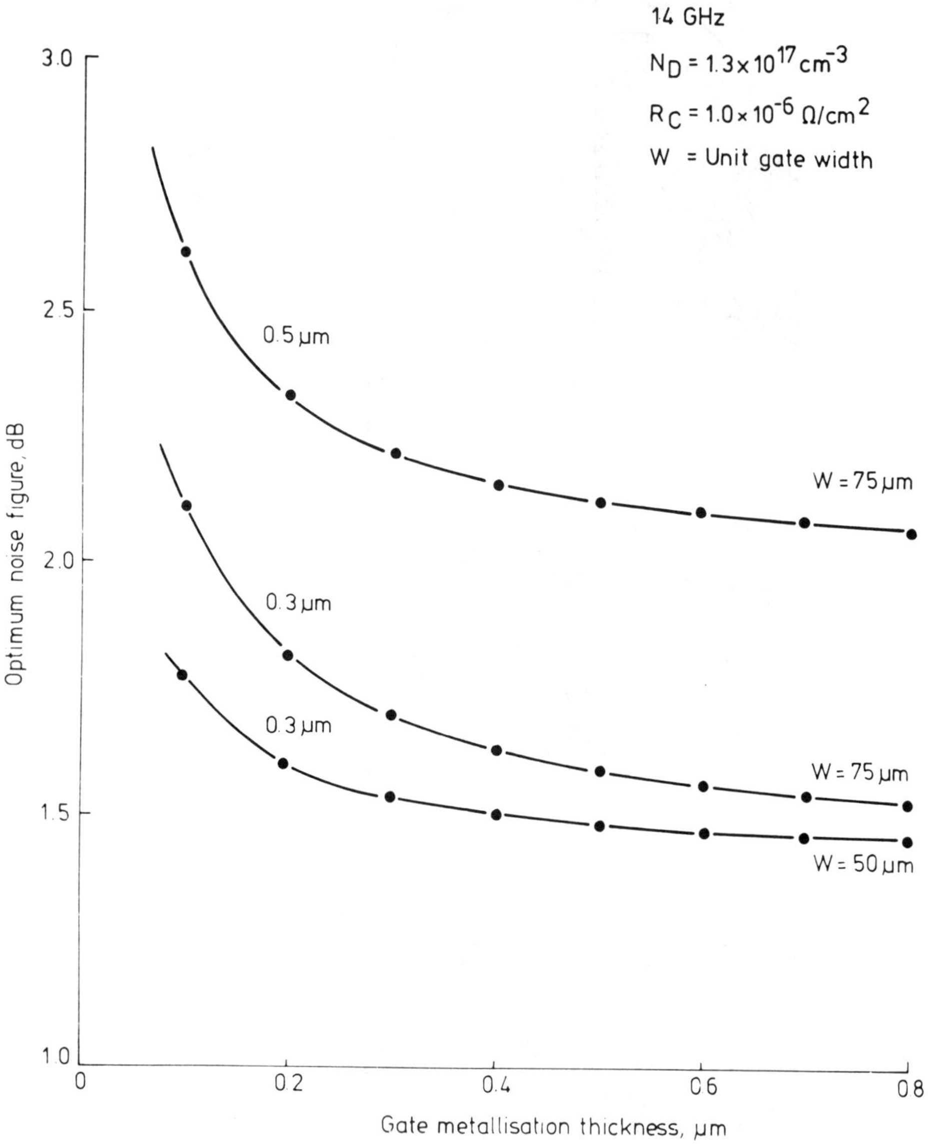

***Figure 2**/Theoretical Variation of Optimum Noise Figure with Gate Metallization Thickness*

R_c = specific contact resistance in $10^{-6}\ \Omega cm^2$

a's and L_{SG}'s are dimensions shown in Figure 1 in microns.

In Equation 1, terms 1 and 2 represent the noise contribution from the gate metal loss, and depend on gate length, unit gate width, metal conductivity and metal thickness. In the devices to be described, the gate metal is aluminum, which has a relatively high electrical conductivity when deposited under the correct conditions, and has been shown not to react with GaAs up to 500°C [3]. This aluminum film forms a Schottky barrier diode to the GaAs surface.

Figure 2 plots the effect of gate dimensions on device noise figure for a material carrier density of $1.3 \times 10^{17}\ cm^{-3}$ at 14 GHz. This shows the pronounced effect on noise figure of reducing the gate length to 0.3 microns while maintaining a constant unit gate width of 75 microns. It can also be seen that a reduction in unit gate width to 50 microns further improves the device performance. For devices of these dimensions, gate metallization thicknesses of around 4000Å are adequate – thicker metallization offers only a small improvement in performance though it may improve device reliability.

Term 3 in Equation 1 is the source series resistance contribution of the material between source and gate electrodes. This expression highlights the importance of the etched channel technology and shows the effect of sideways etching on the resistance R_{S3}. Figure 3 shows the relationship between R_S, R_{S2} and R_{S3} and active layer thickness. There is obviously an optimum layer thickness for minimum total series resistance. This graph is derived by assuming that when etching the channel, the sideways etching (L_{SG_2}) is equal to the etch depth. R_{S3} decreases as the initial layer thickness is reduced, while R_{S2} increases. If an n^+ contact layer is used, the dominant resistance is that due to undercutting during the channel definition process.

Term 4 in the equation relates to the contact resistance and spreading resistance R_{S1} in Figure 1. It can be seen from Figure 4 that, providing the specific contact resistance is better than $10^{-5}\ \Omega cm$, then improving the quality of the ohmic contact will only have a very small effect on the total series resistance. The particular device geometry chosen for the 0.3 micron device described later in this paper was conceived to optimize terms 1 through 4.

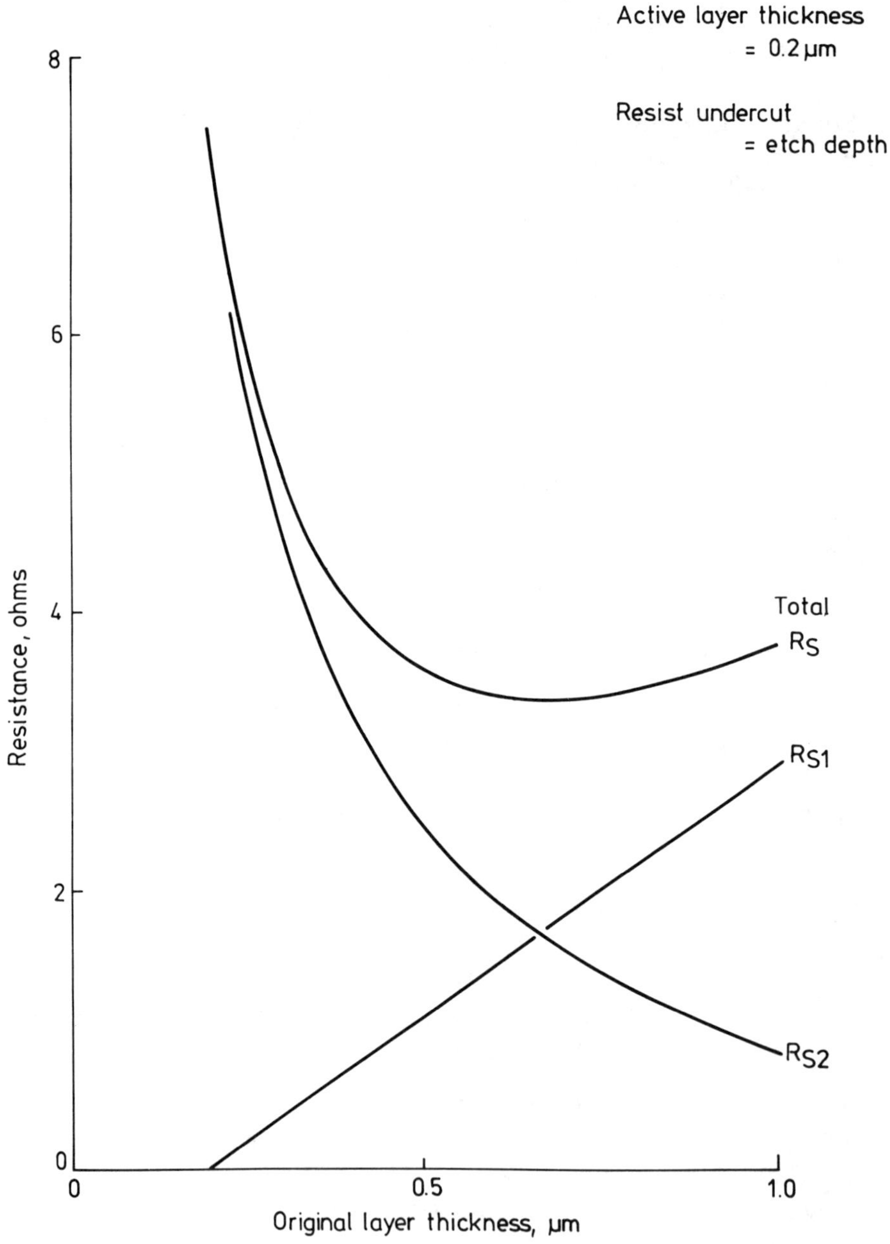

Figure 3/Source Resistance vs. Layer Thickness

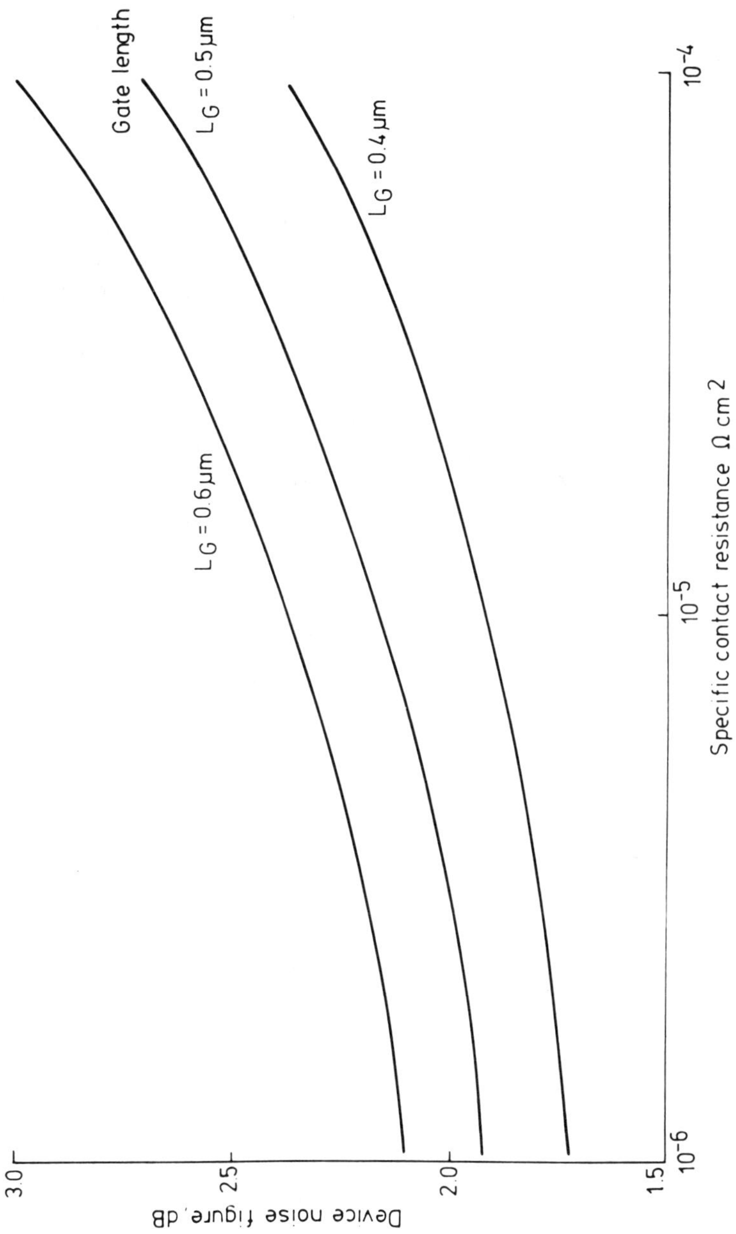

***Figure 4**/Variation of Device Noise Figure with Specific Contact Resistance*

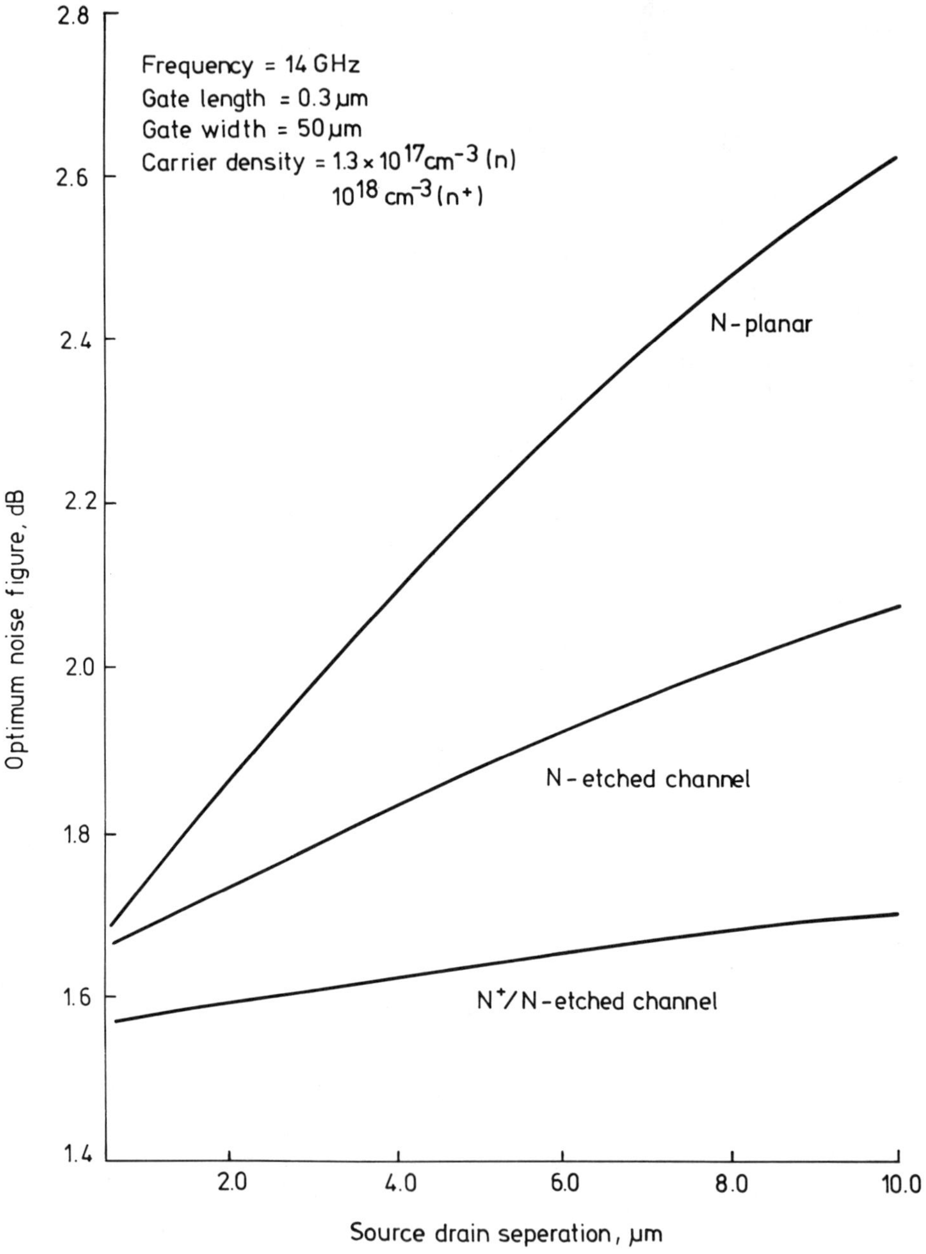

***Figure 5**/Variation of Device Optimum Noise Figure with Source Drain Spacing*

Figure 5 shows the variation of noise figure with source drain spacing (source series resistance) for planar and etch channel FETs. It is clear that the planar (self aligned) device relies heavily on its performance from a close source to gate separation. The n^+ contacted etched channel device has a noise figure almost independent of source to gate separation. This advantage is now being realized by all the major GaAs FET manufacturers.

3. PRACTICAL DEVICE DESIGN AND FABRICATION

Figure 6 shows the device design chosen for the 0.3 micron gate length device. The total gate width is 200 microns, made up of two cells each with two gates 50 microns wide. The source and drain pads are overlaid in order to reduce spreading resistance and to reduce the possibility of ohmic contact migration. The aluminum gate is contacted using a multimetal overlay to prevent interaction with the gold bond wire.

Material Requirements and Characterisation

The material used for these devices is grown by the arsenic trichloride vapour epitaxial process developed originally by Knight et al. [4]. The reactor is equipped with two doping lines to allow n^+/n structures to be grown without a pause in the growth sequence. The layers are assessed for their suitability for low noise FET fabrication through the following characterization routine.

1. Carrier density profile using a mercury Schottky barrier.
2. Buffer resistivity tests in darkness and under a standard illuminator.
3. Mobility profiles using the test structure shown in Figure 7. This consists of a standard Van der Pauw clover leaf with a Schottky diode over the central active area. The layer is successively depleted by back biassing the Schottky diode and Hall measurement taken at each point. Suitable peripheral circuitry allows an immediate plot of mobility through the layer to be determined.

3.1 Device Fabrication Technology

Processing begins by metallizing the back face of the epitaxial slice with a Au/Ge alloy. This facilitates the soldering down of the final transistor chip into the circuit or package. Mesa regions are then defined by conventional photolithography and a chemical etch used to isolate adjacent device areas. The electrical characteristics measured between mesa areas gives a further measure of the quality of the buffer layer. Source and drain contact areas are defined in photoresist by an evaporation and 'float off' process. An In/Au/Ge alloy is used and low resistance contacts are ensured by sintering at about 400°C in purified hydrogen. The gate area is defined by an electron beam lithography process described below. The developed channel region is chemically etched until the monitored source-drain saturation current reaches a predetermined value. This value is directly related to the final saturation current of the device. At this stage aluminum is evaporated to define the gate diode. Excess metal is 'floated off'. An overlay metallization pattern is defined on the gate bonding pad area consisting of chromium then gold. This is clearly seen in Figure 6. To obtain good Schottky diode quality a short sintering process is carried out.

3.2 Electron Beam Lithography

Machine Description

The machine used for electron beam exposure (Figure 8) is a Cambridge Stereoscan which has been modified by addition of an LaB_6 compatible gun, a beam blanking unit and a 2 x 2 inch X-Y stage. Patterns are generated by the 'vector scan' technique, which requires the pattern structure to be broken down into elemental rectangles. The co-ordinates of each rectangle are digitised and fed into 12 bit memories via a paper tape reader. Co-ordinate data is fed sequentially to scan control logic which 'fills in' the area bounded by the co-ordinates. The beam stepping rate for each rectangle is pre-programmed from data held in separate memories. This facility allows correction for 'proximity' and 'size' effects, ensuring optimum exposure throughout the pattern. The final output is converted into scan waveforms which are taken to the scan coils. The beam can be addressed to any one of 4096 positions along each orthogonal axis, and with a field size of 0.5 mm, a minimum linewidth of 0.3μm can be obtained.

In processes where pattern-to-wafer alignment is required, registration marks are included in the pattern field in such a position as to coincide with metal alignment marks on the wafer. The pattern marks can be scanned repetitively and displayed on a VDU screen. The video

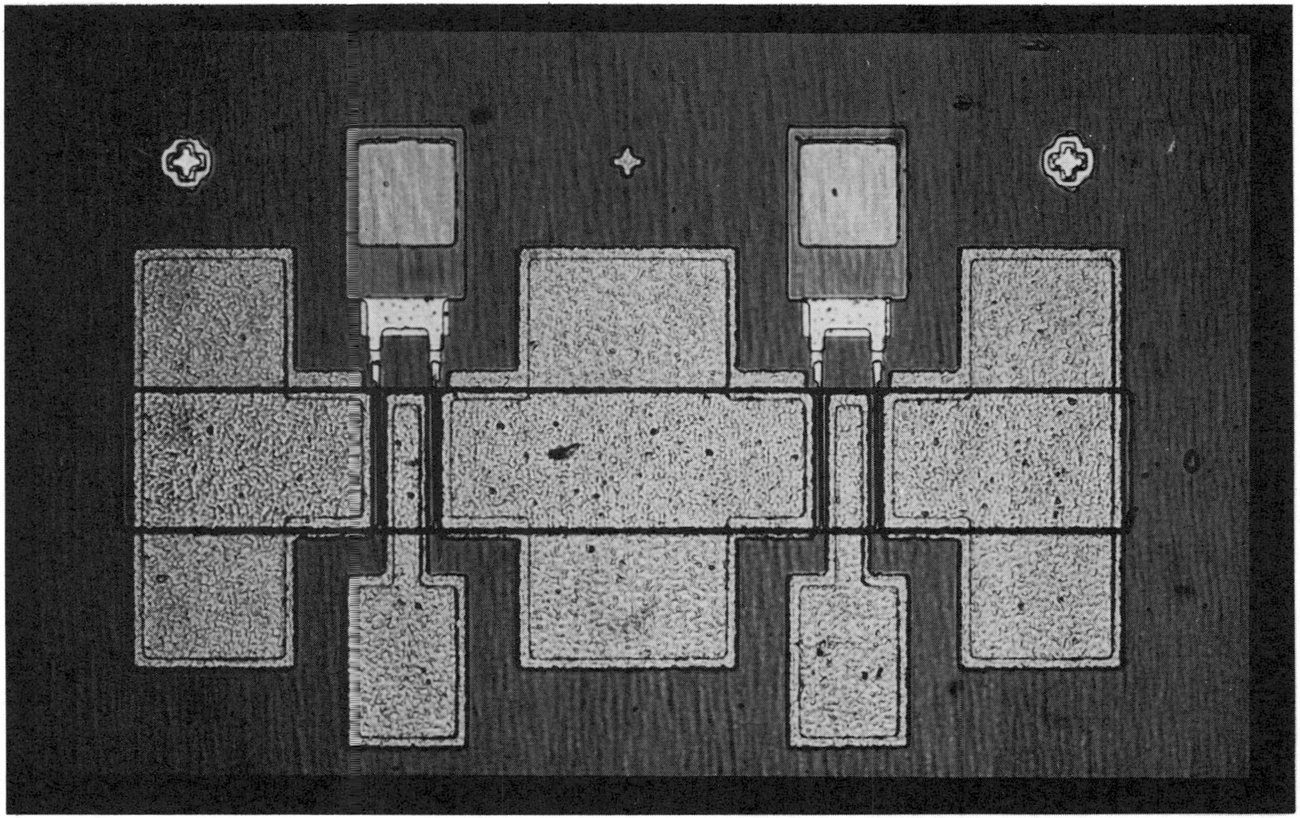

Figure 6/*0.3 μm Gate Length GaAs FET*

signal from the metal alignment marks is superimposed on the VDU display, and the video image is aligned with the mark on the screen by electromagnetic deflection of the beam. Using this technique, an estimated ± 0.1 μm alignment accuracy is achieved over a 0.5 mm field.

Electron Beam Technology

Efficient exposure of sub-micron geometry patterns requires some knowledge of the charge density profile caused by scanning the beam over the resist. Figure 9 shows the variation of line half width with stepping rate for single line scan patterns, and illustrates the technique used to obtain approximately charge profiles. By plotting such profiles side by side and calculating the resultant distribution, profiles for multiple line scan patterns are obtained. A computer program has been devised which will perform the task automatically.

Run-to-run reproducibility of pattern linewidth becomes more dependent on developing procedure as linewidth is reduced, and for

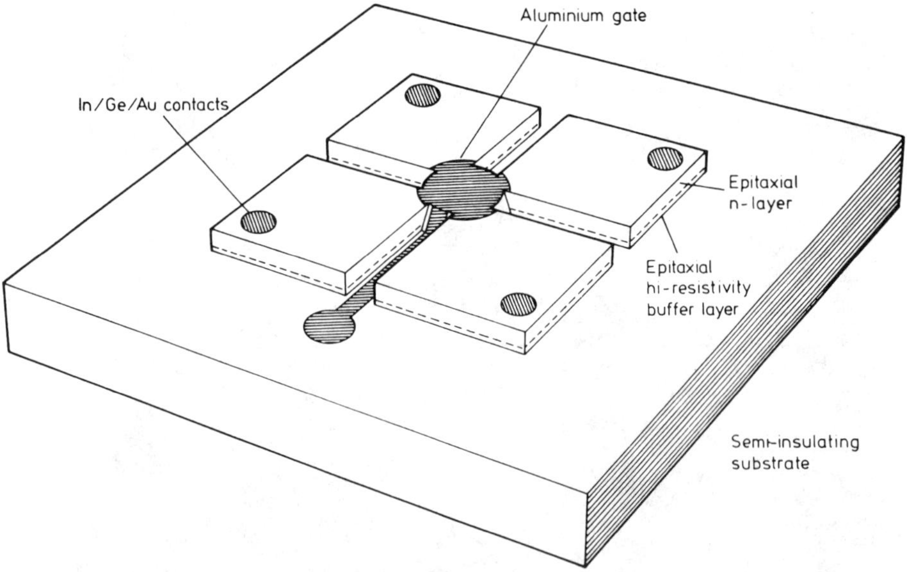

Figure 7/Schematic Illustration of Gated Hall Structure

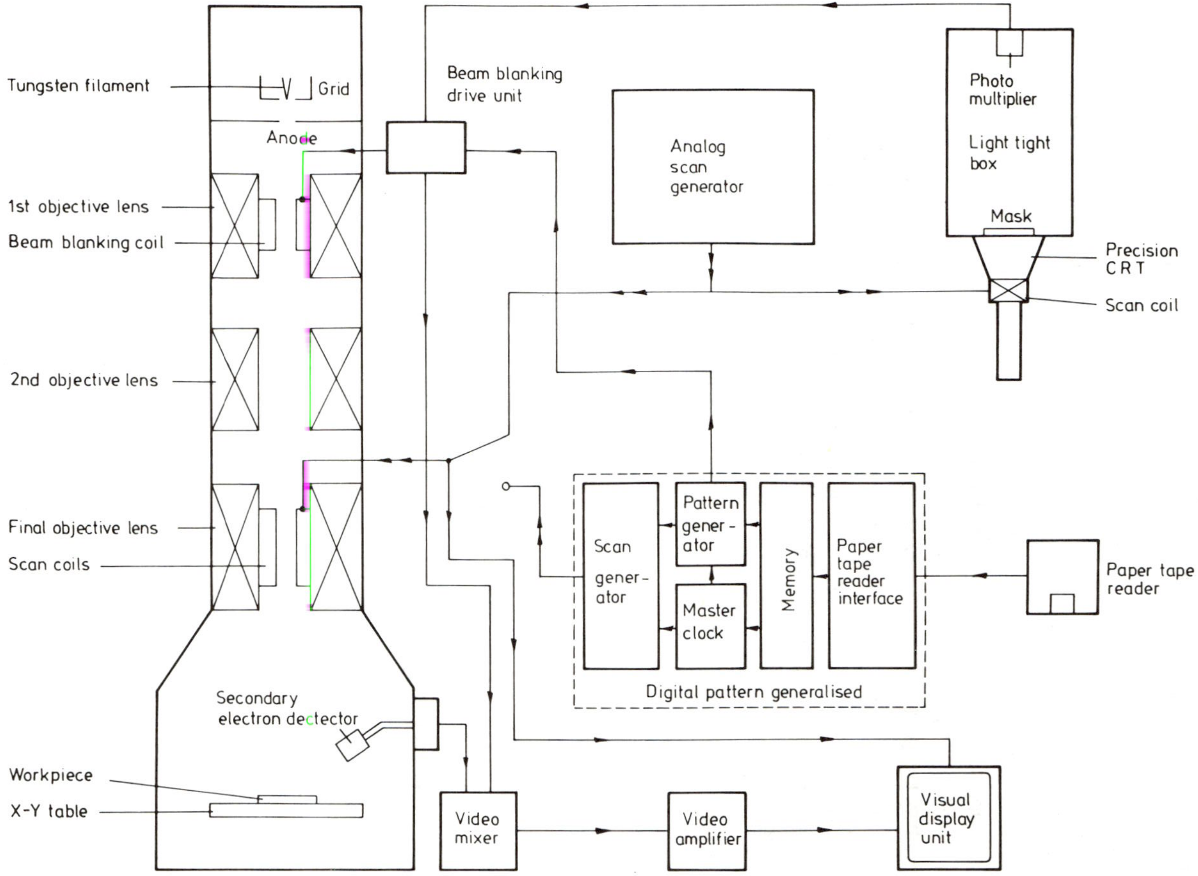

Figure 8/EBMF System

0.3 μm lines, close control of developing time is essential. The patterns themselves are not clearly visible under an optical microscope, and it is necessary to use a developing aid, known as an 'exposure wedge' (Figure 10). The exposure wedge, first reported by Grobman and Speth of IBM[5], is a resist pattern built up from 25 vertical stripes. Each stripe is given a different exposure dose such that the whole pattern becomes a rectangle across which the exposure dose increases linearly. During developing, the higher dose regions are dissolved away, and the position of the line separating dissolved from undissolved regions can be read off against a scale incorporated into the pattern. The optimum development for a given pattern is determined by trial exposure runs, and is maintained throughout all subsequent exposure runs by ensuring that the same wedge scale reading is obtained.

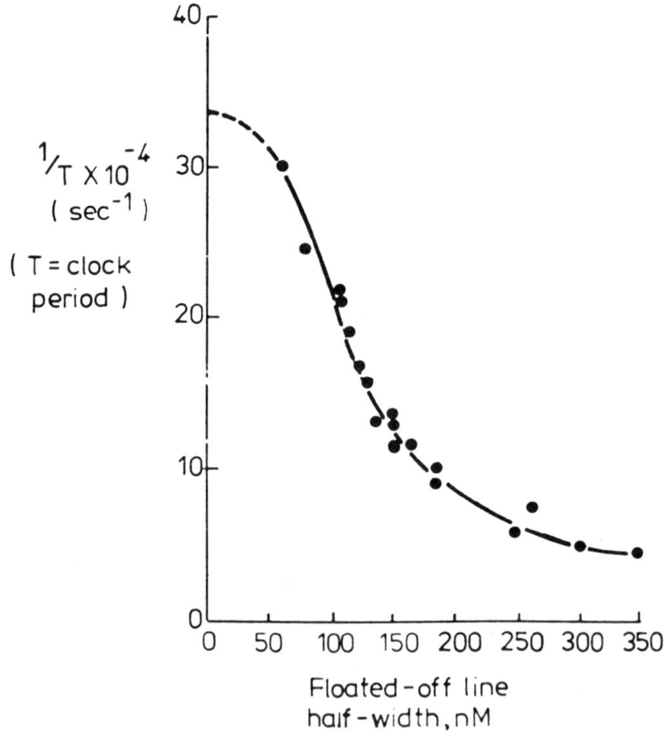

Figure 9/Variation of Line Half-Width with Stepping Rate for Single Line Scans

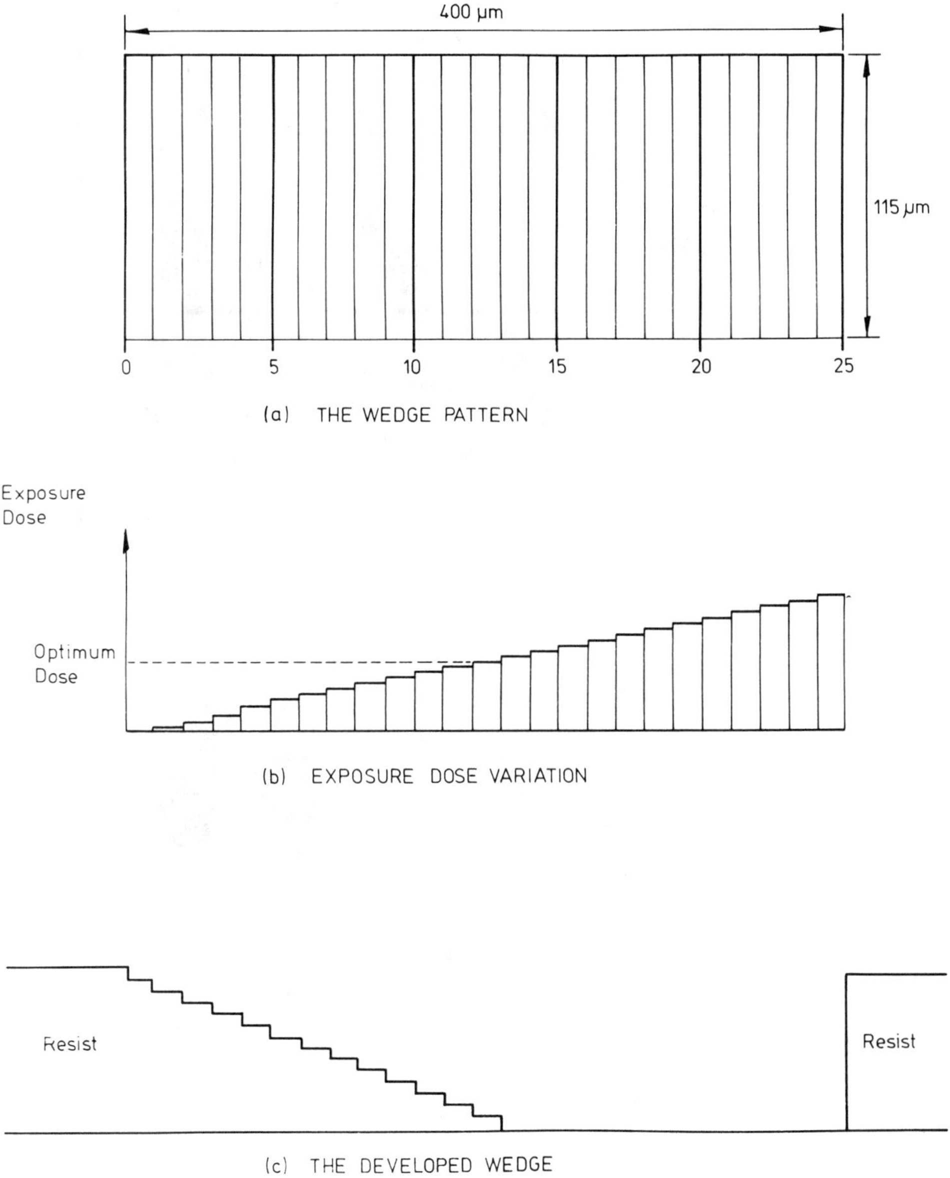

Figure 10/The Exposure Wedge

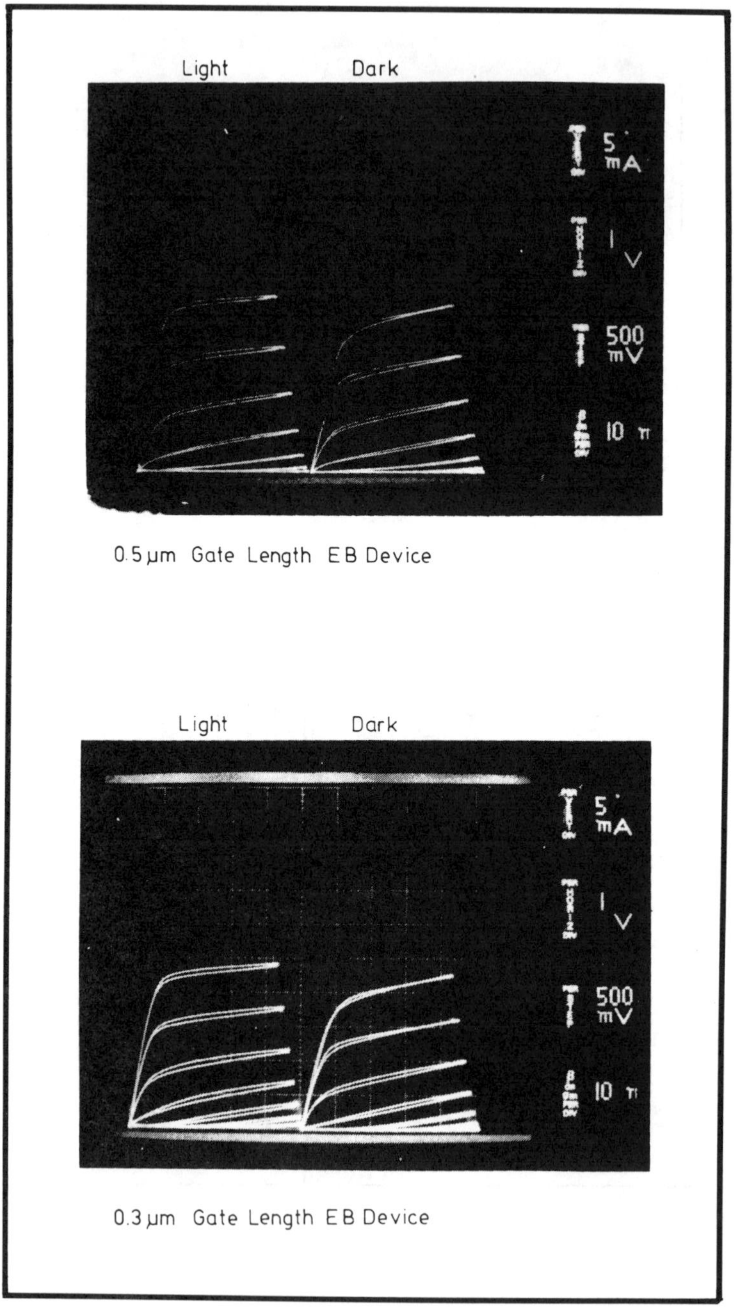

Figure 11/DC Characteristics of Electron Beam Devices

Resist Technology

The resists currently being used are poly-methylmethacrylate (pMMA) and Shipley AZ1350. Of the two, AZ1350 is less convenient to use due to its UV sensitivity. It also shows a smaller margin between optimum exposure dose and cross linking dose than pMMA. This margin can be improved by extended developing, thereby reducing the optimum exposure dose, but this can lead to thickness loss in the unexposed resist. Overall adhesion and sensitivity of AZ1350 are superior compared to pMMA, and AZ1350 has the advantage that it allows both electron beam and photolithographic processing to be carried out on the same resist layer.

4. DC CHARACTERIZATION

Initial d.c. selection is performed by observation of the I-V characteristics of the FETs on a curve tracer/probe station. Typical characteristics of 0.5 micron and 0.3 micron gate length devices are shown in Figure 11. Measurement of transconductance (g_m) provides verification of layer carrier density, while observation of the pinch off behaviour provides qualitative information regarding the active layer/buffer layer interface. On a sample basis the gate/source breakdown voltage and forward diode characteristic is observed as a check on diode quality.

Selected devices also undergo an analysis as described by Howel and bechtel [6] and Butlin et al. [7]. The DC saturated and unsaturated transfer characteristic is made at 5 volts and 50 mV drain bias respectively. From the transfer characteristic at 50 mV, a graph of total source-drain resistance, R_{SD}, versus the function $1/1-\sqrt{V_g/V_p}$ can be drawn where V_g is the applied gate bias and V_p is the pinch off voltage obtained by extrapolating the linear region of the transfer characteristic to zero current. Both V_g and V_p include the built in barrier potential. Extrapolation of the linear region of the graph to the resistance axis yields an intercept resistance ($R_S + R_D$) the total parasitic source and drain resistance, while the slope of the linear region gives the open channel resistance R_i beneath the gate electrode. If the gate region is centrally placed in the source-drain gap, then half the total resistance, $R_S + R_D$, gives the parasitic source series resistance – an important parameter in determining the RF performance of the device. An example of these results is shown in Figure 12.

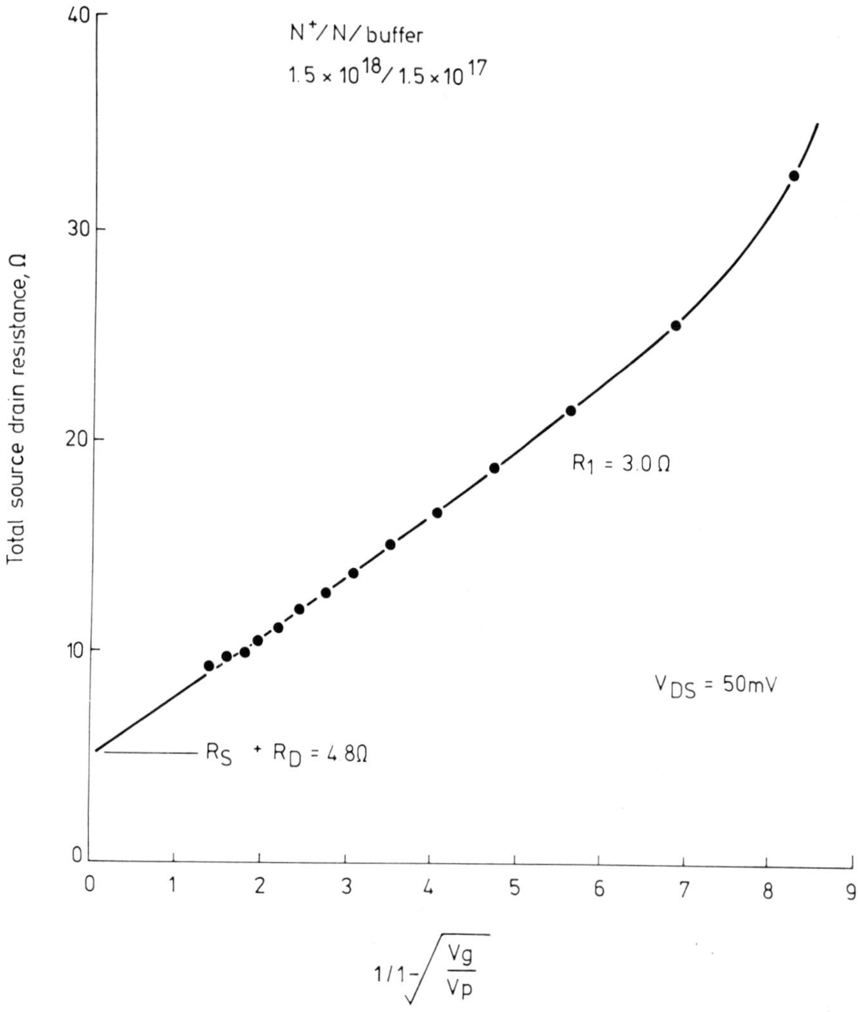

Figure 12/Source Drain Resistance as a Function of Gate Bias

Figure 13 relates these measured resistances to the cross section of the FET. Knowing the resistance of the layer beneath the gate metallization, measurement of the gate length and channel dimensions at the bottom of the etched channel allows the parasitic resistance of the undercut region to be calculated. This calculation shows that 60% - 90% of the source resistance can be attributed to the undercut region of the gate channel. The magnitude of the resistance is dependent upon starting layer thickness and the quality of the photolithography process.

This information provides useful data to the device technologist, and is also invaluable feedback data to the material growth team.

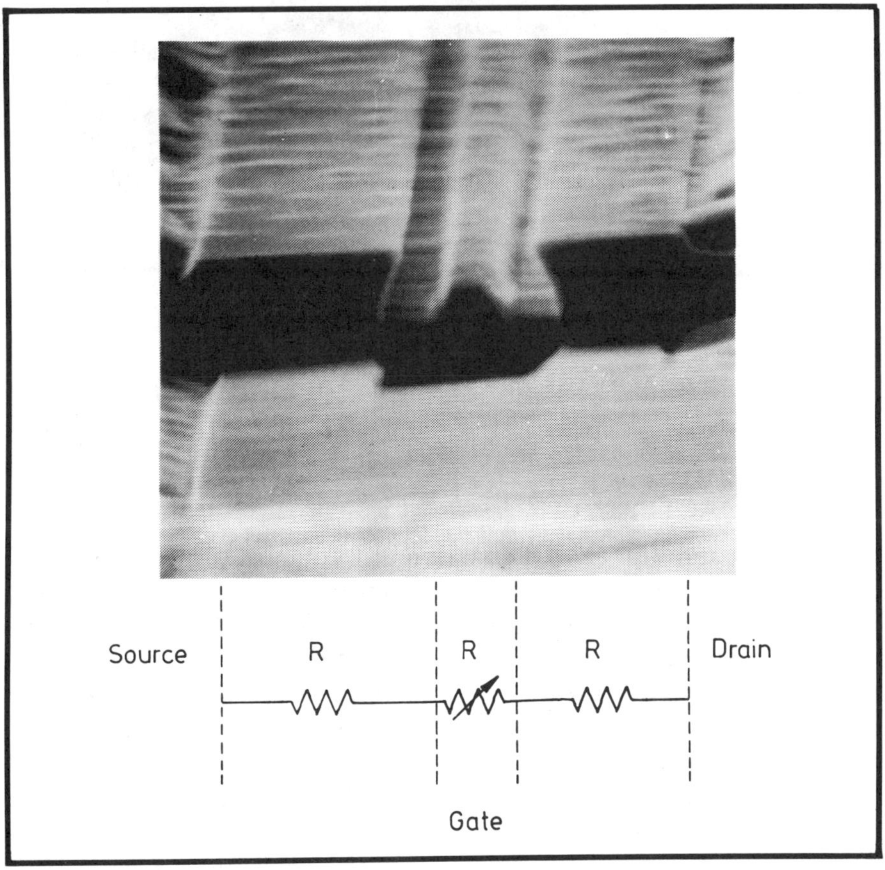

Figure 13/SEM Photograph of FET Channel Region

Figure 14/New J Band Test Fixture

5. RF CHARACTERIZATION

Noise figure and gain characterization at frequencies to 18 GHz is carried out in a relatively simple test fixture, Figure 14. The transistor chip is mounted on to a stud fixed through the alumina substrate of the microwave circuitry. The source gate and drain are contacted using multiple bond wires. Lead inductance, particularly at the source, has a deleterious effect on performance, and multiple bonding is vitally important for good gain and noise performance. In the test fixture chosen tuning for minimum noise is achieved by sliding tuning slugs along the 50 ohm transmission line that connects the device to the measurement circuitry. These form variable shunt capacitor elements with a capability of tuning, narrow band, to most devices.

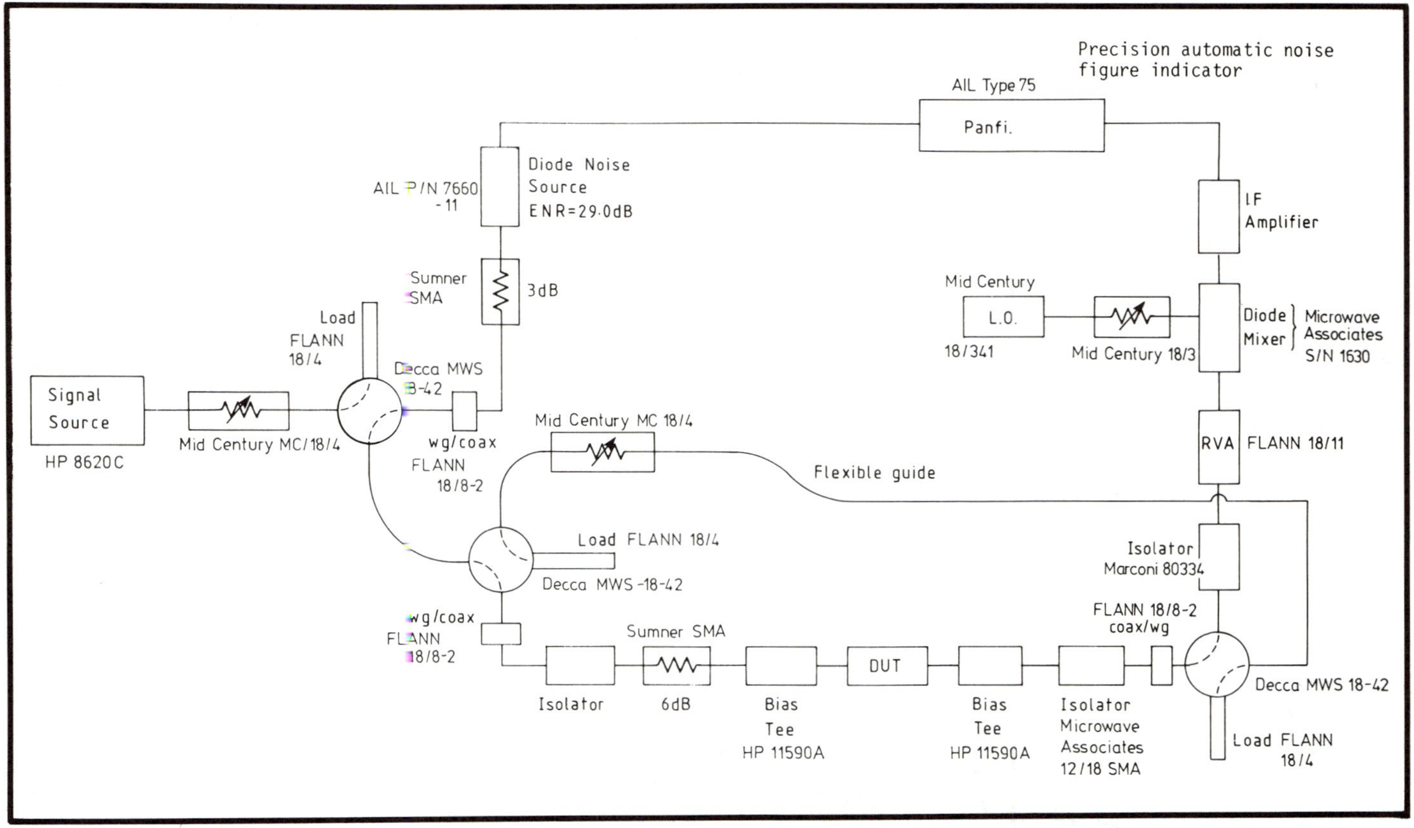

Figure 15/*Noise Measurement System*

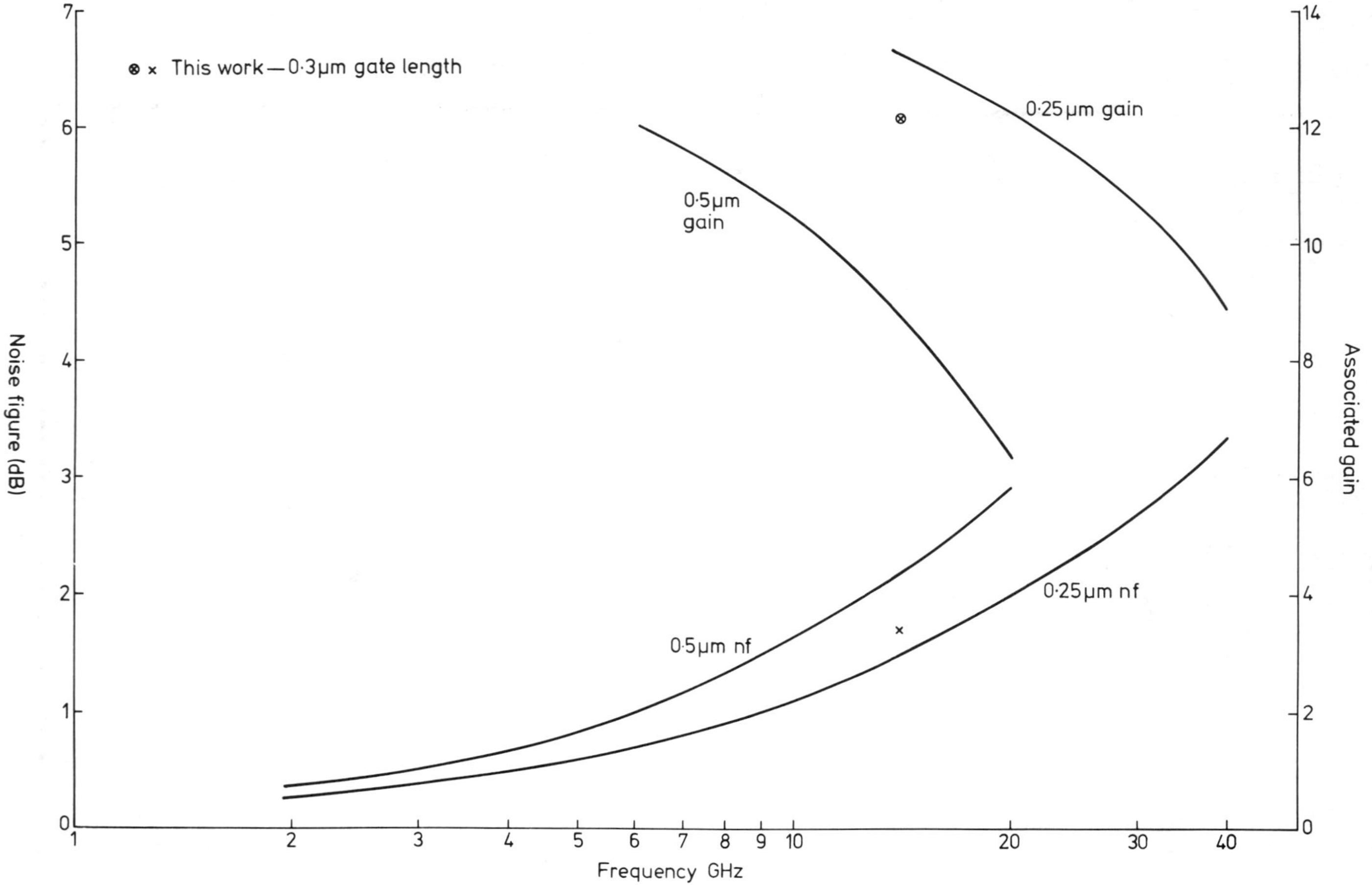

Figure 16/*Noise Figure and Associated Gain Vs. Frequency Predicted and Achieved*

A schematic of the noise and gain measurement equipment is shown in Figure 15. Waveguide components are used wherever possible. Rotary wave guide switches direct the signal paths from the bypass to test lines. The variable attenuator in the bypass line allows the loss of this line to be matched to the loss in the test channel. This allows the input power level to the device and the frequency of the measurement to be checked and corrected without removing the device under test. The system is calibrated using a hot-cold noise generator, though solid state noise sources are used for routine measurement. The sensitivity of the system can be improved by adding a low noise amplifier before the mixer.

5.1 RF Results – Theory and Practice

The solid curves shown in Figure 16 have been derived from the Pucel and Fukui papers and show the advantages – particularly at the higher frequencies – of using short gate length structures. Practical results for 0.5 micron gate length devices from laboratories around the world fit this curve extremely well. The best result for the 0.3 micron gate length devices described in this paper are included in Figure 16. This result, a 1.76 dB noise figure and a 12.2 dB associated gain at 14 GHz, is extremely close to that predicted for devices with the chosen geometry.

The same 0.3 micron devices have also been assessed at 28 GHz. The device was mounted directly into a 50 ohm microstrip circuit formed in quartz. Waveguide to microstrip transitions connected the circuit to the waveguide measurement equipment. A best noise figure of 4.5 dB was measured with a maximum gain of 7 dB. This value of noise figure is rather higher than predicted in Figure 16, but this is almost certainly due to the non-optimised test fixture.

6. SUMMARY

It is predicted that improvements in noise performance and associated gain can be achieved by reducing the gate length of GaAs FETs. This work has demonstrated practically that considerable improvements in these two parameters can be obtained by utilizing devices with gate lengths as short as 0.3 microns. Electron beam lithography is required to achieve such small dimensions, and although the technique used requires that each device be processed individually, the overall yield of working devices is high (80%). The technology for fabrication of the device is given in some detail in Section 3 and, as well as being designed to realise high performance devices, consideration to reliability has been given. Early devices produced to this design but with a different technology failed on long term life testing by void formation in the gate metallisation. The present device technology has almost completely removed this failure mechanism, and the devices now live considerably longer and fail through apparent parametric changes of the material in the channel region.[8]

ACKNOWLEDGEMENTS

The authors wish to thank the directors of Plessey Research (Caswell) Ltd., Allen Clark Research Centre for giving permission to publish this paper. They are also happy to acknowledge the support of the Procurement Executive Ministry of Defence sponsored by DCVD, and also the International Telecommunications Satellite Organisation "INTELSAT".

REFERENCES

[1] Pucel, R.A., et al., *Advanced Elect. Phys.*, 38:195, 1975.

[2] Hewitt, B.S., et al., *Electron. Letters,* 12:309, 1976.

[3] Kim, H., et al., Proc. Symp. on GaAs and Related Compounds. London: Inst. of Physics & the Physical Society, 1975.

[4] Knight, J.R., et al., *Solid State Electronics,* 8:179, 1965.

[5] Grobman and Speth, Microcircuit Engineering Conference Proceedings, Cambridge, England, 1978.

[6] Hower, P., and Bechtel, G., *I.E.E. Trans.,* ED20, 1973.

[7] Butlin, R.S., et al., Proceedings of 6th European Microwave Conference, Microwave Exhibition and Publishers Ltd., Sevenoak, Kent, England, 1976.

[8] Butlin, R.S., private communication.

Low Noise GaAs FETs

F. Hasegawa
Institute of Material Science, Tsukuba University,
Sakura Mura, Niiharigun, Ibaragi Prefecture, Japan.

1. DEVICE STRUCTURE AND PERFORMANCE

1.1 Basic Structure and Device Pattern

The basic GaAs FET structure is simple (as shown in Figure 1(a)). It has a thin active epitaxial layer on a semi-insulating substrate. Normally a high resistive buffer layer is grown prior to the growth of the active layer to improve crystal quality of the active epitaxial layer [1]. Two ohmic contacts are provided as the source and drain electrodes, and a Schottky barrier gate metal is made between them.

Although the basic structure is simple, the dimensions and parameters of the GaAs FETs are difficult ones to achieve. As theoretical analysis indicates, the smaller dimension gives a higher maximum frequency of oscillation, as with other microwave devices. The gate length is one of the most important parameters of GaAs FETs, and should be less than 1 μm to give enough gain and a noise figure at microwave frequencies. Other important parameters are the source and gate parasitic resistances. Since the source resistance is greatly affected by the source and gate spacing, several kinds of self alignment methods are tried and developed to precisely control this spacing at less than 1 μm^2 [4]. The thickness of the active epitaxial layer is also very small, normally 1000 Å to 2000 Å with a carrier concentration of about 2×10^{17} cm^{-3}. The thickness of the high resistive buffer

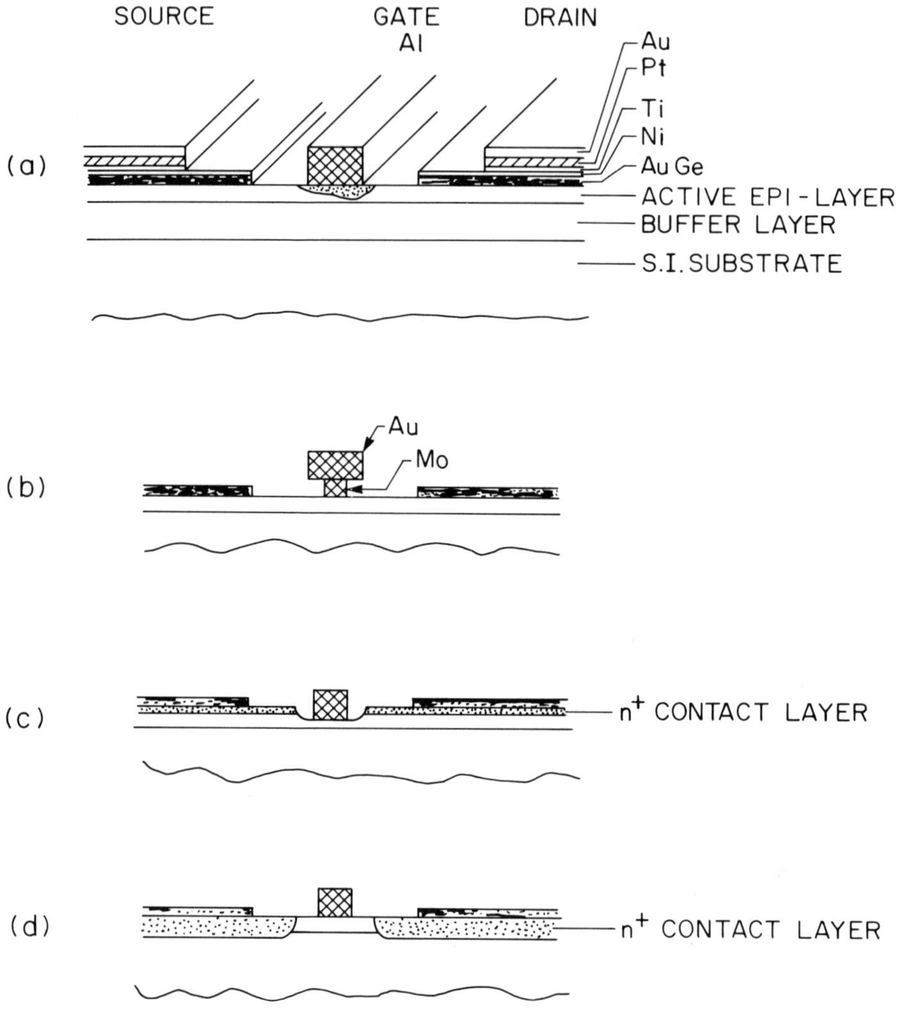

Figure 1/*Fundamental structures of low noise GaAs FETs.*
(a) Flat type FET;
(b) Improved gate structure;
(c) Recessed structure;
(d) FET with a selective n^+ contact layer.

layer is 3 ~ 5 μm. The carrier concentration should be less than $1 \times 10^{14}\ cm^{-3}$.

A typical electrode metallization is listed in Figure 1(a). Au Ge/Ni is the most popular ohmic contact to GaAs [5]. The Au is a capping layer for wire bondings, and Pt (platinum) is used to prevent interdiffusion between Au and Au Ge/Ni.

Aluminum is one of the most popular metals for the gate [6]. The gate metal should have a high conductivity, a good adhesion to GaAs, a high reliability, and must not react with GaAs, at least up to 400°C. There is some discussion of refractory metals, such as TiPt Au, as the preferred gate metal from a reliability point of view, but this is not yet clear.

In order to improve the performance of low noise GaAs FETs, it is necessary to reduce the source and gate parasitic resistances, as well as to reduce the gate length, as discussed in the previous section. Some of the examples of the improved low noise FET structures are shown in Figure 1(b), (c) and (d). Figure 1(b) shows an improved gate structure with a low gate resistance as well as a short gate length. It has a thick gold layer on top of a refractory Schottky metal such as Mo [7] or W [4]. The gold layer is also used as a mask of self aligning ohmic contact formation. Figures 1(c) and (d) show the structures with a reduced source resistance. Figure 1(c) shows a recessed gate structure with a surface n^+ contact epitaxial layer [8]. Even though there is no surface n^+ contact layer, the recessed gate structure is very effective in reducing the source resistance if the recessed depth is deep enough. The structure shown in Figure 1(d) can be made by a selective ion implantation or selective epitaxial growth [9, 10]. The structure itself is an ideal one, but it is difficult to control the actual spacing between the source n^+ contact layer and the gate unless some kind of self aligning technique is developed.

Examples of top views of commercially available low noise GaAs FETs are shown in Figures 2(a), (b), and (c). Figure 2(a) shows a 1 μm gate FET [6]. It has one gate pad and two gate fingers, each of which has 150 μm length. Since the gate pad is on the semi-insulating substrate, a parasitic capacitance of the gate pad is not serious. This is one of the advantages of GaAs devices on a semi-insulating substrate.

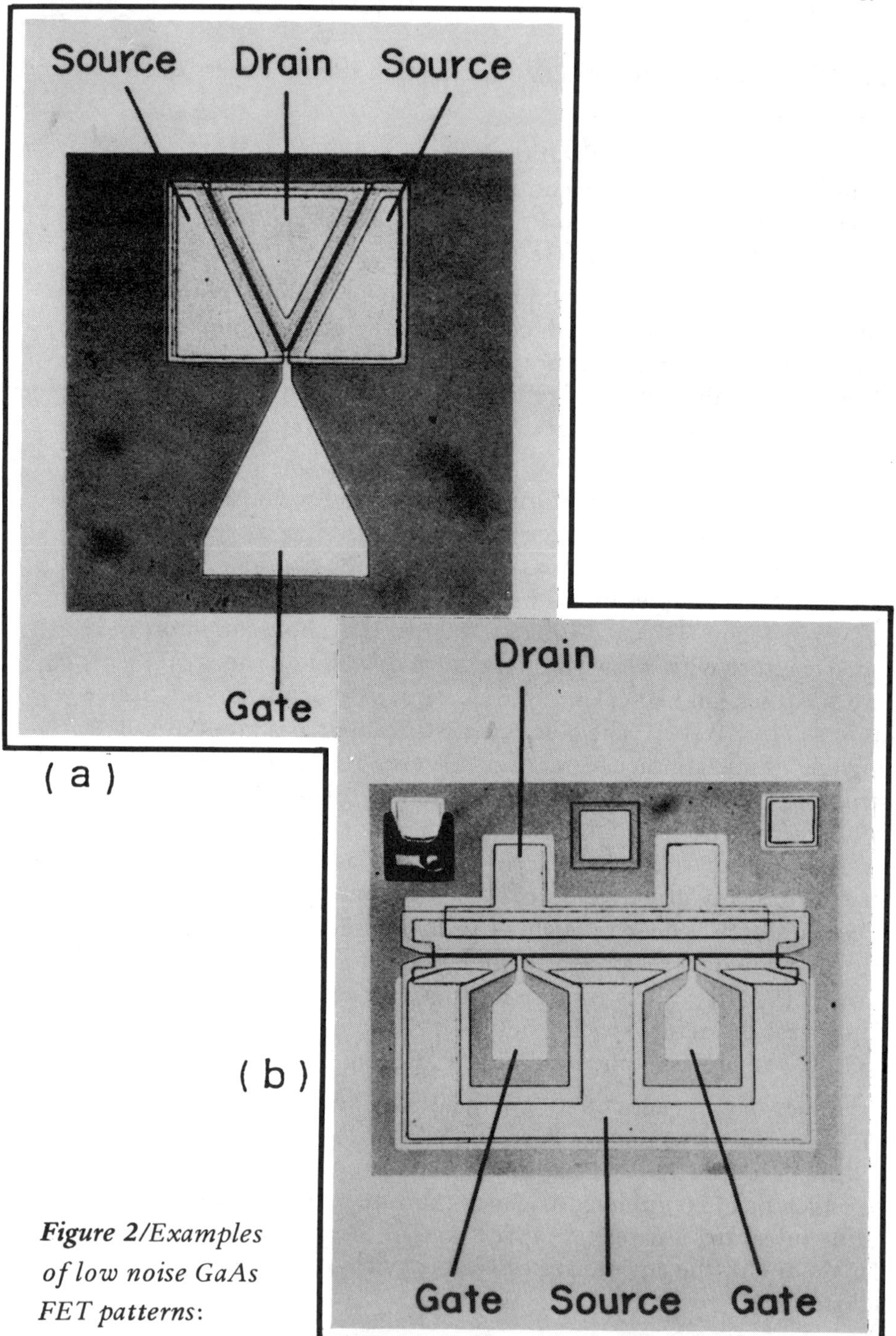

***Figure 2**/Examples of low noise GaAs FET patterns*:

(a) 1 μm gate FET with 150 μm unit gate width (NE244);
(b) 0.5 μm gate FET with two gate pads (NE388);

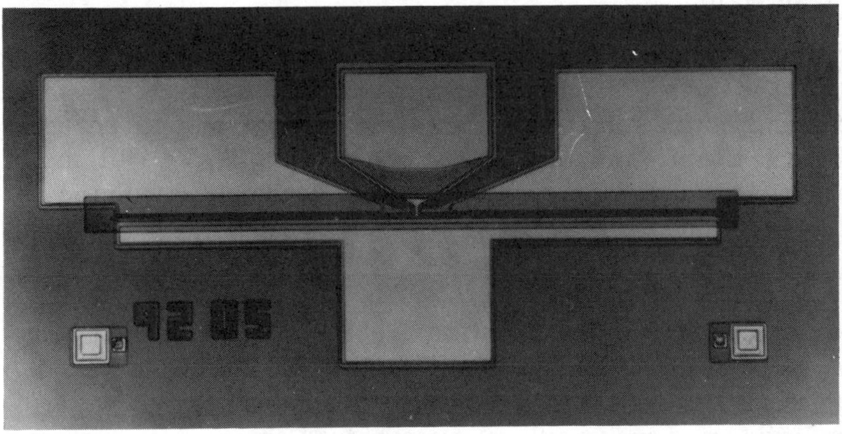

(c)

Figure 2/Examples of low noise GaAs FET patterns:
(c) 1 μm gate FET with 500 μm gate width (HFET 1101).

Figure 2(b) shows a 0.5 μm gate FET [6]. Two gate pads are prepared in this case in order to reduce the gate resistance. If the gate pad or signal feeding point is doubled, with the same total gate width, the overall gate resistance is reduced to one fourth, because the length of unit gate width (gate finger length) is reduced by one half, and these shortened unit gates are connected in parallel. In the case of a 0.5 μm gate low noise FET NE 388 shown in Figure 2(b), the total gate width is 280 μm, the gate metal is aluminum and the thickness is a half micron. If the gate pad is one (the unit gate width is 140 μm), the overall gate resistance would be about 4 Ω. In the actual device, since there are two gate pads, the overall gate resistance is about 1Ω. This is small enough compared with the source resistance of the same device, which is about 5Ω.

In general, then, if there are n gate pads, the overall gate resistance is reduced to n^{-2}.

Another example of a one gate pad low noise FET is shown in Figure 2(c). The total gate width is 500 μm, therefore, the unit gate width is 250 μm. However, if the specific gate resistance is reduced enough by adopting the gate structure like the one shown in Figure 1(b), the gate resistance can be suppressed sufficiently compared with the source resistance.

1.2 Experimental Design Optimization

The performance of low noise GaAs FETs is greatly affected by the doping profile or crystal quality of the epitaxial layer. Although the device pattern, such as the total gate width and number of gate pads can be designed analytically, an experimental optimization is necessary for the doping profile and analysis of the buffer epitaxial layer. In this section, the effect of the high resistive buffer layer, the optimum doping density, the effect of the doping steepness at the interface and the effect of the depth of the gate recessed region will be discussed.

Effect of the high purity buffer layer

Since bulk GaAs has many defects and many impurities, a high purity or a high resistive epitaxial buffer layer is commonly grown prior to the epitaxial growth of active layers.

An example of improvements of the active epitaxial layer by the buffer layer is shown in Figure 3 [1]. The dotted line with triangles in Figure 3 shows an electron mobility profile of the epitaxial layer grown directly on a Cr-doped semi-insulator substrate. The mobility gradually decreases toward the substrate, and it is much lower at the interface than at the surface, because of the diffusion of defects or impurities from the substrate. On the other hand, the electron mobility of the active layer grown on a high purity epitaxial buffer layer is constant or increases slightly at the active layer-buffer interface.

Current-voltage characteristics of the GaAs FETs made on two kinds of wafers with and without buffer layer are shown in Figure 4 [1]. The FET without the buffer layer normally shows a hysteresis or a kinking (increase of the saturation current at a certain drain bias), which the FET with the buffer layer does not show at all. Transconductance of the FET with the buffer layer also seems to be higher than that of the FET without the buffer layer.

For these reasons, and because, in the case of low noise FETs, the current is confined to the vicinity of the interface, the GaAs FETs with a buffer layer normally give a better low noise performance than those without the buffer layer.

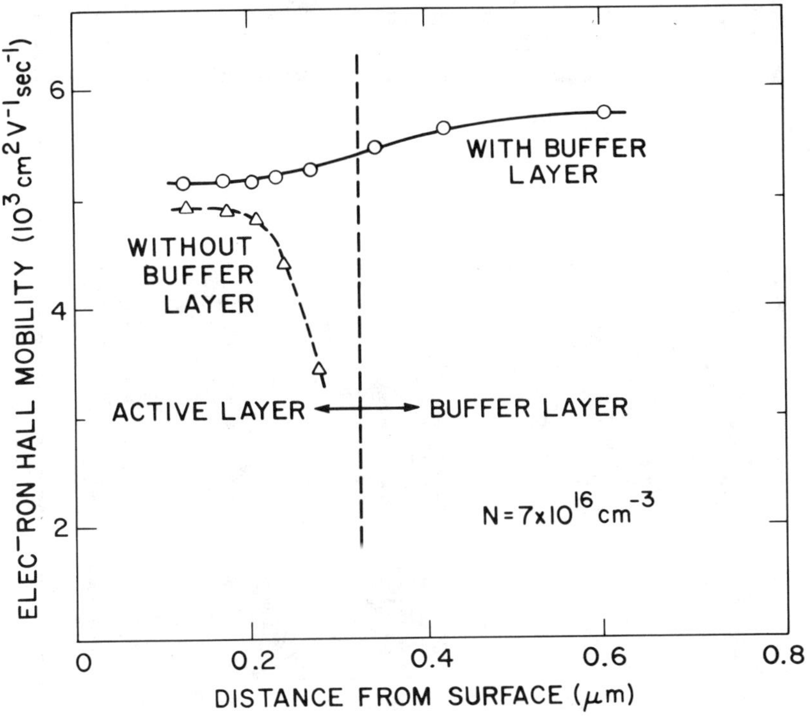

Figure 3/Electron mobility profiles of active epitaxial layers grown on a high resistive buffer epitaxial layer (solid line) or grown directly on Cr-doped substrate.

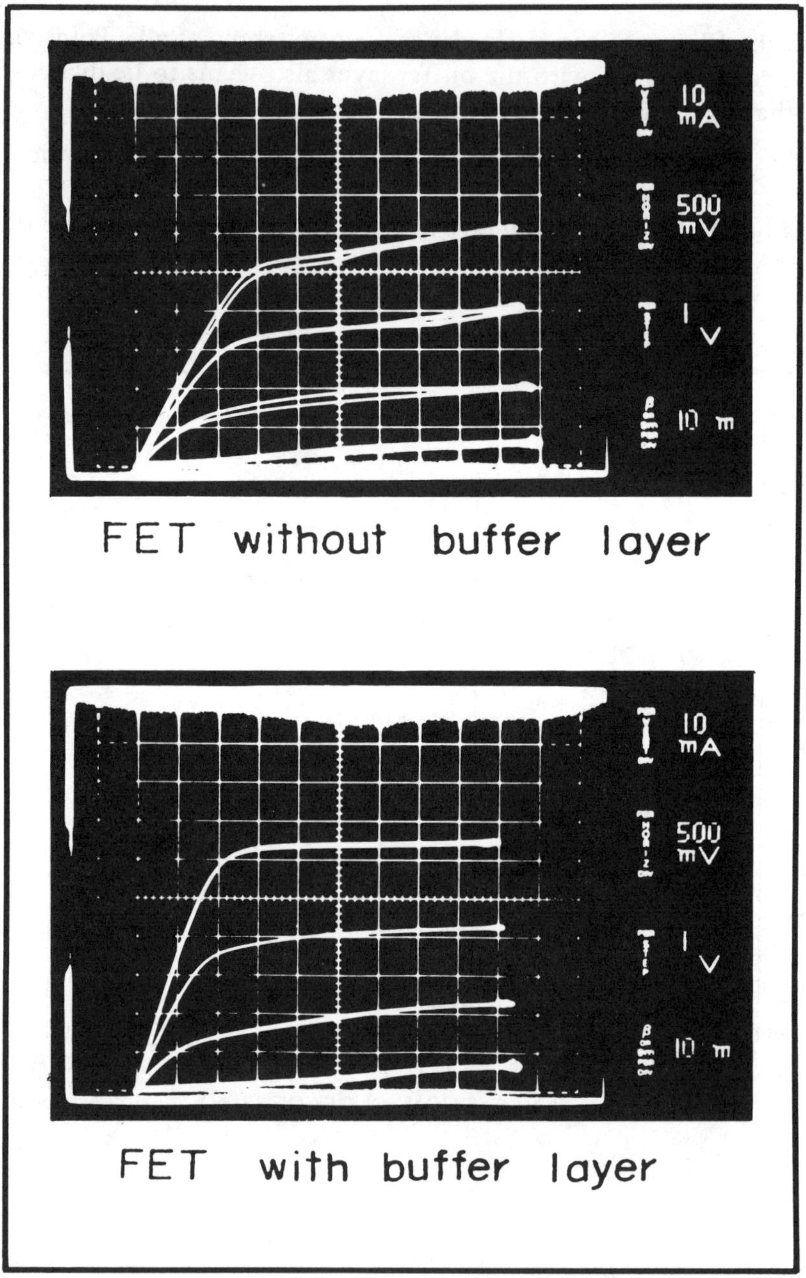

Figure 4/I-V characteristics of the FETs with and without buffer layer.

Optimum doping density

A shorter gate length gives a higher cut off frequency, therefore half micron gate low noise FETs are very popular recently. In order to have a normal FET operation, the thickness of the active layer must be small enough compared with the gate length. In addition to this, theoretical analysis indicates that a higher doping density gives a higher gain. However, the doping density, and therefore the thickness of the active epitaxial layer, is limited by the gate breakdown voltage and by the difficulty of obtaining an ideal doping profile for a very thin active layer with a high doping density.

Dependence of low noise performance on the doping density of the active layer is shown in Figure 5 [6]. The FETs are 0.5 μm gate FETs as shown in Figure 2(b). The noise decreases with an increase in doping density, and reaches the minimum value at around 2.5 x 10^{17} cm^{-3}. The associated gain increases with increase of the doping density and seems to saturate at around 3 x $10^{17} cm^{-3}$. The deterioration of the noise figure at a higher doping density is probably due to deterioration of the sharpness of the doping profile at the interface. Therefore, the optimum doping density might move to a higher doping density than 2.5 x 10^{17} cm^{-3} in the future if epitaxial growth technology progresses.

Dependence on the steepness of the doping profile at the interface

As theoretical analysis predicts [11], a lower drain bias current gives a lower noise figure until the decrease of the transconductance at the interface pushes it up. Normally the minimum noise figure is obtained at the drain bias current I_{DS} of about one tenth of the drain saturation current I_{DSS}, as shown by a solid line in Figure 6. However, if the doping profile is not ideally steep at the interface, the noise figure starts to go up at a higher drain current, as shown by the dotted line in Figure 6, and the minimum noise figure becomes larger even though the noise figure at a higher drain current is almost the same for both cases.

One of the clearer examples is seen in Reference 12. Three wafers with different doping profiles (Figure 7) were prepared by a vapor phase epitaxial growth or an ion implantation of either Si or Si and Mg into a high purity epitaxial buffer layer. The Si implanted layer has a long doping tail into the buffer layer. This doping tail can be

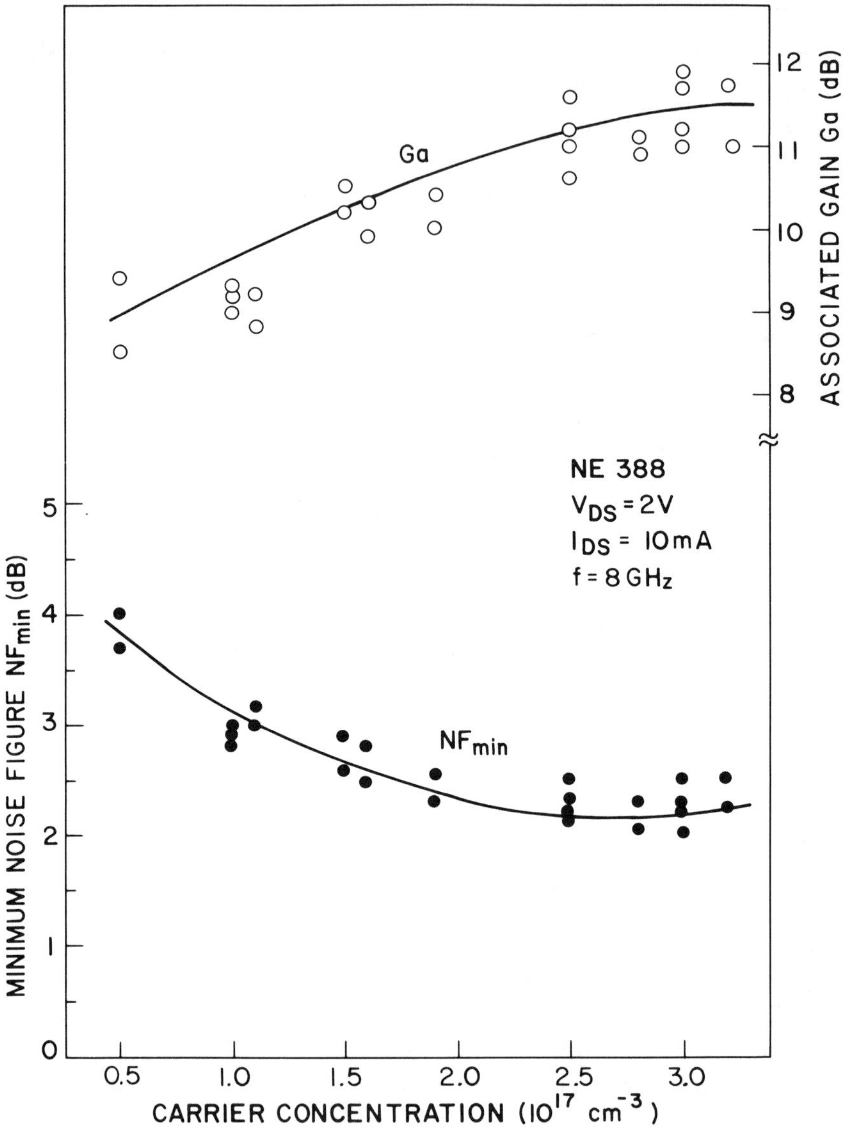

Figure 5*/Dependence of the minimum noise figure on the doping density of the active layer. Devices: 0.5 μm gate FETs. Frequency: 8 GHz. Biases: $V_{DS} = 2V$, $I_{DS} = 10$ mÅ.*

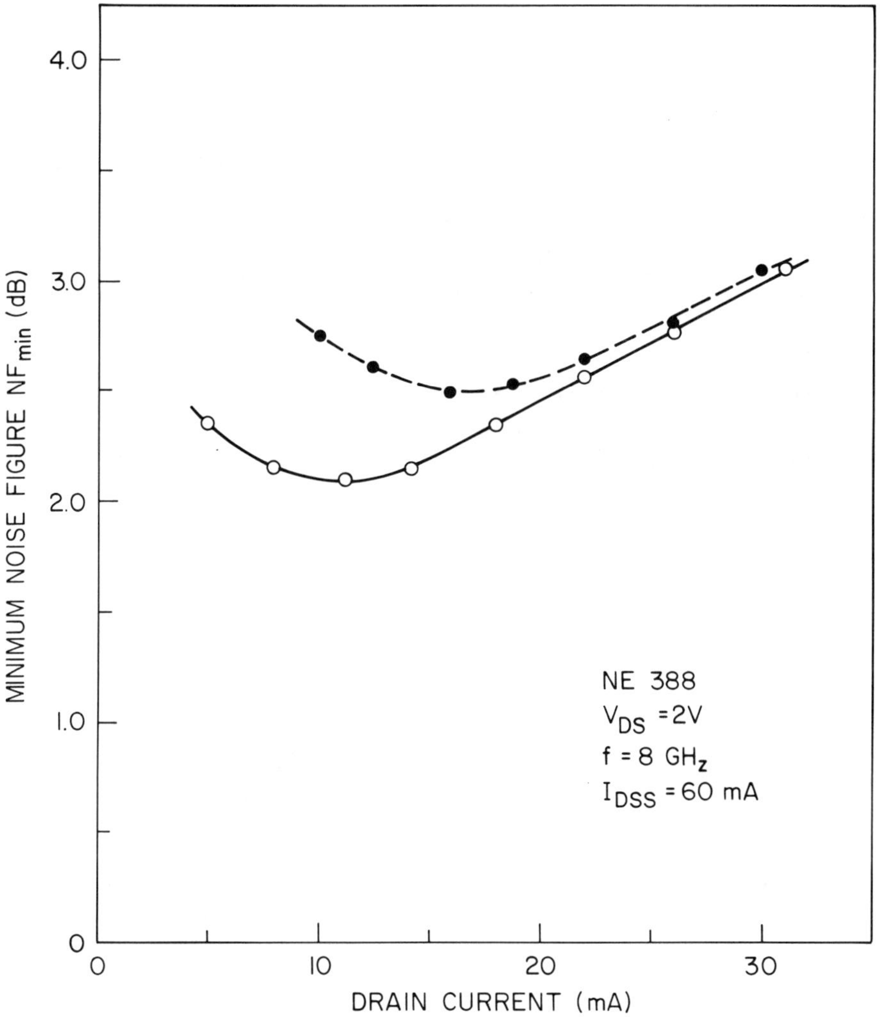

Figure 6*/Dependence of the minimum noise figure on the drain bias current. The dependences are different for different doping profiles at the interface.*

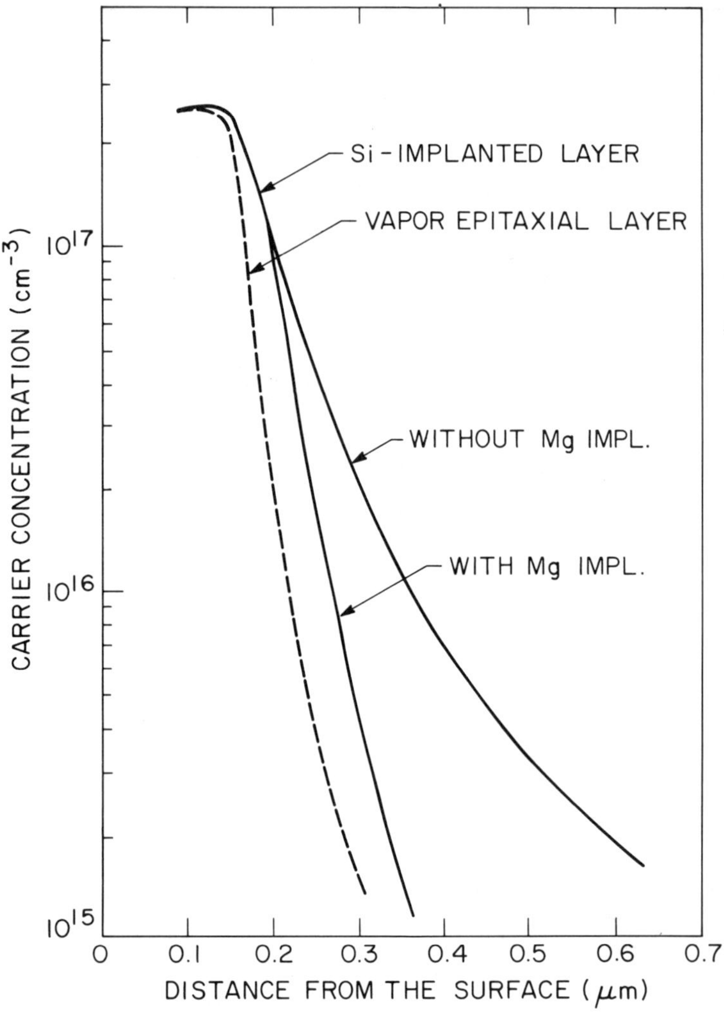

Figure 7*/Doping profiles of the active layer with different profile steepnesses at the interface:* ***(a)*** *Vapor grown epitaxial layer;* ***(b)*** *Si-implanted layer;* ***(c)*** *Si and Mg-implanted layer.*

compensated for by Mg implantation at this region, but, still, the steepness of the doping profile at the interface is worse than that for the vapor grown epitaxial layer.

The performance of low noise GaAs FETs made in the same way on these three kinds of wafers is summarized in Table 1. The drain saturation currents I_{DSS} are almost the same, but the pinch off voltage V_p is the largest and the transconductance g_m is smallest for the Si implanted device because of its doping tail. The minimum noise figure was obtained at the drain current of 20 mA for this device, compared with 10 mA for other devices. Its values of 3.2 dB at 8 GHz and 3.7 dB at 12 GHz are larger than those of the FETs made on the vapor phase epitaxial layer by nearly 1 dB. This difference is mainly due to the difference in the doping steepness at the interface.

Parasitic source and gate resistance

The parasitic source and gate resistance are the largest noise sources of GaAs FETs as described in the previous section. Therefore, reduction of these parasitic resistances is as important as shortening of the gate length.

The gate resistance can be reduced by using two gate pads or by adopting the gate structure of Figure 1(b). One of the most effective methods of reducing the source resistance is recessing the gate region as shown in Figure 1(c). It is effective even though there is no surface n^+ contact layer [13].

Figure 8 shows the decrease of source resistance with increasing depth of the recessed region on a FET without the surface n^+ contact layer. The device is a 0.5 μm gate FET as shown in Figure 2(b). The carrier concentration of the active layer is about 2.5×10^{17} cm^{-3}; therefore, the thickness of the active channel layer under the gate is only about 0.1 μm. The source resistance can be estimated by measuring the potential drop between the source and the gate, flowing a small forward gate current [4].

The source resistance for the zero recessed depth in the figure means the source resistance of the flat type FET without any recess at the gate region. Probably due to the surface depletion layer, this source resistance of about 6 Ω is much higher than that expected from the resistivity and the dimension (gate to source spacing 1 μm, total gate

TABLE I

DC Parameters and Microwave Performance of Low Noise GaAs FETs Made on the Wafers With Different Doping Profiles

CHANNEL	SUBSTRATE	IMPL. ION	I_{DSS} (mA)	V_P (V)	G_m (mΩ)	I_{DS} (mA)	8 GHz		12 GHz	
							NF(dB)	Ga(dB)	NF(dB)	Ga(dB)
I / I	Epi - GaAs	Si	53	5.4	20	20	3.2	10.2	3.7	9.0
I / I	Epi - GaAs	Si + Mg	53	2.4	22	10	2.5	10.7	2.9	9.2
Epi		—	50	2.0	24	10	2.2	11.2	2.8	10.0

width 280 μm). The experimental results indicate that about 400Å of GaAs surface is depleted due to a high surface state density for the GaAs layer with a carrier concentration of 2.5 x 10^{17} cm^{-3}. This is more than one third of the active layer for normal flat type low noise FETs.

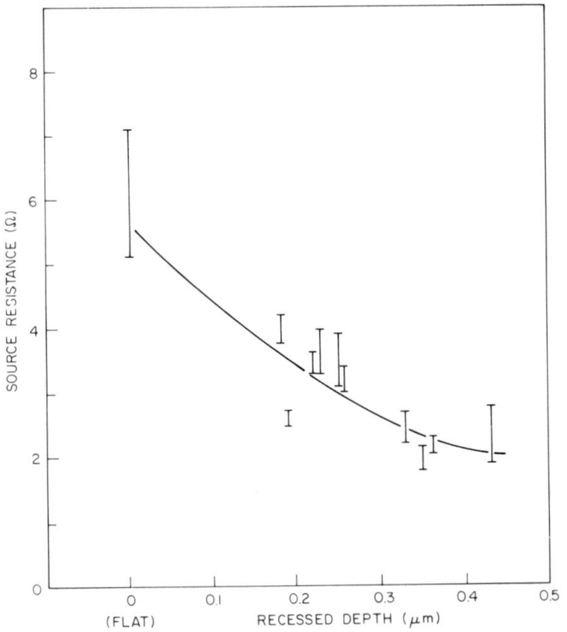

Figure 8/Dependence of the source resistance on the recessed depth. The device pattern is shown in Figure 2(b). The total gate width is 280 μm.

Because of this fact, the source resistance of the low noise FET can be reduced more effectively than expected by recessing the gate region. The source resistance can be reduced to less than half of that of the flat type by recessing more than 0.3 μm (Figure 8). A theoretical estimation indicates that, even though there is a 0.2 μm surface n^+ layer, the reduction of the source resistance is only from 2Ω to 1.6Ω for about a 0.4 μm recess.

The gate resistance of a typical 0.5 μm gate low noise FET with two gate pads is about 1Ω for a 0.5 μm thick aluminum gate.

1.3 Microwave Performance

Typical noise figures of 1 μm gate FETs, 0.5 μm FETs and 0.5 μm gate deeply recessed structure FETs are summarized in Figure 9. The noise figure can be improved by about 0.7 dB at 12 GHz by reducing the gate length from 1.0 μm to 0.5 μm. This noise figure is further

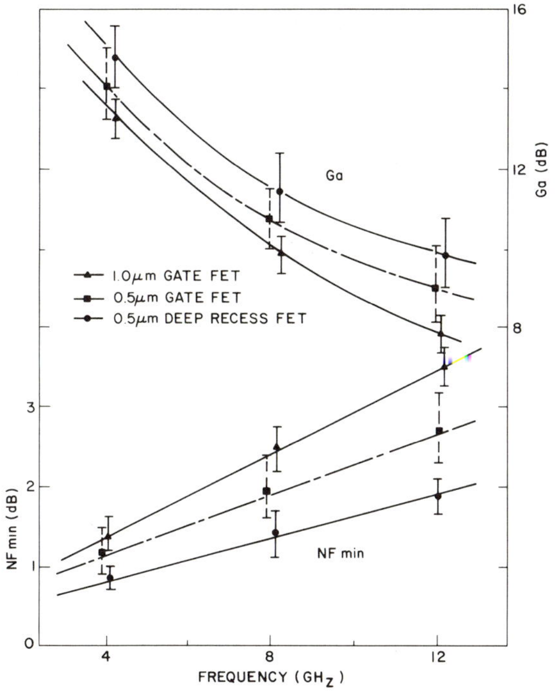

Figure 9/Minimum noise figures and associated gains of different patterns or structures low noise GaAs FETs: 1 μm gate FETs (Figure 2(a), NE244); 0.5 μm gate FETs (Figure 2(b), NE388); 0.5 μm gate FETs with a deeply recessed gate.

improved by more than 0.5 dB at 12 GHz by reducing the source resistance from about 5 Ω to about 2 Ω. Associated gains are also improved by about 0.5 to 1.0 dB with the improvement of the gate length or the source resistance.

The best noise figures obtained for the 0.5 μm gate FET with a deeply recessed gate are 0.7 dB at 4 GHz, 1.2 dB at 8 GHz and 1.68 dB at 12 GHz, with associated gains of 15.5 dB, 12.5 dB and 10.7 dB respectively [13]. Other examples of best performance reported in the literature up to 1979 are 2.0 dB at 14 GHz [8] and 2.4 dB at 18 GHz [4].

Figure 10 is the summary of noise figures of one or two stage amplifiers constructed using 1 μm gate or 0.5 μm gate FETs. One of the things to be noted is that the noise figures of the amplifier, as well as the FET itself, degrade linearly against the frequency, in contrast with the bipolar transistors where the noise figure increases parabolically against the frequency. Therefore, the noise performance at every frequency can be estimated from the noise figure at a certain frequency.

A low noise amplifier with 2.2 dB at 12 GHz has been developed [14] utilizing the 0.5 μm gate FET with a deeply recessed gate. Therefore, it would seem possible that, in the near future, amplifiers with the noise figure of 4 dB at 20 GHz and 6 dB at 30 GHz might be developed.

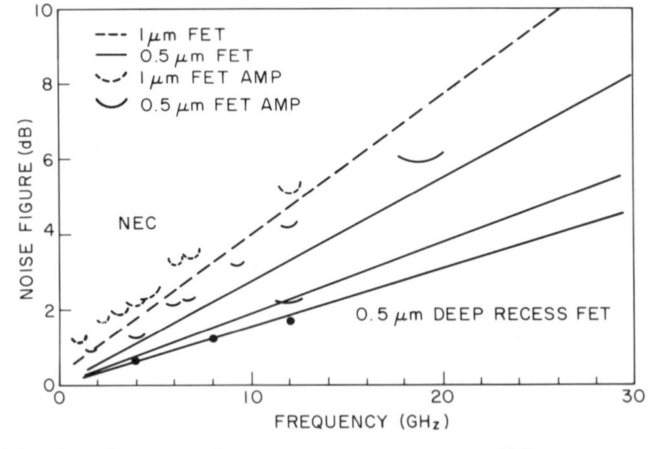

***Figure 10**/Noise figures of one or two stage amplifiers constructed with different types of low noise GaAs FETs.*

REFERENCES

[1] Nozaki, T., M. Ogawa, H. Terao and H. Watanabe, *Inst. Phys. Conf. Ser. No. 24,* Chapt. 2, p. 46, 1975.

[2] Napoli, L.S., J.J. Hughes, W.F. Reickert, and S. Jolly, *RCA Review,* 24:608, 1973.

[3] Baudet, D., M. Binet, and D. Boccon-Gibod, *IEEE Trans. Microwave Theory and Technique,* MTT-24:372, 1976.

[4] Wholey, J. and M. Omori, *Inst. Phys. Conf. Ser. No. 45,* 1979, Chapt. 4, p. 270.

[5] Ohata, K. and M. Ogawa, *12th Annual Proc. IEEE Reliability Physics Symp.,* Las Vegas, p. 278, 1974.

[6] Ogawa, M., K. Ohata, T. Furutsuka, and N. Kawamura, *IEEE Trans. Microwave Theory and Techniques,* MTT-24:300, 1976.

[7] Takahashi, S. et al., *Proc. 8th Conf. Solid State Devices,* Tokyo, p. 115, 1977.

[8] Buthin, R.S. et al., IEEE Int. Electron Devices Meeting, *Dig. Tech. Papers,* p. 136, 1978.

[9] Ohata, K., T. Nozaki, and N. Kawamura, *IEEE Trans. Electron Devices,* ED-24:1129, 1977.

[10] Fukuta, M. et al., *IEEE Trans. Microwave Theory and Technique,* MTT-24:312, 1976.

[11] Pucel, R.A., D.J. Masse, and C.F. Krumm, *IEEE Journal of Solid State Circuits,* SC-11:313, 1976.

[12] Nozaki, T. and K. Ohata, *Proc. 8th Conf. on Solid State Devices,* Tokyo, p. 111, 1976.

[13] Ohata, K., H. Ito, F. Hasegawa, and Y. Fujiki, *IEEE Trans. Electron Devices,* ED-27:1029, 1980.

Chapter 4

Power GaAs FETs

Introduction

This chapter discusses power GaAs FET design and performance and represents the most voluminous chapter in the book. The editors are of the opinion that power GaAs FETs will be one of the most important microwave devices of the 1980's and, therefore, require extensive coverage. Power GaAs FETs are revolutionizing the design of both analog and digital radio communication systems and will be used for active devices in power amplifiers for communications satellites where they will provide life and performance advantages over TWT's. This chapter features contributions from four leading research, development, and manufacturing organizations. Each has a rather different design approach which shows the divergent points of view on obtaining a ' state of the art'' device.

I. Drukier discusses a complete design approach to power GaAs FETs. This contribution is distinct from the others in the chapter because it focuses on the "flip chip" approach to power GaAs FETs. But this is only one element in the design principles outlined. Drukier includes a self aligned gate to minimize gate alignment errors and also to minimize source-drain spacing, a gold based refractory metal system for source, drain and gate metallization, and a multilayer epitaxial wafer to isolate the active layer from the substrate. The advantages of flip chip mounting are demonstrated by a 2-3 db gain advantage over upright mounted devices. Drukier reports impressive performance up to frequencies as high as 20 GHz. Most impressive is 1 watt with 6.3 dB of associated gain at 15 GHz with a power added efficiency of 18%.

F. Hasegawa discusses in detail his design approach to achieving high drain breakdown voltage for power GaAs FETs and concludes that a "graded recess" structure is optimal for achieving high breakdown voltage. Hasegawa considers the gate breakdown voltage, drain bias current and voltage, gate finger width, thermal resistance and matching networks. He reports very impressive microwave performance of 25W at 6 GHz and 6W at 10 GHz.

H.M. Macksey presents data and design principles for GaAs power FETs operating at frequencies from 8 to 20 GHz. Macksey uses an e-beam defined Ti/Pt/Au gate which is unique among power GaAs FET designers. Macksey discusses dependence of power and gain on active layer doping and thickness and observes that a gate "recess" greatly increases device output power and concludes that an optimum gate recess depth is in the 400 Å to 800 Å range. Further data is presented on the effect of gate length on power at various frequencies. Macksey sees 1-1.5 μm gates at 8 GHz, .75-1 μm gates at 10 GHz and 0.5 μm gates at 15 GHz as optimum. Macksey compares 2.0W at 15 GHz and 1.0W at 20 GHz against the currently achieved best microwave data.

In the last contribution W.C. Niehaus et al. discuss their approach to the design of power GaAs FETs. They have design principles which are significantly different from the other three contributions. For example, the doping of the active layer is between 3-7 x 10^{16} cm^{-3} compared to $\geqslant$ 1 x 10^{17} cm^{-3}, a four layer epi structure is used, and gate widths are 500 μm for operation at 4 GHz. Niehaus' results suggest a weak dependence of power on gate width up to 12 GHz. Devices are designed to operate a higher drain bias and burn-out voltages of 40-50 V_{SD} are observed. The key to this is claimed to be N+ "drain ledge" geometry as described in their contribution. Niehaus et al. (like Macksey) state that a recessed gate structure is essential to good rf performance and discuss the dependence of gate drain avalanche on the total charge present in the channel. The model for this is discussed in terms of "lateral spreading" of the depletion layer. The final major difference is the use of a "thin" GaAs substrate [30 - 50 μm compared to 100 - 125 μm] and a plated heat sink for good thermal dissipation. Impressive microwave performance of 18.5W at 4.4 GHz and 8.6W at 7 GHz is reported.

Power GaAs FETs

I. Drukier

Microwave Semiconductor Corporation,
Somerset, New Jersey.

1. INTRODUCTION

Power GaAs FETs have been extensively developed in the past few years [1, 3]. Many different approaches to their fabrication have been used with varying amounts of success. This paper describes the processes used by MSC to fabricate power GaAs FETs, their microwave performance and some aspects of reliability.

Our basic approach to GaAs FET design is to minimize parasitics while maintaining a high degree of device (wafer to wafer) uniformity. Beginning with a multiple wafer epitaxial reactor, attention to uniformity and repeatability are paramount. Some key features of this approach are:

1. Multilayer epitaxial wafers
2. Sintered ohmic contacts
3. Self-aligned gate
4. Refractory gate metallization
5. Flip-chip mounting

These features will be briefly summarized now and discussed in detail in subsequent sections.

A multiple layer epitaxial structure has many advantages. Of primary importance is a high resistivity undoped buffer layer. This isolates the active region from the substrate and substantially improves carrier mobility. After the active region, a contact layer is grown.

A major source of non-uniformity in conventional devices is gate alignment errors. With a self-aligned gate, precise alignments are no longer required and device uniformity is greatly improved. A secondary feature of self-alignment is that the source to drain spacing is almost as small as the gate itself. This reduces the parasitic source and drain resistances to the bare minimum.

Another feature peculiar to MSC's approach is a gold based refractory gate metal system. This allows ease of bonding and no Al-Au diffusion failure mechanisms.

Finally, by flip-chip mounting of transistors, both the thermal resistance and the common source inductance can be reduced by half. A lower thermal resistance increases reliability by lowering the device operating temperature, whereas reduction of the source inductance improves performance.

The approach we have taken gives both good performance and uniformity. The following sections detail our processing approach and the rationale for our approach.

EPITAXIAL MATERIAL

Even the best processing cannot improve bad material. It is therefore imperative that considerable attention be devoted to the quality of the epitaxial material used to make devices. All our GaAs layers are grown by the $Ga/AsCl_3/H_2$ CVD system. The MSC vapor-phase epitaxial reactor system is shown schematically in Figure 1. It combines the best features of several generations of such vapor phase reactors and is designed to produce GaAs epitaxial wafers for FETs to tight tolerances with good yield.

The major features of the MSC reactor are:

1. A moveable furnace with six independently-controlled temperature zones and accurate temperature gradients;
2. A well protected reactor tube;
3. Dual gallium source boat located in a carefully controlled temperature zone;

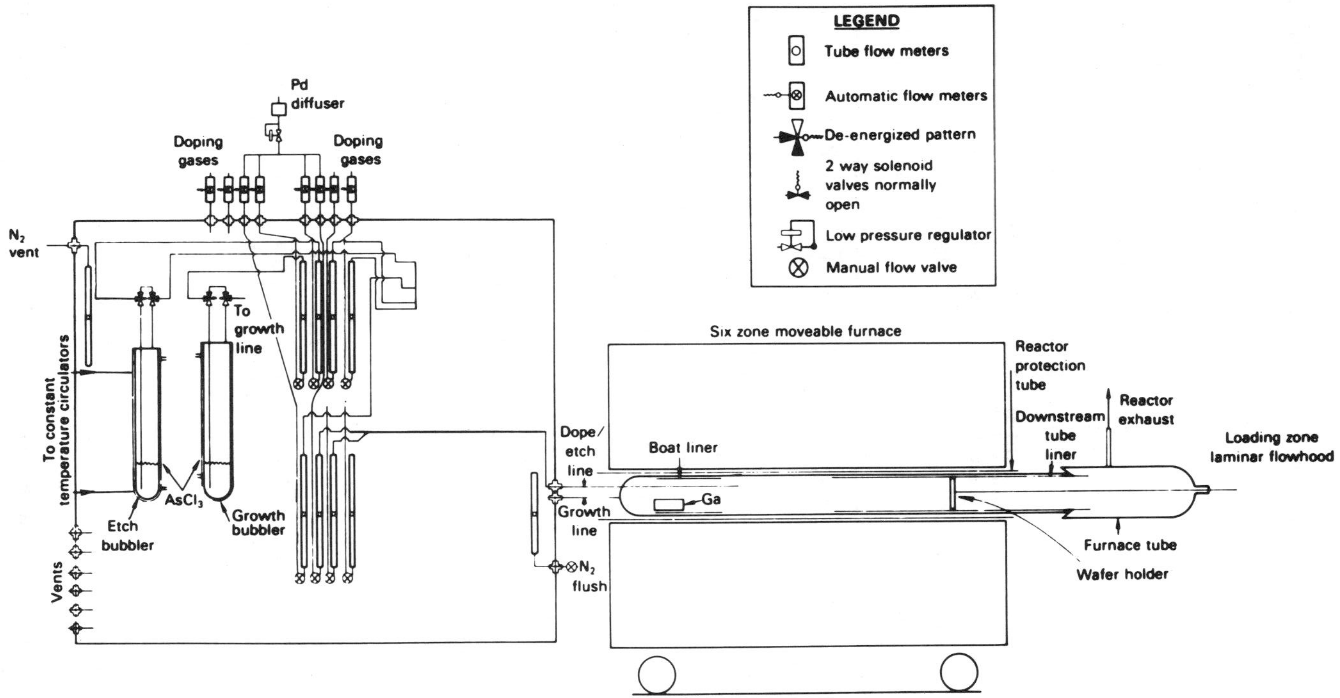

Figure 1/Schematic of epitaxial reactor.

4. Downstream tube liner for convenience in loading and cleaning;
5. A multi-wafer holder which can be accurately positioned;
6. And a precisely controlled automatic gas-handling system enclosed in a sealed, nitrogen flushed plastic box.

All the hydrogen used in the reactor is purified by diffusion through palladium membranes. The reactor is housed in a clean room and the loading of the reactor is done from a laminar flow bench.

Once deposition is initiated, status displays provide process step number, time remaining before step is complete, time programmed for upcoming steps and controls activated during the current step.

This highly sophisticated reactor system, with its precise growth controls, has been a crucial factor in achieving the degree of control required to reproducibly grow GaAs FET wafers to meet our stringent specifications.

All wafers begin with a 5μm thick high resistivity buffer layer. Using a conventional C-V profiler, the doping in this layer appears to be less than 10^{13} cm^{-3}. While a buffer may not be needed for all substrates, it does eliminate the effect of variability in substrate quality. Furthermore, it appears to increase by 15-20% the Hall mobility measured in the active region.

After the buffer, an active region of 0.7 μm thickness is grown. Its doping is typically 1×10^{17} cm^{-3}, with a Hall mobility of 4800 cm^2/V·sec. Since this layer is subsequently tailored in thickness by processing, its uniformity, but not its absolute thickness, is important.

Finally, a highly doped ($> 10^{18}$ cm^{-3}) contact layer is grown. This layer is important, more for consistency and reliability than performance. This layer can act as a buffer zone between metallurgical changes in the ohmic metallization and the active region. There is also some evidence that a contact layer can improve source to drain breakdown voltage [3, 5].

3. FABRICATION

Figure 2 illustrates schematically our fabrication procedure. Processing begins with the multilayer epitaxial wafer. Only conventional photolithography techniques are used throughout and no critical alignment steps are needed. The process is as follows:

1. The wafer is metallized using Au/Ge/Ni. Ohmic contact is made by sintering at 400-450°C in a dry ambient.
2. The mesa is defined and formed (Figure 2A).
3. The gate channel is formed. The source to drain current is monitored and etching is stopped when the optimum current is reached. There is sufficient undercut to allow self-aligning (Figure 2B).
4. Ti/W/Au is deposited to form the Schottky gate. Since the gate channel is undercut, no shorting occurs (Figure 2C). (At this point an optional dielectric coating can be applied.)
5. Gold is selectively electroplated on the source pads so devices can be flip-chip mounted (Figure 2D).

Once the wafer has been completed all devices are automatically dc tested and the results recorded. Devices can then be packaged.

A key row shown in Figure 3 has been incorporated into the photomasks. Among the various parameters that can be monitored on every wafer are:

1. Ohmic contact resistance (relative)
2. Schottky diode characteristics
   a. Barrier height
   b. Ideality factor
   c. Leakage current
3. Impurity density – to verify epitaxial growth area data
4. Mobility using a FATFET
5. Gate resistance
6. Pinchoff as a function of gate length

As an example of the data available, Figure 4 shows a FATFET [6] mobility profile. It shows high mobility rising into the buffer-active interface. While the absolute rise may be an artifact, it does indicate good quality material. By careful study of this parameter, as well as the others available, tight control can be maintained over the fabrication processes.

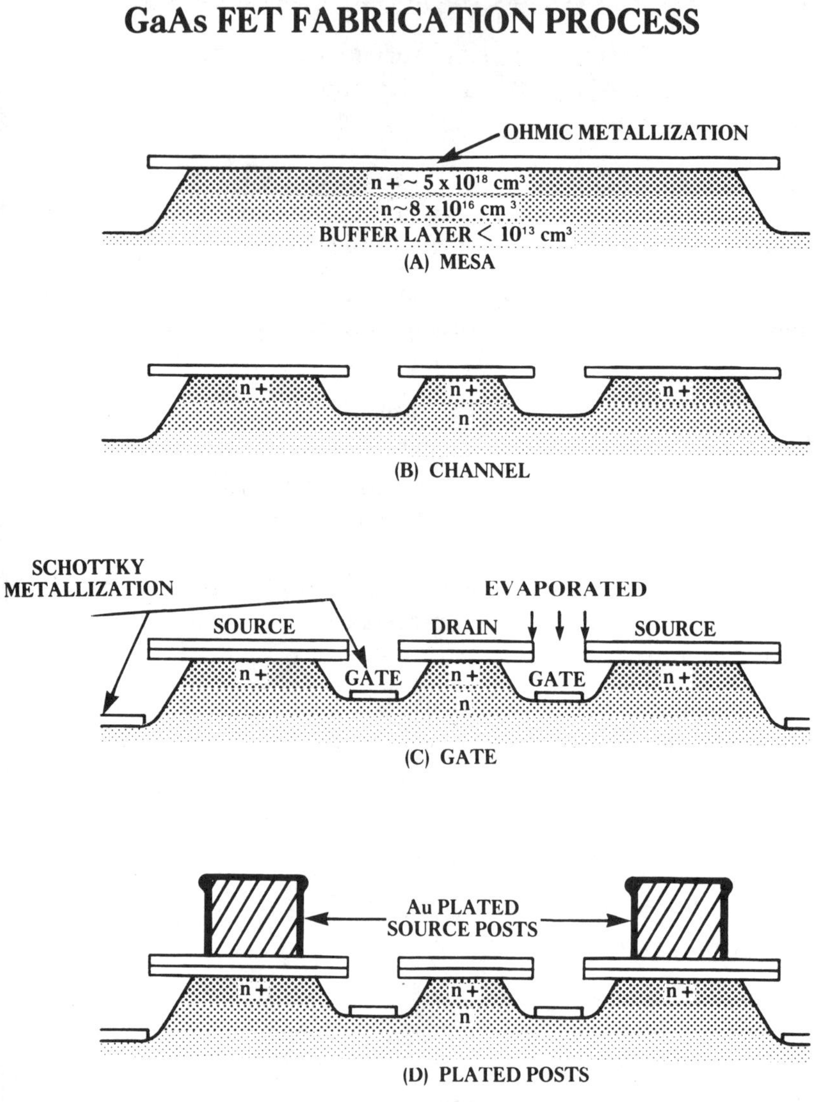

Figure 2/GaAs FET fabrication process

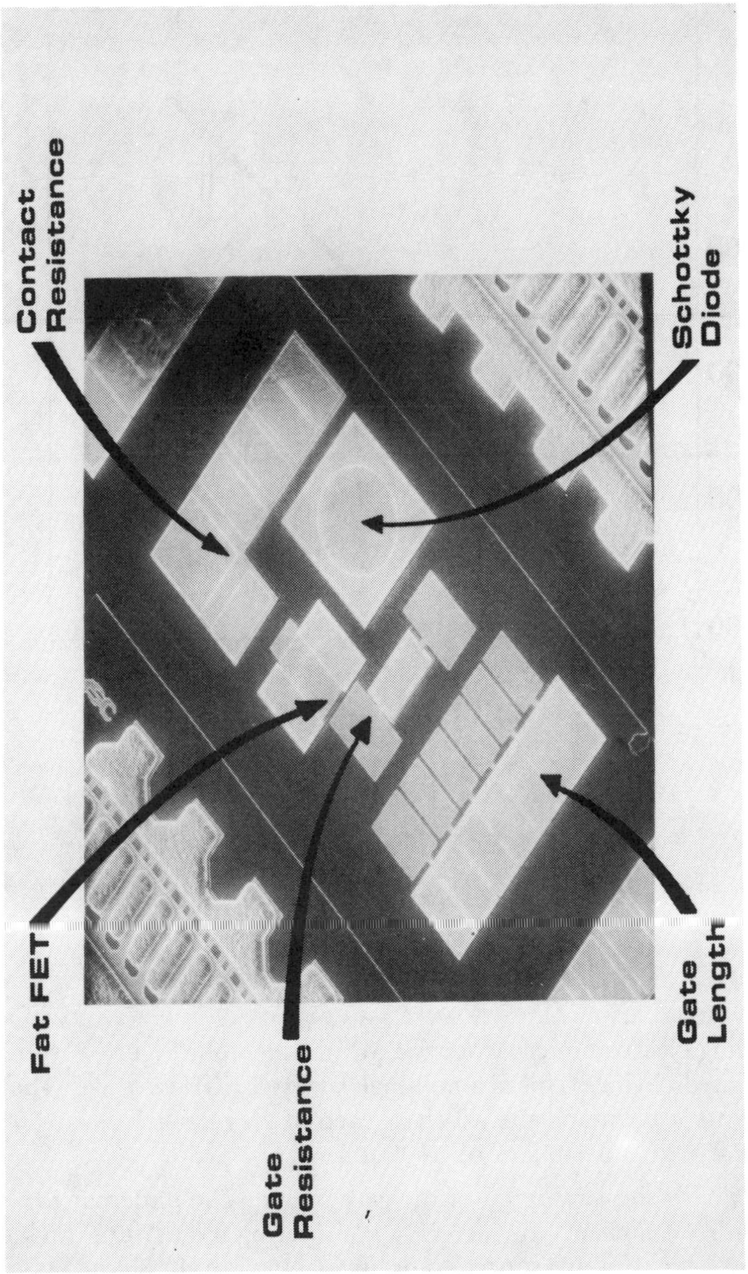

Figure 3/Key row pattern

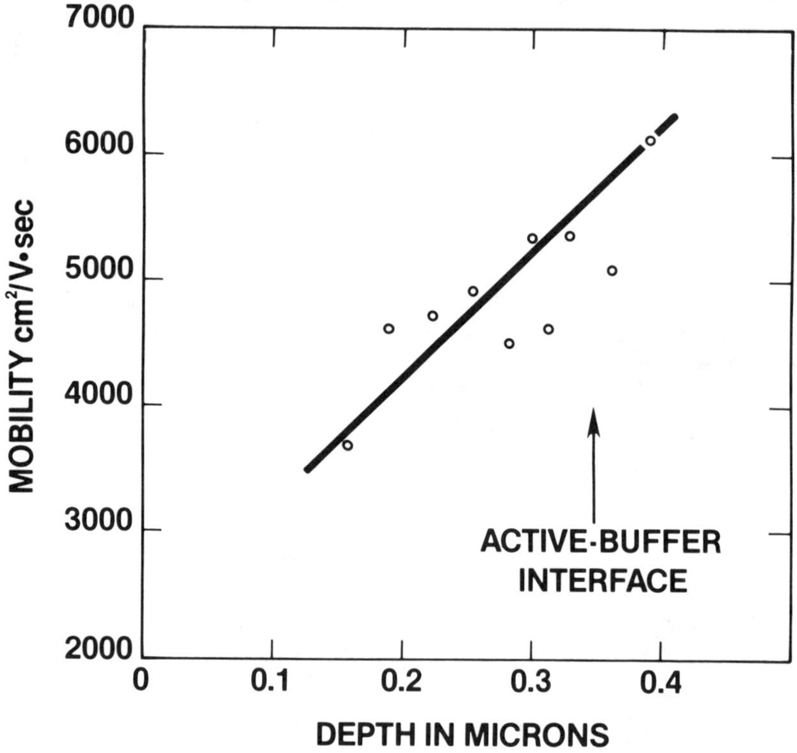

Figure 4/FATFET mobility profile

4. DEVICE GEOMETRY

A scanning electron microscope (SEM) photograph, Figure 5, shows a completed device. This is a nominal 1 watt 15 GHz device. The gate length is 1 μm and the width of each of 24 gate stripes is 100 μm, for a total source periphery of 2400 μm.

The SEM also shows a detail of the gate region. The undercut for self-aligning can clearly be seen. By considering the 0.5 μm thick contact layer, the effective source to drain spacing is only 2 μm. This tight packing allows excellent high frequency performance.

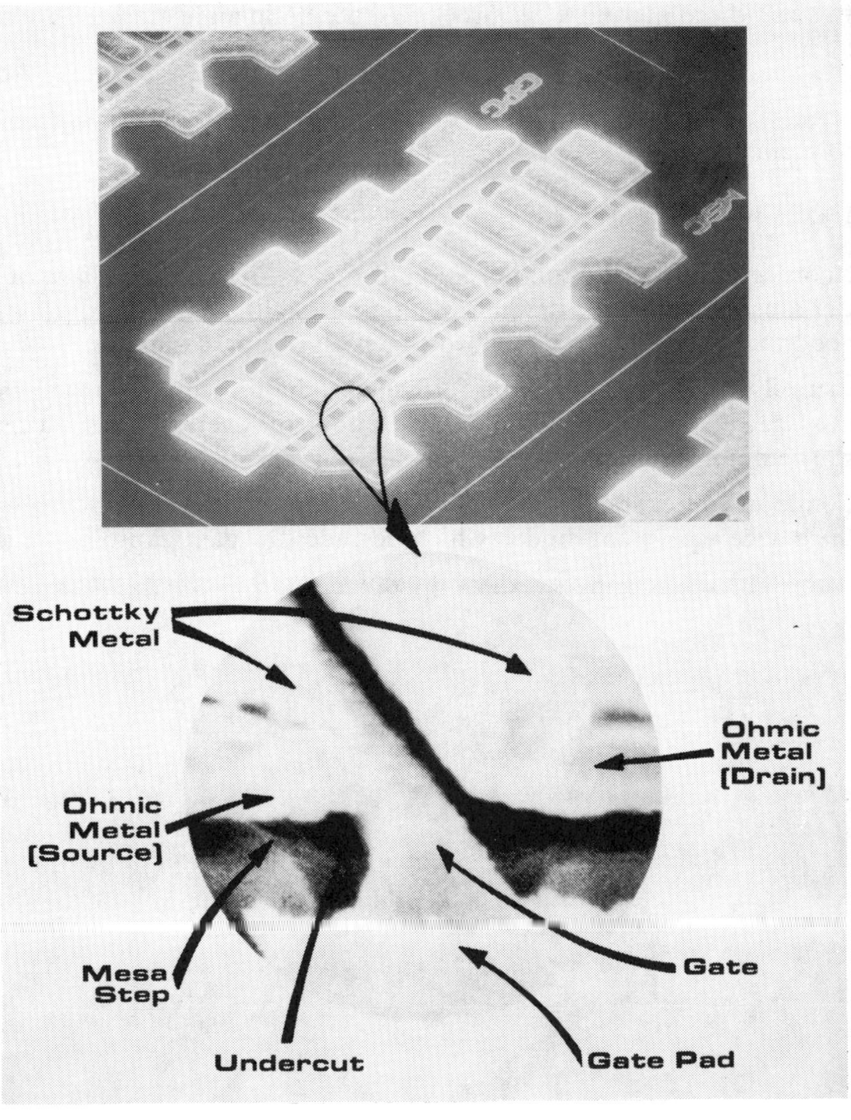

Figure 5/Device cross-section

In addition to plating the source pads, the drain and gate pads are also plated for ease of bonding. Compared to a C-band device with a periphery of 16 x 150 μm, the gate resistance is reduced by half. The effect of this change at high frequencies will be discussed in the performance section.

5. DEVICE ASSEMBLY

In order to utilize the full potential of the GaAs FET, a package or carrier must not limit the device's inherent capability. It should provide a good thermal path and allow broadband device performance.

Critical to our good device performance is the flip-chip mounting of the GaAs FET. This technique lowers both the thermal resistance and the common source inductance of the transistor.

Common source inductance appears as a parasitic input resistance in the device equivalent model, which degrades the gain, noise figure and

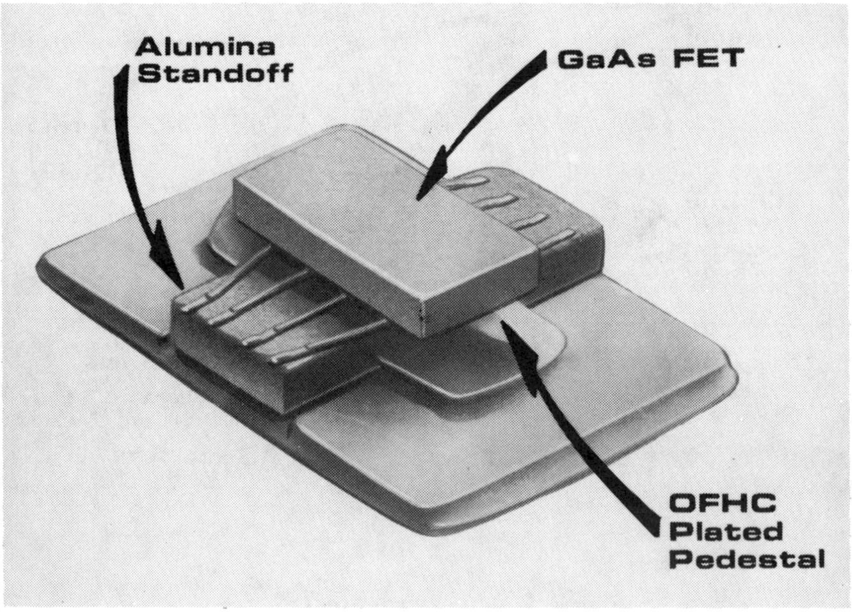

Figure 6/Chip carrier mounted GaAs FET

power output. Furthermore, our technique of source post fabrication, as detailed earlier, insures uniform common inductance for each part of the transistor. This should be contrasted to wire-bond source grounding where small differences in wire length introduce non-uniformities in local device operation, which will, in turn, degrade overall performance. This effect is especially pronounced in large devices capable of high output powers. Experimental results have shown that flip-chip mounting improves gain by 2-3 dB over upright mounted devices.

A chip carrier with a 1 watt X-band device is shown in Figure 6. Note that the wire bonds to the drain and gate are short. This minimizes inductance and makes circuit matching easier. The package base is gold-plated OFHC copper for good thermal heatsinking. Reduction of the thermal resistance improves the gain and output power, but, perhaps even more importantly, increases the operating lifetime.

Measurements on standard production devices have shown the infrared hot spot (IR) thermal resistance of a flip-chip mounted device to be comparable to a silicon bipolar transistor. For example, a one watt GaAs FET has a thermal resistance of 17°C/W, which is identical to a one watt bipolar. Figure 7 shows an IR scan of a one watt device. The uniformity of device temperature is indicative of a good thermal mount.

The results of a random thermal resistance sampling of 500 production devices are shown in Table 1. It clearly indicates the thermal uniformity of devices mounted in this way.

TABLE I

IR Scan Data

Device	Power Output	Thermal Resistance Meas.	Thermal Resistance Std. Dev.
88000	50mW	38.0	6.5
88001	.25W	31.6	1.6
88002	.5W	24.3	1.7
88004	1.0W	16.8	2.0
88012	3.5W	6.7	0.6

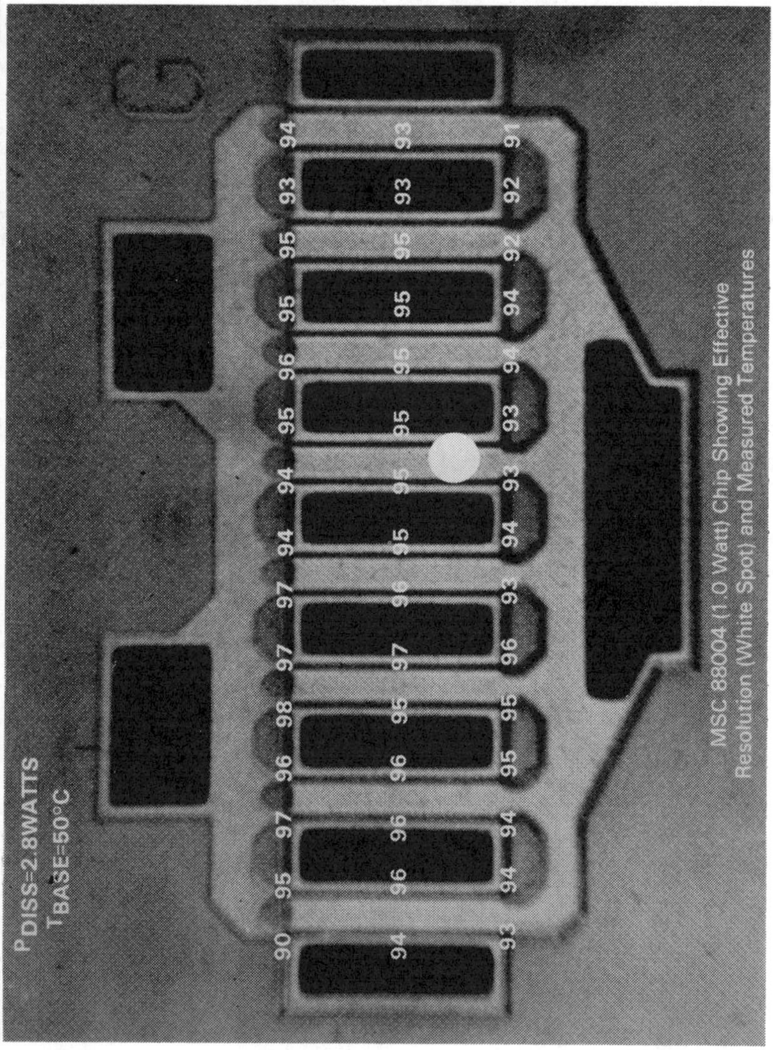

Figure 7*/MSC 88004 (1.0 Watt) chip showing effective resolution (white spot) and measured temperatures*

6. DEVICE PERFORMANCE

The test circuits for power measurements in C- and X-band use microstrip lines on polished alumina substrates in a coaxial system. Above 12 GHz, fused quartz is used instead of alumina. At even higher frequencies (> 18 GHz), measurements are made in a waveguide system. For additional tuning during measurements, either double slug tuners in coax, or E-H tuners in waveguide are used.

When measuring power GaAs FETs, it is important to understand the tradeoffs between output power, efficiency and gain. Any one of these can be optimized at the expense of the other two. For example, Table 2 gives typical values of a single 1 x 2400 μm device optimized for each parameter.

TABLE II.

Performance Tradeoffs at 8 GHz

Output Power (watts)	Gain	Efficiency
1.26	6.5 dB	25%
0.75	5.5 dB	50%
0.7	9.0 dB	15%

What is nominally a moderate efficiency one watt device can also be considered a high efficiency 0.75 watt device or a high gain 0.7 watt device. These tradeoffs make characterizing GaAs FETs rather difficult. Depending on the parameter of interest, the device may be made to appear better than it would be in a general application.

Table 3 shows a summary of MSC device results. As the table shows, excellent performance from 2 GHz through 20 GHz can be obtained using the fabrication procedure described above. Power outputs of 2.5 watts at 15 GHz and efficiencies of 40% at 8 GHz indicate the broad range of performance that devices can reach. In pulsed operation, outputs greater than 10 watts [7] have been achieved. Furthermore, these results, while certainly optimized, are by no means unique to one wafer. Comparable results at all frequencies have been obtained from numerous wafers.

TABLE III.

Performance of GaAs FETs

Freq. (GHz)	P_{OUT} (watts)	Gain @ P_{OUT} (dB)	ηPA @ P_{OUT}	Gate Length (μm)	Source Periphery (μm)
2	2.0	6.5	55%	1.7	6000
4	8.0	6.3	30%	1.0	2 x 8500
8	4.0	5.5	40%	1.0	2 x 4800
15	2.5	4.0	16%	1.0	2 x 2400
18	1.0	3.5	17%	1.0	2400
20	.50	4.0	12%	1.0	1200
10 (Pulsed)	6.0	4.0	—	1.0	6000
8	10.1	4.0	—	1.0	2 x 2600

TABLE IV.

Device Geometry Comparison at 15 GHz

Type (μM)	P_{OUT} (dBm)	Gain @ P_{OUT} (dB)	ηPA %
16 x 150	29.5	5.6	13.5
24 x 100	30.05	6.1	18.7

It is particularly interesting to note that power scaling for flip-chip devices is better than that reported for upright devices. Table 3 indicates that if two pellets are combined in one carrier or package, twice the single pellet output power was obtained. This is due, in part, to the low and uniform source inductance of the flip-chip mounting technique.

Reduction of the gate finger width for higher frequency performance has resulted in improved gain for the same total device periphery. Using 2400 μm as the standard device size, two devices were fabricated, one with sixteen gate fingers, the other with twenty-four. The gate resistance of the 24 finger devices is 40% of the 16 finger design.

Both devices were tested at 15 GHz and the results are shown in Table 4. While both devices performed well, the 24 gate unit gave somewhat more gain and efficiency. In addition, the 24 gate device gave more consistent results from wafer to wafer.

7. RELIABILITY

Reliability is critical to the acceptance of power GaAs FETs. Several papers recently addressed the question of device reliability [8, 10]. Temperature storage, dc bias, and rf bias tests were conducted. The most severe stress condition is the rf bias test; however, results of the dc and rf tests agree rather well and only the dc tests will be discussed here.

Unscreened production devices were dc biased to normal operating conditions at elevated temperatures. Data was taken up to 250°C and the failure criterion used was a 1 dB change in power output at 6 GHz. Figure 8 shows the Arrhenius plot that was obtained from the data. An activation energy of 1.85 eV was calculated. At 120°C an extrapolated MTTF is 2 x 10^7 hours.

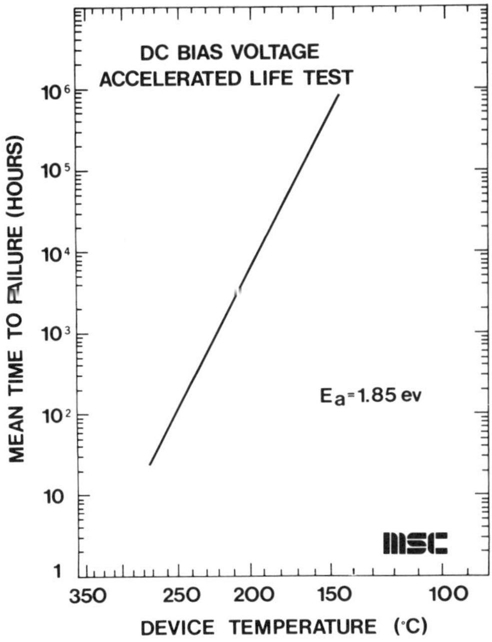

Figure 8/DC bias voltage accelerated life test

Gate metal changes appear to be the dominant failure mechanism, but a full understanding of the exact nature of these changes has not yet been obtained.

Since the tested devices had gold-based gate systems, neither Au-Al migration nor the rf gate current limitations reported in aluminum devices [11] affect the MTTF.

8. CONCLUSION

Using a self-aligned flip-chip device, high power outputs can be achieved from GaAs FETs at frequencies up to 20 GHz.

Flip-chip mounting gives excellent thermal resistance and low uniform common source inductance. Low operating temperature is necessary for reliable performance.

Uniform common source inductance allows excellent power scaling without any external matching. This is especially important at Ku-band and above.

Reliability results for refractory gate devices indicate that today's devices are ready for the needs of modern systems.

ACKNOWLEDGMENTS

The success of any development effort utilizes the talents and labors of many individuals. I wish to acknowledge the GaAs epitaxial work of Dr. J. Thompson, the microwave characterization of Mr. P. Wade, and reliability studies of Mr. J. Silcox, Jr. Also, thanks are due to Mr. P. York, Sr.

REFERENCES

[1] Drukier, I., P.C. Wade, and J.W. Thompson, European Microwave Conference, Brighton, England, 1979.

[2] DiLorenzo, J.V., and W.R. Wisseman, *IEEE Trans. on MTT,* MTT-27:367, 1979.

[3] Niehaus, W.C., et al., *Gallium Arsenide and Related Compounds* (St. Louis), p. 271, 1976.

[4] Fukuta, M., et al., *IEEE Trans. on MTT,* MTT-24:312, 1976.

[5] Furutsukia, T., T. Tsuje, and F. Hasegawa, *IEEE Trans. Electron. Dev.,* ED-25:563, 1976.

[6] Pucel, R.A. and C.F. Krumm, *Electronics Letters,* 12:240, 1976.

[7] Wade, P.C. and I. Drukier, ISSCC, San Francisco, 1980.

[8] Drukier, I. and J.F. Silcox, Jr., European Microwave Conference, Brighton, England, 1979.

[9] Drukier, I., and J.F. Silcox, Jr., International Reliability Physics Symposium, San Francisco, p. 150, 1979.

[10] Cohen, E.D. and A.C. Macpherson, International Reliability Physics Symposium, San Francisco, p. 156, 1979.

[11] Macpherson, A.C., K.R. Gleason, and A. Christou, Proc. IEEE Advanced Techniques in Failure Analysis Symposium, Los Angeles, 1978.

Power GaAs FETs

Fumio Hasegawa

Institute of Material Science, Tsukuba University,
Sakura Mura, Niiharigun, Ibaragi Prefecture, Japan.

1. BASIC DEVICE STRUCTURE

Power GaAs FETs need a high drain breakdown voltage and a high drain saturation current. The drain to source breakdown voltage greatly depends on the basic FET structure such as the cross section of the channel region. The gate breakdown voltage is closely related to the carrier concentration of the active layer and the pinch off voltage, and therefore the drain saturation current per unit gate width. In this section, the dependence of these parameters on the basic device structure will be discussed.

1.1 Drain Breakdown Voltage

Various types of basic GaAs FET structures are shown in Figure 1. The structure shown in Figure 1(a) is a flat type FET. Most low noise GaAs FETs are of this type, and are widely used for practical applications in commercially available devices [1]. There have been attempts to develop power GaAs FETs in this flat type [2], but it was found that this type of GaAs FET exhibits quite low drain breakdown voltage and is difficult to operate stably. Therefore, various types of FET structures, shown in (b), (c), (d), and (e) in Figure 1, were investigated for effectiveness, ease of fabrication, and reproducibility.

The structure (b) has selective n^+ contact regions prepared either by epitaxial growth [3] or by ion implantation [4] under the source and drain ohmic contact layer. This structure was first introduced by Fukuta et al. [3] and has proved very effective. But the fabrication process is not easy and reproducibility seems to be too low.

The structure (c) [5, 6] is a version of (b), but is more reproducible simply because the n^+ contact layer is grown successively on the n type active epitaxial layer and, therefore, the transition is smoother. The gate Schottky contact is prepared by etching off the surface n^+ contact layer on the active region. However, the preparation of the n^+-n-n^- triple epitaxial layer is not very productive by the present (1979) technology of epitaxial growth.

It was found that, as far as the drain breakdown voltage is concerned, it can be greatly improved by simply recessing the active region as shown by the structure (d) [7] in Figure 1. The drain breakdown voltage is as high or higher than that of structures (b) or (c) and can be reproducibly obtained if the recessed depth is enough. However, detailed investigation of the light emission from the device revealed that the structure (d) still has an anomalous high field outside of the gate region.

Light emission from the GaAs FET was first observed by Wemple et al. [5] and Mimura et al. [8], and was extensively investigated using the various types of FETs shown in Figure 1, by Yamamoto et al. [9], and Furutsuka et al. [10]. Some of the results are shown in Figure 2. It was found that a strong light emission is observed at the drain ohmic contact and the active layer interface for the flat type FET (Type A). In this case, the light emission is observed from the drain bias of 2 or 3 V and increases with the increase of the drain bias until the device breaks down at about 8 to 10 V. This light emission is greatly reduced by introducing the n^+ contact layer or the recessed structures, as shown in (b), (c), and (d) of Figure 1. However, the light emitting spots of these structures are not at the drain ohmic contact-active layer interface; they are at the drain n^+ contact layer-n active layer interface, or at the drain edge of the recessed region.

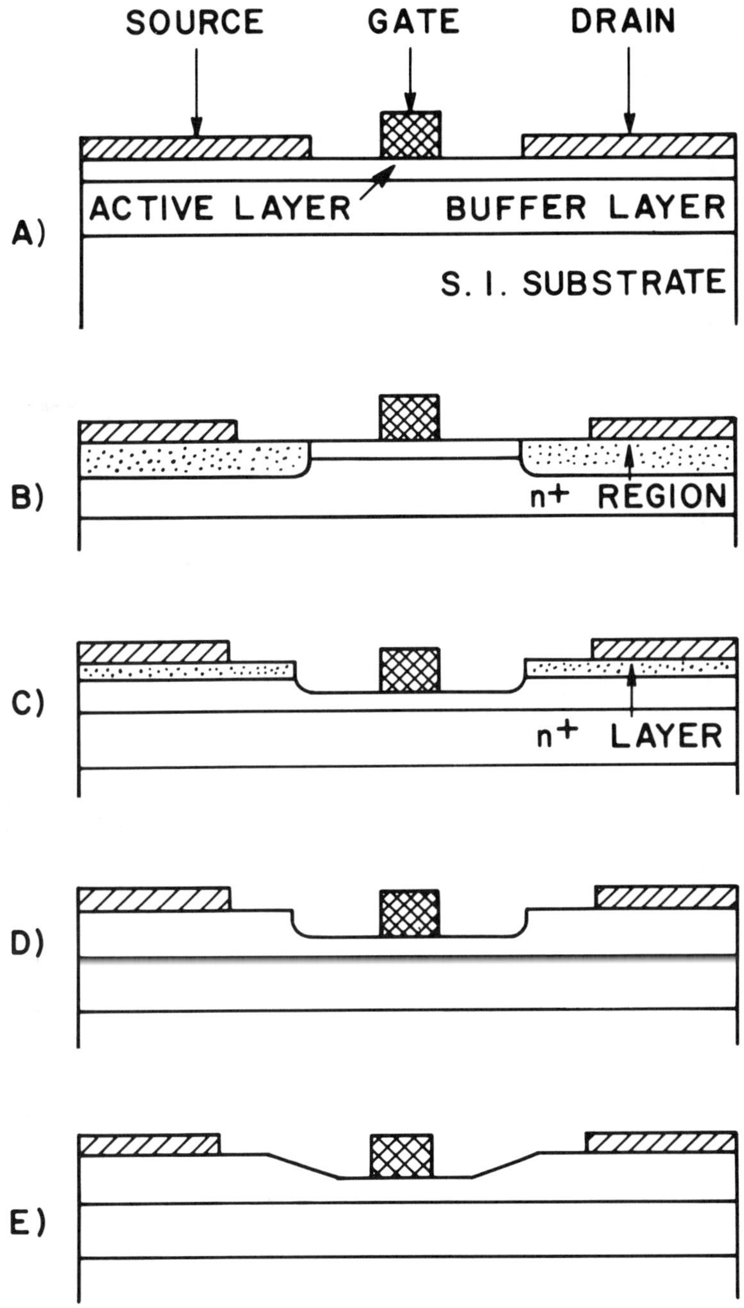

Figure 1/Basic FET structures used in power GaAs FETs.

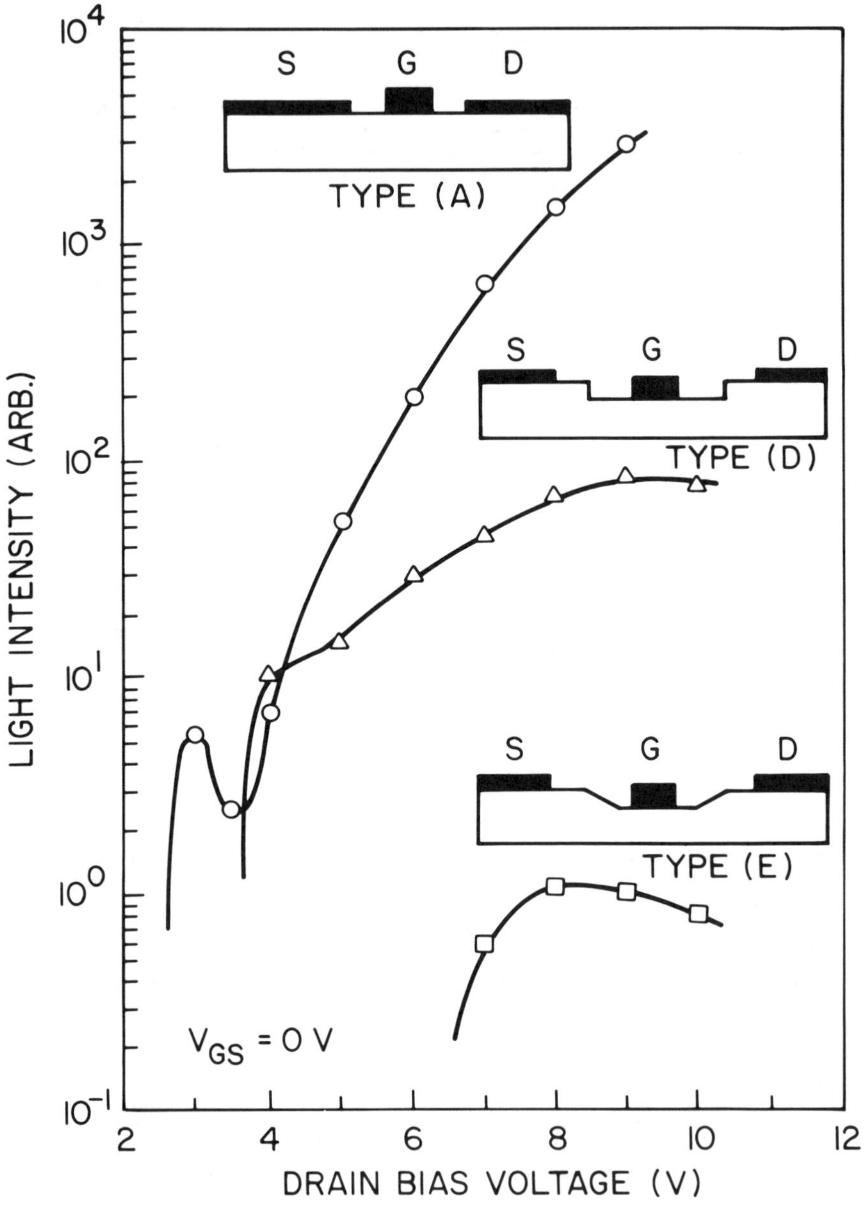

Figure 2/*Light emission intensity versus drain bias voltage for various GaAs FET structures. The gate bias is zero.*

These results indicate that if there is an abrupt change of the carrier concentration or of the cross section of the active layer of the GaAs FETs, an anomalous high field concentrates at that point, probably due to the negative differential mobility in GaAs. Actually, no light emission or abnormal low drain breakdown voltage is observed on a GaAs FET with a p-type active layer [8]. A strong light emission is observed and the drain breakdown voltage is very low for the flat type structure (a) because holes are injected from the drain ohmic contact, probably due to the anomalous high field at the interface.

By analogy with the Gunn diodes, the change of the carrier concentration or the cross section must be smooth to avoid the field accumulation. Therefore, a graded recess structure in which the thickness of the active layer changes gradually and smoothly from the gate to the drain, as shown in Figure 1 (e), was proposed and has been developed [10, 11]. The SEM photograph of the cross section of the developed graded recess structure device is shown in Figure 3. The

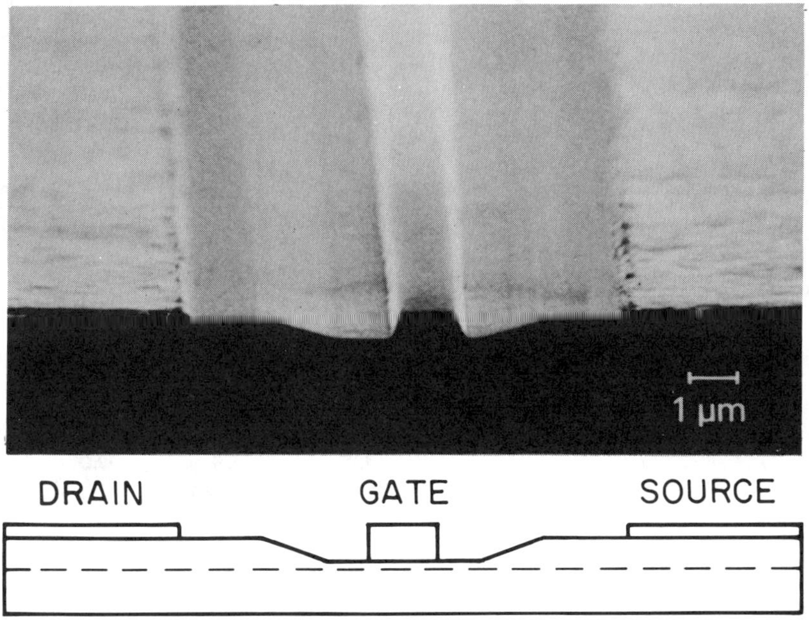

Figure 3/A SEM photograph of the graded recess FET structure.

light emission from this graded recess structure is less than that from the FETs with the structures (b) or (c) in Figure 1 by about two orders of magnitude, as shown in Figure 2. The light emitting spot is not identified. Since the important thing is the smooth change of the carrier concentration or the cross section, any structures which satisfy this condition will be acceptable as long as they have an easy fabrication process and good productivity.

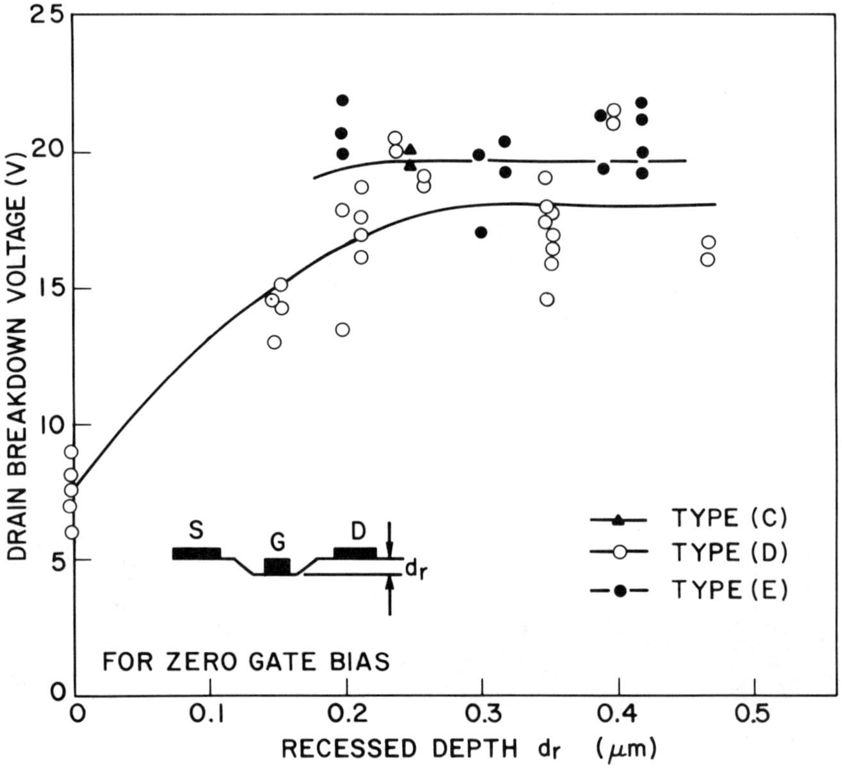

***Figure 4**/Drain breakdown voltage versus the recessed depth of the gate region for zero gate bias.*

Figure 4 shows dependence of the drain breakdown voltage at zero gate bias on the recessed depth for the structures (c) (d) and (e) in Figure 1. The carrier concentration of the active layer is about 1×10^{17} cm^{-3} and the pinch off voltage is about 5 V. The drain breakdown voltage increases from about 8 V for zero recessed depth (the flat type structure) to about 17 V by recessing more than 0.2 μm. This recessed depth is almost the same thickness as the active layer under the gate. If the thickness or the carrier concentration of the epitaxial layer of the drain (anode) region is more than twice the thickness of the active layer under the gate, the anomalous accumulation of the field can be avoided.

The drain breakdown voltage of a device with enough recessed depth does not depend on whether it has a surface n^+ contact layer or not. The graded recess structure seems to give a slightly higher drain breakdown voltage than that of the abrupt recess structure.

For the pinch off condition, the drain breakdown voltage does not depend on the basic device structure such as the recessed type or the n^+ contact layer, but it seems to depend slightly on the effective spacing between the source and the drain. An example of the experimental results is shown in Figure 5. The carrier concentration of the active layer is about 1×10^{17} cm^{-3} and the gate length is about 1.2 μm.

The drain breakdown voltage at near the pinch off condition is much higher (20 ~ 40 V) than that at zero gate bias (15 ~ 25 V). Although it has a large scattering, probably due to the difference between the buffer layer, substrate and the surface condition, it has a slight tendency to increase with the increase in effective source and drain spacing.

The drain breakdown voltage, especially that at near pinch off condition, greatly depends on the carrier concentration of the active layer, and increases rapidly with the decrease of carrier concentration.

1.2 Gate Breakdown Voltage

Breakdown phenomena of the Schottky gate of the GaAs FET are very complicated. Breakdown depends on the carrier concentration of the active layer, the pinch off voltage [12], the gate to drain spacing [1], the surface condition, and also on whether the drain bias is applied or not [9].

The gate breakdown voltage increases with a decrease in the carrier concentration, and with a decrease of the pinch off voltage and, therefore, the maximum drain saturation current. An example of the experimental results is shown in Figure 6. Devices with the same symbol are made from the same wafer. It should be noted that higher gate breakdown voltages are obtained than those expected from the bulk breakdown voltage when the pinch off voltage is low. This is probably because the depletion layer reaches the substrate and spreads toward the source or the drain region. The dependence on the gate-drain spacing is also probably due to the same cause.

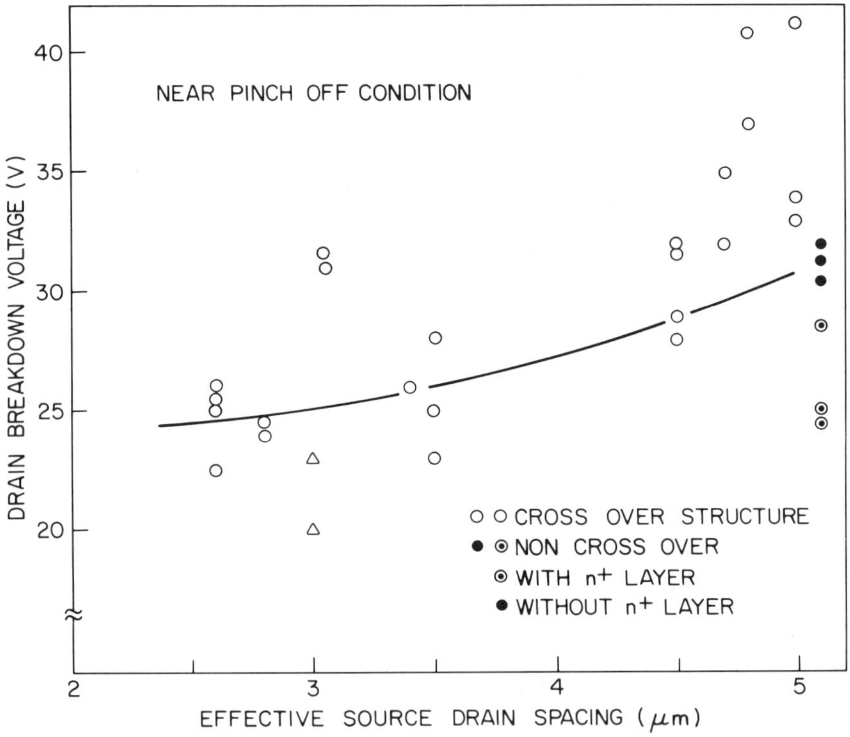

Figure 5/Drain breakdown voltage plotted against the effective source drain spacing for near pinch off condition. The data were obtained for various wafers.

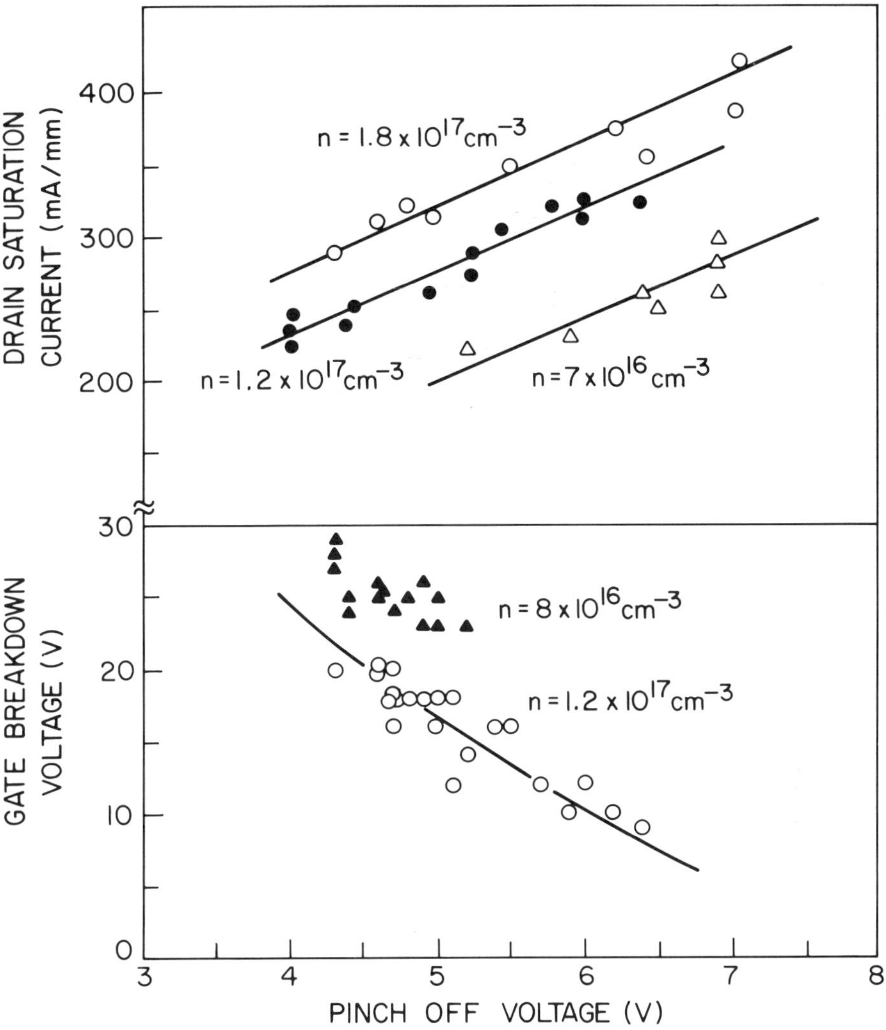

Figure 6*/Dependence of the gate breakdown voltage and the drain saturation current per unit gate width (1 mm) on the pinch off voltage. Parameters are carrier concentrations of the active epitaxial layer.*

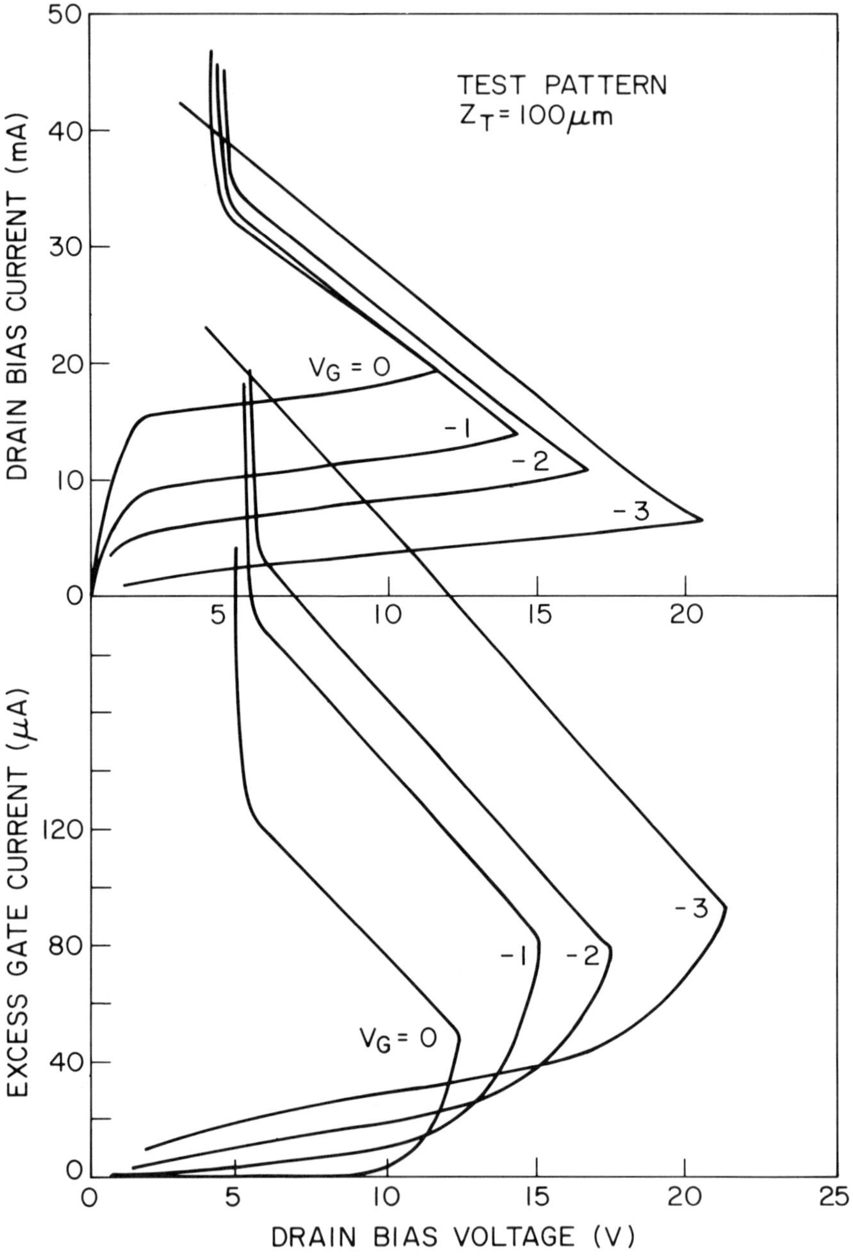

Figure 7/Dependence of the excess gate current on the drain and gate bias.

The gate to drain breakdown voltage often increases with the increase of the negative gate bias and, therefore, with decrease of the drain current [9, 13] as shown in Figure 7. This is quite different from the case of Silicon FET, where the gate breakdown voltage is always constant and is limited by the avalanche breakdown voltage at the gate edge. This difference is inferred to be due to the dipole layer [14] built up by the negative differential mobility in GaAs.

1.3 Drain Bias Current and Voltage

The RF output power from the unit (1 mm) gate width of the power FET is proportional to the DC input power per unit gate width. The drain bias current per mm gate width is normally one third to a half of the maximum drain saturation current if the device is operated in A class, and is determined by the carrier concentration of the active layer and the pinch off voltage. Both the carrier concentration and the pinch off voltage are limited by the gate breakdown voltage simply because the pinch off voltage can not be greater than the gate breakdown voltage. Furthermore, a too low gate breakdown voltage does not give proper transistor characteristics.

Experimental results show that a gate breakdown voltage of at least 10 V is necessary for a normal FET operation. If the carrier concentration of the active layer is about 1×10^{17} cm^{-3}, which is necessary to obtain enough gain at C band, the pinch off voltage should be less than 6 V and the drain saturation current should be less than 300 mA/mm gate width, as shown in Figure 6. Therefore, the drain bias current of 100 mA to 150 mA/mm is a reasonable range for the normal operation of the practical C band power GaAs FETs.

The drain bias voltage should be less than half of the drain breakdown voltage at near the pinch off condition, as described in the previous section (principle of operation). For the C band power GaAs FET whose carrier concentration is about 1×10^{17} cm^{-3}, the drain breakdown voltage at the pinch off condition is 20 to 40 V; therefore, the reasonable drain bias voltage for the practical power GaAs FETs (not for top data) is about 10 V, and the maximum is 15 V.

From the above values, examples of the DC input powers (P_{DC}) and RF output powers (P_{out}) can be estimated as follows, by assuming 30% power added efficiency (η_{add}):

$$P_{DC} = 100\text{ mA} \times 10\text{ V} = 1\text{ W/mm}$$
$$150\text{ mA} \times 10\text{ V} = 1.5\text{ W/mm}$$
$$150\text{ mA} \times 15\text{ V} = 2.25\text{ W/mm}$$

Assuming

$\eta_{add} = 30\%$, 4 dB Associated Gain

$$\frac{P_{out}}{P_{in}} = 2.5$$

$$\eta_{add} = \frac{P_{out} - P_{in}}{P_{DC}} = \frac{(2.5 - 1)\,P_{in}}{P_{DC}} = \frac{1.5\,P_{in}}{P_{DC}} = 0.3$$

$$P_{out} = 2.5\,P_{in} = 0.5\,P_{DC}$$

$$\begin{aligned} P_{out} &= 0.5\text{ W/mm} \quad \text{for } P_{DC} = 1\text{ W/mm} \\ &= 0.75\text{ W/mm} \quad \quad \quad \; = 1.5\text{ W/mm} \\ &= 1.12\text{ W/mm} \quad \quad \quad \; = 2.25\text{ W/mm} \end{aligned}$$

Furthermore, the drain and gate breakdown voltages can be improved (and, therefore, the RF output power per mm gate can be improved), by decreasing the carrier concentration of the active layer, but the power gain decreases with decrease of the carrier concentration. At least $1 \times 10^{17}\text{ cm}^{-3}$ is necessary for C or X band power GaAs FETs.

2. POWER FET PATTER DESIGN

Since the RF output power per mm gate width is limited (as described in the previous section), the total gate width of the power FET should be as large as possible in order to obtain a high RF output power. Therefore, to save the chip area, the device pattern inevitably becomes interdigitated, as shown in Figure 8. Two fundamental parameters of the pattern must be determined: the gate finger width (the unit gate width) and the gate to gate spacing. The former influences the power gain and the latter determines the thermal resistance.

2.1 The Gate Finger Width (The Unit Gate Width)

If the gate finger width Zu in Figure 8 is too long, the input RF signal attenuates along the gate due to the gate resistance and the

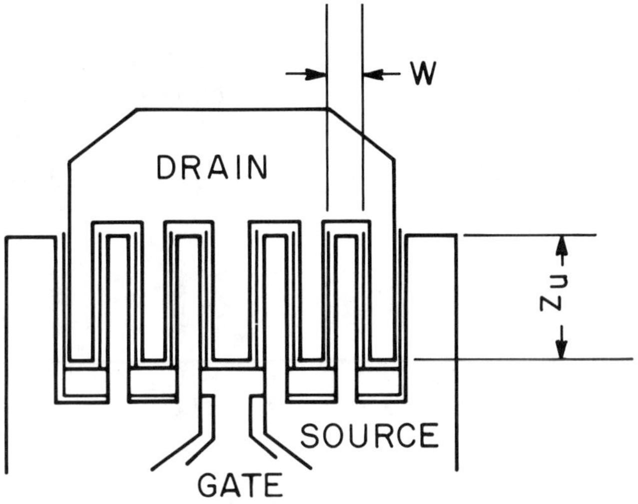

Figure 8/*Interdigitated power FET pattern. Zu = gate finger width or unit gate width; W = gate to gate spacing.*

gate capacitance, and the total power gain or the output power deteriorates. By considering the gate a parallel transmission line, the input R F signal along the gate (Z) can be expressed as follows:

$$V = V_o e^{-\alpha z} (\cos \beta z + j \sin \beta z) \tag{1}$$

where α and β are propagation constants of a parallel transmission line. They are expressed as follows:

$$\alpha = \frac{\omega C_{gs}}{\sqrt{2}} \left\{ -\frac{l_g}{C_{gs}} + \sqrt{\left(\frac{l_g}{C_{gs}}\right)^2 + \left(\frac{\omega l_g}{g_s} + \frac{r_g}{\omega C_{gs}}\right)^2} \right\}^{1/2} \tag{2}$$

$$\beta = \frac{\omega C_{gs}}{\sqrt{2}} \left\{ \frac{l_g}{C_{gs}} + \sqrt{\left(\frac{l_g}{C_{gs}}\right)^2 + \left(\frac{\omega l_g}{g_s} + \frac{r_g}{\omega C_{gs}}\right)^2} \right\}^{1/2} \tag{3}$$

(assuming $g_s \gg \omega C_{gs}$, $l_g \cdot g_s \gg r_g \cdot C_{gs}$)

where ω is angular frequency, C_{gs} is gate source capacitance, l_g is gate inductance, g_s is gate source conductance and r_g is gate resistance. The R F input power density (P_{in}) dissipated per unit gate width can be expressed as follows:

$$P_{in} = \frac{V_o^2}{r_{in}} e^{-2 \alpha z} \cos^2 \beta z \tag{4}$$

Assuming that the intrinsic gain (g) is constant along the gate, the output power (p_{out}) per unit gate width is as follows [15] :

$$p_{out} = g\, p_{in} = \frac{gV_o^2}{r_{in}} e^{-2 \alpha z} \cos^2 \beta z \tag{5}$$

The total output power P_{out} is the integral of Equation 5,

$$P_{out} = \frac{g\, V_o^2}{r_{in}} \int_o^{z_u} e^{-2 \alpha z} \cos^2 \beta z dz \tag{6}$$

For the ideal transistor without the transmission loss and the phase shift, the output power $P_{out,\,o}$ is as follows:

$$P_{out,\,o} = \frac{gV_o^2}{r_{in}} \tag{7}$$

The ratio $P_{out} / P_{out,\,o}$ expresses the overall gain degradation due to the transmission loss. Numerical calculations indicate that the transmission loss is very sensitive to the gate junction capacitance. Unfortunately, the gate capacitance changes with the gate bias and it is difficult to determine accurately when the R F signal swing is too large. An example of the calculated gain degradation at 6 GHz is shown in Figure 9 by changing the gate capacitance as a parameter [16] . The gate inductance l_g and the gate resistance r_g are assumed to be 1.4 pH/μm and 4 x 10^{-2} pF/μm, respectively. The gate capacitance is changed from 0.3 x 10^{-3} pF/μm to 1.5 x 10^{-3} pF/μm.

These values correspond to those of the C band power GaAs FET with an Aluminum 1 μm length gate (the thickness is 0.5 μm). With a gate capacitance of 0.3 x 10^{-3} pF/μm, the gate finger width will be more than 250 μm if the gain degradation is less than 1 dB at 6 GHz, whereas if the C_{gs} is 1.5 x 10^{-3} pF/μm, it is about 140 μm.

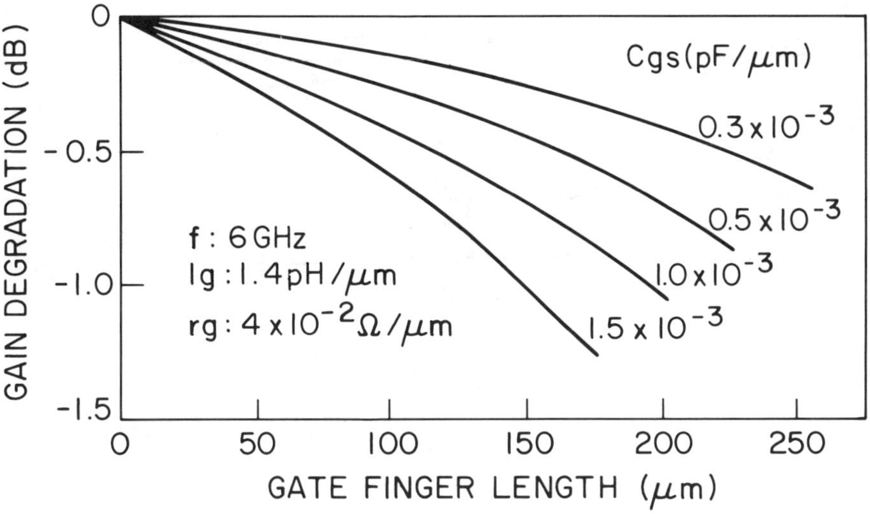

Figure 9/Calculated gain degradation versus gate finger width.

Gate finger widths ranging from 50 μm to 250 μm have been tried by the various laboratories [2, 3, 6]. However, since the wafer, the process, or the structure itself is different, it is difficult to estimate the transmission loss at the gate. Higashisaka et al. have made several test pattern devices with different gate finger lengths on the same wafer by the same process [16]. The gate finger width was changed from 85 μm to 250 μm for the same total gate width (3 mm). The devices are crossover structure C band FETs with gate length of about 1.3 μm. The microwave performance on three different gate finger width devices are summarized and compared in Figure 10. No explicit degradation of the microwave performance

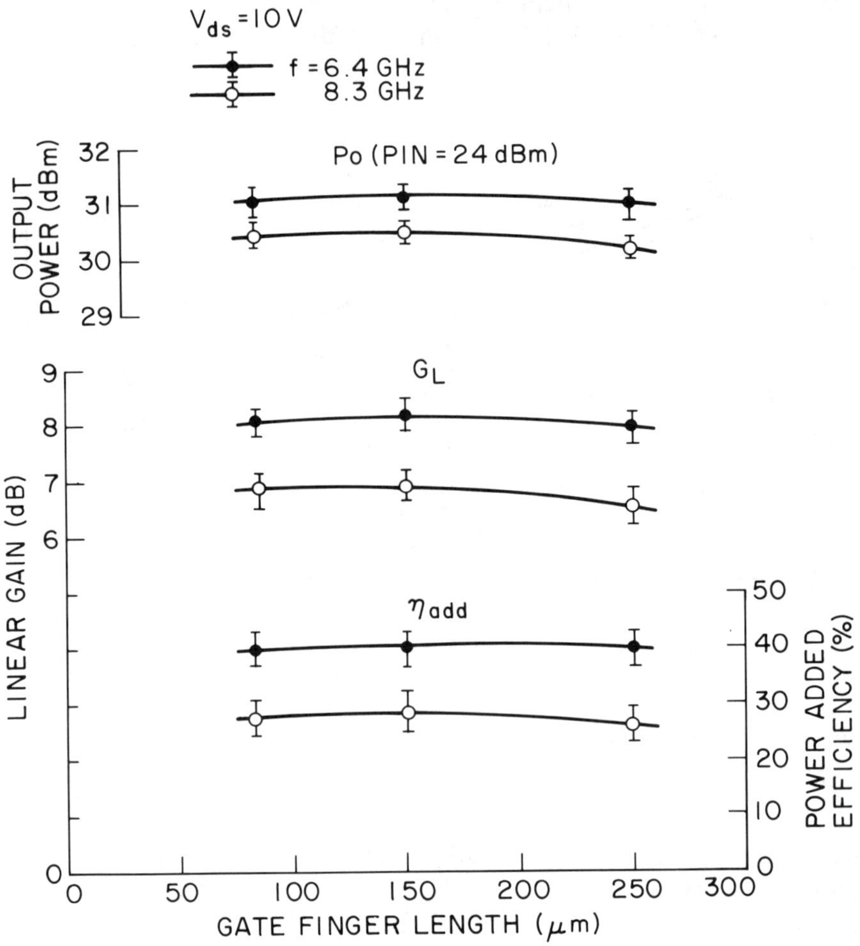

***Figure 10**/Experimentally obtained dependence of the linear power gain, output power and power added efficiency on the gate finger width.*

due to transmission loss at the gate can be seen until the gate finger width reaches about 250 μm at 6 GHz and about 200 μm at 8 GHz. This experimental result is slightly different from that expected from the calculation shown in Figure 9. This may be due to the effects of other factors such as the physical device width, the thermal resistance, or parasitic capacitances. This experimental result indicates, at least, that microwave performance is not very sensitive to the gate finger width.

2.2 Thermal Resistance

For power GaAs FETs, it is very important to suppress the temperature rise of the active region to improve both reliability and performance. Thermal resistance (R_T) of the FET is determined primarily by: 1) the spacing between each active region (B), 2) the gate finger width (Zu), and 3) the thickness of the substrate (H). Although an actual heat flow pattern of GaAs FETs is very complicated, it can be modeled as shown in Figure 11 [15]. Main assumptions of this modeling are as follows:

1) The copper heat sink is semi-infinite,
2) Heat does not flow into within 45° from the surface,
3) When two heat flows intersect, the combined heat flow becomes parallel and flows into the heat sink.

An example of the calculated thermal resistance is shown in Figure 12 as a function of the spacing between each active region (the drain, source width B) for different total gate width Z_T (the source periphery). The thermal resistance is very sensitive to the spacing between each active region (B), and depends on the gate finger width, especially when it is less than 100 μm. The substrate thickness should be as thin as possible but it is limited by handling capability.

Some of the experimental results are shown in Figure 12. Although they are slightly higher than those predicted, they agree fairly well with the calculated ones.

As described in the following section, there are several device structures or chip mount configurations which reduce the thermal resistance, such as flip chip mounting [17, 18] or a via hole connection [19]. The heat flow is more complicated for these structures and no theoretical or experimental comparison has yet been made.

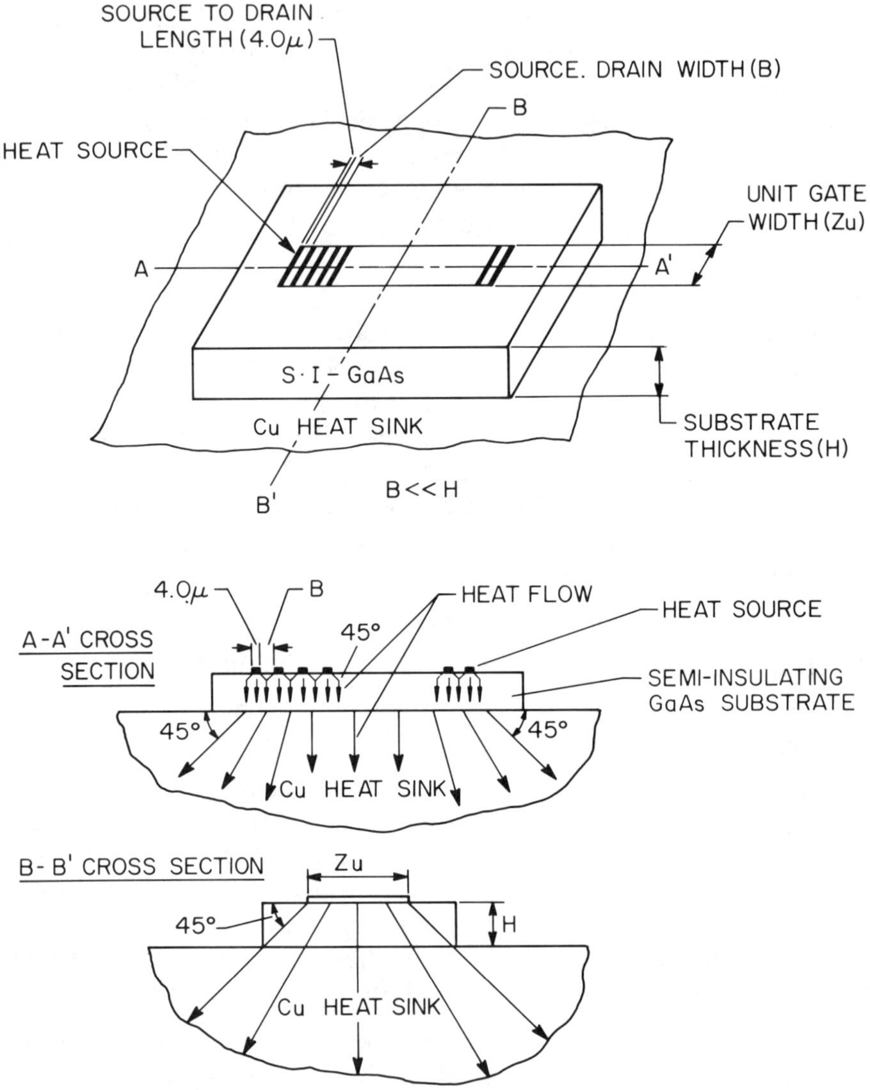

***Figure 11**/Model of the heat flow pattern of the power GaAs FETs.*

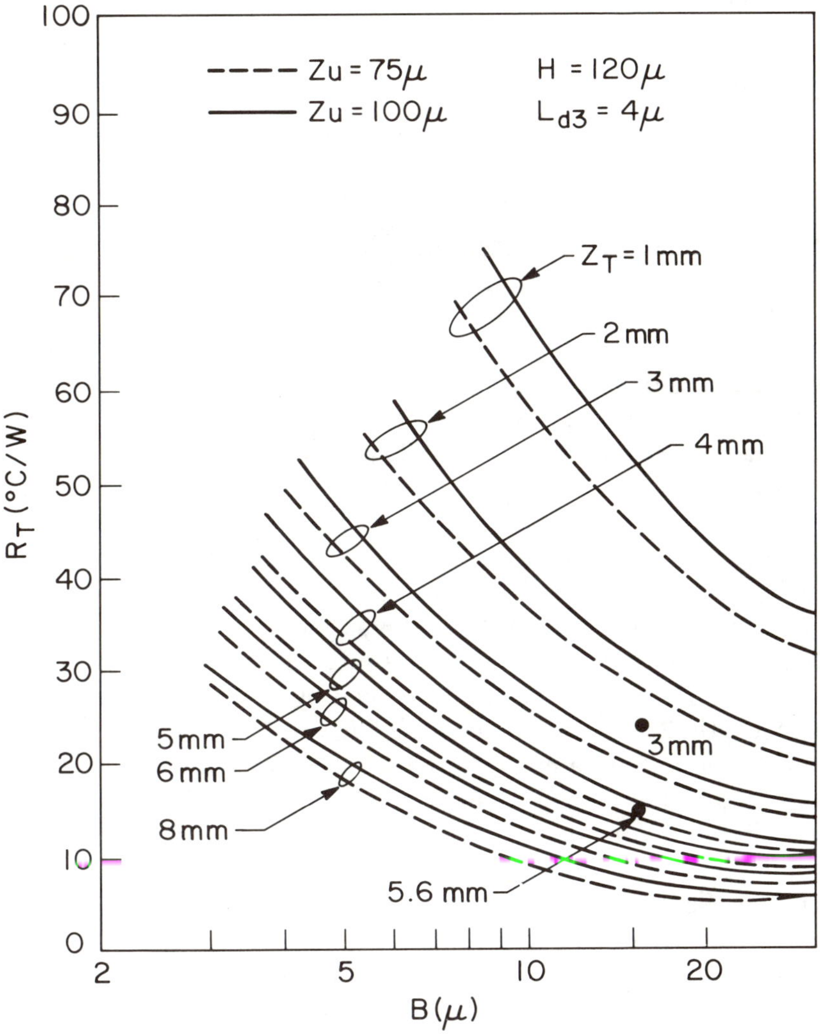

Figure 12*/Calculated thermal resistance against the gate to gate spacing.*

2.3 FET Patterns and Structures

The power GaAs FET structures can be roughly classified as follows:

- Crossover
  - dielectric isolation
  - air gap isolation
- Non-crossover
  - wire bonding construction
  - flip-chip configuration
  - via hole connection

Each of these structures has its own advantages and disadvantages.

Two examples of the crossover structure with a dielectric isolation are shown in Figure 13. A CVD (Chemical Vapor Deposition) SiO_2 is normally used as the dielectric. This structure can give a high packing density (total gate width per unit area) and a simple mounting process due to a smaller bonding pad number, but needs a sophisticated fabrication process including the low temperature deposition of SiO_2 and the metal-metal crossover. The C band device developed by Fujitsu (Figure 13a) has a 100 μm gate finger width, about 14 μm gate to gate spacing and 13 mm total gate width in a 2.1 mm x 0.5 mm chip [20]. The C-band device developed by NEC has 200 μm gate finger length, 25 μm gate to gate spacing and 15 mm total gate width in a 2.2 mm x 0.65 mm chip [16]. It is estimated that the latter has about 15% higher packing density and about 20% lower thermal resistance than the former, simply because it has a longer gate finger width and a larger gate-gate spacing. The source of this structure device can be grounded by a sheet grounding method [13] or a side metallization of the chip.

An air gap isolation or air bridge structure [21, 22] is shown in Figure 14. It is normally made by a thick gold plating on a very thick photoresist to make the bridge after removal of the photoresist. Since the electrodes are isolated by the air, the capacitance at the crossover point is much less than that of SiO_2 isolation. However, it has been reported [23] that the capacitance at the SiO_2 crossover is less than parasitic capacitances at the gate or drain pads, and that removal of that capacitance does not affect the microwave performance very much.

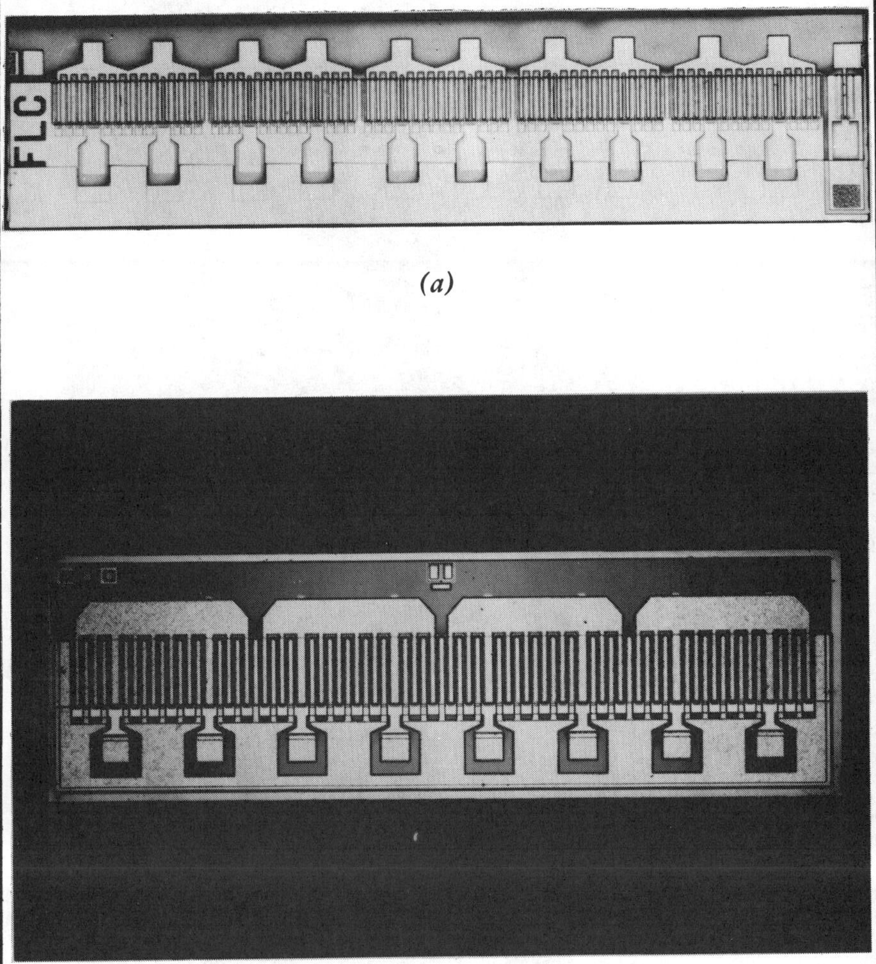

(a)

(b)

Figure 13/Dielectric isolation crossover structure FET: (a) Fujitsu FET; (b) NEC FET.

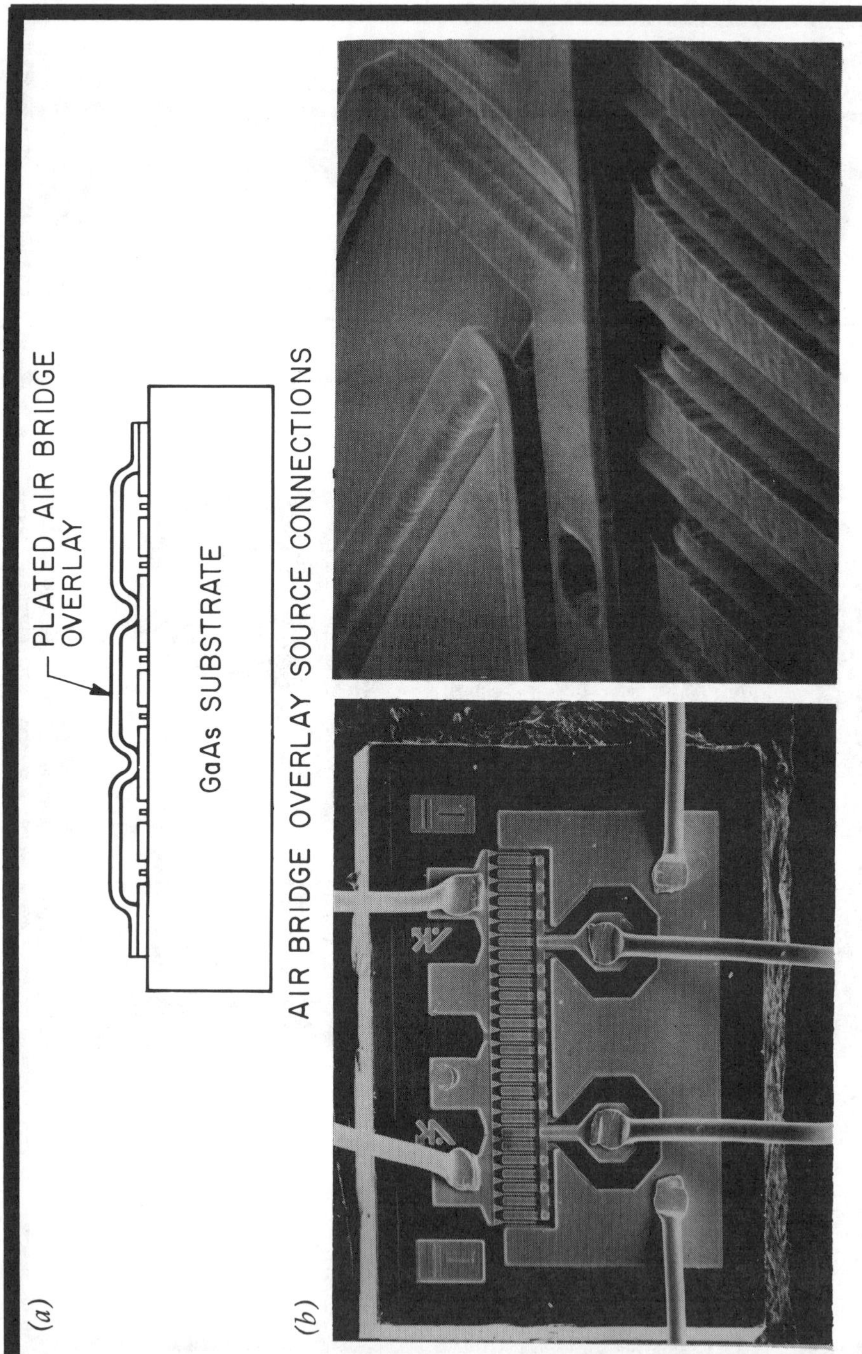

Figure 14*/Air gap isolation crossover structure FET:* ***(a)*** *Sketch of cross section;* ***(b)*** *SEM photographs.*

Wire bonding constructions shown in Figure 15 allow a simple processing technology but make device assembly more difficult and may have increased parasitics. This structure has been adopted by many laboratories because of its simplicity; a 9.6 mm gate width device at 9 GHz [24] and a 24 mm device at 4 GHz [25] have been successfully demonstrated.

A flip-chip configuration utilizes a thick gold plated layer on source pads, as shown in Figure 16. This structure was first introduced by RCA [17] and now also used by MSC and Mitsubishi [18]. It has a low thermal resistance, low parasitic source inductance and low parasitic capacitance, but needs a very sophisticated thermocompression bonding technology.

Figure 17 shows the cross section of a FET with via hole source connections through the substrate, as well as a plated heat sink [19, 22]. This structure has a low source resistance and a low thermal resistance, but requires a complicated fabrication process which involves handling of a very thin, fragile GaAs wafer.

It is not yet known which structure is optimum for practical power GaAs FETs. It will depend on the processing technology. The optimum structure will change as the device technologies progress.

3. MICROWAVE PERFORMANCE

In order to characterize power GaAs FETs, it is necessary to have a matching circuit. When the total gate width is small and the input impedance is not very low, an ordinary microstrip matching circuit can be applied. However, when the device size is increased and the input impedance is decreased, it becomes difficult to obtain a good matching and, therefore, a good microwave performance. A high power GaAs FET has to be characterized as a kind of single stage amplifier.

3.1 Matching Circuit

In order to increase the output power, the total gate width and hence the device size must be increased because the output power per unit gate width is limited as described in a previous section. The input impedance of the power GaAs FET decreases with increasing device size, and reaches about 1 Ω for a 6 GHz 3 W device and to about

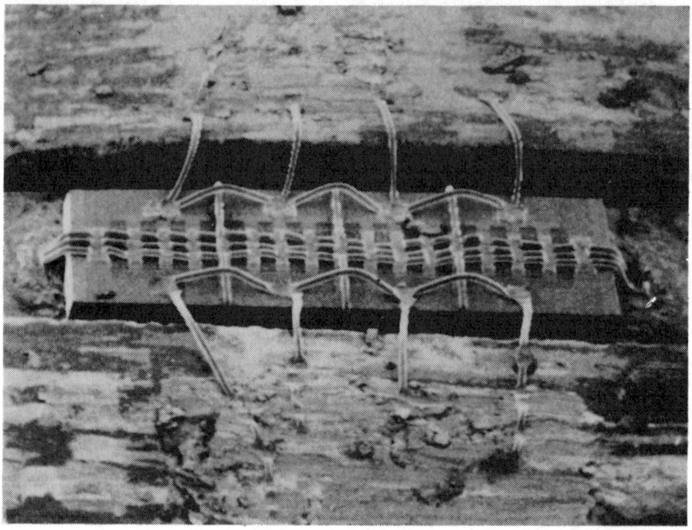

(a)

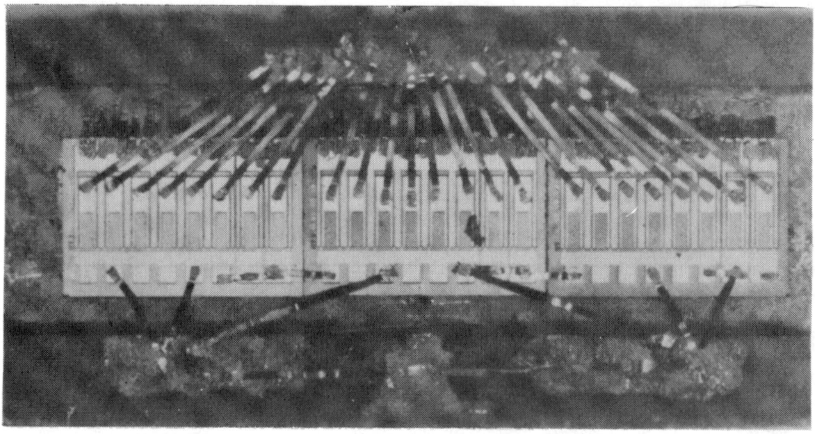

(b)

Figure 15*/Wire bonding constructions:*
(a) Texas Instruments FET (4.8-mm gate width);
(b) Bell Laboratories FET (24-mm gate width).

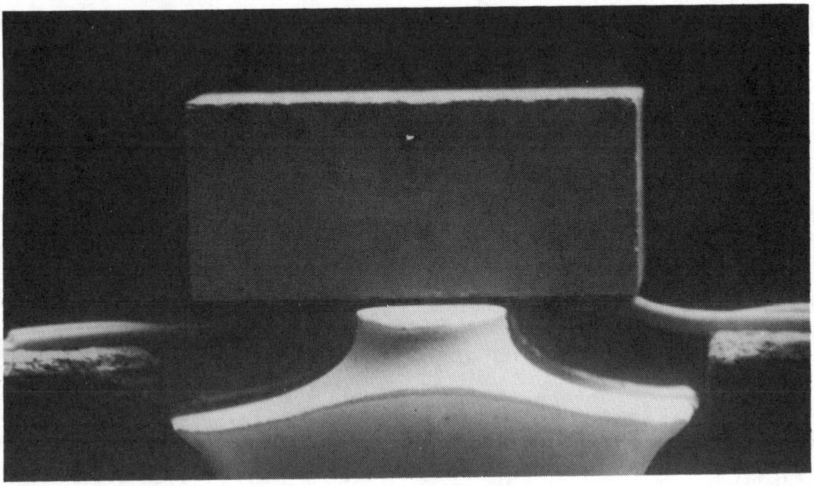

(a)

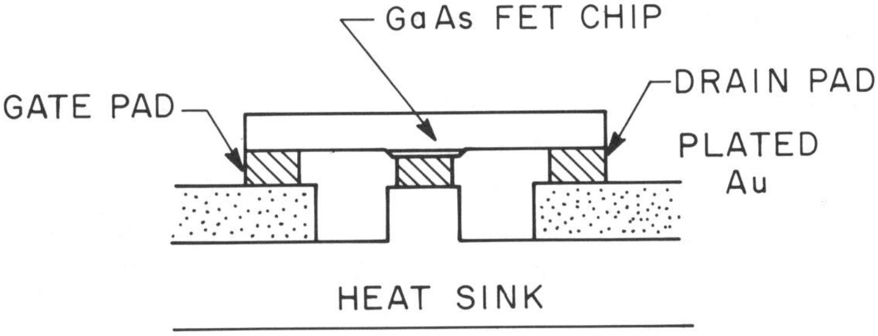

(b)

Figure 16/*Flip-chip mounted FETs:*
(a) MSC flip chip mounted FET;
(b) Mitsubishi flip chip mounted FET — gate and drain pads are also thermocompression bonded.

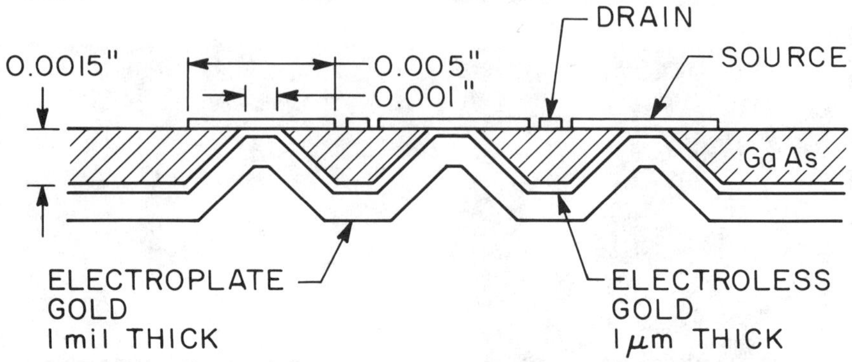

Figure 17*/Via hole source connection structure.*

0.25 Ω for a 6 GHz 15 W device [26]. The physical device width also increases to about 5 mm or even to 10 mm for 6 GHz 10 W class devices. Therefore, it is very difficult to get a good impedance matching by a microstrip circuit, especially outside of the package.

One of the methods used to avoid this kind of difficulty is an internal matching network [27]. An example of the internal matching network developed for C band power GaAs FETs is shown in Figure 18. The input impedance matching is performed by a two section low pass filter type Chebyshev transformer consisting of lumped elements. The bonding wires are inductors. The capacitances are formed on a high dielectric constant ceramics. Since the output impedance is much higher than the input impedance, the output impedance matching is performed by a semi-distributed form in this case. This particular device has a total gate width of 36 mm, a physical device width of 10 mm and can deliver about 15 W at 6 GHz [26].

Another difficulty in getting a good impedance matching to a high power GaAs FET stems from the fact that the optimum load impedance changes with the input power level. Therefore, the output impedance matching has to be determined by taking into account this change [27]. Figure 19 shows an example of the change of the optimum load impedance with increase of the input power level for a device which already has a matching circuit. The device is a C band FET with 16.8 mm gate width, which can deliver about 7 W at 6 GHz. At a small signal level, the optimum load impedance differs from 50 Ω quite a lot, but it converges to 50 Ω as the input power level increases.

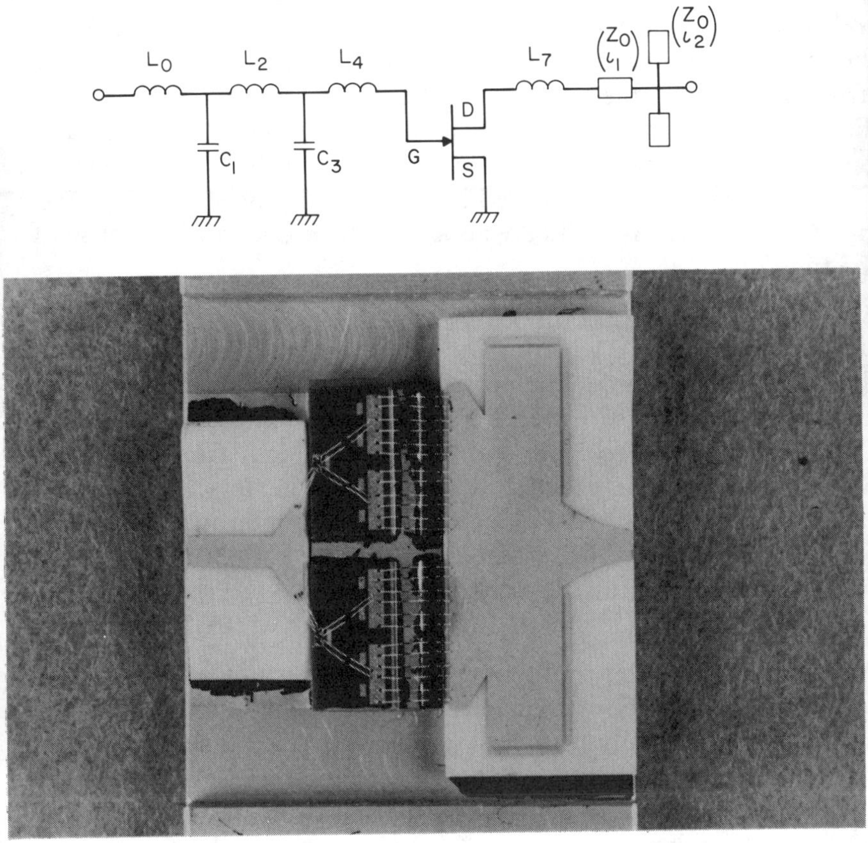

Figure 18/Photograph of an internally matched 6 GHz band power GaAs FET (33.6 mm gate width).

3.2 Microwave Performance

The microwave power of GaAs FETs has increased year by year very quickly. A rough sketch of the change in output power level is shown in Figure 20. Top data reported by various laboratories (open symbols) and catalogue data from various manufacturers (solid symbols) in the spring of 1979 are also plotted in the figure. Although the device structure, size and the laboratory where the devices were made are different, the output power (normally with 4 dB or 3 dB associated gain) is proportional to the reciprocal of the square of the frequency (f^{-2}). Another thing to be noted is that output powers of commercially available GaAs FETs are about a half or one third of those of the top data.

Examples of the output power versus input power characteristics are shown in Figure 21 [16]. The fundamental FET chip is the one shown in Figure 13(b). The impedance matching is performed by lumped elements as shown in Figure 18, combining two or four chips. The linear gain of 7 to 5 dB, and the power added efficiency of 30 to 25% are obtained at 6 GHz, depending on the device size. These values can be improved by reducing the gate length or the source resistance and power added efficiencies of more than 30% can be obtained, especially at lower frequencies (such as 4 GHz) or smaller size devices (such as less than 10 mm total gate width).

As described in the previous section, the output power should be increased almost proportionally to the drain bias voltage. This is proved experimentally for a low frequency device as shown in Figure 22(a) [13]. However, it is not always the case. For higher frequencies, the output power often saturates at the drain bias of 10 to 14 V, as shown in Figure 22(b) [3], even for the device made in the same way as the one whose output power does not saturate until more than 20 V at a lower frequency. It is reported that this power saturation is due to temperature rise of the active region [28], but it can not be explained quantitatively (or even qualitatively) by the temperature rise. The phenomenon is not yet well understood.

The output power of power GaAs FETs increases almost linearly with increases of the total gate width, while the linear power gain decreases slowly with increase of the total gate width. The maximum total gate width is limited by the following factors: i) yield of the device, ii) the device uniformity, iii) difficulty of the impedance matching, iv) decrease of the power gain, and v) physical device size.

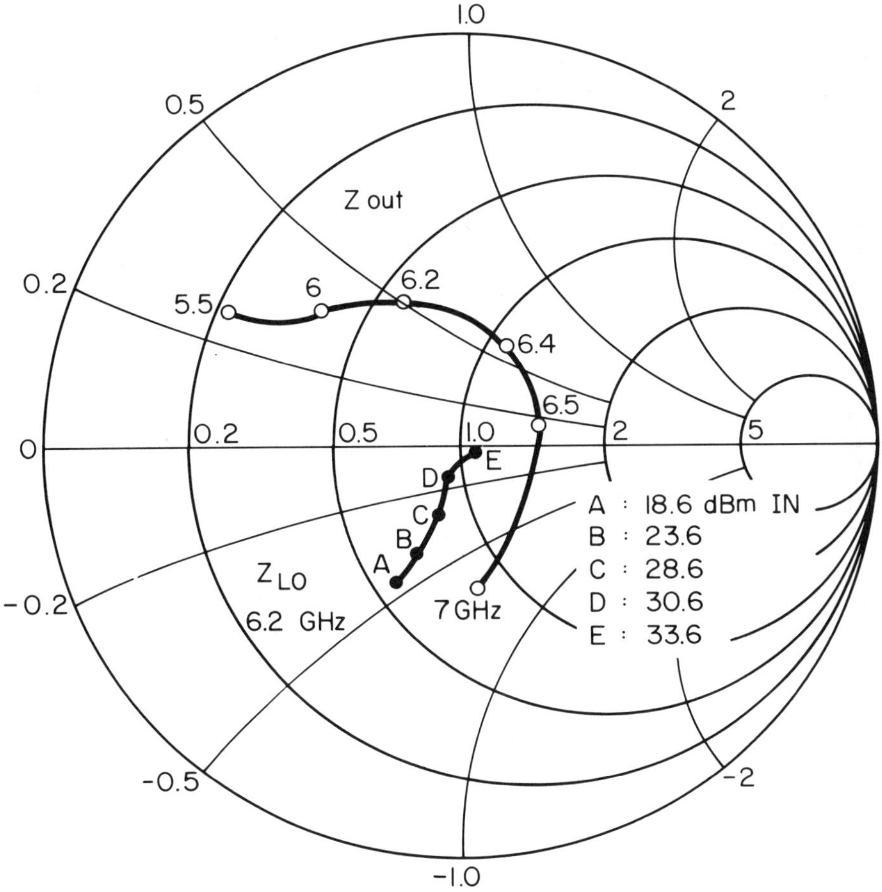

Figure 19*/Small-signal output impedances and large signal optimum load impedances of an internally matched FET (16.8 mm gate width device).*

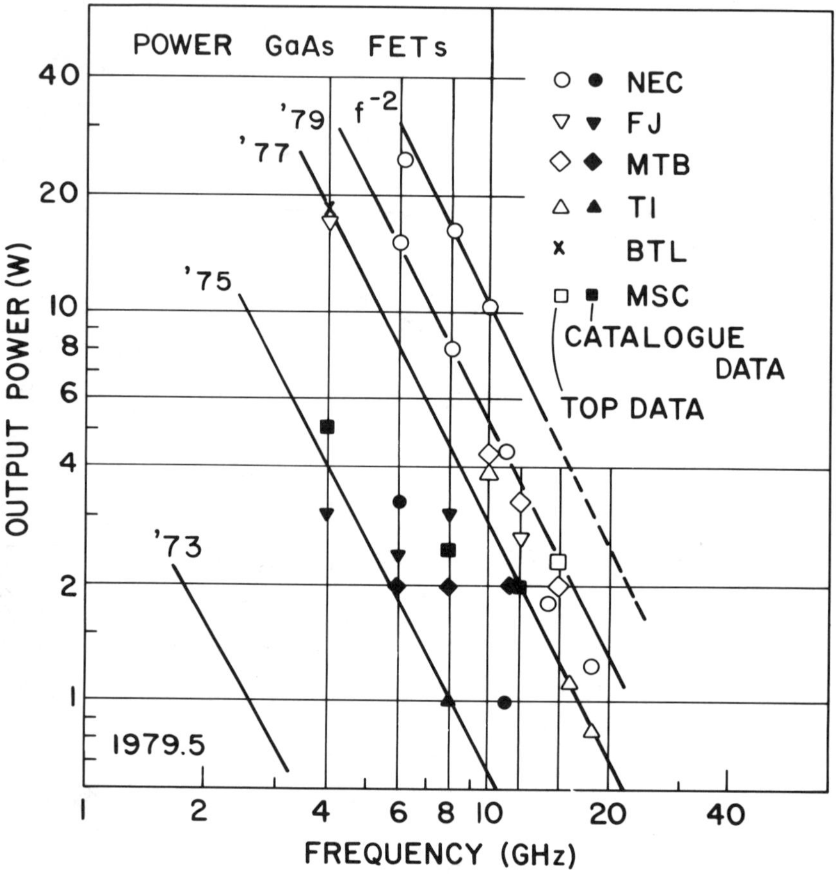

Figure 20/Output powers versus frequency: progress of the output power, top data (open symbols) and catalogue data (solid symbols).

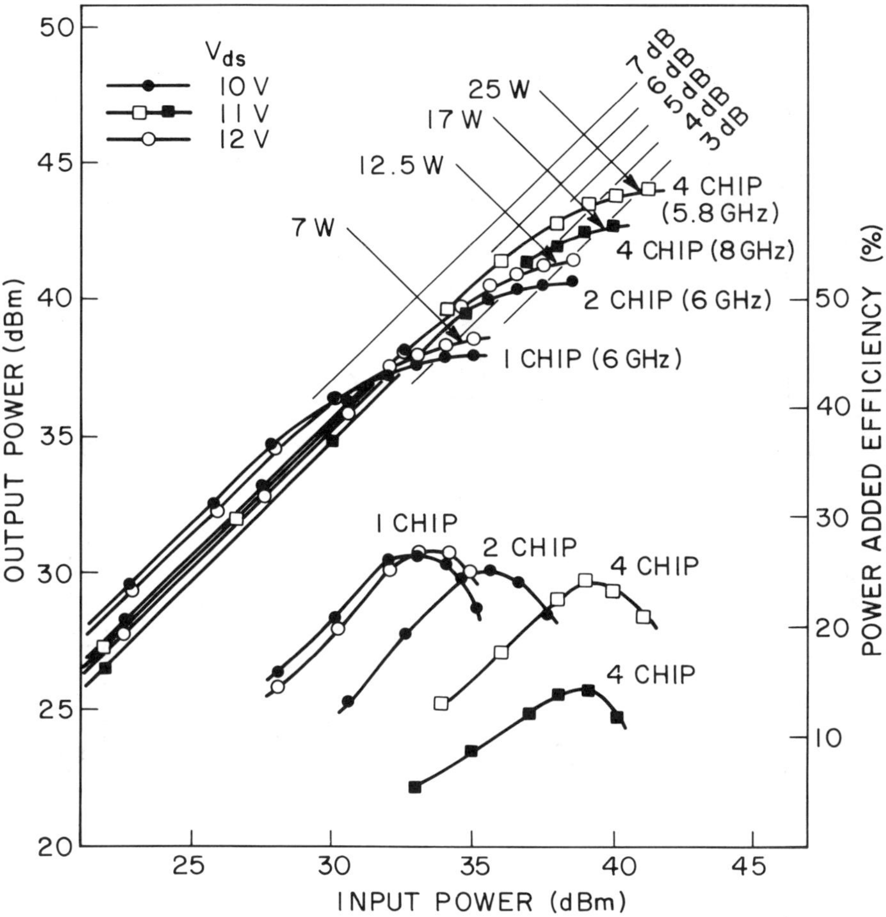

Figure 21*/An example of power transfer characteristics of C band power GaAs FETs.*

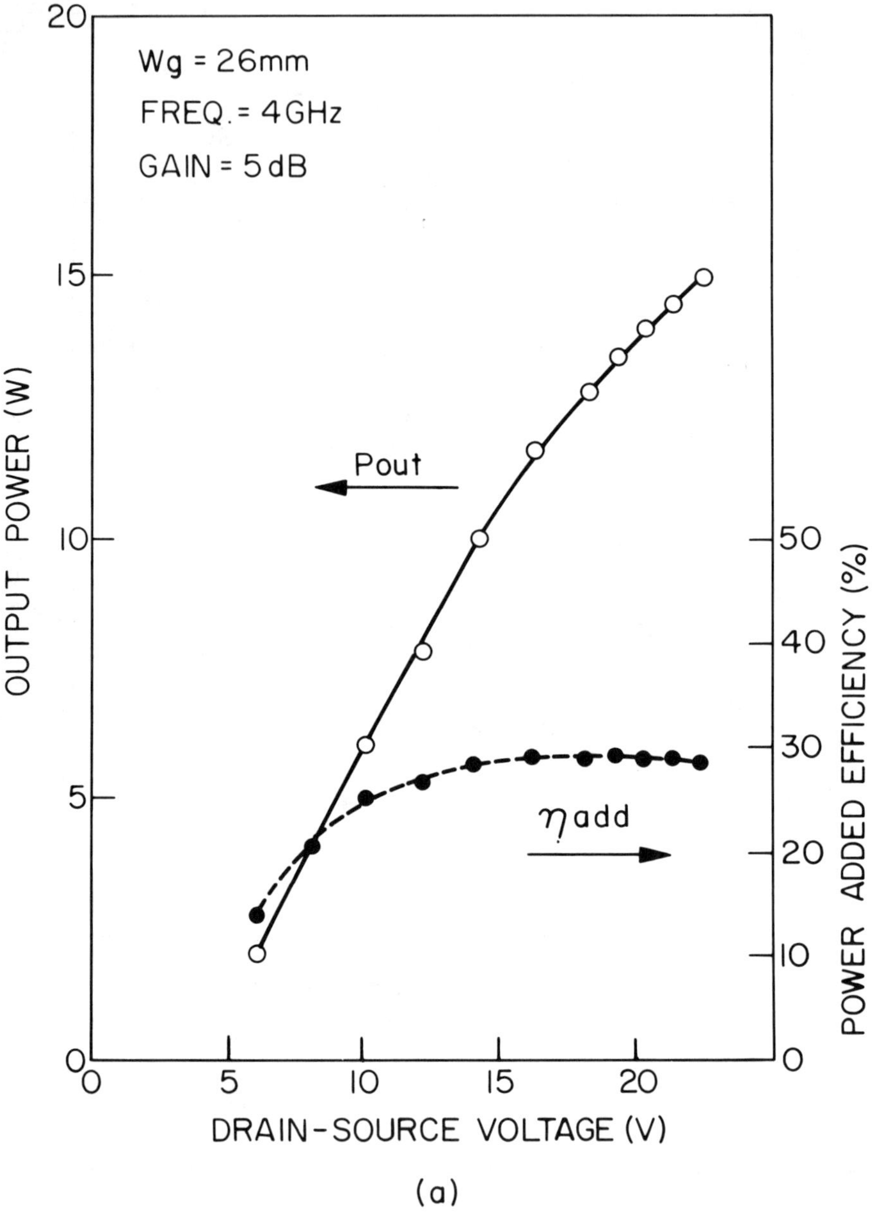

Figure 22*/Dependence of the output power on the drain bias voltage: **(a)** at 4 GHz; **(b)** at 8 GHz.*

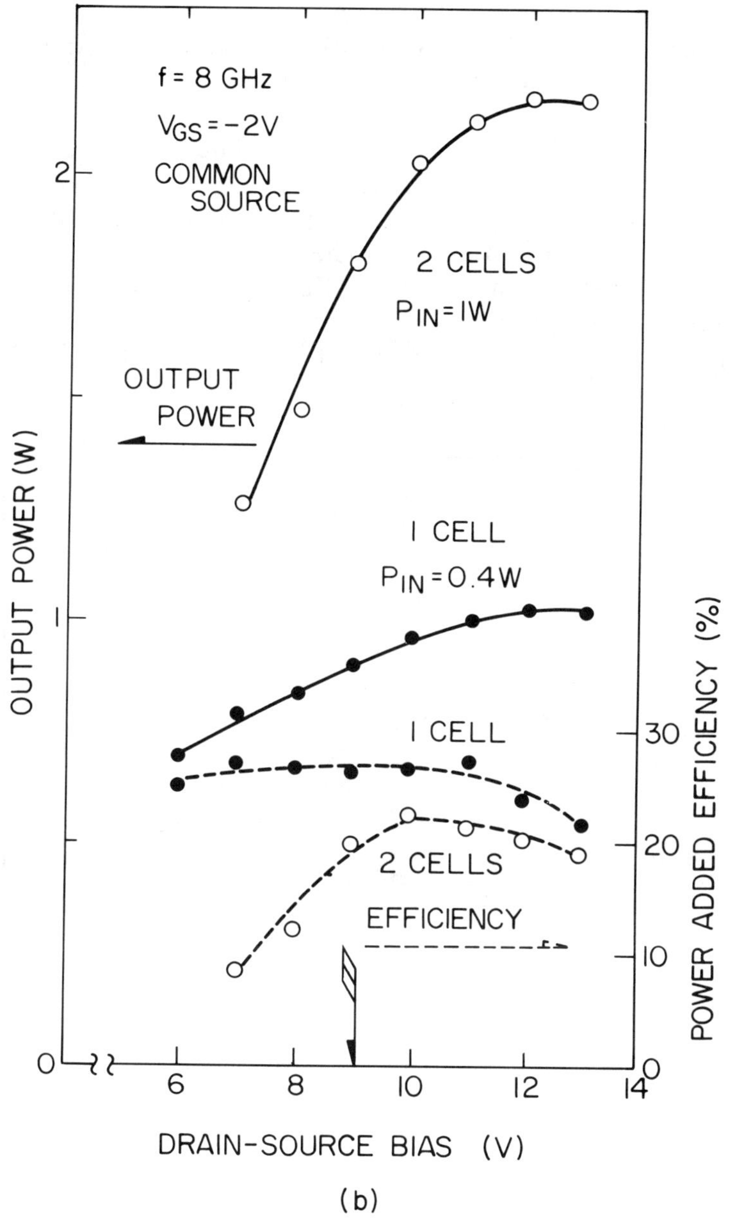

Figure 22/(continued)

Figure 23 shows examples of the relation between the output power, linear power gain and the total gate width. For the same basic device, the output power increases almost linearly against the total gate width, and reaches 25 W at 6 GHz for C band devices and about 6 W at 10 GHz for X band devices [29]. The 6 GHz 25 W device has a total gate width of 60 mm, an input impedance of about 0.15 Ω and a physical device width of 10 mm. The actual device is almost the same as the one shown in Figure 18. Unless a one micron length gate continues 60 mm without any cut, unless the pinch off voltage is reasonably uniform within 16 unit cells or in 4 chips, and unless the

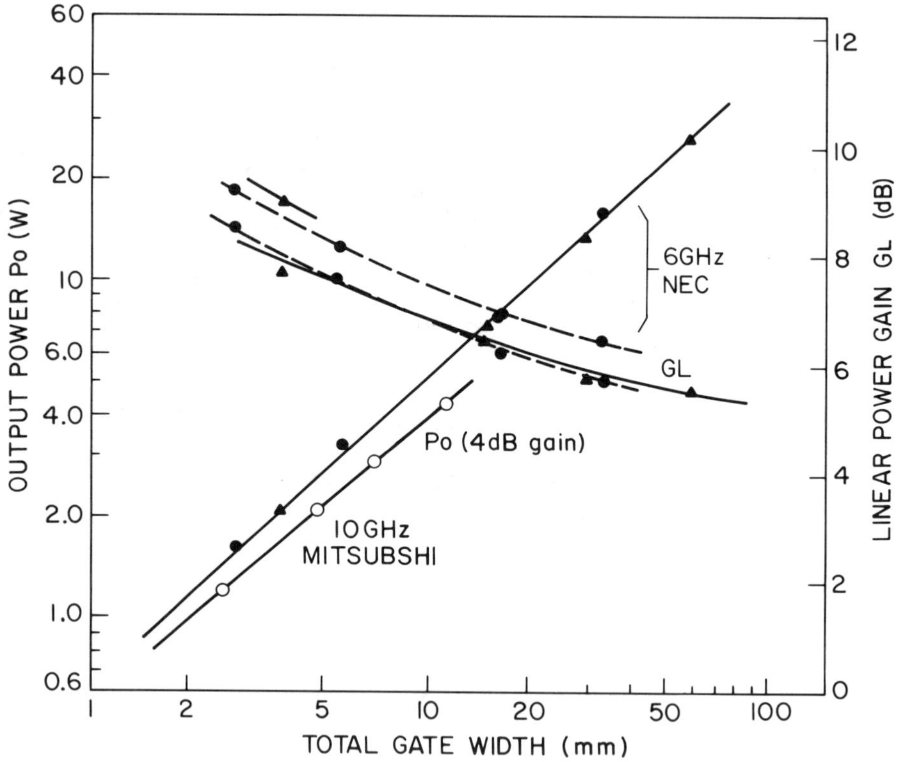

Figure 23/Dependence of the output power and the linear power gain on the total gate width.

phase of input microwave signal to each cell is uniform for 10 mm width, this performance can not be achieved. Further increase of the device size is not very easy, because the lengths of the bonding wires acting as inductors are limited by the operating frequency. It is necessary to increase the packing density (total gate width per unit chip width) without any deterioration of power gain or thermal resistance for further increase of the output power. For lower frequencies, a higher output power can be expected according to $p \propto f^{-2}$, because the physical device size can be increased and therefore the gate finger width and also the packing density can be increased.

The decrease in the linear power gain due to increase of the device size is not very steep if a proper matching circuit is provided. It is only about 0.5 dB for doubling of the total gate width. This degradation of the gain can be easily compensated by reducing the gate length or the source resistance.

For commercially available devices, the device size is mainly limited by the yield and uniformity of the device and the matching circuit – in other words, by productivity. Therefore, output powers of commercially available devices are about one third or a half of the top data for 1979 shown in Figure 20. This discrepancy will be decreased in the near future through progress in the fabrication technologies.

REFERENCES

[1] Ogawa, M., K. Ohata, T. Furutsuka and N. Kawamura, *IEEE Trans. Microwave Theory and Tech.,* MTT-24:300, 1976.

[2] Fukuta, M., T. Mimura, I. Tajumura, and A. Furumoto, IEEE Int. Solid-State Circuits Conf., *Dig. Tech. Papers,* p. 84, 1973.

[3] Fukuta, M., et al., IEEE Int. Solid-State Circuits Conf., *Dig. Tech. Papers,* p. 166, 1976.

[4] Stoneham, E., T.S. Tan, and J. Gladstone, *IEEE Int. Electron Devices Meeting Dig. Tech. Papers,* p. 330, 1977.

[5] Wemple, S.H., and W.C. Niehaus, *Gallium Arsenide and Related Compounds* (St. Louis), p. 262, 1976.

[6] Niehaus, W.C. et al., *Gallium Arsenide & Related Compounds,* p. 271, 1976.

[7] Hasegawa, F. et al., *IEEE Int. Solid State Circuit Conf., Dig. Tech. Papers,* p. 118, 1978.

[8] Mimura, T., H. Suzuki and M. Fukuta, *Proc. IEEE (Lett.),* 65:1407, 1977.

[9] Yamamoto, R., A. Higashisaka, and F. Hasegawa, *IEEE Trans. Electron Devices,* ED-25:567, 1978.

[10] Furutsuka, T. et al., *Electron. Lett.,* 15:417, 1979.

[11] Higashisaka, A. et al., *11th Conf. (Int.) on Solid State Devices,* Tokyo, Paper No. B-1-6, 1979.

[12] Higashisaka, A., private communication.

[13] Fukuta, M. et al., *IEEE Trans. Microwave Theory and Tech.,* MTT-24:312, 1976.

[14] Engelmann, R.W.H. and C.A. Liechti, *IEEE Trans. Electron. Devices,* ED-24:1288, 1977.

[15] Aono, Y., A. Higashisaka, T. Ogawa, and F. Hasegawa, *Japan. J. Appl. Phys.* vol. 17, Suppl. 17-1:147, 1978.

[16] Higashisaka, A., Y. Takayama, and F. Hasegawa, *IEEE Trans. Electron. Devices,* ED-11:104, 1975.

[17] Drukier, I., et al., *Electron. Lett.,* 11:104, 1975.

[18] Mitsui, Y., et al., *European Microwave Conference Tech. Dig.*, p. 272, 1979.

[19] D'Asaro, L.A., J.V. DiLorenzo, and H. Fukui, *IEEE Int. Electron. Devices Meeting Dig. Tech. Papers*, p. 370, 1977.

[20] DiLorenzo, J. and W. Wisseman, *IEEE Trans. Microwave Theory and Tech.*, MTT-27:367, 1979.

[21] Anderson, J.R., M. Omori, and F. Cooke, *IEEE Int. Electron. Devices Meeting Dig. Tech. Papers*, p. 133, 1978.

[22] Hewitt, B.S., et al., *European Microwave Conference Tech. Dig.*, p. 265, 1979.

[23] Aono, Y., et al. in annual spring meeting of IECE, *Japan. Tech. Dig.*, paper No. 353, 1979.

[24] Macksey, H.M., T.G. Blocker, and F.H. Doerbeck, *Electron. Lett.*, 13:312, 1977.

[25] Wemple, S.H., et al., *Electron. Lett.*, 14:175, 1978.

[26] Honjo, K., et al., *1979 Int. MTT-S Microwave Symposium Tech. Dig.*, p. 289, 1979.

[27] Honjo, K., Y. Takayama, and A. Higashisaka, *IEEE Trans. Microwave Theory and Tech.*, MTT-27:76, 1979.

[28] Macksey, H.M., et al., *IEEE Trans. Electron. Devices*, ED-24: 113, 1977.

[29] Mitsui, Y., et al., *Electron. Lett.*, 15:461, 1979.

GaAs Power FET Design

H. M. Macksey
Texas Instruments Incorporated, Central
Research Laboratories, Dallas, Texas

ABSTRACT

During the last few years Texas Instruments has employed a number of different designs for GaAs power FETs operating at frequencies from 3 to 25 GHz. Many experiments have been conducted to optimize devices for operation at each frequency of interest. The present review paper is based on the results of those experiments. The device and material parameters which one must choose in the design of a GaAs Power FET are described along with some of the many trade-offs that must be made.

1. INTRODUCTION

At Texas Instruments, the work on GaAs power FETs has been at frequencies of 3 GHz and above. The goal of the device design effort has been to produce devices having the maximum output power consistent with useful gain ($\geqslant$ 4 dB) and high efficiency ($\geqslant$ 20%). The approach at TI and throughout the industry is to operate several device cells in parallel, each containing several gate stripes. The performance of GaAs power FETs may therefore be improved by increasing the output power, gain, and efficiency of the single cell device and/or by increasing the cell combining efficiency and total number of cells.

An example of an X-band GaAs FET is shown in Figure 1, a photograph of a slice during processing. Each device has a total gate width of 4.8 mm contained in four cells. Multiple cells are employed because a single defect destroys only one cell; the remaining cells may be used in lower power applications. In addition, and probably more importantly in a research environment, the effects of cell combination and very large gate widths can be separated from the different problems of improving the performance of a small gate width device.

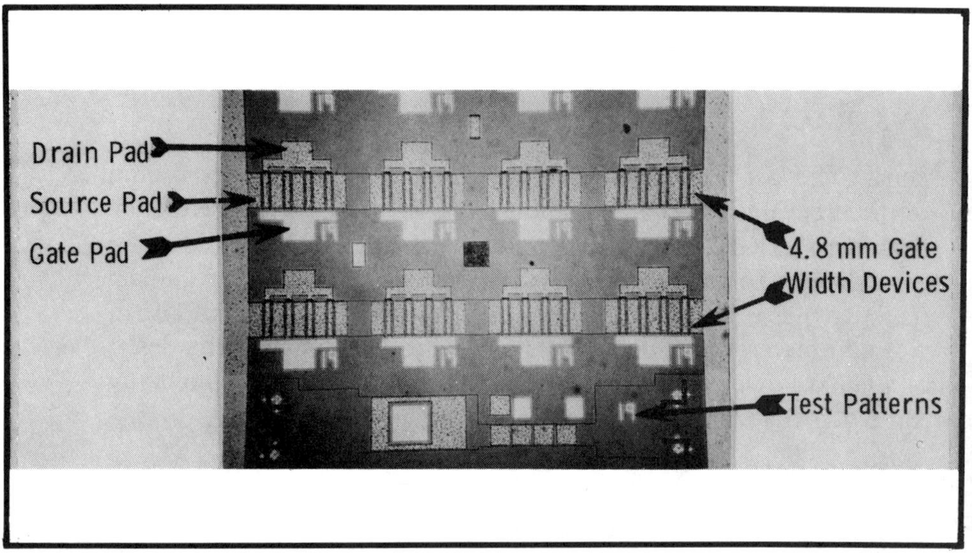

***Figure 1**/Photograph of 4.8 mm gate width GaAs FET slice during processing.*

All of the devices to be described were fabricated by the same basic process [1, 2]. Source and drain ohmic contacts are made of alloyed AuGe/Ni with a 5 μm separation. The gates are defined exclusively by electron beam lithography [2, 3] and e-beam evaporated Ti/Pt/Au is employed as the gate metal. The slice is etched immediately prior to placing it in the gate metal evaporator in order to recess the gates below the epitaxial surface and reduce the device current to the correct value. A layer of Ti/Au is also evaporated on the source and drain to increase conductivity and the bonding pads are plated with 5 to 10 μm of Au to facilitate bonding. The active area is protected by a 0.4 μm layer of silicon nitride. The source pads of some devices are connected by a plated Au air bridge.

The sections below describe the design of individual cells for optimum performance and the combination of several of these cells to obtain high output power. This is followed by a specific example to illustrate the many design factors and their trade-offs. Lastly, the best microwave performance data obtained to date at Texas Instruments is presented, along with the designs of the devices producing it.

2. SINGLE CELL DESIGN

2.1 Active Layer Doping Level and Thickness

All GaAs slices used for power FETs have a buffer layer followed by the active layer. The buffer layer is undoped ($n \leqslant 10^{14}$ cm^{-3}) and is 1 to 3 μm thick. The optimum active layer doping level depends on the intended operating frequency. Measurements on many devices have demonstrated that approximately one watt output power per mm of gate width can be obtained from optimally designed devices between 4 and 16 GHz [4]. However, the higher frequencies require higher doping levels. The type of experiment conducted [5] to determine the optimum doping level is illustrated in Figure 2. Here the optimum doping level for GaAs power FETs operating at 15 GHz is determined. Output power is measured as a function of an intentionally varied I_{dss} (saturated drain current at zero gate voltage) for identical devices fabricated on slices with different active layer doping levels. Clearly, there is a doping level which maximizes output power; at 15 GHz it is about 1.6×10^{17} cm^{-3}. The optimum I_{dss} can also be determined from Figure 2; for $n = 1.6 \times 10^{17}$ cm^{-3} it is 100 mA to

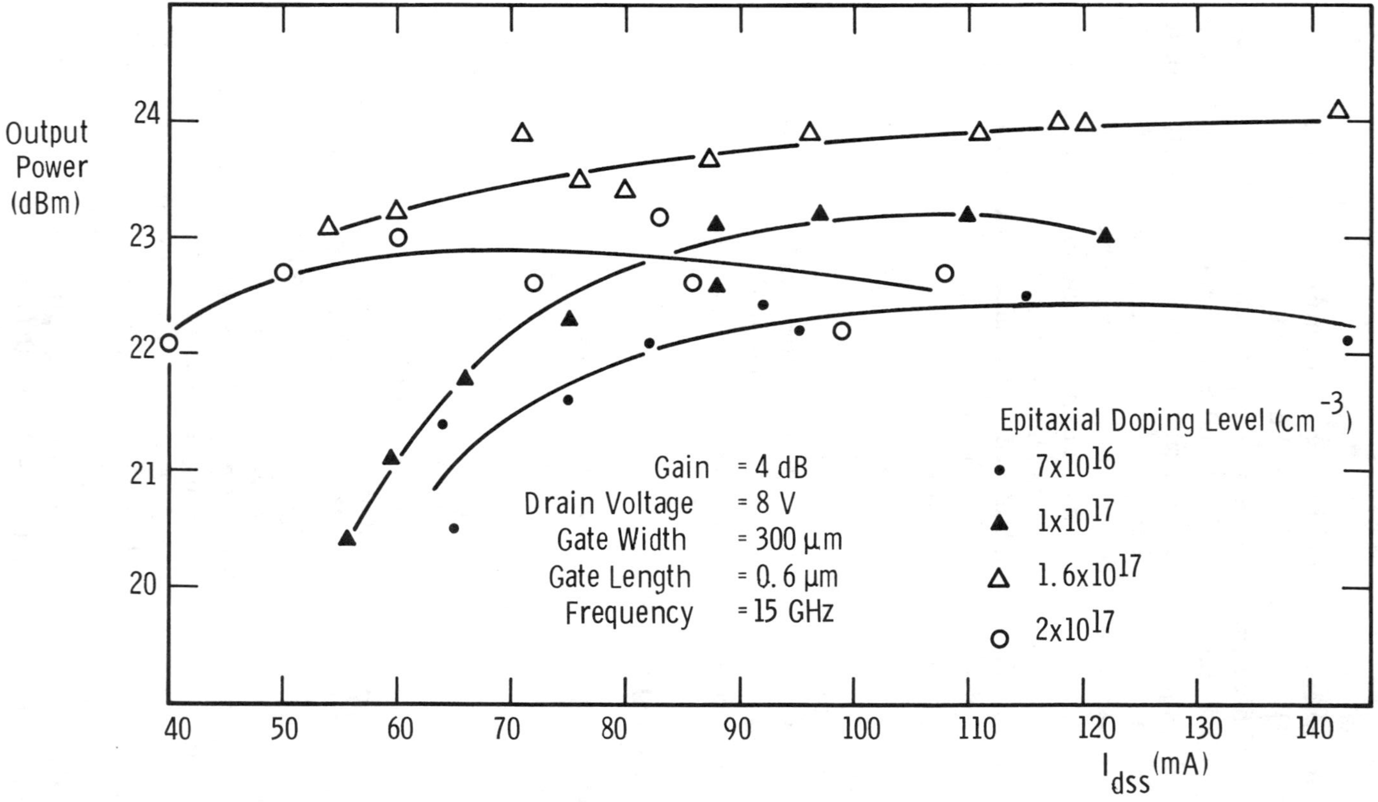

Figure 2*/GaAs FET output power at 15 GHz as a function of I_{dss} for devices from four slices with different active layer doping levels.*

120 mA (330-400 mA per mm gate width). At higher values of I_{dss} output power increases only slightly but efficiency drops (not shown in Figure 2). Similar experiments at other frequencies show that at 10 GHz the optimum is $n \sim 1.2 \times 10^{17}$ cm^{-3} and at 8 GHz $n \sim 8 \times 10^{16}$ cm^{-3}. There is less data at 18 GHz, but it appears that the optimum is near $n = 2 \times 10^{17}$ cm^{-3}. In all cases the optimum I_{dss} is 300-400 mA per mm gate width. Examination of the data also shows that the choice of doping level is more critical at higher frequencies.

The data for Figure 2 was taken at a constant drain voltage of 8 V. The higher Schottky barrier breakdown voltage with more lightly doped material allows those devices to produce significantly higher output powers at higher drain voltages. However, for best efficiency and reliability, a drain bias of 8 V is usually employed. It has been shown that GaAs FETs fail at the drain contact [6,7] ; the failure voltage is practically independent of epitaxial doping level since it doesn't directly involve breakdown of the gate Schottky barrier.

It is, therefore, reasonable to specify the same safe operating voltage for devices with different doping levels but the same geometry. If the device design were changed to improve resistance to failure at the drain contact – for example, by adding n^+ layers under the ohmic contacts – the devices could be operated at a higher drain voltage. This would change the optimum values of the doping level and thickness. The optimum active layer doping level would decrease somewhat, as would I_{dss}, since higher I_{dss} devices have lower Schottky barrier breakdown voltages. Many laboratories are currently working on schemes to increase the drain voltage capability of GaAs power FETs.

2.2 Gate Recess

Recessing the gate greatly increases device output power. An experiment conducted [5] to optimize the gate recess depth for high gain and output power at 15 GHz used a slice having nearly the optimum active layer doping level for operation at that frequency. A uniform slice of $n = 1.7 \times 10^{17}$ cm^{-3} was broken into four pieces that were processed so devices on the different pieces had the same value of I_{dss} but different gate recess depths. The measured doping profiles within the channel were identical, while the gate recess depths varied

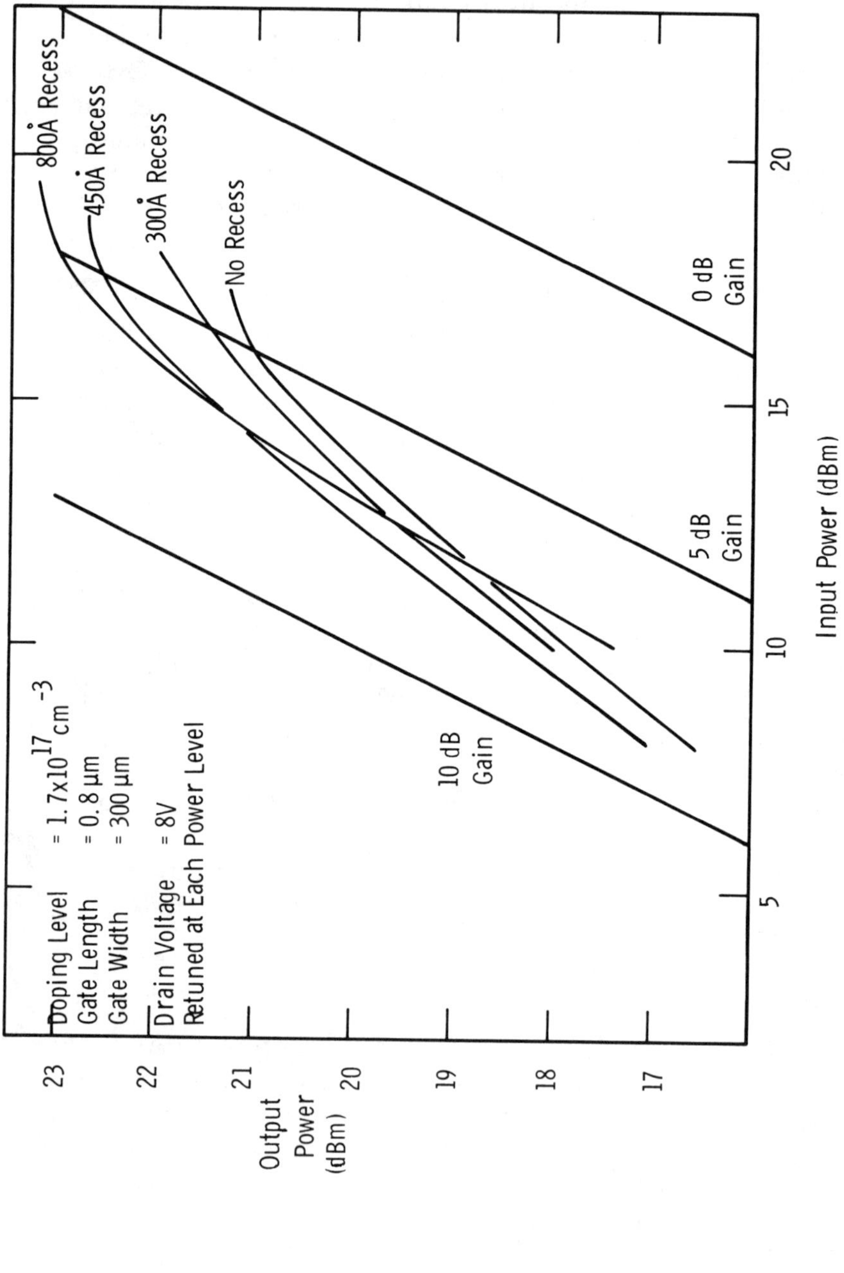

Figure 3*/GaAs FET output power at 15 GHz as a function of input power for devices having different gate recess depths fabricated on the same slice.*

from zero to 800Å. As expected, a larger gate recess correlated with a smaller saturation voltage and a larger transconductance due to the reduced parasitic source-gate resistance.

The microwave performance as a function of gate recess depth was also measured. Figure 3 shows the output power with 8 V drain bias as a function of the input power at 15 GHz for typical devices with each recess depth. Figure 3 shows that the devices with larger gate recess depths had higher output powers at the 4 dB gain level, but they also had lower small signal gains. Although it is not shown in Figure 3, when the input power was reduced sufficiently, the 300Å and no-recess devices had higher gains than the 450Å devices. Also not shown are the results with a device from another slice with the same doping level but 1600Å gate recess. The recess depth was so large that the small signal gain was only 4 to 5 dB, and the device was not useful. Clearly there is an optimum gate recess depth that maximizes the power at whatever gain level one wishes to operate. Since power devices are typically operated at the 4 dB to 8 dB gain level, for $n = 1.7 \times 10^{17}$ cm^{-3} the optimum gate recess depth is in the 400Å to 800Å range. In general, the recess depth should be 25% to 50% of the channel thickness below the gate. Measurements at other frequencies also support this rule.

The present GaAs FET models are not yet sufficiently sophisticated to completely explain the influence of gate recess depth on microwave performance. Other channel modifications have also been employed [8, 9] and there is considerable work in this area.

2.3 Gate Length

The gate length is the major parameter determining device gain and thus operating frequency, since gain decreases at about 6 dB per octave frequency increase. In general, it is desirable to have the longest gate which permits sufficient gain at the operating frequency. Devices with shorter gates have higher gain but are more difficult to fabricate and have slightly lower output powers. Figure 4 is a plot of output power at 8 GHz as a function of input power for two 1200 μm gate width devices from the same slice which have different gate lengths. All parameters except gate length are identical. The shorter gate device has higher output power with high gain but at gain levels less than about 6 dB the larger gate device has higher output power. The choice of gate length depends on the device gain required.

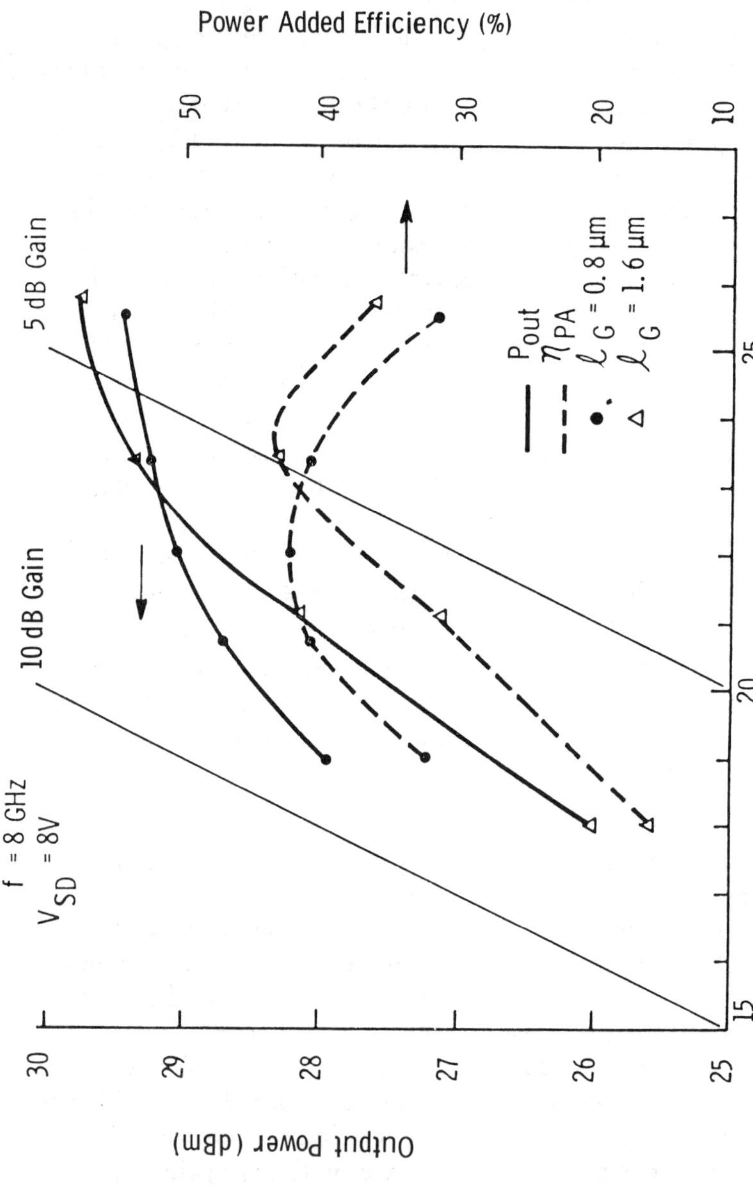

Figure 4*/Dependence of output power at 8 GHz on input power for different gate length GaAs FETs fabricated on the same slice.*

GaAs FETs should be operated at or just below power saturation to obtain the highest power and efficiency. This occurs when the RF input power is increased to a level at which the device gain is ~ 6 dB lower than when it is tuned for maximum small signal gain. Typically, power FETs are operated with 5-6 dB gain. With standard designs this necessitates the use of 0.75 to 1 μm gates at 10 GHz, ~ 1.5 μm gates at 8 GHz, and ~ 0.5 μm gates at 15 GHz. However, these requirements are greatly influenced by other factors. For example, a very low source lead inductance design could have high enough gain to permit the use of longer gates; devices operated at the 8 dB gain level would require shorter gates.

2.4 Gate Finger Width

If the individual gate fingers of a GaAs FET are made too wide, the microwave signal will suffer excessive attenuation and phase shift while propagating down the gate stripes, and the gain will be degraded. This is particularly true for shorter gate lengths. However, it is desirable to make the gate fingers as wide as possible on a power device since that will permit the use of a smaller chip for a given total width. Calculations and measurements in our laboratory demonstrate [10] that for ~ 1 μm gates 0.4 μm thick, gain degradation becomes severe for gate finger widths greater than approximately 150 μm. The results of calculations for a typical X-band structure with 1 μm gates are shown in Figure 5. One surprising result is that the gain degradation is practically independent of frequency up to ~ 30 GHz for finger widths of interest. With 0.5 μm gates, the calculations indicate significant degradation with 75-100 μm fingers, but the measurements were complicated by different source lead inductances of the structures being compared [10].

2.5 Source-Drain Spacing

Devices with a 3 μm source-drain spacing have been compared with otherwise identical devices from the same slice having a 5 μm spacing. Due to the reduced parasitic resistance, the 3 μm devices have slightly reduced saturation voltages and slightly higher transconductances than the 5 μm devices. There are also differences in the microwave performance. Measurements at 15 GHz demonstrate that the 3 μm devices have somewhat lower small signal gains, but higher output powers than the 5 μm devices. Unfortunately, they were much more

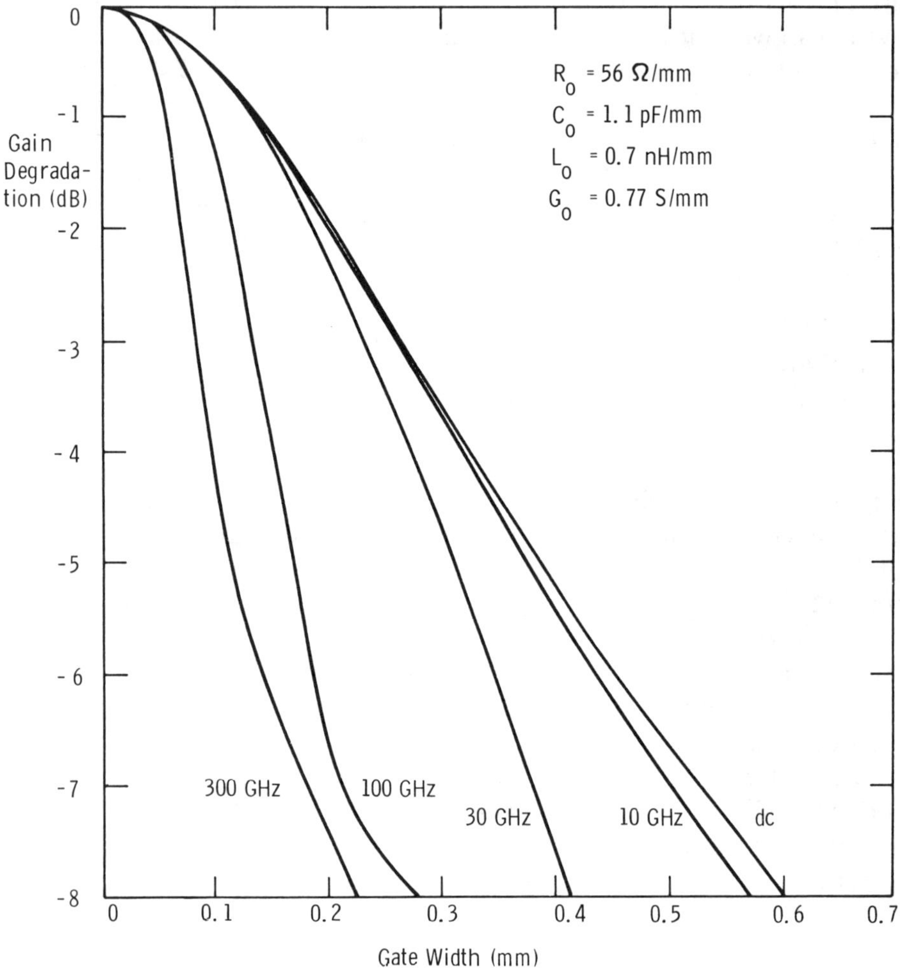

Figure 5*/GaAs FET gain reduction at several frequencies as a function of gate finger width for a typical device structure.*

likely to fail with 8 V drain bias. It is interesting to note that reducing the source-drain spacing has the same effect as recessing the gate: it reduces gain and increases output power. With a conventional structure one should employ a 5 μm spacing for reliable operation. However, if the drain voltage capability is increased (e.g., by employing n^+ regions under the ohmic contacts) it may be desirable to increase the output power by reducing the source-drain spacing.

3. CELL COMBINING

After the microwave performance of individual cells is optimized, it is necessary to efficiently combine them in parallel in order to obtain high output power. It is possible that the requirements of high combining efficiency may require changes from the single cell design. For example, in order to reduce the multicell chip to a reasonable size, it may be necessary to shrink the source pads to the extent that they can't be contacted by bonding, so some type of overlay structure would be necessary.

Several requirements must be met in order to efficiently combine multiple cells: the device thermal resistance must not be excessive; it must be possible to excite each cell uniformly, i.e., the device must not be so large that it is a significant fraction of a wavelength from one end to the other; the source lead inductance must not be excessive; and it must be possible to match the device impedance.

3.1 Thermal Resistance

Calculations and measurements in our laboratory and elsewhere demonstrate that, for a typical GaAs FET chip thickness of $\sim$ 100 μm, the lateral heat flow is negligible at distances beyond the chip thickness and is not very large beyond about 30 μm. Consequently, since cell dimensions are typically several hundred microns, there is little additional heating of multicell chips beyond that obtained with a single cell.

However, the chip must be in good contact with a heatsink capable of removing that generated heat. Typically, soldering to an Au-plated Cu block is employed. A device temperature rise of 50°C - 100°C is usually observed with 100 μm thick chips. Since device gain degrades at $\sim$ .01 dB per °C temperature rise, a reduction of chip thickness from 100 μm to 50 μm would not generally be worth the extra effort involved in handling the thinner chips since only a few tenths dB gain improvement would be observed. For operation in a high temperature environment where device reliability is critical, it may be worthwhile to use thinner chips for the sake of the lower temperature and consequently longer MTTF. Device temperature may also become important if, in attempting to reduce chip size, the gates become closer than ~ 30 μm. One should then consider very carefully the effect of temperature on performance and perhaps employ thinner chips.

3.2 Uniform Excitation

Typically, a GaAs power FET is in a microstrip circuit and must be no larger than the width of the microstrip input and output lines – otherwise, excessively long bond wires will be required for the end cells. The unequal bond wire lengths to the center and end cells will degrade microwave performance. This imposes a limit of ~ $\lambda/4$ at the operating frequency (in the microstrip material) on the maximum chip size. At 10 GHz with microstrip circuits on alumina this is ~ 2 mm.

Figure 6 is a SEM photograph of a 4800 μm GaAs power FET with all pads connected by bond wires. When all four cells are treated as one device in matching impedance, it was found necessary to interconnect the four gate pads and four drain pads as shown. Eliminating these wires decreased output power by ~ 50%. Apparently, they function to equalize the microwave signal at each cell. It is not yet clear whether such interconnections are necessary when the individual cells are matched separately (lumped element matching).

3.3 Source Lead Inductance

Even when the individual cells are interconnected as in Figure 6, the 10 GHz small signal gain of the four-cell device is degraded about 2 dB when compared with a single cell device. The reason is that, with the structure in Figure 6, as more cells are added the source lead inductance per unit gate width increases, degrading gain. When four individual one-cell chips are bonded so that source leads may be grounded between them, and source lead inductance per unit gate width is independent of the number of cells, the gain degradation is reduced to ~ 0.7 dB. About half of this degradation is on the input and half on the output; it has not been possible to reduce it further with the usual bond wire interconnections between cells.

Figure 6/SEM photograph of a conventionally bonded four-cell GaAs power FET.

If additional cells are added, the gain degradation increases rapidly, so usually no more than four cells are employed. It is not yet known whether this limit on the number of cells also holds true when the device cells are individually matched.

Figure 7 shows a more recent device design than that shown in Figure 6. Here it was possible to bond wires to ground between the four cells of the chip. This required that the chip be mounted on a plated heatsink and the GaAs etched away between cells. This further necessitated the use of 25 to 50 μm thick GaAs instead of 100 μm so the etched area was not too wide. An extra advantage was that the chip temperature was somewhat reduced. A further consequence of this design was that the etched areas took up such a large part of the chip that the source pads had to be shrunk in width to maintain the chip size and wires could no longer be bonded to them. Consequently, the air bridge source interconnection scheme seen in Figure 7 was developed.

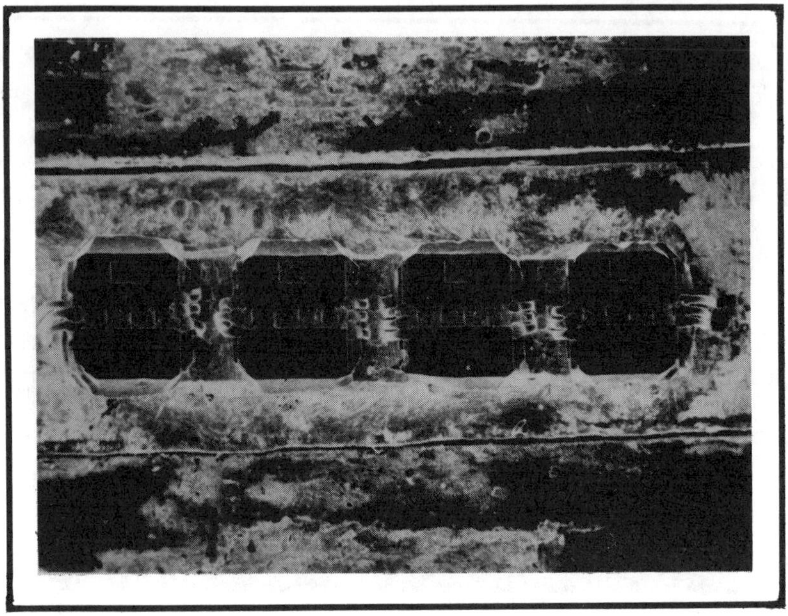

***Figure 7**/SEM photograph of a four-cell GaAs power FET which has a plated heatsink and air bridge source pad interconnection.*

This scheme has proven to have higher yield and greatly simplified bonding compared with the stitch bonded source pad design. Since the processing of the plated heatsink design is more complicated, takes longer, and has lower yield than the conventional process, there is a yield-performance trade-off. For further source lead inductance reduction, one may employ etched vias to the source pads [11], an additional complication.

3.4 Device Impedance Matching

GaAs FET input and output impedances decrease as gate width increases and eventually become too low to match. The input impedance is the most difficult to match. The simplest impedance matching technique is to resonate the input and output capacitances with the gate and drain bond wires and use a simple impedance transformer to match the real part of the impedance. When the cell gate width is larger than a certain amount, the input capacitance can no longer be resonated with physically realizable inductances and more complicated circuits must be employed. This limit corresponds to ~ 1200 μm gate width at 10 GHz. The real part of the impedance of a 4800 μm gate width device is ~ 1Ω, which is about the minimum that can be readily matched at X-band with a simple distributed technique.

Another approach is to use lumped element impedance matching. With this approach, lumped inductors and capacitors are used to transform the impedance of each individual cell to a higher value so that the impedance of the multi-cell device is higher and easier to match. The limits on the number of useable cells or the total gate width which can be matched using this technique have not yet, been established. With four-cell 4800 μm gate width devices, narrow band performance is similar to the distributed technique.

4. SPECIFIC DEVICE DESIGN

As an example to illustrate the device design factors described earlier and the trade-offs necessary, consider the design of a FET capable of 1.5 W output power at 10 GHz with 6 dB gain and 30% power-added efficiency. Further, assume that the fabrication process must be relatively simple and the yield high so that high volume production is possible. In addition, the device must be highly reliable.

The requirements of simplicity and high yield eliminate the use of n^+ contacts and plated heatsinks, since they will not be required to meet the performance goals. The 1.5 W power level can be attained with a 2400 μm gate width device. A larger gate width would permit higher output powers, but is not advisable since efficiency at the 1.5 W level would be lower. Two cells of 1200 μm gate width each will be employed. Four 600 μm cells would have lower source lead inductance and higher gain, but the bonding complexity would be higher and the yield lower. Since the gain requirement can be met with 2 cells, the simpler two-cell design is preferred. One might also consider employing a larger gate width device operated at lower drain voltage (5 to 6 V instead of 8 V) to obtain 1.5 W output power with similar efficiency, since the reliability would be higher under lower voltage operation. However, the reliability is acceptable with 8 V operation, and the gain is lower for lower voltage operation. In addition, the yield of larger gate width devices is lower, their impedance is more difficult to match, and circuit designers generally prefer high voltage low current devices to low voltage high current devices. Going the other direction – to smaller gate width and higher drain voltages – would cause problems with device reliability unless the processing was made considerably more complicated, e.g., by introducing n^+ areas under the ohmic contacts.

A gate length of ~ 0.8 μm is probably necessary to meet the 6 dB gain requirement. If there are no sophisticated lithography techniques available so that this gate length may be routinely achieved, then it is necessary to employ larger gate lengths and increase gain by reducing source lead inductance. One must trade off gate length against design complexity to obtain the combination which meets the gain requirement with highest device yield.

The active layer doping level should be ~ 1.2×10^{17} cm^{-3} and the thickness such that I_{dss} ~ 800 mA for the 2400 μm device. To obtain the best combination of gain and output power, the gate recess depth should be ~ 1000Å. The gate finger width should be ~ 150 μm in order to obtain the best trade-off between number of fingers and gain degradation due to signal propagation along the finger. The source-drain spacing should be ~ 5 μm to obtain high yield and reliability with simple processing (e.g., no n^+ contacts).

A chip thickness of ~ 100 μm offers the best trade-off between thermal resistance and ease of handling. The source pads should be 50 - 75 μm in width. Smaller pads are not necessary since the chip size does not approach ~ 2 mm with 2400 μm total gate width. In order to reduce bonding complexity and source lead inductance without severely increasing process complexity, an air bridge source interconnect can be employed.

It is clear that the design of a GaAs power FET is a complex process involving many trade-offs based on performance specifications and other factors (such as device cost, yield, and reliability) which may not be specified. In the design described above, many of the trade-offs are not based on exact quantitative data but on educated guesses, since the data does not yet exist. Other trade-offs were not even described due to space requirements.

Since GaAs power FETs are continually being improved, the "correct" choices will probably change before sufficient data is accumulated to make all decisions simply a matter of finding a point on a graph.

5. MICROWAVE PERFORMANCE

Over the last few years, GaAs power FETs with many different parameter values have been fabricated and tested at Texas Instruments. The best microwave performance achieved is summarized in Table I. This table includes highest output powers at several frequencies, highest output power per unit gate width, highest output power with high gain, and highest efficiency. The low source lead inductance configuration data is from devices where source wires may be grounded between cells, either by employing separate chips or a plated heatsink.

Almost all of the data from 8 to 12 GHz was obtained from one or more cells of devices similar to that shown in Figure 1. These devices had four cells of 1.2 mm gate width each and source pads connected by bond wires as in Figure 6. The gate length was 0.7 to 1 μm, the source-drain spacing 5 to 6 μm, and the gate finger width 150 μm. The active layer doping level was 8 x 10^{16} cm^{-3} to 1.2 x 10^{17} cm^{-3}. The gate recess depth was 0.1 to 0.2 μm and I_{dss} was in the range of 300 to 400 mA per mm gate width. The 6.4 mm gate width chip had 200 μm gate fingers and, consequently, 1.6 mm gate width cells, but was otherwise identical. The 9.6 mm gate width device was just eight 1200 μm cells in parallel. The device with 2.5 μm gate length had two cells with four 150 μm fingers each but was otherwise identical.

Data at 15 GHz and 16 GHz was taken using devices similar to that shown in Figure 7. These devices had cells of 0.6 mm gate width each and source pads connected by an air bridge. The devices were on integral plated heatsinks. The gate length was 0.7 μm, the source drain spacing 5 μm, and the gate finger width 75 μm. The active layer doping level was 1.5 x 10^{17} cm^{-3} to 2 x 10^{17} cm^{-3}. The gate recess depth was 0.05 to 0.1 μm and I_{dss} was in the range 300 to 400 mA per mm gate width. The device producing the 18 GHz data had four cells of 300 μm gate width each, a 3 μm source-drain spacing, and source pads were connected by bond wires. The other parameters were the same as the air bridge devices. The 20 GHz data was taken on the recently developed "π-gate" devices [12].

The results of Table I will probably improve considerably in the future due to further reduction of the source lead inductance, increasing drain voltage capability, and increasing gate width that can be impedance-matched.

TABLE I

Best GaAs Power FET Performance

Frequency (GHz)	Output Power (W)	Gain (dB)	Power-Added Efficiency (%)	Gate Length (μm)	Total Gate Width (mm)	Drain Voltage (V)	Number of Cells
8	5.1	5.0	35	1.0	6.4	13	4
8	3.9	7.0	39	0.8	4.8*	10	4
8	1.7	4.0	37	2.5	1.2	18	2
9	5.3	4.2	19	1.0	9.6	13	8
10	4.0	3.6	21	1.0	4.8	11	4
10	3.9	6.0	39	0.8	4.8*	10	4
10	0.87	6.0	45	0.7	1.2	8	1
12	3.2	4.0	25	0.7	4.8	9	4
15	1.8	4.0	21	0.7	2.4*	8	4
16	1.6	4.0	20	0.7	2.4*	8	4
18	0.85	4.0	24	0.5	1.2*	10	4
20	1.0	4.0	17	0.5	2.7	8	2

*Low source-lead inductance configurations

ACKNOWLEDGEMENTS

The author would like to thank G.E. Brehm, H.Q. Tserng, and W.R. Wisseman for many helpful discussions during the course of the experiments described here. He would also like to thank W.R. Frensley for the calculations in Figure 5, T.G. Blocker and R.C. Vail for the electron beam lithography, and F.H. Doerbeck for the epitaxial layer growth. For technical assistance he thanks T.B. Brandon, L.P. Graff, R. Kung and A.J. Stinedurf.

REFERENCES

[1] Macksey, H.M., et al., *IEEE Trans. Electron Devices,* ED-24: 113, 1977.

[2] Macksey, H.M., T.G. Blocker, and F.H. Doerbeck, *Electronics Letters,* 13:312, 1977.

[3] Blocker, T.G., H.M. Macksey, and F.H. Doerbeck, *J. Vac. Sci. Technol.,* 15:965, 1978.

[4] DiLorenzo, J.V. and W.R. Wisseman, *IEEE Trans. Microwave Theory and Techniques,* MTT-27:367, 1979.

[5] Macksey, H.M., F.H. Doerbeck, and R.C. Vail, *IEEE Trans. Electron Devices,* ED-27:467, 1980.

[6] Fukuta, M., K. Suyama, H. Suzuki, Y. Nakayama, and H. Ishikawa, *IEEE Trans. Microwave Theory and Techniques,* MTT-24:312, 1976.

[7] Wemple, S.H., and W.C. Niehaus in *Gallium Arsenide and Related Compounds* (Inst. Phys. Conf. Ser. 33b), p. 262. Bristol: Institute of Physics, 1977.

[8] Furutsuka, T., A. Higashisaka, Y. Aono, Y. Takayama, and F. Hasegawa, *Electronics Letters,* 15:417, 1979.

[9] Macksey, H.M., D.W. Shaw, and W.R. Wisseman, *Electronics Letters,* 12:192, 1976.

[10] Frensley, W.R. and H.M. Macksey, *Proceedings of 7th Biennial Cornell Electrical Engineering Conference,* p. 445. Ithaca: Cornell University, 1979.

[11] D'Asaro, L.A., J.V. DiLorenzo, and H. Fukui, *Intl. Electron Device Meeting Technical Digest,* p. 370. New York: IEEE, 1977.

[12] Macksey, H.M., H.Q. Tserng, and S.R. Nelson, 1981 International Solid State Circuits Conference, *Digest of Technical Papers,* p. 70. New York: IEEE, 1981.

GaAs Power FETs Device Design Principles, Fabrication Processes and Material Technology

W. C. Niehaus
Western Electric, Reading, PA.

S. H. Wemple
L. A. D'Asaro
H. Fukui
J. C. Irvin
H. M. Cox
J. V. DiLorenzo
J. C. M. Hwang
W. O. W. Schlosser
Bell Laboratories, Murray Hill, New Jersey.

1. INTRODUCTION

There has been rapid progress in power GaAs field effect transistor (GaAs FET) technology in recent years [1-9], with major emphasis on the 2-20 GHz microwave band. In this chapter we will summarize various aspects of this new technology as developed at Bell Laboratories (BTL). Important considerations reviewed herein are: material technology, fabrication processes, design principles, including both channel and metallization parameters, and performance achievements. The interplay between design variables is complex and it is therefore not surprising that a variety of device layouts and channel designs have evolved in different laboratories in the quest for ever improved microwave performance.

2. MATERIALS TECHNOLOGY

BTL power GaAs FETs are fabricated using multi-layer epitaxy which presently consists of four distinct layers sequentially grown without interruption in an $AsCl_3$—Ga—H_2 chemical vapor deposition (CVD) reactor. These layers are grown on Cr-doped semi-insulating GaAs substrates prepared by either Bridgeman or Czochralski techniques. First, a semi-insulating buffer, several microns thick, is grown using a two-step $CrO_2Cl_2/AsCl_3$ Cr-doping technique [10]. This is followed by a thin (< 1 μm) undoped buffer which serves to reduce the chromium level in the n-type active layer. Third, a sulphur doped active layer (typically doped to $3 < n(\times 10^{16}\ cm^{-3}) < 7$ and $0.6 < x\ (\mu m) < 0.9$ thick) is then grown. This is followed by a n^+ ($\approx 10^{18}\ cm^3$) thin (0.25 μm) contact layer.

Early in the GaAs FET development, devices were made from active layer epitaxy grown directly on Cr-doped substrates. An undoped buffer-active layer structure grown in a two-bubbler $AsCl_3$—Ga—H_2 CVD system [11] was later added, which eliminated looping and light sensitivity of the device current-voltage (I-V) characteristics and improved consistency of device performance [1]. Undoped buffer layer characteristics were found to be dominated by acceptor out diffusion, however, so that their properties were still substrate-dependent [11]. A technique for doping with chromium during growth in order to make the semi-insulating buffer independent of the substrate was then developed to alleviate this problem [10]. Detailed discussion of n and n^+ layers is contained in subsequent sections.

3. EPI-SIDE PROCESSING

The six major processing steps in making the FET chip are based on positive photoresist and optical lithography using contact printing to define the resist pattern. The process sequence is as follows:

Mesa Isolation. Mesas are formed by ion milling away the unwanted n^+ and n epitaxy using photoresist as a milling mask.

Ohmic Contact Formation. The source and drain ohmic contacts spaced 8 μm apart are defined in photoresist. The contact is formed by evaporating a 400 Å AuGe eutectic from a 12% Ge-88% Au (by weight) source, followed by 1000 Å silver and 1250 Å gold overlays.

The unwanted layers are removed by standard lift-off techniques and the metals are alloyed in an oxygen-free atmosphere such as He or H_2 to form an ohmic contact.

n^+ Drain Ledge. Using photoresist as an etch mask, the n^+ layer is removed between source and drain to form a recessed channel structure while leaving a 2 μm n^+ ledge at the drain contact edge (Figure 3d). The GaAs is etched, and S-D saturation current, I_{DS}, is monitored in order to reduce the current to 300-400 mA/mm. For avalanche control I_{DS} is monitored in order to reduce the current to 300-400 mA/mm. For avalanche control I_{DS} is generally kept below 350 mA/mm [12].

Gate. A 1 μm thick, 1.0-2.5 μm long Al gate is defined using photoresist as a lift-off medium. The gate is recessed in its own notch by chemically etching the GaAs to reduce I_{DS} to a preselected value (typically 250 mA/mm) prior to Al evaporation by E-gun.

Thick Gold Overlay. After the Al gate has been defined, a Ti-Pt-Au (0.2 μm, 0.2 μm, 1.0 μm) overlay is added to the source, drain, gate busbar and gate bonding pads. This 1.4 μm metal thickness is difficult to form by a simple lift-off procedure. Therefore the following combination of lift-off and ion milling is used: first, the standard photo-resist lift-off layer is put in place; second, the Ti-Pt-Au layers are evaporated from a source heated by E-gun; third, a second resist layer, which is the anti-pattern of the first resist layer, is defined; fourth, the Au and Pt layers are ion etched using this second resist level as a milling mask, and, finally, the first resist layer is spray dissolved, thus defining the Ti layer by lift-off.

Passivation. As the final step in epi-side processing, the wafer is coated with plasma-deposited silicon nitride. Windows are then opened to the device bonding pads using plasma etching and appropriate photoresist masking.

Chip Thinning, Separation, and Mounting in the Microwave Package. After the epi-side processing is complete, the substrate is thinned to 30-50 microns, plated with 1 mil of Au in an electro-less bath, and laser scribed to form separated GaAs chips [13]. Finally, the device is bonded in a microwave package using an AuSn preform, and the source, drain, and gate electrodes are wire bonded to the package. In some cases, source via connections are made through the GaAs substrate to the back metal as part of the substrate thinning and

Au plating chip formation process.

Figure 1 shows an SEM photograph of the device channel after the epi-side processing is complete. Note that the gate has a tab as it enters the 1 μm high mesa step to facilitate step coverage. The dielectric had been moved from this device to avoid electrostatic charging while taking the SEM photograph. A vestige of the dielectric remains in the channel on the drain pad showing the opening in the dielectric for a drain bond wire. We will defer the discussion of S-D channel dimensions until a later section.

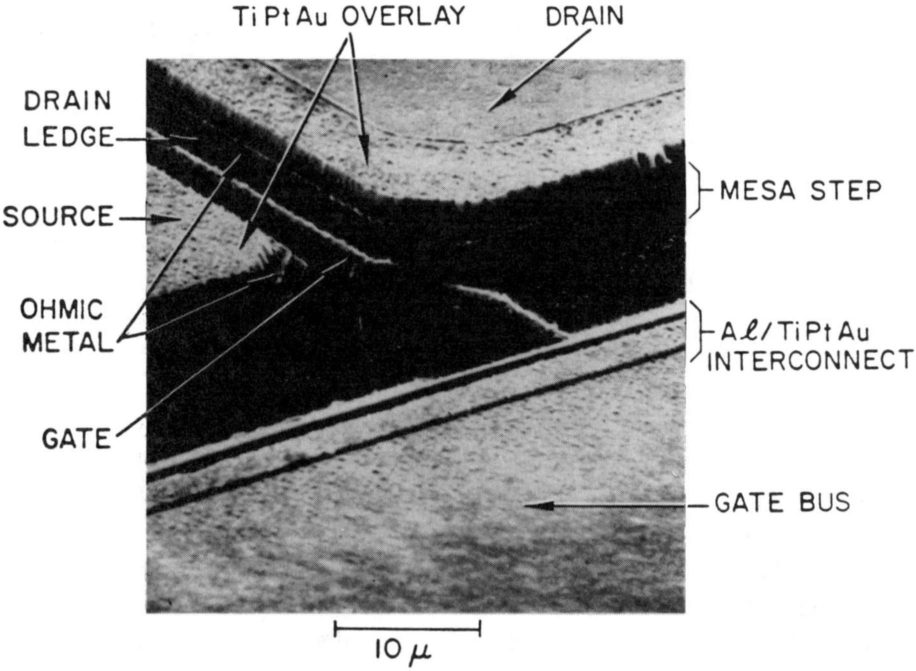

***Figure 1**/Shallow angle SEM photograph of the mesa edge of a single device cell at the point where the gate is stepped onto the mesa. The AuGe ohmic contact, Al gate, and thick TiPtAu overlay metallizations are identified. The drain ledge in the S-D channel can also be seen. Note the dielectric layer has been removed to make this photograph but the residues from the plasma etch process indicate the original drain bonding window opening in the SiN.*

4. POWER FET DESIGN PRINCIPLES – OVERVIEW

The power FET can be considered as a simple current switch. The maximum power capability, P (watts), is related to the static drain $I_D - V_D$ as (see Figure 2):

$$P = P_w \times W = \frac{1}{4} I_F (V_{SD} - V_K) \times W \ , \qquad (1)$$

where I_F represents the maximum drain current in A/mm under forward gate bias, V_F; V_K is the knee voltage as defined in Figure 2, and W is the total device width (usually in millimeters). It is commonly assumed that V_{SD} is limited by the gate-drain bulk avalanche voltage, V_{GD}, as depicted in Figure 2, according to the expression

$$V_{SD} < \frac{1}{2} (V_{GD} - V_p + V_K) \ , \qquad (2)$$

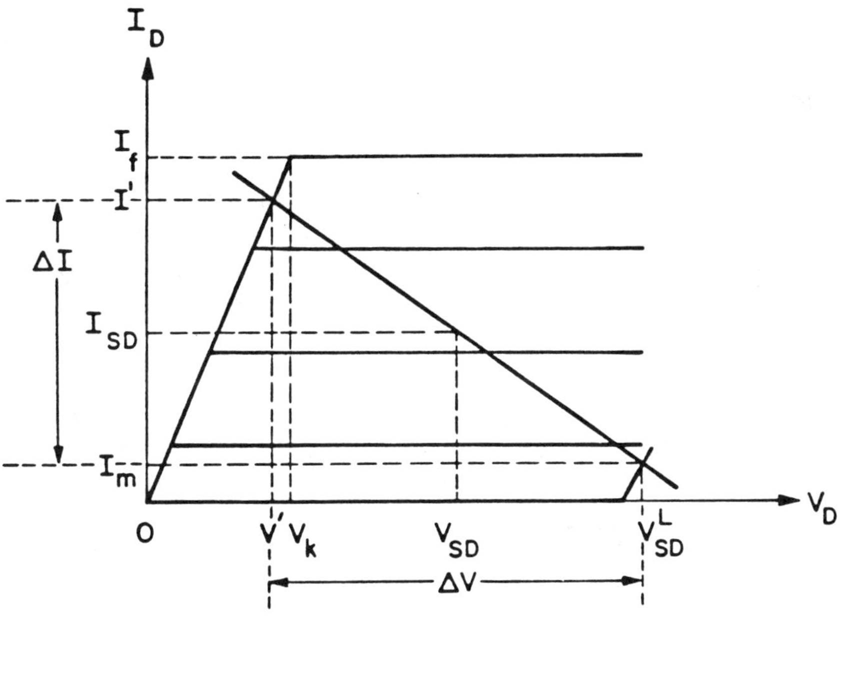

Figure 2/Schematic drain characteristics for a power GaAs FET showing the $I_D(V_D)$ characteristics important for producing power.

where V_p is the pinch-off voltage. As discussed below, Equation (2) is inappropriate for the channel design used at BTL. The device channel design which maximizes the output power per unit cell width, P_w, trades off the need for high source-drain currents, I_{DSS} and I_F, with the need for high source-drain bias, V_{SD}. The elements of the BTL power GaAs FET design developed to meet these requirements can be grouped into three main categories:

Source-Drain Channel Design. The power FET geometry includes: i) A 10^{18} cm^{-3} doped n^+ ledge at the drain contact in order to achieve high source-drain burnout capability, ii) a recessed gate structure which takes into account the active layer doping and surface zero bias depletion in order to minimize the charge in the S-D channel and thereby suppress the onset of gate-drain avalanche, iii) proper choice of active layer doping, and iv) a Si_3N_4 dielectric coating which greatly extends device life and shields the metallizations from chemical attack.

Device Metallurgy. A reliable device must utilize source, drain, and gate metallizations which are compatible with the high current densities present in power FETs. In choosing the three basic metallization systems, i) AuGe-Ag-Au S-D ohmic contacts, ii) Al gates, and iii) thick Au used in combination with Ti and Pt barrier layers, we have considered the effects of electromigration in both the Al and the Au as well as intermetallic formation between Au and Al. In general, current densities in the Au overlay are kept below 2×10^5 A/cm^2.

Chip Layout. An optimum cell paralleling scheme required for high power (i.e., large W) must account for complex interplay between the principles for best thermal and best microwave design. A proper layout to reduce channel operating temperatures and metal electromigration must include: i) the electrode dimensions, ii) gate-to-gate spacing, iii) individual gate finger width, and iv) substrate thickness. For the latter dimension we generally use 30-50 μm. Microwave performance is usually assumed to depend on i) gate length, ii) gate finger width, and iii) the specific interconnect pattern. In particular, we have investigated both through-the-chip source via connections and multiple wire bonding to a source bus bar located at the chip edge. Our results to date do not support a strong dependence on gate length or finger width up to 12 GHz operation.

5. CHANNEL DESIGN FOR MAXIMUM OUTPUT POWER AND RELIABILITY

5.1 Recessed S-D Channel Design for Increased dc Bias Voltage Capability

One of the keys to increasing RF output power is to increase its dc burnout capability. Early experiments [14-16] implicated the drain contact as the initiation site for instantaneous burnout. Several laboratories have modified the drain contact configuration by introducing either an n^+ inlay structure [5], an n^+ contact layer [1, 15, 16], or a simple recess structure [15] and observed improved S-D bias capability. Experiments in our laboratory [17] and elsewhere [18] show that instantaneous burnout is ultimately limited by dc input power and not voltage or current separately, with thermal runaway occurring when some local temperature reaches about 550°C [17]. We have chosen to optimize the device channel design for I_{dss} values near 200 mA/mm corresponding to ~ 100 mA/mm optimum S-D bias current under RF operating conditions. Our standard devices operate at V_{SD} = 14V.

Before proceeding further it is important to distinguish short-term (or instantaneous) S-D burnout from long-term S-D failure [17]. Simply put, instantaneous burnout occurs during initial application of dc bias, whereas long-term failures occur during normal operation or during accelerated bias-temperature stress aging [19-21]. The channel design we have chosen is based on the short-term burnout data displayed in Table 1.

TABLE I

Typical S-D Bias (Input Power) Capabilities As a Function of Drain Geometry

	Material	Drain Geometry	S-D Bias	Input Power
i)	n-layer	planar	10-20V	1-2 watts/mm
ii)	n-layer	pedestal, 2μ output space	25-35	2.5-3.5
iii)	n^+/n	pedestal, 4μm output space	25-35	2.5-3.5
iv)	n^+/n	ledge, drain only	$40\text{-}50^+$	$4\text{-}5^+$
v)	n^+/n	ledge, source and drain	$40\text{-}50^+$	$4\text{-}5^+$

These data were obtained from 6mm wide devices, biased with 600 mA S-D current with 28 dBm RF input power. Five basic channel designs have been studied at TBL (see Figure 3):

(i) Planar drain contact – No n^+ Surface Layer (Figure 3A). Used until 1976, this structure was typically capable of 10-20 V S-D bias (1-2 W/mm). In this case the bias capability of the device depended weakly on S-D spacing [14].

(ii) and (iii) n^+ Drain Pedestal (Figures 3B and 3C): Introduction of the n^+ surface layer and its subsequent removal by chemical etching using the metals themselves as an etch mask resulted in 25-35 volt S-D bias (2.5-3.5 W/mm). Extending the G-D output distance from 2 μm to 4 μm did not improve either the G-D avalanche or S-D burn-out capabilities [1].

(iv) and (v) n^+ Drain Ledge (Figures 3D and 3E): Introducing an extra lithographic step to define a drain contact ledge increased the bias capability to the 40-50+ V range (4-5 + W/mm). This dc input power capability in excess of 5 watt/mm is equal to or superior to those reported elsewhere [22] for 1 mm to 1.5 mm devices fabricated by Texas Instruments (4.5 watt/mm), Dexcel (3.8 watt/mm), M.S.C. (4.8 watt/mm) and NEC (2.9 watt/mm). Addition of a ledge to the source contact made no further improvement, thus reinforcing the idea that the drain was the initiation site for catastrophic failure. It is clear that the n^+ contact layer, especially when used in combination with a drain ledge geometry, suppresses S-D failure mechanisms and makes possible higher S-D voltage operation.

Extensive experiments in this laboratory involving changes in ohmic contact details show that the drain contact plays an important role in device resistance to long-term burnout. That data will be discussed in detail in a separate publication [23]. An additional long-term burn-out mechanism which involves surface chemical changes has been suppressed by using Si_3N_4 passivation rather than, for example, SiO_2 [17].

5.2 Recess Gate Structure – Control of Gate-Drain Avalanche

Up to this point our design discussion has focused on catastrophic source-drain failure mechanisms as limiting the dc input power (and hence RF output power) of the GaAs FET. The gate plays only a secondary role in the burnout process by its control of S-D currents;

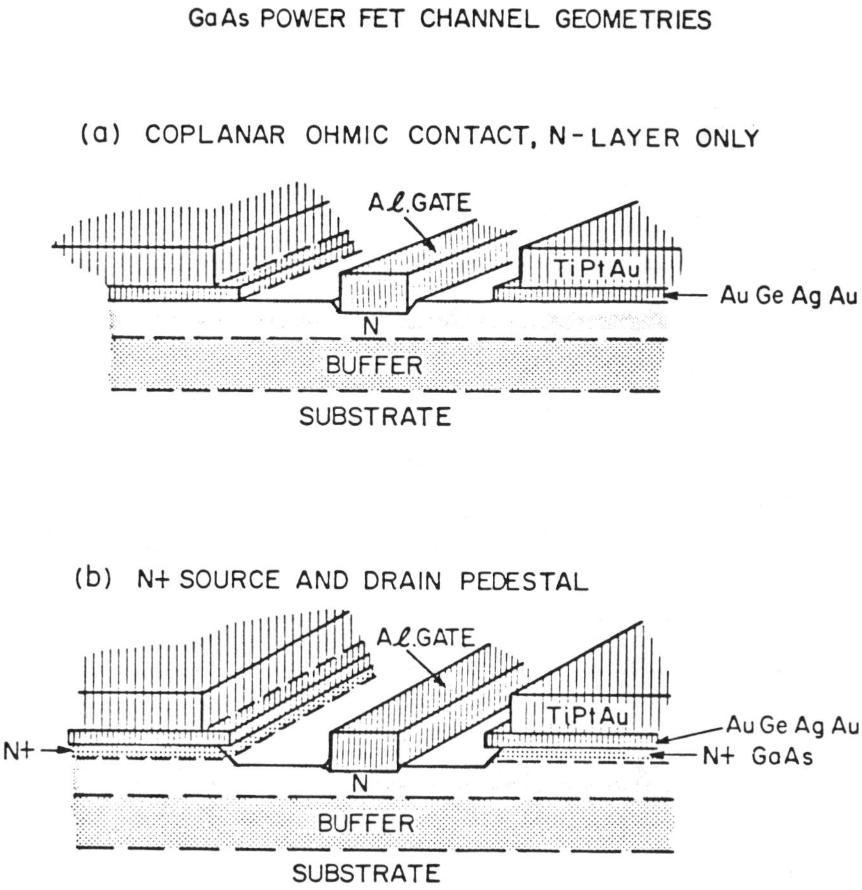

Figure 3*/Schematic representation of various channel designs investigated at Bell laboratories. All designs have recessed gate structure with*

(A) a planar drain edge contact without an N+ GaAs contact layer.

(B) an N+ pedestal drain contact and 2 μm output space.

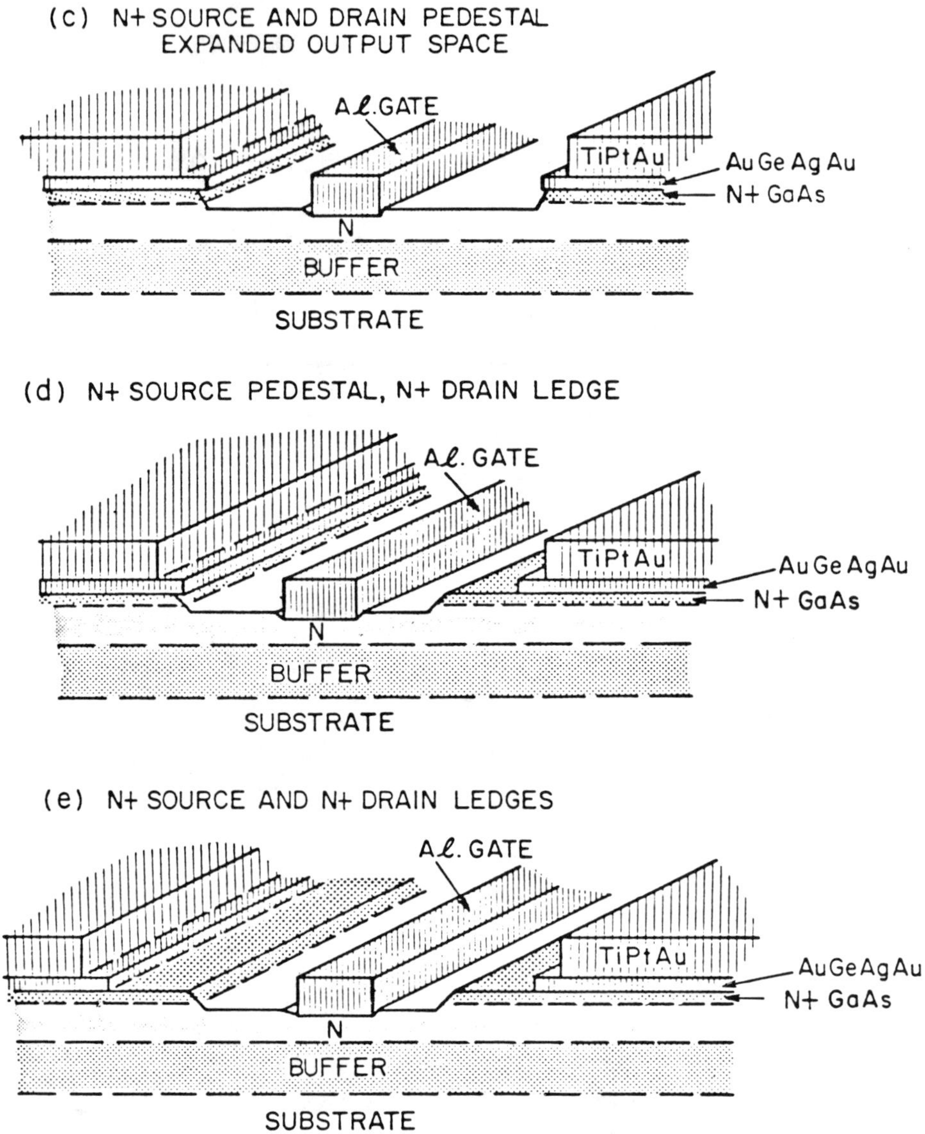

***Figure 3**/Schematic representation of various channel designs investigated at Bell Laboratories. All designs have recessed gate structure with*

(C) *an N+ pedestal drain contact with a 4 μm output space.*

(D) *an 2 μm N+ drain ledge geometry.*

(E) *an 2 μm source and a 2 μm drain N+ ledge geometry.*

however, the gate does play an important role in controlling the output power since the onset of gate-drain avalanche imposes a fundamental constraint on voltage and/or current swings at the drain.

Wemple and Niehaus [12] have shown that gate-drain avalanche voltages factors of 2-3 in excess of those predicted by one dimensional bulk theory can be achieved. This can be understood in terms of two dimensional, lateral spreading of the gate depletion region towards the drain at large G-D bias if a suitable channel design is employed. Specifically, the desirable consequences of lateral spreading are enhanced if the undepleted epitaxial charge per unit area, Q_u, in the gate-drain output space is minimized. This undepleted charge [in units of 10^{12} cm^{-2}] is calibrated in terms of saturated S-D current using $I_{DS} = 176\ Q_u$. Here I_{DS} is in mA/mm units, and the factor 176 reflects an effective saturated drift velocity in GaAs of 1.1×10^7 cm/sec.

The optimum channel design must: 1) provide minimum undepleted charge in the G-D output space of the transistor in order to improve G-D avalanche, 2) maximize the charge under the gate since the maximum current under forward gate bias, I_F, is proportional to this charge, and 3) take into account the depletion of charge at the GaAs surface due to band bending associated with mid-gap pinning of the Ferni level. These design principles require the lowest undepleted charge in the device output space consistent with the desired RF current swing under the gate. This leads to a design which places the gate in a notch slightly shallower than the surface zero-bias depletion depth, a_0. For $n = 10^{17}$ cm^{-3}, $a_0 \approx 0.1$ μm, and for $n = 4 \times 10^{16}$ cm^3, $a_0 \cong 0.16$ μm.

Experimentally we find the bulk avalanche model is appropriate for $Q_u > 2.6$, but the thin film lateral spreading model is appropriate for $Q_u < 2.3$. In the thin film case there is no statistically significant dependence of avalanche voltage on doping level over the range 3×10^{16} $cm^{-3} < n < 2 \times 10^{17}$ cm^{-3}. This experimental result is consistent with the lateral spreading model but it is, of course, at odds with the one-dimensional avalanche theory which predicts that $V_{gd} \sim N^{-1}$. Our transistors with $4 < n \times 10^{16}$ $cm^{-3} < 7$ are designed for: 1) $2.0 < Q_u < 2.3$ corresponding to saturation currents, prior to gate etching, of about 350 mA/mm and 2) an under-gate epi thickness corresponding to a 250 mA/mm saturation current prior to Al deposition. This recipe results in $I_{ds} \approx 200$ mA/mm after the Al gate has been defined by lift-off. Figure 4 shows a typical pulsed $I_D - V_D$ curve [100ns pulses at a 50 Hz rate] for a 6 mm FET with this channel design. Notice the nonpinchable drain current beginning at $V_{SD} \approx 14$V. This is due to gate avalanche starting at $V_{GD} \approx 21$V. In the classic FET described by Equations (1) and (2) and Figure 2, the gate-drain voltage swing would be limited to about 21V due to a rapid increase in G-D current. In contrast, an FET designed using lateral spreading principle can realize additional voltage swing and a further increase in output power since the avalanche current curves coalesce and saturate in a way not predicted by bulk avalanche theory. Indeed, in this thin film design, avalanche imposes a minimum drain current rather than a maximum drain voltage as expected classically. For example, the minimum S-D current at 30 volt bias, I_m (30), reaches 67 mA/mm in the example of Figure 4. The minimum drain current at V_{SD} = 30V is a convenient avalanche figure of merit for devices biased at 14 V where the maximum RF swing approaches 26 V.

Devices fabricated according to the above design principles typically (at least at 4GHz) show increasing output power with S-D bias up to

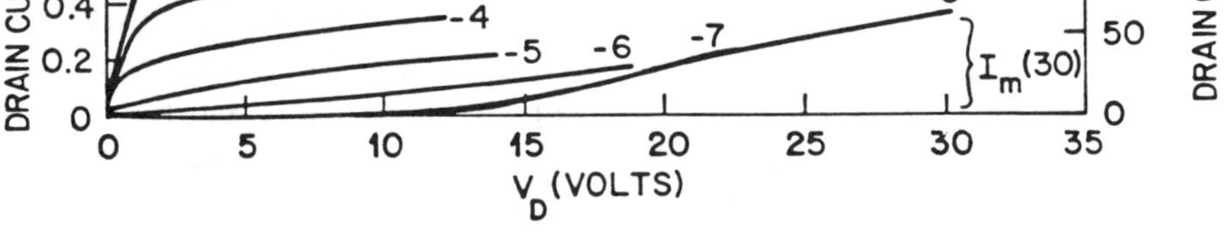

Figure 4/Pulse source-drain I-V of typical 6 mm periphery FET with undepleted charge $Q_u = 2.0 - 2.3 \times 10^{17}$ cm^{-2}. The gate is dc biased and 100 ns voltage pulses at a 50 Hz repetition rate have been applied to the drain.

25 V. Nevertheless, the power per millimeter of gate width, P_w, remains for all but the best devices below 0.9 W/mm (BEST = 1.2 W/mm). As a result, the 40-50 volt S-D bias capability still outstrips the ability to suppress G-D avalanche. Gate-drain avalanche remains the major cause of output power saturation as a function of drain voltage in our FETs.

Whereas the I_m (30) value can be easily adjusted to any value between 0 and 200 mA/mm simply by controlling Q_u, there still remain excursions between 30 and 80 mA/mm for nominally the same channel design. Two processing factors appear to be associated with this variability [12]: first, the gate lift-off formation process itself plays an important role. The rough edges of the gates as formed by lift-off are a likely cause for this wafer-to-wafer and even device-to-device variability within a wafer. For a fixed G-D bias, the rough gate edge corresponds to regions of enhanced electric field ($E \sim V/r$) and result in localized regions reaching the critical field for avalanche faster than predicted by the lateral spreading model discussed above. For devices with $Q_u < 2.3$, applying a gate-drain bias sufficient to yield a total gate current per unit gate-width of 1-2 mA/mm causes white light emission from discrete points on the drain side of the gate edge [12]. At times, very obvious physical defects such as tweezer scratches occur in the Al gate. Under gate drain bias this defect is the site of local microplasmas [24]. Such defect-microplasma correlations tend to confirm the enhanced field concept. In our laboratory, FETs fabricated with PtTiPtAu gates seem to yield a smoother gate edge and a subsequent slightly higher G-D voltage and lower I_m (30) [ΔI_m (30) ≈ -20 mA/mm] for a given FET channel design. At this point, the gate data is limited and is being pursued further, although the general principle that smooth gate edges are desirable for avalanche control seems valid.

Application of a dielectric passivation layer is the second process which degrades G-D avalanche. Although the effect is variable and not easily controlled, the reduction in V_{GD} can be caused by either 1) the dielectric introducing approximately 0.2×10^{12} cm^{-2} positive surface charge, thus increasing Q_u, or 2) the dielectric deposition altering the gate edge roughness and hence the avalanche field. Recent experiments show that it is the dielectric itself and not the accompanying thermal cycling which causes avalanche degradation.

5.3 Active Layer Doping

Suppression of S-D catastrophic burnout and G-D avalanche control are independent of active layer doping over the range N = 2 - 20 x 10^{16} cm^{-3}. The BTL design employs a 0.6-0.9 μm thick, 3 - 7 x 10^{16} cm^{-3} doped active layer. The lower doped, thicker active layer is used to facilitate control of GaAs etching required to meet the charge minimization design goals discussed above.

5.4 Chip Passivation

The preferred passivation dielectric for our devices is ~ 4000Å of plasma deposited $Si_3N_4H_x$ [25]. These devices display significantly improved resistance to long-term burnout [17, 19, 20] when compared with devices coated with SiO_2 [21]. It has been proposed [17] that chemical instability of the "natural" GaAs oxide plays a major role in long-term burnout and that any surface treatment, such as a Si_3N_4 layer, that minimizes oxide formation would improve burnout resistance.

6. DEVICE METALLURGY

The Au-Ge-Ag-Au ohmic contact used by BTL is similar to that used elsewhere in that it offers a smooth morphology, low resistance contact (0.5 - 2 x 10^{-6} ohm-cm^2) [1] and remains stable under high temperature aging. The majority of other laboratories also use AuGe based ohmic contacts [5, 6, 26-29]. Nickel [6, 28-29] or some refractory metal [26-28] is frequently used as a replacement for silver and serves the same purpose in keeping a smooth contact morphology, which is important for achieving the high quality, fine line lithography required in the steps to follow.

On the other hand, there is not universal agreement on the choice of Al as the most desirable gate metallization. Others, like BTL, are investigating composite gate metallization systems [30, 31] which employ such metals as W, Ti, or Pt, as the Schottky contact. For Al, Irvin et al. [32] and others [29, 30] have catalogued several failure mechanisms, i.e., Au-Al phase formation, electrolytic corrosion, and electromigration. Fukui et al. [19, 20] have reported that 6 mm power devices with Al gates, in conjunction with a Ti/Pt barrier placed between the Al and the Au have operated in excess of 3,000 hours at 250°C channel temperature with 14 V S-D bias and 28 dBm 4-GHz input power. They show neither electrical degradation nor cosmetic changes in the Al gates. At this point we see no reliability-related reason to depart from the Al gate technology.

A one micron thick gold overlay serves both: 1) to limit the current density in each source finger to less than 2 x 10^5 A/cm^2 in order to suppress Au electromigation [33] and 2) to facilitate bonding of the device in the microwave package. The intermediate Ti and Pt layers prevent the formation of Al-Au intermetallics [34] at the Al gate-gold bonding pad interconnect.

7. CHIP LAYOUT

7.1 Thermal Design

Thermal design of a power FET is a complex interplay of substrate thickness, gate-to-gate spacing, and gate finger width [35, 36]. Most power FETs under development [1, 5, 6, 9], including those developed at BTL, are designed so heat is removed through the GaAs substrate. An alternate arrangement (RCA, MSC, and Mitsubishi) involves flip-chip bonding of the source pads to the heat sink [37]. The lowest thermal impedances are achieved with normal backside mounting provided the chip is thinner than about 3 mils [35].

For the BTL designs with 30-50 μm thick substrates, the typical normalized thermal impedance is 30-40°C-mm/W, or about 6°C/W for the 6 mm device [35, 36]. This design generally yields a channel temperature rise less than 50°C above ambient with the device biased at 14V and 0.6A in order to delivery 4-5 watts RF power at 630 m W RF input power.

As implied above, gate width is a factor in thermal design since end cooling is more effective for narrow gates than for wide gates. Here the quantity of interest is the ratio of gate width to substrate thickness. When this ratio is large, as for 500 μm gate fingers, end cooling is entirely negligible. This would not be true, however, for 100 μm wide gates so that for a given thermal impedance value, narrow gate designs permit thicker substrates. When end cooling can be neglected, as for 500 μm gate fingers, our thermal impedance design goal can be achieved simply by maintaining the gate-to-gate separation equal to or greater than the substrate thickness. For some of our transistor designs this simple rule is automatically satisfied because other, non-thermal constraints require it. Thus, 125 μm (or even 75 μm) drain bonding pads on the mesa in conjunction with a source stripe length (the short dimension) dictated by a 2 x 10^5 A/cm^2 current density (electromigration) constraint, yield an average gate-to-gate spacing requirement considerably larger than the 30-50 μm value permitted by our substrate thickness.

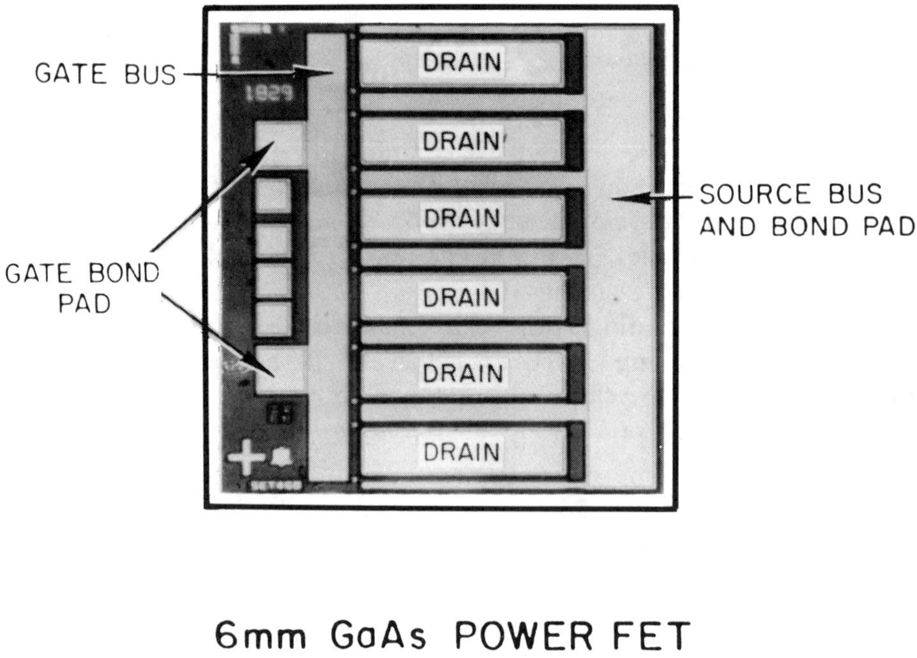

Figure 5*/Photograph of a 6 mm GaAs FET. The 6 mm device has twelve, 500 μm wide cells. The gate and source electrodes are fed from a common busbar to the left and right, respectively, of the chip. The non-integrate chip design employs six drain pads on the device mesa.*

7.2 Gate Paralleling

Figure 5 shows a photograph of our 6 mm device used at 4 and 7 GHz. The chip has twelve, 500 μm wide gate fingers fed by two gate bonding pads. The source stripes connect to a common busbar, which is wire bonded to the package ground over the edge of the chip. There are six, 5 mil by 20 mil drain pads located on the device mesa. The overall average gate-to-gate distance in this design is 95 μm. Each drain is individually bonded to the package drain tab with 2 mil Au tapes. Using this basic approach, various combinations of gate widths (125 μm, 250 μm, 500 μm) and number of gates (2, 4, . . ., 64) have been used to give device widths ranging from 0.5 mm to 32 mm. Figure 6 shows a 3-chip, 24 mm packaged device.

A through-the-chip source via design has also been developed [13]. Such a design offers the advantages of: 1) lower source lead inductance, 2) reduced packaging costs, and 3) more efficient use of GaAs especially for gate finger widths narrower than 500 μm.

We have made extensive measurements on source via devices at frequencies up to 12 GHz and have found significant gain advantage (several dB at 12 GHz) for frequencies of 7 GHz and above. At 4 GHz there seems to be little if any advantage, and even at 7 GHz via source connections do not improve saturated power. Although the interplay between frequency, device size, and the source grounding scheme seems to be relatively complex, our motivation for using source vias in X-band devices is strong.

8. MICROWAVE PERFORMANCE

8.1 Power

Increasing the number of device cells in order to increase the total device size, W, provides a straightforward procedure for increasing output power. The best BTL 24 mm device (3 chips) has given 18.5 W of power with 3 dB gain at 4.4 GHz [3] while at 7 GHz a 16 mm unit has achieved 8.6 watts with 3 dB gain. In general, though, decreasing cell combining efficiency limits the expected benefits of larger and larger gate periphery. This scaling loss depends not only on total size but also on frequency and power compression. In the case of saturated power (defined at the 3 dB gain

Figure 6/*Photograph of a packaged 24 mm GaAs FET.*

level), data shown in Figure 7 display our observed trends. Note that there is no significant scaling loss up to 8 mm widths at either 4 or 7 GHz. For 16 mm devices, on the other hand, some power degradation occurs amounting to approximately −0.5 dB at 4 GHz and − 1.2 dB at 7 GHz. Modest scaling problems with 16 mm devices are also revealed by the "best" 4 GHz data points which were obtained with V_{SD} = 20 V. These "best" results fall near the state-of-the-art 1 W/mm value.

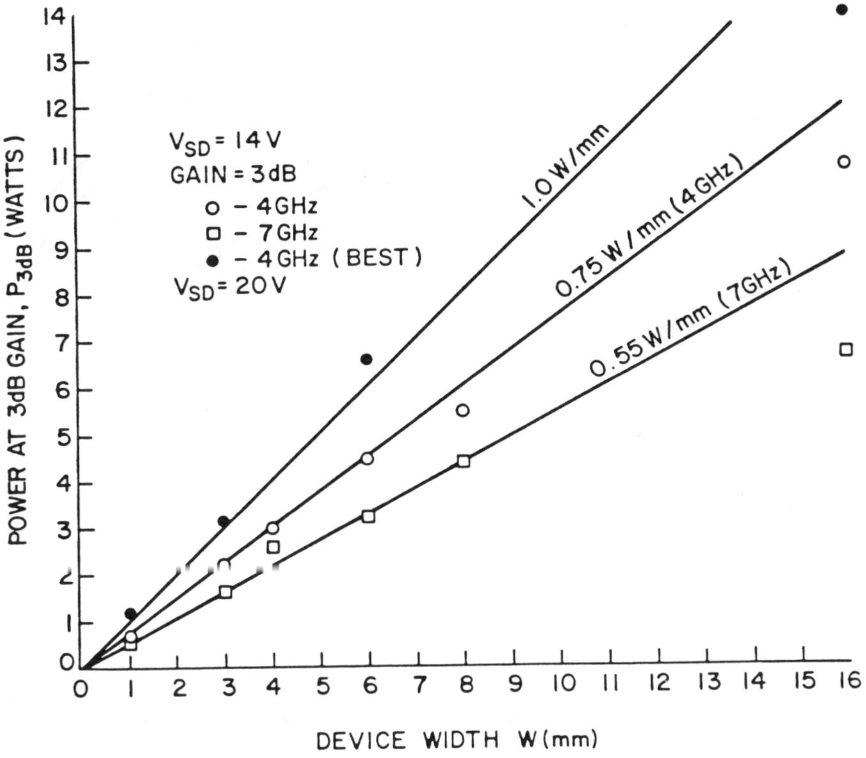

Figure 7/Power at 3 dB gain, $P_{3\,dB}$ (watts), as a function of device width, W_T (mm), at 4 and 7 GHz.

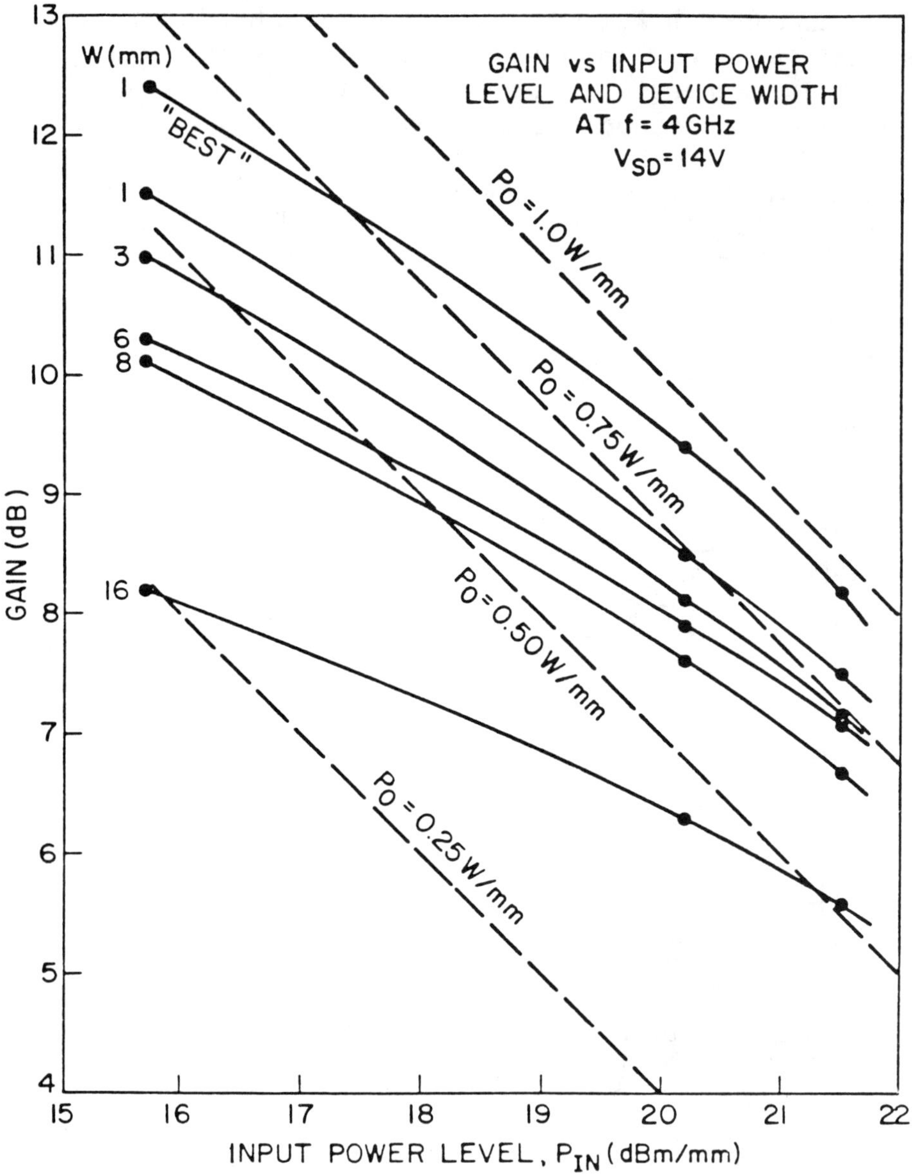

Figure 8/*Gain, G(dB), as a function of input power, P_{IN} (dBm/mm), and total device width, W_T (mm), at 4 GHz. The device cell widths is 500 μm.*

According to Figure 8, the saturated power associated with increasing the operating frequency from 4 to 7 GHz falls on the average nearly 1.3 dB for devices with W $\leqslant$ 8 mm, whereas for 16 mm devices the loss is closer to 2 dB. We have found, however, that for specific 8 mm or smaller devices this power loss can range anywhere from 0 to 2 dB. The actual value does not correlate with total periphery (W = 1-8 mm); gate finger width (125-500 μm); or gate length (1.0-2.5 μm). The latter two results suggest that, at least for saturated power, little if any penalty is paid at 7 GHz by using a 500 μm x 2 μm gate finger design.

8.2 Gain

Turning now to gains at lower input compression, we show in Figures 8 and 9 the average experimental results at 4 and 7 GHz along with our "best" gains for W = 1 mm devices measured at 14 Volts. Higher gains can be observed at higher S-D voltages. The selected power input range corresponds to large signal operation appropriate to typical power devices with output power, P_0, in the 0.25-1.0 W/mm range. Each data point encompasses a spread of measured values extending over approximately 1 dB. At fixed input power we conclude that there is a clear trend toward reduced gain with increasing size. Surprisingly, 16 mm devices operated at 7 GHz seem to show better gain scaling than they do at 4 GHz even though, as noted earlier, exactly the opposite result is observed for saturated power.

We now ask what design variables are at our disposal which can be used to improve gain values presented as Figures 8 and 9. First, it should be noted that the observed gain loss between 4 and 7 GHz generally lies between 1 and 2 dB. This result is only slightly larger than the saturated power value noted above and is much smaller than the 4.9 dB loss predicted by small signal theory (i.e., gain ~ f^{-2}). It is doubtful, therefore, that design directions deduced from small signal analysis will be applicable to power FETs. We have found, for example, that performance at 7 GHz is independent of gate finger width up to at least 500 μm and gate length up to at least 2.5 μm. At even higher frequencies (12 GHz) we have found that gains in excess of 10 dB can be achieved with 2 μm gates (W = 1 mm). It is our contention that gate length and gate width limits for X-band operation remain to be established.

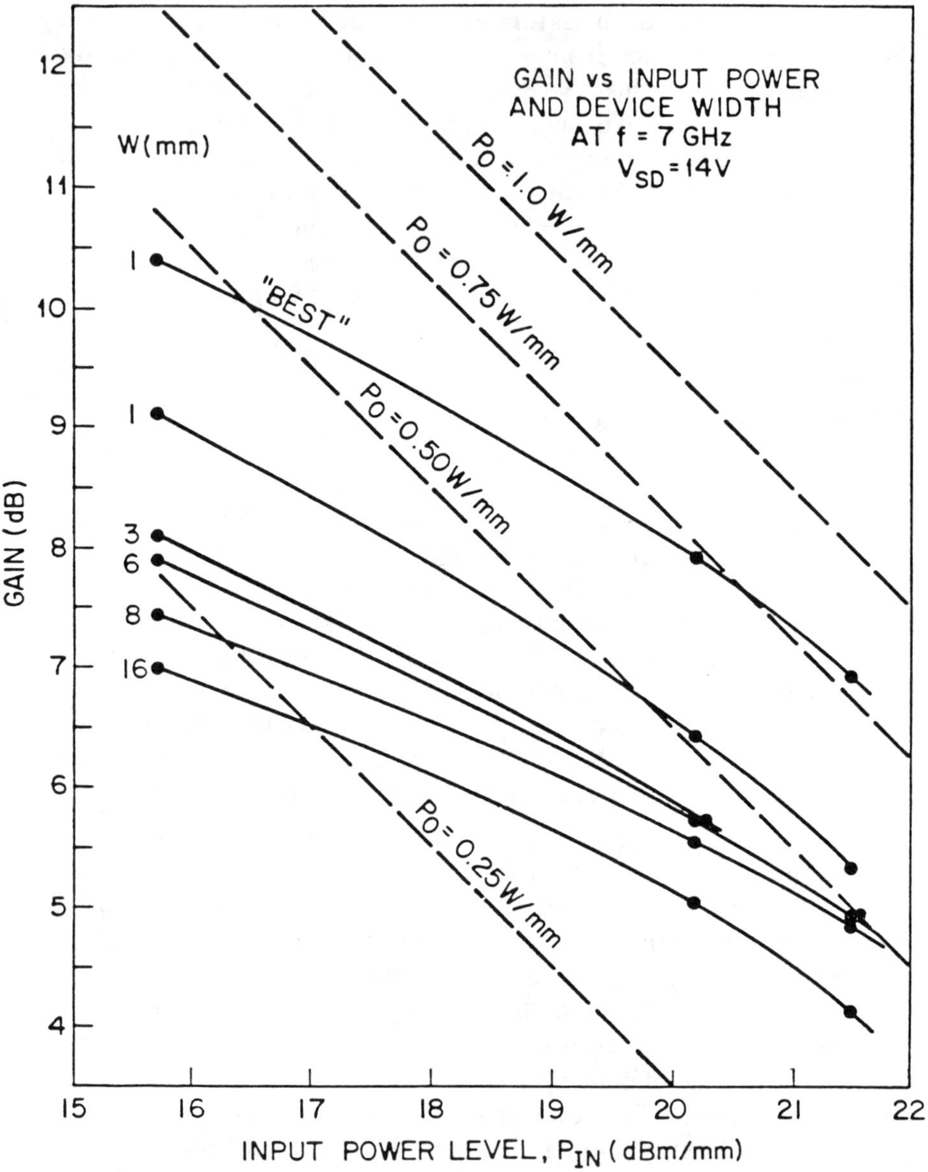

Figure 9/Gain, G(dB), as a function of input power, P_{IN} *(dBm/km), and total device width,* W_T *(mm), at 6 GHz. The device cell width is 500 μm.*

9. SUMMARY

We have reviewed GaAs materials technology, fabrication processes, S-D channel geometry, metallurgies and the chip thermal design factors important to power FETs. Throughout the discussion we have emphasized that suppression of catastrophic S-D burnout originating at the drain contact and control of G-D avalanche are the key elements for maximum microwave power. Burnout power capability can be increased by using a recessed S-D channel design. Our particular design, which employs an n^+ contact layer with a 2 μm ledge at the drain, has achieved 4 - 5^+ W/mm dc burnout capabilities. The RF output power can be increased by suppressing the onset of G-D avalanche. The proper channel design to do this is based on the following concepts: 1) the epitaxial charge in FET G-D output space should be minimized as a means of achieving avalanche voltages higher than 1-dimensional bulk theory; 2) the epi charge in this G-D output space must be consistent with the charge needed under the gate for maximum RF current swing; and 3) the gate should be recessed in a gate notch to a depth comparable to the surface depletion depth. We find experimentally that the two-dimensional lateral spreading model applies for $Q_u < 2.3 \times 10^{12}$ cm^{-2}, independent of active layer doping ($2 \leqslant N \times 10^{16}$ cm^{-3}) $\leqslant 20$. To better control the etching procedure when tailoring the channel design we have chosen to use the lower end of the doping range. Since we operate our devices at about 100 mA/mm S-D current and 14 V S-D bias, we are led to a channel design with G-D epi thickness corresponding to 350 mA/mm and saturated S-D current at zero bias of about 200 mA/mm. With this design, output power typically continues to increase with S-D bias up to 25V, but non-ideal processing factors such as rough Al gate edges or silicon nitride surface charge still cause avalanche to limit the microwave performance.

We have stressed the importance of proper chip design to avoid current densities in excess of 2×10^5 A/cm^2 and temperature rises greater than 50°C. We briefly reviewed our choice of AuGe - ohmic contact, Al-gate, and Ti-Pt-Au-thick metal overlays metallizations as well as silicon nitride as the preferred passivation layer.

Finally, in the microwave section we presented data showing that the saturated power per unit gate width at 4 and 7 GHz is:

1. Independent of gate length for $1 < l_g$ (μm) < 2.5.
2. Independent of gate finger width for $125 < W_f(\mu m) < 500$.
3. Independent of total gate width, W_T, up to 8 mm and suffers a 0.5 dB loss at 4 GHz and 1.2 dB loss at 7 GHz in going to 16 mm widths.
4. Reduced only 1.3 dB when comparing 7 GHz to 4 GHz results for 1-8 mm devices and 2 dB for the 16 mm devices.

As a result, we concluded that microwave performance at 4 and 7 GHz does not suffer by using devices with 500 μm gate widths and 2 μm gate lengths.

We observed that gain for low RF compression levels at 4 and 7 GHz for low RF input levels:

1. Shows a general trend for reduced gain ($\approx$ 2 dB) with increased size ($1 \leqslant W_T$ (mm) $\leqslant 16$). The noted exception was the 4 GHz, 16 mm result which had an overall 3.5 dB drop compared to the 1 mm device.
2. Is reduced less than predicted by small signal theory.
3. Is independent of gate finger width over the range, $125 < W_f$ (μm) < 500.
4. Seems independent of finger widths to 500 μm to total widths up to 8 mm.

The optimum design for a power FET is, indeed, a complex interplay of many factors. These BTL state-of-the-art devices: 1) operate at relatively low bias currents (100 mA/mm) and high bias voltages (14 V), 2) have large gate dimensions (2 μm x 500 μm), and 3) are fabricated from relatively low doped active layers, ($3 < N$ (x 10^{16} cm^{-3}) < 7). Generally, the FET design reported by others: 1) employs a high current (200 mA/mm), low voltage (10V) bias condition, 2) has short gate dimensions (1 μm x 200 μm) and 3) uses $\geqslant 10^{17}$ cm^{-3} doped active layers. On the other hand the BTL devices are similar to some other devices in their S-D recessed structures, metallurgy, and chip design.

ACKNOWLEDGMENTS

The authors would like to acknowledge G.E. Mahoney, W.R. Robertson, P.F. Sciortino and J.R. Velebir for their processing expertise, J.A. Seman for mask design, J.P. Beccone, D. E. Iglesias, F. Magalhaes, D.J. Talley and V.M. Wagner for their packaging and testing support, and the Model Lab personnel for their attention to detail which make a material and device development effort such as this possible.

REFERENCES

[1] W.C. Niehaus et al., *Gallium Arsenide and Related Compounds (St. Louis) 1976.* (Inst. Phys. Conf. Series No. 33b) pp. 271-280. London: Adam Helger/The Institute of Physics,1977.

[2] W. Niehaus, Fourth Annual European Microwave Conference, Vienna, April 1977.

[3] S. Wemple et al., *Electronic Letters,* 14:175, 1978.

[4] J.V. DiLorenzo, W.R. Wisseman, *IEEE Transactions on Microwave Theory and Techniques,* MTT-27:367, 1979.

[5] M. Fukuta et al., *IEEE Trans. Microwave Theory and Techniques,* MTT-24:312, 1976.

[6] H.M. Macksey et al., *IEEE Trans. Electron Devices,* ED-24: 113, 1977.

[7] F. Hasegawa et al., *ISSCC Technical Digest,* 118, 1978.

[8] H.M. Macksey et al., presented at 1978 Workshop on Compound Semiconductor Microwave Materials and Devices (WOCSEMMAD), February 1978.

[9] M. Fukuta et al., *IEEE Trans. Electron Device,* ED-25:559, 1978.

[10] H.M. Cox, J.V. DiLorenzo and L.A. D'Asaro, Paper presented at the Gallium Arsenide IC Symposium, Lake Tahoe, Nevada, Sept. 1979.

[11] H.M. Cox, J.V. DiLorenzo, *Gallium Arsenide and Related Compounds (St. Louis) 1976.* (Inst. Phys. Conf. Ser. 33b). London: Adam Helger/The Institute of Physics, 1977.

[12] S.M. Wemple, W.C. Niehaus, *IEEE Trans. on Electron Devices,* ED-27, no. 6:1013, 1980.

[13] L.A. D'Asaro, J.V. DiLorenzo, and H. Fukui, *IEDM Technical Digest,* December 1977, pp. 370-371; and *IEEE Trans. Electron Devices,* ED-25:1218, 1978.

[14] S.H. Wemple and W.C. Niehaus, *Gallium Arsenide and Related Compounds (St. Louis) 1976* (Inst. Phys. Conf. Series No. 33b), pp. 262-270. London: Adam Helger/The Institute of Physics, 1977.

[15] T. Furutsuka, T. Tsuji and F. Hasegawa, *IEEE Trans.*, ED-25: 563, 1978.

[16] R. Yamamoto, A. Higashisaka, F. Hasegawa, *IEEE Trans. Electron Devices*, ED-25:567, 1978.

[17] S.H. Wemple et al., in press.

[18] K. Morizane, M. Dosen, Y. Mori, in *Gallium Arsenide and Related Compounds (St. Louis) 1978* (Inst. Phys. Conf. Ser. No. 45), p. 287. London: Adam Helger/The Institute of Physics, 1978.

[19] H. Fukui et al., IEDM Technical Digest (Late News), Washington, DC, December 3-5, 1979. *IEEE Transactions on Electron Devices*, in press.

[20] H. Fukui et al., in *18th Annual Proc. Reliability Physics*, 151, 1980.

[21] A.S. Jordan, J.C. Irvin, and W.O. Schlosser, *18th Am. Proc. Reliab. Phys. Symp.*, Las Vegas, 123, 1980.

[22] H.M. Macksey, L.W. Joy, *RADC-TR-80-124 (Rome Air Development Center Interim Technical Report)*, April 1980.

[23] W.C. Niehaus, in press.

[24] A.R. Tretola, J.C. Irvin, *J. of Appl. Physics*, 39:3563, 1968.

[25] J.C.M. Hwang, in press.

[26] K. Chino, Y. Wada, *Japan J. of Appl. Phys.*, 16:1823, 1977.

[27] A. Christou, K. Sleger, Sixth Biennial Conference on Active Microwave Semiconductor Devices and Circuits. Cornell University, Aug. 16-18, 1977.

[28] K. Ohata, M. Ogawa, *12th Annual Proc. Rel. Phys. Symposium (IEEE)*, 278, 1974.

[29] D.A. Abbott, J.A. Turner, *IEEE Trans. on Microwave Theory and Techniques*, MTT-24:317, 1976.

[30] R. Lundgren, *RADC-TR-78-213*, Oct. 1978.

[31] M. Macksey, F. Doerbeck, R. Vaele, *Trans. on Electron Devices*, ED-27:467, 1980.

[32] J.C. Irvin, A. Loya, *Bell System Technical Journal*, 57:2823, 1978.

[33] A.T. English, K.L. Tai, P.A. Turner, *J. Appl. Phys.*, 45:3747, 1974.

[34] S.P. Muraka et al., *J. of Electrochemical Society*, 125:156, 1978.

[35] S.H. Wemple, and H.C. Huang, contribution in this volume.

[36] H. Fukui, in *1980 IEDM Technical Digest,* Washington, D.C., December 8-10, 1980.

[37] I. Drukier et al., *Electron. Lett.*, 104, 1975.

Chapter 5

Thermal Design of Power GaAs FETs

Introduction

In this chapter, S.H. Wemple and H.C. Huang discuss the theoretical and experimental approaches to thermal design of power GaAs FETs. Wemple and Huang discuss various design options associated with gate finger width, gate-to-gate spacing, substrate thickness and the two ways to mount power GaAs FETs, the "normal" backside mounting and the "flip-chip" mounting techniques. The measurement of thermal resistance by emitted infrared radiation, liquid crystal techniques, and electrical techniques are discussed in detail. They conclude that the most versatile and useful technique for component manufacturers is the electrical technique.

The authors of this chapter point out that the thermal design principles for power GaAs FETs are rather complex. An issue of much debate among FETs designers is whether "flip-chip" or backside mounting techniques give lower thermal impedance. The conclusion reached in this chapter is that a thin (25-50μm) substrate backside mounted gives the lowest thermal impedance, but requires a technology that necessitates handling rather fragile GaAs wafers. Gate-to-gate spacing is the critical design parameter but there are constraints such as current density in source or drain stripes which may influence or override gate-to-gate spacing.

Thermal Design of Power GaAs FETs

S. H. Wemple
Bell Laboratories, Murray Hill, New Jersey.

H. Huang
RCA Laboratories, Princeton, New Jersey

1. INTRODUCTION

As with any power device, the ultimate limits on the performance of power GaAs FETs are imposed by thermal considerations. In this chapter we review both theoretical and experimental approaches to thermal design with emphasis on the various trade-offs and layout options available to the transistor designer. Our intent is to be comprehensive in the sense that full three-dimensional solutions to the heat flow equation are presented in simple graphical form. The thermal conductivity temperature dependence is included exactly using the Kirchhoff transformation technique described by Joyce [1]. These computed results display various design options associated with gate finger width (the long dimension), gate-to-gate spacing, substrate thickness, "normal" backside mounting, and "flip-chip" mounting. It is worth noting at the outset that there is an unavoidable adjustable parameter in the theory, *viz.*, the heat source dimension, d, in the direction of source-drain current flow. Although, as we shall show, thermal impedance depends rather weakly on d, it is not legitimate, although it is common practice, to equate d with the metallurgical gate length. Our approach is to adjust d by fitting the theory to measurements on a specific transistor layout and then to use this value for all subsequent calculations. Comparison with experiment is then made using thermal impedance values reported in the literature for a variety of transistor designs including "normal" and "flip-chip" configurations. The experimental techniques described include electrical, infrared, and liquid crystal methods.

Although it is not our purpose to present here an exhaustive treatment of the reasons for keeping the channel temperature as low as possible, it is self-evident that, by doing so, both performance and reliability benefits can be achieved. For example, the transistor transconductance, g_m, is proportional to the saturated electron drift velocity, V_s, and the latter quantity varies approximately as the reciprocal of the absolute channel temperature, T, [2] i.e., $g_m \sim T^{-1}$. More importantly, perhaps, any thermally activated degradation or failure mechanism will be minimized by keeping the channel temperature as low as possible. As an illustrative example, we note that reducing the channel temperature from 120°C to 100°C yields a factor of 4.9 improvement in median life for a process having an activation energy of 1 eV.

For reliability assessment involving accelerated temperature aging, it is important to know the actual channel temperature with reasonable accuracy so that realistic extrapolations to normal operating conditions can be made. In this area we feel that the technique described herein for including the temperature-dependent thermal conductivity will be of value.

This paper is organized as follows. First, we present both simplified thermal models and "exact" three-dimensional computer calculations. Measurement techniques which include optical, liquid crystal, and electrical methods are described and, finally, we compare theory with experiment and make a detailed evaluation of the important thermal design parameters.

2. ANALYSIS

2.1 Analytical Solutions

As an introduction to the subject of heat flow from thin line sources, it is useful to begin with an infinitely long planar line source having a narrow dimension, 2d, (or, equivalently, a half-cylinder source of diameter, d) placed on an infinite slab of thickness C (see Figure 1). Cooke [3] has noted that an analytical solution to this, or any multiple line problem, can be extracted directly from the microwave stripline literature [4, 5], since the thermal and electrical problems obey the same partial differential equation [3, 6]. Straightforward comparison of terms yields the following general result:

$$\theta = k^{-1} Z/Z_0 \ , \tag{1}$$

where θ is the thermal impedance of a unit line length in °C-mm/W, k is the substrate thermal conductivity in W/°C-mm, Z is the characteristic impedance for the analogous electrical line, and Z_0 (=377 ohms) is the characteristic impedance of free space.

For a single infinite line, Equation 1 becomes [4, 5]

$$\theta = (2k)^{-1} K_1(p)/K_1(s) \ , \tag{2}$$

where K_1 is the complete elliptic integral of the first kind, and

$$p = \sinh\ [\pi(2d)/4C] \ , \tag{3}$$

$$s = \tanh\ [\pi(2d)/4C] \ . \tag{4}$$

Note that (2d) equals the narrow dimension of the heat source. Equation 2 can be reduced to a simple logarithmic form [6, 7] when, as is generally the case, $d \ll C$,

$$\theta = (\pi k)^{-1} \ln \frac{|8C|}{|\pi d|} \tag{5}$$

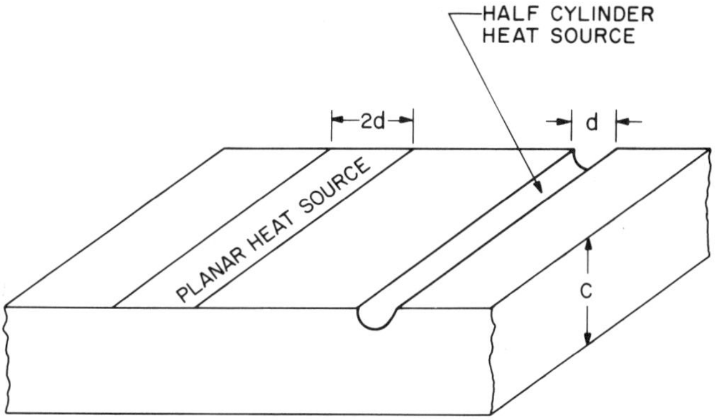

Figure 1/Schematic of equivalent planar and half-cylinder line heat sources on an infinite dielectric slab of thickness C.

This expression is accurate to better than 10% even for the extreme case when d = 4 μm and C = 25 μm. We have plotted Equation 5 in Figure 2 taking k^{-1} = 26.3 °C-mm/W corresponding to the thermal resistance of GaAs at 60°C [8]. Figure 2 serves to illustrate the relatively weak logarithmic dependence on heat source dimensions and substrate thickness associated with a single narrow infinite line. The dimension, d, in Equation (5) is either half the stripe width for a flush surface source [4] or the diameter of a half-cylinder source penetrating a distance d/2 below the surface [5]. In actual transistors, of course, all lines are of finite length so that end cooling may be substantial and should not be neglected. Furthermore, multiple gates are often placed in parallel, leading to a reduction in the heat

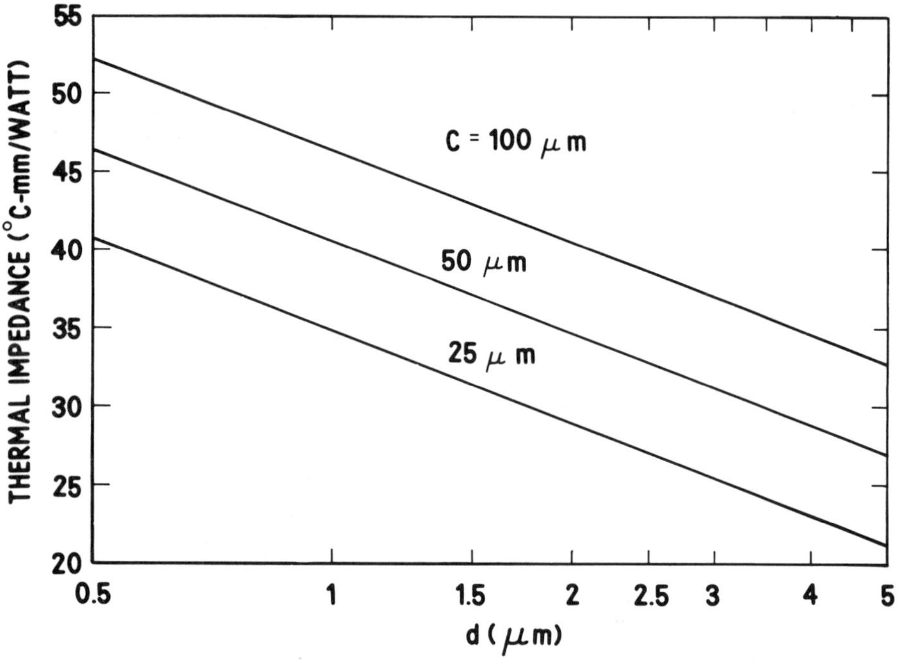

Figure 2/Calculated thermal impedance based on Equation (5) as a function of heat source dimension, d, for the infinite lines shown in Figure 1. The quantity, C, is the substrate thickness in μm.

transfer area from the semi-infinite slab available for a single gate to the average gate-to-gate spacing G. We consider this more complicated situation below, but at this point it is worth noting that the logarithmic dependence on the heat source dimension, d, shown in Figure 2, remains essentially unchanged in the more complicated multi-gate geometry. Only the thermal impedance scale and substrate thickness dependence are altered.

Before turning to the three-dimensional computer calculations, we first present thermal conductivity data for GaAs and then show how the relatively strong conductivity temperature dependence can be accounted for exactly without solving a non-linear partial differential equation. The thermal conductivity curve k(T) shown in Figure 3 has been taken from the measurements of Maycock [8]. In Figure 3, the $T-T_0$ scale applies to the temperature-dependent thermal conductivity analysis which we now consider. In an actual FET chip, particularly during accelerated temperature aging, the temperature varies between the channel value and the heat sink value so that it is not initially clear what average thermal conductivity is appropriate. This apparent difficulty is easily circumvented using the Kirchhoff transform technique described by Joyce [1]. As is well-known, the true temperature distribution satisfies the nonlinear heat flow equation

$$\nabla \cdot k \; \nabla T = 0 \, , \tag{6}$$

whereas the foregoing analysis, and the computer solutions described below, satisfy only the constant-conductivity form of Equation (6), i.e.,

$$\nabla^2 \tau = 0 \, . \tag{7}$$

Joyce has shown, however, that exact solutions to Equation (6) can be obtained from solutions to Equation (7) by using the following transform:

$$\tau(T) = T_0 + k_0^{-1} \int_{T_0}^{T} k(T)\, dT \, , \tag{8}$$

where T_0 and k_0 refer to the heat sink temperature and associated GaAs thermal conductivity, respectively. Substitution of the k(T) data shown in Figure 3 into Equation (8) yields the $\tau(T)$ graphs also

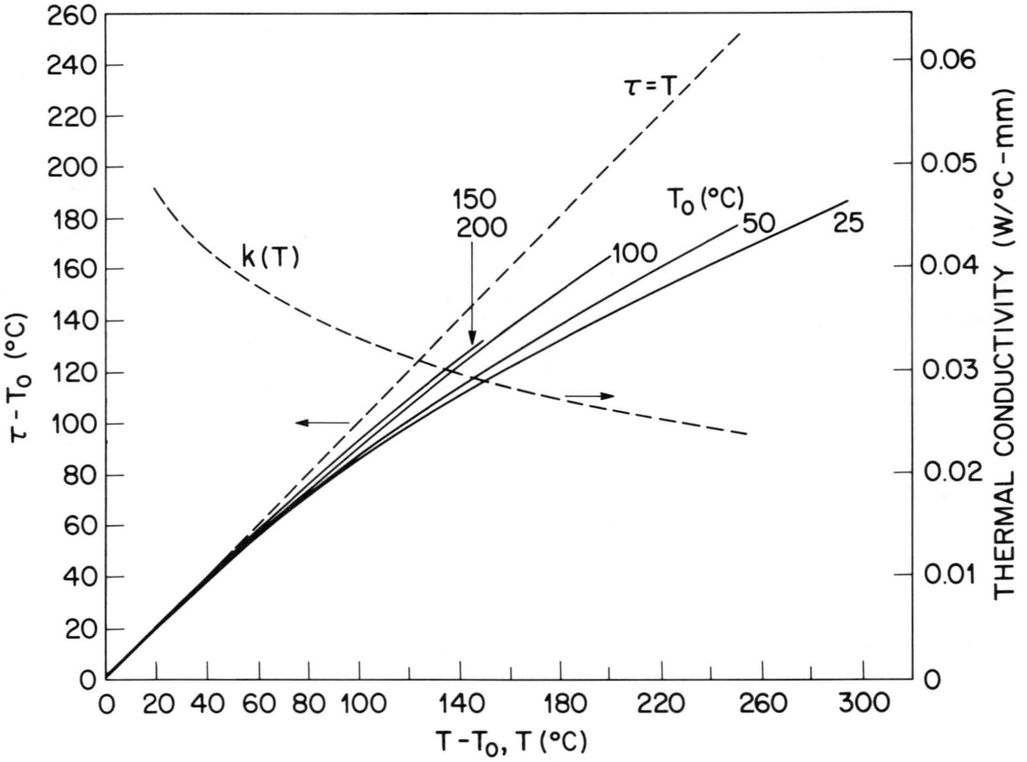

Figure 3/*Thermal conductivity (k) of GaAs (Reference 8) as a function of temperature T(°C). Also shown is the linear temperature rise* (τ-T_o) *based on Equation (7) as a function of the nonlinear (true) temperature rise* (T-T_o) *based on Equation (6) for various ambient temperatures* T_o*(°C).*

displayed in Figure 3. From these curves, it is a simple matter to obtain the exact channel temperature, T, from the constant-conductivity result, τ. For example, a computed temperature rise of 100°C actually corresponds to a true temperature rise of 120°C when T_0 = 50°C.

2.2 Computer Solutions

Because a realistic power transistor layout generally includes many gates having finite widths (the long dimension) and either equal or unequal spacings, any quantitative thermal analysis must resort to a full computer solution to Equation (7). The geometry of interest is shown in Figure 4. Here we show only four gates of a multi-gate device. In general, the gates may or may not be equally spaced, and both possibilities are analyzed in detail. We assume that the end gates (or channels) are sufficiently remote to allow a solution restricted by symmetry to a slab of width G, where G is the average gate-to-gate spacing. For the equal gate spacing case, it is clear that the solution can be restricted to a slab of width G/2.

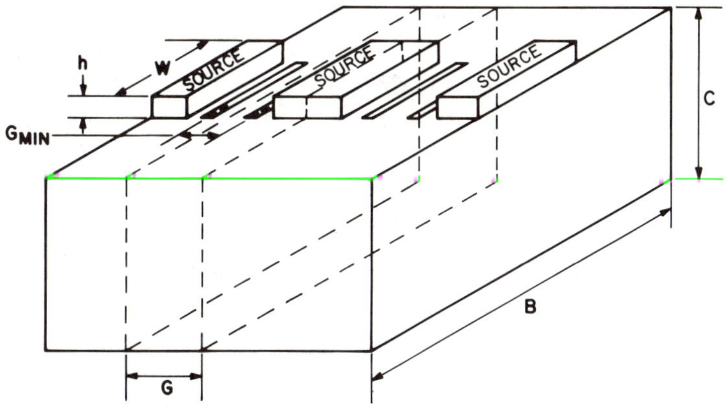

Figure 4/Schematic illustration of transistor metallizations and definitions of dimensions used in computer calculations. The drain metallizations (not shown) fall between each gate pair.

There are two thermal design options that yield viable alternatives for extracting heat from the channel regions. These are (1) provide a thin chip and extract heat through the GaAs and out the bottom, and (2) provide thick top surface metallizations, as shown in Figure 4, and extract heat out the top. Under these conditions the chip need not be thinned. The latter approach is often called flip-chip mounting since it is structurally advantageous to attach the source bonding posts to a rigid heat sink located underneath the chip. In Figure 4 we show the common flip-chip arrangement which extracts heat only through the rectangular source posts. The drain metallizations (not shown) are thin and remain both electrically and thermally isolated from the heat sink. It should be noted that the heat flux generated in the channel region must first traverse the GaAs to the sources, then conduct through the Au source posts, and finally enter the Cu heat sink. The latter spreading resistance, as noted in the literature [9], can be an important contributor to the overall thermal impedance of the flip-chip design.

Solutions to Equation (7) have been obtained for both flip-chip and backside mounting configurations using a versatile thermal program developed by Dietzel et al. [10]. The inputs include any combination of rectangular heat sources and heat sinks located on the top surface of a rectangular slab of dimension G x B x C (see Figure 4). For our calculations, we have made the reasonable assumption that all surfaces not in contact with a heat sink are thermally insulated. Thus, for the flip-chip case, no heat is extracted through the bottom, whereas all the heat passes through this surface and into the Cu heat sink in the backside mounting geometry. It has been further assumed that the thermal impedance for the latter process can be neglected.

The single adjustable parameter on all our calculations is the heat source dimension d (Figure 1). This quantity has been fixed by comparing calculations for a specific backside mounted 6 mm periphery Bell Laboratories FET with the measurements of Fukui [11]. These calibration measurements, which were made using the forward gate current method described in the next section, are shown in Figure 5, along with computed results for three different values of d. The best fit is obtained when d = 4 μm, and this value, which is double the gate length, has been used in all subsequent calculations. Note that this result can be interpreted either as a semi-cylindrical heat source with a diameter of 4 μm or an 8 μm flat heat source. The Bell Laboratories FET has a gate length of 2 μm and a source-to-drain spacing of 8 μm. Clearly, the thermal pattern in the channel cannot be viewed as a simple flat heat source coincident with the gate metallization. Note that a value d = 2 μm rather than 4 μm yields about 5°C-mm/W increase in θ (see Figure 2).

Computed thermal impedance values are presented in Figures 6-8 as a function of average gate-to-gate spacing, G. These graphs have been computed using k = 0.038 W/°C-mm corresponding to a heat sink temperature of 60°C. If other heat sink temperatures are desired, then it is merely necessary to adjust the thermal impedance scale by the associated reciprocal thermal conductivity ratio. For example, according to Figure 3, k = 0.045 W/°C-mm at T_0 = 25°C, leading to a 16% *reduction* in the thermal impedance scale of Figures 6-8. If the computed temperature rise falls into the nonlinear region shown in Figure 3, then the graphs presented there can be used as outlined above to extract the additional temperature rise associated with the thermal conductivity temperature dependence.

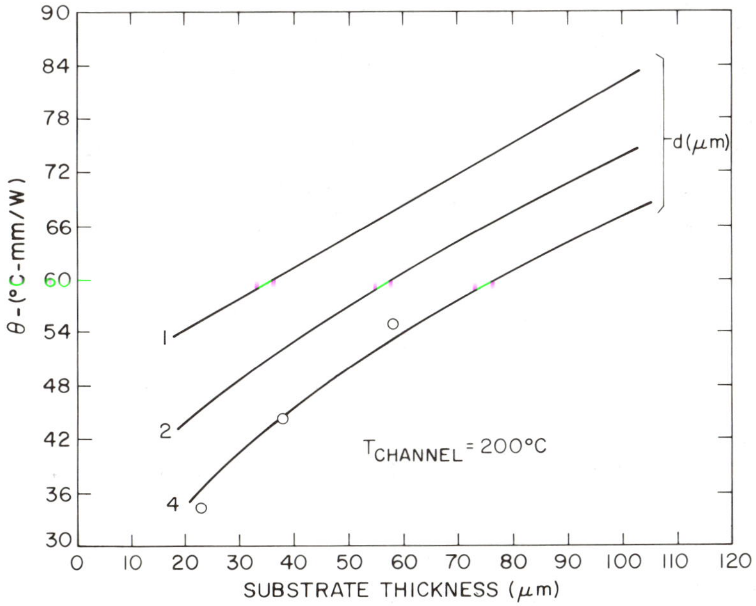

Figure 5*/Calibration data used to adjust the heat source dimension, d, in the computer calculations. Data taken from Reference 11.*

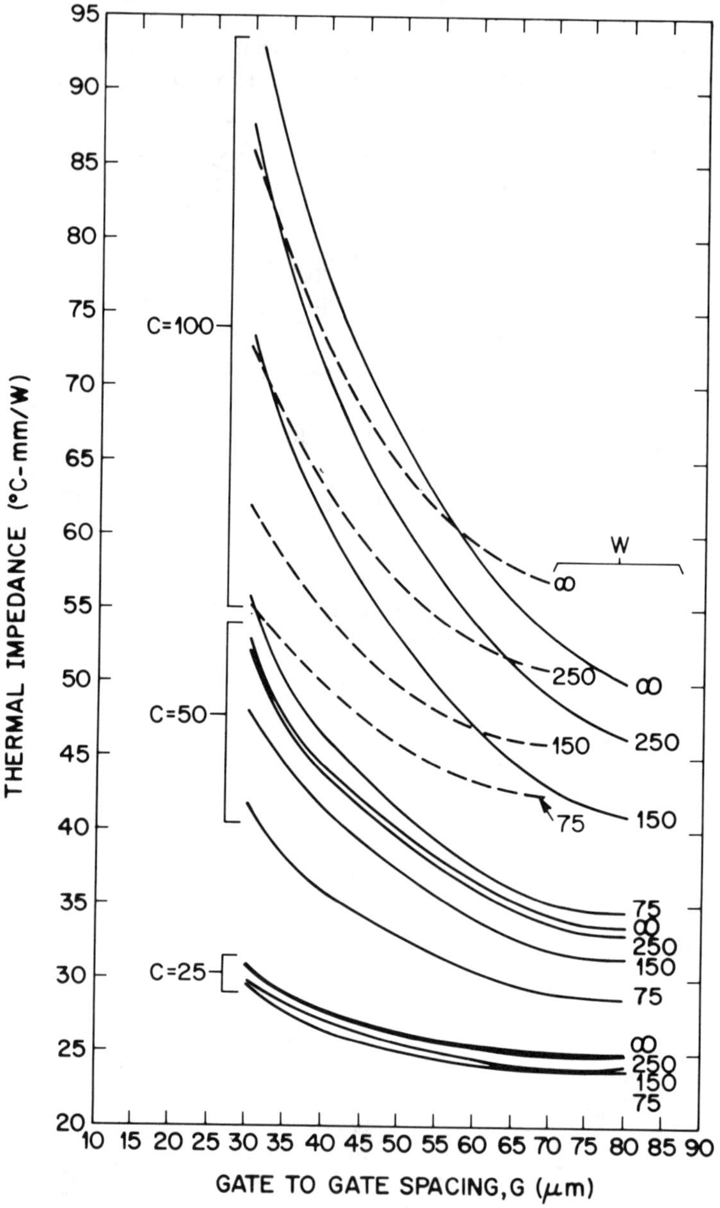

Figure 6/Thermal impedance as a function of uniform gate to gate spacing (G) for various substrate thicknesses (C) and gate finger widths (W). Backside mounted (solid lines) and flip-chip (dashed lines) configurations are included. For these graphs k = 0.038 W/°C-mm.

Figure 6 applies to an equal gate separation geometry both for flip-chip (dashed lines) and backside mounting (solid lines) configurations. In the flip-chip case, we have assumed that the chip thickness is 250 μm, the Au source posts are 20 μm thick, and the Cu heat sink is 500 μm thick. The final results depend very little on these nominal values. Figures 7 and 8 present unequal gate separation cases for backside mounting and flip-chip geometries, respectively.

To illustrate the use of these graphs, we consider a backside heat sink with a 50 μm thick substrate, 150 μm finger width, and evenly spaced gates separated by 35 μm. In addition, we will assume 2 W/mm power dissipation at a 100°C heat sink temperature. From Figure 6, we see that the nominal thermal impedance is 45°C-mm/W. Correcting for the thermal impedance value at 100°C (0.033 W/°C-mm) yields θ = 52°C-mm/W(= .038/.033 x 45) or a temperature rise of 104°C. The final correction for thermal conductivity temperature dependence taken from Figure 3 then gives a predicted 115°C temperature rise at the center of the channel.

After reviewing experimental techniques in the following section, we will return to a comparison of experimental results with the predictions of Figures 6-8.

3. THERMAL RESISTANCE MEASUREMENT TECHNIQUES

The thermal resistance $\bar{\theta}$ of an FET is generally determined using the relation

$$\bar{\theta} = \frac{T_c - T_0}{P_{dis}} , \tag{9}$$

where T_c is the measured channel temperature, T_0 is the temperature of the mount, and the power dissipated, P_{dis}, is given by $V_{dc} I_{dc}$ + $P_{in}^{rf} - P_{out}^{rf}$. For comparison with theory, we focus on the normalized thermal impedance $\theta = W_T \bar{\theta}$, where W_T is the total gate periphery in mm, and the units of θ are °C-mm/W (i.e., °C per W/mm of power dissipation). The channel temperature, T_c, in Equation (9) can have several meanings depending upon the purpose of the thermal investigation. For example, T_c can correspond to the channel temperature averaged over all gates; it can be defined at the highest temperature point, i.e., at the middle of the central gate in a multi-gate device; or

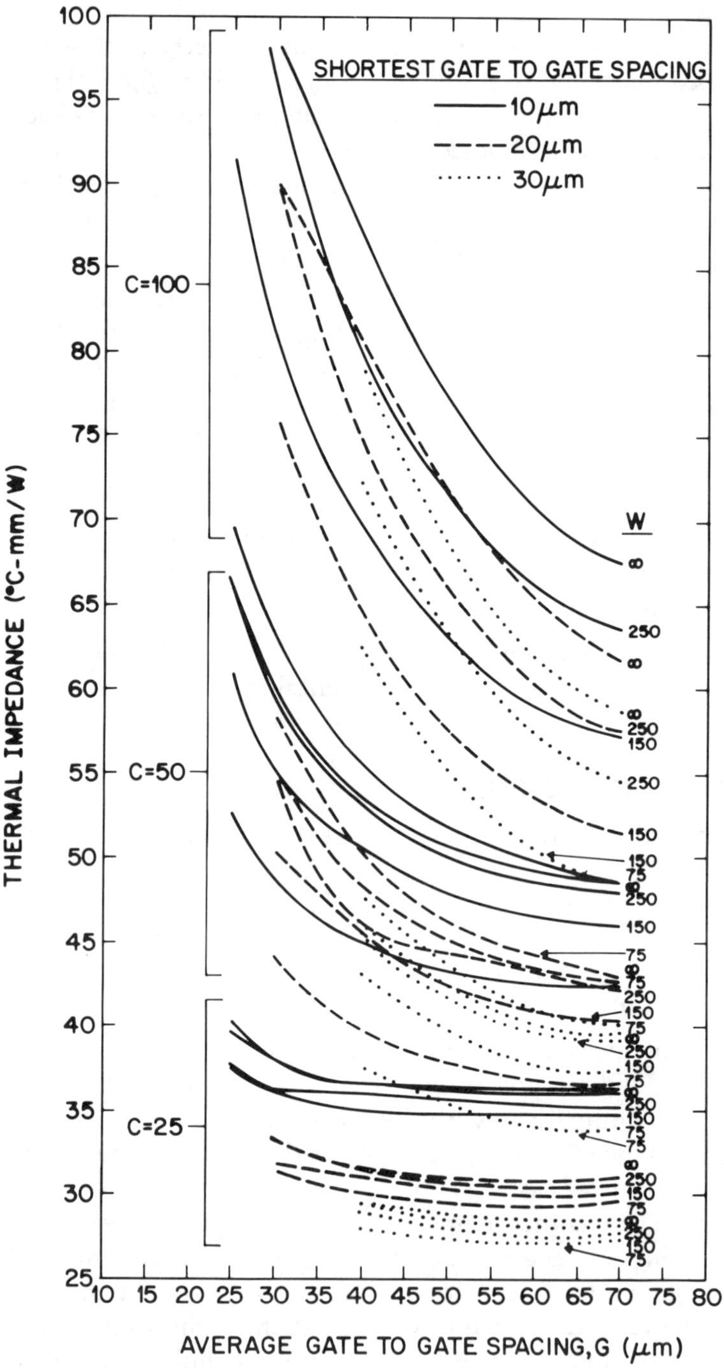
SHORTEST GATE TO GATE SPACING
10μm
20μm
30μm
C=100
C=50
C=25
W
THERMAL IMPEDANCE (°C-mm/W)
AVERAGE GATE TO GATE SPACING, G (μm)
100
95
90
85
80
75
70
65
60
55
50
45
40
35
30
25
10 15 20 25 30 35 40 45 50 55 60 65 70 75 80

T_c can be the temperature at some local hot spot associated, for example, with current nonuniformities induced by processing errors. Although thermal measurements that have the spatial resolution to detect local hot spots may be of interest as a diagnostic tool, our primary interest in this work is in ideal devices which are free of such flaws. The theoretical results presented earlier in Figures 6-8 apply, for example, to the center point of ideal multi-gate devices.

In the remainder of this section, we describe in detail the most commonly used thermal impedance measurement techniques, viz., infrared, liquid crystal, and electrical.

3.1 Infrared (IR) Measurement Technique

The temperature profile of an FET can be measured by detecting the thermal IR radiation emitted from the surface. In a particular implementation [9, 12, 12] an FET is mounted on a computer-controlled motorized X-Y platform. Infrared radiation is then observed at discrete points on the surface with a spatial mechanical resolution of almost 1.6 μm. The temperature measurement begins with a calibration of the emissivity of the FET surface at each point. This is accomplished by uniform heating to several elevated temperatures and scanning along the selected spatial pattern. Having stored all calibration and measurement data in a disc file, the temperature of each point on the surface is then computed using a pre-determined algorithm. Temperature profiles of the device may then be displayed graphically on a hard copy plotter. Since the calibration data are stored in the computer, they may be reused for scanning the same device under different conditions of bias and rf drive.

***Figure 7** (opposite)/Thermal impedance for backside mounted chips as a function of average gate-to-gate spacings (G) for various substrate thicknesses (C), gate finger widths (W), and shortest gate-to-gate spacings (i.e., gate spacings need not be equal). For these graphs* $k = 0.038$ *W/°C-mm.*

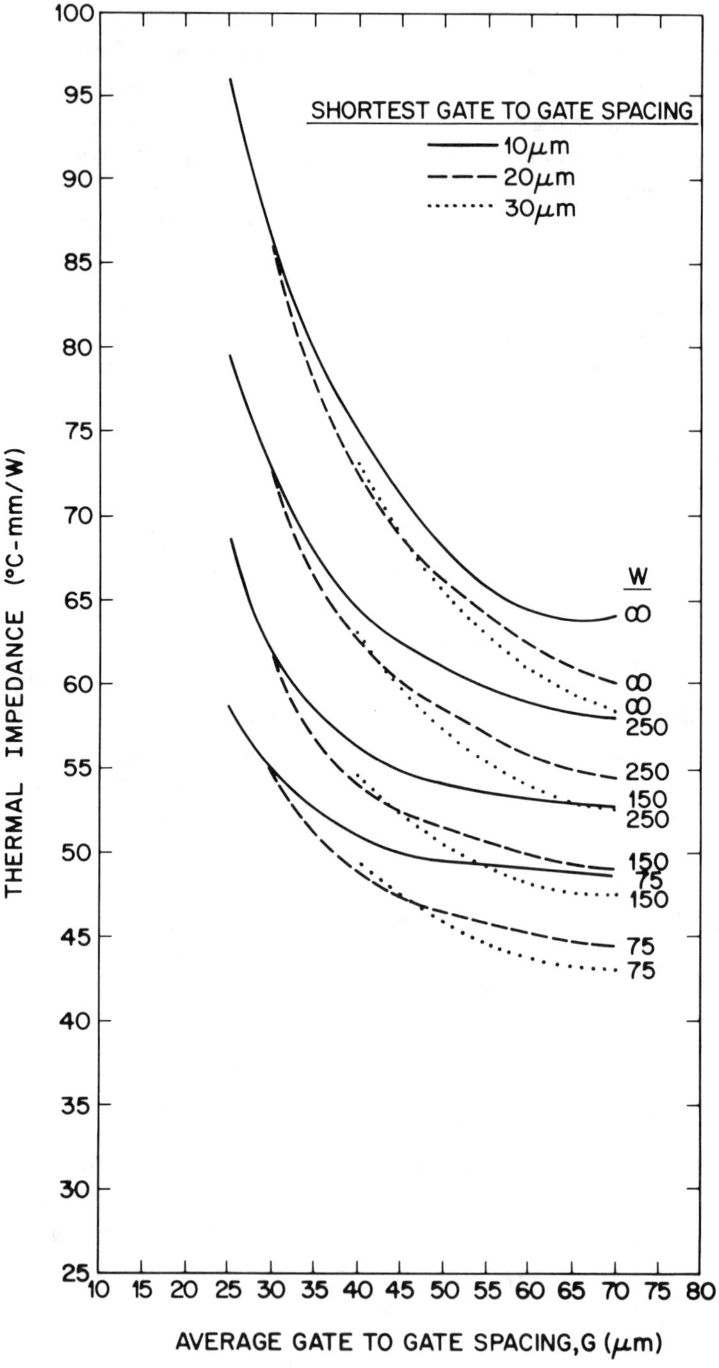
SHORTEST GATE TO GATE SPACING
10μm
20μm
30μm
W
∞
250
150
75
THERMAL IMPEDANCE (°C-mm/W)
AVERAGE GATE TO GATE SPACING, G (μm)
100
95
90
85
80
75
70
65
60
55
50
45
40
35
30
25
10 15 20 25 30 35 40 45 50 55 60 65 70 75 80

***Figure 8**/Thermal impedance for flip-chip configuration as a function of average gate-to-gate spacing (G) for various gate finger widths (W) and shortest gate-to-gate spacings (i.e., gate spacings need not be equal). For these graphs k = 0.038 W/°C-mm.*

The IR measurement technique has both advantages and drawbacks. The spot size resolution is only of the order of 25 μm. This dimension is much larger than the 4-8 μm dimension of the heat source associated with the channel region. For a flip-chip mounted FET, the temperature at the drain (≈12-25 μm long) is approximately the same as that in the channel because there is no heat flow out of the drain. Therefore, the IR measurement should be reasonably accurate for the flip-chip configuration. For the backside mounted FET, on the other hand, the drain temperature is lower than that of the channel. As a result, the temperature measured by the IR technique tends to be lower than the actual channel temperature. The experimental setup, theory of operation, and measurement examples using the IR technique are described below.

The measurement system [13] includes an infrared microscope (Barnes Engineering Co., Model RM-2A) which is shown diagrammatically in Figure 9. Infrared plus visible energy received from the device under test is focused by a reflective objective. A germanium optical filter is used to deflect the visible energy into a separate channel, allowing a clear view of the device under test. The infrared beam to be measured is passed through a chopper and received by an InSb detector cooled to liquid nitrogen temperature (77°K). The output signal from the detector is then amplified and demodulated. The signal at the output of the demodulator gives a measure of the infrared radiation emitted by the device. The optical resolution of the instrument, which is the diameter of the measuring dot, is 10-25 μm, depending on the objective used.

The basic microscope shown in Figure 9 was modified to include: (1) numerical X-Y position control of the substage platform by using a pair of precision stepping motors; (2) a built-in substage heater and temperature controller; and (3) a means for autoranging the sensitivity of the radiometer amplifier. A photograph of the RCA IR system,

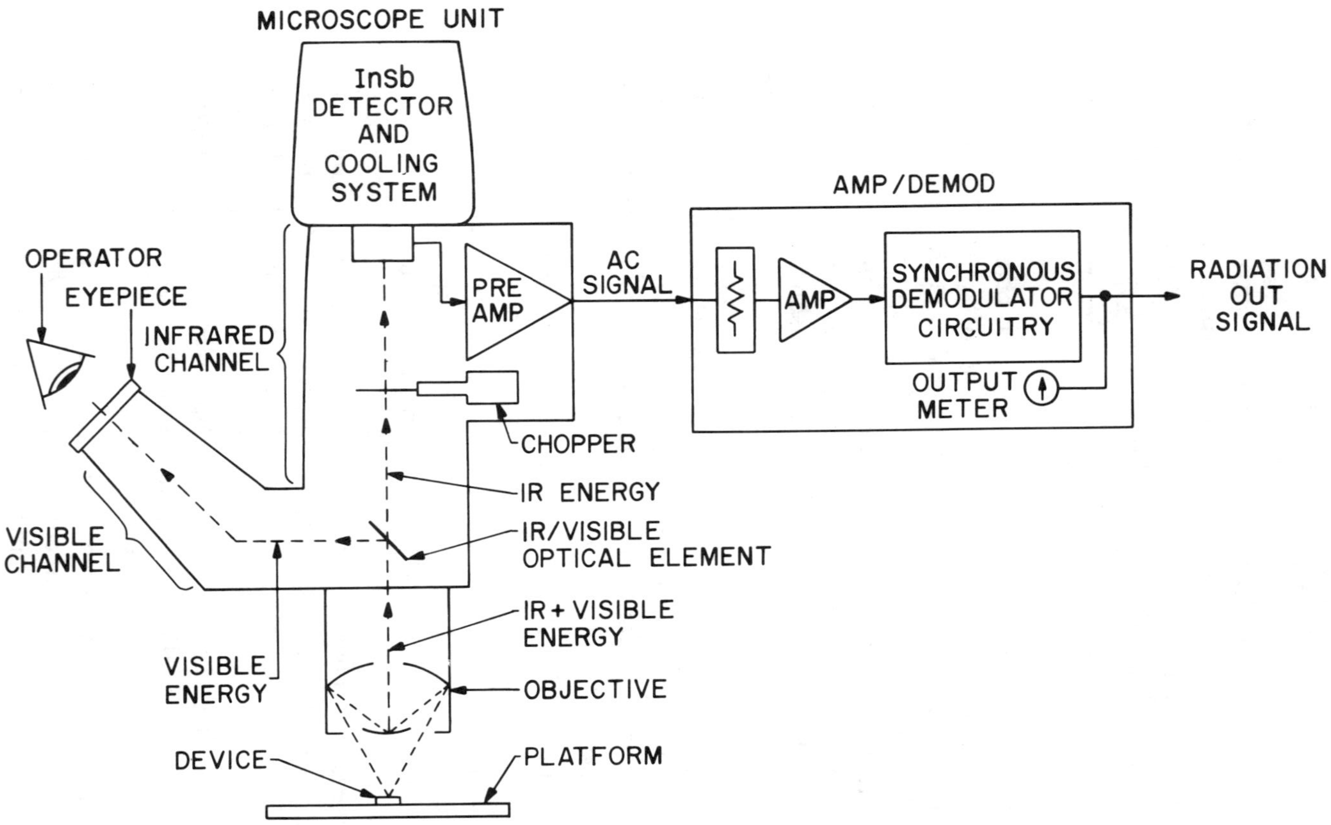

Figure 9/Functional description of infrared microscope.

including its controllers, is shown in Figure 10. The Platform Control Unit (PCU) (left) provides circuitry for motor and heater control with provision for digital readout of relative position offset, while the Radiometer Control Unit (RCU) (right) provides radiometer amplifier control and readout.

A built-in heating element is used to raise the temperature of the top surface of the substage platform up to 200°C. This element is controlled by the PCU. This feature is provided to calibrate the instrument for variations in surface emissivity.

A real-time computer operating system (RTE)* is used to control and monitor the operation of the infrared microscope. A schematic diagram for the automated infrared microscope system is given in

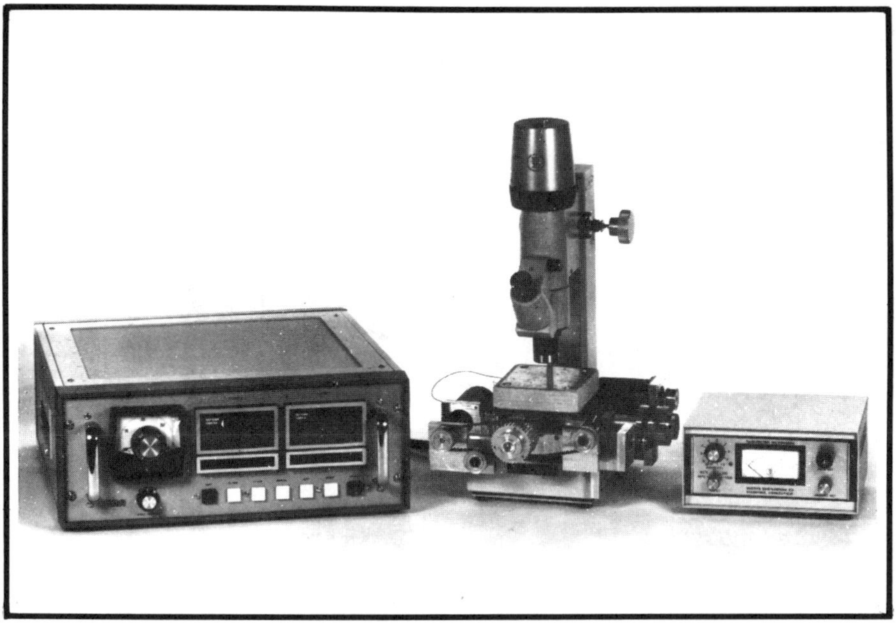

Figure 10*/Photograph of an infrared microscope with the platform control unit (PCU) (left) and the radiometer control unit (RCU) (right).*

*RTE – Real Time Executive operating system – is a product of the Hewlett-Packard Co., Cupertino, California.

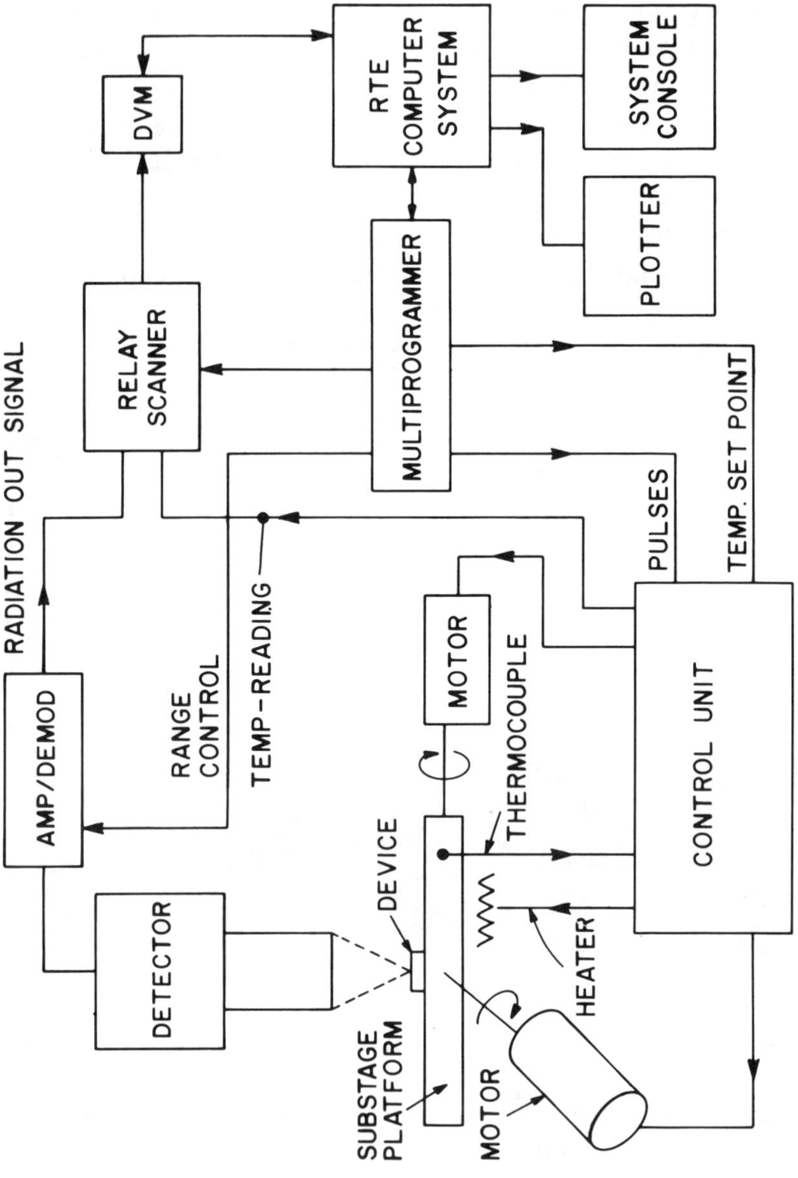

Figure 11/Schematic of automated infrared microscope system.

Figure 11. The multiprogrammer, under the control of the RTE computer, contains specialized hardware for providing: (1) pulse trains for motor translation; (2) a set point voltage level using a D/A converter to control heater temperature; (3) range control of an attenuator in the radiometer amplifier to maintain the output signal within a specified range; and (4) a relay register used as an analog scanner. The computer tracks the position of the substage platform by recording the total number of pulses sent to each motor. Voltages proportional to temperature as sensed by a thermocouple and the radiation output signal are recorded by interfacing the analog scanner to a digital voltmeter.

The temperature profile along the center of the flip-chip FET shown in Figure 12 was obtained using the equipment described above. The peak temperatures correspond to the drain region, and valley temperatures correspond to the source region. The peak temperature of 85°C corresponds to a thermal resistance of this 5-cell (total gate periphery, W_T = 3 mm) FET of 19°C/W (θ = 57°C-mm/W).

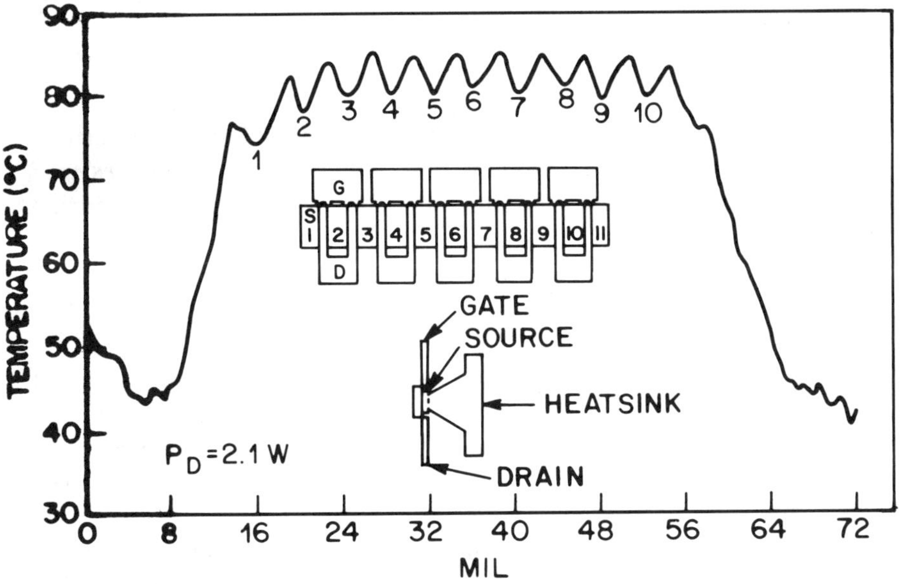

Figure 12/Temperature profile of a 5-cell 4-gate FET.

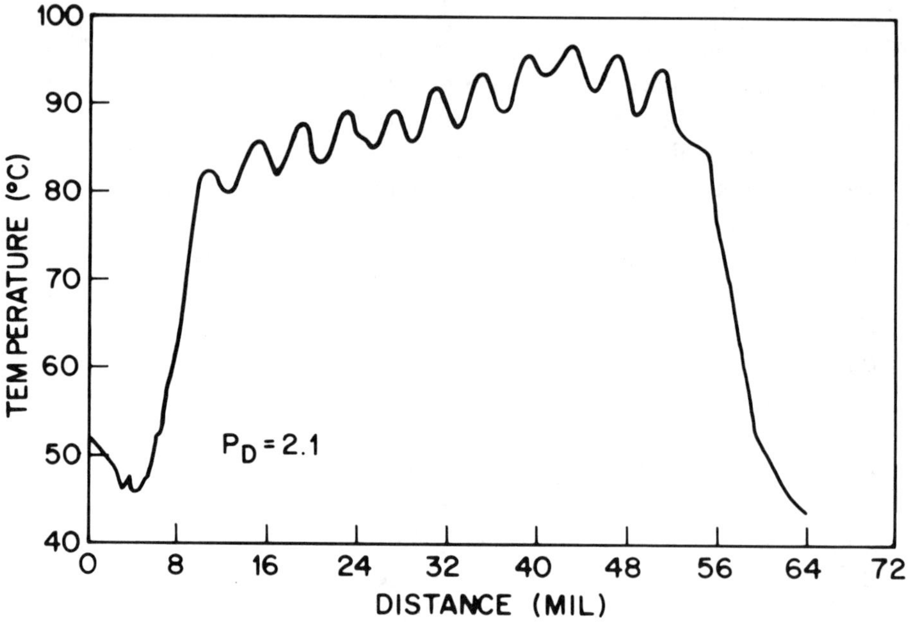

Figure 13/Temperature profile of a high power FET with defective source bonding.

For comparison, the temperature profile of an FET with defective source bonding is shown in Figure 13. Although all the sources are electrically connected, the thermal contact is not uniform, resulting in degradation of the rf performance and reliability. At the hottest region $\overline{\theta} = 24°C/W$ ($\theta = 72°C\text{-mm/W}$).

3.2 Liquid Crystal Measurement Technique

The liquid crystal technique [14] is probably the best method available for measuring the temperature of hot spots on a biased FET. This technique relies on the fact that between the melting point and the isotropic temperature, nematic liquid crystals enter a state in which all the molecules are aligned in the same direction. Above the isotropic temperature the molecules are randomly oriented. This isotropic temperature is characteristic of each liquid crystal material and is accurate to at least ±0.5°C [15]. A number of nematic liquid crystals [16] have isotropic temperatures between 50°C and 200°C and are liquid at room temperature, making them easy to use.

The device temperature is measured by applying a thin layer of liquid crystal to the surface of a mounted and bonded GaAs FET chip to which external bias may be applied. One problem is that if the liquid crystal is too thick, its thermal conductivity will influence the results. The liquid crystal may be applied in a thin layer by first dissolving in a solvent such as methylene chloride or chloroform (~ 10% solution). A drop of this solution is placed on the device and the solvent allowed to evaporate. The liquid crystal will be thicker near bond wires due to surface tension, so for accurate results it is desirable to bond the device such that all bond wires are 100 μm or more from the measurement area. It may be necessary at times to reconstitute the liquid crystal by first flooding with chloroform, followed by draining with a micro-pipette [17].

Following application of the liquid crystal, the device in its test fixture is placed under a microscope objective and is illuminated with polarized light. The light reflected from the device is examined with a rotating analyzer in the light path. The analyzer is rotated so that regions without liquid crystal appear dark, i.e., the analyzer is crossed with the incident light polarizer. When the device covered with the liquid crystal is examined, it doesn't appear dark, however, because the liquid crystal acts as a third polarizer which rotates the plane of polarization of the illuminating light so that a fraction of the reflected light is now able to pass through the analyzer. If a portion of the device is above the liquid crystal isotropic temperature, then the liquid crystal in that region no longer affects the polarization of the light and the surface appears dark. This set-up is shown schematically in Figure 14.

As the power in a device shown in Figure 14 is increased from zero by increasing the drain voltage, the device becomes hotter. When any portion of the surface exceeds the liquid crystal isotropic temperature that part suddenly becomes dark. As the drain voltage is increased further, the dark region expands. By noting the drain voltage and current when the first part of the device becomes dark, the thermal resistance to the hottest part of the device may be determined. This value will be quite accurate because the heat is generated only a few tenths of a micron from the surface, which should be at almost the same temperature as the device channel. As an example, suppose that the first part of the device becomes dark at 5 volt drain bias and 100 mA current, the liquid crystal has a 70°C isotropic tempera-

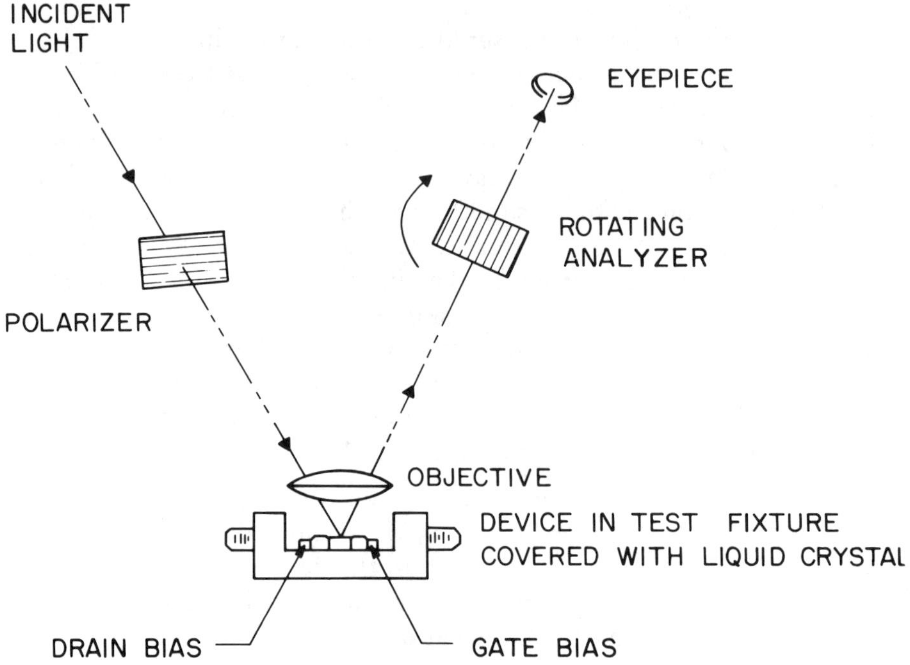

Figure 14*/Schematic diagram of liquid crystal temperature measurement system.*

ture (liquid crystal E7 from BDH Chemicals, Ltd.), and the device heat sink is at 25°C. The thermal resistance to the hottest spot on the device is then 45°C/0.5 watt = 90°C/W.

Using the liquid crystal technique, thermal impedances have been measured for many GaAs power FETs at Texas Instrument Incorporated. Some interesting phenomena are observed with the high spatial resolution (1 μm) associated with this technique. Often the temperature does not uniformly decrease from the center to the ends of a gate stripe, but there are one or more slightly hotter spots. These may be due to slight epitaxial layer or gate recess etch nonuniformities causing regions of slightly higher current and greater power dissipation.

One problem with the liquid crystal method is that the device surface must be accessible so flip-mounted and hermetically sealed packaged devices cannot be checked. In addition, it is more time consuming than other methods, and, while the temperature measurement does not destroy devices, it contaminates them with a liquid crystal which is difficult to remove. Since the method is most accurate when no bond wires are within 100 μm of the area being measured, a specially bonded device, useless for microwave measurements, is often required.

3.3 Electrical Measurement Technique

The IR and the liquid crystal techniques described previously are useful for FET manufacturers because of their capability of measuring temperature profiles, device mounting uniformity, and hot spots. The most versatile and most useful technique for component manufacturers is probably the electrical measurement method described in this section [18].

This technique utilizes the temperature dependence of the FET electrical characteristics as a calibration parameter for obtaining the channel temperature. The most common electrical characteristic used for this application is the forward biased gate voltage at a fixed value of gate current.* The measurement begins with calibration of forward gate voltage as a function of temperature. The FET should be mounted in an amplifier fixture with input and output matching circuits to prevent bias circuit oscillations. The drain and the gate of the FET are connected to specially-designed bias control circuits and the source is grounded. During calibration the drain is isolated from the source by an MOS switch in the power supply; therefore, only the forward gate Schottky barrier circuit to source is characterized. At each case temperature, a pulsed forward gate current from a constant current supply is applied to the gate-source terminal. In a particular implementation [18], the pulse width of the gate current pulse is 10 μs, and the repetition rate is 50 Hz (Figure 15). The magnitude of the forward current corresponds to a forward gate voltage roughly equal to the gate Schottky barrier voltage (i.e., 0.5-0.8 volts). As an example, the forward gate current was chosen to be 7 mA for measuring the FLC-30MA FET (manufactured by Fujitsu, Inc., Tokyo, Japan) which has a total gate periphery (W_T) of 7.28 mm and is rated at 3 W output power in C-band. The corresponding for-

*Commercial apparatus for performing this measurement is manufactured by Sage Enterprises, Mountain View, CA 94043.

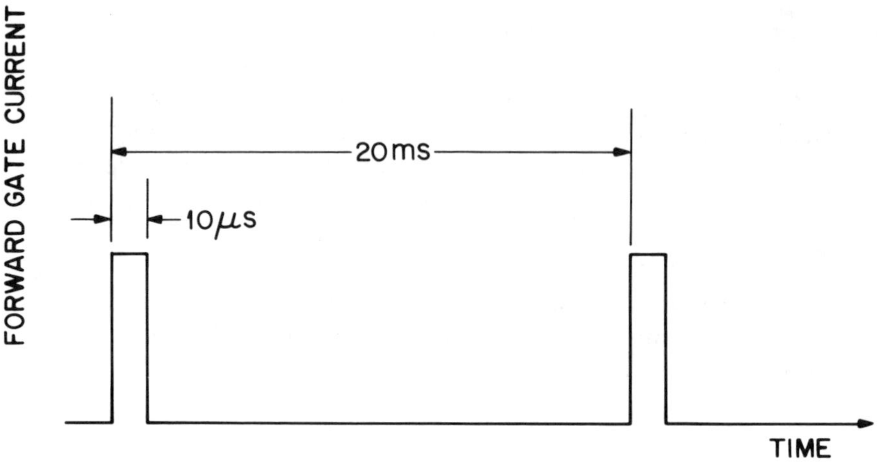

Figure 15/Constant gate current pulse for calibration of electrical thermal impedance apparatus.

ward gate voltage in this case was about 0.6 V. Arranged in this way, the forward gate voltage of the unbiased FET as a function of temperature can be obtained. This forward gate voltage decreased linearly with increasing temperature. A detailed measurement example will be given later in this section.

After the forward gate voltage versus temperature relationship is obtained, the FET is biased to a particular operating point with the mount set at the desired temperature. Figure 16 shows the drain and gate bias pulses. The drain voltage is held constant except for a very short time during which the forward gate voltage corresponding to 7 mA (or some other selected value) forward gate current is measured. In this interval, the drain bias is removed and the gate bias is switched from the desired operating voltage to a constant current source of 7 mA. A comparison of this forward gate voltage with the previously obtained calibration curve yields the FET channel temperature. The thermal resistance corresponding to the particular channel temperature can be easily calculated from Equation (9). It should be recognized that this method determines the average channel temperature, not the peak.

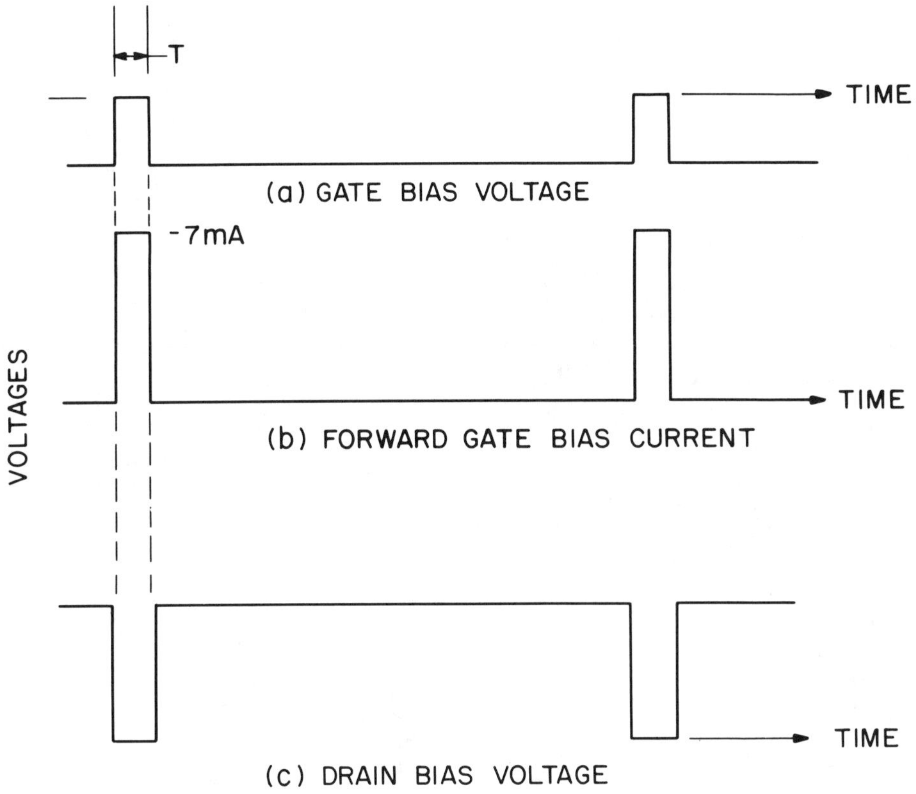

***Figure 16**/Bias condition for thermal resistance measurement. The FET is biased at a desired operation condition as shown in (a) and (c). During the measurement period, T, of about 10 μs, the gate and drain bias voltages are off. The gate is forward biased by a constant current source, (b).*

A schematic diagram of a control circuit for electrical thermal resistance measurement is shown in Figure 17 and the timing diagram for this circuit is shown in Figure 18. The device under test, denoted as DUT, is connected to the control circuit. The drain electrode is connected to VN64GA, as shown in the dashed box 1 of Figure 17, and the gate electrode is connected to the common point of 2N4124 and 2N4126. The VN64GA, shown in dashed box 1, serves as an on/off switch for the external dc drain box bias voltage V_D. The 2N4124

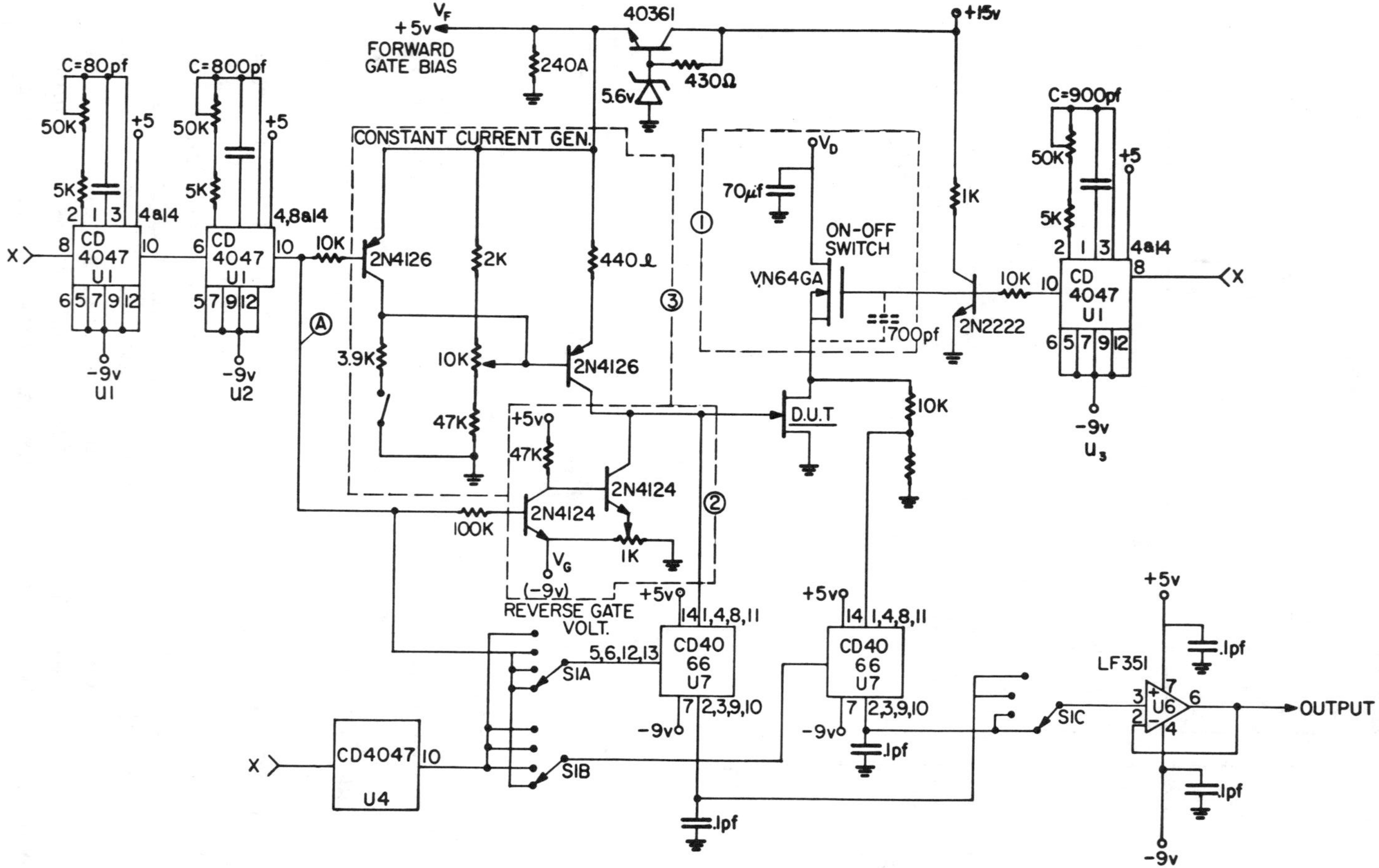

Figure 17/GaAs FET thermal impedance test set.

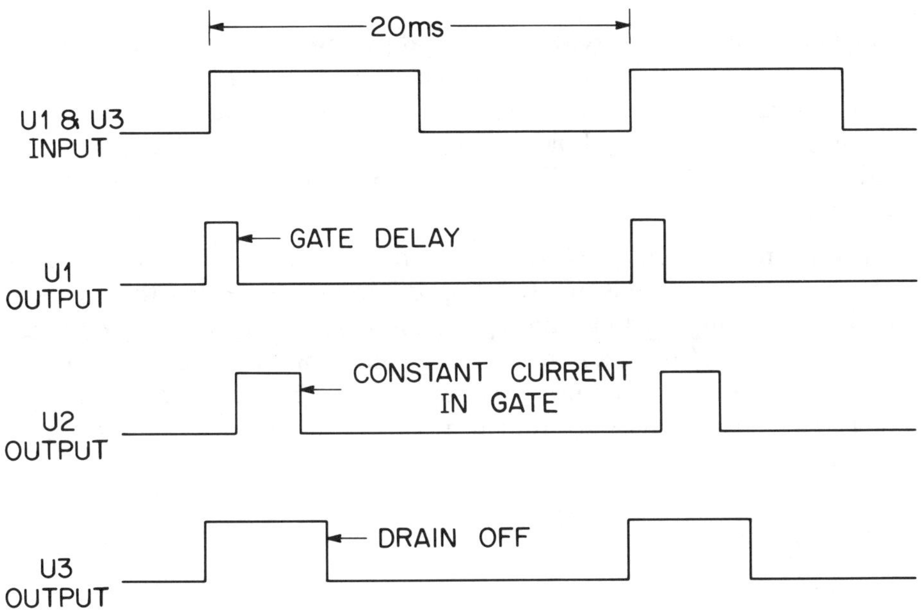

Figure 18/Measurement timing sequence.

shown in dashed box 2, serves as an on/off switch for the external reverse gate bias voltage V_G. The 2N4126 circuit (dashed box 3) is a constant current source to supply the forward gate current. The gate bias voltage is measured by the sample and hold circuit (dashed box 4).

An external square wave generator with pulse width of 10 ms is connected to u_1 and u_3 of the control circuit (Figure 17). A positive input to u_1 and u_3 initiates the measurement cycle. The positive output from u_3 turns off VN64GA, which disconnects the drain bias voltage V_D from the FET. The positive output from u_1 acts as a delay circuit, allowing the drain bias to be removed without changing the gate output. When the output of u_1 begins to turn off, it activates u_2. The positive pulse from u_2 disconnects the negative gate bias voltage V_G and connects the constant current source. The output of u_2 is designed to turn off automatically before u_3 turns off, thus ensuring that the gate voltage be switched on before the drain bias voltage is on. This entire cycle is repeated at a rate of 50 Hz. At each ambient temperature and set of bias conditions, the forward gate voltage corresponding to the 7 mA constant forward gate current is measured through the sample and hold circuit. This forward gate voltage is then compared with the calibration data to obtain the channel temperature.

With this measurement technique, the temperature is measured when the drain bias voltage is off. It is important to be sure that the measurement is completed in a time period much smaller than the thermal time constant of the FET so that the channel has not cooled appreciably. To address this point we note that the thermal time constant of a slab of heat conductor on a heat sink with infinite conductivity is given by [19] :

$$t = \frac{4L^2 C_v}{\pi^2 k} \tag{10}$$

where L is the thickness of the slab, k is the thermal conductivity, and C_v is the heat capacity. For GaAs, $k \approx 0.04$ W/(°C-mm); $C_v \cong$ 1.7 joules/(°C-cm^3); therefore, $t \approx 1.7L^2$, where t is in seconds and L is in centimeters. The thermal time constant is roughly 10 μs for a FET with a 25 μm thick backside mounted substrate. The thermal time constant for a FET with a 100 μm thick substrate is about 170 μs. The Fujitsu FLC-30 FET has a substrate thickness of 100 μm so that a measurement period of 10 μs is only 6% of the thermal time constant. A pulse width near 1 μs may be required for very thin substrates.

As an example, we present results of thermal resistance measurements on Fujitsu FLC-30 FETs [18] . We begin by obtaining the channel temperature calibration charts for each device. The device under test was heated with no drain bias applied. At approximately 25°C intervals, the gate was pulsed on in the forward direction with a constant current source of 7 mA, and the gate forward voltage was recorded. Each device was measured to a mount temperature of 225°C. The gate forward voltage decreased linearly with temperature as shown in Figure 19, for a typical device. Data from each device were stored, and a least square curve fitting method was used to obtain the device calibration curve.

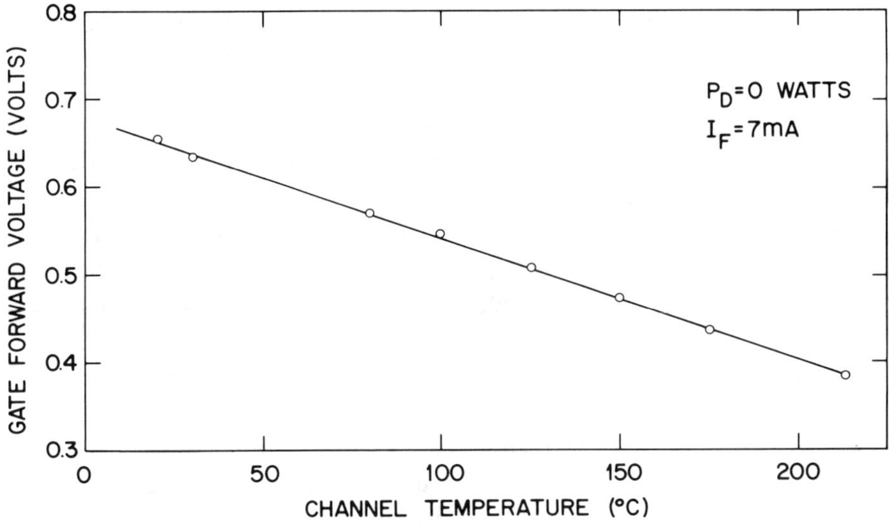

Figure 19/*Channel temperature calibration curve for a Fujitsu FLC-30 GaAs FET (W_T = 7.28 mm).*

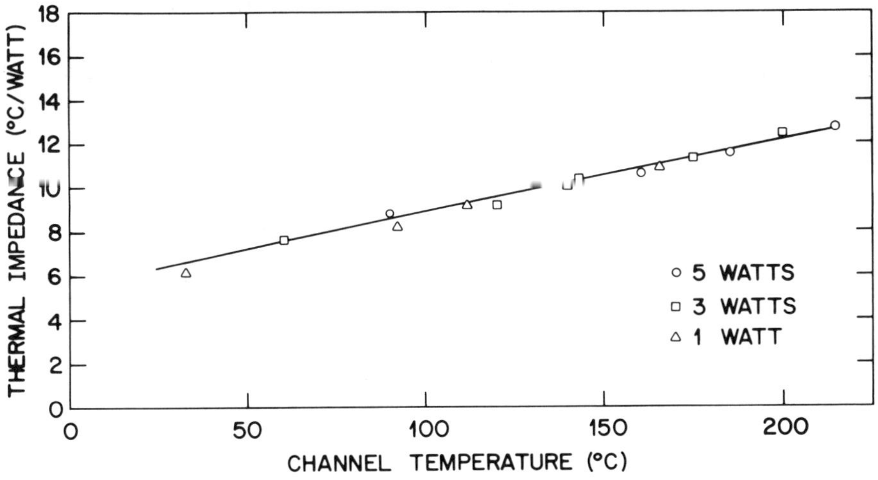

Figure 20/*Dependence of thermal resistance on total input power for Fujitsu FLC-30 GaAs FET.*

After calibration, the device was allowed to cool to room temperature. The device was then biased to a power level of 5 W and reheated. At selected case temperatures, the forward gate voltage was recorded. The thermal resistance was then calculated using Equation (9). Initially, devices were measured during both heating and cooling cycles. No hysteresis effects were observed, and subsequent devices were measured while cooling after the calibration measurement. As shown in Figure 20, the thermal resistance increased linearly with channel temperature over the temperature range 30 to 225°C.

In addition to measuring the thermal impedance at a dc power level of 5 W, several devices were measured at operating levels of 1 and 3 W. There is an advantage in measuring at the highest possible power level as this reduces the uncertainty in the measurement associated with the test method. As shown in Figure 20, the measured thermal resistance is independent of dissipated power over the range 1.0 to 5.0 W. Evidently, for this device (W_T = 7.28mm) dissipated power up to 0.69 W/mm is insufficient to require the complete nonlinear analysis described earlier.

Thermal impedances of 20 FLC-30 FETs (W_T = 7.28 mm) were measured at a mount temperature of 45°C and a dissipated power level of 3 W. In this particular case, the channel temperature ranged from 73°C to 77°C. Therefore, the thermal impedances of these FETs were 9.3-10.7°C/W (68-78°C-mm/W). Similar measurements obtained at Bell Laboratories [11] on WE 131C devices (W_T = 6 mm) gave $\overline{\theta}$ = 6°C/W (θ = 36°C-mm/W).

4. COMPARISON OF THEORY WITH EXPERIMENT

In this section we review the rather limited experimental thermal impedance literature and compare it with the theoretical results displayed in Figures 6-8. Briefly, there are only three available thermal designs for power GaAs FETs. These are: (1) relatively thick ($C \approx$ 100-125 μm) backside mounted chips which have narrow gate fingers ($W \leqslant 150$ μm); (2) thin backside mounted chips ($C < 50$ μm) with finger widths ranging up to 500 μm; and (3) flip-chip configurations. As shown in Figures 6-8, there is no fundamental thermal advantage to flip-chip mounting; rather, it provides an alternative approach to backside mounting. The crossover, where backside mounting be-

TABLE 1

Representative Experimental Thermal Impedance Values for Commercial Power GaAs FETs*

Device	Mounting	C(μm)	W(μm)	W_T(mm)	G(μm)	G_{min} (μm)	$\bar{\theta}$(°C/W)	θ(°C-mm/W)	θ (colc.) Figs. 6-8	References**
BTL (131L)	Backside	40	500	6.0	95	58	6	36	≈35	a
BTL (131J)	Backside	40	125	1.0	95	58	30	30	≈30	a
FUJ (FLC30)	Backside	100	100	7.3	14	14	10	73	≈80	b
NEC (868196)	Backside	125	100	1.4	20	20	77	108	≈100	c
TI (MSX802)	Backside	100	150	1.2	55	30	46	55	≈50	c
Dextel (3651A)	Backside	100	187.5	1.5	63	50	41	62	≈60	c
RCA (16G)	Flip	–	150	2.4	40	30	24	58	≈55	d
MSC (88002)	Flip	–	150	1.2	40	27	46	55	≈55	c
MTB ( )	Flip	–	100	2.4	30		25	60	≈60	c

*The column headings provide design parameters which are defined in Figure 4 except for W_T, which is the total gate periphery in mm. The listed manufacturers are FUJ (Fujitsu), NEC (Nippon Electric), BTL (Bell Laboratories), RCA (Radio Corporation of America), MSC (Microwave Semiconductor Corporation), MTB (Mitsubishi), and TI (Texas Instruments). For comparison with theory, the thermal impedances displayed in Figures 6-8 have been reduced by 5% corresponding to an ambient temperature of about 45°C (k = 0.04 W/°C-mm). The calculated results in some cases require extrapolation beyond the range of Figures 6-8, but this should not detract from the conclusion that theory and experiment are in reasonable agreement.

**References for Table 1:

a. H. Fukui (private communication).
b. Measured at RCA Corporation, Princeton, NJ 08540.
c. H.M. Macksey and L.W. Joy, "Reliability Evaluation of GaAs Power FETs", Interim Technical Report RADC-TR-80-124, April 1980, Rome Air Development Center, Griffiss Air Force Base, New York 13441.
d. Reference 12.
e. Mitsubishi data sheet.

comes advantageous, falls near C = 75 μm. The exact crossover, of course, depends on details of gate spacings and finger widths. Therefore, to obtain the lowest thermal impedances, backside mounting is required. We stress again that not only substrate thickness, but also gate finger width and gate-to-gate spacing are important thermal parameters.

The above qualitative conclusions are fully supported by the experimental results presented in Table I. Note that $\theta \approx$ 70-80 °C-mm/W for the thick-chip backside mounted design, $\theta \approx$ 50-60 °C-mm/W for flip-chip configurations, and θ = 20-40 °C-mm/W for the thin-chip backside mounted case. These results are in reasonable agreement with the three-dimensional theory presented in Section II. Because of experimental uncertainties, and also because chip thicknesses are rarely known with high accuracy, it does not appear possible at this time to ascribe better than 10-20% accuracy to the claimed agreement. For most purposes, however, this result is adequate, and moreover, since the theory clearly delineates the relevant design parameters, we view Figures 6-8 as providing all the information required for thermal design of new metallization layouts. In addition, using Figure 3, the theory provides a simple procedure for including the thermal conductivity temperature dependence in the analysis as required, for example, when undertaking high temperature stress-aging experiments.

What, then, are the underlying thermal design principles for power GaAs FETs? First, the designer must assess his chip thinning capability since, for the lowest thermal impedances, backside mounting is required. The second design choice is gate finger width, W. This dimension is generally based on phase shift and resistance limitations along each gate imposed by microwave considerations, although it is not at all clear at this point that narrow fingers (W $<$ 200 μm) are actually required for operation up to 12 GHz. The final design choice involves the average gate-to-gate spacing, G. We emphasize that this choice is *not* necessarily simply a thermal question since it is possible to envisage thermally acceptable values of G that yield current densities in source and/or drain stripes that are unacceptable from an electromigration viewpoint. Thus, especially for wide gate designs, both thermal and current density requirements must be simultaneously satisfied. At Bell Laboratories it has been found that current densities up to at least 2 x 10^5 A/cm^2 in the Au source and drain stripes yield devices free of Au electromigration problems. A third consideration that may control gate spacings occurs when the drain (or source) fingers serve as bonding pads. In this case, bonding pad size requirements may override *both* thermal and electromigration considerations. When through-the-chip source via connections are used, however, then gate spacings (and chip size) can be reduced to the limits imposed by thermal and/or electromigration constraints.

A final word is in order as to the actual channel temperature expected during rf operation of power GaAs FETs. A typical dc bias point yields 1-1.5 W/mm of dc input power (e.g., 100 mA/mm at 14 V). Depending on the rf power compression point, the net rf output power (i.e., output minus input) usually falls in the 0.3-0.7 W/mm range so that the resulting dissipated power is about 1 W/mm or below. This means that, to a first approximation, the normalized thermal impedances shown in Figures 6-8 are close to the actual maximum temperature rise in the channel. The GaAs FET is therefore a remarkably cool device, able to easily achieve channel temperatures less than 50°C above ambient.

ACKNOWLEDGMENTS

The authors appreciate several useful discussions with H. Fukui, M. Macksey, W.B. Joyce, W. Slusark, Jr., M. Plesher, and E. Finn. In particular, we thank H. Fukui for allowing us to use his thermal impedance data prior to publication, and K.R. Dietzel, J.A. Seman and V.M. Wagner for programming consultation and assistance.

REFERENCES

[1] Joyce, W.B., *Solid State Electronics,* 18:321, 1975.

[2] Ruch, J.G., and W. Fawcett, *J. Appl. Phys.,* 41:3843, 1970.

[3] Cooke, H.F., *Microwaves,* February 1978, p. 55.

[4] Cohn, S.B., *IRE Trans. Microwave Theory and Techniques,* MTT-2:52, 1954.

[5] Oliver, A.A., *IRE Trans. Microwave Theory and Techniques,* MTT-3:134, 1955.

[6] Wang, Y.C., and M. Bahrami, *Int. J. Electronics,* 47:147, 1979.

[7] Abromowitz, M., and I.A. Stegun, *Handbook of Mathematical Functions,* U.S. Dept. of Commerce, National Bureau of Standards Applied Mathematics Series, No. 55, p. 587, 1970.

[8] Maycock, P.D., *Solid State Electronics,* 10:161, 1967.

[9] Huang, H.C., F.N. Sechi, and L.S. Napoli, 6th Biennial Cornell Electrical Engineering Conference, Ithaca, 1977.

[10] Dietzel, K.R., V.L. Hein, and V.D. Lenzi, *TASIC* (Thermal Analysis of Substrates and Integrated Circuits), unpublished BTL program.

[11] Fukui, H., (unpublished work).

[12] Sechi, F.N., B.S. Perlman, and J.M. Cusack, MTT Symp. Digest, 1977, p. 143.

[13] The IR measurement system described in this section was developed by B.S. Perlman, F.N. Sechi, J.M. Cusack, and W.A. Schroeder of RCA Laboratories. Reported in an *Automatic Infrared Microscope Manual Report,* No. 2, July 1977. Prepared for Sandia Corporation, Contract No. 03-6003.

[14] Macksey, M., R.L. Adams, E.N. McQuiddy, Jr., D.W. Shaw, and W.R. Wissman, *IEEE Trans. on Electron Devices,* ED-24: 113, 1977; M. Macksey, private communication.

[15] Stephens, C.E., and F.N. Sinnadurai, *J. Phys. E.,* 7:671, 1974.

[16] For example, the anisotropic liquid ranges for several materials are as follows:

Butyl p-(p-Ethoxyphenoxy-carbonyl)-phenyl carbonate
4,4′-Bis (penyloxy)-azoxybenzene
4,4′-Dibutoxyazoxybenzene
p,p′-Anzoxydiphenetole

These liquid crystals are available from Eastman Kodak.

[17] Buchsbaum, R.J. (unpublished).
[18] The information in this section on "Electrical Measurement Techniques" was contributed by W. Slusark, Jr., M. Plesher, and E. Finn of RCA Corp.
[19] Carslow, H.S., and J.C. Jaeger, *Conduction of Heat in Solids.* Oxford: Oxford University Press, 1959.

Chapter 6

The Reliability of GaAs FETs

Introduction

In this chapter the reported failure mechanisms and projected reliability of low noise and high power GaAs FETs are reviewed. J.C. Irvin reviews the principal failure mechanisms which are burnout, metallization failures, and gradual degradation.

In his comprehensive review, Irvin first discusses the source, drain, and gate metallizations which have been used by various laboratories. He then goes on to discuss reliability mathematics including the parameters of the log-normal distribution, confidence limits, and the impact of censored data.

In a comprehensive review of the reliability of low-noise FETs, Irvin concludes that sufficient data exists to indicate that the low noise GaAs FET is capable of very long life.

Power GaAs FETs are a different matter, however. They operate at much hotter temperatures and much higher bias voltages than low noise FETs. Unlike the low noise FET, the power FET has a major burnout problem. This problem is discussed in detail and it is demonstrated that properly designed power GaAs FETs can be rendered nearly burnout-free.

However, since these devices operate at higher temperatures, aside from the burnout problem, there are the reliability problems of gradual degradation and metallization failures. These, too, are addressed in detail and Irvin concludes that power GaAs FETs can be made reliable if proper attention is paid to the design of the device.

The Reliability of GaAs FETs

J. C. Irvin

Bell Laboratories, Murray Hill, New Jersey

1. INTRODUCTION

The ultimate performance limits of GaAs FETs may be approximately predicted, based only on the fundamental properties of the material and the gate dimensions. The results of such predictions are very enticing and the performance realized to date is encouraging. The eventual usefulness of the GaAs FET depends, however, not only on its performance but also on its reliability. The question of reliability does not lend itself to such ready prediction as do performance figures. Only experiment (usually long, patient, and expensive) can resolve the problems of reliability and, unlike performance generalizations which apply broadly to all GaAs FETs, the results of stress-aging experiments refer only to a specific structure. A slightly different combination of metals or even the same one processed in a different manner can yield entirely different aging properties. Furthermore, all predictions derived from stress-aging results involve an acceleration factor, a factor the accuracy of which cannot be completely assured until field results are available – which may be too late. Fortunately, to whatever extent the metallization employed on a GaAs FET replicates the metallization already used and studied previously on silicon devices, that knowledge may be applied (cautiously) to the GaAs FETs. Also, it is pleasant to note that the present results of the most extensive, published reliability programs to date indicate that properly designed and fabricated GaAs FETs can indeed have very low failure rates.

This chapter summarizes the reliability status of low-noise as well as high-power GaAs FETs, including failure mechanisms and stress-aging data. The failure mechanisms of GaAs FETs may be roughly categorized into two classes: 1) those mechanisms involving the metallization, such as electromigration and interdiffusion; and 2) those taking place in the GaAs itself or on its surface, such as changes in net donor density, trap, and defect formation. Metallization failures appear to be the more common and serious causes of GaAs FET degradation and, therefore, that subject is reviewed in Section 2 of this chapter. This will be followed in Section 3 by a brief summary of the essential reliability mathematics employed here. Section 4 contains a description of the stress-aging facilities appropriate to GaAs FETs. Section 5 discusses the published data and projections on the reliability of low-noise FETs. Section 6 contains a similar review for power FETs.

2. METALLIZATIONS

The properties of an ideal metallization for the gate, source, or drain contacts of a GaAs FET would include: 1) high electrical conductivity; 2) low tendency to electromigration; 3) inertness with respect to its immediate environment including GaAs, passivation, ambient, and its own components; 4) good adhesion to the GaAs surface; 5) accurately controllable penetration into the GaAs, or none at all, during processing; 6) bondability with leads; 7) tractable processing and lithographic qualities. In the case of the ohmic source and drain contacts, the additional requirement of low contact resistance must be met. For the gate metal, a reasonably high barrier height is desirable. No single metal or combination of metals meets all these requirements. The usual solution, especially for source and drain, has been a compromise which optimizes most of the properties at the expense of one or two. Almost invariably, one of the desirable qualities sacrificed is simple, tractable processing.

Low contact resistance is achieved in most source and drain metallization schemes by alloying a thin AuGe film into a n+ GaAs epitaxial (or implanted) layer. Later, a thick layer of Au is deposited to provide high conductivity, bondability and corrosion resistance. However, such a system is unstable with time and temperature; Ga readily diffuses through the AuGe into the Au layer which acts as a sink while Au (and O_2) diffuse inward [1, 2, 3]. As a result, the contact resistance increases and can eventually degrade the performance of the transistor. To prevent the slow and undesirable interdiffusion of Ga and Au, a suitable barrier layer must be interposed. Some of the choices for this barrier are Ag, Cr, Ni, Pt, Ta, and Ti. If this barrier metal is deposited before alloying, it can also help prevent balling of the AuGe and consequent erratic penetration of the latter into the GaAs. The metals listed are not equally effective as a diffusion barrier, Ag apparently being one of the best [3] . If the barrier layer is omitted or is inadequate, the interdiffusion of components and degradation of the contacts proceeds with an activation energy, of 0.8 to 1.8 eV [4, 41] . This relatively high energy assures reasonable contact life if the operating temperature is low enough ($< 100^\circ$C) but also means rather rapid deterioration at elevated temperatures ($> 150^\circ$C).

In any case, however, the drain and source contacts are commonly not the origin of the most acute reliability problems. The gate of the GaAs FET is usually the more failure-prone element. For reasons of fabrication ease, Al was one of the earliest choices for GaAs FET gate metal. The major failure modes of Al are pulse burnout, electromigration, electrolytic corrosion, and Au-Al or Pt-Al phase formation. Though that's a forbidding list of weaknesses, there are adequate "fixes" for all of them, which is why Al has remained the choice for many successful Si devices. Nevertheless, to avoid the frailties of Al, some designers have chosen a refractory metal gate, topped by one or more layers of other metals to improve electrical conductance. Some of these are Ni, Au/Mo, Au/Pt/Ti, Au/Pt/Ti/Pt, Au/Pt/Cr/Ti and Au/W/Ti. The refractory gate systems are not without their own failure mechanisms, however; the failure modes are simply not as well known or understood (and thereby cured) as those of Al. In particular, Lundgren [5] found the Au/Pt/Cr/Ti gates to be as susceptible to burnout as Al; Drukier and Silcox [6] reported fairly rapid destruction by electromigration of a Au/W/Ti gate.

Further discussion of the various metallizations will be given in sections 5 and 6. Details of their failure mechanisms as actually observed in low-noise and high-power GaAs FETs will be described together with the associated lifetime data.

3. RELIABILITY MATHEMATICS

The essence of a reliability study is to determine the cumulative failure distribution of a sample population under some known degree of acceleration and project that distribution to normal (operating) conditions. From the latter distribution, then, the failure rate as a function of time is obtained. The accelerated distribution is found by experiment and may turn out to be of an arbitrary shape not describable by previously developed functions. Fortunately, however, many or most semiconductor devices obey a cumulative failure distribution known as log-normal [7]. The parameters characterizing the log-normal distribution are well defined and graphs permitting the rapid determination of failure rate, given these parameters, are available [8, 9]. Since the difficulties and economics of stress-aging GaAs FETs enforce relatively modest sample sizes, it is important to consider the uncertainties inherent in these statistics. Again, for a log-normal distribution, the confidence limits afforded by different sample sizes have been worked out and published in simple graphical form [9, 10].

In view of the computational convenience of a log-normal distribution, it is tempting to jump to the conclusion that a set of failure data follows this law when the data are insufficient to prove the point. For example, as shown in Figure 1, a set of random numbers plotted on log-normal coordinate axes appears to approximate a straight line up to the 20% failure point (especially if the earliest 2 failures are discarded as freaks), but as the full plot shows, the distribution, of course, is not log-normal [11]. Happily, the few published cumulative failure distributions for GaAs FETs indicate that both gradual degradation (at least in the case of low noise FETs) and burnout (in power FETs) obey a log-normal relationship [5, 11, 12].

3.1 Parameters of the Log Normal Distribution

The cumulative failure function, Q(t), is the probability that a device will fail in a time equal to or less than t, i.e., the fraction of the whole sample or population that has failed up to time t. For a log-normal distribution, Q is given by

$$Q(t) = \int_0^t \frac{1}{t\sigma\sqrt{2\pi}} \exp\left[\frac{-1}{2\sigma^2}(\ln t/t_m)^2\right] dt$$

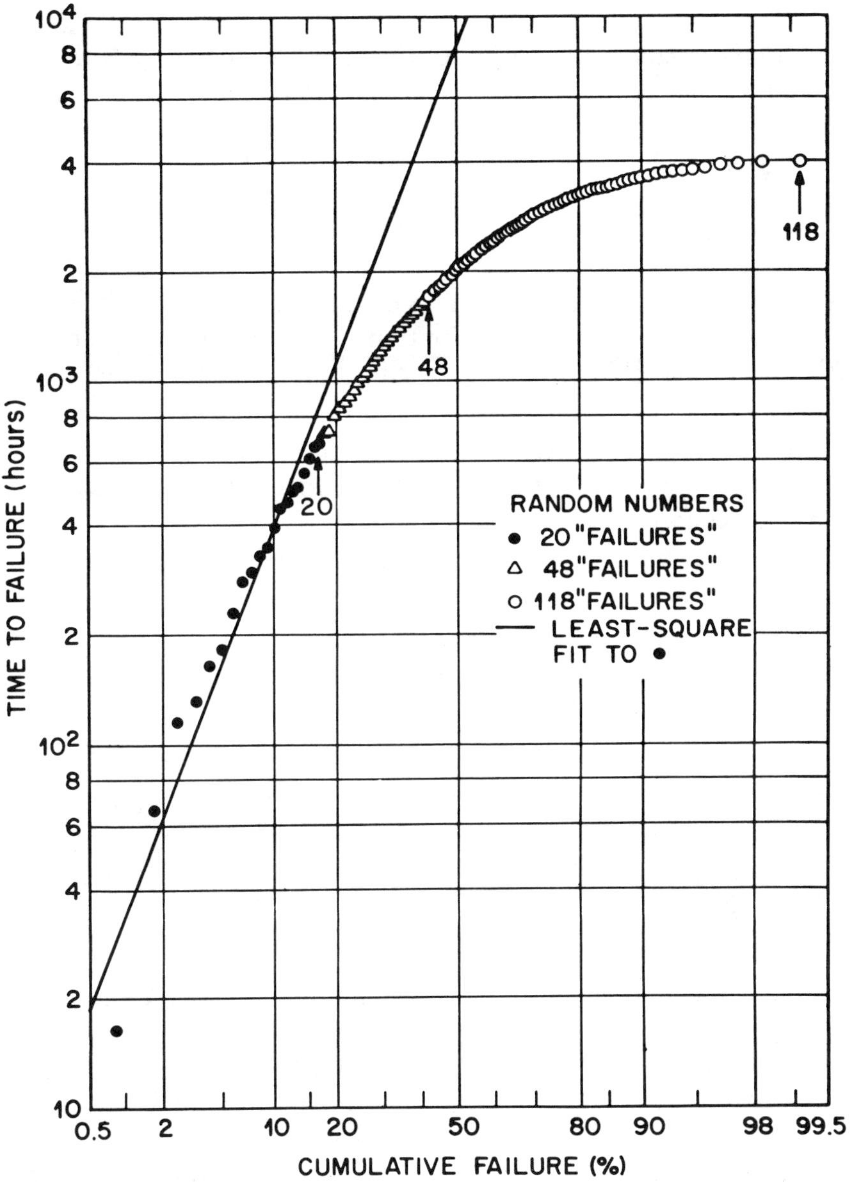

Figure 1/The log normal projection of time to failure versus cumulative failure percent for a sample of random numbers. The times to "failure" were based on 118 pseudorandom numbers between 0 and 1, computer generated and subsequently ordered and scaled to a time domain. The least-square fit to the first 20 of these data points indicates the apparent (but erroneous) log normality of the early failures. (From Jordan et al. [11]).

where σ is the standard deviation of the distribution and t_m is the median lifetime, i.e., the time at which 50% have failed. The standard deviation is readily calculated for a given log-normal distribution using the expression

$$\sigma = \ln(t_m/t(16))$$

where t(16) is the time at which 16% of the population has failed. An example of a log-normal distribution is given in Figure 6.3 in the section on power FETs. For that distribution, $t_m = 1800$ hours and $\sigma = 1.2$.

The failure rate, $\lambda(t)$, is defined such that $\lambda(t)dt$ is the probability that a device will fail in the interval between t and t+dt. It is given by the expression

$$\lambda(t) = \frac{\sqrt{2}\exp\left[\frac{-1}{2\sigma^2}(\ln t/t_m)^2\right]}{\sqrt{\pi}\, t\, \sigma\, \mathrm{erfc}\left[\frac{1}{\sqrt{2}\sigma}\ln t/t_m\right]}$$

where erfc (x) is the complimentary error function

$$\mathrm{erfc}(x) = \frac{2}{\sqrt{\pi}}\int_x^{\infty} \exp(-y^2)\,dy.$$

The failure rate at a given time, t, is completely determined by the two parameters t_m and σ which characterize the failure distribution. Jordan has computed λ for several values of t (5, 10, 20, and 40 years) and presented the results graphically [9]. The graph for the 5-year point is reproduced in Figure 2. The unit of failure rate employed in Figure 2 and throughout this chapter is the FIT which equals one failure in 10^9 device-hours. Note that a failure rate of 1% per 1,000 hours is 10^4 FITs. In general, for a given t_m and σ, the failure rate increases with time to a maximum value and then declines. For very small values of σ, the time at which λ is maximum will correspond roughly to the median life. (If the standard deviation were zero, the failure rate would be zero for $t < t_m$ and then briefly be-

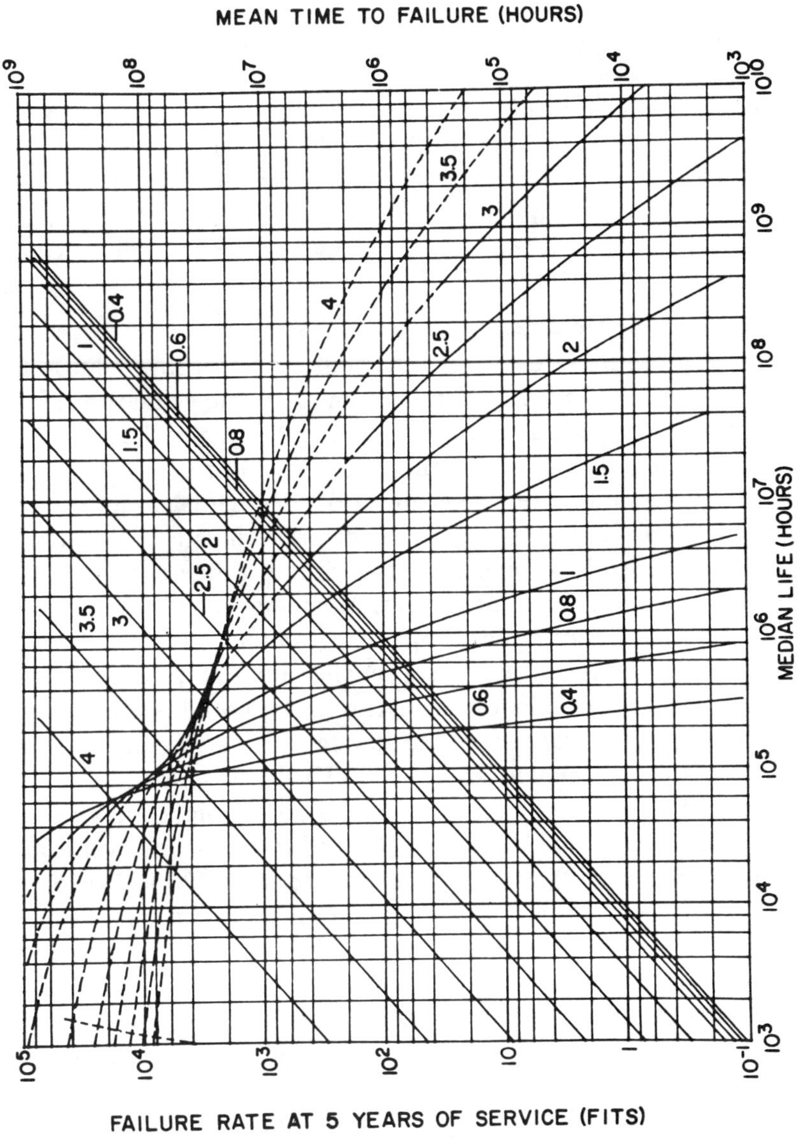

***Figure 2**/A family of curves for the log normal failure rate in FITs (left-hand scale) at 5 yr. of service life vs. median life. The straight lines show the mean time to failure (right-hand scale) vs. median life. Each line, as labeled, is for a different standard deviation (from Jordan [9]).*

come infinite at $t = t_m$ when all the devices failed at once.) Thus, Figure 2 does not necessarily show the maximum λ up to 5 years of service. If the maximum value has occurred before 5 years, the curve is dashed; if the curve is solid the value of λ is still rising at the end of 5 years. Median life and mean time to failure, MTTF, are uniquely related, depending only on σ for a log-normal distribution. The transformation is given by the straight lines and the right-hand ordinate scale in Figure 2.

Two aspects of Figure 2 deserve particular attention. One of these is how rapidly failure rate varies with median life for low values of σ. For example, with $\sigma = 1$ (a common and realistic value for many semiconductor devices), a median life of 1×10^6 hours means a failure rate of about 70 FITs at the end of 5 years. For $t_m = 2 \times 10^6$ hours, however, the failure rate is about 6 FITs. Thus, a factor of two change in median life in this region of the graph means more than an order of magnitude difference in failure rate. Uncertainties in measured median life or in the activation energy, by which the accelerated values are translated to operating temperature, will have, therefore, a greatly magnified effect on the uncertainty in the corresponding failure rate. The other feature to be noted is the rapid dependence of failure rate on σ, the standard deviation of the failure distribution. In fact, median life (or MTTF) alone is *insufficient* to determine failure rate. As a fairly reasonable example, it may be noted that the difference between $\sigma = 1$ and $\sigma = 2$ means over *two* orders of magnitude difference in failure rate (6 vs 700 FITs) for a median life of 2×10^6 hours. Thus, uncertainties in determination of *either* t_m *or* σ have a large effect on the resultant calculation of failure rate. It's also clear how sensitive failure rate predictions are to the *assumption* that the operating temperature failure distribution has the same σ as the one observed at the stress temperature, an assumption not subject to confirmation without years of experience. Fortunately, there is a mitigating factor present, as follows. Suppose that at the termination of a stress-aging experiment, only 10% of the devices have failed. The projection of the full failure distribution and associated t_m and σ would be quite uncertain. However, if the estimated σ is erroneously high, the corresponding estimated t_m will also be high and these two errors have opposite effects on the resulting calculation of failure rate. Thus in this situation, the errors tend to cancel each other.

3.2 Confidence Limits

The parameters of Figure 2, median life and standard deviation, refer to the true values that characterize the actual failure distribution of a very large population. Any stress-aging experiment involves finite samples which, indeed, are frequently very small (10 to 20 devices). Observations on a succession of finite samples will not give identical results but instead a range of values. The extent of this range and the associated confidence that the true value is included in this range are related by formulas well known in statistical theory [13]. Figure 3 shows the range of median lifetimes for a log-normal distribution (as a ratio of the true value, t_{mp}, to the observed value, t_{mh}) within various confidence limits as a function of sample size and of the observed standard deviation, s_h. Thus, if $s_h = 1$ and the sample size is 10, in 9 trials out of 10 the observed median lifetime would range between about 0.52 and 1.8 times the true value of the population (and 1 time out of 10 it would be even farther off). A particularly convenient presentation of confidence limit as applied to log-normal statistics has been given by Jordan [10]. Figure 4 is his graph showing the range of failure rates that may be expected with 90% confidence, at the end of 5 years, as a function of sample size and *observed* median life for an *observed* standard deviation of 1. (Similar curves for other values of observed standard deviation and service life are found in Jordan's publication [10].) Figure 4 shows that if a median life of 10^6 hours (at operating temperature) is projected from stress-aging tests on a sample of 8 units, where the observed standard deviation of the failure distribution is 1.0, then one can predict with 90% confidence only that the failure rate after 5 years would lie somewhere between 1 FIT and 1,000 FITs. This prediction still assumes that the acceleration factor (i.e., activation energy) is known exactly and that the standard deviation does not change with temperature. This degree of uncertainty, present in many small scale reliability studies, is frequently blissfully ignored since to emphasize it often produces great consternation. The only solution is large, very large, samples drawn from a static population. If this is not possible, the investigator should appreciate the inherent limitations from the outset and avoid wasting time on a degree of experimental precision which cannot possibly provide an escape from the tyranny of small numbers.

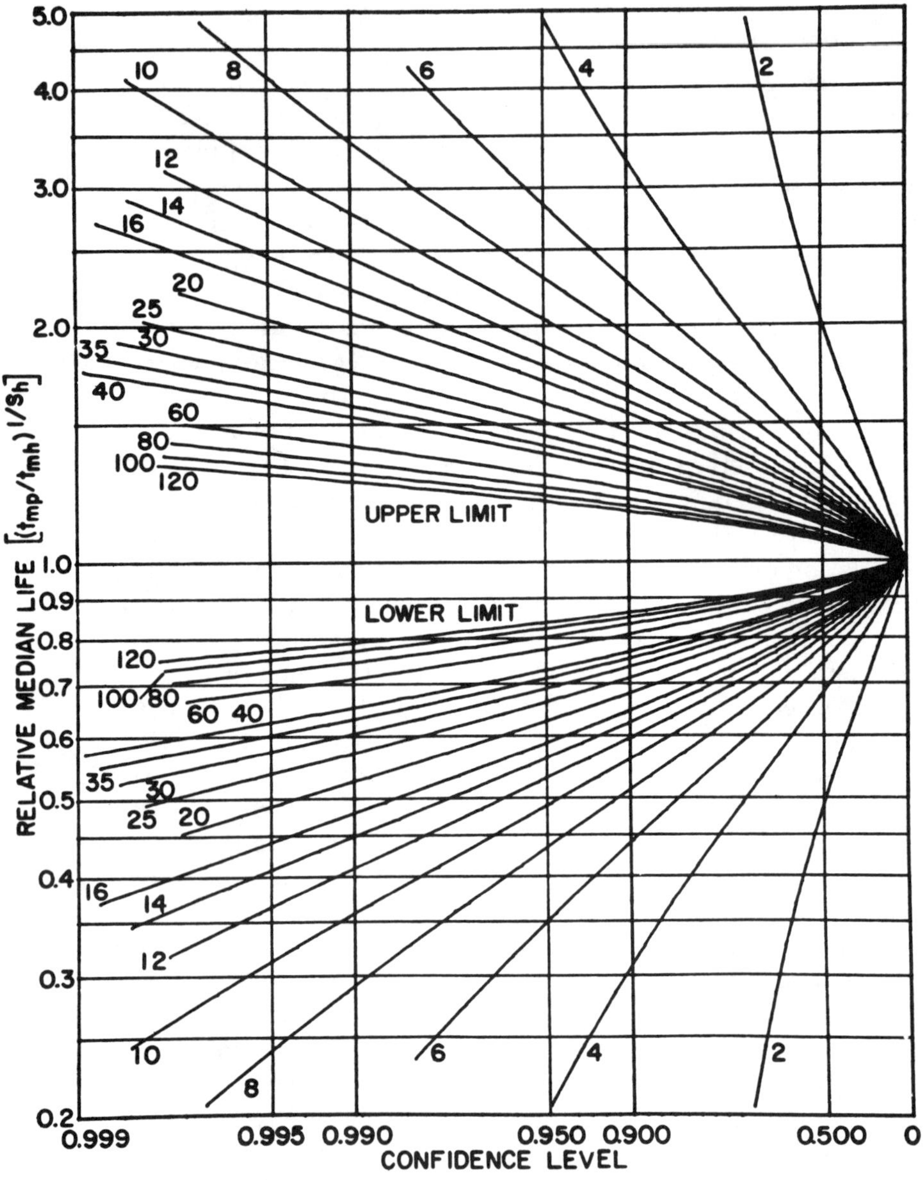

Figure 3/The upper and lower confidence limits of median lifetime as a function of sample size. See text for explanation of symbols (from Jordan [9]).

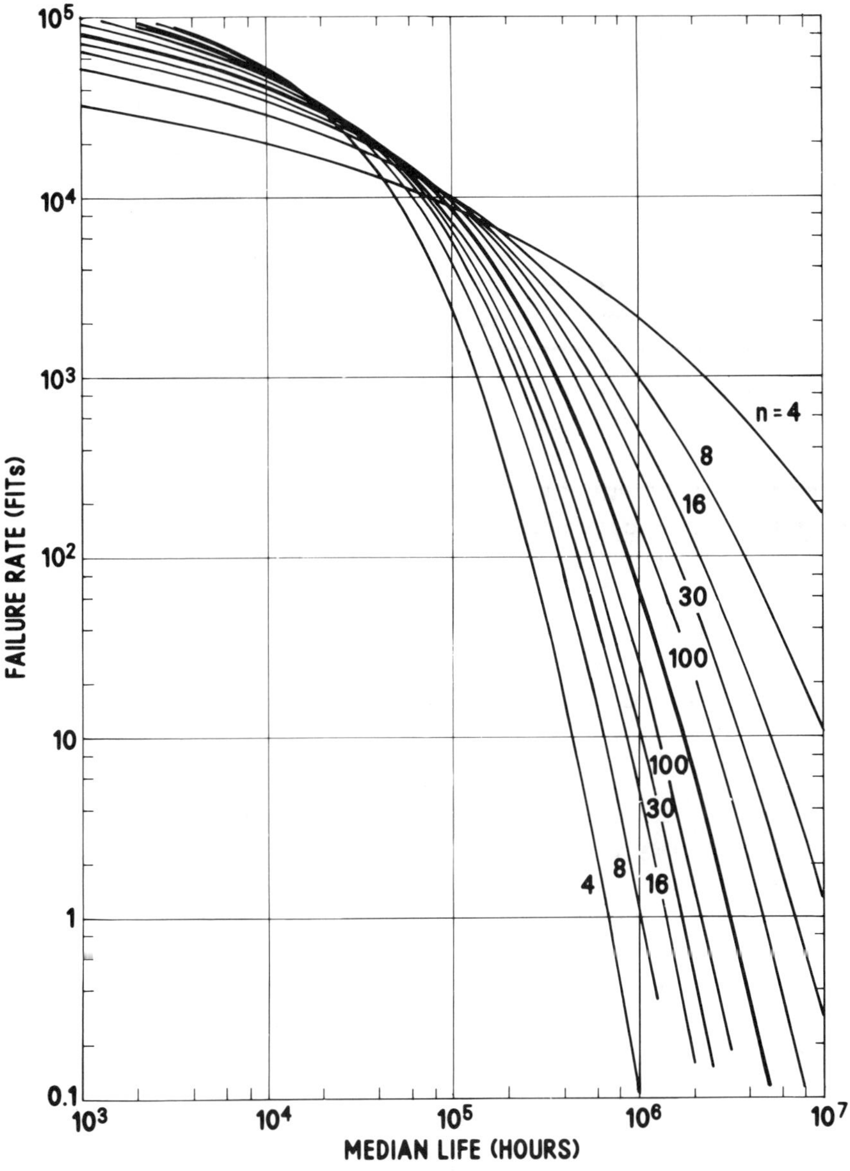

Figure 4/Upper and lower confidence limits of the failure rate at the end of 5 years of a log normal population at the 90% confidence level shown as a function of sample size and the observed median life for an observed standard deviation of 1.0 (Jordan [10]).

3.3 Censored Data

As indicated in section 3.1, the accurate determination of the nature of a failure distribution requires data extending at least to the 40% or 50% failure point, if not further. On some occasions, however, due to limited rack space and the desire to age other samples of different design (or just a different slice), it is necessary to terminate an aging experiment before the cumulative failures have reached 50%. Also accident or some failure mode other than the one under examination may prematurely remove devices from the sample tested. In these cases there may not be any problem in obtaining the needed data if the failure of interest has a precursor that allows its prediction from the available data even though the final event has not yet occurred. However, if the failure lacks a precursor and occurs spontaneously without prior indication, such as a burnout in a power FET, then the data base is incomplete and is said to be "truncated" or "censored." Techniques for the handling of such data have been published [14, 15]. A recent application of these methods to burnout data for GaAs power FETs has been reported by Jordan et al. [11]. Some of the results of that work are given in section 6 of this chapter. For details of the technique one is referred to the publications cited [11, 14, 15].

4. AGING FACILITIES

Most reliability programs have a three-fold purpose: 1) to obtain in a reasonable time span (1 to 2 years) the data necessary to make useful projections of the operational reliability (failure rate) of the device in question; 2) to discover and understand the failure mechanisms so that they might be corrected where necessary; 3) to establish screening procedures to assure that the production device is a faithful replica of the units on which the reliability data are based. (The last-mentioned purpose will not be discussed here.) To accomplish these tasks, the degradation of a large number of devices must be accelerated in such a way as to exercise all the relevant failure mechanisms without introducing any extraneous ones. Or, if the introduction of extraneous mechanisms is unavoidable (artifacts of the testing and aging procedures which would not occur in field usage), then at least they must be recognized as such.

Many failure mechanisms involve the diffusion of atoms and possess, therefore, an Arrhenius type time-temperature dependence, i.e., a reaction rate which is exponentially dependent on the inverse temperature and characterized by an activation energy. These mechanisms are readily accelerated by increasing the temperature. However, new mechanisms can also be introduced at an elevated temperature, especially (but not only) if a phase change occurs. Thus, this means of acceleration must be used judiciously. Most workers in the field have limited their tests of completed GaAs FETs to temperatures of 300°C or less. (Even at this temperature electromigration in Al involves different diffusion processes than at lower temperatures.) In test structures involving only a particular metallization, such as the ohmic contact, aging at a higher temperature may be meaningful.

Most failure mechanisms are also bias dependent. These include burnout, electromigration, some aspects of contact degradation, and trap formation, with its associated effects on gain and noise figure. Thus, high temperature storage without bias can be informative but will not accurately accelerate all the degradation modes. Of course, an operating FET is subject not only to dc bias but also to RF signals. For a low-noise (low-power) device, the RF signal amplitudes are too small to perturb significantly the dc values. However, for the power FET, the RF signals are large enough to more than double the momentary source-drain voltage (over the dc value). It can also cause large net (dc) gate currents, which may be either of forward or reverse polarity, depending on tuning and bias adjustments. Thus, an adequate aging program for power FETs must include some RF fixtures (amplifiers) which are operable to 250° or 300°C. The latter are especially expensive and bulky, however, so the majority of the test data are usually obtained under high temperature, DC biased conditions.

Since a dc biased GaAs FET is prone to oscillate over a wide band of frequencies, and these oscillations can be destructive in magnitude and present an unknown, uncontrollable test condition as well, it is necessary to suppress such oscillations. No completely effective means of quelling oscillations has been reported. Among the methods employed with some success are ferrite beads on the bias leads (which, however, cannot go in the oven), R-C networks on the gate and drain contacts of the mounting fixture, and RF absorbers (Eccosorb) built into the fixture. Figure 5 shows a mount used for dc bias aging of GaAs power FETs at temperatures up to 300°C. One of the RC networks is visible (drain side) as well as some of the Eccosorb material. Several of these fixtures are mounted on a copper block (for uniformity) which is equipped with thermocouple controlled heaters. An alternate procedure is to place the block without heaters in a hot air oven. A significant amount of the heat input to the block is from the dc dissipation of the FETs. If a FET burns out or severely degrades, that dissipation is altered. With individual block heaters, the resultant change of dissipation may be compensated by the heaters, but in an oven, the setting of the oven must be altered. The latter adjustment is not feasible if there are other blocks of fixtures in the oven with varying numbers of burned-out units.

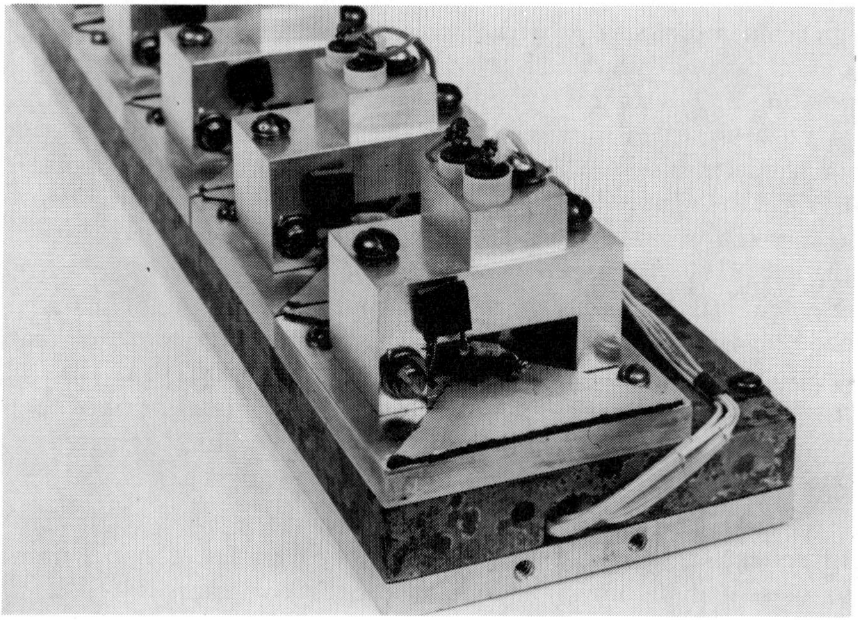

Figure 5*/Photo of dc power FET aging fixture from drain side, showing RC stabilizing network.*

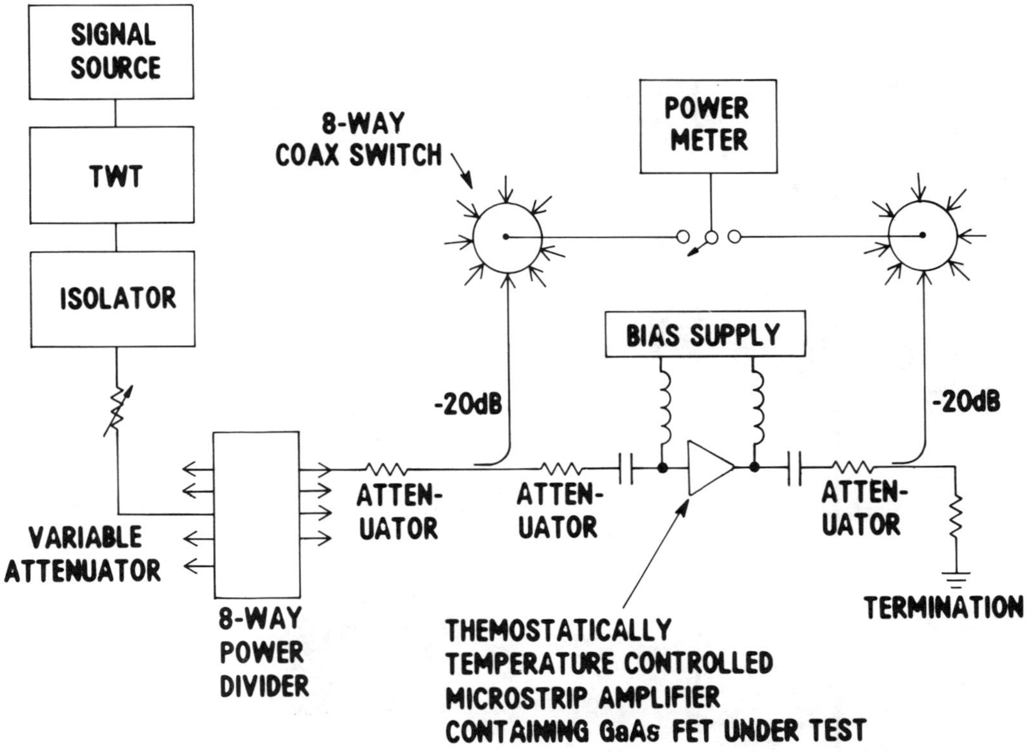

Figure 6/Block diagram of arrangement for RF aging of power FETs.

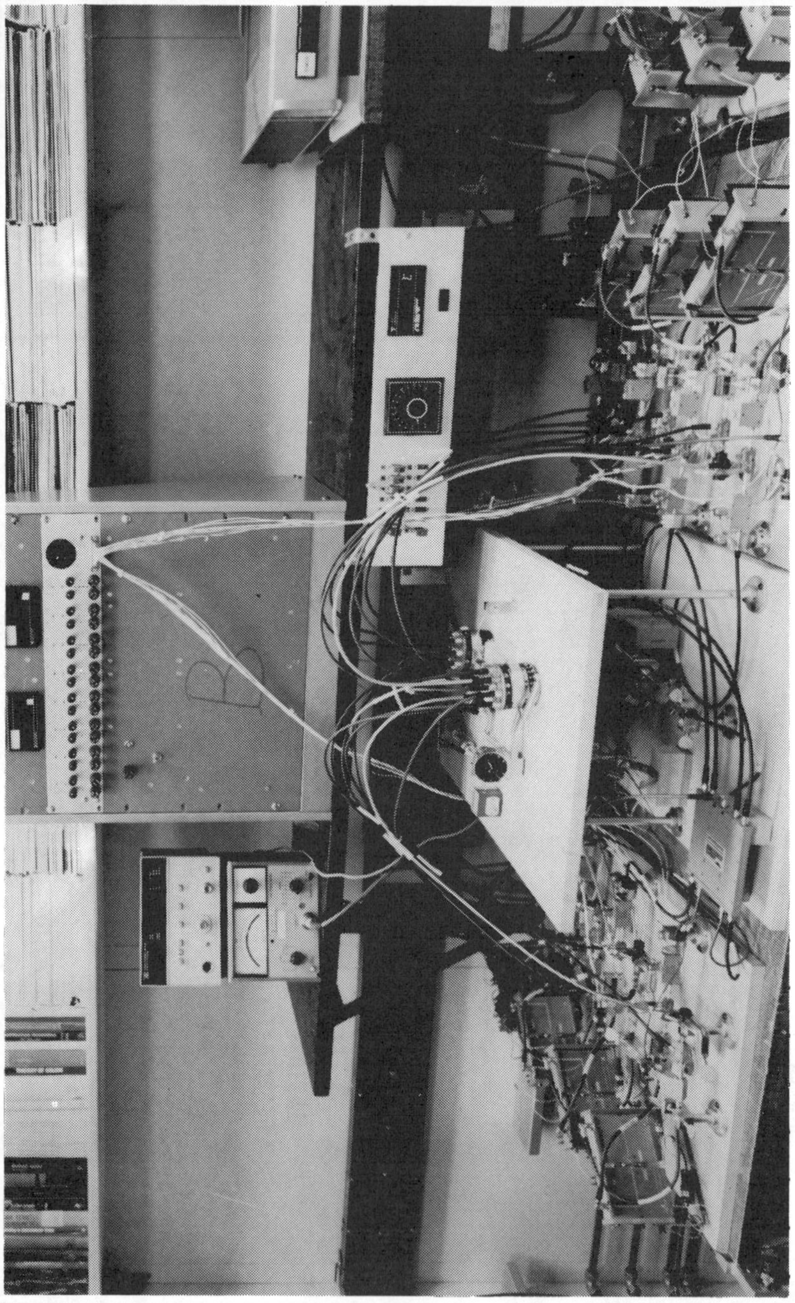

Figure 7/Photo of RF aging set up for eight power FETs.

The block diagram of an RF aging arrangement is shown in Figure 6 and a photograph of the same in Figure 7. The amplifiers are microstrip designs and are individually maintained at the desired temperature with thermocouple controlled heating elements. The RF input and output power, as well as the dc bias values, may be read from appropriate meters at any time (but are not automatically recorded in the system shown). The FETs are soldered into the amplifiers and individually tuned at room temperature by altering stubs in the microstrip circuit. To carefully monitor the degree of device degradation, the amplifier may be periodically cooled to room temperature and the necessary dc and RF parameters measured (without removing the transistor from the amplifier). Such an RF aging facility costs over $1,000 per device position, occupies a good deal of space, and the installation of the FETs is tedious.

If useful stress-aging data could be obtained only under RF conditions, the reliability testing of power FETs would be an extremely arduous and expensive undertaking, especially in view of the necessarily large data base, as demonstrated in section 3. Fortunately, most experience to date [11, 16, 17] indicates that RF aging does not introduce any new failure mechanisms not *potentially* present under dc conditions. As mentioned before, however, large RF signals can produce significant gate currents of either polarity, as contrasted with dc aging, which generally only results in small, reverse gate currents. (The polarity may be significant since only a reverse gate current, that is, an electron flow from metal to GaAs, causes a current density gradient at the tapering throats of the gate stripes in the direction that is apt to produce voids at that point.) Consequently, gate electromigration is more probable in RF aging at elevated temperatures than it is in dc aging at the same temperature and bias voltages. Also, the higher peak gate and drain voltages which occur under large signal RF conditions may produce higher burnout rates than in dc aging unless the devices have been previously screened at a sufficiently high voltage. A suitable compromise between the exorbitant cost and more faithful reproduction of operating conditions that are inherent in RF stress-aging is to do sufficient aging under RF to reveal the failure mechanisms, obtain the more massive data base in dc racks, and correct the latter data with correction factors derived from comparisons of the RF and dc results or from specific investigations of those failure mechanisms, such as gate electromigration, which are not adequately exercised in dc aging.

To accelerate corrosion (electrolytic and otherwise), the above described facilities are not generally very useful, since the high temperatures produce very low water vapor levels. Aging in a high-humidity atmosphere is an effective means of accelerating corrosion; an atmosphere of air at 85% relative humidity and 85°C (hereafter referred to as 85/85) is a common choice. Electrolytic corrosion of Al gates is the failure mechanism of GaAs FETs most likely to be induced by humid atmospheres [12]. However, refractory metal-gold gates are not immune to humidity-induced deterioration; migrated gold resistive shorts (MGRS) have been observed in Si devices [18, 19] as well as in GaAs FETs [20]. To accelerate these two probable mechanisms, the aging of FETs in an environmental chamber must be done with the gates biased.

Thermal and mechanical shock tests, thermal cycling, and bond pull tests are other stresses to which GaAs FETs have been subjected. None of these tests is peculiar to or requires special arrangements for GaAs FETs. The tests and apparatus are the same as those used with silicon devices. There is very little literature on the thermal or mechanical shocking of GaAs FETs; however, Irvin and Loya found no failures from such stresses [12].

5. FAILURE MECHANISMS AND RELIABILITY OF LOW NOISE GaAs FETs

5.1 Burnout

Since the low noise FET has been around several years longer than the power version, there is a larger literature on the reliability and degradation modes of the former [4, 5, 12, 22, 23, 24]. The earliest failure mechanism observed in low noise FETs was gate burnout due to electrical transients. (This problem also vexes the Si MOSFET [25].) Static discharges from personnel or equipment induced transients on leads are the usual sources of these damaging electrical pulses. Studies by various workers have shown that: 1) Al gates are usually more susceptible to such injury than are refractory metals; 2) that pulses in the reverse polarity sense of the Schottky barrier are more apt to cause damage than forward bias pulses; and 3) that the destructive energy threshold varies from approximately 1 to 10 ergs for high voltage pulses (~ 50 v) to 25 to 50 ergs for low voltage (10 v) pulses [5, 22, 23]. Appropriate handling precautions such as grounding of personnel, deionizing air streams, and shielding can prevent this kind of damage to low noise GaAs FETs. This failure mode is not a time-temperature dependent mechanism and, though certainly fatal, it is avoidable if good practices are employed, unpredictable if such practices are not followed, and not a reliability problem in the sense of the present chapter.

Another cause of low noise FET burnout is the uncontrolled oscillation which may be readily induced by RF circuit mismatches and inappropriate bias circuits. These failures, like static-discharge burnout, are avoidable and not a reliability problem to be discussed further in this chapter.

The above mentioned causes, together with lithographic defects and dirt, are the principal causes of burnout in low noise GaAs FETs. Thus, careful fabrication and post-fabrication inspection and careful handling and operating procedures effectively reduce burnout to such rare instances that burnout is not a significant problem in low noise GaAs FETs.

5.2 Catastrophic Failure Mechanisms

The time and temperature dependent failure modes of low noise GaAs FETs, which are therefore acceleratable and eventually amenable to some degree of prediction, may be divided into two categories: catastrophic and gradual. The former are catastrophic in the sense of a sudden, quantum deterioration, distinctly altering both the dc and RF properties of the device. These failures are due to gate destruction by either phase formation, electromigration, field aided migration or electrolytic corrosion. The result of phase formation and electromigration is usually an open gate; the other two mechanisms may produce either an open or a shorted gate. Since low-noise devices usually only have a few gates, the loss of one of them is immediately apparent in both the dc and RF characteristics (no pinch-off and loss of gain). If the device has a multitude of gates (like a power FET) the loss of a single gate may make a relatively small change in the characteristics, but this is not apt to be the case for a low noise FET. The gradual failure mechanisms (which will be discussed in the following section) are those which produce a slow deterioration of noise-figure and/or gain. This RF degradation may or may not be accompanied by a significant change in dc (I-V) characteristics, and is attributable to deterioration of drain and source contacts and of channel properties.

Figure 8 illustrates a case of Au-Al phase formation with consequent formation of voids in the gate structure. This device has Al gates with a Au/Pt/Ti (read Au over Pt over Ti) bonding pad. The Pt and Ti layers, among other functions, were supposed to isolate the Al and Au from each other and thus prevent Au-Al phase formations.

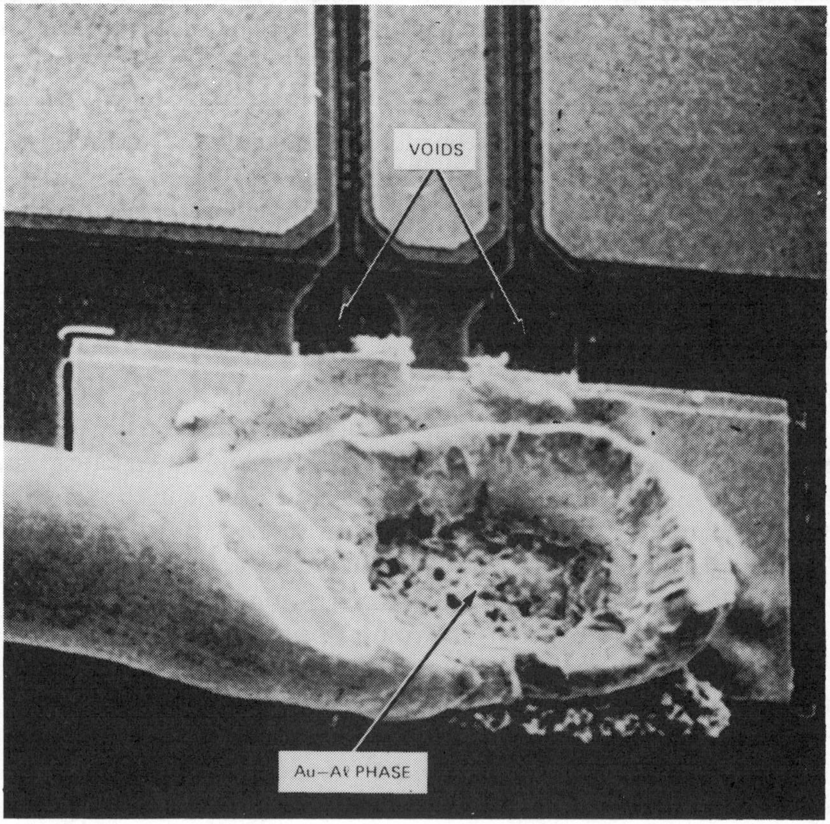

Figure 8*/SEM photo of Au-Al phase at gate-bonding pad of early model low noise FET after aging at 250°C with bias (from Irvin and Loya [12]).*

Likewise, the Ti layer should inhibit Al-Pt phase formation. However, the force of bonding of the Au lead ruined the integrity of the Pt/Ti layers, thus permitting Al-Pt and especially Al-Au interdiffusion. The large supply of gold acts as a sink for the available Al, producing the voids shown. A better structure is shown in Figure 9. In this case, the Au/Pt/Ti bonding pad (not visible) is separated laterally from the Al gates and connected only by a Au/Pt/Ti bridge. This prevents the bonding pressure from directly destroying the Al-Au isolation. However, slight fabricational vagaries, such as redeposition of Au during ion-milling, have nevertheless caused Au-Al contact and the ensuing loss of gate continuity. An even more reliable structure uses Ti only in the pad-to-gate bridge. But this costs gate resistance and an additional mask.

It will be noted that in Figure 9 the voids appear at the point in the gates of maximum current density and also where the current density *gradient* is maximum and is in the direction of the "electron wind" (assuming negative gate bias). The latter facts suggest that the real villain may be electromigration instead of phase formation. Irvin and Loya found evidence that *both* phenomena were active [12]. When devices were aged *without* bias, the median life was 3 to 10 times longer, suggesting an electromigration contribution, but also showing that void formation occurs in the absence of current flow. No voids were ever found (among over 100 inspected examples of this failure mode) in which phase-formation was not also visible.

Since two mechanisms are involved here, an activation energy is not entirely appropriate. However, Irvin and Loya showed that an activation energy around 1 eV is expected for each of the separate mechanisms in this case (taking into account the gate current-temperature dependence) and that a similar value was actually observed for the joint phenomenon. The median lifetimes they observed varied greatly, with many slices showing no tendency at all to this failure mechanism. In the worst case, however, a median life of 94 hours at 250°C (with bias) was observed. Thus, it is vital that design and fabrication techniques be chosen that minimize the possibility of phase formation. Furthermore, since the disease varies from slice to slice, a screening procedure involving the aging of a sample lot from each slice is a worthwhile precaution. Some other reports on gate destruction by the mechanism of phase formation (and its prevention) are those of Huang et al [23] and Lundgren [5].

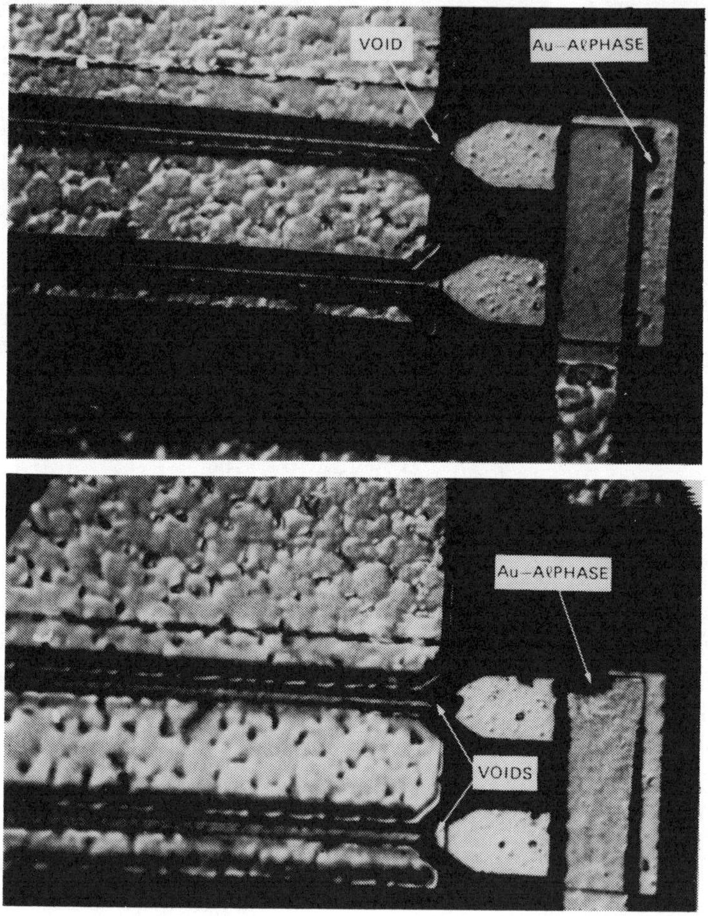

Figure 9*/Optical photos of Au-Al phase and consequent open gates in later model low noise FET after aging at 250°C with bias (from Irvin and Loya [12]).*

Figure 10 shows a case of electrolytic corrosion of the Al gate of a low-noise FET. This device had been aged for 48 hours, uncapped and unpassivated, in an 85/85 atmosphere, with 2 V negative bias on the gate (with respect to the drain, source floating). Destruction of the gate is evident and not unexpected. It is more surprising to find gate corrosion in an hermetically sealed, leak tested package, as shown in Figure 11 for a device aged 1,816 hours ar 85/85. In this case, the corrosion is blamed on a trace of chlorine and water absorbed on the surfaces of the chip or package before sealing. Such instances of electrolytic corrosion due to trace contaminants have been reported for Si devices [19]. One cannot exclude the possibility of a package leak below the detectable limit (10^{-8} std cm^3/sec). However, Irvin and Loya report that chips which had visible traces of photoresist on them had a very high incidence of gate corrosion when sealed and bias-aged at 85/85; hence residual contamination is thought to be the most likely reason for electrolytic gate corrosion in these cases [12].

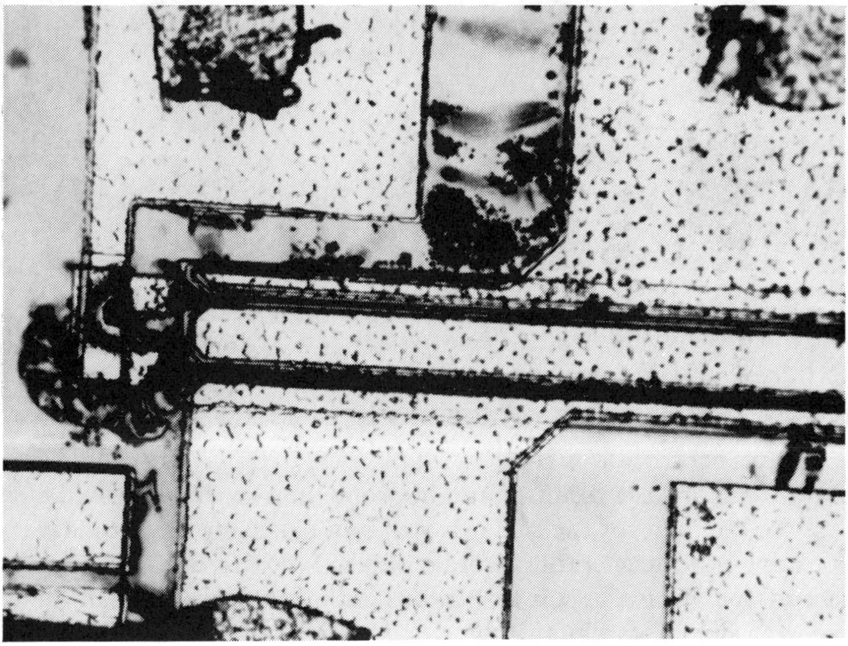

***Figure 10**/Photo of low-noise FET after 48 hours of aging, uncapped in 85°C/85%RH chamber, showing electrolytic corrosion of Al gate (from Irvin and Loya [12]).*

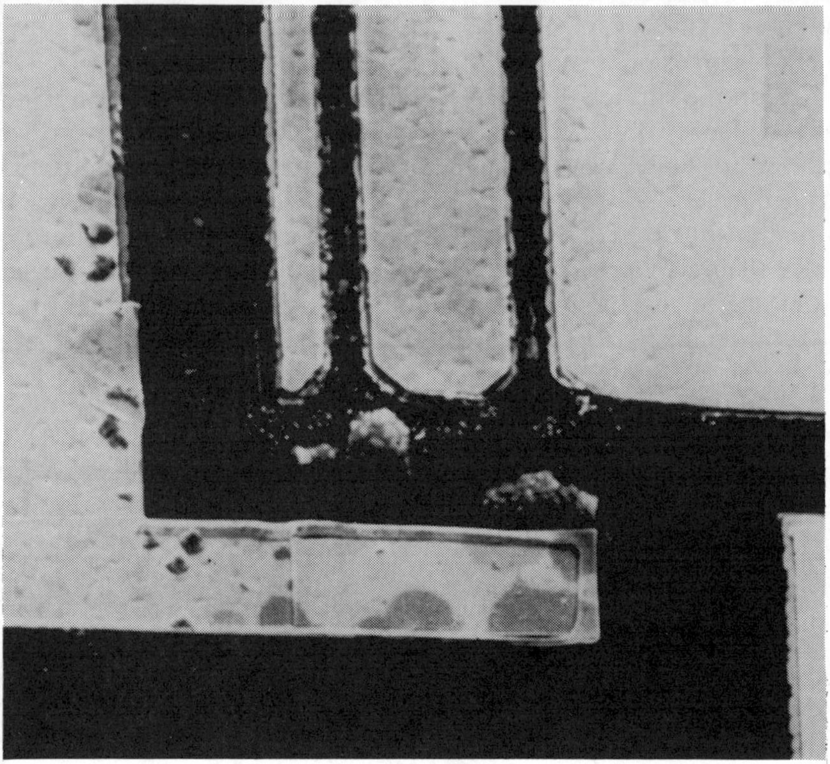

Figure 11/SEM photo of corroded gate of FET after 1816 hours of aging in sealed, leak-tight package. Analysis reveals traces of Cl at corrosion site (from Irvin and Loya [12]).

Electrolytic corrosion is accelerated by increased humidity and temperature mainly as a result of and in proportion to the increased electrical conductivity of the surface. Sbar and Kozakiewicz [26] have determined the acceleration factors of an 85/85 atmosphere with respect to operating ambient for various encapsulations. Though the absolute value of conduction on a GaAs surface may differ from that on Si, Si_3N_4, or alumina, the temperature and humidity dependence are expected to be similar. Irvin and Loya [12] have considered several possible operating circumstances. They calculate, for example, that in an outside ambient of 30°C/25% RH (where an equipment bay operating at 60°C would have a local ambient of 5% RH) the median life for electrolytic corrosion of an unprotected, Al-gate, low-noise GaAs FET would be 7 x 10^5 hours, corresponding to a failure rate of 1,000 FITs after 5 years. For an outside condition of 30°C/100% RH (producing 60°C/25% RH in the apparatus bay) the median life is projected to be 3.5 x 10^4 hours and the failure rate 20,000 FITs. For an unprotected low-noise chip sealed in an amplifier with trace water and chlorine contaminants, the failure rate could be as high as 50,000 FITs. However, in the case of a clean, hermetically sealed device, electrolytic corrosion is effectively prevented and no failures due to that mechanism are anticipated.

Anderson et al. [20], in a comparative study of Al-gate and gold-refractory gate GaAs FETs, also observed rapid electrolytic corrosion of Al gates in humid atmospheres. They found the gold-refractory gates to be less susceptible to failure, though these gates also were eventually degraded by leakage paths due to migrated gold. They did not estimate failure rates.

5.3 Gradual Degradation

While the catastrophic failure mechanisms just described are related entirely to the metallizations and are found in many Si devices as well as GaAs FETs, the gradual degradation mechanisms occur in the GaAs itself. The latter failure modes involve either the diffusion of impurities or defects in the GaAs or at its interfaces or the interdiffusion of GaAs and a contacting metal.

Contact degradation and an associated decrease in I_{DSS} were reported in one of the earliest papers on gradual degradation of GaAs FETs [4]. In that work, the devices were aged without bias, only dc (no RF) properties were monitored, and an activation energy of 1.8 eV was observed. More recent studies have reported activation energies of 0.8 to 1.1 eV, again without bias [24, 41]. The mechanism involved is the interdiffusion of GaAs and the ohmic contact metal as discussed in section 2 of this chapter and studied by various workers [1, 2, 3]. However, in devices utilizing an effective barrier to interdiffusion in the ohmic contacts, the degradation behavior is substantially different. Irvin and Loya [12] studied FETs which employed Ag and Ti Pt barriers between the AuGe ohmic contact metal and the final heavy gold metallization. The results of their study of gradual degradation in low noise GaAs FETs may be summarized as follows:

1) There are significant differences among slices in the way the dc (I_{DSS} and R_S and contact resistivity) and RF properties (noise figure and gain) change upon aging. The variations in the former are probably attributable to the contact alloying cycle, which is considered the least precisely controlled operation in fabrication. Variations in the aging of RF qualities are probably due to traps and defects which form in the active layer or its interfaces. (A recent work by Omori, et al. [24] also attributes a severe degradation of performance to changes in the active layer, with an activation energy of 1.5 eV).

2) There is little correlation between changes of the dc and changes of the RF properties. Radical deterioration of contact resistance (50-fold in one case) was observed with only minor or no change in gain and noise figure. Conversely, the RF properties sometimes degraded without any change in R_S and I_{DSS}.

3) Both dc and RF properties generally degraded more rapidly (by as much as a factor of 5) when aged with bias than without bias. In fact, aging only without bias would lead to erroneously optimistic failure predictions.

4) The median life at 250° with bias, where failure is defined as a 0.2 dB increase in noise figure or 0.8 dB change in gain, was at least 1,500 hours. The activation energy was 0.8 to 1.0 eV. On this basis, a median life at 60°C channel temperature of 10^7 hours is projected, corresponding to a failure rate of less than 2 FITs in 20 years of service.

5.4 Summary of Some Low-Noise GaAs FET Reliability Papers

Seven of the more extensive published studies of the reliability of low-noise GaAs FETs are very briefly summarized in the following paragraphs. (Temperatures are estimated channel temperatures.) The underlined source and drain elements are those present during alloying.

Abbott and Turner [22] (1976): Metallizations studied were Al and Ni gate; <u>AuGe</u>, <u>In/AuGe</u>, <u>Ni/AuGe</u>, and Au/Cr/<u>Pt/AuGe</u> sources and drains. Some units were passivated with SiO. Stresses included voltage spikes to gates, 250°C DC and 100° RF aging, and long duration (50,000 hours) room-temperature RF life tests. Principal failure mechanisms were 1) gate burnout from voltage spikes, Ni gates tolerating about 6 times more energy/spike than could Al; and 2) electromigration (at 250°C, 10^5A/cm^2) of sources and drains with all of the metallizations listed. On the basis of an assumed activation energy of 1 eV, an MTTF of $> 10^7$ hours is predicted at 70°C. No failure distributions, failure rates, confidence limits or sample sizes are given. Changes with aging of RF and dc parameters were not reported.

Huang et al. [23] (1979): Metallizations studied were Al gates with Au/Pt/Cr/Ti bonding pads; Au/Pt/Cr/AuGe drains and sources. Units were passivated with SiO_2. Stresses included voltage spikes to gates and unbiased aging at 265°C and 300°C. Principal failure mechanism was gate burnout from voltage spikes. Extensive data are presented on effect of energy, voltage and polarity of spikes. The formation of intermetallics in the gate structure was specifically investigated and none was found up to 500 hours at 300°C or 1500 hours at 265°C. On this basis, an MTTF $> 10^7$ hours at 100°C is projected. No failure distributions, failure rates, confidence limits, or sample sizes are given. Changes, if any, of RF and dc parameters with aging were not reported.

Irie et al. [4] (1976): Metallizations involved were an Al gate with Au/Pt/Ti bonding pad and an Au/Pt/Ti/Pt/AuGe source and drain. No passivation is mentioned. Stresses applied were voltage spikes to the gates and drains, unbiased aging at 227°C, 259°C, aging at 150°C with reverse gate bias only, and room temperature, 500 hours, 120 mW and 240 mW power burnin. Room temperature drift of dc and RF parameters was monitored and attributed, where appreciable, to wafer quality (traps at the interface). Failure mechanisms observed were 1) burnout of gate or drain when spiked, the drain being more likely to fail than the gate for voltage less than 40 V (reverse polarity on the gate); 2) contact resistance increase during high temperature storage. Only I_{DS} (below saturation), which reflects source and drain contact resistance (among other things), was monitored in the high temperature tests. An activation energy of 1.8 eV was extracted. The effect on RF parameters was not measured but implied from equivalent circuit analysis. An MTTF of 3 x 10^8 hours at 100°C is projected from these data. 116 devices were involved in this study and the sample size in any one test was 18 or 20. No failure distribution, failure rates, or confidence limits are given.

Irvin and Loya [12] (1978): Metallizations studied were an Al gate with Au/Pt/Ti bonding pad and Au/Pt/Ti/Au/Ag/AuGe source and drain. The devices were not passivated. Stress-aging conditions included temperatures of 228°, 258°, and 284°C both without bias and with full operating dc bias, long-term (6,000 to 14,000 hours) dc life testing at 35°, 96° and 188°C, 85/85 environmental chamber with negative gate bias, and temperature cycling under bias. RF and dc parameters were periodically measured as aging progressed. Failure mechanisms observed were electromigration and intermetallic phase formation in the gate structure, electrolytic corrosion of gates, and gradual degradation of channel properties. The measured activation energy of the phase formation/electromigration process was near 1 eV as was also the gradual degradation mechanism. For a hermetically sealed device, a median life (based on measured RF parameters) over 10^7 hours at 60°C is projected, with a corresponding failure rate of 40 FITs after 20 years of service. Failure rates for various non-hermetic situations are also given. About 1,500 devices were involved in this study; failure distributions are presented and confidence limits are discussed.

Lundgren [5] (1978): Metallizations studied were Al and Au/Pt/Cr/Ti gates and Au/Pt/Ti/Pt/AuGe and Au/Pd/Cr/Au/Ni/AuGe sources and drains. The devices were protected variously with SiO or SiO_2 or not at all. The stresses applied included pulsed dc and pulsed and CW RF inputs to the gates and aging with and without dc bias at 200° to 245°C. The failure mechanisms were 1) gate burnout under dc or RF stress, in which the Au/Pt/Cr/Ti gates were not significantly different from Al, 2) rapid deterioration of the Au/Pt/Cr/Ti gates under bias at elevated temperature, and also of certain passivated Al gates (but not unpassivated Al), 3) intermetallic phase formation in the gate structures, and 4) contact resistance increase. The RF parameters of some units were monitored, but failure data were based on dc parameters. An activation energy of about 1 eV was observed. The study involved 230 devices. Failure distributions are presented though no failure rates or confidence limits are given. An operational MTTF in the 10^6 to 10^7 hour range is indicated.

Mizuishi et al. [41] (1979): Metallizations studied were an Au/Mo gate and Au/Ni/AuGe source and drain. The chips were passivated with a "sintered organic material." Stress conditions were storage (no bias) at 175° to 265°C. DC parameters were monitored periodically at room temperature for all devices and correlation of noise figure with drain conductance was established for a limited sample. The failure mechanism observed was ohmic contact degradation with a 0.8-1 eV activation energy. Degraded contacts were analyzed by ion-microspectroscopy and Ni diffusion into the GaAs was observed as well as Ga out diffusion. An MTTF of 10^8 hours at 80°C is projected. The study involved 400 units. No failure distributions, failure rates, or confidence levels are given.

Omori et al. [24] (1980): Metallizations investigated were (primarily) an Au/TiW gate and an Au/TiW/Au/Ni/AuGe source and drain. No passivation is mentioned. The stress aging was with full dc bias, partial dc bias ($V_g = 0$, $I_d \neq 0$), and without bias at temperatures to 300°C. The dc characteristics and S_{21} at 6 GHz were monitored and a 1 dB drop in S_{21} was defined as failure. The failure mechanisms observed were ohmic contact degradation with 1.05 eV activation energy in the partial bias and 0-bias aging, and a degradation of channel properties (decrease in net donor density) when aged with full dc bias. The activation energy of the latter process was 1.5 eV, and ion implanted channels degraded more slowly than VPE layers. An MTTF of 10^6 hours at 150°C was predicted for the better material. The number of devices involved in stress-aging is not mentioned but the total device-hours is stated to be 2 x 10^5. No failure distributions, failure rates, or confidence limits are given.

5.5 The Present Status of Low-Noise GaAs FET Reliability

In sum, these papers report a variety of failure mechanisms, including gate burnout, drain burnout, electromigration, intermetallic phase formation in the gate structure, electrolytic corrosion of Al gates, migrated resistive shorts of Au/W/Ti gates, ohmic contact degradation, and channel degradation. However, with the exception of channel degradation, the means of preventing or at least sufficiently suppressing all these mechanisms have been demonstrated. A failure rate due to channel degradation of less than 2 FITs in 20 years service has been projected.

The study of failure mechanisms in low-noise FETs continues, under both commercial and military sponsorship. Special attention is being given to non-aluminum gate structures in view of the recognized pitfalls of Al. Of course, Al also has certain important advantages, among them facility of fabrication, nonreactivity with GaAs except at relatively high temperature (> 400°C) [27], and the published stress-aging results that show that properly designed and fabricated and hermetically sealed devices have excellent reliability projections [12]. However, the requirement of hermetic sealing (in the absence of some as-yet-undeveloped tenacious, impervious, pinhole-free, low loss passivation) is quite restrictive, since it prohibits the use of unpackaged chips unless the whole amplifier is absolutely free of trace water and halide contamination and can be itself hermetically sealed. So far, no exhaustive reliability study has been reported which would establish the superiority of the non-Al-gate low-noise GaAs FET. In particular, it has not been shown that refractory-metal-gate FETs are unaffected by moisture.

6. FAILURE MECHANISMS OF GaAs POWER FETs

6.1 Metallization Failures

The temperature and current regimes of operation of the power FET are much more severe than those of the low-noise FET. For example, in an equipment bay where the ambient temperature is allowed to rise to 52°C, the maximum channel temperature of a 6-mm (2W) power FET could be 110°C, whereas for a low-noise device the maximum would be only 60°C. Thus, for a 1 eV failure mechanism, such as Au-Al phase formation, the median life of the power FET would be 2 orders of magnitude shorter than for the low-noise FET, if both operated at their respective maximum temperatures. The corresponding failure rates could soar from a negligible 1 FIT at 60°C to a somewhat perturbing 1,000 FITs at 110°C. This does not mean that the same metallization cannot be used in the power FET, only that the precautions taken to isolate Au and Al from each other must be very secure. As subsequent data will show, this is indeed possible.

The dc gate current density of a power FET can be in the range of 10^4 to 10^5 A/cm^2, with peak values each RF cycle at least 10 times higher. In contrast, the current density in the gate of a low-noise FET is about 10^3 A/cm^2 and the RF swings are insignificant. Using an activation energy of 0.5 eV as typical for Al electromigration, and assuming a quadratic current density dependence for void formation in this current density range, the expected median lifetime for failure by electromigration in the gates is shorter by a factor of 10^{-3} to 10^{-5} for the power FET as compared to the low-noise model. As a result, electromigration failure in an Al power FET gate is a real possibility, whereas it's extremely remote in the low noise version (if unassisted by Au-Al phase formation as described in section 5.2). The current density in a power FET source or drain stripe is of the same order or even higher than the gate current densities. Gold is universally used as the final metallization on sources and drains and has a much lower tendency for electromigration than aluminum. This may make up for the higher current density but electromigration failure in a power FET source or drain is also a distinct threat and has been reported by Abbott and Turner [21].

Experimental results confirm these predictions, though the situation is not as bad as might be feared. Observations on SiO_2 protected 6 mm Al-gate power FETs, aged at 260°C (channel) with RF applied, have shown some cases of gate electromigration [17]. A more recent study of SiN protected, Al-gate FETs, however, found no evidence of gate damage (as gauged by pinchoff voltage, gate resistance or capacitance) after 6,000 hours of RF aging at 250°C [16]. In the latter study, forward gate current densities of about 2 x 10^5 A/cm^2 prevailed during most of the aging. In dc electromigration tests, SiN covered Al stripes of the same cross section and length as the FET gates failed with median lifetimes of 200-400 hours at the same temperature and dc current density [17]. The unexpected endurance of the actual FETs may be due to RF effects or to the favorable polarity of the gate current, as mentioned in Section 4.

To avoid the phase formation and electromigration to which Al is prone, much effort is being devoted to the development of power FETs with refractory metal-gold gates. So far, there is no evidence that such structures are actually more reliable than good Al-gate devices. Though some comparisons have found the refractory gates more durable [28], the better Al-gate devices [16, 17] actually have observed median lifetimes at high temperature at least 50 times longer than those reported for refractory metal-gold devices [6, 28]. Figure 12 superposes the observed MTTFs under RF aging of Al-gate units (from unpublished work by Irvin [17]) on the data plot of the

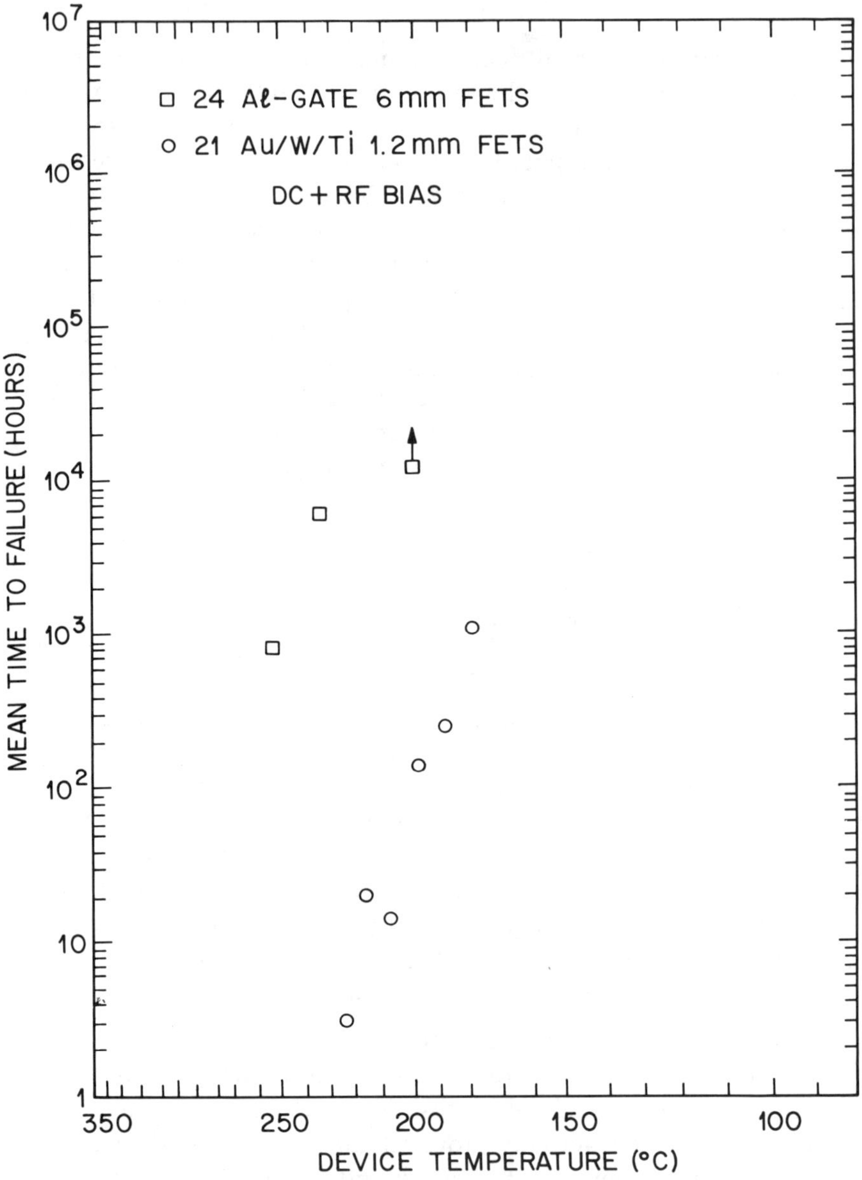

Figure 12/Comparison of MTTFs of Al-gate FETs and Wu/W/Ti-gate FETs under RF aging conditions. One dB decrease in output power is defined as failure. The arrow means that the median failure had not yet been observed at the time shown and exceeds the duration of the experiment. (From Irvin [17] and Cohen and MacPherson [28]).

Au/W/Ti gate results reported by Cohen and MacPherson [28]. The Al-gate devices were 6 mm devices, aged with a 4-GHz drive of 28 dBm or 105mW/mm. The Au/W/Ti-gate FETs were 1.2 mm units and were aged with a 8-GHz input of 100mW or 83mW/mm. Failure for both types of devices was defined as a 1 dB decrease in output power, or burnout if that occurred first. It is evident that the MTTF of the Al-gate units is about 1,000 times longer than that of the Au/Ti/W-gate FETs at 200 to 250°C. It has not been proven, however, that electromigration is the primary failure mechanism in either case.

Figure 13 shows a comparison of stress-aging data for Al-gate and Au/W/Ti-gate power FETs obtained under DC conditions [6, 16]. In this case, the data for the Al-gate units are from Fukui, et al. [16], who defined failure as a 0.2 dB decrease in saturated output power capability; failure criterion for the Au/W/Ti-gate units remained a 1 dB decrease in output power. The former were 6-mm devices and the latter were 2.4-mm devices. In this comparison, the MTTF of the Al-gate units is about 50 times longer than that of the Au/W/Ti-gate devices, ignoring the substantial difference in failure criteria. Drukier and Silcox attributed the failures among the Au/W/Ti-gate FETs to electromigration in the gates. Thus it appears that the expected superiority of the refractory-gold gate has yet to be fully realized. (In both the above comparisons, the Al-gate data, which were originally given in terms of median lifetimes, have been converted to MTTFs using Figure 2. Also, the MTTF of the dc-aged Al-gate FETs, Figure 13 appears shorter than that of the RF-aged Al-gate units, Figure 12, only because of the more stringent failure criterion in the dc-case.

6.2 Burnout

Unlike the low-noise FET, where burnout is a minor and almost entirely avoidable problem, the power FET has a major burnout problem. Though rarely discussed, the difficulty is actually widespread, as off-the-record conversations with workers in the field reveal. Usually, gradual degradation has been the principal focus in reliability, and burnouts are more or less dismissed without comment, even when they exceed 1/3 of the sample [28]. In some cases, especially in devices with a very large number of source-gate-drain fingers, certain failures are misinterpreted as degradation but are actually due to the

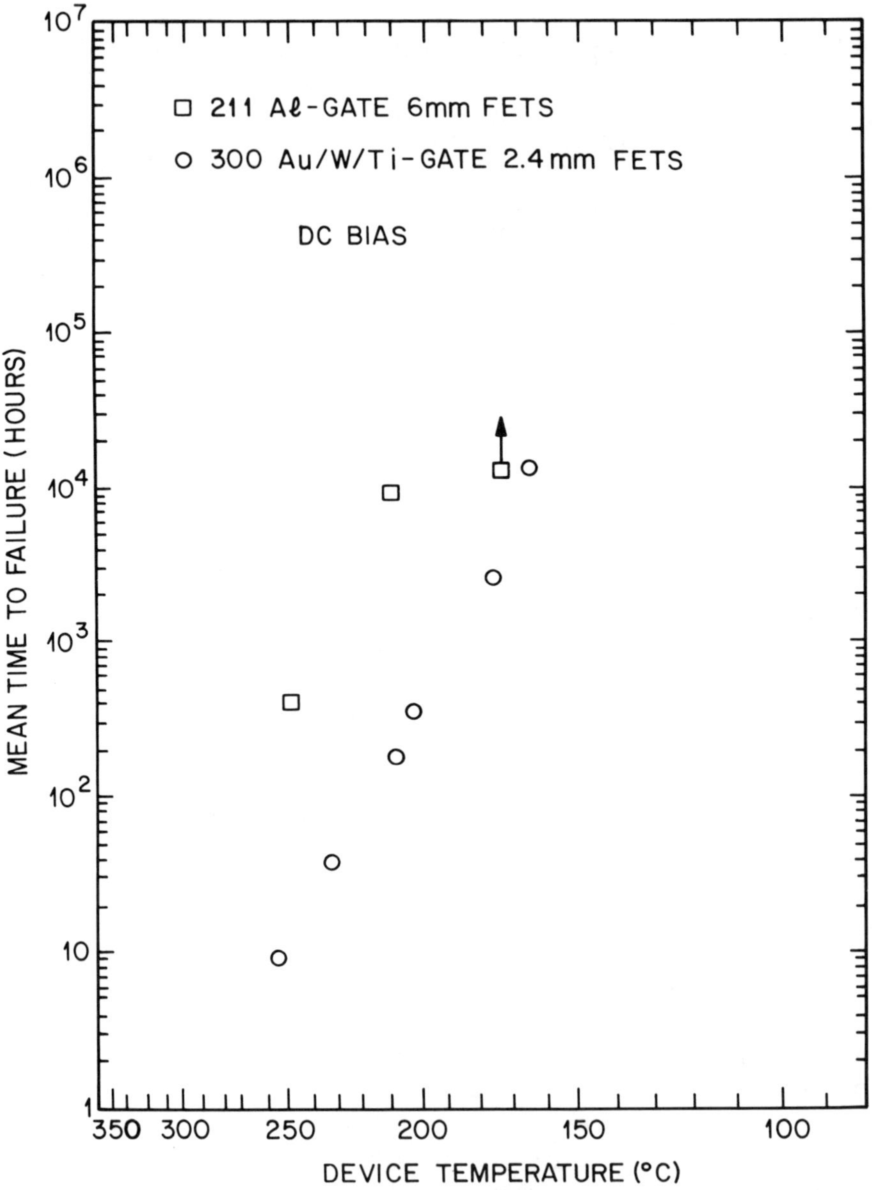

Figure 13*/Comparison of MTTFs of Al-gate FETs and AU/W/Ti-gate FETs under dc aging conditions. For the Al-gate devices, failure is defined as 0.2 dB decrease in output power. For the Au/W/Ti-gate devices, failure is a 1 dB decrease in power. (Fr m Fukui et al. [16] and Drukier and Silcox [6]).*

burnout of one or more fingers. The dead digits are electrically amputated (local vaporization of metallization) and do not short out the remaining live structure, which continues to operate with diminished performance.

Burnout is a thermal runaway process taking place in the buffer or substrate when that normally nonconducting material becomes warm enough for its negative temperature coefficient of electrical resistivity to take over. Morizane et al. [29] showed that this occurs at about 500-550°C. This process is apt to be initiated at the drain contact where nonuniformities and current crowding cause local hot spots. It will occur instantaneously when sufficient drain-source power is applied and dissipated to cause the requisite temperature rise. Wemple et al. [30] showed that burnout during aging or normal operation can occur at more modest and normally "safe" power inputs due to gradual surface changes which produce a high field region at the drain. The resultant power dissipation in that region again produces hot spots and the same thermal runaway ensues.

Only one statistical study of power FET burnout has appeared in the literature [11]. In that study of approximately 500 6-mm, Al-gate GaAs FETs (mostly SiO_2 passivated), aged with dc bias at 170° to 260°C, Jordan et al. verified a log-normal cumulative failure distribution for the burnout of these devices. Figure 14 shows the cumulative failure plot for a group of 119 units aged at 260°C with 14 V, 550 mA drain bias. It was found that both the standard deviation of the distribution and the activation energy varied with temperature. At 260°C, standard deviations (σ) between 1 and 2 were observed, while at 160° and 208°, values of σ nearer 4 were found. The activation energy was around 0.9 eV at 160°C to 208°C and 1.6 eV for the 208° to 260°C interval. This behavior indicates the presence of more than one mechanism. One mechanism has a high activation energy (1.6 eV) and consequently tends to dominate at high temperature. This dominance of one mechanism gives rise to lower values of standard deviation. At lower temperatures, one or more other mechanisms of lower activation energy cause many of the burnouts and the mixture of mechanisms produces a large standard deviation. The high-activation energy burnout mechanism is probably the surface effect described by Wemple et al. [30]. The lower activation energy mechanisms are not identified.

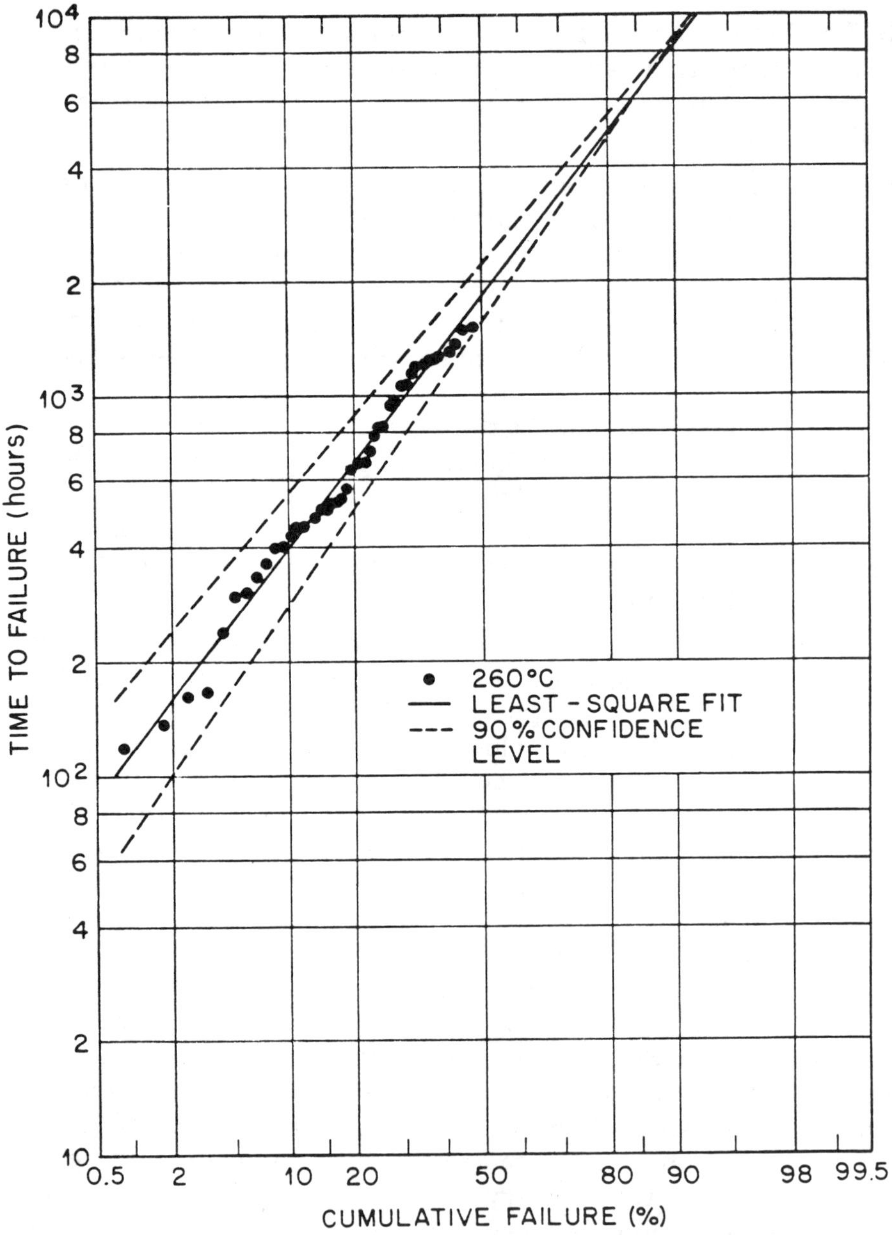

Figure 14*/The log normal projection of time to burnout versus cumulative burnout percent for a group of 116* SiO_2 *passivated power FETs aged under dc bias at about 260°C channel temperature. The solid line is a least square fit and the dotted lines are the upper and lower limits of the 90% confidence level (from Jordan et al. [11]).*

Figure 15 shows the projected burnout rates of the FETs studied by Jordan et al. [11] when operated at 0.53 A, 14 V drain bias in an ambient at 25°C or 52°C (the normal and the maximum temperatures which would be encountered in Bell System microwave repeater stations in the U.S.). The upper and lower confidence limits correspond to the combined effect of 67% confidence limits on the population's median life and standard deviation. It is seen that at a constant 25°C, (channel temperature of 67°C) the expected maximum burnout rate is about 150 FITs, which occurs in the first year and thereafter diminishes. The upper confidence limit would forecast a rate 4 or 5 times higher. At a constant 52°C ambient throughout the life of the device (channel temperature 99°C), the expected burnout

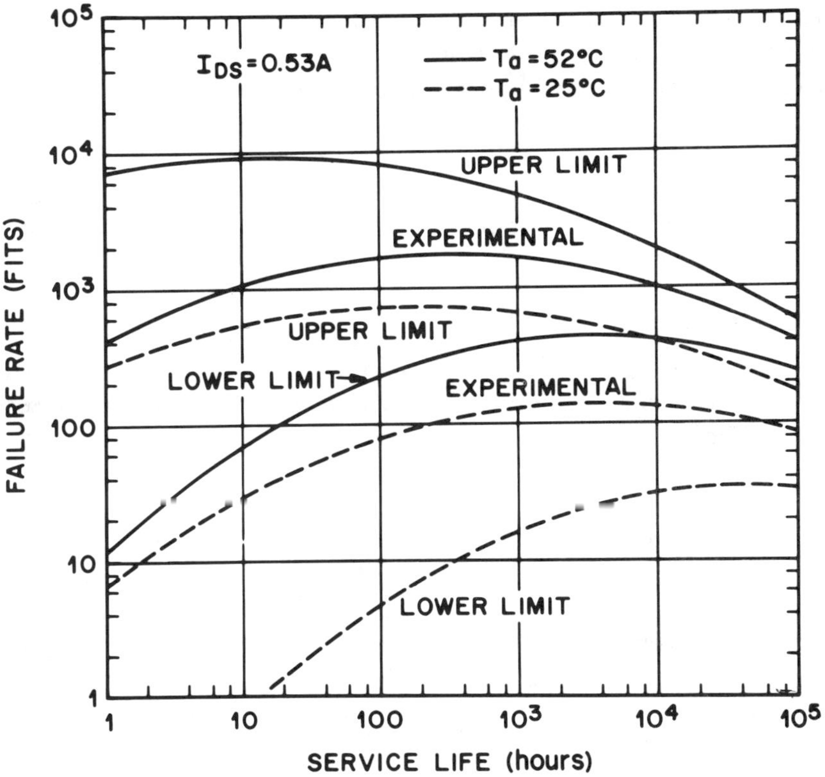

Figure 15/The estimated burnout rate as a function of service life for a type of SiO_2 passivated power FET operating at I_{DS} = 0.53A and at ambient temperatures of 25°C and 52°C. The upper and lower "limit" curves correspond to 67% confidence levels on the populations' median life and standard deviation (from Jordan et al. [11]).

rates would be an order of magnitude larger than at 25°C. (Of course, in no station does such a high ambient temperature exist *constantly.*) If the transistor is operated at a drain current of 0.7 A instead of 0.53 A, the channel temperature will be about 10°C warmer and the failure rates two to three times higher.

Guided by the work of Wemple et al. [30] and extensive stress-aging studies of many design variations [17], a new generation of GaAs power FETs has evolved with high resistance to burnout. A burnout study of this design, which employs aluminum gates and silicon nitride passivation, was included in the work of Fukui et al. [16]. Among 270 FETs and 2 million device-hours of aging, only one burnout occurred. At the time of this writing, that study is the only published investigation which provides a substantial body of data (obtained under dc and RF conditions at temperatures of 175°C to 250°C) confirming a nearly burnout-free GaAs power FET. The very lack of burnouts makes it impossible to determine the activation energy of burnout for these units. However, the most pessimistic assumptions indicate a burnout rate under 500 FITs at 110°C channel temperature and probably much less than 100 FITs, in ten years of service.

6.3 Gradual Degradation

The destruction of the gate metallization, as described in section 6.1, partial burnout in a highly interdigitated device, as mentioned in section 6.2, and increase of contact resistivity, as treated in sections 2 and 5.3, will cause degradation of RF performance in a power FET. Those mechanisms have already been discussed. Two other causes remain: bulk and surface semiconductor effects and gate metal encroachment on the channel. Surface and bulk semiconductor effects were mentioned in section 5.3 as being responsible for the slow degradation of certain low-noise FETs [12, 24] and have also been indicated for power FET degradation [16]. Such surface and bulk effects may be ascribed to 1) a change in the (shallow) donor or acceptor density, 2) traps (i.e., deep levels), and 3) structural defects. Of these three phenomena, traps are the most probable and, in some sense, include the other two. Traps can alter the field distribution, the carrier density, and carrier mobility.

Light sensitivity, hysteresis in the output characteristics, or differences between the dc and pulsed drain current (if not of thermal origin) are indicative of the presence of traps in GaAs FETs. Several workers [31, 32, 33] have investigated traps in GaAs FETs, usually using some form of deep level transient spectroscopy [34]. Meignant et al. [33], report electron traps in the active layer when the latter is grown by v.p.e. at relatively low temperature (650°C) using a GaAs source. They found hole traps, localized at the active/buffer layer interface, in VPE structures grown at higher temperature (750°C) with a Ga source. In all the above mentioned papers the GaAs FETs were unaged. No reports comparing trap densities in GaAs FETs before and after stress-aging or correlating their presence with degradation have been published, though at least one such study has been undertaken [35]. Since traps have demonstrable effects on microwave performance [31], since their movement is stimulated by elevated temperature, electric fields, and carrier recombination [36] (all of which obtain in operating power FETs), and since the channel interfaces as well as buffer and substrate dopants are potential sources of traps, it is to be expected that traps will play a role in the gradual degradation of power FETs. It is noteworthy that recombination assisted propagation of traps was found responsible for Pt-GaAs IMPATT degradation by Ballamy and Kimerling [37].

The interdiffusion of gate metal and GaAs can potentially affect channel properties through two mechanisms. One of these is the advancement of the effective Schottky barrier into the channel, thereby decreasing the available channel thickness. This should be manifested in reduced I_{DSS} and reduced pinch-off voltage. The other mechanism is the dopant action of the diffused metal atoms (or also atoms of oxygen which sometimes reside at the metal/GaAs interface) with resultant alteration of the net donor density in the channel, also reducing I_{DSS}. In a study of the interaction of Al and GaAs in GaAs FET gates [38], no degradation of 3-terminal dc characteristics was found in 24 hour anneals at 275°C, though severe degradation occurred at 450°C. In studies of Al-gate GaAs low-noise FETs [12], median lifetimes of at least 1,500 hours at 250°C were observed and no significant decrease in pinch-off voltage was found. Hence, channel narrowing due to Al/GaAs interdiffusion, though an ultimate possibility, does not appear to be an important reliability limitation for GaAs FETs. The low noise experiments do not rule out the possibility of deleterious doping of the channel, since a decrease in I_{DSS} was sometimes observed [12]. This does not have a first order effect on low noise microwave properties (noise figure and associated gain) but would be expected to degrade the gain and saturated power of a power FET. As will be discussed below, however, the degradation limitations, if any, of the Al gate, are quite acceptable. Research on the interaction of other gate metals with GaAs in actual FET structures has not been published. Of course, numerous studies of the interaction of GaAs and various metals (in non-FET geometries) have been conducted [1, 2, 27]. These indicate several promising candidates from a metallurgical viewpoint, though none without some disadvantages (high electrical resistivity, mismatch of thermal expansion coefficients, propensity to oxidation, etc.). Exploration of the possible undesirable doping properties of these metals in a GaAs FET application has not been reported.

Some details on the nature of the gradual degradation of power FETs are found in the paper by Fukui et al. [16]. Several dc and RF parameters were measured periodically during the course of aging. The behavior of two of these, I_{DSS} and the saturated output power, P_S, (at 4 GHz), is shown in Figure 16. At 250°C (channel temperature), under operational *dc bias but no RF,* the saturated power declined about 5% in 250 hours. The corresponding change in gain (not shown in the figure) was 0.2 dB. The same degree of degradation in P_S took about 6,500 hours at 200°C, whereas at 175°C no measurable change was evident at the time that report was written (8,000 hours). These data give rise to the Arrhenius plot shown in Figure 17. The apparent activation energy of the degradation mechanism is 1.8 eV; in the worst case, taking the possible errors in temperature estimation into account, the activation energy could be no lower than 1.2 eV. Using these values of activation energy, the degradation failure rate (5% decrease in P_S) at 110°C operational channel temperature is predicted to be no higher than 100 FITs at the worst and probably less than 10 FITs in ten years of service.

After special accelerated dc aging (285°C) of similar units, it was found that the gate capacitance and resistance, the pinch-off voltage, and the ohmic contacts had not changed at all, though I_{DSS} and maximum channel conductance had decreased markedly. Detailed analysis of these and other electrical parameters indicated that the essential changes were in the properties of the active channel, namely a decrease of saturated velocity and of low-field mobility. The root cause of this deterioration is not known, but traps are a candidate. It is interesting to note that the degradation of P_S, as shown in Figure 16, does *not* faithfully follow the decline in I_{DSS}. The large decrease of P_S at 500 hours (at 250°C) and subsequent recovery is not due to measurement error, since it has been witnessed repeatedly in groups of units from different slices. There is no such oscillation in

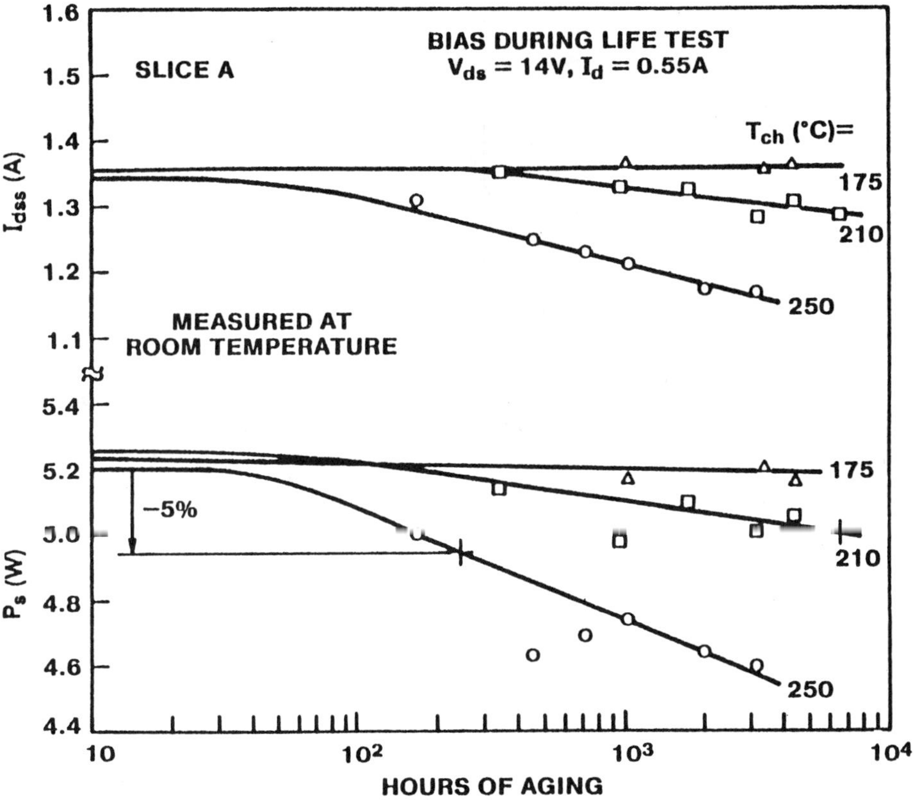

Figure 16/Typical behavior of I_{DSS} and P_S with aging at 250°C, 210°, and 175° (channel temperature) under 14V, 0.55A drain bias, of an improved version of silicon nitride passivated power FET (from Fukui et al. [16]).

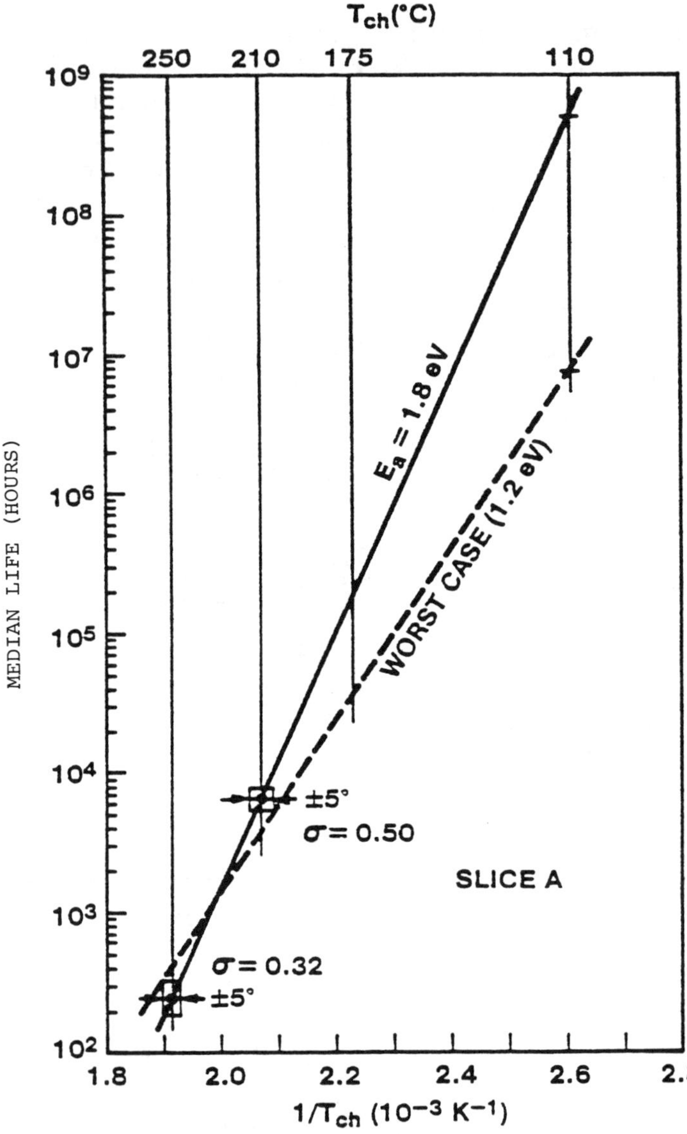

Figure 17/Median life as a function of channel temperature of dc bias aged, silicon nitride passivated power FETs. Failure is defined as a 5% decrease in 4 GHz saturated output power, measured at room temperature. The solid line is based on median values and the dashed line is the worst case derived from the error limits shown (from Fukui et al. [16]).

the value of I_{DSS}, which falls monotonically. This discrepancy in behavior is not presently understood but points to a multiplicity of mechanisms.

It is important to remember that these median lifetimes and failure rate predictions are based on dc-bias aging data. RF-aging data up to 3,000 hours at 250°C (with 28 dBm, 4 GHz drive, as described in section 6.1) of the same types of units which gave rise to Figures 16 and 17, reveal the same degradation rate as under dc conditions. This is demonstrated by Figure 18 for I_{DSS}, which compares the data obtained under dc and RF conditions [16]. More importantly, the saturated power and gain of the RF-aged devices degrade at the same rate as the dc-aged units up to 3,000 hours, at least [39]. This is a very encouraging result. However, the dc data involve far more device-hours of aging than do the RF data. Since dc aging conditions do not fully reproduce actual large signal RF operating conditions, especially with regard to gate currents, further RF aging data are needed for confident predictions of operating degradation rates.

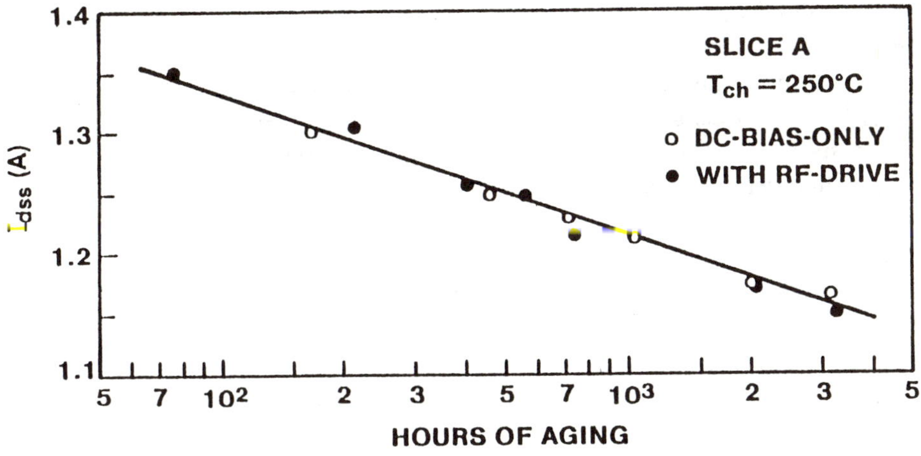

Figure 18/Aging behavior of I_{DSS} for a group of silicon nitride passivated power FETs aged with dc bias only and another group aged under RF drive, both at 250°C channel temperature (from Fukui et al. [16]).

6.4 Summary of Power FET Reliability

In summary, failures in GaAs power FETs may be classified as either due to 1) burnout, 2) metallization weakness, or 3) changes in the bulk and interface properties of the semiconductor. The first of these can be reduced to very low rates through proper design [30] and probable burnout rates below 100 FITs (in 10 years service) have been estimated [16]. In a less effective design, burnout rates (at 110°C channel temperature) of 1,700 to 10,000 FITs are projected [11]. Burnout is highly dependent on details of design and can be quite different for devices of different construction.

Metallization failures take the form of ohmic contact degradation, electromigration in either the source, drain, or gate metal, or, finally, phase formation due to interaction between two or more metals. Contact degradation and phase formation are both the result of interdiffusion which is preventable by the interposition of suitable barriers, physical separation of the intermingling components, or choice of noninteracting metals. Where these precautions are not taken or are inadequate, the resultant failure rate can be catastrophically high and will depend intimately on the details of design and processing. The failure rate is apt to vary from slice to slice, so the stress aging of samples from each slice can detect slices where this failure mechanism exceeds certain bounds. Electromigration is discouraged by use of thick metal layers and a high degree of interdigitation to reduce current densities, good heat sinking, and the avoidance of Al. However, substitutes for Al gates provoke other problems and the best published reliability results to date have been obtained with power FETs employing silicon nitride covered Al gates [16]. The degradation failure rate for these devices (5% decrease in P_S) is estimated at well below 100 FITs (but requires further RF aging for confirmation.)

Burnout and metallization failures are premature failures in the sense that they are avoidable in principal, if not always in fact. If these shoals are avoided, a power FET will founder eventually on degradation of bulk and interface properties. Such changes are usually considered long-term [12, 16], but that is not always the case [24].

TABLE 1

Summary of Time and Temperature Dependent Failure Mechanisms of the GaAs FET

Mechanism and Structure	ML and Aging Condition	Activation Energy	References
Burnout	700 hr/265°C/dc	0.9-1.6 eV	11, 16
Electrolytic Corrosion Al (G)	3 hr (85/85)		12, 20
Electromigration			
Al (G)	168 hr/130°/RF	0.3-0.7	28
	400 hr/260°/RF	0.3-0.7	17
Au (G)	10 hr/260°/dc	0.5-1.0	6
AuGeIn (S, D)	60 hr/250°/dc	1.1	21, 22
Electromigration/AuAl Phase Formation			
Al/Au (G)	94 hr/250°/dc	1.0	12
Gradual Degradation Due to Bulk on Interface Effects			
GaAs (Channel)	30 hr/275°/dc	0.8-1.8	12, 16, 24
Intermetallic Phase Formation			
Al/Au, Al/Pt (G)	94 hr/250°/dc	0.6-1.0	5, 12, 23, 40
Ohmic Contact Resistance Increase			
AuGe/GaAs (S, D)	31 hr/250°/O-bias	0.8-1.8	1, 2, 3, 4, 12, 41

Changes in interface and bulk properties (such as compensation and decreased mobility) may be attributed to formation or redistribution of traps. However, no study of the effect of aging on traps has been published to date.

The principal time-temperature dependent failure mechanisms of low-noise and high-power GaAs FETs are listed in Table 1. The metallizations and the portion of the structure involved are indicated.

The worst reported median lifetimes under accelerated aging are given, along with corresponding aging conditions. A range of activation energies associated with each failure mechanism is shown. This range includes the values obtained from GaAs FET studies as well as those generally reported in the literature. In the case of electromigration, the range is particularly large, since the mechanism is influenced by grain size and therefore deposition conditions, as well as by passivation and any added or incidental impurities. There is no assurance that the values found in one system would be applicable to another. Also, in most cases, the median lifetimes and activation energies have been measured under aging conditions without RF and in some cases even without dc bias; thus the extrapolation of lifetimes under large signal RF operating conditions (i.e., for power FETs) may be presumptive. This table is intended primarily as a guide to the possible failure mechanisms of a given structure and to the computation of a first order, worst case estimate of its median lifetime. The references given in the last column are useful for further illumination of the mechanism in question.

ACKNOWLEDGMENT

The author is indebted to A.S. Jordan for his contributions to section 3 and particularly Figure 3.3. The valuable comments of W.O. Schlosser are also greatly appreciated.

REFERENCES

[1] Christou, A., and K. Sleger, 6th Biennial Conf. on Active Microwave Semiconductor Devices and Circuits, Cornell, 1977.

[2] Sinha, A.K., and J.M. Poate, *Appl. Phys. Lett.*, 23:666, 1973.

[3] Chino, K., and Y. Wada, *Jpn. J. Appl. Phys.*, 16:1823, 1977.

[4] Irie, T., I. Nagasako, A. Kohzu, and K. Sekido, *IEEE Trans. on Microwave Theory and Tech.*, MTT-24:321, 1976.

[5] Lundgren, R., *Final Technical Report*, RADC-TR-78-213, Oct. 1978.

[6] Drukier, I., and J.F. Silcox, 17th An. Proc. Reliab. Phys. Symp., San Francisco, p. 150, 1979.

[7] Peck, D.S., and C.H. Zierdt, Jr., *Proc. IEEE*, 62:185, 1974.

[8] Goldthwaite, L.R., Proc. 7th Nat'l. Symp. Reliab. Quality Control, Phila., PA, p. 208, 1961.

[9] Jordan, A.S., *Microelectronics and Reliability*, 18:267, 1978.

[10] Jordan, A.S., to be published.

[11] Jordan, A.S., J.C. Irvin, and W.O. Schlosser, 18th An. Proc. Reliab. Phys. Symp., Las Vegas, p. 123, 1980.

[12] Irvin, J.C., and A. Loya, *Bell Syst. Tech. J.*, 57:2823, 1978.

[13] Ireson, W.G., *Reliability Handbook*. New York: McGraw Hill, 1966.

[14] Kaplan, E.L., and P. Meier, *Jour. of the Am. Statis. Assoc.*, 53:457, 1958.

[15] Herd, G.R., Proc. 6th Nat'l. Symp. on Reliab. and Qual. Control, p. 217, 1960.

[16] Fukui, H., S.H. Wemple, J.C. Irvin, W.C. Niehaus, J.C.M. Hwang, H.M. Cox, W.O. Schlosser, and J.V. DiLorenzo, 18th An. Proc. Reliab. Phys. Symp., Las Vegas, p. 151, 1980.

[17] Irvin, J.C., (unpublished work).

[18] Wright, J., and E.B. Hakim, 11th An. Proc. Reliab. Phys. Symp., Las Vegas, p. 224, 1973.

[19] Shumka, A., and R.R. Piety, 13th An. Proc. Reliab. Phys. Symp., Las Vegas, p. 93, 1975.

[20] Anderson, W.T., A. Christou, and K.J. Sleger, 17th An. Proc. Reliab. Phys. Symp., San Francisco, p. 127, 1979.

[21] Abbott, D.A., and J.A. Turner in *Gallium Arsenide and Related Compounds (Edinburgh) 1976* (Inst. Phys. Conf. Ser. no. 33a) p. 255. London: Adam Helger/The Institute of Physics, 1977.

[22] Abbott, D.A., and J.A. Turner, *IEEE Trans. on Microwave Theory Tech.,* MTT-24:321, 1976.
[23] Huang, C., F. Kwan, S.Y. Wang, P. Galle, and J.S. Barrera, 17th An. Proc. Reliab. Phys. Symp., San Francisco, p. 143, 1979.
[24] Omori, M., J. Wholey, and J.F. Gibbons, 18th An. Proc. Reliab. Phys. Symp., Las Vegas, p. 134, 1980.
[25] Madzy, T.M., *IEEE Trans. Electron. Devices,* ED-23:1099, 1976.
[26] Sbar, N.L., and R.P. Kozakiewicz, 16th An. Proc. Reliab. Phys. Symp., Las Vegas, p. 161, 1978.
[27] Mukherjee, S.D., D.V. Morgan, M.J. Howes, J.G. Smith, and P. Brook, *J. Vac. Sci. and Technol.,* 16:138, 1979.
[28] Cohen, E.D., and A.C. MacPherson, 17th An. Proc. Reliab. Phys. Symp., San Francisco, p. 156, 1979.
[29] Morizane, K., M. Dosen, and Y. Mose in *Gallium Arsenide and Related Compounds (St. Louis) 1978* (Inst. Phys. Conf. Ser. no. 45) p. 287. London: Adam Helger/The Institute of Physics, 1979.
[30] Wemple, S.H., et al., *IEEE Trans. Electron. Devices,* ED-28:834,
[31] Asai, S., S. Ishioka, H. Kurono, S. Takahashi, and H. Kodera, Supplement to *Jour. of Jap. Soc. of Appl. Phys.,* 42:71, 1973.
[32] Alderstein, M.G., *Electron Lett.,* 12:297, 1976.
[33] Meignant, D., D. Boccon-Gibod, and J.M. Bourgeois, *Electron Lett.,* 15:779, 1979.
[34] Lang, D.V., *J. Appl. Phys.,* 45:3023, 1974.
[35] Chen, F., and J.C. Irvin, (unpublished work).
[36] Kimerling, L.C., *IEEE Trans. Nucl. Sci.,* NS-23:1497, 1976.
[37] Ballamy, W.C., and L.C. Kimerling, *IEEE Trans. Electron. Devices,* ED-25:746, 1978.
[38] Sleger, K., and A. Christou, *Solid-State Electron.,* 21:677, 1978.
[39] Fukui, H., private communication.
[40] White, P.M., B.L. Hewitt, and J.A. Turner, Proc. 8th European Microwave Conference, Paris, p. 405, 1978.
[41] Mizuishi, K., H. Kurono, H. Sato, and H. Kodera, *IEEE Trans. Electron. Devices,* ED-26:1008, 1979.

Chapter 7

GaAs FET Applications

Introduction

Having covered the fabrication technology of GaAs FETs in the previous chapters, we now turn to a discussion of the characterization and microwave applications of these devices. GaAs FETs have been used in almost all microwave applications requiring an active device, making them the true Jack-of-all-Trades in the microwave applications arena.

In the first part of this chapter, Vendelin discusses the characterization of FETs for the dc and RF parameters. He covers the wide range of microwave functions that have been successfully performed by GaAs FETs to date, including their configurations, design principles and performance results. The dual-gate configuration of the FETs has not been explicitly covered in this book; these dual-gate FETs, however, represent a simple extension of the conventional single gate configuration, but with two separate gate electrodes realized between the source and the drain spacing, instead of one gate electrode.

The use of GaAs FETs for microwave amplification in linear, small signal, low noise and power configurations represents the most widespread application of these devices. These applications are covered by Sokolov and Schlosser in the second part of this chapter. After discussing linear amplifier design, they discuss linearity in detail. The general FET circuit aspects are then covered, including the various aspects of microstrip technology, the most common hybrid circuit approach when GaAs FETs are the active devices. The section concludes with a discussion of FET amplifiers including monolithic GaAs microwave amplifiers, which are becoming increasingly important.

The Schlosser and Sokolov contribution to this chapter contains a comprehensive reference list. For this reason, the style of referencing in the text is a departure from the style used throughout the book.

Small-Signal and Nonlinear Applications of GaAs FETs

G. D. Vendelin

Eaton Corporation, Sunnyvale, California.

1. INTRODUCTION

The GaAs FET has become the workhorse of the microwave industry because of the excellent microwave gain achievable from this device. The historical development [1-9] leading to this device is summarized in Figure 1. Today (1981) GaAs FETs are commercially available for low-noise and high-power amplifier applications [10-13]. In addition, these FETs have been applied in other microwave functions which include: oscillators [14-17], mixers [18-19], switches [20-22], attenuators [23-25], modulators [26], multipliers [27-28], and limiters [25].

Historical Development of GaAs FET

1948	-	Germanium Point-Contact Bipolar Transistor
1952	-	Unipolar FET Description
1960	-	Silicon Planar Bipolar Transistor
1965	-	Germanium L-Band Bipolar Transistor
1966	-	Schottky-Barrier Gate Field-Effect Transistor (Ref. 1)
1969	-	Silicon MESFET
1970	-	Silicon and GaAs FET Technology
1972	-	Low Noise GaAs FET Amplifier
1973	-	Commercial GaAs FET
1981	-	Status

Low Noise Performance		High Power Performance	
NF	Frequency	Power (Single Chip)	Frequency
0.7 dB	4 GHz (Ref. 10)		
1.5 dB	6 GHz (Ref. 11)	18.5	4 GHz (Ref. 12)
1.7 dB	12 GHz (Ref. 10)	5.1	8 GHz (Ref. 12)
2.5 dB	18 GHz	1.10	16 GHz (Ref. 12)
		(Four Chips)	
		25 watts	6 GHz (Ref. 13)
		17 watts	8 GHz (Ref. 13)

Figure 1/Historical Development of GaAs FET

The basic principle of operation of the GaAs FET can be explained using Figure 2. Beginning with Ohm's law, the current density between the ohmic contacts (source and drain) of a linear resistor is:

$$i = \sigma E \tag{1}$$

For n-type epitaxial material the conductivity is:

$$\sigma = q\mu N_D \tag{2}$$

and the mobility is related to the drift velocity by:

$$v = \mu E \tag{3}$$

For an epitaxial layer of thickness t and width Z, the total current through the resistor becomes:

$$\begin{aligned} I_{DS} &= iA \\ &= (\sigma E)(tZ) \\ &= qN_D vtZ \end{aligned} \tag{4}$$

As the electric field is increased in the resistor, the velocity of the carriers (and therefore I_{DS}) saturates. Next, a Schottky-barrier gate electrode is fabricated between the ohmic source and drain contacts. With the source-gate potential at zero volts, the drain-source current again has a linear and saturation region (I_{DSS}) with a reduced channel thickness t_o. An increasing negative voltage between gate and source will eventually pinch off the channel and reduce the resistor current to essentially zero. The three-terminal MESFET is a drain-source resistor which has a variable current controlled by the gate-source reverse bias voltage.

The RF amplification occurs when the transistor is dc biased in the saturation region ($V_{DS} > V_{DS}$ sat) and an RF input signal is applied between gate and source (v_{gs}). This signal will modulate the RF current in the channel and result in an RF component of drain current given by:

$$i_{ds} = g_m v_{gs} \tag{5}$$

A large gain occurs for transistors optimized for transconductance per unit gate width or g_m/Z. The device theory shows this parameter will be maximized by dc biasing in the saturation region and minimizing the gate length, Lg, to 0.5 - 1 μm. Further reductions in gate

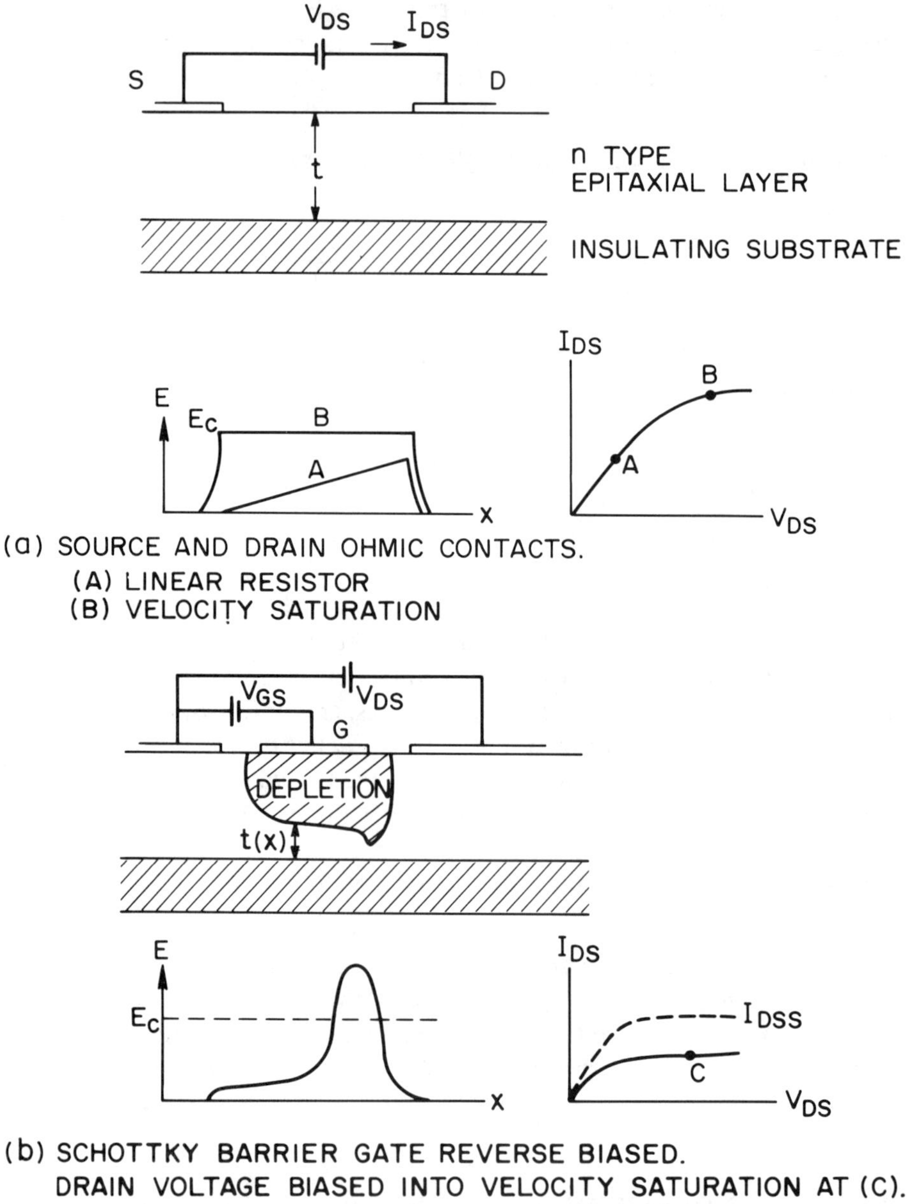

Figure 2/GaAs FET Cross Section

length below 0.5 μm have not increased the microwave gain due to device parasitics, largely the gate resistance.

For present day GaAs FETs, the maximum transconductance is:

$$\frac{g_m}{Z} = \frac{\epsilon v_{sat}}{t} \cong 0.067 \frac{\text{mmho}}{\mu\text{m}} \tag{6}$$

Further improvements in this parameter are expected from other III-V compounds such as InP, InGaAs, and related compounds.

In summary, the field-effect-transistor is a variable resistor with the voltage at the gate controlling drain current at the load. Amplification occurs when the input gate-source RF voltage produces a large output current at the drain-source terminals. The microwave circuit designer must characterize the transistor at the optimum dc bias point, usually resulting in a two-port s-parameter description of the device. Next, the engineer must design input and output matching circuits which maximize the overall rf gain of the device.

This chapter will describe the important dc and RF parameters of typical GaAs FETs. Using two-port network theory, the design approach for GaAs FET amplifiers is presented in detail. Next the additional applications of this device are briefly described and summarized. In the second part of the chapter, the design considerations for high power GaAs FET amplifiers will be presented.

2. DC CHARACTERISTICS OF GaAs FETs

The curve-tracer display of a small-signal GaAs FET is given in Figure 3. The important dc parameters are:

I_{DSS} — The maximum drain-source current which flows in the saturation region. Typical bias point for measurement is:

$V_{GS} = 0$, $V_{DS} = 5$ volts

g_m — The ratio of $\Delta I_{DS}/\Delta V_{GS}$ which is approximately a constant in the saturation region of the transistor ($V_{DS} > V_{DSsat}$).

V_p — The pinchoff voltage is the value of negative gate-source voltage required to reduce the drain current to zero ($\sim 10\ \mu$A).

These parameters are approximately related by $I_{DSS} \cong g_m V_p$.

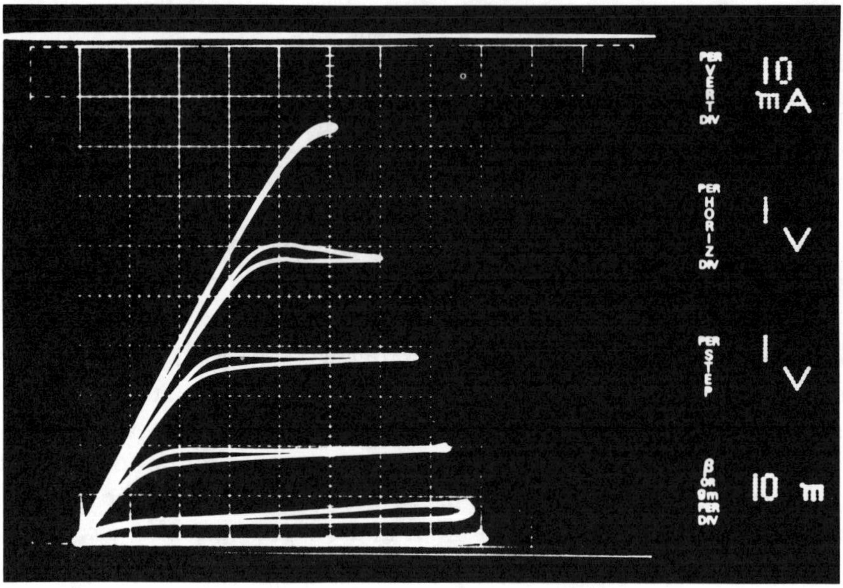

Figure 3/I_{ds} vs. V_{ds} for GaAs FET

The curve tracer display may be viewed as three different regions:

(1) The linear region at low V_{DS} where the device has not yet reached saturation. This region is important for mixer applications and other nonlinear effects (note the anomaly in names). The linear region is characterized by carriers which follow a linear relation between voltage and current (i.e., Ohm's law). When V_{DS} reaches $V_{DS(sat)}$, the channel is pinched off at the drain edge so that the current can no longer increase (ideally).

(2) When the channel is pinched off, the current is in the saturation region where the current is essentially independent of V_{DS}. For short channel devices, such as GaAs FETs, the drain-source electric field is sufficient to reach another "saturation" level where the carriers can no longer increase in velocity. This saturation region occurs at $E > 3$ kv/cm. For microwave GaAs FETs this voltage is ~ 1-2 volts. Most amplifier devices are operated in the saturation region, particularly the small signal amplifiers. The

saturation region extends between the linear region and the burnout region. In the saturation region the carriers travel at the saturated drift velocity.

(3) The burnout region or high field region occurs at large V_{DS}, typically 10-20 volts, depending upon drain current and the device technology. Considerable effort is needed to extend this burnout voltage to a higher value for both higher power and improved reliability. For small-signal low-noise applications, the burnout region is less significant since the optimum RF performance occurs at low fields.

The burnout mechanism is related to both temperature and electric field. Minimum channel temperature is required for device reliability, with an upper limit of 175°C recommended for power GaAs FETs. Before explaining the high-field burnout mechanism, a more detailed description of the channel current at high field (Figure 2b) is required.

The drain current in the saturation region is given by (Eqn. 4):

$$I_{DS} = qn(x)v(x)t(x)Z \tag{7}$$

where n(x) is the density of conduction electrons, which is N_D, the doping density of the epitaxial layer at low electric field.

Referring to Figure 4, increasing the drain voltage above V_{DSsat} results in an electron accumulation layer at the drain edge of the channel. Since current continuity in the channel is required ($I_{DS}(x)$ = constant), the channel may be divided into two regions:

(1) The linear region of the channel where:

$E(x) < Ec$

$v(x) < v_{sat}$

(2) The saturation region of the channel where:

$E(x) > Ec$

$v(x) = v_{sat}$

In the saturation region, t(x) is reducing because of the increasing depletion layer due to the ohmic drop in the channel. Since v(x) is constant, n(x) must increase to give current continuity. Thus an electron

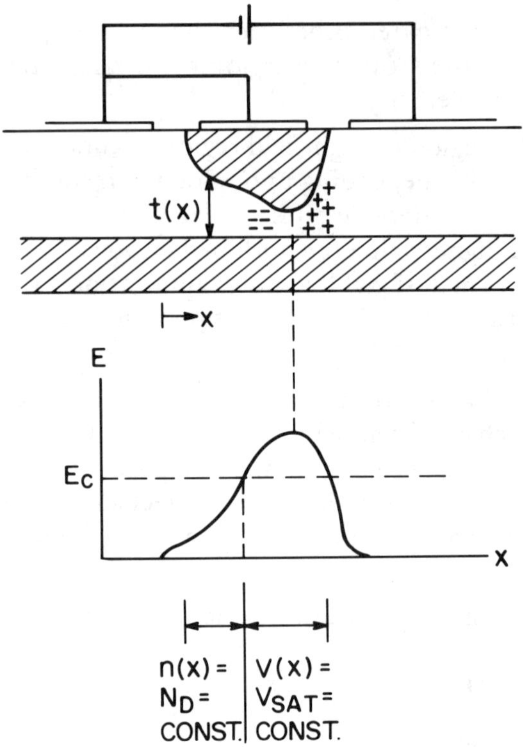

Figure 4/GaAs FET with Velocity Saturation

accumulation layer occurs at the drain edge of the channel which is balanced by a partial electron depletion as the electric field drops below E_c at the drain ohmic contact. A dipole layer is formed in the channel that extends beyond the drain end of the gate. This dipole layer gives rise to a small capacitance effect in the device model ($C_{dc} \cong 0.02$ pf) which has a negligible effect on the RF model. However, the formation of the dipole layer at the substrate-epitaxial layer in a high electric field region is very significant to device limitations near burnout.

The high density of accumulated electrons may become trapped in the substrate or at the channel/buffer-layer interface. Light emission has been observed at voltages (drain-source) as low as 3-4 volts, when the gate bias is 0 volts [29-30]. Simultaneous with the light emission, there is a flow of excess gate current, which is probably a collected hole current from the drain region [29]. When the gate current becomes too large, burnout occurs.

Typical burnout characteristics of the high power FET are related to the turn around in the drain-source I_{DS}-V_{DS} characteristic as shown in Figure 5. This phenomenon is also related to the light emission behavior [30].

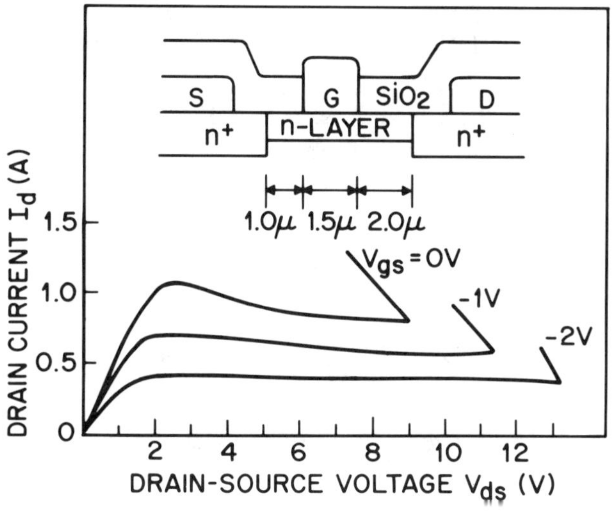

Figure 5/High Field Effects in GaAs FET

Other dc parameters of interest for GaAs FETs are the gate-source and gate-drain breakdown voltages and the forward bias gate-source and gate-drain curves. Typical performance is shown in Figure 6. A large forward diode current is related to the "ruggedness" or burnout resistance of the gate.

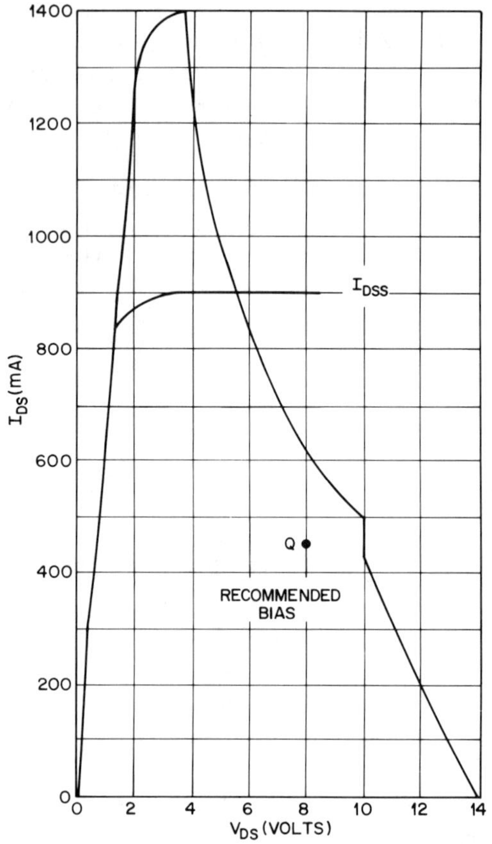

Figure 6/Safe Operating Region of GaAs FET (Z = 1500 μm).

The recommended dc operating point for various RF applications is given in Figure 7. The low noise bias point occurs at low voltages and low current (.15 I_{DSS}). The high gain point is near I_{DSS} and may be sensitive to the amount of light in the amplifier. The maximum drain to source voltage is given by

$$V_{SD}{}^{L} = V_{GD} - V_p \tag{8}$$

where V_{GD} is the gate-drain avalanche voltage and V_p is the pinch-off voltage. For maximum Class A output power with minimum distortion, the recommended bias point is 0.5 I_{DSS} and 0.5 $V_{SD}{}^{L}$.

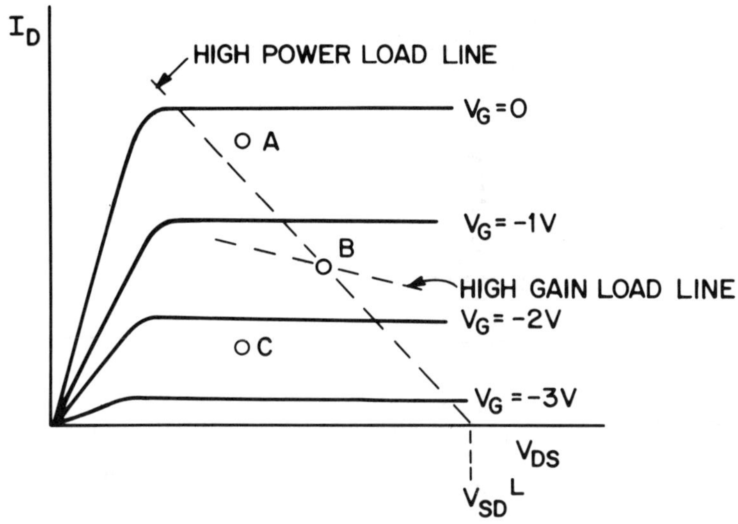

A. HIGH GAIN OPERATION
B. MIN DISTORTION OR HIGH POWER OUTPUT STAGE
C. LOW NOISE OPERATION

Figure 7/Recommended DC Operating Points

Typical bias circuits for common-source devices are given in Figure 8. The first circuit requires two power supplies of opposite polarity. This provides the advantage of minimum source inductance and, therefore, maximum microwave gain. The circuits with source resistors are widely used because of the built-in transient protection. As the drain supply voltage is applied, the gate-source junction is simultaneously reverse biased by the source resistor. A 50 pF bypass capacitor is usually sufficient for RF bypassing and transient protection. The circuits with source resistors offer a simpler dc supply circuit but also dissipate dc power and therefore reduce amplifier efficiency. Thus, for high power amplifiers, the positive and negative supplies are recommended with a sequencing circuit which reverse biases the gate-source junction before applying the drain bias. This sequencing is required because V_{DS} applied at I_{DSS} may result in burnout.

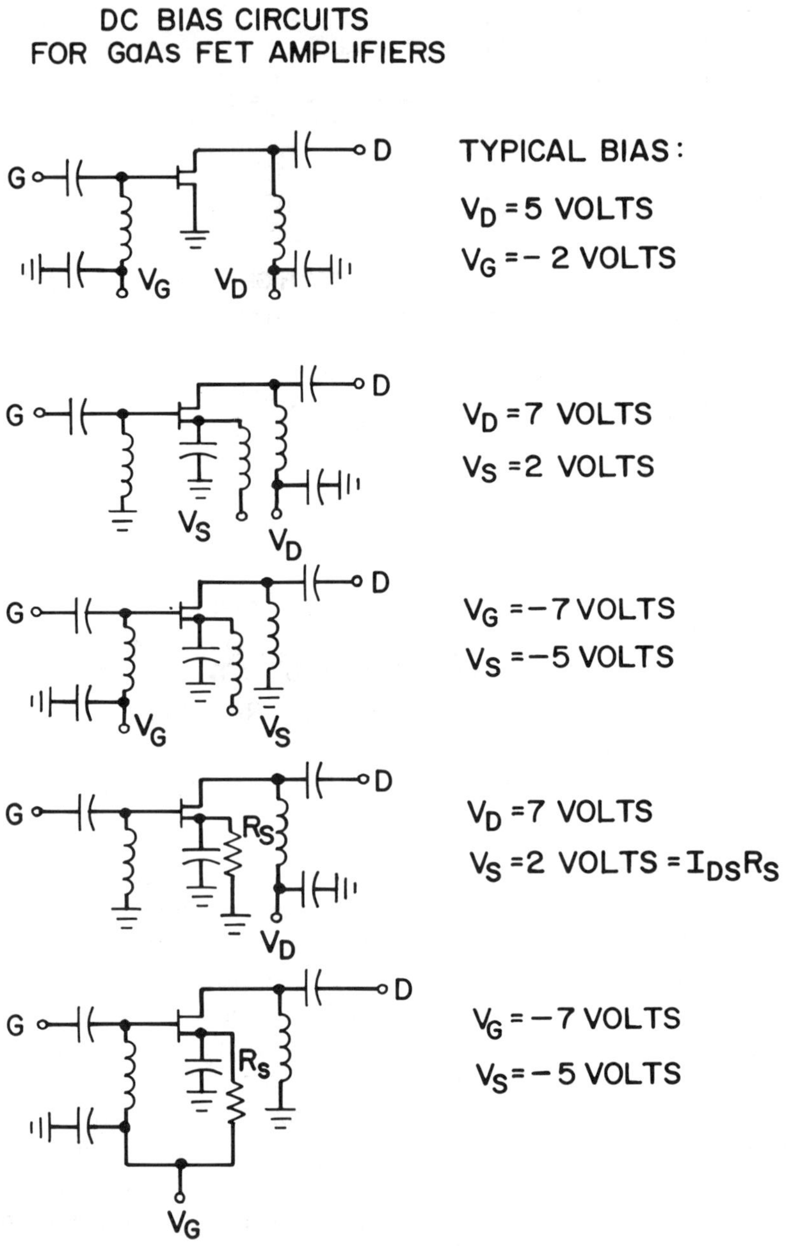

Figure 8/*Bias Circuits for Common-Source GaAs FET*

In addition to the single-gate GaAs FET, the dual-gate FET [23-28, 34-37] is a useful transistor for many applications. A cross-section of a dual-gate transistor is shown in Figure 9, where the transistors are dc biased in series, which normally offers the best gain in the pentode-pentode region. The dc properties of this transistor are determined by the relative bias at each terminal. There are four distinct

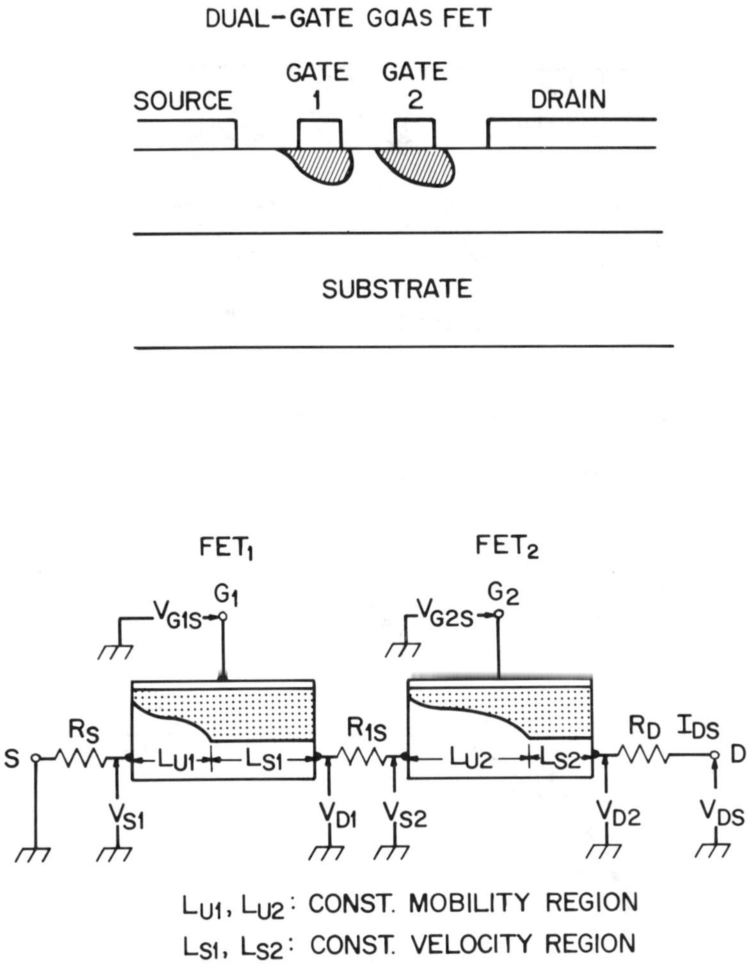

Figure 9/Dual-Gate GaAs FET [37]

regions of operation, depending on the bias at each gate junction:

1) Triode-Triode
2) Triode-Pentode
3) Pentode-Triode
4) Pentode-Pentode

where current unsaturated region is the triode operation and current saturated region is the pentode region.

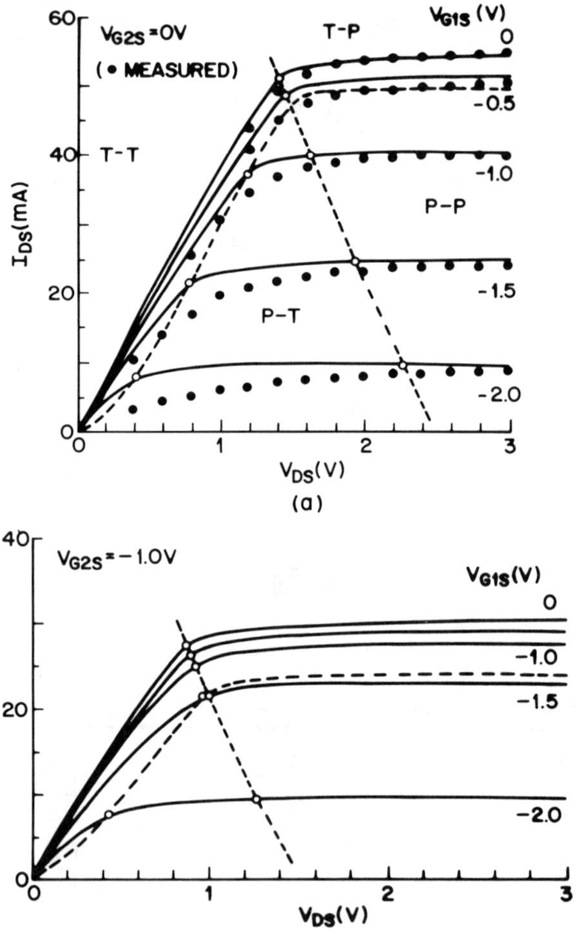

Figure 10/I_{DS} vs. V_{DS} for Dual-Gate GaAs FET [37]

The location of the four dc regions is given in Figure 10. The I_{DSS} of the transistor is defined where both gates are zero biased with respect to the channel. This would require

$V_{G1S} = 0$ volts

$V_{G2S} \cong 1.7$ volts

where the value of the second gate bias is chosen for zero bias between gate and channel. The g_m of both gates is important, but normally the first gate is optimized for this parameter. The pinch-off of each gate has the same definition as that for the single-gate FET.

3. RF CHARACTERISTICS OF GaAs FETs

Microwave GaAs FETs are usually characterized in the common-source configuration by the two-port S-parameters. Either a two-port or a three-port S-parameter set may be used to characterize the RF performance (Figure 11). Since the transistors are usually used in the common-source configuration, the S-parameters are normally measured for amplifier designs. For the single-gate FET, two-port S-parameters are measured. For the dual-gate FET, three-port S-parameters or two-port S-parameters with the second gate rf grounded are measured.

The S-parameters are preferred for microwave characterization for several reasons. First, a 50 ohm termination at input and output is relatively easy to provide since a 50 ohm transmission line terminated in 50 ohms provides an accurately determined termination at all frequencies. Secondly, the 50 ohm terminations usually give stable operation at all frequencies (no oscillations). Third, an extensive amplifier design theory using S-parameters has been developed for the design engineer. Finally, microwave instruments for measuring two-port S-parameters have been available for more than 15 years.

S_{21} is a simple measure of the gain of the transistor. It is given at low frequencies for the single-gate FET by:

$$S_{21} = -2g_m Z_o \tag{9}$$

In a 50 ohm system, a transconductance of 50 mmhos gives:

$$S_{21} = -100(.05) = -5$$

which is a power gain of 14 dB. Higher gain will be achieved by designing the amplifier for gain and stability.

For the dual-gate FET, the gain is affected by the transconductance of both gates by

$$y_{21} = \frac{y_{21}^{1} y_{21}^{2}}{y_{22}^{1} + y_{11}^{2}} = \frac{g_{m1} g_{m2}}{y_{22}^{1} + y_{11}^{2}} \tag{10}$$

For maximum gain, the transconductance of both stages should be optimized. The dc bias at the second gate is often used for gain control (AGC).

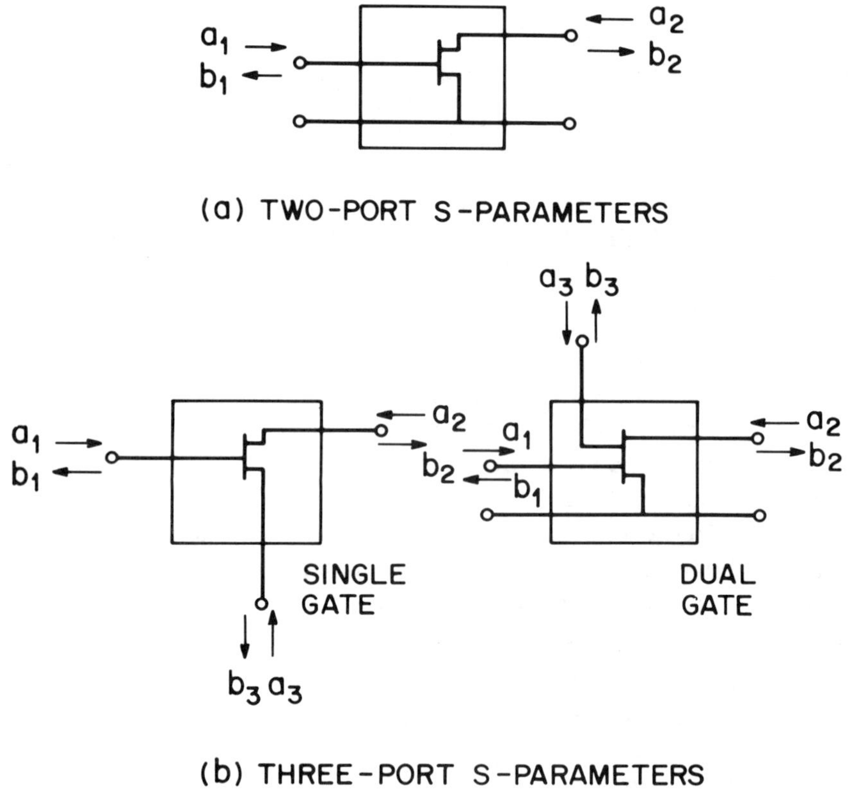

Figure 11/ Two-Port and Three-Port S-Parameters

The measurement of common-source s-parameters and the use of the indefinite scattering matrix will give either the common-gate or common-drain S-parameters. The three possible configurations for single-gate FETs are summarized in Figure 12. This argument only applies to the basic GaAs FET chip. If the bonding inductances are included in the measurement, it will usually be necessary to measure the two-port S-parameters in the configuration in which it will be used with the corresponding bonding or packaging parasitics.

In practical GaAs FET circuits, the two-port common-source S-parameters are usually measured. The reason is that this configuration gives the largest small-signal gain in an amplifier. For oscillator applications, the common-gate S-parameters should be measured or calculated since S_{22} is normally greater than 1 over a wide frequency range in this configuration. This parameter set is a sensitive function of the common-gate inductance, which should be carefully controlled for the measurement. The three-port S-parameters are rarely measured.

For measurement convenience, the S-parameters are normally measured in a 50 ohm transmission line system. The GaAs FET chip S-parameters are usually measured in a 50 ohm microstripline system. (The reference plane for the start of the input gate inductance and the start of the output drain inductance (see Figure 12) must be accurately determined by either a short circuit or through-line measurement. The transitions to the microstrip (from coaxial measurement system) must be either characterized or have a negligible effect on the measurement. The measured common-source S-parameters must include all of the bonding inductances since these parasitics will be present when the chip is bonded in the amplifier. Normally, all bonding inductances are kept at a minimum length ($\sim$ 15 mils) in order to maximize the bandwidth of the circuit.

Typical S-parameters of a small signal GaAs FET are shown in Figure 13 (a) and (b). The common-source and common-gate S-parameters are for the DXL 2501A chip with $Z = 500\ \mu m$ and $L_g = 1\ \mu m$ over a frequency range of 2-18 GHz at the bias for maximum gain. The S-parameters are relatively insensitive to current, except S_{21}, which increases with increasing current. A simple equivalent circuit for the GaAs FET is useful for wideband amplifier design. An equivalent circuit is given in Figure 14 for this transistor at bias for high gain. The time delay term in the transconductance current generator is related to the gate length L_g and is approximately given by:

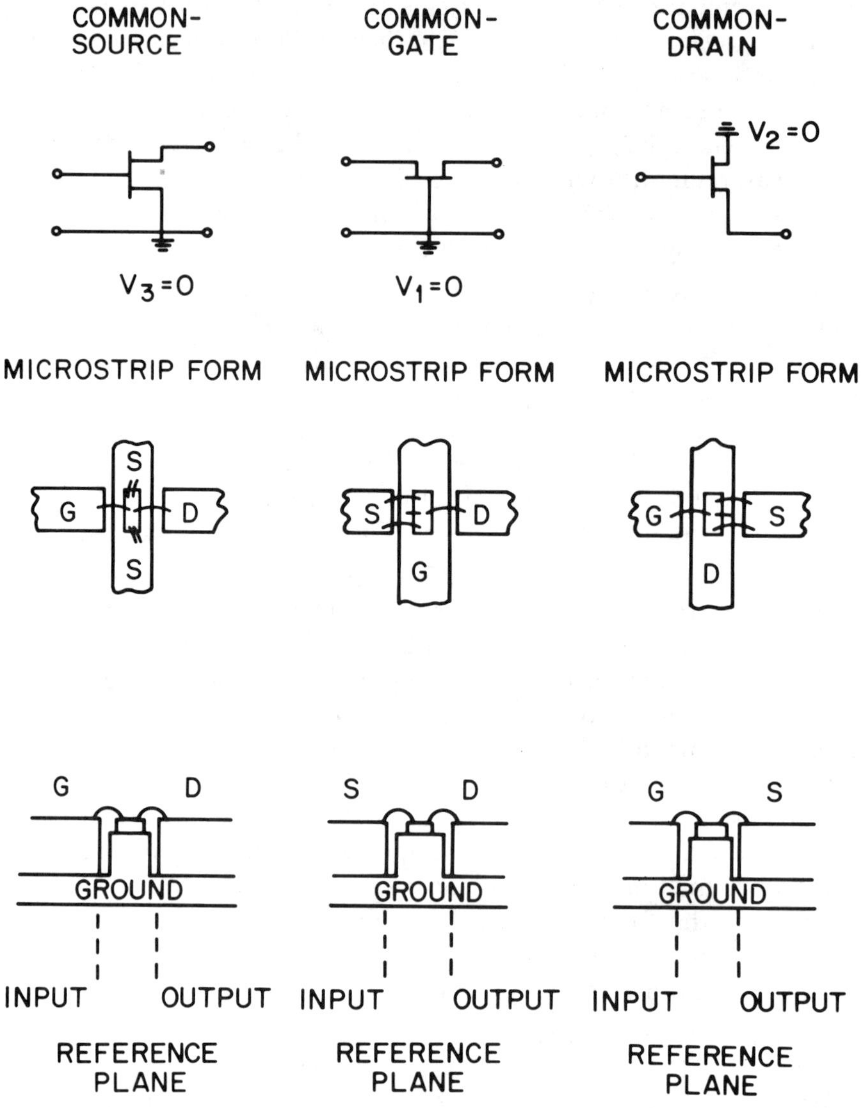

Figure 12/GaAs FET Chip Circuit Configurations.

current generator is related to the gate length Lg and is approximately given by:

$$\tau_D \simeq \frac{L_g}{v_{sat}} = \frac{10^{-4}\ \text{cm}}{10^7\ \text{cm/sec}} \simeq 10\ \text{psec} \quad (11)$$

The common-gate S-parameters of Figure 13 are for minimum gate inductance. The input port is relatively well matched in a 50 ohm system since

$$S_{11} \simeq Z_o g_m \quad (12)$$

in the common gate configuration. Since S_{22} is greater than unity for frequencies above 6 GHz in Figure 13, an oscillator can easily be achieved at the output drain port. The S_{22}' of the oscillator is a sensitive function of the termination at the source port and is given by

$$S_{22}' = S_{22} + \frac{S_{12}S_{21}\Gamma_s}{1 - S_{11}\Gamma_s} \quad (13)$$

where Γ_s is the reflection coefficient of the source termination. Additional details of oscillator design are presented in a later section.

The S-parameters for a packaged small signal GaAs FET are also given in Figure 13 for a typical microstripline package. Generally, the package parasitics lead to more dispersion or spread in the S-parameters and therefore circuits with narrower bandwidth. If the package is lossless, the maximum available amplifier gain and minimum noise figure are usually unchanged.

The primary advantages of considering packaged GaAs FET circuits are

(1) Simpler assembly techniques (low temperature soldering iron);

(2) Selection of device before mounting in circuit, which is particularly important in low noise amplifiers where the lowest noise figure transistor should be in the first stage;

(3) Guaranteed measured performance provided by device supplier;

(4) High reliability screening of packaged transistors is possible.

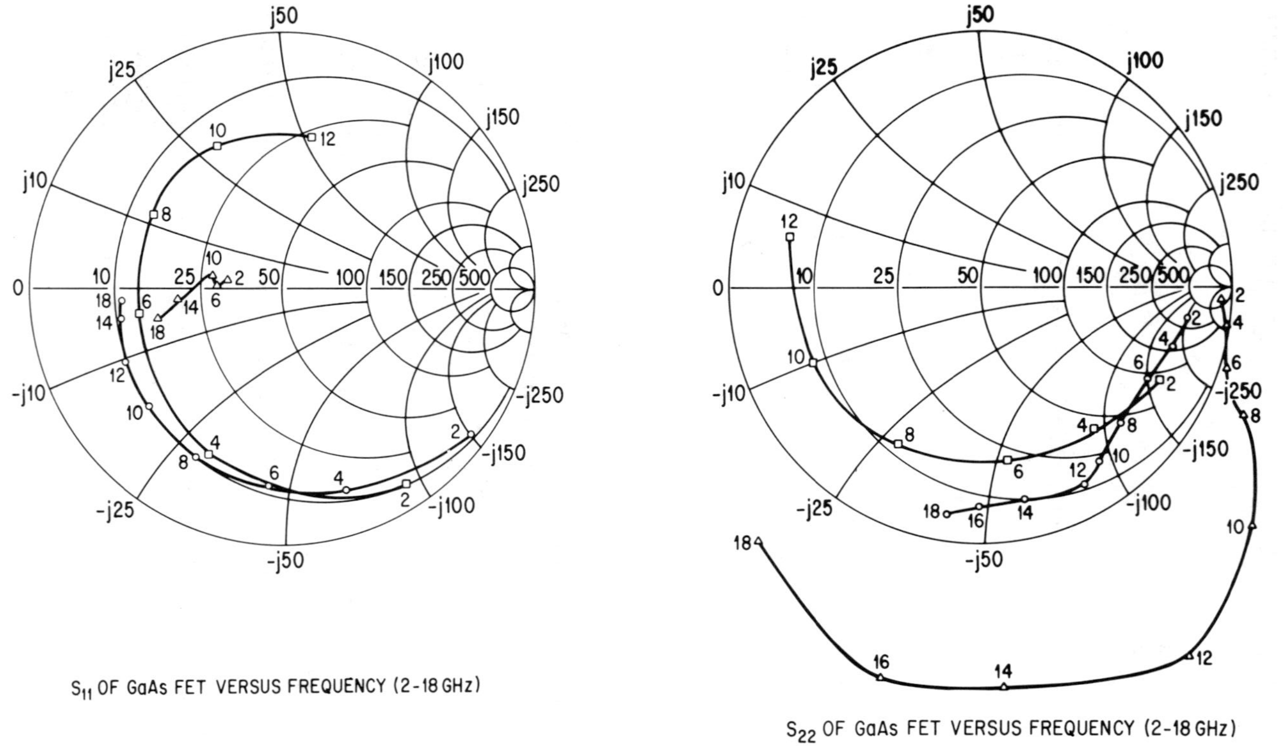

Figure 13 */S-parameters for common-source DXL 2501A chip, ○, and common-gate DXL 2501A chip, △, at* $V_{DS} = 3.5V$, $I_{DS} = 0.9\ I_{DSS}$*; common-source HFET-1101 packaged, □, at* $V_{DS} = 4.0V$, $I_{DS} = I_{DSS}$ *(coordinates in ohms).*

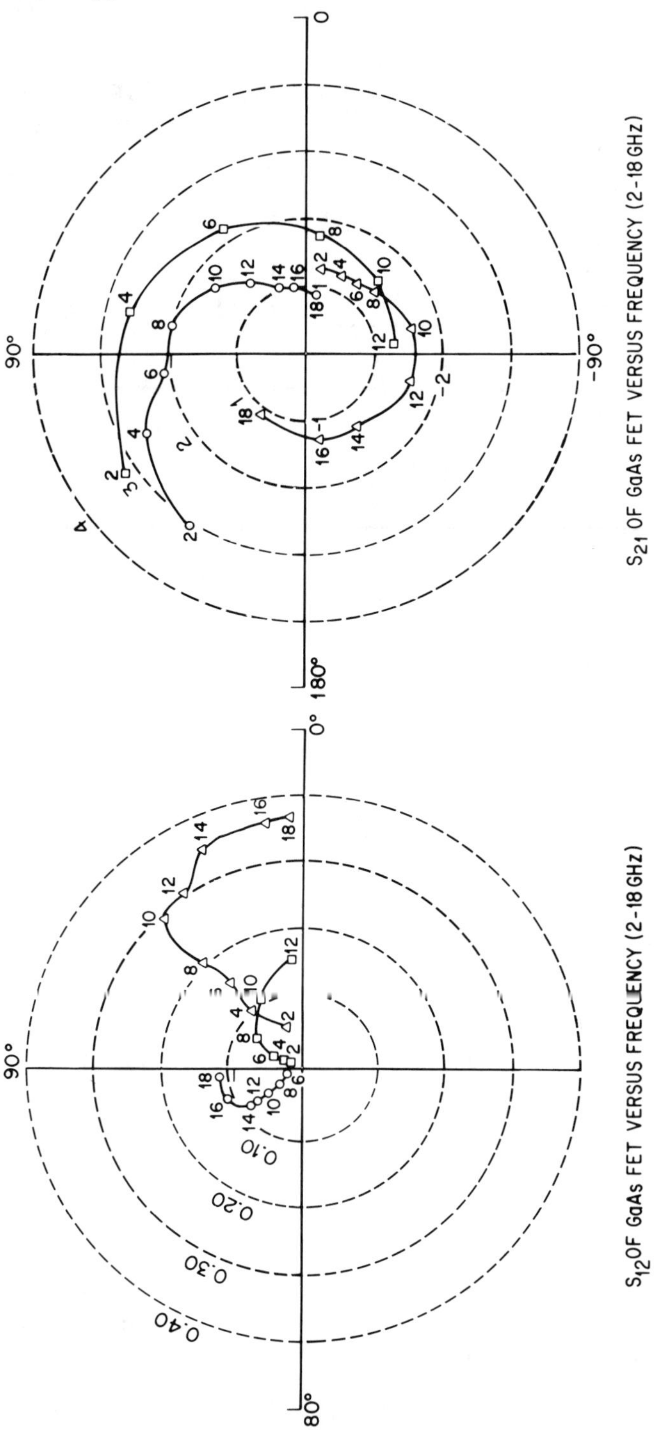

Figure 13/S-parameters *for common-source DXL 2501A chip,* ○, *and common-gate DXL 2501A chip,* △, *at* $V_{DS} = 3.5V$, $I_{DS} = 0.9\ I_{DSS}$; *common-source HFET-1101 packaged,* □, *at* $V_{DS} = 4.0V$, $I_{DS} = I_{DSS}$ *(coordinates in ohms).*

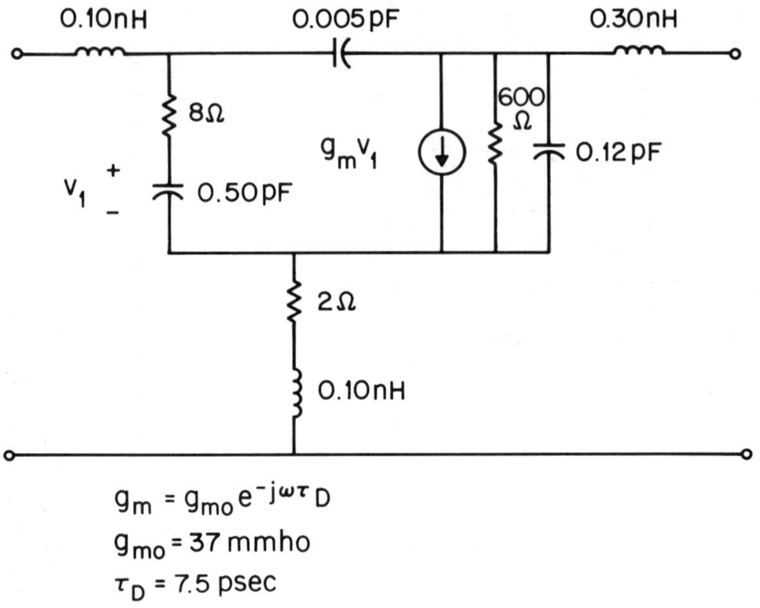

Figure 14*/Equivalent Circuit for DXL 2501A Chip at $V_{DS} = 3.5$ volts and $I_{DS} = 0.9\ I_{DSS}$.*

For low noise designs, the four noise parameters must be supplied as data or measured by the user. These parameters are (at each frequency):

F_{min} = minimum noise figure

G_{on} = $Re(Y_{on})$

B_{on} = $Im(Y_{on})$

R_n = parameter giving increase in noise figure vs. detuning from minimum noise figure

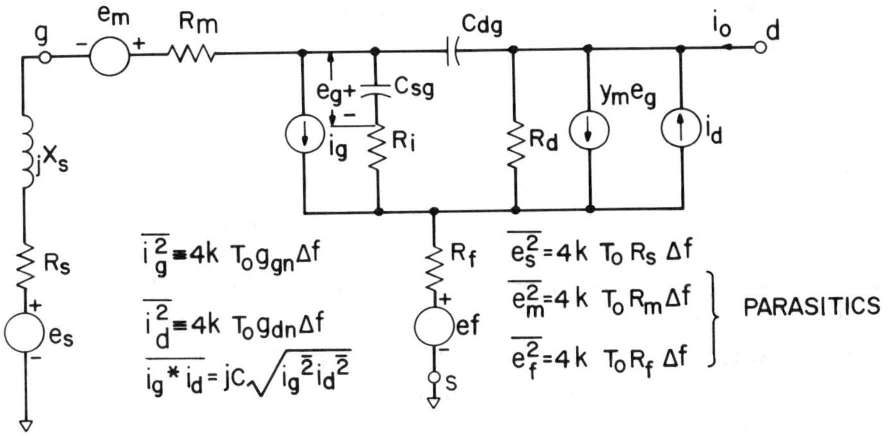

Figure 15/ Noise Equivalent Circuit of GaAs FET

These four parameters are related by

$$F = F_{min} + \frac{R_n}{G_S} |Y_S - Y_{on}|^2 \tag{14}$$

where

Y_S = source admittance

$Y_{on} = G_{on} + jB_{on}$ = optimum source admittance required for $F = F_{min}$

The noise equivalent circuit [31] for the GaAs FET is given in Figure 15. The intrinsic noise sources are:

(1) i_d, caused by thermal noise of the hot electrons in the channel and diffusion noise in the velocity saturated region of the channel.

(2) i_g, caused by the thermal channel noise being coupled to the input circuit through C_{gs}, the "induced gate noise".

The noise figure is a function of the two internal noise sources and the complex correlation of these generators. The extrinsic resistors R_m and R_f also add thermal noise to the overall transistor noise. From published correlations between noise figure measurements and theory [32-33], the extrinsic resistors add 1-2 dB to the noise figure. The primary bias dependence of the noise figure is due to increasing drain current, which increases the i_d generator due to dipole layers drifting to the drain region at the saturated drift velocity. Thus, the minimum noise figure occurs by minimizing the length of the velocity saturated drift region. Notice that the effect of intervalley scattering has not been included in the noise theory, but this effect is minimized by operating at low drain-source voltage. The application of these noise parameters will be discussed in the amplifier design section.

For large signal or power applications the small-signal S-parameters have limited usefulness. In this case, the output circuit (and input circuit) are tuned for maximum power out leading to a set of large-signal tuning parameters which are a function of the transistor, the bias, and the frequency of operation. The details are covered in the second part of this chapter.

In amplifier designs, the input circuit is usually matched for low noise figure and the output circuit is usually matched for high power both at the expense of gain. If maximum gain is required, both input and output circuits are optimized for maximum gain while maintaining stability at all frequencies (no oscillations). The details of this design procedure will follow.

For dual-gate amplifier designs, the termination at the second gate will have an important effect. A capacitive ground will normally give stability and reduce the dual-gate amplifier design to the similar single-gate amplifier design. An inductive termination in the second gate can be used to peak the gain.

4. APPLICATIONS OF GaAs FETs

4.1 Amplifiers

Small-signal Design [38-41]

The design of small-signal amplifiers may be explained using Figure 16. The common-source S-parameters of the transistor are provided over the microwave frequency range at the suggested bias point or measured by the designer. Single stage circuit design consists of: (1) finding an input matching circuit to give $S_{11}' = 0$ and stability at all frequencies, and (2) finding an output circuit which simultaneously gives $S_{22}' = 0$ and stability at all frequencies. Usually, several circuits are possible and the logical choice is dictated by size, physical realizability and a topology which allows dc biasing of the transistor.

A useful approximation for narrow-band amplifiers is to assume the $S_{12} = 0$. This gives a simple form for maximum available gain called the maximum unilateral transducer gain.

$$G_{Tumax} = \frac{|S_{21}|^2}{(1 - |S_{11}|^2)(1 - |S_{22}|^2)} \tag{15}$$

which consists of three terms:

$|S_{21}|^2$ — the transistor transducer power gain, the power gain between a 50 ohm source and a 50 ohm load;

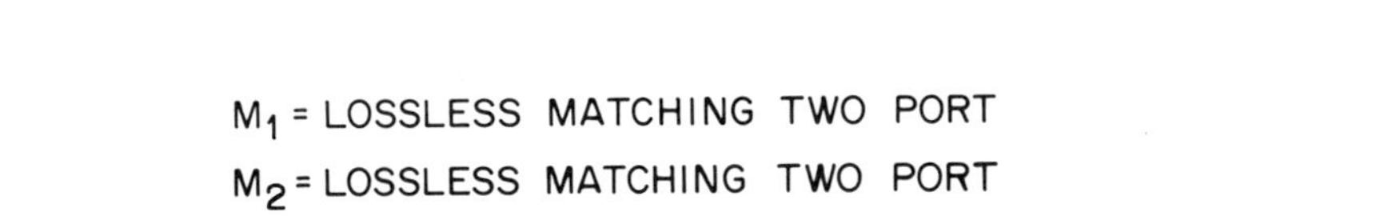

Figure 16/Amplifier Design with S-Parameters

$(1 - \|S_{11}\|^2)^{-1}$	the increase in gain due to matching the input port;
$(1 - \|S_{22}\|^2)^{-1}$	the increase in gain due to matching the output port.

Although the G_{Tumax} may be about 1-2 dB lower than the true maximum available transducer power gain, the approximation is useful because it provides insight into the circuit operation at each port.

Before considering the various definitions of power gain, the stability factor k must be calculated using the equation

$$k = \frac{1 + |S_{11}S_{22} - S_{21}S_{12}|^2 - |S_{11}|^2 - |S_{22}|^2}{2\,|S_{12}S_{21}|} \qquad (16)$$

for all frequencies of interest, i.e., at all frequencies where gain is significant and oscillations are possible. The significance of the stability factor k is that for a transistor to be unconditionally stable for all possible load and source terminations, k must be greater than 1.

The necessary and sufficient conditions for a one-port which contains a single transistor are:

$$k > 1 \qquad (17)$$

$$|D| < 1 \qquad (18)$$

where

$$D = S_{11}S_{22} - S_{12}S_{21} \qquad (19)$$

For $k < 1$, the stability circles will indicate the source and load terminations which give rise to device instability, i.e., $|S_{11}| > 1$ or $|S_{22}| > 1$.

The location of the stability circles is given by:

$$\text{Center } C_i = \frac{(S_{ii} - (S_{11}S_{22} - S_{21}S_{12})S_{jj}^*)^*}{|S_{ii}|^2 - |S_{11}S_{22} - S_{21}S_{12}|^2} \qquad (20)$$

$$\text{Radius } r_i = \frac{|S_{12}S_{21}|}{|S_{ii}|^2 - |S_{11}S_{22} - S_{21}S_{12}|^2} \qquad (21)$$

where i = 1, j = 2 for the input impedance plane;
i = 2, j = 1 for the output impedance plane.

Usually the location of these circles will track the S_{11}^* or S_{22}^* parameter vs. frequency.

Typical stability circle plots are given in Figure 17, along with the appropriate S_{11} and S_{22} plots. When the stability circles are outside the input and output impedance plane, the transistor is unconditionally stable for all possible terminations ($k > 1$). When the circles intersect the impedance plane, the terminations must not present instability, particularly at low frequencies. Usually a short circuit termination at low frequencies will guarantee stability and an open circuit termination creates instability.

In Figure 16, three useful definitions of power gain are given. The transducer power gain is the power delivered to the 50 ohm load divided by the power available from the 50 ohm source and is given by:

$$G_T = \frac{P_{del}}{P_{avs}} = f(\Gamma_S, S, \Gamma_L)$$

$$= \frac{1 - |\Gamma_S|^2}{|1 - S_{11}'\Gamma_S|^2} |S_{21}|^2 \frac{1 - |\Gamma_L|^2}{|1 - S_{22}\Gamma_L|^2} \tag{22}$$

$$S_{11}' = S_{11} + \frac{S_{12}S_{21}\Gamma_L}{1 - S_{22}\Gamma_L} \tag{23}$$

$$S_{22}' = S_{22} + \frac{S_{12}S_{21}\Gamma_S}{1 - S_{11}\Gamma_S} \tag{24}$$

The power gain is the power delivered to the load divided by the power incident on the transistor and is given by:

$$G = \frac{P_{del}}{P_i} = f(S, \Gamma_L) = \frac{1}{1 - |S_{11}'|^2} |S_{21}|^2 \frac{1 - |\Gamma_L|^2}{|1 - S_{22}\Gamma_L|^2} \tag{25}$$

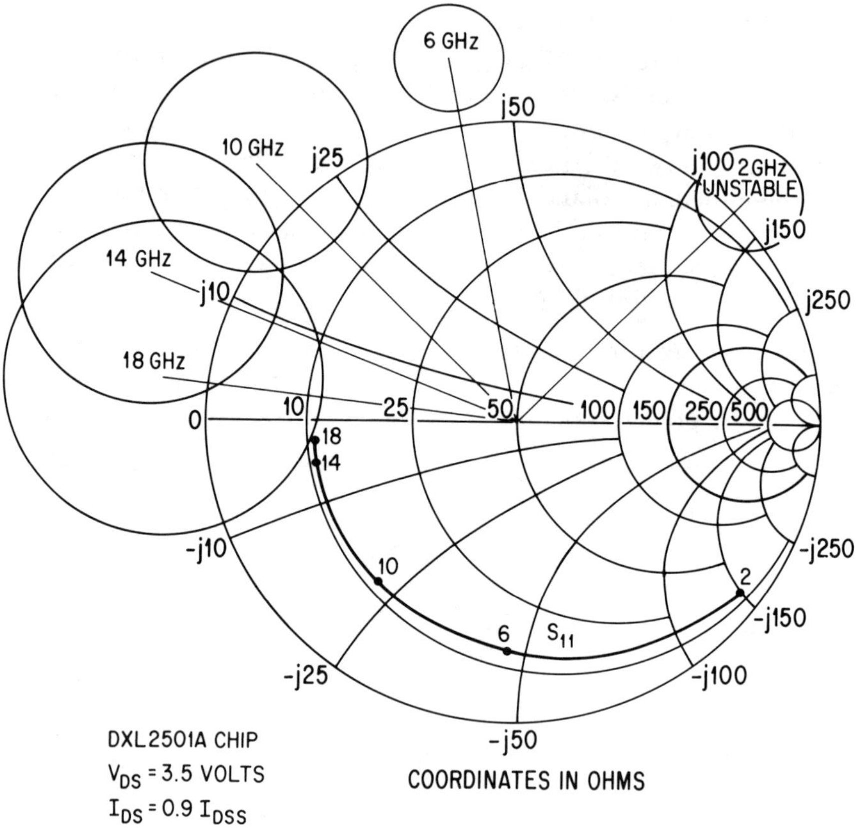

(a) INPUT STABILITY CIRCLES

Figure 17/GaAs FET Input and Output Stability Circles.

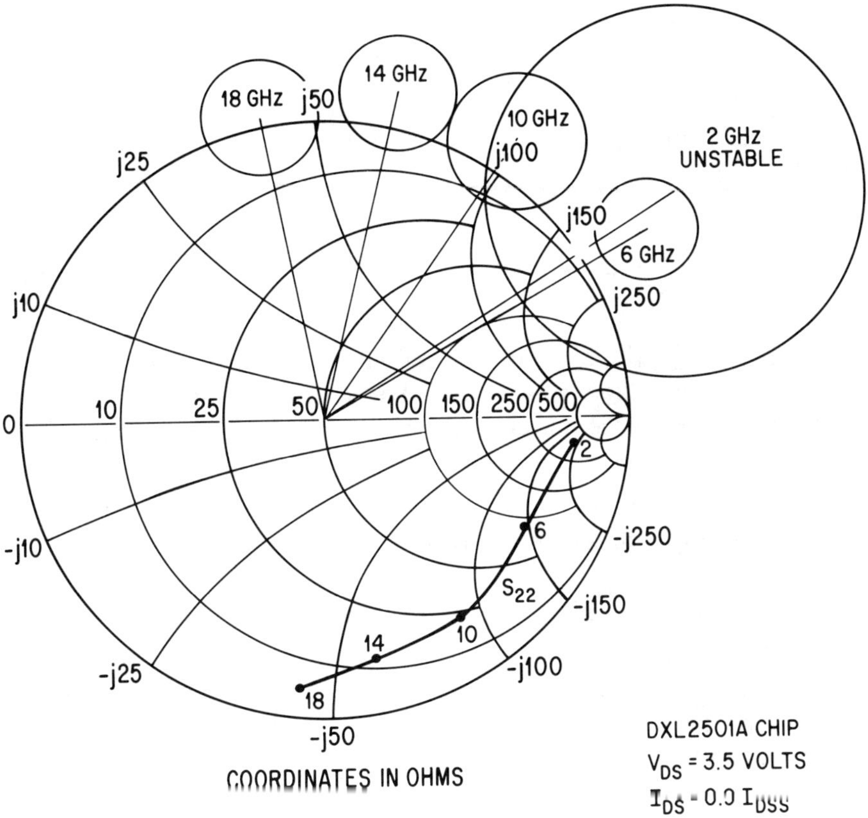

(b) OUTPUT STABILITY CIRCLES

Figure 17/GaAs FET Input and Output Stability Circles.

Since $P_i \leqslant P_{avs}$ the power gain G is always greater than or equal to the transducer power gain. The power gain is a function of the load match and is the transducer power gain when the input is conjugately matched ($S_{11}' = 0$).

The available power gain is the power available at the output divided by the power available at the source and is given by:

$$G_A = \frac{P_{avo}}{P_{avs}} = f(\Gamma_s, S)$$

$$= \frac{1 - |\Gamma_S|^2}{|1 - S_{11}\Gamma_S|^2} |S_{21}|^2 \frac{1}{1 - |S_{22}'|^2} \quad (26)$$

Since $P_{avo} \geqslant P_{del}$ the available power gain G_A is also greater than or equal to the transducer power gain. The available power gain is a function of the source match and is the transducer power gain when the output is conjugately matched ($S_{22}' = 0$).

When the unilateral assumption ($S_{12} = 0$) is applied to the transducer power gain, the unilateral transducer power gain is:

$$G_{Tu} = \frac{P_{del}}{P_{avs_{S_{12}=0}}} = f(\Gamma_S, S, \Gamma_L)$$

$$= \frac{1 - |\Gamma_S|^2}{|1 - S_{11}\Gamma_S|^2} |S_{21}|^2 \frac{1 - |\Gamma_L|^2}{|1 - S_{22}\Gamma_L|^2} \quad (27)$$

When $S_{11} = \Gamma_s^*$ and $S_{22} = \Gamma_L^*$ this reduces to G_{Tumax} given earlier by Equation 15.

If the transistor is unconditionally stable ($k > 1$) a simultaneous conjugate match at the input and output port can be found which will give $G_T \max = G_A \max = G \max$. From Equations 23 and 24:

$$S_{11}' = S_{11} + \frac{S_{12}S_{21}(S_{22}')^*}{1 - S_{22}(S_{22}')^*} \quad (28)$$

$$S_{22}' = S_{22} + \frac{S_{12}S_{21}(S_{11}')^*}{1 - S_{11}(S_{11}')^*} \tag{29}$$

Solving simultaneously gives:

$$\Gamma_{ms} = \frac{B_1 \pm \sqrt{B_1^2 - 4|C_1|^2}}{2C_1} \tag{30}$$

$$\Gamma_{m1} = \frac{B_2 \pm \sqrt{B_2^2 - 4|C_2|^2}}{2C_2} \tag{31}$$

$$B_1 = 1 + |S_{11}|^2 - |S_{22}|^2 - |D|^2 \tag{32}$$

$$B_2 = 1 + |S_{22}|^2 - |S_{11}|^2 - |D|^2 \tag{33}$$

$$C_1 = S_{11} - DS_{22}^* \tag{34}$$

$$C_2 = S_{22} - DS_{11}^* \tag{35}$$

$$D = S_{11}S_{22} - S_{12}S_{21} \tag{36}$$

In practical GaAs FET amplifier designs, a first order circuit may be derived by assuming S_{12} is zero and designing the input match for $\Gamma_s = S_{11}^*$ and the output match for $\Gamma_L = S_{22}^*$. The approximate gain is estimated from G_{Tumax}. The exact solution given by Equations 30 and 31 is usually applied when computer resources are available. The maximum available gain which occurs for the simultaneous conjugate match is

$$G_{Amax} = MAG = G_{Tmax} = G_{max}$$

$$= \frac{|S_{21}|}{|S_{12}|}\,(k \pm \sqrt{k^2 - 1}) \tag{37}$$

where $k > 1$ and a negative sign is chosen for $B_1 > 0$.

Another definition of "power gain" is the maximum stable gain given by:

$$MSG = \frac{|S_{21}|}{|S_{12}|} \tag{38}$$

This gain exists regardless of the k value and can be taken to be the highest possible gain which can be achieved before instability occurs. If the two port is resistively loaded until k = 1, the MAG of this two port is the MSG given above.

One final form of power gain is the maximum unilateral power gain which is given by:

$$U = \frac{\frac{1}{2}\left|\frac{S_{21}}{S_{12}} - 1\right|^2}{k\left|\frac{S_{21}}{S_{12}}\right| - Re\left(\frac{S_{21}}{S_{12}}\right)} \tag{39}$$

This is the power gain which occurs when lossless feedback produces $S_{12}' = 0$ and matching, gives $S_{11}' = 0$ and $S_{22}' = 0$. This power gain is larger than any of the others and is often quoted by transistor manufacturers. The frequency where unilateral gain becomes zero is f_{max} which is also called the maximum frequency of oscillation. The f_{max} is typically about 60-100 GHz for transistors with usable gain below 20 GHz.

The many forms of power gains are summarized by plotting the gain responses for the DXL − 2501A transistor chip described earlier in Figure 18. The transistor transducer power gain is $|S_{21}|^2$, the MAG is given by Equation 37, the MSG by Equation 38, the G_{Tumax} by Equation 15, U by Equation 39, and k by Equation 16.

The amplifier designer has the task of selecting M_1 and M_2 so that the amplifier transducer power gain approaches MAG or MSG of the transistor. This condition will give $S_{11}AMP = S_{22}AMP = 0$. If a significant bandwidth is required, the amplifier transducer gain at the high frequency band edge is chosen as the design target across the band.

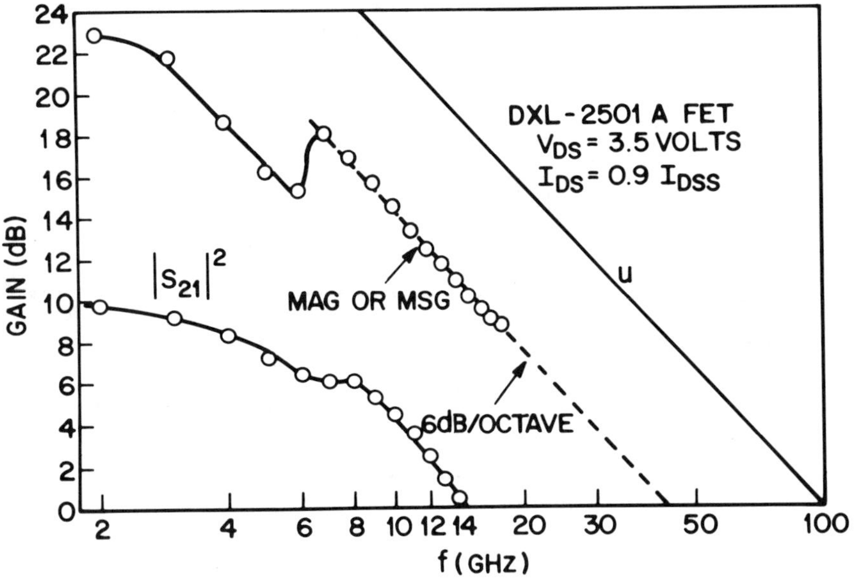

Figure 18/Gain vs. Frequency for DXL2501A Chip

For a single stage amplifier, a corresponding mismatch loss vs. frequency must be incorporated at the input and/or output port. A design example using the DXL – 2501A chip at 8-12 GHz will illustrate the procedure. Using the S-parameters at 0.9 I_{DSS}, the gains were plotted in Figure 18. Since the MSG is 12.3 dB at 12 GHz, a reasonable target is 11 dB across 8-12 GHz. An amplifier design which achieves this goal is given in Figure 19. Only two matching elements are required on the input and three matching elements on the output. Distributed microstrip transmission lines were used for matching elements in this frequency range. The circuit is built on 25 mil alumina substrates and meets the design goal.

Low Noise Design

The design of low noise amplifiers requires both S-parameters and noise parameters, which at each frequency amounts to 12 numbers.

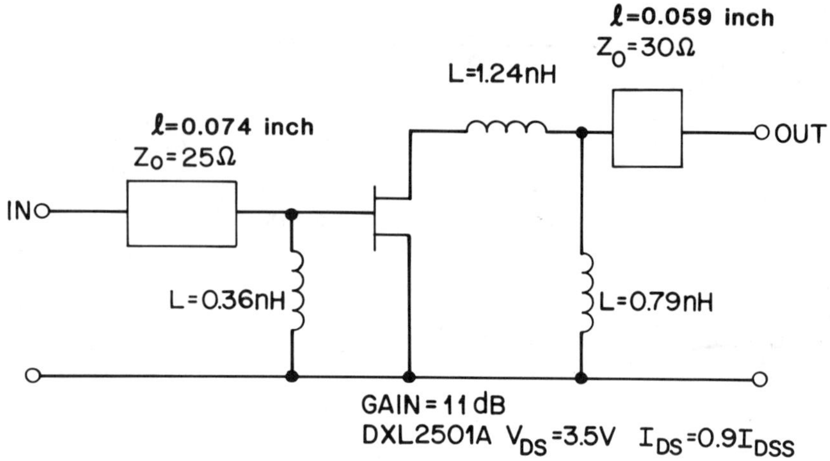

Figure 19/8-12 GHz GaAs FET Amplifier Design Using DXL2501A Chip

The noise parameters were previously defined in Equation 14. The design of single and multiple stage low noise amplifiers can be explained with Figure 20.

A design procedure follows:

(1) Design M_1, the input matching circuit, to present Γ_{on} to the input port of the transistor. This is a passive circuit design problem which can usually be solved with two matching elements at a single frequency. Note that a low value of S_{11} will not necessarily occur at the input port of the amplifier. Attempts to modify the transistor characteristics to give both minimum noise figure and minimum input VSWR have not been successful in practice since gain is usually degraded.

(2) Design M_2 for maximum available gain, i.e., a conjugate power match at the output port. After M_1 is found, the $S_{22}'^*$ is calculated using Equation 24.

(3) Investigate for amplifier stability at all frequencies. Since there are multiple solutions to the amplifier design, a different topology may be required until unconditional stability is found. If instability remains a problem, a resistor may be required (preferably in M_2) to guarantee stability, but this will increase the noise figure.

The noise figure of a multiple stage amplifier is given by:

$$F_{tot} = F_1 + \frac{F_2 - 1}{G_{A1}} + \frac{F_3 - 1}{G_{A1} G_{A2}} + \ldots \quad (40)$$

The minimum overall noise figure can be found by minimizing F_1 and F_2 since additional terms are negligible if the available gain of the first stage (G_{A1}) is about 6 dB or more.

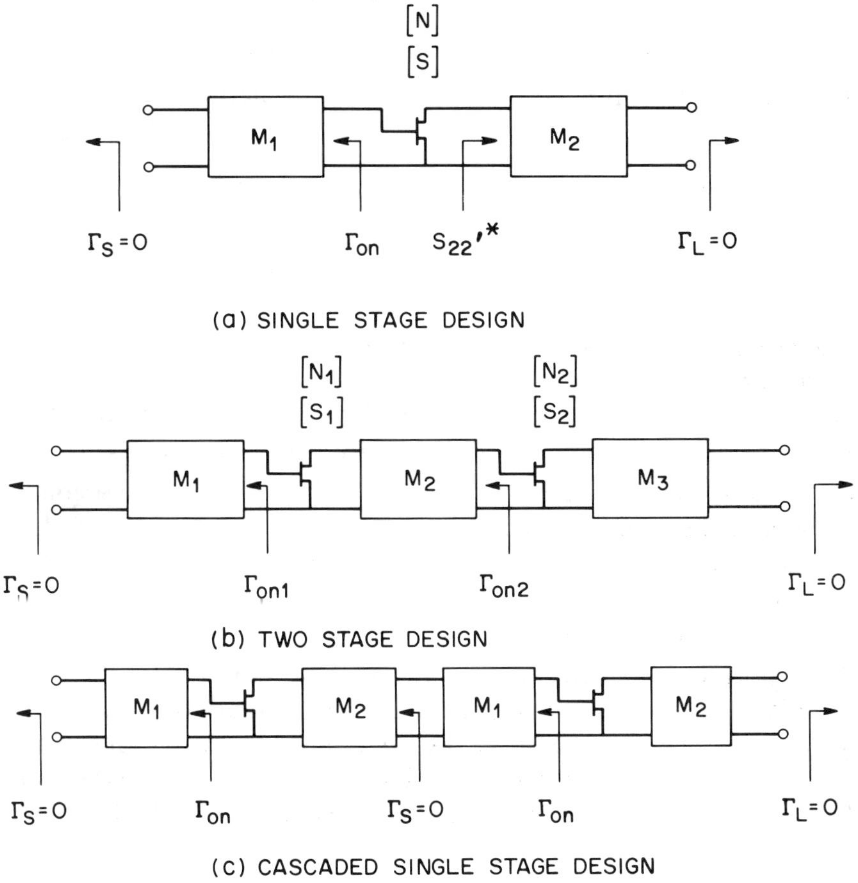

Figure 20/Low Noise Amplifier Design

An important special case is the minimum possible value of F_{tot}, which occurs for an infinite cascaded chain with $F_1 = F_2 = F_3 = \ldots$, which gives

$$F_{totmin} = F_1 + \frac{F_1 - 1}{G_{A1}} + \frac{F_1 - 1}{G_{A1}^2} + \ldots$$

$$= (F_1 - 1)(1 + 1/G_{A1} + 1/G_{A1}^2 + \ldots) + 1$$

$$= (F_1 - 1)\left(\frac{1}{1 - 1/G_{A1}}\right) + 1 \qquad (41)$$

This equation yields noise measure M which is defined by:

$$M = \frac{F_1 - 1}{1 - 1/G_{A1}} \qquad (42)$$

$$F_{totmin} = M + 1 \qquad (43)$$

The noise measure gives a lower limit which is considered a figure of merit for the transistor. In practice F_{min} is not achievable, but the noise figure (M + 1) can be approached by low noise amplifier design.

For a multiple stage amplifier (Figure 20b) the noise design should be extended to find an interstage circuit which minimizes the second stage contribution. When considering bandwidth and stability, this can become a difficult design problem. A possible solution is shown in Figure 20c, a cascade of single stage designs. This allows the selection of the lowest noise stages for the amplifier front end.

An important consideration in multiple stage amplifiers is the overall stability. The two-port stability factor k is no longer adequate to guarantee amplifier stability. Now the interstage may produce oscillation at either transistor. Introducing a resistor may solve the problem but increase the overall noise figure. In summary, the low noise design procedure is:

1) Design M_1 for $\Gamma_{on\,1}$;
2) Design M_2 for conjugate power match;
3) Cascade two stages and examine stability of each stage at all frequencies.

If a significant bandwidth is required, an exact solution to the problem does not exist. A practical compromise usually requires concentrating on the noise figure at the upper frequency and flattening the overall gain in the interstage and output circuit. Since an exact solution for Γ_{on} may not be possible, the noise circles in the input impedance plane should be plotted to find the increase in noise figure.

These equations become:

$$C_i = \frac{\Gamma_{on}}{1 + N_i} \tag{44}$$

$$r_i = \frac{1}{1 + N_i} \sqrt{N_i^2 + N_i(1 - |\Gamma_{on}|^2)} \tag{45}$$

$$N_i = \frac{F_i - F_{min}}{4R_n/Z_o} |1 + \Gamma_{on}|^2 \tag{46}$$

A low value of R_n will minimize the effect of detuning from the optimum noise match.

If a tradeoff between noise figure and gain is required, the bias dependence of noise figure is important. Generally, an increase of 1 to 2 dB in noise figure is expected as I_{DS} is increased from 0.15 I_{DSS} to full I_{DSS}.

Wide Band Design [42-49]

For the design of wide band amplifiers (typically one octave or more), special consideration must be given to achieving the highest gain at the upper band edge while simultaneously flattening the gain throughout the band (Figure 21). Several design approaches are available, depending upon the amplifier application.

For a single stage amplifier, an input match may be required for low noise or an output match may be required for high power. Thus, the mismatch for gain flattening must be provided at the opposite port. Notice a simultaneous input match, output match, and flat gain is not possible without adding resistors in the circuit (or using the balanced amplifier approach described in the next section).

For wideband amplifier design, the conditions for broadband imped-

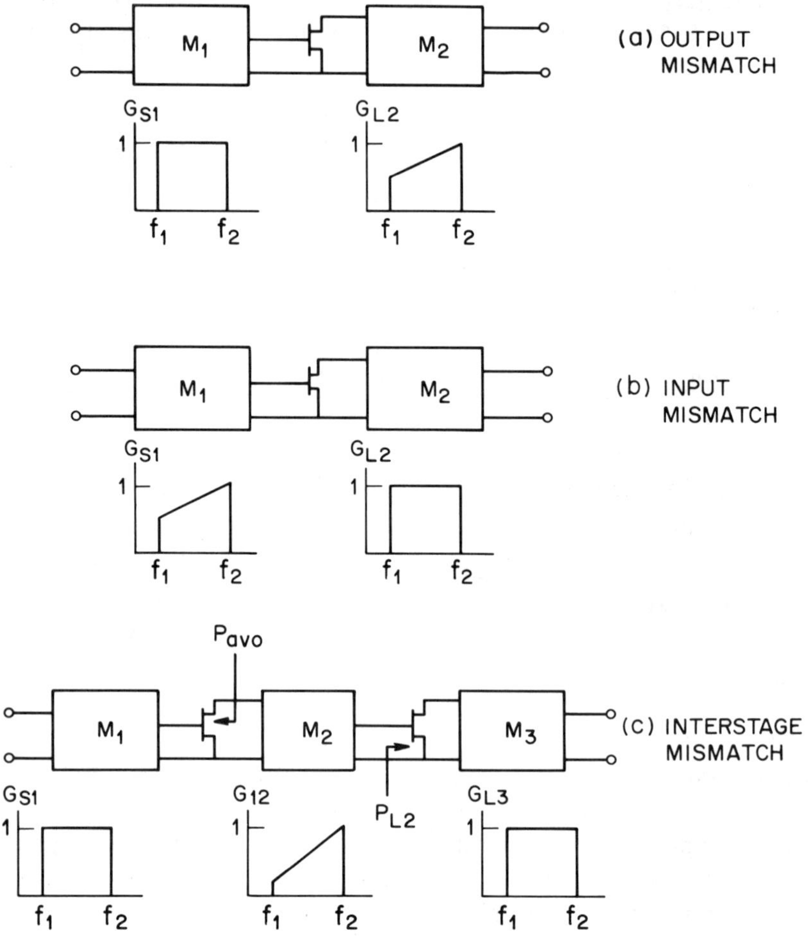

Figure 21/Wideband Amplifier Design with Lossless Elements

ance matching will limit the amplifier performance. This limitation is generally referred to as Fano's limit [50], and one of the four possible cases is considered below (see Figure 22):

$$\int_0^{\infty} \ln |1/\Gamma| \, d\omega \leqslant \pi/R_1C \tag{47}$$

$$Q_{LOAD} = \omega R_1 C \tag{48}$$

$$\int_{\omega_a}^{\omega_b} \ln |1/\Gamma| \, d\omega \leqslant \pi/R_1C \tag{49}$$

$$|\Gamma| \geqslant e^{-\pi/R_1C(\omega_b - \omega_a)} = e^{-\pi(Q_{ckt}/Q_{LOAD})} \tag{50}$$

$$Q_{ckt} = \frac{\omega}{\omega_b - \omega_a} \tag{51}$$

For a narrow band circuit, Q_{ckt} is large and the achievable Γ is very low. But for a wide band circuit, Q_{ckt} may be low and Q_{LOAD} is prescribed by the transistor. The best possible match is given by Equation 50 , which is Fano's limit. The other three possible cases are parallel R-L, series R-C, and series R-L for the load. The Fano limit applies to all cases.

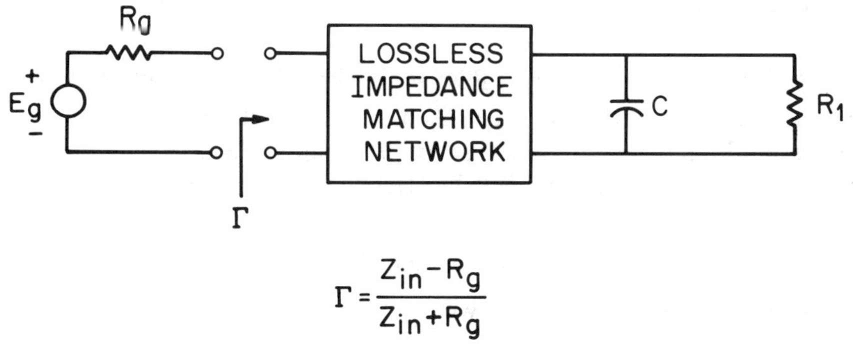

Figure 22/Broadband Impedance Matching

Since the transistor has parasitic capacitances, the broadband matching circuits must absorb these parasitic elements. If a broadband match is designed at both input and output, the circuit will give a 6 dB/octave gain rolloff due to the transistor gain characteristic.

The design procedure for wide band amplifiers is the following:

(1) From S-parameters determine the transistor device model.

(2) Set the objective of each matching circuit depending upon the application (Figure 21), and the gain-bandwidth limitations due to the transistor model.

3) Design each matching network and check the total amplifier response over the operating bandwidth.

(4) Readjust matching elements for acceptable gain flatness.

Several design examples are available in the literature [41-48] . From a practical point of view, a minimum of reactive elements should be used and resistors should be avoided. A large gain slope in the interstage only (e.g., 12 dB/octave) will be difficult to realize over a wide band. Usually some gain sloping in all matching networks is recommended. Finally, amplifier stability must be checked at all frequencies for all stages.

Practical Balanced Designs

The most common approach to GaAs FET amplifier design is the balanced approach [51] shown in Figure 23. After the amplifier design is verified, two amplifiers are connected at the ports of a 90 degree, 3 dB hybrid, usually a Lange coupler [52] . This approach allows the input and output amplifier ports to be mismatched, but the combined amplifier will appear matched if the stages are balanced.

Another balancing approach uses two in-phase Wilkinson power dividers and a pair of 90° phase shifts, also shown in Figure 23. This approach is simpler to build but has less bandwidth inherent in the 90° phase shifters.

Several other advantages of the balanced approach are:

(1) 3 dB increase in output power;

(2) Stages may be cascaded with minimum interstage effects, i.e., ripple;

(3) Excellent amplifier stability;

(4) A minimum amount of tuning labor is required.

The disadvantages are:

(1) Two transistors are required, thus doubling the transistor cost;

(2) Bandwidth is limited by the Lange coupler, which is usually greater than an octave, or the 90° phase shifter;

(3) Coupler loss is about 0.3 dB.

In modern microwave integrated circuit factories, the balanced approach is by far the most common design. Single section couplers have given excellent bandwidths for amplifier stages including 2-8 GHz, 4-12 GHz, and 6-18 GHz designs.

Another practical approach to achieving amplifier mismatch with good overall matching is the use of isolators. However, larger size and higher cost has limited the application of this approach.

When a balanced amplifier is designed for wide bandwidth, the first step is the coupler design, which is indicated in Figure 24. The bandwidth of back-to-back couplers will usually be the limitation on bandwidth. As shown in Figure 24, a coupling factor of about 1.5 to 2.5 dB will give the largest bandwidth with the least ripple. A 3 dB coupler does not give the best bandwidth.

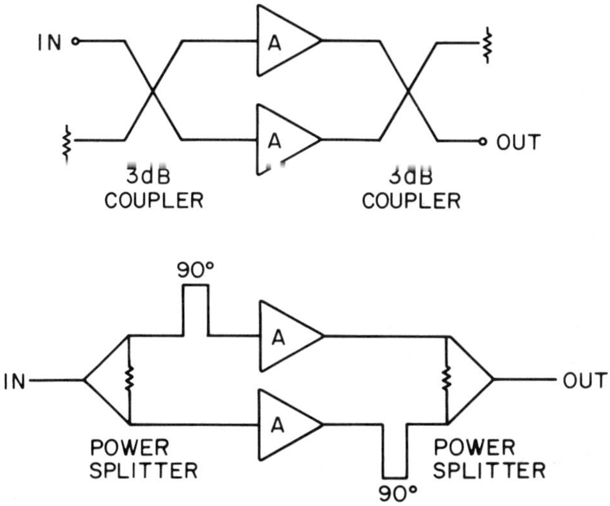

Figure 23/Balanced Amplifier Designs

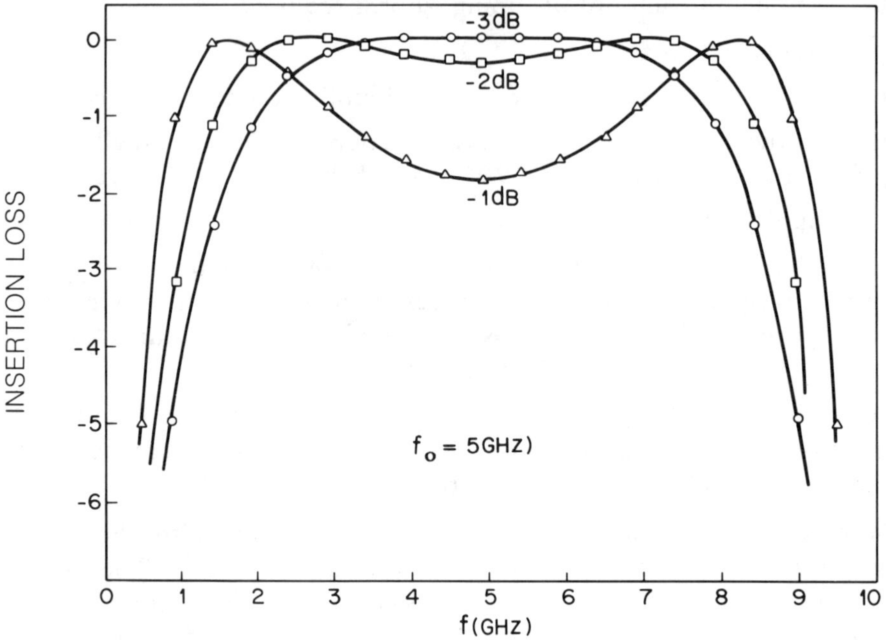

Figure 24/Cascaded Coupler Performance over 1-10 GHz

4.2 Oscillator Design [14-17, 41]

Oscillator and amplifier designs are closely related for two reasons. First, the same GaAs FET transistor at approximately the same dc bias can be used for both applications, except for the configuration. Common gate is preferred for oscillator design and common source for the amplifier design. In addition, nearly equivalent power levels and equivalent bias conditions are required for both designs.

The second reason for the close relationship is the use of S-parameters for both designs, which is illustrated in Figure 25. In amplifier design, two matching structures are added to the two-port to give a maximum transducer power gain and a low input and output reflection coefficient. In oscillator design, the same two-port (or one which is appropriately modified with feedback) can be designed to deliver nearly the same output power to the 50 ohm load. The load will receive approximately the same power and the transistor is dc biased

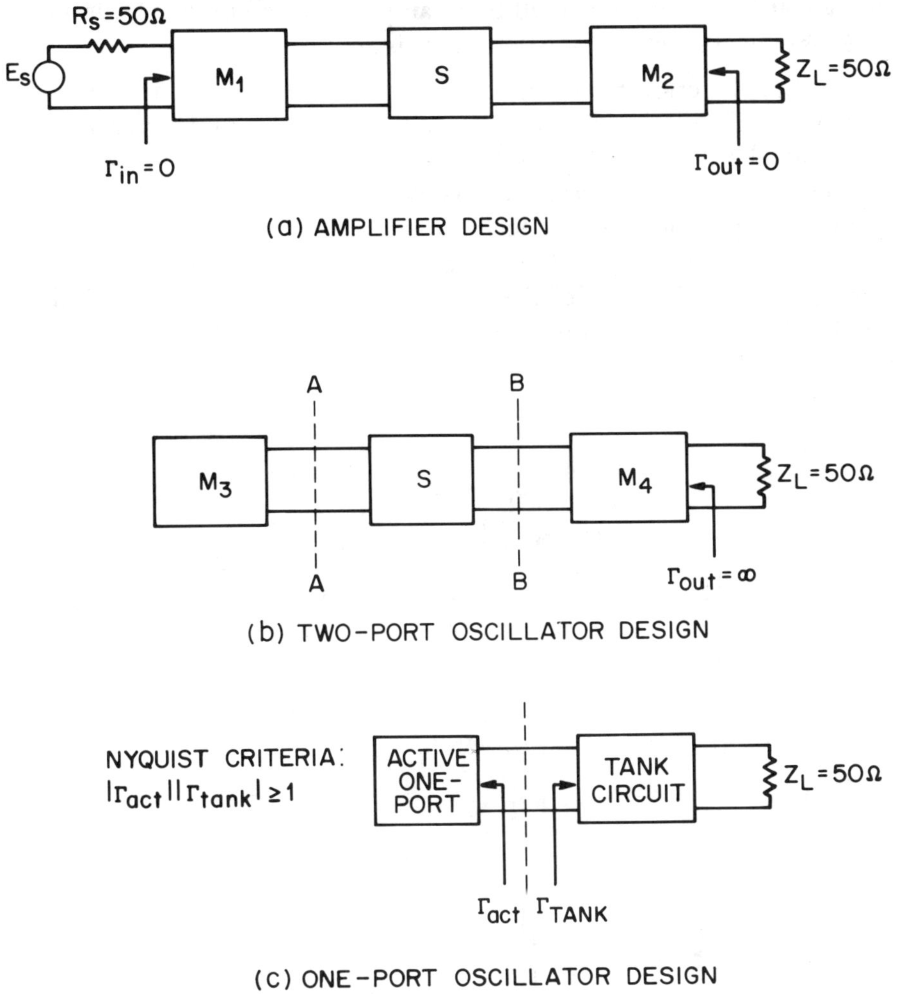

Figure 25/GaAs FET Oscillator Design

to approximately the same operating point. Since the load does not know if it has been connected to an amplifier or an oscillator, in both cases the output reflection coefficient is zero.

A possible objection to the use of two-port S-parameters in oscillator design is the large signal nature of oscillator operation. However, small signal S-parameters have been successfully applied to many oscillator designs [14-17] with results which correctly predict the frequency of operation. As the technology improves, additional modifications to the small signal S-parameters will improve the oscillator design procedures. The basic two-port approach to the oscillator problem has been clearly established.

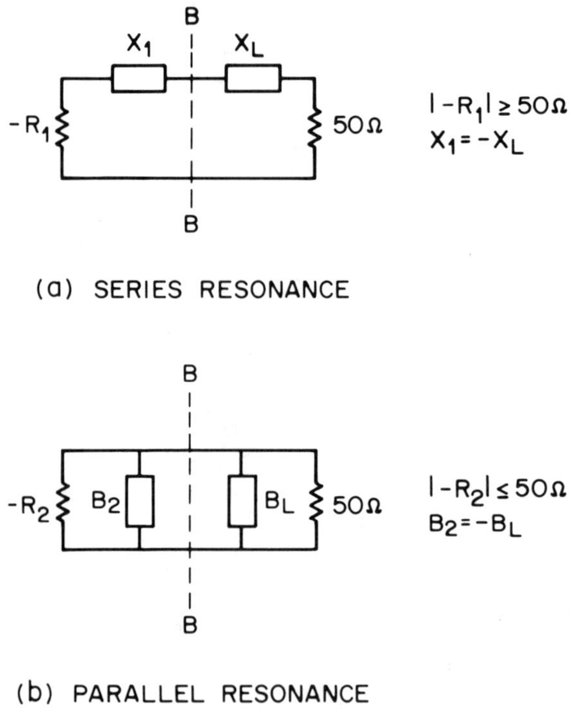

Figure 26/Oscillator Resonance

The condition for oscillation is the Nyquist criteria which must exist for reference plane BB at the output of the active two-port. The two-port must be terminated by a lossless termination M_3 and possibly lossless feedback elements to satisfy the Nyquist criteria at the output port BB. This output port could be the drain of the FET (with either the source or gate as the ground terminal), but any terminal of the transistor could deliver power as an oscillator. After achieving the Nyquist criteria at port BB, the remainder of the circuit design is illustrated in Figure 26 for both the series and parallel resonance conditions.

A useful theorem [17] for oscillator design is that which states that if the Nyquist criteria is satisfied at port AA, it is also satisfied at BB and vice versa. Thus at oscillation:

$$S_{11}'\Gamma_1 = 1 \tag{52}$$

$$S_{22}'\Gamma_2 = 1 \tag{53}$$

(see Figure 27).

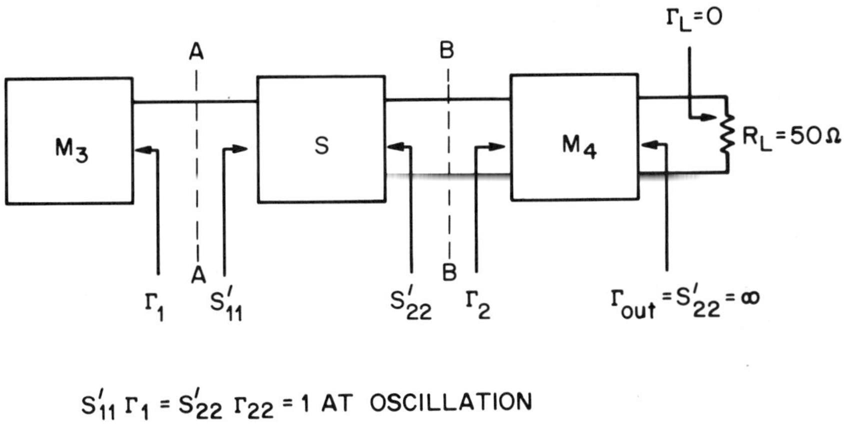

Figure 27/Simultaneous Oscillation at Both Ports

For any two-port the necessary and sufficient conditions for unconditional stability are:

$$k > 1 \tag{17}$$

$$|D| < 1 \tag{18}$$

$$D = S_{11} S_{22} - S_{21} S_{12} \tag{19}$$

For this two-port at this frequency, oscillation is impossible. Two choices are possible: add feedback or change configurations. Usually the oscillation designer uses the common-gate configuration since this gives the condition over a wide frequency range:

$$|S_{22}| > 1$$

Recall the two-port S-parameters are measured with 50 ohm generator and load terminations, but an oscillator will operate with only a 50 ohm load termination. Thus, if $|S_{22}| > 1$ with a 50 ohm termination at the generator port, then the condition for oscillation can be satisfied with a lossless termination M_3 at the input port.

The condition for oscillation will exist at both ports simultaneously. The oscillation condition is:

$$|S_{11}'|\ |\Gamma_S| \geqslant 1 \tag{54}$$

$$|S_{22}'|\ |\Gamma_L| \geqslant 1 \tag{55}$$

The matching circuits M_3 and M_4 must be designed to satisfy Equation 54 or 55.

4.3 Mixers [18, 19, 53-54]

The FET is useful for mixing because of conversion gain and low noise figure. In microwave mixers, a strong local oscillator frequency and a small RF signal are simultaneously applied to the terminals of a nonlinear device to produce a low frequency intermediate frequency. In a normal diode mixer the diode is forward biased to produce a nonlinear resistor. The resulting conversion circuits typically give 6-10 dB conversion loss when optimized.

Both single and dual-gate FET mixers have been reported with similar low noise figures and conversion gain [53]. The nonlinearity which produces conversion gains is the rapid change in g_m as the current is cut off. For amplifiers, a constant transconductance is desired for linear power gain; for mixers, a rapidly changing transconductance is required for conversion gain. For simplicity, the single gate mixer analysis will be considered. Some possible mixer circuits are suggested in Figure 28 where usually the IF appears at the drain. The drain current may be written as

$$i_D = i_0 + \frac{\partial i_D}{\partial e_{in}} e_{in} + \frac{\partial^2 i_D}{\partial^2 e_{in}} \frac{e_{in}^2}{2!} + \ldots$$

$$= i_0 + g_m e_{in} + \frac{\partial g_m}{\partial e_{in}} \frac{e_{in}^2}{2} + \ldots \qquad (56)$$

When the input signal (at gate-source) is the sum of a local oscillator component and a signal component,

$$e_{in} = e_g \cos \omega_1 t + e_0 \cos \omega_0 t \qquad (57)$$

the drain current becomes:

$$i_D \simeq i_0 + g_m e_{in} + \frac{\partial g_m}{\partial e_0} \frac{1}{2} (e_g \cos \omega_1 t + e_0 \cos \omega_0 t)^2 \qquad (58)$$

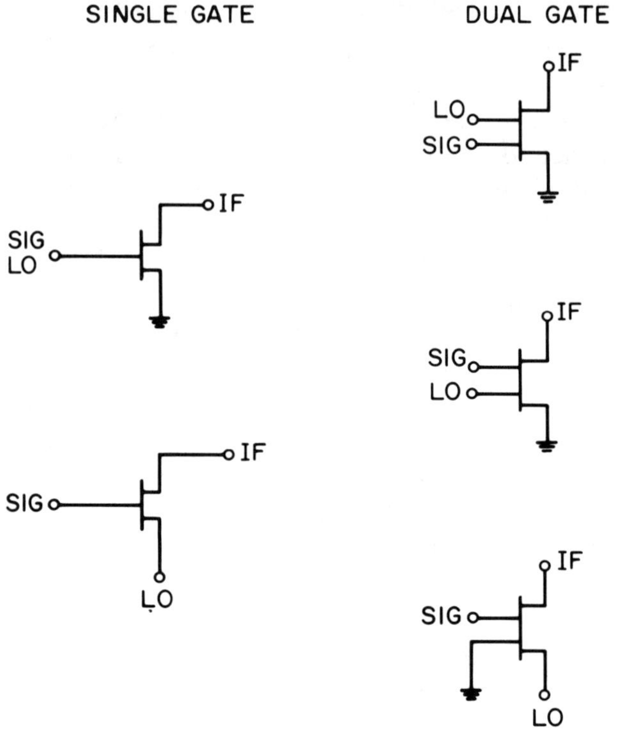

Figure 28/GaAs FET Mixer Configurations

The intermediate frequency term is

$$\frac{\partial g_m}{\partial e_0} \; \frac{1}{2} \; e_g e_0 \; \cos\,(\omega_0 - \omega_1)t$$

The conversion conductance is defined as

$$g_1 = \frac{\partial i_D\,(\omega_1)}{\partial e_g(\omega_1)} = \frac{\partial g_m}{\partial e_0} \; \frac{e_0}{2} \tag{59}$$

This conductance is optimized by providing sufficient local oscillator power and dc biasing near the pinch off voltage which gives

$$g_1 \simeq \frac{g_m}{3} \tag{60}$$

Calculating the maximum available amplifier power gain gives the required insight for high mixer conversion gain. For the simplified amplifier case

$$P_{avs} = \frac{e_s^2}{4R_g} \tag{61}$$

$$P_{out} \simeq \frac{g_m^2 v_g^2}{4} R_d \tag{62}$$

The maximum value of input voltage at the gate-source capacitor is

$$v_{g\,max} = \frac{1/\omega_1 C}{1/\omega_1 C + R_{in}} \frac{e_s}{2} \simeq \frac{e_s}{2\omega_1 C R_{in}} \tag{63}$$

Thus the maximum available gain becomes

$$MAG \cong \frac{g_m^2}{4\omega_1^2 C^2} \frac{R_d}{R_{in}} \tag{64}$$

The same analysis can be applied to mixer conversion gain. The input is tuned for the maximum signal voltage at the gate-source capacitor. The output is tuned for maximum output current at the intermediate frequency. The mixer available conversion gain becomes

$$G_c = \frac{P_{out}(\omega_i)}{P_{in}(\omega_s)} = \frac{i_D^2(\omega_i)}{4} \overline{R}_d \frac{4R_g}{e_s^2} \tag{65}$$

$$v_g = \frac{e_s}{2\omega_1 \overline{C}\, \overline{R}_{in}} \tag{66}$$

$$g_1 = \frac{i_D(\omega_i)}{v_g} \tag{67}$$

$$G_c \simeq \frac{g_1^2 v_g^2}{4} \overline{R}_d \frac{4\overline{R}_{in}}{v_g^2 4\omega_1^2 \overline{C}^2 \overline{R}_{in}^2} \tag{68}$$

$$G_c \simeq \frac{g_1^2}{4\omega_1^2 \overline{C}^2} \frac{\overline{R}_d}{\overline{R}_{in}} \tag{69}$$

The bar terms $\overline{C}$, $\overline{R}_d$, and $\overline{R}_{in}$ refer to the time average value since the local oscillator signal will vary these parameters from the small-signal value. Several observations can be made from Equation 69. The conversion gain will be dependent on the impedance transformation $\overline{R}_d/\overline{R}_{in}$ which can be optimized with bias and local oscillator drive. High transconductance is required, but biasing at best power gain (where g_m is maximum) would give about 10 dB lower conversion gain compared to the optimum power gain because of Equation 60. Also a 6 dB/octave gain rolloff is expected for conversion gain. The optimum device for power gain at low bias current would appear to give the best mixer conversion gain.

The FET mixer must be designed with the following objectives:

(1) conversion gain;

(2) low noise figure;

(3) wide dynamic range;

(4) signal-local oscillator isolation;

(5) low intermodulation distortion;

(6) wide bandwidth.

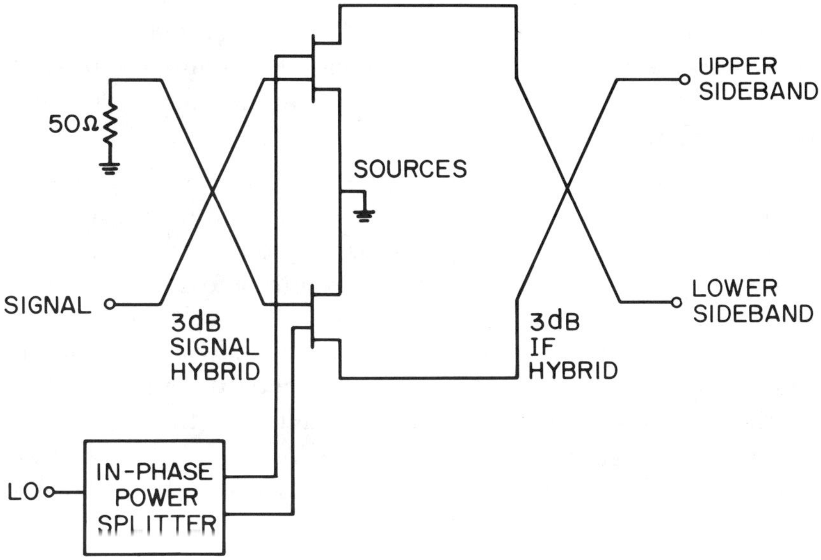

Figure 29/Image-Reject Balanced Mixer Using Dual-Gate FETs

Often a balanced (2 transistors) or doubly balanced (4 transistors) mixer is used to satisfy these additional requirements. An example of a balanced mixer which also offers image rejection is given in Figure 29. These circuits generally use 3 dB hybrids, baluns, and power splitters to simultaneously apply the local oscillator and signal voltages.

4.4 Switches [20-22]

Microwave switching is usually accomplished by series or shunt diodes in a transmission line system. The PIN diode is usually chosen for low insertion loss in the on state and high isolation in the off state. The two-terminal device is chosen for low cost. The GaAs FET can be applied to the application of switching. Largely because of its high cost, this transistor has not been economically applied to switching. However, the possibility of GaAs FET switches may be attractive in monolithic circuits.

The GaAs FET switch is shown in Figure 30. In the on state the circuit is simply an amplifier with gain over the desired operating band. In the off state, the control bias at the gate is adjusted for zero current, i.e., the channel is pinched off. The resulting loss is approximately the $|S_{12}|^2$ of the FET, modified by the mismatch of the amplifier matching circuits. The design of the broadband GaAs FET switch follows the concepts presented in the S-parameter design of small-signal amplifiers. For the off-state, the S-parameters of the off bias condition must be measured. An off state isolation of -20 dB is easily achieved from small signal GaAs FETs below 10 GHz. If low VSWR is required, the balanced amplifier configuration could be used.

Because the dual-gate GaAs FET has a very low value of S_{12}, this transistor can be used for a high isolation switch, i.e., high isolation in the off state. The same result can be achieved by building a two-stage amplifier for a switching application.

The speed of GaAs FET switches is extremely fast because of the low values of input capacitance. The input switching speed can be estimated using an approximate formula for speed

$$\tau \simeq R_{in} C_{in} \tag{70}$$

For the DXL2501A chip shown in Figure 14, this gives $\tau_{in} \cong 5$ psec.

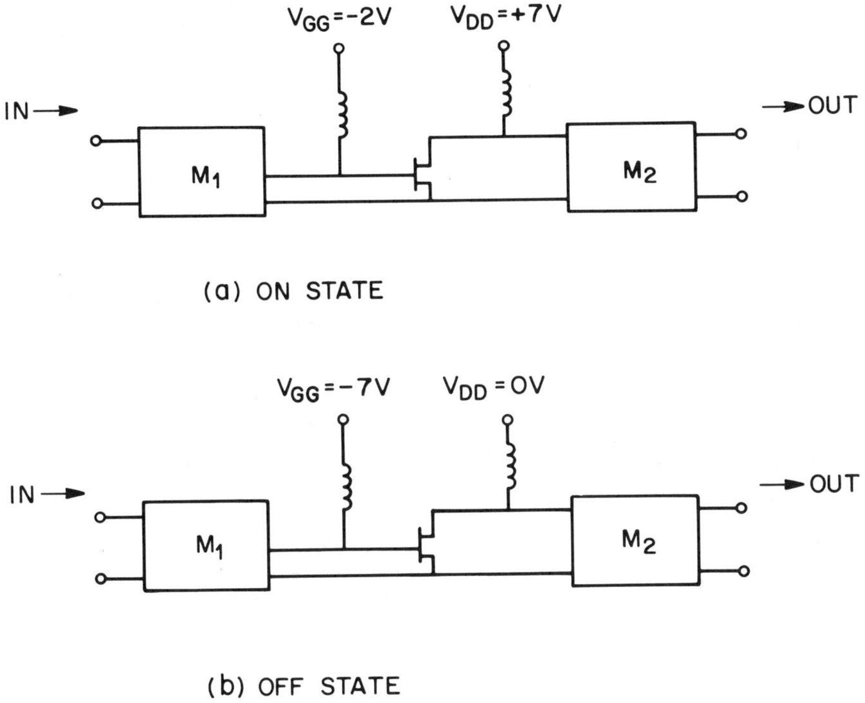

Figure 30/GaAs FET Switch Configuration

Switching speeds of the order of 10 psec have been reported for GaAs FETs in logic gates [55].

Some work has been reported on GaAs FET devices designed for microwave switching [21-22]. For example, by parallel resonating the transistors, a very low dc dissipation, low insertion loss switch has been reported. This SPDT switch had an insertion loss of 0.7 dB with 28 dB isolation at 10 GHz. There was no dc bias applied to the drain terminal. The on condition corresponded to 0 voltage applied to the gate with the channel essentially open. The off condition corresponded to -4 to -8 volts applied to pinch off the channel. The equivalent circuits and design performance of the series and parallel GaAs FET switches are given in Figure 31. The resonated GaAs FET SPDT switch is shown in Figure 32 [21].

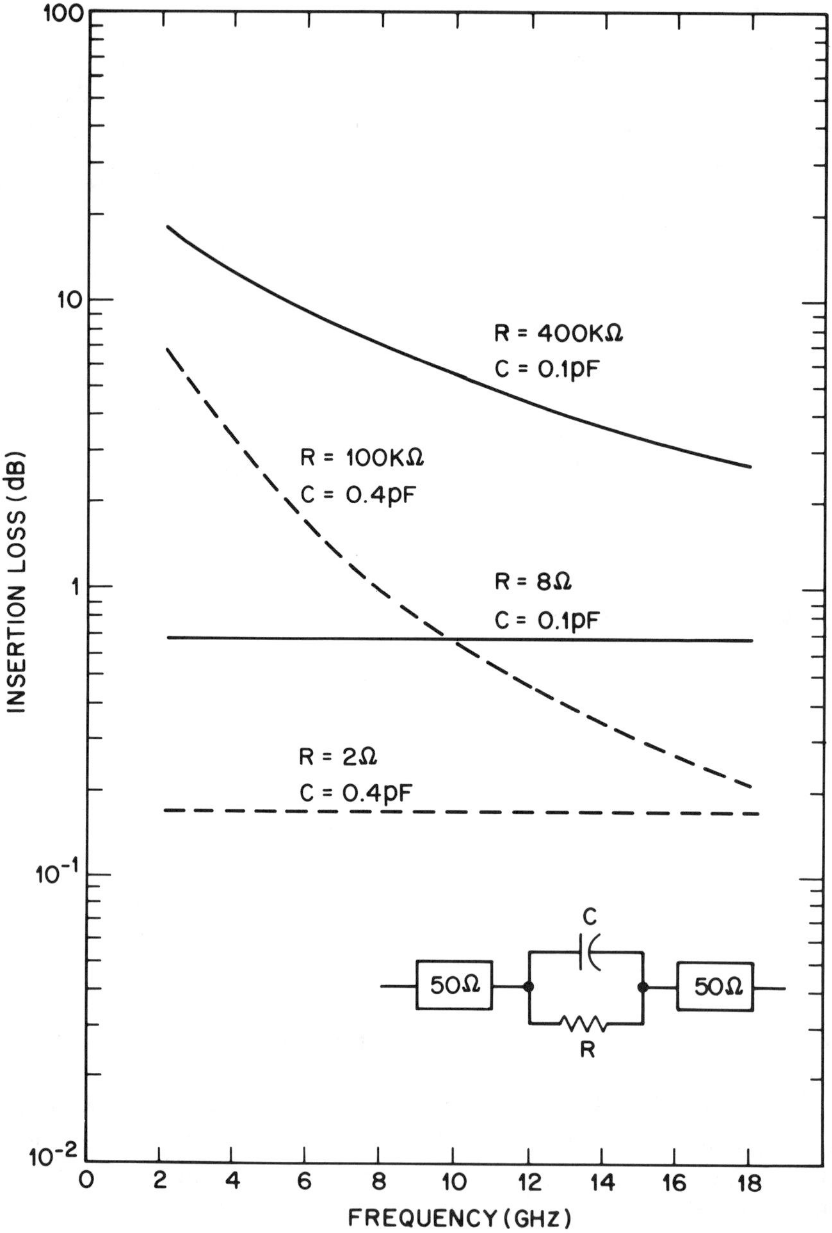

(a) Series Switch

Figure 31*/Equivalent Circuits of GaAs FET Switches [21]. Solid lines correspond to 300 μm device and dashed lines correspond to 1200 μm device.*

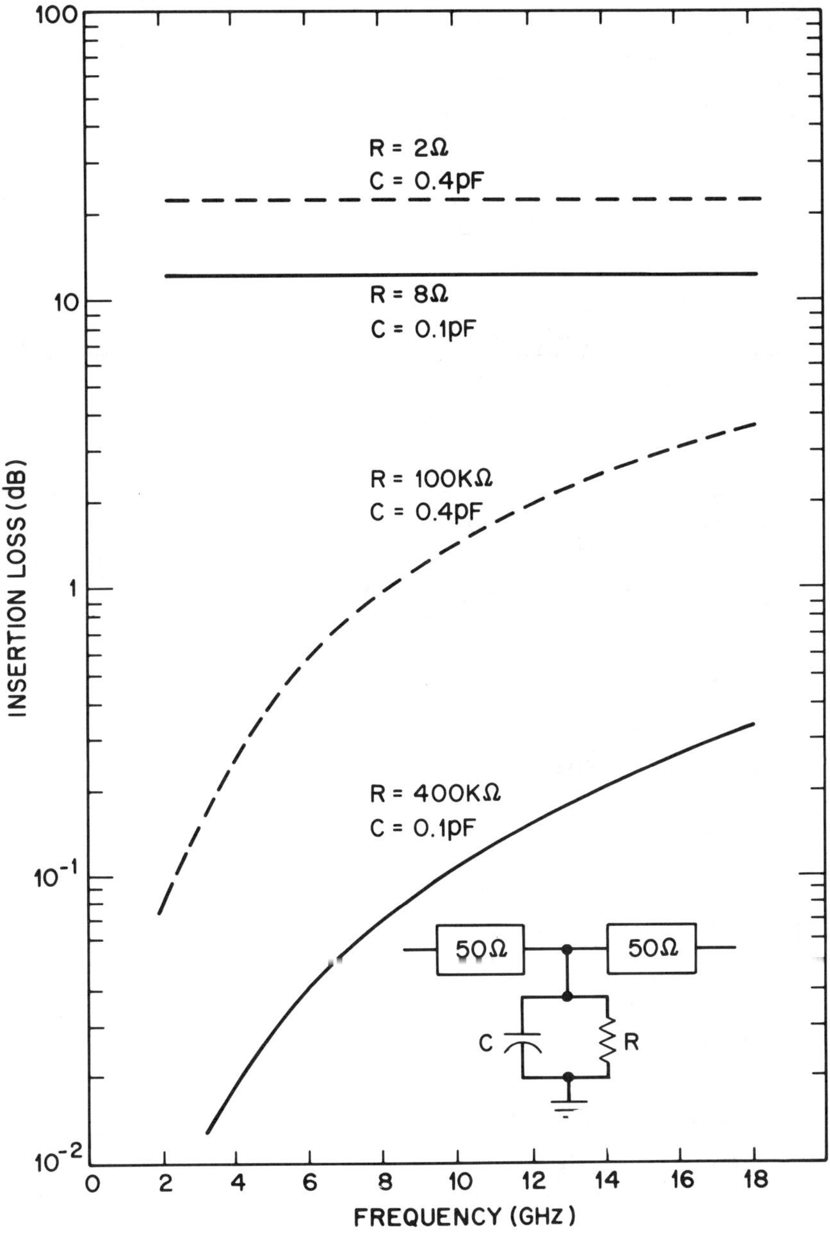

(b) Parallel Switch

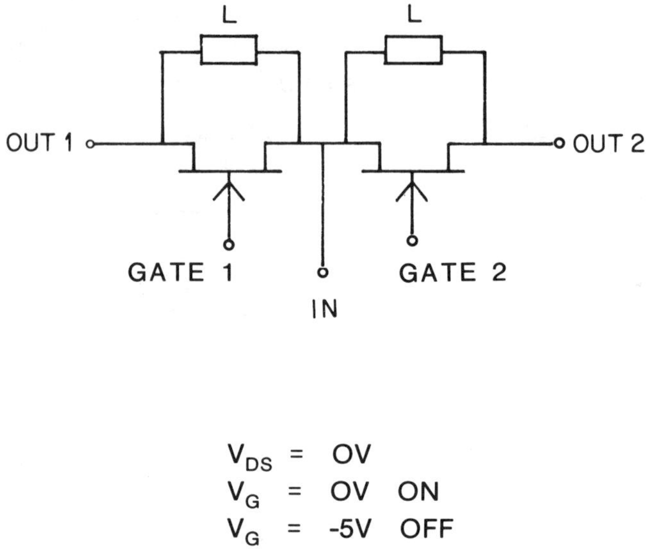

***Figure 32**/Resonated GaAs FET Switch-SPDT [21]*

The dual gate GaAs FET has also been applied to X and Ku-band switching [22]. Both SPST and DPDT structures have been built with over 30 dB isolation at 10 GHz. These switches were designed equivalent circuit models derived from S-parameter data.

4.5 Attenuators [23-25]

An attenuator is either a fixed or variable value of loss in a microwave transmission system. It should present a low VSWR or low mismatch to the system. Usually this function is also performed by PIN switches but the GaAs FET is a useful attenuator in some applications (e.g., monolithic circuits). Fixed attenuators are designed by methods described in the previous section for microwave switches in the off-state.

A variable attenuator is a useful component in a microwave system since the gain can be effectively adjusted by a dc control voltage. The design of the broadband variable-gain attenuator using a dual-gate GaAs FET is shown in Figure 33. Here the RF termination at gate 2 is an important consideration in the overall amplifier design. Using a simple two stage amplifier design, this amplifier can give a gain of 20 dB or a loss of 40 dB by varying the voltage at gate 2 from +1.0 volts to -2.0 volts over an octave bandwidth.

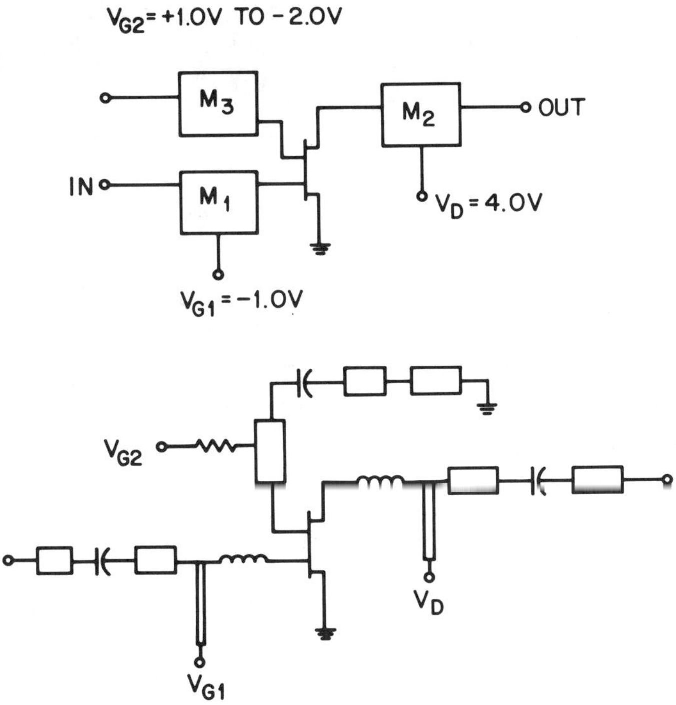

Figure 33/Broadband Variable Gain GaAs FET Attenuator [24]

4.6 Modulators [26]

A modulator is a three-port network where a low frequency (IF port) is added to a carrier (LO port) to produce an amplitude modulated signal (RF port). A modulator is simply a mixer circuit which is operating in reverse. The low frequency IF signal is being upconverted to the RF band. Since the GaAs FET has the nonlinear properties required for mixing, it also has the characteristics needed for a modulator or frequency upconverter.

The dual-gate GaAs FET offers the possibility of high speed AM modulation. By pulsing the bias of the second gate, the microwave signal can be modulated very quickly, with switching speeds ~ 100 psec already reported for an 8 GHz amplifier. The circuit for the pulse-amplitude modulated signal is given in Figure 34, where no RF matching was used. The circuit was built using 50 ohm coplanar lines on a sapphire substrate.

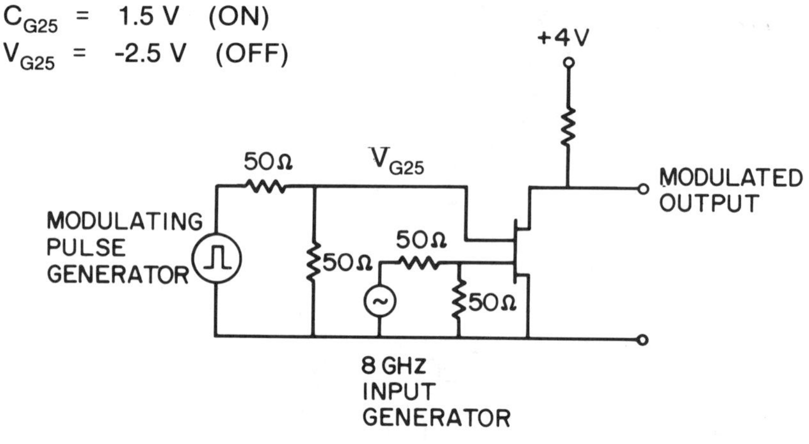

Figure 34/Dual-Gate GaAs FET Pulsed Amplitude Modulator Configuration [26].

For a mathematical description of the modulator, assume that the transistor has a linear dependence on transconductance given by

$$g_m = Kv_{bias} \tag{71}$$

The modulation signal is applied to the bias terminal such that

$$v_{bias} = v_{dc} + v_m \cos \omega_m t. \tag{72}$$

Since the low frequency voltage gain is

$$A_v = -g_m R_L \tag{73}$$

the input and output signal are given by

$$v_{in} = v_s \cos \omega_s t \tag{74}$$

$$\begin{aligned} v_{out} &= -g_m R_L v_{in} = -K(v_{dc} + v_m \cos \omega_m t) R_L v_s \cos \omega_s t \\ &= -Kv_{dc}(1 + m \cos \omega_m t) R_L v_s \cos \omega_s t \end{aligned} \tag{75}$$

where $m = v_m / v_{dc}$ is the modulation index.

Thus, the output signal is amplified and amplitude modulated if the transistor has the proper transconductance curve.

4.7 Multipliers [27-28]

The nonlinearity of the GaAs FET may be applied to both mixers and frequency multipliers. Generally there are four nonlinear components in the GaAs FET: the input C_{gs}, the g_m, the output resistance R_o, and the output capacitance C_{ds}. Under the large signal drive, these parameters will change in the transistor model, which will lead to frequency conversion either by mixing or harmonic generation. Frequency multiplication is usually done by low-loss varactor diode circuits. The GaAs FET offers the possibility of frequency multiplication with gain.

An example of a dual-gate GaAs FET doubler is given in Figure 35. At high drive (+15 dBm) this circuit can give 0 dB conversion loss with 14 GHz output as a doubler. By designing for tripling, a conversion loss of −6 dB was found at 18 GHz with an input drive of +10 dBm. The nonlinearity of the output resistance is the multiplying element for this circuit.

Using the balanced doubler circuit of Figure 36, an output power of +16 dBm has been achieved across 18-26 GHz (K-Band) with about 1 dB conversion loss. At 40 GHz, −7 dB conversion gain has been achieved with an output power of +5 dBm.

The operation of the dual-gate multiplier (Figure 35) can be understood as follows. The drain-source dc voltage is 6 volts, so the drain of the first FET is at 3 volts. At this low bias voltage and at a high drain current, the first gate can be overdriven to give two nonlinearities:

1) Drain current clipping at high drain currents

2) Forward bias gate current from the second gate which is dc biased at 2.5 volts.

The RF loads at all three ports and the dc bias levels at three external terminals have been optimized for the best frequency conversion.

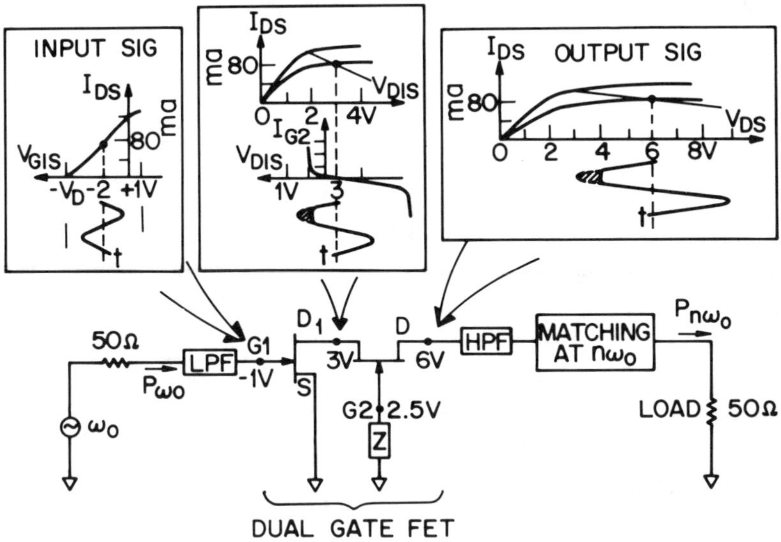

Figure 35/Dual Gate GaAs FET Multiplier [28]

The measurements of the optimum output load for doublers suggests a load of

$$\frac{Z_L - Z_o}{Z_L + Z_o} \simeq S_{22}^* \tag{76}$$

which implies the output conductance is the nonlinear element in the dual-gate doubler.

The balanced dual-gate doubler simply allows both FETs to contribute harmonics in the proper phase to emphasize the second harmonic.

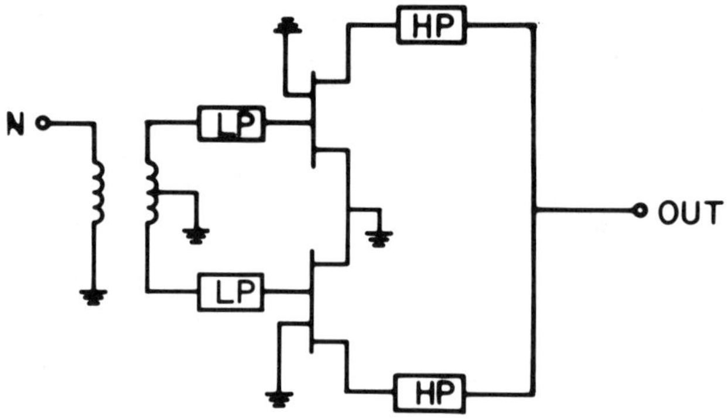

Figure 36/Balanced Dual Gate GaAs FET Doubler

4.8 Limiters [25]

A limiter is simply a two port network which limits the output power level to a constant value at high input power levels. The required power transfer curve is given in Figure 37. The GaAs FET amplifier is an excellent limiter. When the output power level reaches the 1 dB compression point, the output is beginning to limit and the signal may have some AM/PM conversion. As the input level is increased further, the output power will usually increase only 1-2 dB higher than the 1 dB compression point. This is the region of hard limiting,

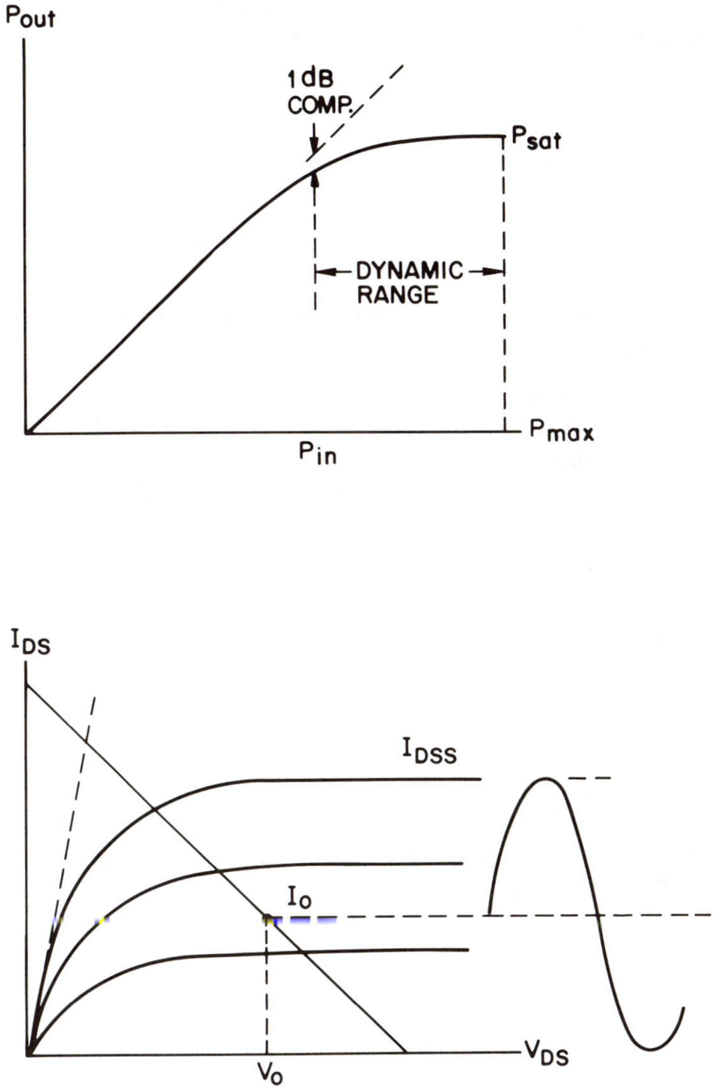

Figure 37/Limiting Operation of GaAs FET

and the dynamic range can be 20-30 dB of input power change. The limitation is the input power which will permanently damage the transistor. Assuming the maximum Class A power of the transistor is the power for output limiting (see Figure 37), the output power is

$$P_{max} \simeq \frac{V_o I_o}{2} \tag{77}$$

which is determined by the I_{DSS} of the transistor, the maximum safe operating voltage, and the rf load line of the amplifier.

The dual-gate GaAs FET is also an excellent limiter [56]. It offers the design advantages of lower power output limiting, a sharper saturation knee, and the ability to set the output power level by changing the dc bias point. The design of these limiters requires the use of three-port S-parameters for the design of wide-band stable amplifiers.

5. SUMMARY

Some of the present applications of GaAs FETs have been reviewed in this chapter. The present trend in engineering is the application of this technology to monolithic integrated circuits [57, 58]. When the technology matures, millimeter-wave integrated circuits [59] will be built using GaAs FET technology. Some exciting results have already been reported at 40 GHz for amplifiers [60] and 69 GHz for oscillators [61]. As the computer technology advances into the range of microwave speeds, the superior performance of this GaAs FET technology will undoubtedly become an important factor in future electronic designs.

REFERENCES

[1] Mead, C.A., *Proc. IEEE,* 54:307, 1966.

[2] Hooper, W.W. and W.I. Lehrer, *Proc. IEEE,* 55:1237, 1967.

[3] Wolf, P., *IBM J. Res. Develop.,* 14:125, 1970.

[4] Drangeid, K., R. Sommerhalder, and W. Walter, *Electron. Lett.,* 6:228, 1970.

[5] Baechtold, W. and P. Wolf, *Solid-State Electron.,* 14:783, 1971.

[6] Baechtold, W., W. Walter, and P. Wolf, *Electron. Lett.,* 8:35, 1972.

[7] Baechtold, W., *IEEE J. Solid-State Circuits,* SC-8:54, 1973.

[8] Baechtold, W. et al., *Electron. Lett.,* 9:232, 1973.

[9] Liechti, C.A., *IEEE Trans. MTT,* MTT-24:279, 1976. (Includes detailed bibliography up to 1976.)

[10] Ohata, K., H. Itoh, F. Hasegawa, and Y. Fujiki, *IEEE Trans. Electron. Dev.,* ED-27:1029, 1980.

[11] Fukui, H., et al., *IEEE Trans. Electron Dev.,* ED-27:1034, 1980.

[12] DiLorenzo, J.V. and W.R. Wisseman, *IEEE Trans. MTT,* MTT-27:367, 1979.

[13] Higashisaka, A., Y. Takayama, and F. Hasegawa, *IEEE Trans. Electron Dev.,* ED-27:1025, 1980.

[14] Maeda, M., K. Kimura, and H. Kodera, *IEEE Trans. MTT,* MTT-23:661, 1975.

[15] Abe, H., et al., *IEEE Trans. MTT,* MTT-26:156, 1978.

[16] Johnson, K., *IEEE Trans. MTT,* MTT-27:217, 1979.

[17] Basawapatna, G.R. and R.B. Stancliff, *IEEE Trans. MTT,* MTT-27:379, 1979.

[18] Pucel, R.A., D. Masse, and R. Bera, *IEEE Trans. MTT,* MTT-24:351, 1976.

[19] Kurita, O., and K. Morita, *IEEE Trans. MTT,* MTT-24:361, 1976.

[20] Garver, R.V., *IEEE Trans. MTT,* MTT-27:523, 1979.

[21] McLevige, W.V. and V. Sokolov, *IEEE Trans. Electron Dev.,* ED-28:198, 1981.

[22] Vorhaus, J.L., W. Fabian, P.B. Ng, and Y. Tajima, *IEEE Trans. Electron Dev.,* ED-28:204, 1981.

[23] Maeda, M. and Y. Minai, *IEEE Trans. MTT,* MTT-22:1226, 1974.

[24] Goel, J.A. and H.J. Wolkstein, *Electronic Letters,* 14:167, 1978.

[25] Rosen, A., H.J. Wolkstein, J.A. Goel, and R.J. Martarese, *RCA Rev.,* 38:253, 1977.

[26] Liechti, C.A., *IEEE Trans. MTT,* MTT-23:461, 1975.

[27] Chen, P.T., C.T. Li, and P.H. Wang, *IEEE Int. Microwave Symp. Dig. Tech. Papers,* 1978.

[28] Chen, P.T., C.T. Li, and P.H. Wang, *IEEE Trans. MTT,* MTT-27:411, 1979.

[29] Yamamoto, R., A. Higashisaka, and F. Hasegawa, *IEEE Trans. Electron Dev.,* ED-25:567, 1978.

[30] Fukuta, M. et al., *IEEE Trans. Electron Dev.,* ED-25:559, 1978.

[31] Statz, H., H.A. Haus, and R.A. Pucel, *IEEE Trans. Electron Dev.,* ED-21:549, 1974.

[32] Baechtold, W., *IEEE Trans. Electron Dev.,* ED-18:97, 1971.

[33] Baechtold, W., *IEEE Trans. Electron Dev.,* ED-19:674, 1972.

[34] Turner, J.A., et al., *Electronic Letters,* 7:661, 1971.

[35] Asai, S., F. Murai, and H. Kodera, *IEEE Trans. Electron Dev.,* ED-22:897, 1975.

[36] Ogawa, M., K. Ohata, T. Furutsuka, and N. Kawamura, *IEEE Trans. MTT,* MTT-24:300, 1976.

[37] Furutsuka, T., M. Ogawa, and N. Kawamura, *IEEE Trans. Electron Dev.,* ED-25:580, 1978.

[38] Liechti, C.A. and R.L. Tillman, *IEEE Trans. MTT,* MTT-22: 510, 1974.

[39] Neidert, R.E. and H.A. Willing, *IEEE Trans. MTT,* MTT-24: 342, 1976.

[40] Hornbuckle, D.P. and L.J. Kuhlman, *IEEE Trans. MTT,* MTT-24:338, 1976.

[41] Vendelin, G.D., *Design of Amplifiers and Oscillators by the S-Parameter Method.* New York: John Wiley and Sons, 1982.

[42] Tucker, R.S., *IEEE Trans. MTT,* MTT-21:322, 1973.

[43] Medley, M.W. and J.L. Allen, *IEEE Trans. MTT,* MTT-27: 784, 1979.

[44] Pitzalis, O. and R.A. Gilson, *IEEE Trans. MTT,* MTT-19:381, 1971.

[45] Mokari-Bolhassan, M.E. and W.H. Ku, *IEEE Trans. MTT,* MTT-25:837, 1977.

[46] Mokari-Bolhassan, M.E. and W.H. Ku, *IEEE Trans. MTT,* MTT-25:848, 1977.

[47] Mellor, D.S., *Computer-Aided Synthesis of Matching Networks for Microwave Amplifiers,* Stanford Electronics Laboratories SEL-75-005, (Ph.D. Thesis), 1975.

[48] Ku, W.H. and W.C. Petersen, *IEEE Trans. Circuits and Systems,* CAS-22:523, 1975.

[49] Mellor, D.J. and J.G. Linvill, *IEEE Trans. MTT,* MTT-23: 1013, 1975.

[50] Fano, R.M., *J. Franklin Inst.,* 249:57, 1960 and 249:139, 1960.

[51] Kurokawa, K., *Bell Syst. Tech. J.,* 44:1657, 1964.

[52] Lange, J., *IEEE Trans. MTT,* MTT-17:1150, 1969.

[53] Stitch, J.E. and P.N. Robson, *Proc. IEEE,* 61:399, 1973.

[54] Cripps, S.C. et al., *IEEE Int. Microwave Symp. Dig. Tech. Papers,* 1977.

[55] Van Tuyl, R.L. and C.A. Liechti, *IEEE ISSCC Dig. of Tech. Papers,* 1974.

[56] Hamilton, R.J. Jr., *Int. Microwave Symp. Dig. of Tech. Papers,* 1979.

[57] Special Issue on Advances in Monolithic III-V Devices and Circuits, *IEEE Trans. Electron Dev.,* ED-28:133, 1981.

[58] Pucel, R.A., *IEEE Trans. MTT,* MTT-29:513, 1981.

[59] Mao, S., S. Jones, and G.D. Vendelin, *IEEE Trans. MTT,* MTT-16:455, 1968.

[60] Watkins, E., H. Yamasaki, and J. Schellenberg, *IEEE ISSCC Dig. of Tech. Papers,* 198, 1982.

[61] Schellenberg, J.M., H. Yamasaki, and D.W. Maki, *IEEE Int. Microwave Symp. Dig. Tech. Papers,* p. 328, 1981.

Circuit Aspects of Power GaAs FETs

W. O. Schlosser

Bell Laboratories, Murray Hill, New Jersey.

V. Sokolov

Honeywell Corporate Technology Center
Minneapolis, Minnesota

1. LINEAR AMPLIFIER DESIGN

The design of GaAs FET amplifiers is identical in principle to that of bipolar and similar to that of reflection amplifiers (Impatt, Gunn). It has, therefore, been extensively treated in the literature even before the advent of GaAs FETs. The GaAs FET does pose a somewhat more difficult problem than the bipolar transistor because the reflection coefficients are greater and the accuracy and loss of the transformation networks are thus more critical.

It is, of course, impossible to treat the subject in any depth in this chapter. We have, therefore, attempted to summarize the essential steps in linear amplifier design and refer to the pertinent literature for details. Although this discussion is limited to the linear case, the methods are applicable to transistors operating in the nonlinear regime, as we shall see a little later. Special requirements on distortion, efficiency or power output will be discussed in Sections 2-4.

One can break amplifier design into the following steps:

a. Check if the gain-bandwidth requirements are compatible with limitations imposed by the device parameters.
b. Determine the source and load impedance from the transistor scattering parameters or from direct impedance measurements.
c. Synthesize the matching networks from these impedances either using a closed form theory or a computer aided design (CAD) approach.
d. Decide on the circuit technology to be used and realize the circuits.

We will now discuss the first three steps. Step d will be extensively discussed in Section 5.

1.1 Performance Limitations

Gain Limitations
The strategy in determining the maximum gain is different if the transistor is operated in a linear or nonlinear regime, i.e., if the required output power is small or large compared to the saturated power, i.e., the maximum power capability of the device. As a rule of thumb, one can assume that saturation effects start to play a significant role at 10 dB below the saturated power, assuming the transistor is biased close to the bias conditions for maximum power.

If the transistor is *linear* one can use the S-parameters to determine the maximum gain. This topic has been extensively treated in the literature [69]. We have summarized the most important equations in Table 1 for easy reference. If the transistor is unconditionally stable ($K > 1$) the maximum available gain is well defined (Table I). Only one word of caution — the maximum available gain determined in this way depends on the difference between almost equal quantities (e.g., $1 - |S_{11}|^2$). It is only reliable if the S-parameters are very accurately known, especially in the case of large devices with $|S_{11}| \approx 1$. If the device is only conditionally stable no maximum gain can be defined and the gain is determined from the tradeoff between it and the stability margin.

In the *nonlinear* case the situation is much more difficult. Although work is underway [177, 180, 243, 100-102, 145] to develop an S-parameter representation for large signal operation, the problem has not been solved in general. (We will discuss this topic in more detail in Section 1.2.) The only strategies then available are to use (a) the small signal maximum available gain as an upper bound since the gain decreases with increasing power level and (b) to use narrowband tuneable test fixtures to experimentally determine the gain limit.

Bandwidth Limitations

The circuit synthesis problem is complicated by the fact that the devices have a built-in reactance, e.g., the input impedance of the FET chip looks, in general, like a series RC combination, whereas the output can be represented to first order by a parallel RC combination (Figure 1). In packaged devices the situation is even more complicated because even well designed packages add at least a series inductance. Bode first showed [23] that the design of an impedance transformer with a given accuracy is possible only over a limited frequency range if such a built-in reactance is present. The theory of transformer synthesis under reactive constraints and its bandwidth limitations has subsequently been considerably refined [60, 109, 246]. The following equations from Ku and Petersen [109] describe the frequency gain-loss trade-offs due to this reactive constraint if the maximum available gain is frequency independent:

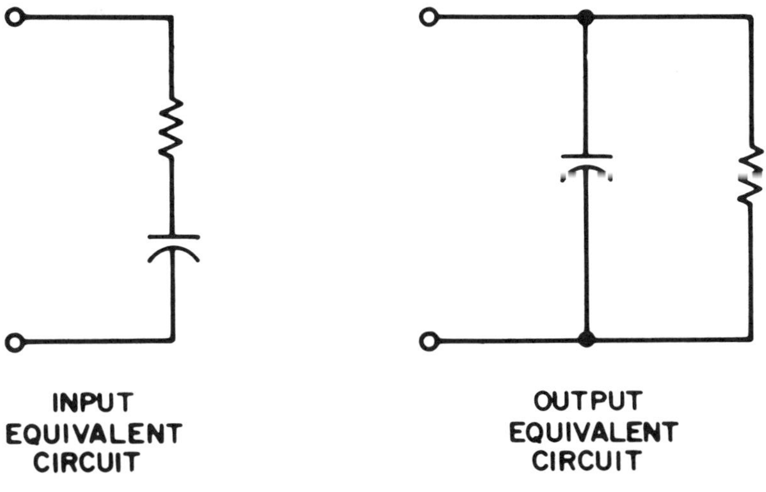

Figure 1/Equivalent circuits of input and output of an idealized GaAs FET chip.

$$\frac{\Delta f}{f_h} = \frac{1}{f_h RC \ln(1/1-k)} \quad \text{output} \tag{1}$$

$$\frac{\Delta f}{f_h} = 4\pi^2 RCf_h \frac{1}{\ln(1/1-k)} \quad \text{input} \tag{2}$$

where the gain-loss is 10 log k, f_h is the upper and Δf the difference frequency. The case of general gain-slope has been treated by Ku and Petersen as well [109]. The resulting expressions are more complicated than the ones shown above and therefore this discussion is confined to the results. In Figures 2a and 2b the frequency difference normalized to the upper frequency limit is plotted as a function of the normalized time constant RCf_h. A 0.5 dB gain degradation and a 0 and 6 dB gain slope per octave were assumed. It is interesting

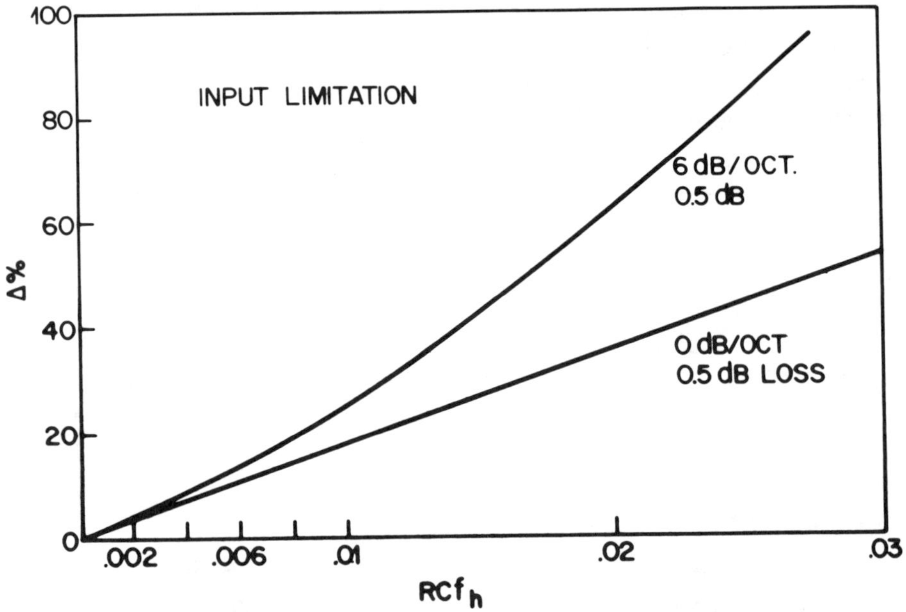

Figure 2/Bandwidth limits in percent of the upper frequency f_h due to reactive constraints.
a) Input for a gainslope of 0 and 6 dB/octave and .5 dB loss of gain.

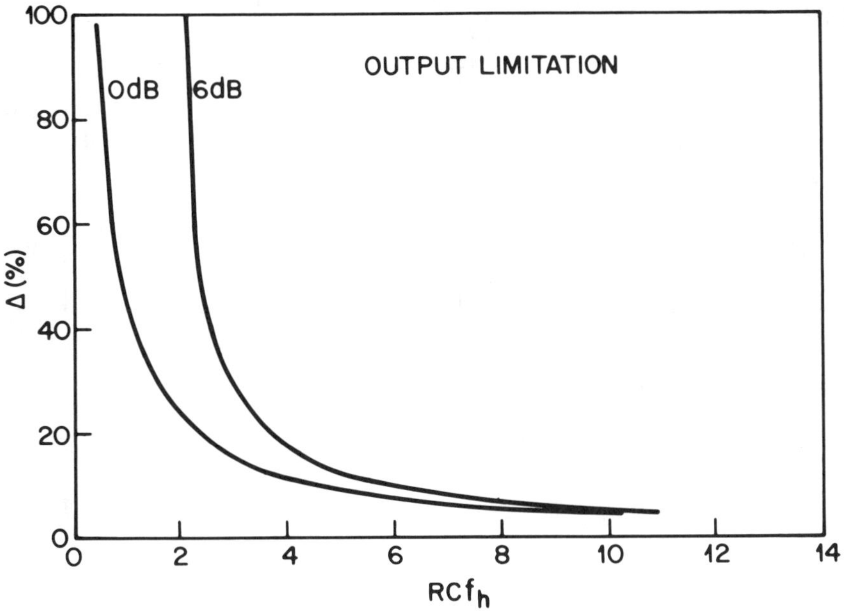

Figure 2/Bandwidth limits in percent of the upper frequency f_h due to reactive constraints.
b) Output for a gainslope of 0 and 6 dB/octave and .5 dB loss of gain.

to see that the frequency independent gain restricts the bandwidth more than the 6 dB slope. Figure 2b shows that the output limitation is a very steep function of the time constant. It is, therefore, mandatory to keep the parasitics low enough to prevent the output from being the limiting factor.

As mentioned before, the capacitive reactance for very large devices is relatively small and the series inductance dominates the input impedance behavior at high frequencies. If one then represents the input impedance by a series combination of L and R the output curve in Figure 1.2b can be used by setting $f_h \cdot L/R$ instead of RC f_h.

For easy reference we constructed charts both for the 0 dB and 6 dB/octave case to give the frequency limitation for the most likely parameter values. Figure 2c covers the 0 dB case, both for series (input) RC and parallel (output) case. To use it, one determines the quantity RCf_h (f_h is the upper frequency limit), and connects the origin with this quantity either on the top or right hand side of the

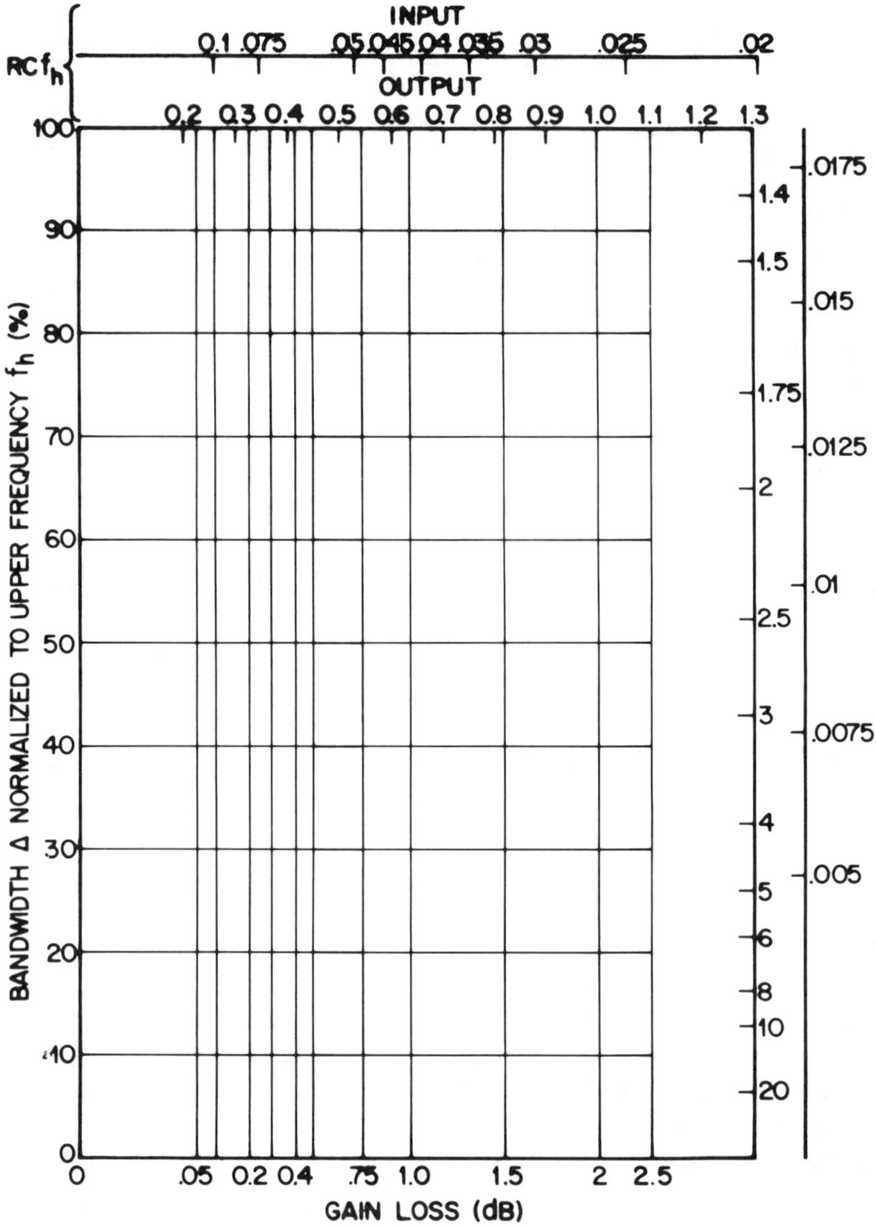

Figure 2/Bandwidth limits in percent of the upper frequency f_h due to reactive constraints.
c) Nomogram to determine the bandwidth limits for 0 dB/gainslope for arbitrary gainloss at input and output.

figure. The resulting line gives the relation between gainloss and bandwidth. The values for input (series) RCf_h are marked on the outer lines on the top and right hand side, the output (parallel) values cover the inside. Figures 2d and 2e are applicable for the 6 dB case. For gain slopes between 0 and 6 dB per octave, it is generally sufficient to use these extreme cases and interpolate, since the limits

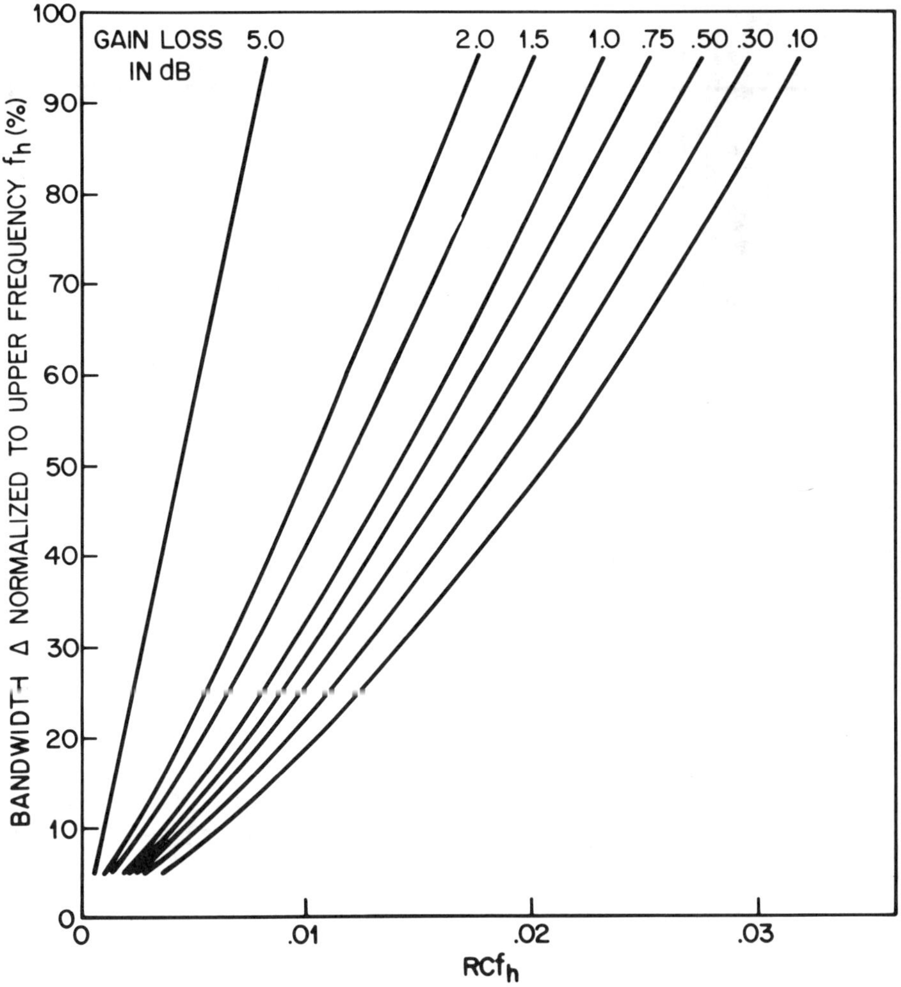

Figure 2/Bandwidth limits in percent of the upper frequency f_h due to reactive constraints.
d) Input for a range of gainlosses for a 6 dB/octave gainslope.

can be taken as approximate only. These limitations give overly optimistic answers sometimes, since the synthesis allows ideal transformers, which cannot always be realized at microwave frequencies [235].

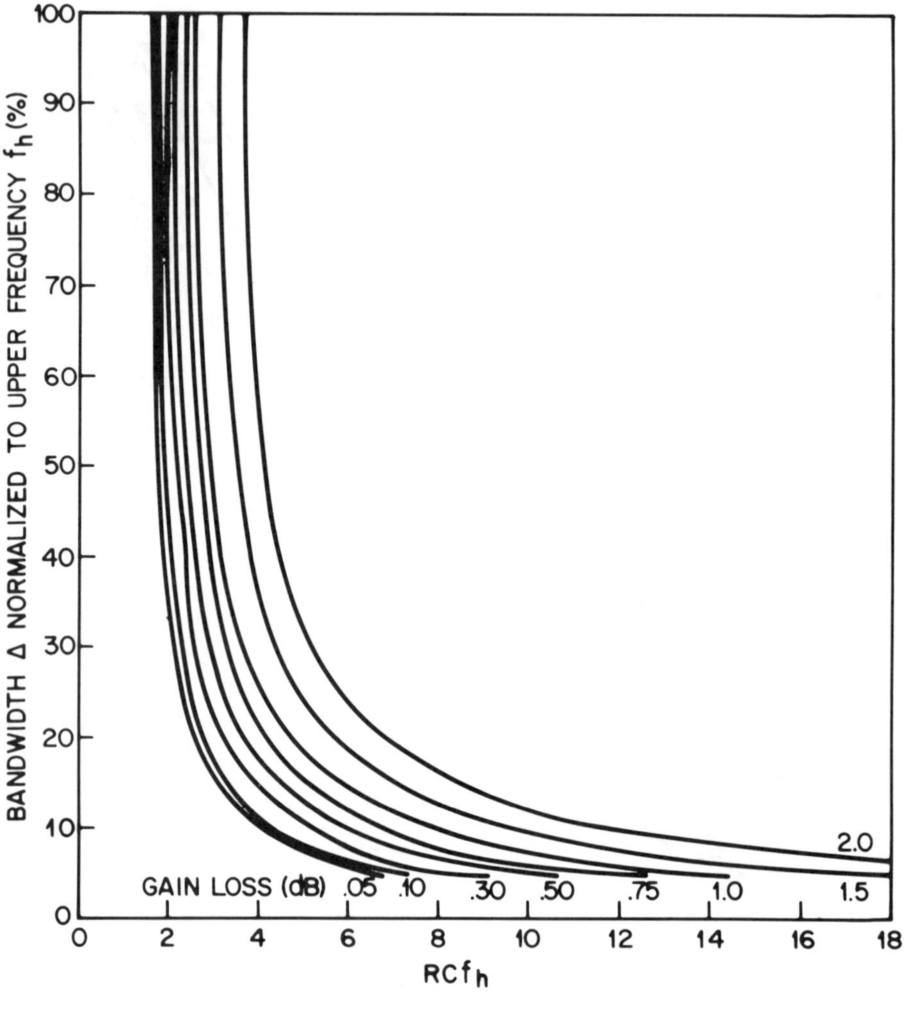

Figure 2/Bandwidth limits in percent of the upper frequency f_h due to reactive constraints.
e) Output for a range of gainlosses for 6 dB/octave gainslope.

1.2 Determination of Source and Load Impedance

The synthesis problem for transistor amplifiers at microwave frequencies consists of the synthesis of two in general lossless two-port networks which provide the match at input and output, as shown in Figure 3. At low frequencies it is possible to change the scattering parameters of the transistor itself by introducing feedback. This is generally not done at microwave frequencies because it is very difficult to realize technologically with proper phase control. However, the emerging techniques of in-package tuning and on chip matching (see Section 5) may alter this limitation. Recently negative feedback has been used to improve the wideband impedance properties of GaAs FETs and impressive results have been achieved. Ulrich [227] reported on an amplifier spanning 100 MHz to 6 GHz and Niclas et al. [153-156] built an amplifier from 350 MHz to 14 GHz. These are the widest bandwidths achieved to date with GaAs FET amplifiers.

Since, in general, the transistor is not unilateral, i.e., $|S_{12}|$ is nonzero, the synthesis of the input and output network is not independent. This two-port synthesis problem has not been solved. It is, therefore, customary to regard the transistor as unilateral, which is usually a good assumption but can lead to problems, especially in the case of conditionally stable devices. For the unilateral case one can then determine a transistor input and output impedance which is used in the subsequent synthesis of the circuit. In the linear case a subsequent

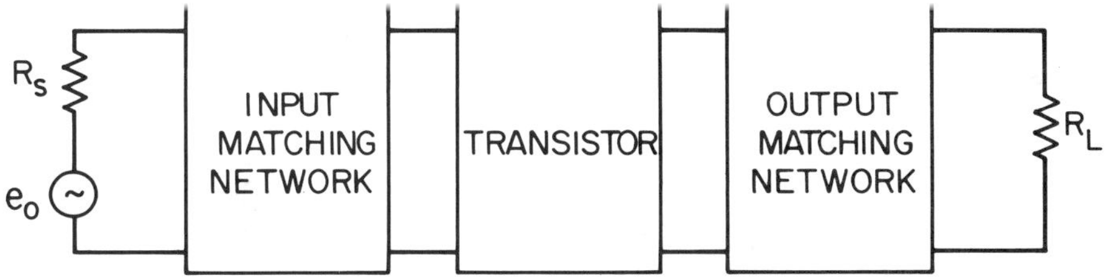

Figure 3/Schematic representation of a transistor amplifier.

TABLE 1

	Unilateral	Unconditionally Stable	Conditionally Stable
Stability Factor $K = \frac{1 + \|\Delta\|^2 - \|S_{11}\|^2 - \|S_{22}\|^2}{2\|S_{12}\| \cdot \|S_{21}\|}$	$K \geqslant 1$	$K > 1$	$K \leqslant 1$
Regions of Stability	—	—	center: $r_{si} = \frac{C_i^*}{\|S_{ii}\|^2 - \|\Delta\|^2}$ radius $= \rho_{si} = \frac{S_{12}S_{21}}{\|S_{ii}\|^2 - \|\Delta\|^2}$
Maximum Gain	$MAG = \frac{\|S_{21}\|^2}{(1 - \|S_{11}\|^2)(1 - \|S_{22}\|^2)}$	$MAG = \frac{S_{21}}{S_{12}} (K \pm \sqrt{K^2 - 1})$	
Constant Gain Circles for Gain loss – ratio G $= G_i/G_{imax}$			

Center: Location r_{gi}:	$\frac{G\, S_{ii}^*(1 - \lvert S_{ii}\rvert^2)}{1 - \lvert S_{ii}\rvert(1 - G)}$	$\left\lvert \frac{G}{1 + D_iG} \right\rvert C_i^\dagger$
Radius ρ_{gi}	$\frac{\sqrt{1 - G} \cdot (1 - \lvert S_{ii}\rvert^2)}{\lvert 1 - S_{ii}\rvert^2(1 - G)}$	$\frac{(1 - 2K\, \lvert S_{21}S_{12}\rvert\, G + \lvert S_{12}S_{21}\rvert^2 G^2)^{1/2}}{1 + D_iG}$
Optimum Load Reflection Coefficient	$r_i = C_i^\dagger \frac{B_i \pm \sqrt{B_i^2 - 4\lvert C_i\rvert^2}}{2\lvert C_i\rvert^2}$	

Definitions: $\Delta = S_{\ 1}S_{12} - S_{12}S_{21}$

$D_i = \lvert S_{ii}\rvert^2 - \lvert\Delta\rvert^2$

$B_i = 1 + \lvert S_{ii}\rvert^2 - \lvert S_{jj}\rvert^2 - \lvert\Delta\rvert^2$

$C_i = S_{i} - \Delta \cdot S_{jj}^*$

if $i = 1$, $j = 2$

if $i = 1$, $j = 1$

In the equation for MAG the plus sign is used for B_i negative and the minus for B_i positive.

In the equation for Optimum Load Reflection Coefficient the plus sign is used if B_i is negative; minus signs are used if B_i is positive.

complete analysis of the amplifier circuit based on the S-parameters is used to assess the deviation from the initial assumption. If necessary, a second synthesis iteration step is added (usually a computer optimization) to optimize the total amplifier. If the transistor is operating in the nonlinear regime, the S-parameters are not available. In that case one has no other choice than to determine the load and source impedance. One can then proceed in an identical way both for the linear and nonlinear case. In the nonlinear case, however, the gain variations which are calculated from the linear case will be inaccurate since, in addition to mismatch losses, the compression will play a role. Furthermore, the last analysis step is not possible unless the large signal S-parameters are known [178].

We will now discuss the determination of source and load impedance from the transistor characteristics. Again we must distinguish between the linear and nonlinear case. In the *linear* case we can use the previously mentioned S-parameter methods. This topic has been covered [69] so that it is not necessary to discuss it any further. The necessary equations are summarized in Table 1. In the *nonlinear* case, as mentioned in the previous section, these methods break down. The primary effect is on S_{21} and S_{22} whereas S_{11} and S_{12} are relatively unaffected even at high power levels. This is understandable because in the nonlinear case the higher power capability is achieved by changing the loadline compared to the small signal case (Figure 4). As a consequence, one can use to a first approximation the input impedance determined for a small signal amplifier even for highly compressed power amplifiers. The variation is mainly caused by the change in output impedance in conjunction with the nonzero S_{12}.

A number of methods are being used to determine the output impedance under nonlinear, i.e., high power conditions. One can of course use the load pull method where the device is tuned with variable tuners to the desired gain value. The impedance can then be determined by calculation or direct measurement. This method is time consuming since it has to be done on a narrow band basis separately for each frequency and requires a well controlled circuit environment which does not add undesirable parasitics. Despite these shortcomings, the method usually produces reliable results. It has been used up to 12 GHz with relatively small tuner losses ($<$ 1 dB). It is dangerous to subtract tuner losses from the results, because it is very diffi-

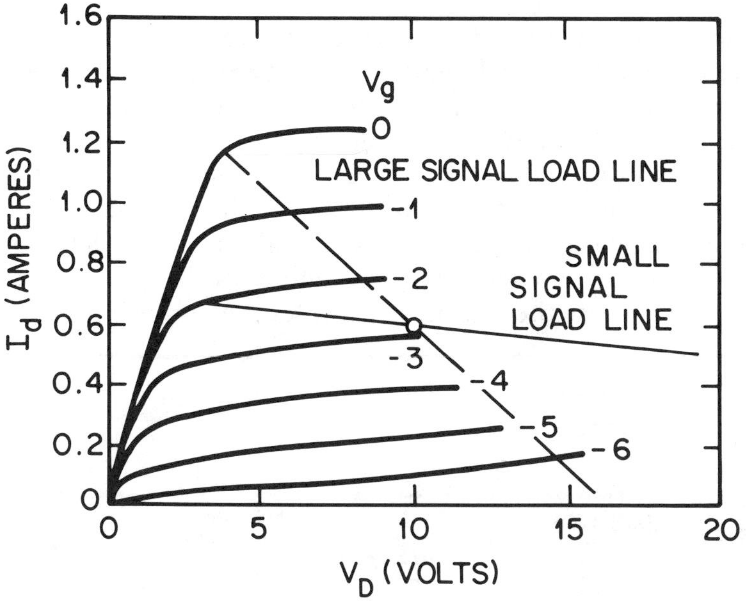

Figure 4/Typical I-V characteristics with small and large signal load-lines.

cult to reliably determine those losses. For production applications, where a large number of measurements are necessary, a fully automated tuning setup has been reported by Cusack et al. [42].

A second method [211] by Takayama (Figure 5) operates on a similar principle, but instead of using tuners at the output, it feeds some of the input power back into the output. This incident signal can be varied by phase and amplitude and thus arbitrary load impedances can be generated. The advantage is the elimination of output tuners which do not allow orthogonal tuning and which create reproducibility problems. There is a problem in determining the extracted power if the output reflection coefficient is close to one as it is apt to be in unmatched large devices. This method lends itself particularly well to computer control, since it does allow orthogonal tuning.

An attempt has been made to extend the conventional S-parameter characterization to high power [118]. The S-parameters were measured with high incident powers. In the case of the FET this has not been successful. The reason is relatively easy to see. The output is measured with a 50 Ω load impedance. The limitations of the output power are, however, not simply related to the mismatch (this is the basic assumption underlying all S-parameter treatments) but depend nonlinearly on the load-impedance. Consider Figure 4 for example, where two load lines are shown with the I-V characteristics. In case of loadline (a), corresponding to small signal operation, the output power is limited by forward drive capabilities of the gate, whereas in the loadline (b) case the avalanche provides the power limit.

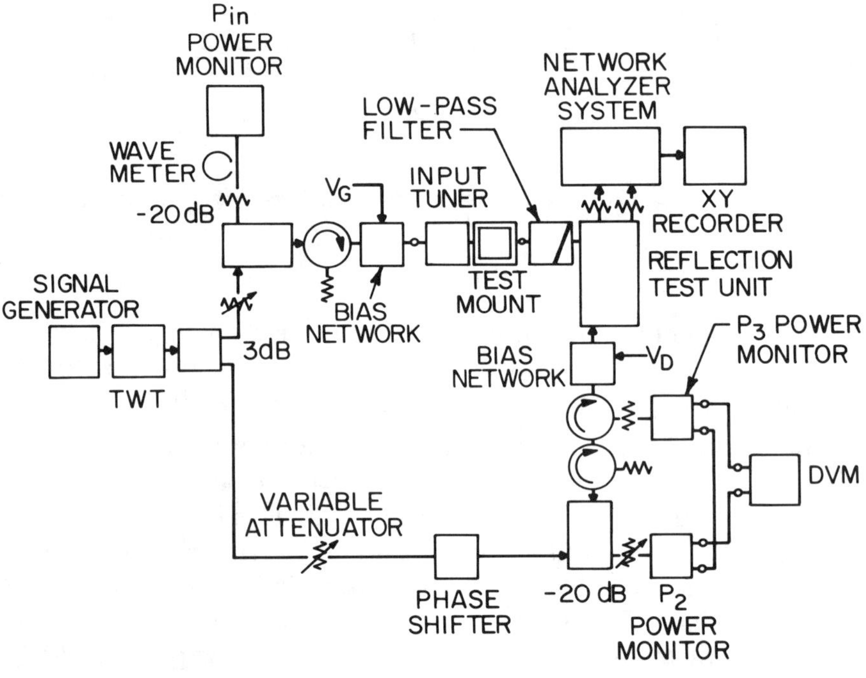

Figure 5/Load pull test setup (Takayama).

A somewhat different and potentially more successful approach has been taken by Rauscher, Willing et al. [243, 177, 178] who use a standard small signal equivalent circuit as a starting point. They postulate that some of these quantities are power dependent and determine their values from S-parameter measurements at different bias conditions. They show good agreement between measurement and theory for one device type, which unfortunately, operates at very low power levels (~ 200 mW/mm). It will have to be seen if this method is applicable to a wide range of devices which show different causes for power saturation (e.g., avalanche which almost surely was not present in the test devices at 6V source drain bias). There is, furthermore, the question of how the standard equivalent circuit can serve to explain the weak frequency dependence of gain at high compression levels.

1.3 Determination of Matching Networks

After measuring input and output impedances, both have to be converted into equivalent circuits. Usually fairly simple circuits with not more than 4 elements are adequate. It is often advantageous to use a circuit optimization routine to determine the parameters. Before attempting the synthesis of the matching networks, it is advisable to calculate the stability regions using the S-parameters both at operating frequencies and at low frequencies. This applies equally for low noise and power amplifiers. These stability regions serve a) to ensure stability in the operating range and b) allow the design of appropriate bias structures which prevent oscillations at low frequencies.

If one assumes that the input and output impedances are adequately represented by an equivalent circuit, one is able to design the matching networks. We will describe the common approaches in degrees of increasing complexity:

(a) The simplest case is that of a device which has frequency independent gain and purely resistive impedance. For this case simple lossless lumped low- or band-pass structures have been extensively investigated. The low pass structure is the easiest to realize. Tables have been published which give element values for Chebyshev [130], elliptic [182], and maximally flat [40] return characteristics. These tables include the relationships between numbers of elements, bandwidth and gain variation. One of the most important conclusions is that, in general, a relatively small number of elements (4 or 6) is adequate and larger numbers provide only marginal improvements [120, 130]. These lumped element values can easily be converted into distributed elements if necessary. For the conversions, Richards transformations can be used. Since these conversions are only approximate, it is usually necessary to check the resulting distributed network with a circuit analysis program and adjust parameters if necessary.

(b) The more complicated case of a device impedance including reactances, can very often be solved by using the previously mentioned circuits and incorporating the reactances as part of the circuit. This works in general for the output since it can be modeled adequately, as shown in Figure 1. The same is true for the input of large transistors where the capacitive reactance can be neglected compared to the lead and package inductance. In all other cases, the matching problem has not been reduced to a simple tabulated design and one has to use standard synthesis techniques [23, 60, 246]. The most efficient strategy is to use a computer aided technique.

(c) Since in practice the gain of a transistor is a function of frequency, this gain variation has to be taken into account in broadband amplifier design. In general one uses a flat network for one port and corrects for the gain slope at the other port.

In contrast to the previously mentioned synthesis problems (Section 1.2), which have been adequately treated by the Bode-Fano-Youla theory some time ago, the gain roll off problem has been treated only recently. A number of treatments are available which give adequate results. Ku and co-workers [105-111] have developed an exact synthesis for distributed networks. Previous work by Mellor and Linvill [134] describes a method to synthesize lumped networks with gain slopes of multiples of 6 dB per octave. Other gain slopes are approximated by linear combinations of these slopes. Carlin and Kamiak [31] use a piecewise linear approximation method in the real frequency domain.

All methods, except the initially mentioned simple ones which utilize tables, are quite complex and require that the designer carry out a relatively large number of algebraic calculations. Most recent papers on amplifier synthesis thus report that extensive use has been made of computers. The circuit engineer with no specific interest in the details of circuit theory has increasingly utilized computer aided design (CAD), without needing to penetrate the complex basis for these methods.

The CAD methods fall into two categories: (a) the previously described methods which are computerized for easy evaluation and (b) an optimum seeking routine. Option b looks very tempting at first, since it involves only the choice of a realizable topology on the part of the designer. All the rest can be left to the computer. Unfortunately, the optimization problem is highly nonlinear, with generally many local minima. Secondly, the number of variables is generally quite large. For relatively narrowband amplifiers (~ 10% bandwidth) it is usually adequate to start with a single frequency design and then use the optimization method to achieve adequate bandwidth. In the broadband case, this approach results in uneconomically long computations and unreliable optimizations. The computer aided synthesis then proceeds along very similar lines to standard amplifier synthesis except that a higher degree of complexity is possible and an overall optimization step can be added. The procedure is typically as follows:

1. Determine input and output impedance and find equivalent circuit over the appropriate frequency range.
2. Choose an appropriate realizable circuit structure and carry out a one-port synthesis using either tables or a program based on the above mentioned methods (e.g., Ampsyn, Cadsyn by Compact Engineering). The result is a network which has to be checked for realizability. If after the subtraction of the built-in reactance the network is not realizable, one has to repeat the synthesis. One can use the freedom of choice in the poles of the network to achieve the appropriate structure.
3. In case of a linear transistor, the complete equivalent circuit including the newly designed network is then analyzed using a straightforward analysis program (e.g., Compact). This shows whether the interactions due to feedback change the overall response and if further optimization is necessary.
4. If further optimization should be needed, the above mentioned optimization routines are appropriate.

In summary, although great progress has been made in amplifier synthesis, the procedure is far from automatic and requires a good deal of user interaction and common sense. Otherwise it is possible to waste time and money without arriving at a good circuit.

2. POWER AMPLIFIER DESIGN

Power amplifier design is considerably more complex than that of linear amplifiers and has not yet been fully theoretically treated. The reasons are two-fold:

a. In general, the gain performance is prescribed not only for a specific power level but for a range of power as well. Therefore, additional efforts have to be made to determine the appropriate impedances which give an adequate compromise. Linear synthesis methods are then employed to determine the appropriate networks. The prediction of gain-variations becomes very difficult and the standard follow-up by a general optimization program is not possible because the device information is not available.

b. The transistor operates in a nonlinear regime, i.e., the S-parameters depend on the power level. So far no adequate S-parameter characterization at high powers has been found although good progress is being made (see Section 1). In the standard linear synthesis procedures the gain ripple is ascribed to the mismatch loss. In power amplifiers an additional gain variation is introduced because the mismatch variations change the compression levels and therefore the gain.

In this section we will discuss the choice of bias point, the influence of the output impedance on gain compression and methods to choose both bias point and impedance to approximate a prescribed gain compression curve. In the last section we give a brief summary of the various realizations of power amplifiers.

2.1 Determination of dc Operating Point

The power output and gain of a device is determined by a number of operating parameters which are under the control of the amplifier designer. To achieve an optimum design it is mandatory to know the influence of these parameters.

Drain bias voltage

The bias voltage for maximum power output can be easily determined from dc considerations (Figure 6). It is in the middle between the knee voltage V_k and the avalanche voltage, V_A (i.e., the voltage at which the gate-drain avalanche currents starts contributing significantly to the drain currents). In practice, the maximum is not very sharp, because the avalanche current does not always rise very sharply with voltage. The sharpness of the rise depends on the particular transistor design. An increase in bias voltage generally results

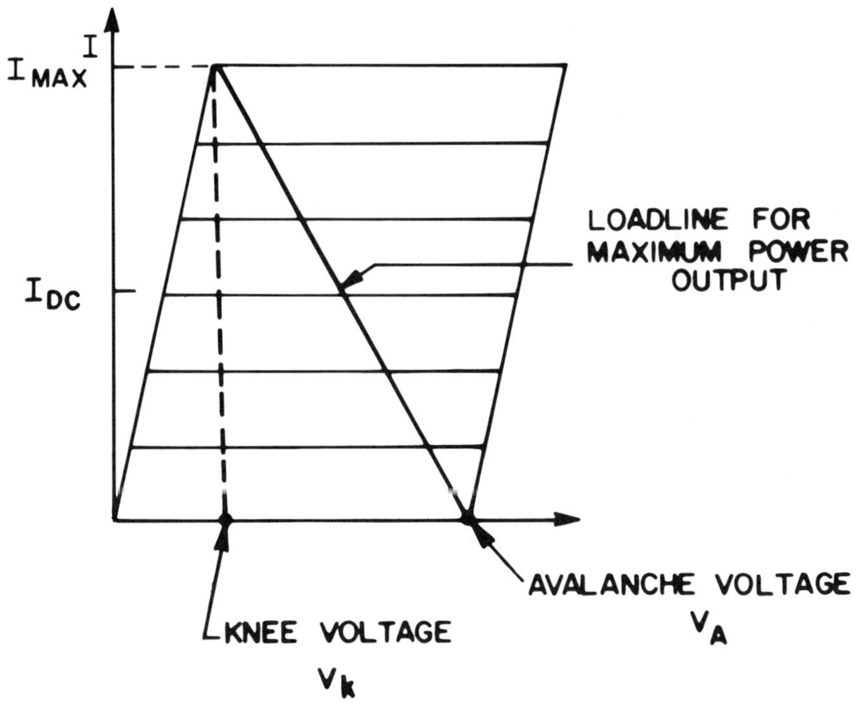

Figure 6/Idealized I-V characteristic with load line for maximum power output.

in an increase in power gain even at low compression levels until heating and/or avalanche start to limit the performance. The small signal gain does not necessarily behave this way. For reliable operation, it is generally advisable to operate at a considerable margin below the burnout-voltage, which can degrade as a function of time. The avalanche can vary considerably even among devices of the same type.

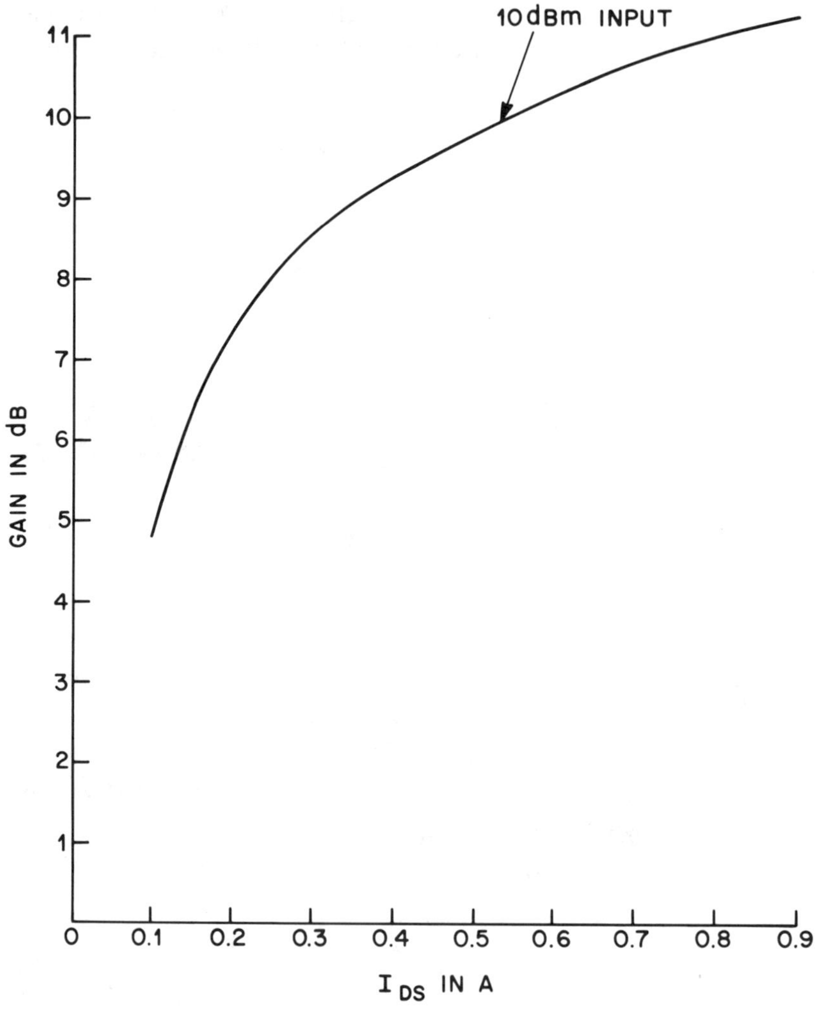

Figure 7*/Gain at 10 dBm input levels as function of drain current for a WE6 mm transistor at 4 GHz.*

Drain current
The small signal gain increases with current (Figure 7). This is simply a consequence of the increase in g_m with current. The optimum bias current for small signal operation is I_{dss}. One could of course bias at I_{max}, i.e., a value greater than I_{dss}, to maximize this gain. This is generally not done to avoid gate current or in some devices burnout.

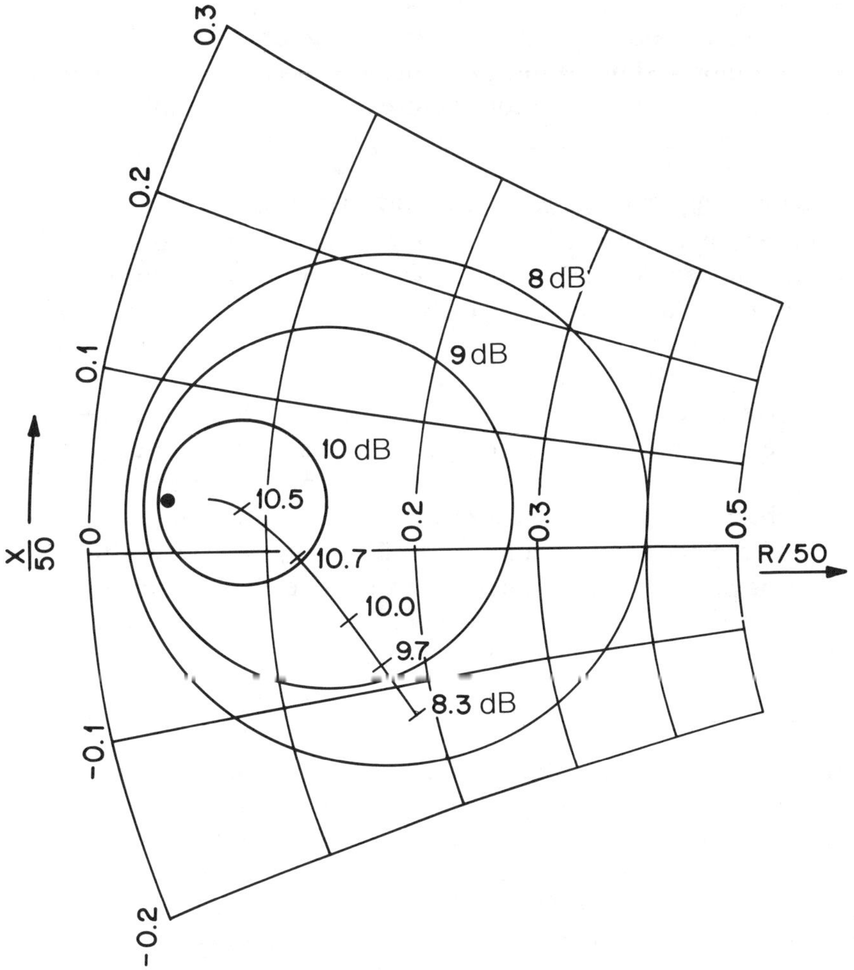

***Figure 8**/Optimum output impedance with associated gains for different input levels (solid line) at 4 GHz for a WE6 mm device at constant bias (14V, 600 mA). The circles are the small signal constant gain circles.*

For power operation the optimum bias point converges towards $I_{max}/2$ (Figure 6). Note that I_{max} is not related to I_{dss} by a fixed factor but that the ratio I_{max}/I_{dss} depends on the channel characteristics. It has to be determined separately for different types of transistors. At intermediate power levels, one has to find a compromise between gain and power output. For powers close to the saturated power, there is a tendency to self bias, i.e., the RF incident on the gate causes rectification and – depending on the power supply impedance – pushes the operating point close to optimum. This is generally accompanied by sizable gate currents. It is advisable to check the manufacturers specifications before operating an amplifier at high gate currents, since they can degrade reliability.

2.2 Gain Compression and Output Impedance

As discussed in Section 1, the output impedance varies considerably as a function of power. Figure 8 shows the optimum output impedance at various power levels for a packaged Western Electric 131 6 mm transistor at 4 GHz. The bias point was held constant. It shows that besides a variation in real part, which is predicted from the dc characteristics, there is a reactance variation as well. As a consequence, even for a fixed bias point, the transistor should be retuned for different power levels to achieve maximum performance. The details of this particular impedance variation are a function of the transistor and the packaging employed. In Figure 9, we have plotted a number of gain compression curves at the same dc operating point. Note the tradeoffs that can be made between 1 dB compression power and gain. This figure shows that it is meaningless to quote a 1 dB compression power for a device. This quantity only has meaning for an amplifier.

In amplifier design we have found the following procedure helpful: determine the optimum gain for a number of input power levels at the optimum bias point. In Figure 10 we have done this for an average 131 Western Electric 6 mm device at 4 GHz. It is obviously advisable to generate this plot for typical devices to maintain good yield in amplifier production. After generating these plots for a family of devices, it is easy to determine the gain budget at nominal power and assign devices of proper size to the various stages of a multistage amplifier, starting with the last stage and working backwards.

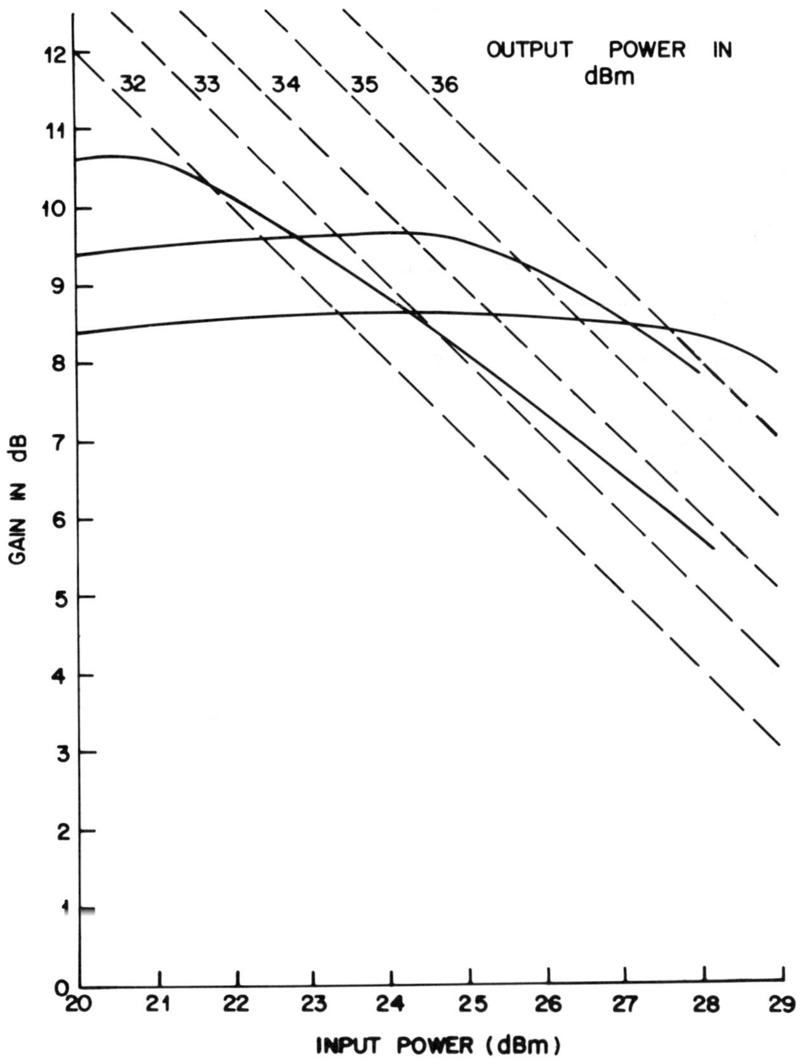

Figure 9/Gain as a function of input power for constant bias for three different tuning conditions at 4 GHz. The device is the same as in Figure 8.

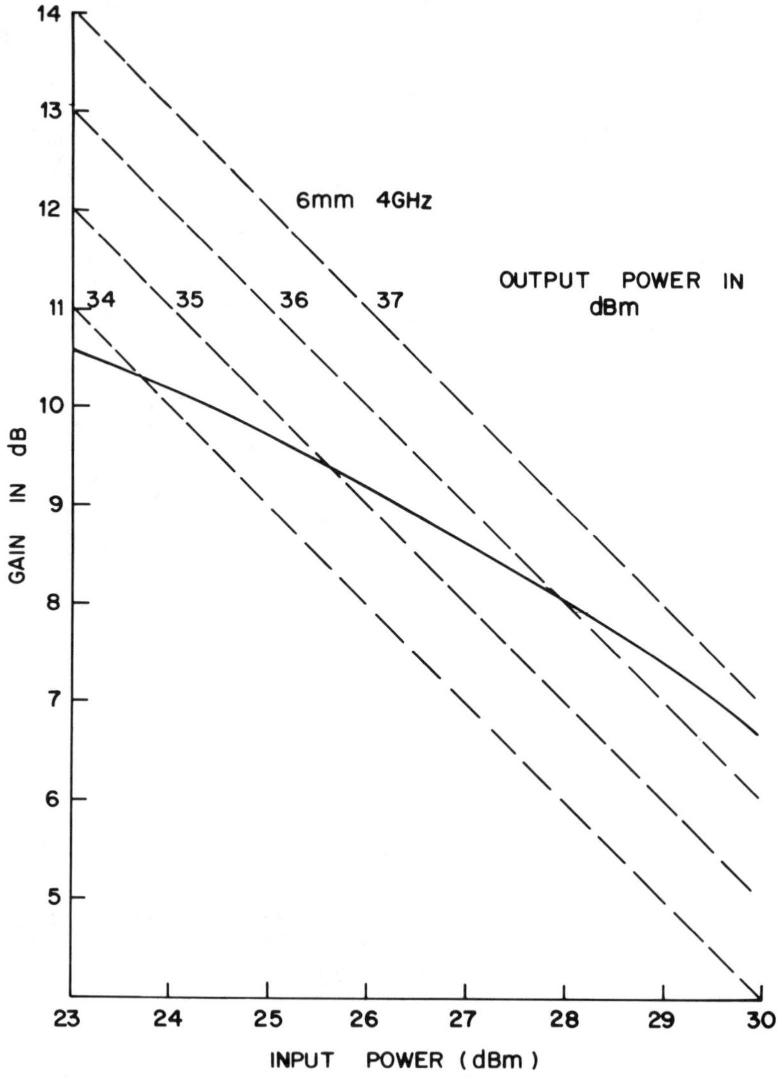

Figure 10/Optimized gain with respect to tuning and bias as function of input power for the same device as in Figure 8.

For actual amplifier-design a compromise has to be found between small signal and large signal performance. This can be done on a purely experimental or a partly theoretical basis. We will discuss both approaches.

As pointed out above, ripple predictions from linear theory tend to be wrong because of the gain compression in power operation. Usually this effect tends to reduce the ripples. Therefore, this problem has not found too much attention and the amplifiers are generally experimentally readjusted to make up for deviations from theoretical predictions.

The problem of gain flatness has been treated on a semi-empirical basis by de Koning et al. [47] and Kotzebue [100-102]. We will briefly summarize the method here. It is based on the fact that the large signal gain contours can be approximated by circles. One can then express the gain at a particular input power by:

$$G_{LS} = G_{max} - \frac{K_1}{G_L} |Y_L - Y_{LOPT}|^2 \qquad (3)$$

where G_{max} is the maximum gain at this input power if terminated by the optimum load admittance Y_{LOPT}. Y_L is the load admittance, G_L its real part, and K_1 a constant which has to be determined experimentally from load pull measurements. Note that this expression is identical to the noise figure as a function of input impedance. The circle parameters are given by the following expressions:

$$\text{center } M_g = \frac{S_0}{1 + \beta}$$

$$\text{radius } R_2 = \sqrt{\frac{\beta}{1 + \beta}\left[1 - \frac{|S_0|^2}{1 + \beta}\right]}$$

$$\text{with } S_0 = \frac{Y_c - Y_{LOPT}}{Y_c + Y_{LOPT}} \text{ and } \beta = \frac{G_{max} - G_{LS}}{K_1} \frac{|1 + S_0|^2}{4Y_c}$$

$$\text{with } Y_c = \frac{1}{50\,\Omega}$$

One can now draw small signal and large signal constant gain circles on the same plot and find an impedance at which both gains are identical (Figure 11). This kind of load impedance will insure that the amplifier has a flat gain vs drive characteristic. Putting the load impedance between optimum large signal gain and the edge of the Smith chart increases the difference between small and large signal gain, whereas moving towards the center of the chart tends to reduce this difference. The optimum load impedance is not the point for

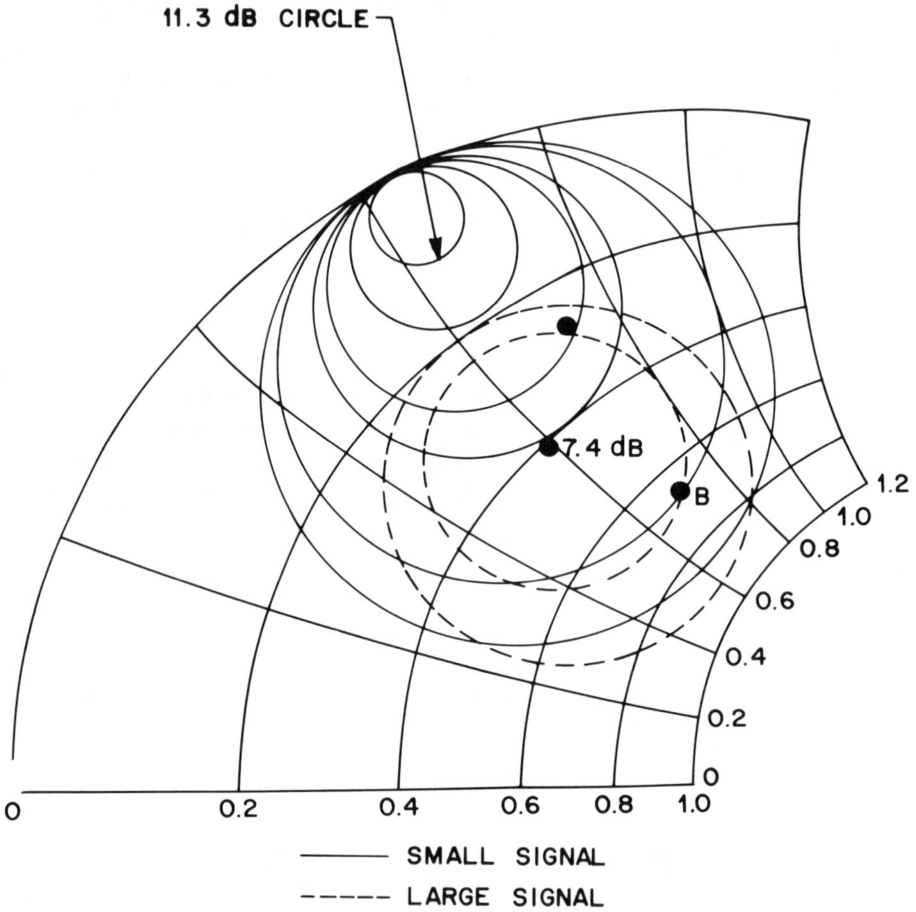

Figure 11/Small signal and large signal constant gain circles (deKoning et al.)

best gain flatness. This is consistent with the observation that the optimum large signal impedance and the impedance for optimum linearity are different [188] (see Section 3). This type of interpretation can explain the frequently observed gain expansion as well. Consider Figure 8, where we have plotted the small signal constant gain circles and the variation of maximum gain impedance at various power levels. Note that at the impedance corresponding to a 9.7 dB large signal gain, the small signal gain is 9.2 dB. The measured gain compression curve shows indeed this amount of expansion. The gain expansion is then mainly a consequence of the reactance variation of the optimum output impedance. The degree to which large and small signal impedances differ can be very different depending on the transistor design. Recently a result has been published [184] by Sakane et al. in which small signal and large signal impedance were identical. This is generally true if the bias is maintained considerably below I_{DSS}.

The determination of the impedance for a prescribed compression characteristic can be done on a purely experimental basis as well. The following test-setup has proven helpful (Figure 12). The key item is the use of an amplitude modulated signal. In our case the modulation was applied with a PIN modulator before the power amplifier. A signal proportional to the amplitude of the input power is detected at point B and used as the "X" axis of the gain display on a network analyzer. In this way a gain versus input power display is generated. It allows the direct adjustment of the gain compression characteristics by bias and impedance changes.

2.3 Amplifier Realizations

Besides the problem of nonlinear design, the power amplifier often poses additional problems. The device size must increase as power increases and, therefore, impedance levels become low enough to make circuit losses non-negligible. If one assumes that a 1 mm transistor has an input resistance between 3-6 Ω, one can expect seeing influences from circuit losses starting at gatewidths of 6-10 mm. As present saturated power capabilities range from .4 to .7 W/mm for commercial units, this means that above the 3-5 W level careful attention must be paid to circuit loss. A combination of techniques must be used to overcome the gain loss due to matching difficulties. A number of different techniques have been employed:

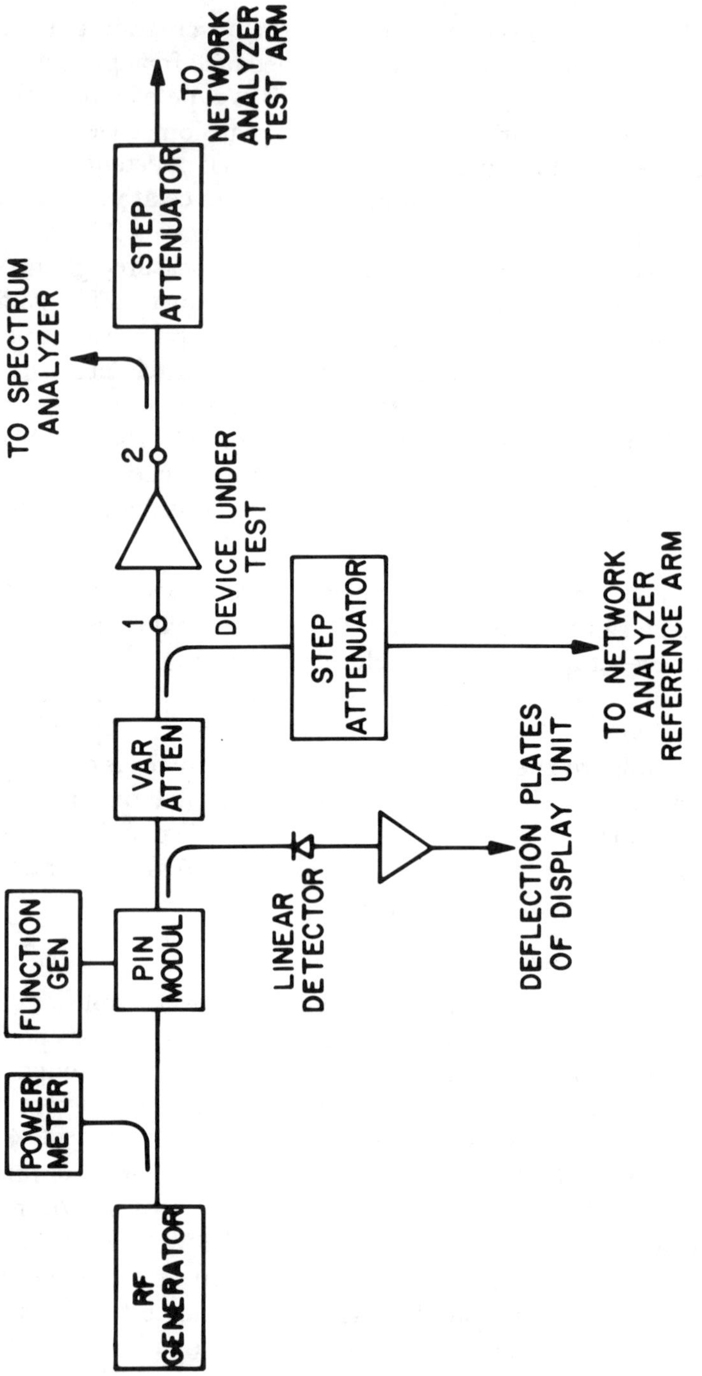

Figure 12/Gain compression measurement setup.

Paralleling of Stages
The classic approach of paralleling two identical amplifier modules using 90° hybrids usually has the purpose of fully matching the amplifier to avoid either the tuning problems inherent in cascading stages or the necessity for isolators [114]. It has, of course, the added advantage of doubling the power without increasing the matching problems. It does, however, require well matched amplifier modules; this exacts a certain penalty in selecting devices and fine tuning of the modules. This approach is widely followed. Results for amplifiers with up to 10 W over relatively narrow bands and 1 W over very wide bands using commercial devices have been reported (see Tables 2 and 3). The coupler losses are typically in the .2 to .5 dB range and thus are negligible if the gain of the combined stages is high enough. They do have noticeable impact on the efficiency of low gain stages (see Section 3 for curves of efficiency vs. loss).

If it becomes necessary to increase the level of combining, the combining losses very soon become prohibitive. It is then advantageous to utilize combining schemes like, for instance, the Wilkinson type which have low loss and are easier to fabricate [154, 219] than the Lange coupler. A very high degree of combining has been demonstrated by Cohen et al. [36] by using a radial combiner. A combining efficiency of 90% was achieved with 12 individual amplifier modules. The bandwidth extended from 7.7 to 9.9 GHz and was limited by the amplifier modules. Much wider bandwidths are predicted for the radial combiners.

Chip Combining
In this approach, in-package matching networks transform the device impedance to a higher level and the paralleling is accomplished at this higher impedance level. In essence, this approach is similar to the one previously mentioned using Wilkinson couplers except that tuning is done both before and after the paralleling. This approach has been able to match very large chip sizes (up to 67 mm [81, 213]) and produce powers up to 25 W at 6 GHz.

Push Pull Technique
In effect, a push pull technique series combines the device impedance instead of paralleling it. This concept, which is common at low frequencies, has been recently demonstrated by Sokolov et al. [199] at X-band frequencies.

TABLE II

Power Amplifier Results With Output Powers Less Than 4W

First Author	Reference #	Date	f_h Upper Freq.	Δf %	Gain Variation dB	Max Power W	Output Stage: # of Elements	Output Stage: Total Device Size (in mm)	Technology
Aihara	5	79	12.1	3	1	1	2	12	Single ended
Arai	8	76	6.3	3.5	1.	1.	NA	NA	Single ended
Goel	66	76	8.4	6.	1.	.16	2	1.2	3 dB hybrids
Hornbuckle	82, 83	78, 76	6.2	67.	2	1.1	3	4.5	3 way combiner, Lossy network
Kadowaki	93	79	12.1	3.3	1	2	1	4.8	Single ended
Niclas	152, 153	78	18.	33	1.4	.20	2	.6	3 dB hybrid
Niclas	154	79	18	39	1.	.500	NA	NA	Comparison of Wilkinson & Hybrid 4-way combining
Niclas	151	78	18	33	1.2	.1	1	.300	Single ended
	155	80	14	98	2.4	.02	1	.8	Single ended, feedback
Pucel	174	79	10.4	11	1	.4	1	NA	Monolithic
Schroeder	188	78	4.2	7	1	2.5	1	6	Single ended
Sechi	189	76	5.1	17	1.	1.1	2	3.6	3 dB hybrid
	190	77	4.3	16	NA	1.1	1	3.	Single ended
Sone	202	79	14.5	4	1.	1.1	1	1.	Single ended
Tserng	220	76	9.8	6.	1.	1.	1	2.4	Single ended

TABLE III

Power Amplifier Results With Powers More Than 4W

First Author	Reference #	Date	f_{U} Upper Freq.	Δf %	Gain Variation dB	Max Power W	Output Stage: # of Elements	Output Stage: Total Device Size (in mm)	Technology
Cohen	36	79	10	23	1.	10	12	28.8	Radial combiner
Dornan	50	80	4.0	7.5	.5	10	4		3 dB hybrids
Drago	52	78	3.9	7.	1.	5.6	2	10.4	3 dB hybrids
Higashisaka	74	80	6.3	11	1	25	4	60	Internally matched
Ho	78	79	4.2	12.	(?)	10.	4	28.8	3 dB hybrids
Ho	79	78	4.2	14	1.	7.	3	15.6	3 way hybrid
Honjo	81	79	6.2	6.5	1	20	8	67.2	Internally matched + 3 dB coupler
			6.[illegible]	11	1	15	4	33.6	Internally matched
Honjo	80	79	7.[illegible]	42	~2.	4.4	2	11.2	Internally matched (5 dB gain)
Mitsui	142, 143	80	10.[illegible]	8.6	1	10	4	28.8	Internally matched
Sone	203	80	18.[illegible]	6.4	1	1	2	6	Internally matched
Takayama	213	80	5.[illegible]	8.3	1	25	2 x 4	121.6	Couplers + internally matched
Tsai	219	77	5.[illegible]	15.	1.	5.5	NA	NA	Split 90° Comb. (4 parallel)
Tserng	222	78	7.[illegible]	12	.5	8	2	19.2	Internally matched/3 dB hybrid
Yokuchi	249	78	4.[illegible]	14	1	6	2	NA	3 dB hybrid

2.4 Practical Results

Tables 2 and 3 provide an overview of the types of amplifiers which have been reported to date (mid-1980). We split them arbitrarily into power outputs below and above 4 W.

The lower power category (Table 2) is usually either optimized for gain or bandwidth. Consequently, the power output per mm transistor-width is typically in the .2-.3 W/mm category. There is a 50/50 split between hybrid approaches and single-ended amplifiers reflecting the fact that units with adequate power are commercially available and matched pairs may be more difficult to procure than single units. The majority of bandwidths are below 16%. The few *wideband* approaches deserve special attention because they deviate from the classical reactive matching concept. Hornbuckle et al. [82, 83] achieve a 67% bandwidth by using a lossy network, which is designed to equalize the gain and improve the match. The other ultra-wideband result by Niclas et al. [155, 156] broadens the device response by introducing negative feedback.

The higher power table (3) shows the increasing difficulties of matching to large transistors. All approaches use either in-package matching or a combining scheme. The relative power outputs are higher than in the lower power case (up to .5 W/mm) because these stages are optimized for power, not gain. Almost all bandwidths are in the 10-15% range, with only one exception. In that case a 42% bandwidth is covered although with a minimum gain of only 3 dB. Almost 50% of all amplifiers in this category are internally matched.

In conclusion:

1. The reactive tuning schemes are relatively easily applicable up to approximately an octave bandwidth. For larger bandwidths these methods are still usable but at a considerable penalty. Lossy networks and feedback networks become attractive at these bandwidths.
2. Combining schemes are unavoidable for large power outputs. In fact, almost all power records to date have been achieved by combining chips directly with combining networks. At low power outputs the advantages of combiners are presently not as clear-cut. In particular, the need for matched transistor pairs is a serious disadvantage of this approach.

3. EFFICIENCY

In some applications like satellite amplifiers, high efficiency is of paramount importance. It is, furthermore, clear that a substantial increase in efficiency allows the transistor to be operated at lower temperatures and that, therefore, the reliability can be increased. We will first discuss the tradeoffs to be made in the choice of the operating conditions (Class A or B) and then show how the efficiency is related to transistor properties. This will give us simple design principles for high efficiency amplifiers.

We will now consider the consequences of operating the transistor in a Class B mode. By definition this means that the current at low input levels is essentially zero. The dc gate voltage is $\approx V_p$ (pinch-off voltage) and the gate voltage swing is $2(V_B + |V_P|)$ where V_B is the built-in voltage. The power dissipation in the input circuit is proportional to the gate voltage swing:

$$P_{in} \sim 4(V_B + |V_P|)^2$$

In Class A operation the dc gate bias is typically $1/2(|V_P| - V_B)$ and the total swing will then be $V_B + |V_P|$. The input power is up to 6 dB higher in Class B operation than in Class A for the same output, which means, of course, that the Class B gain is approximately 6 dB lower than Class A gain. In Figure 13 we have plotted some measurement results which demonstrate this point. The data were taken on a WECO-3 mm device, operated at 14V source drain bias at 4 GHz. The efficiency numbers are noted at every experimental point. The lowest curve corresponds almost to Class B operation as evidenced by the steep gain drop at low power levels. The experimental results clearly show the considerable drop in gain as one moves into Class B operation. Note that the efficiencies are close to the theoretical Class B efficiencies. In general, it will not be desirable to run FETs in this mode since the gains are not high enough to sustain this loss, but a compromise between gain loss and increased efficiency can be made. One has to furthermore keep in mind that the gain reduction in itself reduces the power added efficiency according to a factor $\eta = (1 - 1/g)$ which is plotted in Figure 14.

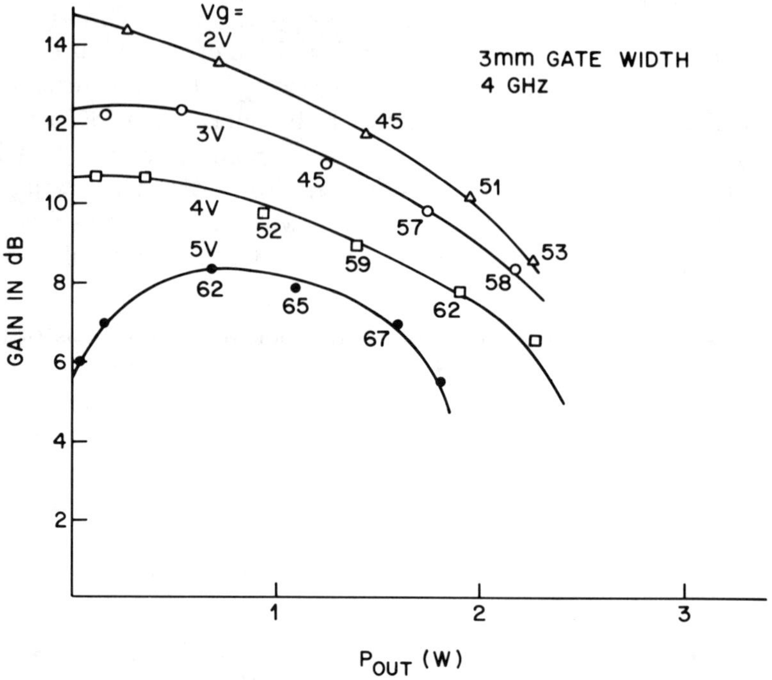

Figure 13*/Gain as function of output power for various gate voltages. The device is a WE 3 mm device. The number at the experimental points shows the power added efficiency.*

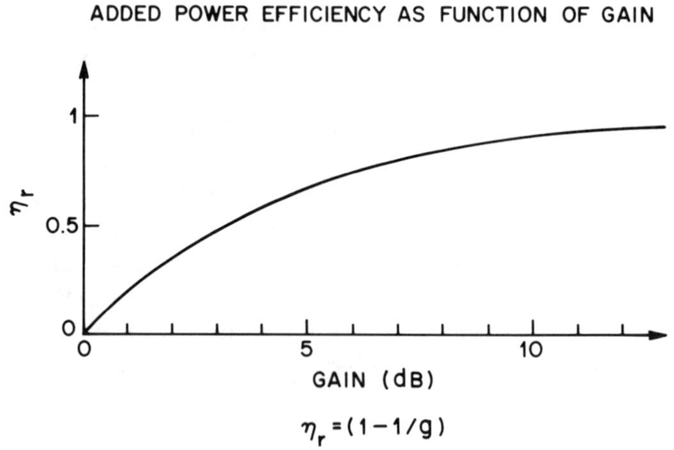

Figure 14*/η, the contribution of gain to the power added efficiency.*

Class B operation has an effect on saturated power output as well. The reason is that the increased gate voltage reduces the available voltage swing at the drain. The maximum voltage between gate and drain is limited by the avalanche breakdown voltage V_A. The maximum power output can then be calculated from:

$$P_B = \frac{I_{max}}{8} \left(|V_A| + V_B - 2|V_P|\right. \tag{4}$$

whereas Class A power output is given by

$$P_A = \frac{I_{max}}{8} \left(|V_A| + V_B - |V_P|\right. \tag{5}$$

The difference of $I_{max}(V_P/8)$ can be a substantial quantity. This trend is evident in the experimental data shown in Figure 13.

Let us now turn to the transistor parameters affecting efficiency. In Figure 15 we have plotted a typical I-V characteristic. In the saturated class A mode, the transistor is switched along the loadline between the points ($V_{sat,}$ I_{max}) and V_A, I_m) (see Figure 6). The bias points for this condition are:

$$V = \frac{1}{2} (V_A + V_K) \text{ and } I = \frac{I_{max} + I_m}{2} . \tag{6}$$

The avalanche voltage V_A is generally not well defined (e.g., Figure 15) and it is often desirable to operate at a fixed bias-voltage determined by other considerations, e.g., reliability. We can represent the efficiency by:

$$\eta = \frac{1}{2} \frac{I_{max} - I_m}{I_{max} + I_m} \cdot \frac{(V - V_K)}{V} \cdot \eta_r \tag{7}$$

There is an additional influence due to the nonlinear nature of the $V_g - I_d$ characteristic, but it is generally small. The key point of this expression is that compared to ideal values, the voltage swing is reduced by V_K and the current swing by I_m. This expression can be separated into a voltage efficiency η_V and a current efficiency η_I The voltage term

$$\eta_V = \frac{V - V_K}{V} \tag{8}$$

is easy to evaluate. V_K values range from 1.5 to 2V and bias voltages from 7 to 14V. The corresponding values for η_V range from 89 to 70%, which means that this effect alone limits the maximum efficiency in Class A to ~ 45%. The strategy to maximize η_V is obviously to find transistors with minimal V_K and maximum operating voltage. Differences in V_K, however, are generally very small.

The second term

$$\eta_I = \frac{I_{max} - I_m\ (V)}{I_{max} + I_m\ (V)} \tag{9}$$

describes the efficiency loss due to avalanche current. To demonstrate that this is the dominant variable the following experiment was

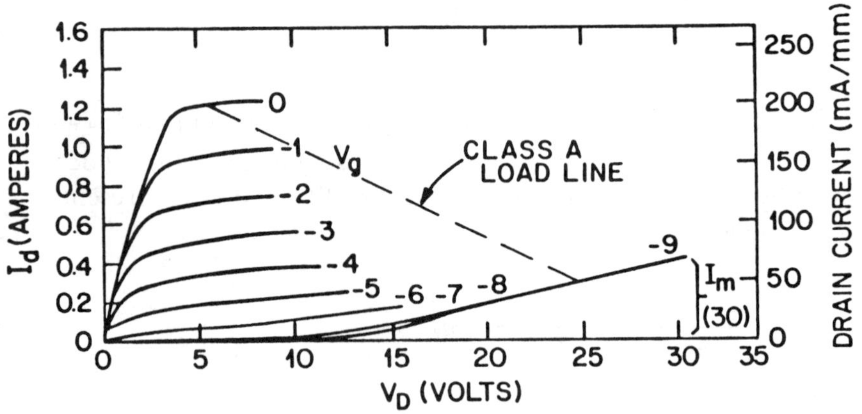

Figure 15/Typical I-V characteristic of a 6 mm WE transistor with a class A load line.

carried out [240]. A number of FETs with different avalanche properties were operated at 14V and a fixed bias current of 500 mA.

The RF input power was 28 dBm. Figure 16 shows the power added efficiency as a function of I_m at 30V. The efficiency drops by 50% when $I_M(30)$ reaches the 500-600 mA range. This result is consistent with expectations, particularly when it is noted that the maximum drain voltage is ~ 26V and $I_m(26) = .75 \cdot I_m(30)$ (see Figure 15).

If gate drain avalanche provides the main efficiency degrading mechanisms, then the key to higher efficiency is clearly a reduction in avalanche current by device design or selection, or a reduction in bias-voltage, which is exactly the opposite to the maximization of voltage efficiency as previously mentioned.

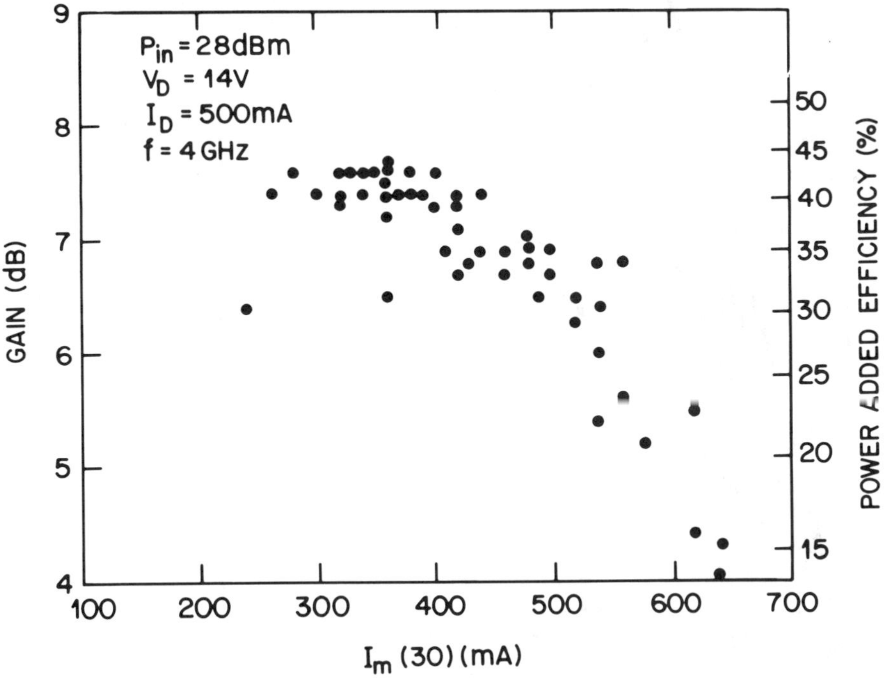

Figure 16/Gain at 500 mA, 14V as function of avalanche current I_M at 30V for different 6 mm WE transistors.

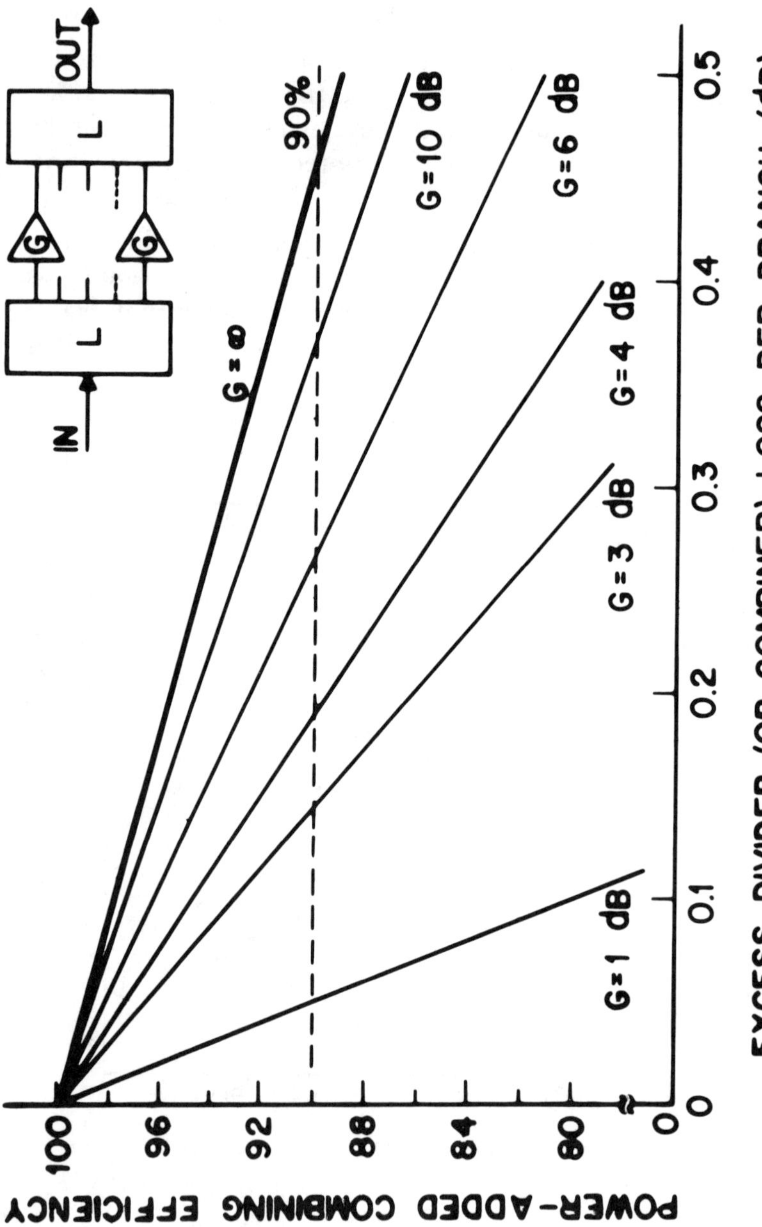

Figure 16/Added power combining efficiency vs. combiner loss with amplifier gain as a parameter.

As noted in Section 2.3, the coupler losses in the frequently used power combining schemes have considerable impact on the power added efficiency. To quantify this effect assume a n-way combiner as in Figure 16 with divider loss L and amplifier stages with gain G. G and L are expressed as ratios rather than dB; therefore, L is a positive number greater than 1. It is assumed that the divider loss L is equal to the combiner loss L_{out} or $L_{in} = L_{out} = L$, where L represents the loss through any one branch of the divider (or combiner) that is in excess of the ideal power split loss. Let P_{in} be the input power to the individual amplifiers. Then the total power added by the n amplifiers, P_A, is

$$P_A = P_{in}(G - 1)n\,. \tag{10}$$

The power added by the n-way power combined amplifier, P_c, is

$$P_c = n\left[\frac{P_{in}G}{L} - P_{in}L\right] = nP_{in}\left[\frac{G - L^2}{L}\right], \tag{11}$$

if P_{in} to the amplifier module is assumed to be the same as without the loss L. The power-added combining efficiency η_{PC} is then

$$\eta_{PC} = \frac{P_c}{P_A} = \frac{G - L^2}{L(G - 1)} = \frac{1/L - L/G}{1 - 1/G}\,. \tag{12}$$

Figure 16 shows curves of η_{PC} versus L with G as a parameter. Both G and L are expressed in dB in the figure for convenience of interpretation. The curves indicate the advantage of combining higher gain amplifier modules.

4. LINEARITY

In this section we will first review the characterization methods for linearity. This is necessary, since the GaAs FET does not behave like an ideal third order device. A basic knowledge of the underlying assumptions in linearity description is, therefore, useful. The topic has been thoroughly reviewed by Heiter [72]. We will briefly summarize his results and state the assumptions and simplifications pertinent to our application.

4.1 Characterization Methods

All the discussions are restricted to odd order distortion, since for most applications the bandwidth is less than an octave. This situation may change because very broadband amplifiers can be made with GaAs FETs. In that case almost none of the presently available data on linearity can be used.

Let us now consider a two port terminated at input and output as indicated in Figure 17. Under these conditions, the output voltage $e_0(t)$ can be related to the input voltage $e_I(t)$ as follows:

$$e_0(t) = \sum_{m=0}^{\infty} a_n e_I{}^n (t - t_n) \ .$$

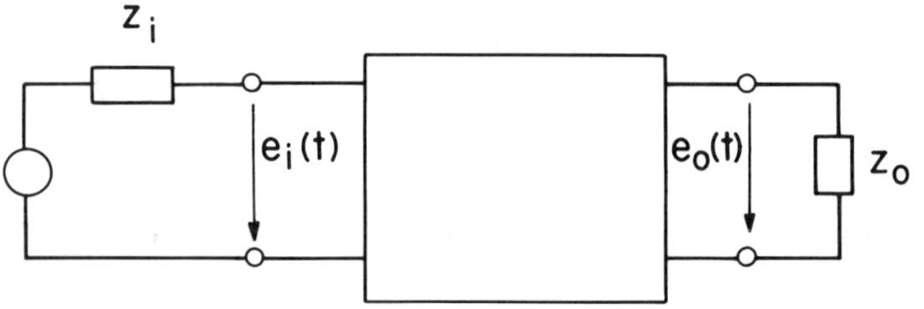

Figure 17/Schematic of a nonlinear two port.

Note that all terms have different time delays in addition to the amplitude components a_n. If one carries out a spectral decomposition for sinusoidal inputs, these time delays are transformed into phase angles. Although we are only considering inband components in this spectral analysis, it has to be recognized that the out of band components can have influence. Consider the following simple example: with the input voltage $e_I = \cos \omega t$, the device-characteristic can be described by

$$e_0 = C_1 e_I + C_3 e_I^3 \ . \tag{13}$$

e_0 can then be written as:

$$e_0 = (C_1 + \frac{3}{4} C_3) \cos \omega t + C_3 \cos 3 \omega t. \tag{14}$$

Even if we do not consider the third harmonic, we have to recognize that the amplitudes of the third harmonic and the third order distortion term at the base frequency are related. Should the impedances in two measurements be the same at the operating frequency but differ at the harmonic, the compression characteristic will be different.

There are a number of different ways to characterize the nonlinearity behavior. The simplest way is the measurement of the 1 dB compression power P_{1dB}. It is only well defined if the device is compressive, i.e., if the gain always decreases with increasing power. This is not always the case for GaAs FETs. To the first order, P_{1dB} is not sensitive to phase distortion. It is, therefore, not necessarily a representative measurement of distortion. The single tone measurement can be refined to give detailed information about the phase and amplitude distortion by using a bridge measurement. This is particularly useful for device diagnostics and to measure the AM-PM conversion which is of considerable practical interest.

The two tone test is the most popular of the distortion tests. It uses two closely spaced frequencies and has thus a spectrum which resembles the real application better than the single tone test but is still relatively easy to perform. With proper filtering, the fundamental output and the distortion products are measured separately. Two quantities are generally used to describe the nonlinearity. The first one is the intercept point. It is determined by plotting fundamental power output (from one of the tones) and the power in the modulation product P_{2A-B} as a function of input power (Figure 18). The intermodulation (IM) product will be at frequency 2A − B or 2B − A if A and B are the frequencies of the two tones. If the nonlinearity is dominated by the third order term, P_{2A-B} will have a 3:1 slope. The power at which the P_A and P_{2A-B} intersect is called the intercept point P_I. Another measure which is frequently used is the carrier to intermod ratio C/I. It has the disadvantage that it is not constant as a function of power even for a third order device and, therefore, it must always be specified at a particular power level. This, however, is an advantage in the case of the GaAs FET which is not a third order device. Here the intercept power can be a very misleading quantity, if one does not make absolutely sure one is operating in the third order range. Under the assumption of pure third order behavior, C/I and P_I are simply related as shown in Table 4, where we have summarized all the relationships between the various quantities.

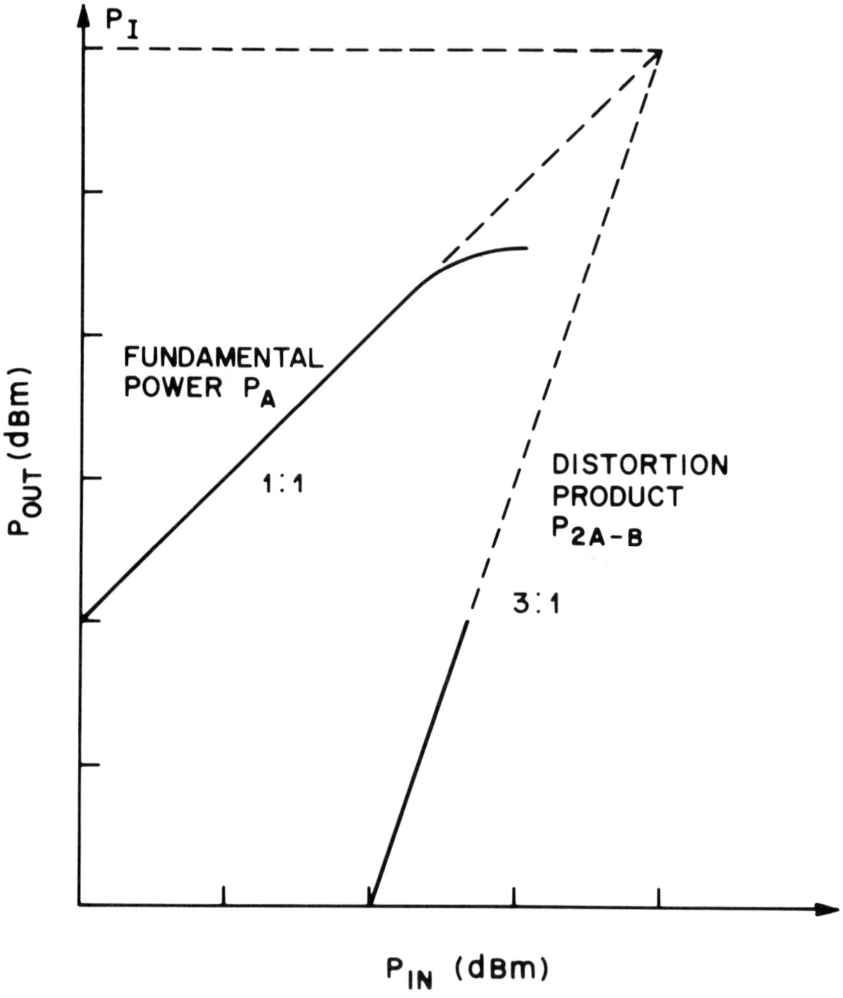

Figure 18/Definition of the intercept point.

The last test which we will discuss is the three tone test. It utilizes three closely spaced frequencies. The signal is more realistic at the expense of more complicated testing. The measured quantity is M_{A+B-C}, the ratio of intermodulation power to the cube of the power output at a single tone (Table 4). M_{A+B-C} is power independent for pure third order devices and low enough power levels.

When comparing the various tests, one must remember that the peak voltage levels in 2 and 3 tone experiments are higher than the ones corresponding to average power levels (see Table 4). This effect has considerable influence for power levels close to saturation.

4.2 FET Device Behavior

One would expect three major contributors to the device nonlinearity in the normal operating range:

(1) The input capacitance. The capacitance varies as a function of gate voltage and, therefore, it is reasonable to assume that it acts like a varactor and adds noticeably to the nonlinearities.
(2) g_m is a nonlinear function of the gate voltage and it should therefore contribute to the nonlinearity.
(3) From S-parameter measurements one knows that S_{22} varies with gate voltage and thus is another potential contributor.

A number of workers have made measurements to determine the relative importance of the nonlinearities [3, 226, 188, 190, 207, 208, 212] and there appears to be almost general consensus that the input capacitance contribution is negligible compared to the g_m and output impedance contributions. For linear amplifier design, it is therefore necessary to design the output network for linearity, whereas the input network can be used for other purposes. Other measurements done on much smaller devices at low power levels [122] show an influence of the gate capacitance nonlinearity. At this point it is not clear if the discrepancy is due to differences in operating level or the device structures.

If the device is operated close to its compression point, two additional nonlinearities will play a role:

(a) The saturation current limitation as V_g comes close to the built in voltage, and

TABLE IV

Conversion Chart for the Various Measures of Distortion for Pure Third Order Devices

	Symbol	Type of Measurement	Average Power	Peak Power	Definition
1 dB compression point	P_{1dB}	Single tone	P_A	P_A	See Fig. 18
Intercept point	P_I	Two Tone	P_A+3	P_A+6	See Fig. 18
Carrier to Intermod ratio	C/I	Two Tone	P_A+3	P_A+6	$P_{A+B-C}{}^{-3P_A}$
Third order Intermodulation coefficient	M_{A+B-C}	Three Tone	$P_A+4.77$	$P_A+9.54$	$P_{A+B-C}-3P$

Comments:

All quantities are either in dB or dBm

P_A is the power in one tone. The powers in all tones are assumed to be equal.

†The conversion of P_{1dB} and P_i or C/I is only valid if the distortion has zero phase angle.

P_{1dB} and M_{A+B-C} can be converted into each other if the phase angle of the distortion is known (see Heiter [72]).

(b) The drain current nonlinearity caused by V_g reaching pinch-off or high drain voltages driving the transistor into the avalanche region.

To minimize the latter two components, one has to bias the transistor to maximize the saturated power, which corresponds to a current level of 1/2 I_{MAX}. For good linearity, it is furthermore advantageous to use as large a transistor as possible since, for third order behavior, the $C_I \sim 20 \log w_g$, where w_g is the gatewidth. In comparing IM performance of different devices it is necessary to normalize with respect to size and power capability. Unfortunately, not all reports do that and it is thus sometimes very difficult to assess reports on performance improvements.

TABLE IV (cont.)

P_{1dB}	P_I	C/I	M_{A+B-C}
	$P_I - 10.6$†	$P_A + \frac{1}{2} C/I - 12.1$†	See Text
$P_{1dB} + 10.6$†	—	$P_A + \frac{1}{2} C/I - \frac{3}{2}$	$-\frac{1}{2} M_{A+B-C} + 3$
$2P_{1dB} - 2P_A + 24.2$†	$2P_I - 2P_A + 3$	—	$-M_{A+B-C} - 2P_A + 9$
—	$-2P_I + 6$	$-2P_A - C/I + 9$	—

A number of authors have observed that the optimum match for linearity does not coincide with that for maximum power gain [191, 188, 207, 208]. In Figure 19, which is copied from Schroeder et al. [188], constant gain and intermod circles are plotted to indicate the tradeoffs that have to be made between linearity and power output. This is consistent with the observation made in Section 2, that the impedance for maximum power output does not achieve the best gain flatness. To a first order the methods described there to achieve gain flatness can be utilized to design low distortion amplifiers.

Another very important point has been made by Higgins [75] and by Strid and Duder [207, 208] who showed that the intermodulation product (C/I) does not follow a 2:1 slope, but that the slope varies as a function of input level (Figure 20). This indicates that the FET is, in general, not a third order device. The characterization by the intercept point (Table 4) can be misleading since it depends on this property. Since the slope in general is greater than 2:1, the C/I can be improved much more by power backoff than in pure third order devices (e.g., TWTs). (See Figure 21 from Schroeder et al. [188].)

There have been a number of attempts to model the nonlinearity behavior. One approach has been to determine the bias dependence of

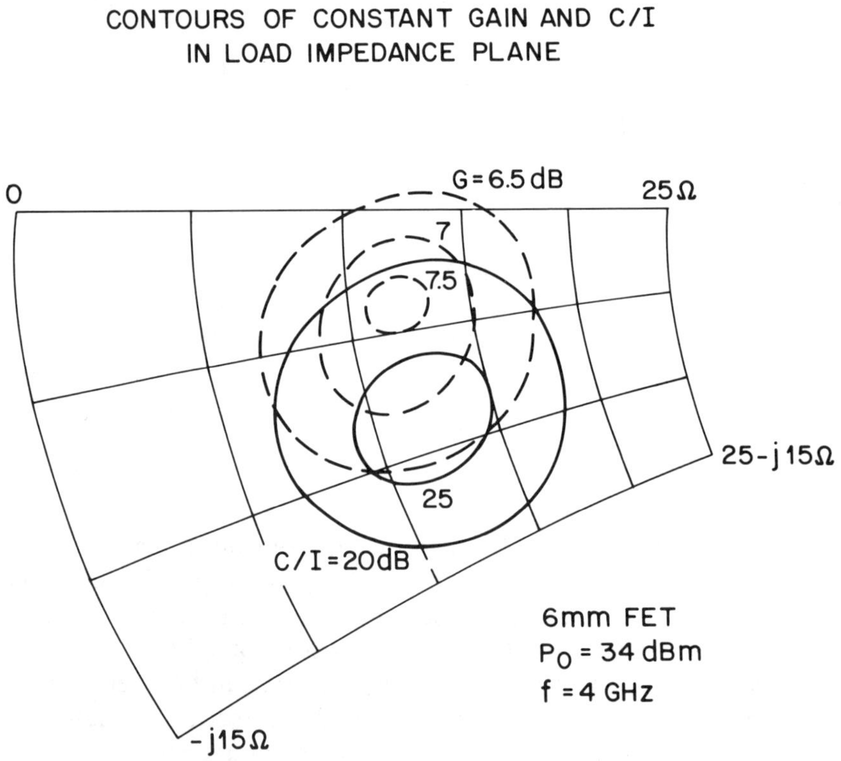

Figure 19/Contours of constant gain and C/I in load impedance plane (Schroeder).

the S-parameters [137, 138, 243] and use this information to predict the nonlinearities. Minasian [137, 138] uses this method to model a transistor operating in a high gain condition ($V_g \approx 0$). Since he assumes a load of 50 Ω, the output impedance nonlinearity does not

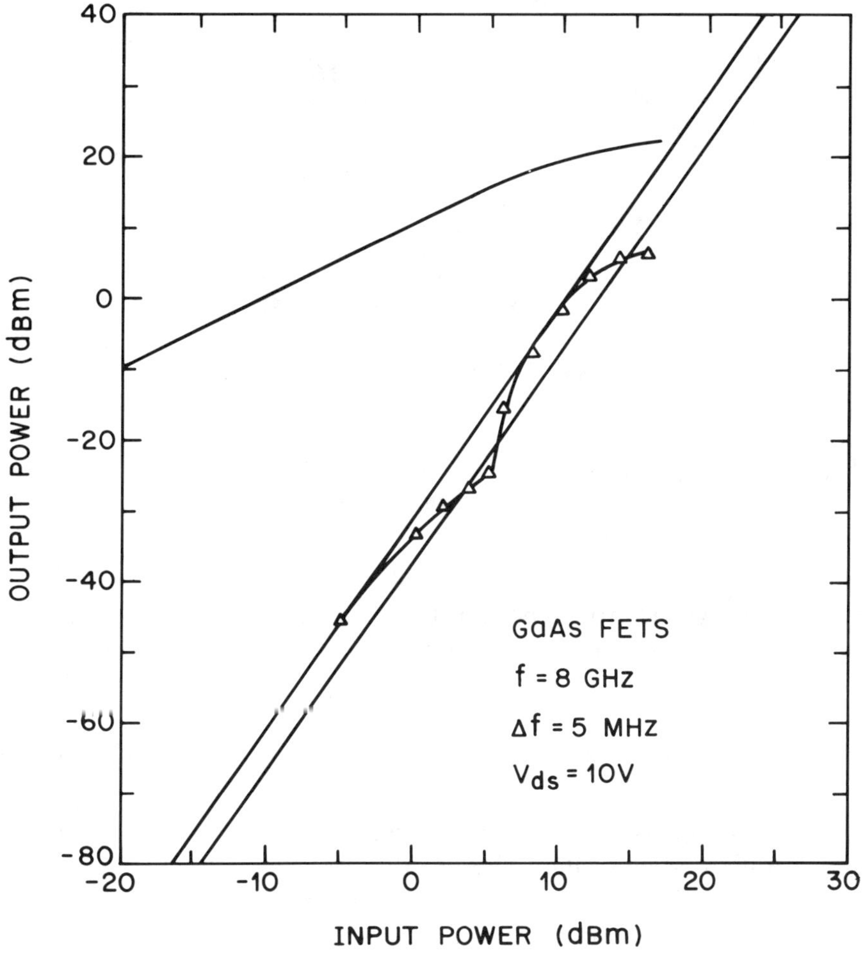

***Figure 20**/Output power and intermodulation level as function of input power (Higgins).*

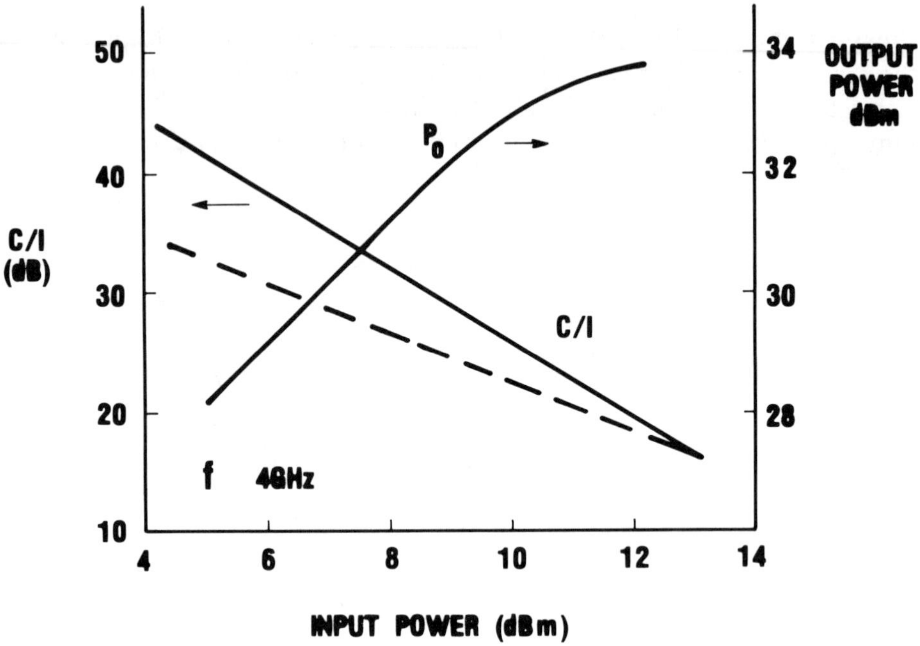

Figure 21/Output power and intermodulation level as function of input power for a 6 mm device (Schroeder).

contribute significantly. The resultant distortion dependence on P_{IM} is third order, which may indicate that the deviation from this behavior is attributable to the interaction of g_m and G_D nonlinearities. Willing et al. [243] calculate the distortion up to very high saturation levels with realistic matches. In that case a deviation from third order behavior is observed.

A different approach was taken by Tucker et al. [226]. They employ a unilateral equivalent circuit, identify the most important sources of nonlinearities and develop a closed form expression for IM in terms of third order parameters. The free parameters can be determined by comparison with IM measurements. This method gives a dependence of IM on output match, but is at present restricted to third order distortion.

The optimization of the transistor parameters with respect to linearity has been treated by three different workers. Pucel [169] minimizes the g_m contribution. He derives an expression for g_m in terms of the doping profile and then minimizes the third order IM by choosing either a $1/x^3$ or clump profile. Williams and Shaw [242] follow a similar path and experimentally generate an exponential doping profile. They claim a ~ 6 dB improvement over a similar unit with a flat doping profile. Higgins [75, 77] models the distortion taking g_m and output conductance variation into account. He then uses a computer generated doping profile to improve the IM. His experimental data indicate approximately a 6 dB improvement over a flat profile. In both cases the information given is unfortunately not sufficient to compare the results with other devices.

4.3 Practical Results

It is very difficult to compare practical results for the following reasons:

(a) Very often intercept points are given without showing that the device or amplifier indeed follows a third order behavior;
(b) Many results are given for amplifiers with multiple stages and the contribution from the various stages is not given; and
(c) Device size and power capability are not always given.

Nevertheless, Table 5 gives a compilation of some typical results. This table shows approximate levels of linearity and is not intended as a comprehensive summary. The normalized intercept power with respect to device size varies by only 4 dB amongst the various devices of different construction, size, and frequency. The higher frequency results tend to be a little lower, most probably due to the generally lower power output of the devices at the higher frequencies. One can make a crude comparison of the intercept power (note the caveats in Table 4) and the 1 dB compression point. This results in a prediction of 1 dB compression powers of 100-250 mW/mm which is consistent with the observations made in Section 2.

4.4 The Design of Highly Linear Amplifiers

A number of conclusions can be drawn on the improvement of amplifier linearity from the preceding discussion.

TABLE V

Name of First Author	Reference #	$P_{1\,dB}$ dBm	PI dBm	Dev. size mm	PI dB/mm	f GHz	Amp. Dev.	Date	Dev. Manuf.
Arai	9	40	48	20.8	34.8	4	D	79	Fujitsu
Fukuta	63	26	37.5	5.2	30.	6	D	76	Fujitsu
Honjo	81	38	45.	16.8	33	6.2	D	79	NEC
Schroeder	188	34	41.5	6	34.	4	A	78	BTL
Tsai	219	30	38.5	2.4	34.7	4.5	A	78	Fujitsu
Tserng	221	25	33.5	2.4	30	9	A	76	TI
Williams	242	20	31.	.6	33.2	9	A	78	TI

(1) The output match determines the linearity. The same methods which have been used for improving gain compression characteristics are applicable to linearity.
(2) For third order behavior, C/I is proportional to 20 log w_g where w_g is the gate width for constant relative power output or 20 log P_{SAT}. This means that linearity at a given power level can be improved by using higher power transistors.
(3) For a given transistor size, linearity can be improved by power back-off. For third order behavior of an FET, the carrier to intermod ratio (C/I) follows approximately the rule

$$C/I = C_1 + 20 \log (P_{1dB}/P) \tag{15}$$

with C_1 between 20 and 24 dB. In general, FETs show a more rapid improvement. For good linearity, it is thus desirable to operate far below the saturated power level. This is the opposite strategy to the one pursued if optimum efficiency is desired. In applications where both quantities are of importance, a careful tradeoff has to be made between them.
(4) The bias point should be chosen for maximum power capability. Therefore, the current should be near $(I_{max} - I_{av})/2$. The linearity increases with increasing voltage as well. Figure 22 shows a strong decrease of C/I for a voltage increase. This improvement is only partially explainable in terms of increased power output. Rectification can take place at the gate at relatively high saturation levels which in turn leads to selfbiasing effects. The linearity behavior can be strongly affected. Thus the gate bias impedance should be tailored to give the best linearity.
(5) In a multistage amplifier all stages can contribute to the intermodulation. If one assumes that all intermodulation products add randomly, the C/I product can be calculated from

$$\frac{1}{C_{TOT}} = \sum_{n=1}^{N} \frac{1}{C_n} \tag{16}$$

where C/I of the overall amplifier is log C_{TOT} and $C_n = 10^{C/I}$ of the nth stage. If all the voltages of the distortion products add in phase, the total inter-modulation is larger and given by:

$$\frac{1}{C_{TOT}} = \left[\sum_{n=1}^{N} \frac{1}{\sqrt{C_n}} \right]^2 \quad (17)$$

In reality, the overall C/I will fall between these two linearity values.

Up to this point, the focus has been on the influence of the transistors on linearity. There are, however, circuit measures which can be taken to change the linearity. One traditional way is to utilize negative feedback. If the gain is reduced by G_R, the C/I is improved by $2G_R$ for a third order device. At present the difficulty with this approach is the lack of gain in power devices, especially at higher frequencies.

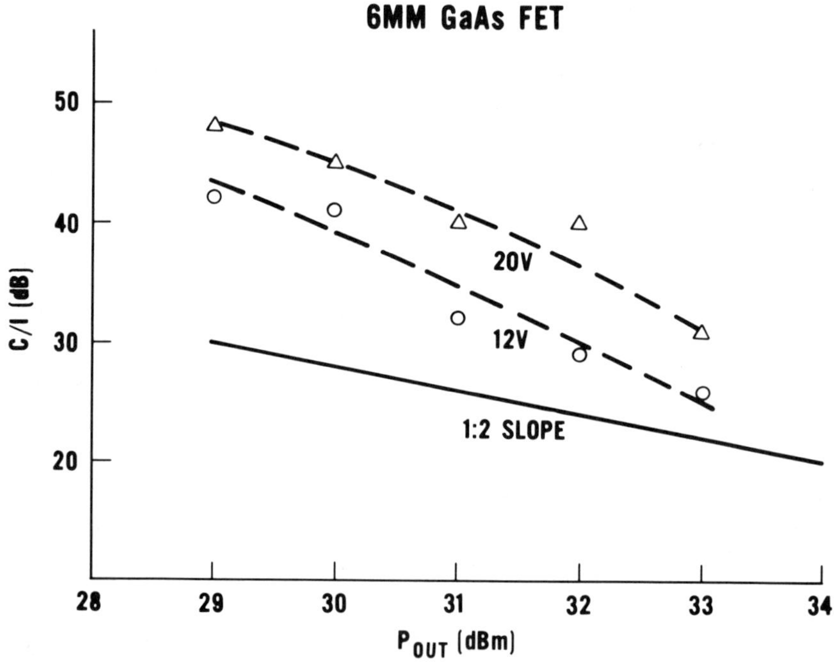

Figure 22/Intermodulation level as a function of output power for different drain voltages.

Two other methods can be employed for very high linearity applications. In predistortion, one assumes that the nonlinearity of the device is known and time invariant. A negative error is injected which then cancels the nonlinear contribution. This works well for a pure third order device. It has not, however, been demonstrated for more complex situations such as the GaAs FET.

Another approach is feetforward which separates the error signal, amplifies it separately and then injects it to cancel the original error. This can generate improvements of up to 25 dB in C/I. However, this is achieved at the expense of considerable circuit complexity and requirements for phase accuracy. A more moderate 10 dB improvement relaxes the requirements for high accuracy. So far this concept has not been utilized in FET amplifiers but it is to be expected that the emerging integration methods make it feasible for high linearity applications.

5. FET CIRCUIT TECHNOLOGY

Microwave circuit technology appropriate to FET amplifier development is presented in this section. The first sub-section gives a summary of microstrip technology. The second sub-section considers the microstrip/FET interface and compares the use of lumped and distributed circuit elements in typical FET circuits. Finally, the last sub-section presents several representative examples taken out of the technical literature of FET amplifier technology ranging from microstrip circuits using discrete FETs to totally monolithic FET amplifiers fabricated on GaAs substrates.

5.1 Summary of Microstrip Technology

The use of microstrip technology for the interconnection of microwave solid state devices and passive circuit components is a common approach to the development of microwave integrated circuits (MIC). Three terminal devices – including GaAs FETs – can be conveniently connected in chip or packaged form to a planar microstrip circuit since both "center" and ground conductors of microstrip transmission lines are readily accessible.

Such a line in its basic form is shown in Figure 23. It consists of a top strip conductor and a bottom ground plane separated by a suitable dielectric substrate. For nearly all practical applications, it can be assumed that the propagating mode is a transverse electric and magnetic mode (TEM). The characteristic impedance of microstrip depends on the width to height ratio, W/H, of the top strip conductor (where H is its height about the ground plane), and the relative dielectric constant of the substrate material. Since a microstrip transmission line consists of a mixed dielectric medium, i.e., air and the substrate material (assumed here to be non-magnetic, $\mu = \mu_0$), the imaginary component of the propagation constant, β, is given in terms of an effective relative dielectric constant, ϵ_{eff}, as

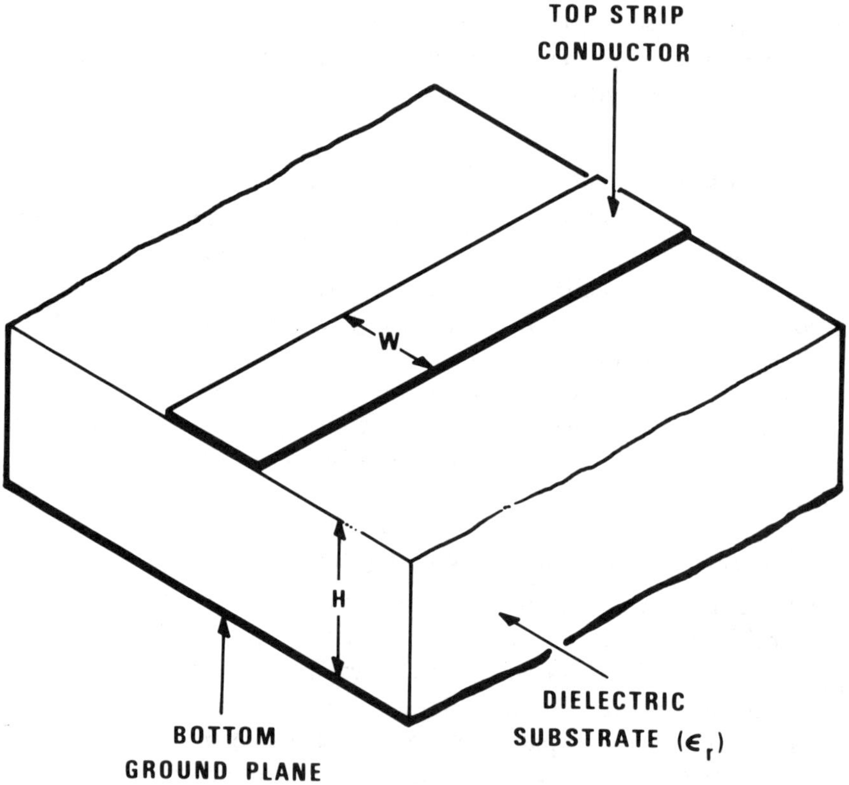

Figure 23/Basic form of microstrip transmission line.

tive dielectric constant, ϵ_{eff}, as

$$\beta = \frac{2\pi}{\lambda_g} = \frac{\omega}{v_{ph}} = \omega\sqrt{\epsilon_0 \mu_0} \cdot \sqrt{\epsilon_{eff}} \tag{18}$$

where λ_g is the guide wavelength, v_{ph} is the phase velocity, ω the radian frequency and $(\epsilon_0 \mu_0)^{-1/2}$ the phase velocity of light in air. Because of the mixed dielectric, ϵ_{eff} is always less than (or equal to in the case of an "air" microstrip line) ϵ_r, the relative dielectric constant of the substrate material, and is a function of W/H. Extensive characterization of such lines in terms of W/H and ϵ_r can be found in the literature for commonly used substrate materials (see, for example, References 187, 170, 186). Figure 24 shows the dependence of the microstrip characteristic impedance Zo, on the width to height ratio, W/H, for three substrate materials: quartz, alumina, and chrome-doped semi-insulating GaAs (ϵ_r = 3.78, ~ 9.7 and 13.0 respectively).

In most cases, (including those to be considered in this chapter) where good quality substrates are used, RF losses occurring in the top strip and ground plane conductors dominate the dielectric losses within the substrate. Thus, in terms of the transmission line parameters, the microstrip line can be approximately characterized by the three quantities, R, L, and C, the resistance, inductance, and capacitance per unit length (G, the conductance per unit length is assumed negligible). The complex propagation constant, γ, and the characteristic impedance, Zo, of the line are then given by

$$\gamma = \sqrt{(R + j\omega L)(j\omega C)} = \alpha + j\beta \tag{19}$$

$$\beta \cong \omega\sqrt{LC} \tag{20}$$

$$Zo = \sqrt{\frac{R + j\omega L}{j\omega C}} \cong \frac{L}{C} \tag{21}$$

where α is the attenuation constant and the approximations in Equations 20 and 21 are based on the fact that $\omega L/R \gg 1$ for most practical situations.

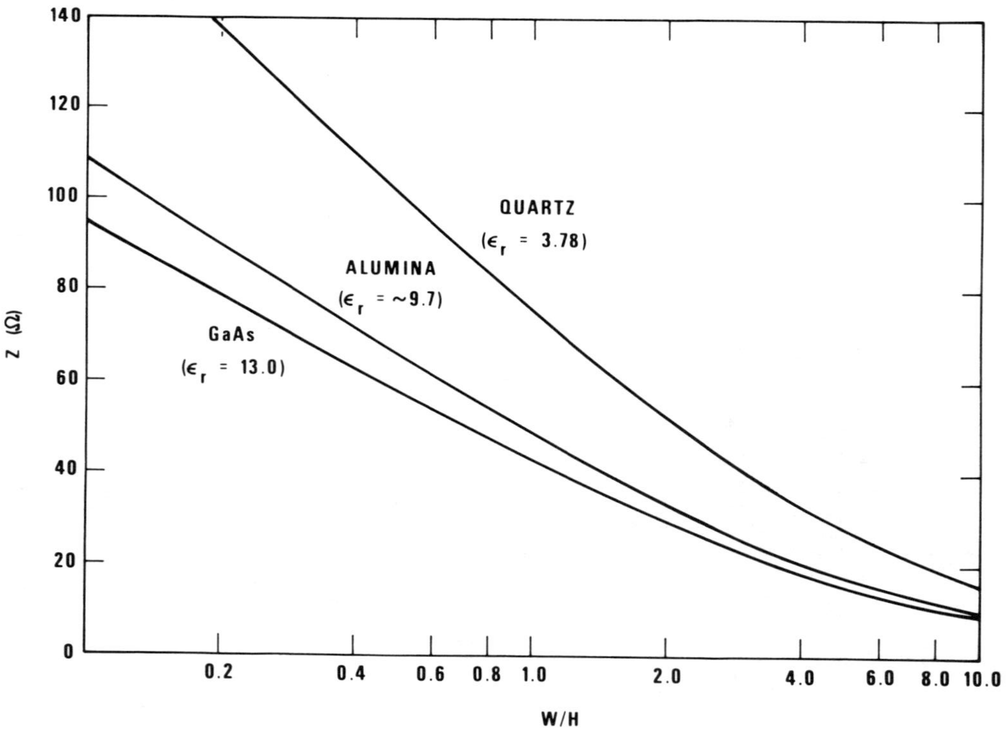

Figure 24/*Characteristic impedance of microstrip as a function of W/H for three substrate materials, Quartz ($\epsilon_r = 3.78$), Alumina ($\epsilon_r \sim 9.7$), and GaAs ($\epsilon_r = 13.0$).*

The unloaded Q for the line is inversely proportional to the attenuation per guide wavelength and is given by

$$Q_u = \frac{\beta}{2\alpha} = \frac{\pi}{\lambda_g \alpha} \quad (\alpha\lambda_g \text{ in nepers})$$

$$= \frac{20}{\ln 10} \cdot \frac{1}{\lambda_g \alpha} \quad (\alpha\lambda_g \text{ in decibels}) \tag{22}$$

From Equation 19 it can be shown that $\beta/2\alpha = Q_0 + (1/4Q_0) +$ higher order terms in Q_0^{-1} where $Q_0 = \omega L/R$. Since values for Q_0 are in the several hundreds range for typical microstrip, a simple expression for the unloaded Q factor is

$$Q_u = \frac{\omega L}{R} \tag{23}$$

It is important to note that L and R, and therefore Q_u, depend on the geometrical configuration of the top strip conductor and its height above the ground plane. They do not depend on the substrate dielectric constant. This should not suggest, however, that for a given microstrip characteristic impedance, the choice of substrate, or substrate thickness, has no influence on the conductor losses. Indeed for the same characteristic impedance, a microstrip line fabricated on a thicker substrate of the same material will result in less losses due to a wider top strip conductor. For this case, ϵ_{eff} is the same as are β and L (from Equations 18, 20, and 21), but the resistance per unit length, R, for the microstrip on the thicker substrate is less due to a wider strip.

The theoretical dependence of conductor attenuation, α_c, on the microstrip characteristic impedance, Z_0, is shown in Figure 25 in terms of the normalized attenuation factor, $A_c = \alpha_c(H/R_s)$, for two substrate materials, quartz and alumina. H is the substrate thickness and R_s the skin resistance given by $R_s = \sqrt{\pi f \mu_0 \rho}$ where ρ is the conductor resistivity of the metal. The curves are calculated from Equations 44 and 45 of Reference 61 and include the effects of conductor thickness, t. In Figure 25 the assumed values are t = 7.5 μm for all curves and H = 0.25 or 0.64 mm as indicated.

The substrate surface finish also affects the losses in the metallization. A rough or unpolished surface results in higher metal losses because of a longer effective current path. Experimentally it has been found [228] that if the ratio of the root mean square of the surface roughness to the value of the skin depth is less than about 0.1, then the additional losses due to this phenomenon may be considered negligible. Where rms surface roughness approaches the skin depth, the relative attenuation factor (i.e., the attenuation of the microstrip on the rough substrate as compared to the attenuation of the microstrip fabricated on a smooth substrate) is about 1.5. At 10 GHz the skin depth for gold is about 0.8 μm. Commercially available polished alumina, for example, can be obtained having an rms surface roughness of about 0.05 μm. Substrates of this quality can therefore be used for microstrip through X-band and part of Ku-band without incurring significant metal losses due to surface roughness. For higher frequencies, another commonly used substrate material is fused quartz. From the point of view of surface finish, it has the advantage over alumina of being a softer, glass-like material and therefore amenable to very high polish (its disadvantage, however, is that it is considerably more fragile than alumina).

Metallization is applied to the dielectric substrates either by thick film (essentially a silk screening process using suitable conductive inks which are screened and fired onto the dielectric) or thin film techniques (metal is vacuum evaporated or sputtered onto the substrate and then selectively etched to form the circuit pattern using standard photo-resist and photo-lithographic techniques.) For thin film circuits, the metallization consists of a thin layer of an appropriate metal to obtain good adhesion to the substrate (e.g., chrome or nichrome ~ 200Å thick) followed by a thick layer (several microns) of a good conductor such as copper ($\rho = 1.7 \times 10^{-6}$ Ω-cm) or gold ($\rho = 2.44 \times 10^{-6}$ Ω-cm). Gold is very often used because of its resistance to tarnishing. The thickness of the top layer is usually several skin depths to minimize loss. Since skin depth, which is given by $\delta_s = \sqrt{\rho/\pi f \mu_0}$, is inversely proportional to the square root of frequency, more metal is required at the lower frequencies. Thick film can be a cost effective approach for large volume production of circuits suitable for the lower microwave frequencies extending possibly up to lower X-band. Its disadvantages over thin film fabricated circuits include higher loss and significantly poorer pattern resolution due to the porous nature of the resultant metallization. At X-band, typical attenuation factors (for a 50Ω line on 0.63 mm substrates) are 0.24 dB/cm for a thick film line versus 0.09 dB/cm for thin film. Line and space resolution is limited to about 120 μm and 80 μm for thick film and about 20 μm for each case for thin film. For these reasons thin film technology is used almost exclusively for X-band and higher frequencies.

Numerous other low-loss dielectric materials besides quartz and alumina are used for specialized microstrip circuit applications. These include the ferrites (required for non-reciprocal circuits as encountered in MIC circulator coupled reflection amplifiers), beryllia (valued for its high thermal conductivity and often used in microwave bipolar transistor microstrip amplifiers), spinel, and sapphire. For applications requiring high permittivity dielectrics ($20 \leqslant \epsilon_r \leqslant 150$), rutile, the various titanates [117], and other more exotic materials are used. For monolithic applications high resistivity silicon and semi-insulating, chrome-doped GaAs must also be considered. To focus attention on those substrates which are important for GaAs FET circuits, the discussion to follow will restrict itself to four substrate materials: quartz ($\epsilon_r = 3.8$), alumina ($\epsilon_r = 9.7$), sapphire ($\epsilon_{r\parallel} = 11.5$, $\epsilon_{r\perp} = 9.6$), and chrome-doped semi-insulating GaAs ($\epsilon_r = 13$) [38].

Table 6 lists the microstrip electrical characteristics for the four substrate materials. Alumina is the most popular because it is especially suited for an MIC assembly and production environment. It is relatively inexpensive, a strong and durable ceramic with a relative dielectric constant ranging from about 9 to 10.2 depending on the manufacturer and the specified purity of the material. Round holes or holes with square corners can be easily pre-punched by the manufacturer while the material is still in its "green" state, prior to firing.

Ground and polished substrates having an rms surface roughness of as little as 0.05 μm are commercially available. Sapphire is the same material as alumina except that it is single crystal aluminum oxide.

An advantage of sapphire over alumina for MIC applications is that it has well defined and repeatable electrical characteristics and its surface can be polished to have an optical grade finish. The crystal, however, is anisotropic, having a relative dielectric constant parallel to the optical axis (C-axis), $\epsilon_{r\parallel}$, which is higher than the relative dielectric constant in the plane perpendicular to this axis, $\epsilon_{r\perp}$ [161].

Consequently, microstrip on sapphire is usually fabricated such that the C-axis is perpendicular to the ground plane. Assigning the y-axis to be along the C-axis, the principal relative dielectric constants for sapphire are $\epsilon_{r\perp} = \epsilon_x = \epsilon_z = 9.4$ and $\epsilon_{r\parallel} = \epsilon_y = 11.6$. For microstrip design purposes, a new equivalent relative dielectric constant, ϵ_{req}, which is a function of W/H, can be defined for sapphire which accounts for the anisotropy [161]. In Table 6 the value of $\epsilon_{req} = 11.1$ is the relevant value for a 50Ω microstrip line (W/H = 0.86) fabricated on sapphire oriented as described above. Fused quartz is much more fragile than alumina, but it has an advantage at high frequencies (10 GHz $\leqslant$ f $\leqslant$ 30 GHz) because of its lower relative dielectric constant, $\epsilon_r = 3.8$. This results in larger circuit dimensions (approximately a factor of 1.5 greater in the guide wavelength when compared with alumina) and lower loss. Relative to alumina, the latter characteristic is due to wider strips for the same characteristic impedance (see Figure 25), and a smoother surface, since quartz can readily be polished to have an optical grade finish (0.03 μm rms surface roughness). Finally, for monolithic circuits, GaAs is also a satisfactory microstrip substrate. It is a single crystal material on which may be fabricated both active and passive devices. Typically, GaAs FETs are fabricated in regions of highly conducting epitaxial material, while the passive components are fabricated directly on the chrome-doped semi-insulating substrate. Bulk dc resistivity for good quality semi-insulating substrates is of the order of 70,000Ω-cm.

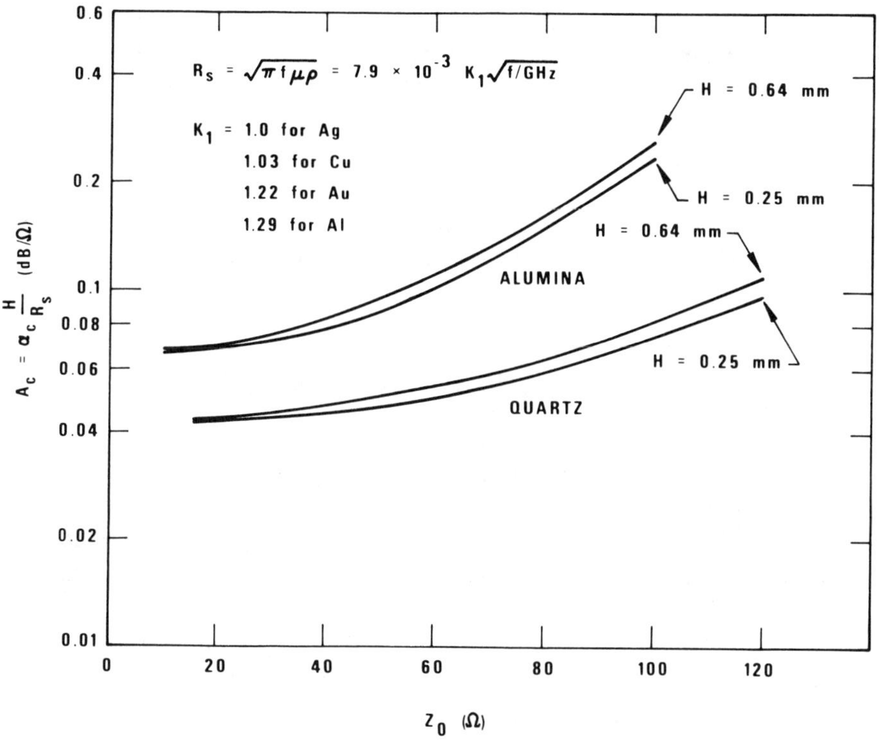

Figure 25/Normalized conductor attenuation factor as a function of microstrip characteristic impedance for Alumina and Quartz substrates.

For design purposes it is important to establish the range of realizable impedance values and the loss associated with microstrip fabricated on the various substrates. For reference, Table 6 shows the W/H ratio, the effective relative dielectric constant (for a 50Ω microstrip line), and the physical length for a quarter wave section of line at 10 GHz. In the case of sapphire the equivalent relative dielectric constant of 11.1 was used to arrive at the W/H ratio and the effective dielectric constant (using Equations 16 and 23 of Reference 187).

The approximate limits on the range of practical characteristic impedances achievable with microstrip can be determined by consideration of the substrate thickness, the ratio of the strip width to guide wavelength, and the metallization loss.

From a loss point of view, a thicker substrate is desirable to minimize metal losses. However, for a given frequency of operation and substrate dielectric constant, an upper limit on substrate thickness is used to prevent excessive radiation loss and the propagation of extraneous, undesired modes. For example, to prevent the propagation of the TE plane surface wave (TE version of ordinary TM Zenneck wave) over the dielectric substrate, the thickness must be limited to [129].

$$H_{max} = \frac{\lambda_0}{4\sqrt{\epsilon_r - 1}} \qquad (\lambda_0 = \text{free space wavelength}). \tag{24}$$

For alumina, $H_{max} = 0.25$ cm at 10 GHz. In practice, substrate thicknesses are often considerably less than this limit in order to prevent excessive radiation loss from open circuits like those encountered in open circuited resonant tuning stubs. This is particularly significant in the case of ordinary unshielded microstrip. It has been shown theoretically by Lewin [121] and experimentally by Denlinger [48] that the radiation resistance of a simple microstrip half wavelength resonator is approximately given by

$$R_r \approx (h/\lambda_0)^2 F(\epsilon_{eff}) \tag{25}$$

where $F(\epsilon_{eff})$ is a rapidly decreasing function of the effective dielectric constant. From the empirical formula of Denlinger, an approximate criterion for preventing excessive radiation loss (less than 25% of the power will be radiated) for microstrip resonators is that the ratio (h/λ_0) be less than 0.01 and that the substrate's relative dielectric constant be greater than 2.5. In terms of the substrates listed in Table 6, it is desirable, therefore, that the substrate thickness be no greater than 0.3 mm for 10 GHz operation, especially in the case of quartz. Van Heuven [228] contends that the relationship between radiation and frequency is not as simple as indicated in the previous formula and that a resonant microstrip line can radiate not only from its ends but along its entire length, especially when the resonator length approaches an integral multiple of half-wavelengths in free space. Consequently, the Qs of unshielded microstrip resonators may have maxima and minima as a function of frequency. Such microstrip resonators must be appropriately shielded to prevent radiation. The use of substrates much thinner than 0.2 mm is discouraged because of the resulting thin lines, excessive metal losses, and the mechanical fragility of the material. An exception is GaAs, however, where the monolithic circuits require thin substrates for adequate heat sinking of the active devices (discussed in detail later in this chapter). For these reasons the substrate thicknesses listed in Table 6 are limited to between 0.125 mm and 1.25 mm for the range of frequencies 5-15 GHz.

With the substrate thickness established, consideration can now be given to the practical lower limit (wide strips) on the characteristic impedance, Z_0, for the various substrates in Table 6 as a function of frequency and substrate thickness. This limit comes about because the transverse dimension of the strip becomes a significant fraction of the guide wavelength as the strip is made wider in order to decrease Z_0. This can result in the excitation of unwanted transverse resonances (similar to adding a parallel stub) and can distort the normal longitudinal excitation. From a designer's point of view it becomes difficult to predict the circuit behavior. For example, in the simple but often encountered case of impedance matching with $\lambda/4$ transformers, if the width of the transforming section is equal to the quarter guide wavelength, then the etched microstrip pattern becomes a square. It is not clear whether to treat such a circuit as a $\lambda/4$ transformer, a lumped capacitor shunting to ground, or as a loaded transverse resonator (a $\lambda/4$ section of line with open circuited ends is equivalent to a simple R-L-C series resonant circuit). In practice, especially when matching to low impedance solid state power devices, this lower limit on Z_0 is often approached. To correct for some of the semi-lumped behavior which was not taken into account in the circuit design, empirical tuning (e.g., with metallized ceramic tuning chips) may be used to optimize the circuit performance. Generally, however, from the point of view of predictable distributed circuit design, such low Z_0's which result in etched circuit patterns with aspect ratios close to unity should be avoided if possible. In Table 6 an arbitrary criterion that the strip width be no wider than 0.8 of $\lambda_g/4$ is used to establish the practical lower limit on the characteristic impedances for different frequencies and substrate thicknesses. The formulas in Reference 187 were used for these calculations. As shown

TABLE VI

Microstrip Characteristics

Substrate Material	Quartz	Alumina	Sapphire	Gallium Arsenide
	Fused SiO_2	99% Al_2O_3-Ceramic	Al_2O_3-Single Xtal	Cr Doped Semi-Insulating GaAs - Single Xtal
ϵ_r*	3.8	~9.7	Anisotropic $\epsilon_{r_1} = 9.4$, $\epsilon_{r_{11}} = 11.6$	13.0
tan δ	<0.0001	0.0001	<0.0001	0.0004
ϵ_{eff}*	2.97	6.64	7.4	7.9
W/H*	2.1	0.96	0.86	0.78
$\lambda_g/4$ @ 10 GHz* (mm)	4.34	2.90	2.74	2.67
Q @ 10 GHz	420*	300*	380**	70-100*
$a\lambda_g = 8.686\ \pi/Q$ (dB)	0.065	0.091	0.072	~0.27
R_a (μm)	0.05	0.05	0.005	~0.05
H(mm)	0.5	0.64	0.51	0.2
M&T	Ni-Ph, Au 1000Å, 7μm	Cr, Au 200Å, 7.5 μm	Cr, Au, Cu plate 200Å, ~3.0 μm total	Au ~1.5 μm

ϵ_r = Relative dielectric constant
ϵ_{eff} = Relative effective dielectric constant
tan δ = Dielectric loss tangent
W/H = Width to height ratio
λ_g = Guide wavelength
Q = Quality factor

$a\lambda_g$ = Attenuation per guide wavelength
R_a = Rms surface roughness
M&T = Metallization and thickness

*For $Z_o = 50\Omega$
**For $Z_o = 20\Omega$

TABLE VI

Microstrip Characteristics
(Continued)

Approximate Upper and Lower Limits for Realizable Microstrip Characteristic Impedance Z_o, as a Function of Substrate Material, Height, and Frequency

Substrate Material		Quartz			Alumina			Sapphire			Gallium Arsenide		
Z_o Lower Limit (Ω)†													
frequency (GHz)		5	10	15	5	10	15	5	10	15	5	10	15
H (mm)	0.125	3.7	7.1	10.1	3.8	6.7	9.4	3.6	6.6	9.3	3.6	6.5	9.2
	0.250	7.1	13	18	8.6	12	16.4	6.6	11.7	15.9	6.5	11.5	15.6
	0.50	13	23	31	12	20	26	11.6	19.6	25.9			
	1.25	27			24	36	47	23	35	43			
Z_o Upper Limit (Ω)††													
H (mm)	0.125		137			91			85			79	
	0.250		165			109			103			95	
	0.50		193			128			120				
	1.25		229			152			144				

† $Z_{o_{lower}}$: $W = 0.8 \cdot (\lambda_g/4)$

†† $Z_{o_{upper}}$: $W = 25\ \mu m$

in the figures of Table 6, low impedances are more difficult to achieve for the higher frequencies and the thicker substrates. Note that approximately the same impedance levels are achievable with fused quartz as with alumina for the same substrate thickness, even though there is a factor of 2.5 between the two relative dielectric constants. From Table 6 it is seen that at 10 GHz the lower limit on Z_0 for an alumina substrate, 0.25 mm thick, is 12Ω. For example, a single section λ/4 transformer having a 12Ω characteristic impedance will transform from 50Ω to 2.9Ω. To put this in perspective, large power FETs can have input impedances significantly lower than this value. For these cases, lumped and semi-lumped circuit elements (to be discussed later) are used in impedance matching networks to transform to these lower levels, often in conjunction with partial distributed circuit matching.

For the upper limit on Z_0, the dominant factors are line resolution and metal losses. The criterion used in Table 6 is that the strip width be no smaller than 25 μm. Much smaller lines are difficult to resolve with standard photomasks and etching techniques. Furthermore, the metal losses become excessive. It is easier to achieve high impedances for thick substrates than for thin substrates. For monolithic GaAs circuits, where the substrate thickness may be as small as 0.1 mm, this is a significant limiting factor for the realization of high Q and high impedance transmission lines. On 0.1 mm thick semi-insulating GaAs, for example a strip 25 μm wide corresponds to only a 74Ω line. Depending on the thickness of the conductor metallization, the Q of such a line may be only in the 20-50 range.

Finally, Table 6 shows typical experimental loss factors for microstrip expressed in terms of Q and the attenuation per guide wavelength. The data for quartz was obtained from measurements of open circuited shielded 50Ω microstrip resonators [228]. The loss data for alumina was obtained from measurements of a 23 cm, 50Ω, microstrip meander line [186]. Both cases make use of highly polished substrates metallized with a thin adhesion layer followed by a layer of gold several skin depths thick. The advantage of fused quartz in terms of lower loss is apparent (see also theoretical data shown in Figure 25). The conductor losses for microstrip on sapphire are similar to those of microstrip on alumina for the same substrate thickness, metallization, and surface roughness, since the relative dielectric constants of the two materials are nearly the same. However, sapphire can be obtained with a highly polished surface. The loss data for sapphire in Table 6 was obtained from an average value of several measurements of Q-factors of unshielded open-circuited 20Ω microstrip resonators in the frequency range of 7.9-12.4 GHz [94]. The Q factor for a 50Ω microstrip line on a 200 μm GaAs substrate is an approximate experimental figure interpolated from Q-factor data obtained from several laboratories engaged in monolithic GaAs circuit development.

5.2 The FET/Microstrip Interface

5.2.1 General

It is relatively straightforward to interface GaAs FETs with standard microstrip circuitry. Figure 26a shows a typical FET chip connected to microstrip, while Figure 26b shows the corresponding configuration for a FET in packaged form. As shown in Figure 26a, the chip is mounted directly onto the metal surface (the carrier plate) which is an extension of the microstrip ground plane. Usually backside metallization is applied to the substrate side of the chip so that it can be soldered to the carrier plate ensuring a low thermal resistance path to the metal heat sink. Since the chrome-doped substrate is an electrical insulator, RF grounding must be realized either with bond wire connections from the common terminal pad to the carrier plate or, in some cases, with plated through holes which are prefabricated in the FET chip and directly connect the back metallization with the common terminal (refer to the FET design and fabrication section of the chapter). For common source operation using discrete FETs, the source pad (or pads in the case of power FETs) must be connected to the carrier plate with minimum inductance, while the gate and drain pads are connected via bond wires to the input and output microstrip circuits respectively. Another mounting configuration is to attach the FET to the carrier plate with the topside metallization facing down such that the source pads are directly in contact with special ground posts fabricated on the carrier plate. This "flip-chip" technique minimizes source lead inductance and is sometimes used in the fabrication of power FETs [53, 142, 143]. In the case of monolithic FET circuits, proper RF grounding is an especially significant problem and will be discussed later in the chapter.

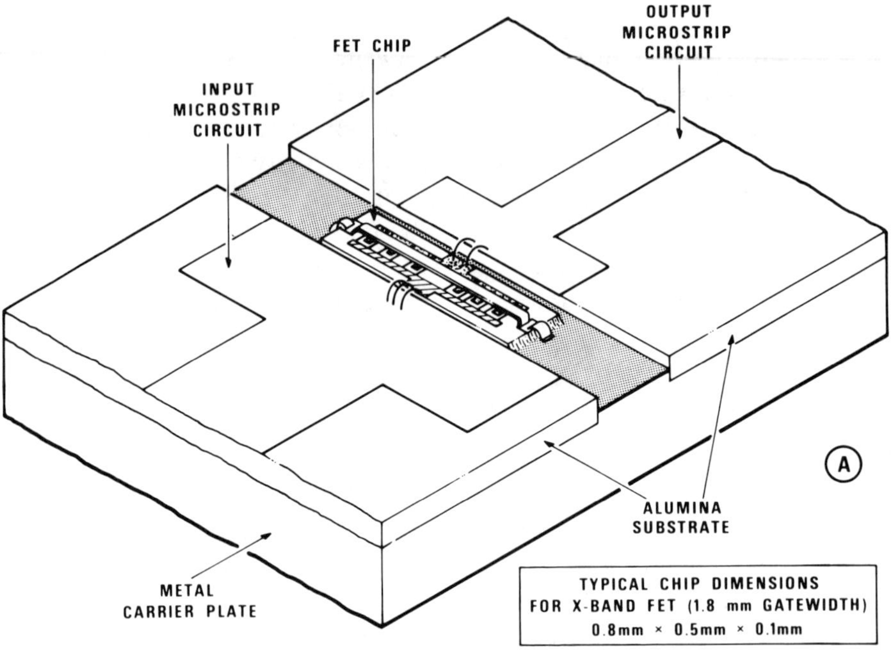

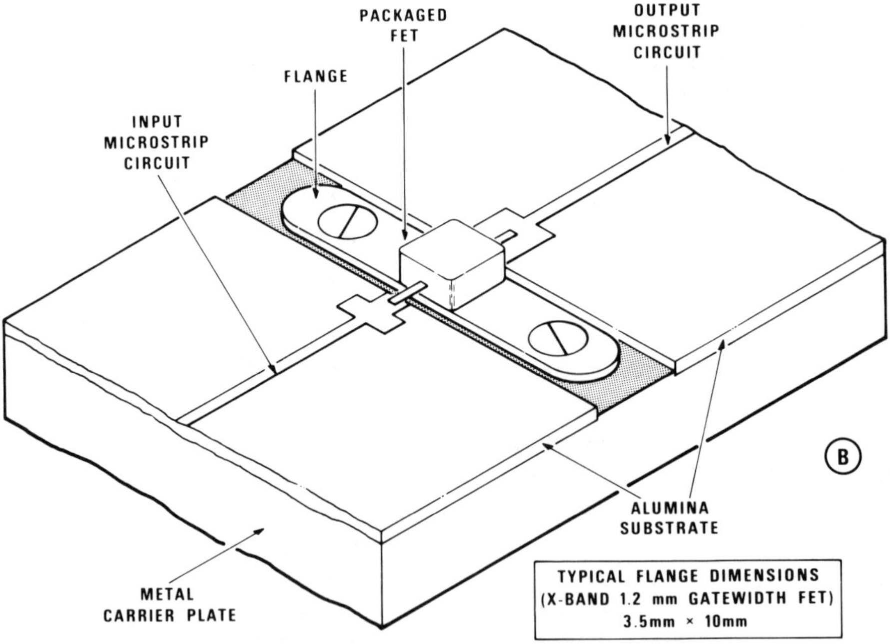

Figure 26/Mounting and connection of FET to microstrip
a) FET in chip form; b) Packaged FET.

The bond wire used to connect the FET to the microstrip circuit is usually much shorter than a wavelength and can be considered for all practical purposes to behave as a relatively high Q lumped inductance. The exact value of the self inductance depends on several factors including the diameter and length of the wire, its shape, its proximity to other neighboring bond wires, and its location with respect to the ground plane. Indeed, it is often difficult to assess the bond wire inductance in an actual FET circuit. Nevertheless, guidelines for bond wire inductance values are available [215, 149]. Terman [215] gives formulas for several specific cases of self inductance of isolated round straight wires and straight wires in proximity to a ground plane. These can be used to estimate the inductance of a particular bond wire configuration. For typical MIC FET applications a rough approximation for the self inductance of a straight, long piece of 25 μm diameter bonding wire far removed from the ground plane (i.e., far enough that the self inductance of the wire is dominant and that mutual inductance contributed by the ground return path is negligible) is about 0.8 nH/mm. Note that, in reality, the inductance per unit length for a finite piece of wire is not a constant but rather a slowly increasing function of the length. From Terman, for example, the high frequency limit (where the skin depth is much smaller than the diameter of the wire) for the self inductance, L, of an isolated straight wire 1 centimeter long is given by

$$L = 2l \left(\ln \frac{4l}{d} - 1 \right) \text{nH} \tag{26}$$

where d is the diameter of the wire. At high frequencies an estimate of the quality factor for a 25 μm round gold wire can be calculated from

$$Q = \frac{\omega L}{R_{ac}} \quad \frac{\omega L(\pi d\delta)}{\rho l} = \frac{\omega L \pi d}{R_s l} \tag{27}$$

where L is the inductance, R_{ac} is the ac resistance, ρ is the resistivity, δ is the skin depth at the radian frequency ω, and R_s is the skin resistance. Note that since $\delta \ll d$, R_{ac} can be considered to be the resistance between the two ends of a pipe having a wall thickness δ. Estimating L from Equation 26, the Q factor for a 0.25 mm length of gold bond wire at 10 GHz is calculated from Equation 27 to be about 86.

From the above discussion it is seen that the input and output bond wires for the case of discrete devices introduce lumped inductance elements into the FET circuit topology. Even at 10 GHz, a 25 μm bond wire only 0.25 mm long results in an inductive reactance of 12Ω. Although this reactance could be reduced by using larger diameter wire, ribbon wire, or multiple bond wires, in practice the bond wire inductance is purposely incorporated as part of the input and output impedance matching network. Often the bond wire inductance is used to resonate the input (or output) capacitive reactance and transform the impedance of the FET to a real value. A $\lambda/4$ transformer printed in microstrip can then complete the resistive transformation to 50Ω. Such simple schemes may be used for cases involving impedance matching to relatively small gatewidth FETs (impedance transformation ratio $\leqslant 5$) and for narrow bandwidths ($\leqslant 15\%$). Obviously, more sophisticated networks are required for broadband applications or for impedance matching to large gatewidth FETs such as encountered in power amplifiers. In most cases, however, lumped circuit elements as well as distributed circuits are used in the circuit topology. Packaged FETs offer an advantage in that the chip and the bond wire connections are fixed inside the hermetically sealed package. This allows the manufacturer to provide a more exact RF characterization of the device in terms of the S-parameters with respect to the external package terminals. However, the package introduces additional reactive parasitics, e.g., series inductance and shunt capacitance, which limit the intrinsic broadband capabilities of the device.

Sometimes the package may contain partial or complete impedance matching to reduce or eliminate, respectively, the impedance transformation ratio necessary in the external circuit. Before giving examples, however, a brief discussion of the use of lumped and distributed circuit elements in typical FET applications is presented next.

5.2.2 Lumped Versus Distributed Circuit Elements in FET Impedance Matching Networks

Introduction

It is difficult to generalize whether one should use distributed or lumped circuit elements for implementation in a given impedance matching network, especially in light of the fact that both circuit forms are employed in typical FET amplifiers. In a production environment, for example, the choice of the circuit configuration is made on the basis of the physical reproducibility of the network and the repeatability of the corresponding circuit performance. As discussed in the previous section, lumped inductive elements are naturally introduced into the network via bond wire connections from the FET to the microstrip circuit. The equivalent circuits of packaged devices have a lumped capacitance which is associated with the package parasitics. More interesting is the fact that it is now possible to obtain relatively high Q lumped capacitors which are physically small enough to behave as lumped circuit elements even as high as 20 GHz [204]. Two fabrication types of such microwave chip capacitors are commercially available. One type is manufactured as a silicon MIS (metal-insulator-semiconductor) device. The second type is fabricated from high dielectric constant, low loss ceramic materials such as barium titanate. Quality factors for these capacitors range from 50 to 100 at X-band frequencies, with the ceramic capacitors having somewhat higher values. It is, therefore, quite practical to construct various forms of L-C lumped element matching networks using such capacitors and bond wire inductance. Advantages of lumped element matching networks include circuit compactness and ability to match to low impedance levels. Lumped or semi-lumped circuit elements are also important in monolithic applications because of their space-saving attributes. Circuit reproducibility, however, is somewhat greater in networks which employ mostly distributed elements.

To assess more quantitatively the characteristics of distributed and lumped circuit elements in typical FET microstrip applications, a comparison of some specific examples of impedance matching networks can be made.

Loss Considerations.

For any specific FET circuit design it is important to check the impedance matching networks for losses, especially where low impedance levels are involved, as in power amplifier applications. For these cases, current densities are high and conductor losses can become significant. In general the circuit designer should check each case for losses as well as for the other important considerations and not depend solely on either the lumped or the distributed element approach for the realization of the desired network.

To begin with, one can compare the losses associated with a given matching network in terms of the typical Q values of the individual elements. Figure 27 shows an often used configuration for matching between real impedances R_o and R_{in} using lumped elements. An important feature of this T network is its low pass characteristic which allows dc bias to be transferred through it in the case of active devices. Assume, for example, as in Figure 27, that this network is used to impedance match to the input equivalent circuit of a power FET having an input resistance R_{in}, from a source resistance R_o. Typically, $R_{in} < R_o$, where R_o is 50Ω. Further, let $R_o = nR_{in}$, where n is the impedance transformation ratio. Note that in Figure 27 the gate capacitance of the intrinsic FET is resonated with bond wire inductance to bring the input impedance to a real value, R_{in}, at the center band frequency, f_o. The circuit efficiency for power transfer through the network at f_o can be approximated in a fairly straightforward manner in the case of small losses, i.e., for reasonable values of the element Qs($Q \geqslant 10$). In this case the input and output currents I_1 and I_2 in Figure 27, can be related by the approximation that the input power to the matching network equals the power delivered to the load.

$$P_{in} = |I_1|^2 \cdot R_o \doteq |I_2|^2 \cdot R_{in} = P_{load} \tag{28}$$

Associated with each of the reactances, X_1, X_2, and X_3, is an equivalent dissipative resistive element such that the element quality factors are given by

$$Q_L = \frac{|X_1|}{R_1} = \frac{\omega L_1}{R_1} = \frac{|X_2|}{R_2} = \frac{\omega L_2}{R_2}\,; \quad Q_c = \frac{|X_3|}{R_3} = \frac{1}{\omega C R_3} \tag{29}$$

where the Qs of the inductors are assumed to be the same. The total dissipative loss in the network is given by

$$P_{diss} = |I_1|^2 R_1 + |I_2|^2 R_2 + |I_1 - I_2|^2 R_3 \qquad (30)$$

Normalizing the network's reactances to R_o such that $x_1 = X/R_o$, $x_2 = X_2/R_o$, $x_3 = X_3/R_o$, the general form for the loss in the T impedance matching network may be shown to be

$$P_{diss} = P_{in}\left[\frac{1}{Q_L} + \frac{1}{Q_c}\right] \cdot \left[|x_1| + n|x_2|\right] \equiv P_{in}\left[\frac{1}{Q_L} + \frac{1}{Q_c}\right] B(n, \theta) \qquad (31)$$

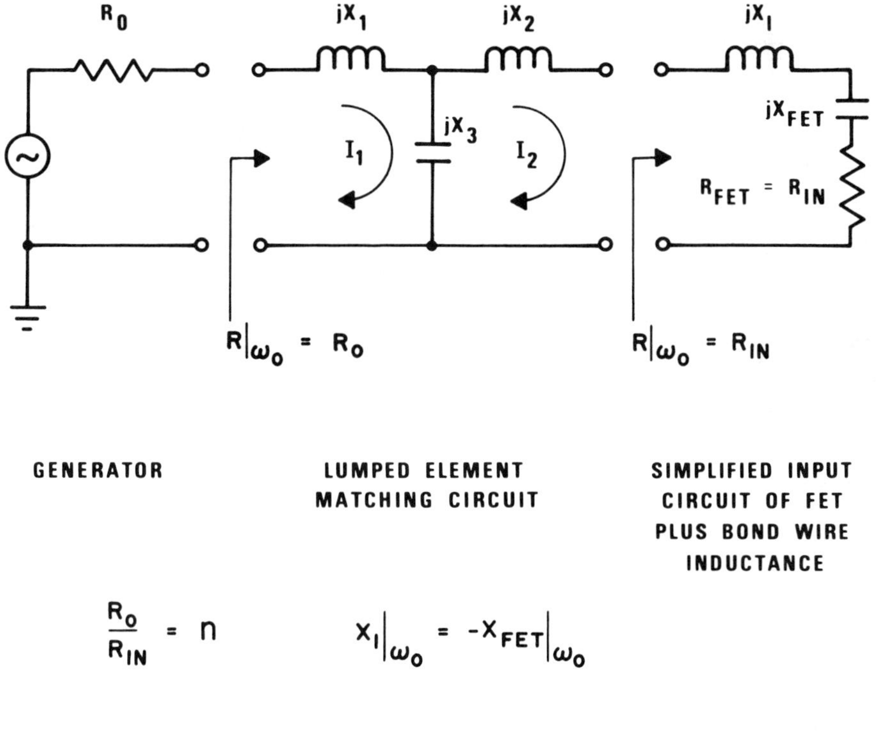

Figure 27/Example of lumped element impedance matching to simplified input equivalent circuit of FET.

where θ is the phase angle between the input and output currents, I_1 and I_2, or simply the transfer phase angle, and $B(n,\theta)$ is a loss factor which depends on the impedance transformation ratio, and the transfer phase angle. It may be shown [59] that the element values are uniquely determined if θ and n are specified, i.e., x_1, x_2, and x_3 are functions of only θ and n and are given by

$$x_1 = \frac{1 - \sqrt{n}\cos\theta}{\sqrt{n}\sin\theta}$$

$$x_2 = \frac{\sqrt{n} - \cos\theta}{n\sin\theta}$$

$$x_3 = \frac{-1}{\sqrt{n}\sin\theta} \tag{32}$$

It follows, therefore, from Equation 31 that for given quality factors, Q_L and Q_c, the network dissipative loss is likewise a function of the impedance transformation ratio, n, and the phase shift parameter, θ. Figure 28 shows the loss factor, B, plotted for values of $1 \leqslant n \leqslant 100$ and $\cos^{-1} n^{-1/2} \leqslant \theta < 180°$. The dashed line shows the limiting case where $x_1 = 0$ and the T network degenerates to an L matching network. Note that minimum loss occurs for the L section and for low transformation ratio. To compare on a direct basis with this circuit's distributed counterpart, let $\theta = 90°$. For this case $x_1 = x_2 = -x_3 = 1/\sqrt{n}$ and the circuit corresponds to the critically coupled symmetric T matching network. From Equations 31 and 32, the dissipative loss becomes

$$P_{diss} = P_{in}\left[\frac{1}{Q_L} + \frac{1}{Q_c}\right]\frac{1+n}{\sqrt{n}} \qquad (\theta = 90°) \tag{33}$$

The circuit efficiency, η_{ckt} is defined in terms of the loss by the equation

$$\eta_{ckt} = 1 - \frac{P_{diss}}{P_{in}} \tag{34}$$

For a typical microwave lumped element matching network using bond wires and chip capacitors, assume $Q_L = 86$, $Q_c = 75$, and $n = 10$. From Equations 33 and 34 the circuit efficiency at center frequency f_o is calculated to be $\eta_{ckt} = 91.3\%$. This corresponds to a 0.39 dB attenuation of the signal as it passes through the network.

Suppose that the matching circuit of Figure 27 were to be realized using a microstrip distributed, single section $\lambda/4$ transformer etched on an alumina substrate. The losses associated with a distributed $\lambda/4$ impedance transformer can be related to the attenuation constant,

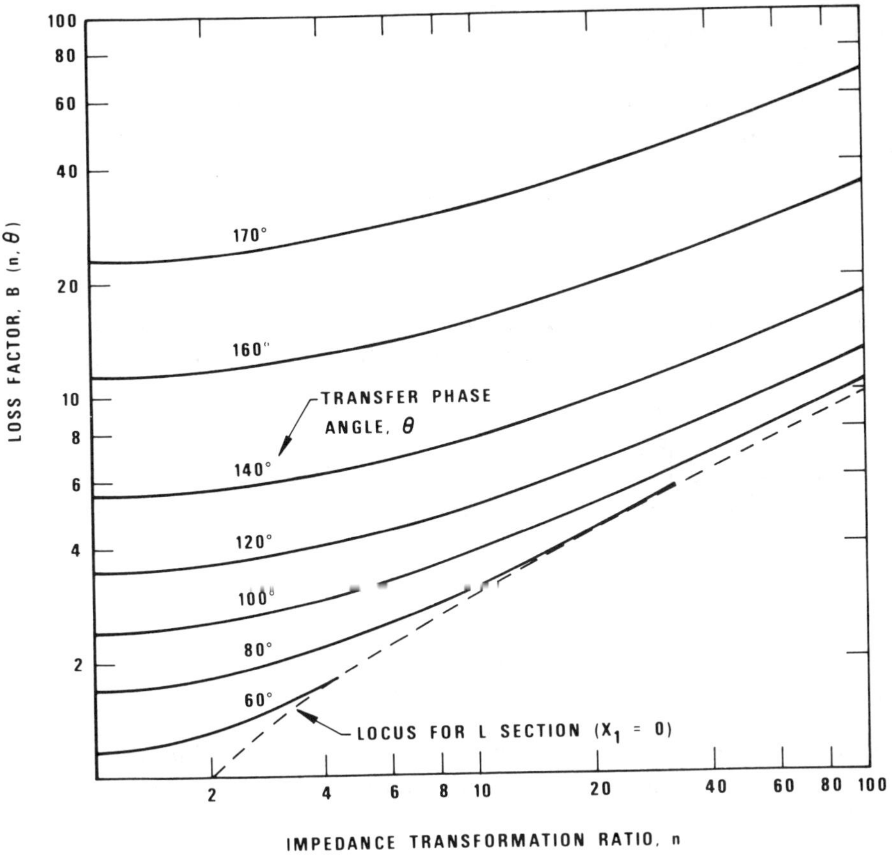

Figure 28/Loss factor, $B(n, \theta)$, for lumped element T matching network.

α, of the transmission line and the voltage standing wave ratio, S, which exists on the quarter-wavelength section. The VSWR is dependent in turn on the impedance transformation ratio, n. Losses on a line supporting a standing wave are greater than on a matched or "flat" line. For the former case, the power loss is the sum of that loss caused by the incident and reflected waves. A new attenuation constant, α_s, can be derived which is the product of the ordinary attenuation constant, α, and a function of the voltage standing wave ratio, S [175]. For a transmission line supporting a standing wave, the transmitted power decrease along the line is therefore expressed by

$$P(l) = P_{in}\, e^{-2\alpha_s l} \tag{35}$$

where P_{in} is the incident power, l the length along the line, and α_s, the attenuation constant for a mismatched line given by [175]

$$\alpha_s = \alpha \frac{S^2 + 1}{2S} \tag{36}$$

For a single section $\lambda/4$ transformer, which transforms from R_{in} to R_o as in Figure 27 (where $R_{in} < R_o$), S on the quarter wave section is

$$S = \frac{R_o}{\sqrt{R_{in} R_o}} = \sqrt{\frac{R_o}{R_{in}}} = \sqrt{n} \tag{37}$$

From Equations 35 and 36 the power dissipated along the quarter-wavelength section is found to be

$$P_{diss} = P_{in} - P = P_{in}(1 - e^{-\alpha_s \cdot \lambda/2}) \cong P_{in}(\alpha_s \cdot \lambda/2) \tag{38}$$

where the approximate equality is made assuming small losses. From Equations 22, 36, 37, and 38 the dissipated loss for the $\lambda/4$ single section transformer is

$$P_{diss} = \left[\frac{\pi}{4Q_u} \cdot \frac{n+1}{\sqrt{n}}\right] P_{in} \tag{39}$$

(A more general form for losses in a multisection transformer can be found in Reference 131, Section 6.14.) Note the similarity of Equation 39 to the expression for the losses in the lumped equivalent of a $\lambda/4$ transformer given in Equation 33. Assume the same impedance transformation ratio, n = 10, is required and that the circuit is to be fabricated on alumina. The required characteristic impedance of the quarter wavelength section is just $\sqrt{50 \times 5} = 15.8\Omega$. From Table 6 one can see that a 0.25 mm or thinner substrate is required for 10 GHz operation to ensure that the width of the quarter wavelength section be less than 0.8 $\lambda q/4$. The unloaded Q for such a line on a 0.25 mm polished alumina can be estimated to be about 190. (From Table 6 one can scale the Q = 300 value for a 50Ω line on a 0.64 mm alumina substrate to a 15.8Ω line on a 0.25 mm substrate using the equations in Reference 187). From Equations 39 and 34 the circuit efficiency is calculated to be 98.6% which corresponds to only 0.06 dB attenuation of the signal passing through the transformer section.

Although the distributed circuit gives less loss for this particular case than the lumped equivalent (for the assumed element Q values) it is seen that because of the low impedance required, the realization of the distributed circuit results in an etched circuit pattern whose dimensions approach the limiting values appropriate for the given frequency of operation as indicated in Table 6. If either a higher frequency or a greater transformation ratio were required, the lumped element network, or another form of a distributed element circuit would be required. A more practical solution may dictate the use of both types of transforming circuits, dividing the total transformation ratio between the lumped and distributed networks. This is typically done in power FET amplifiers where the impedance transformation ratio may be 50:1 or greater.

Although the above specific example was applied to the input equivalent circuit of the FET, the same arguments would apply to the output circuit at a single frequency, f_o, if the T-matching network were used to match to the output FET resistance whose capacitive reactance was resonated with bond wire inductance at f_o. At other frequencies, a computer calculation of the losses for the specific matching network and equivalent load configuration may be necessary.

Generally, losses become significant whenever impedance levels are low and comparable to the ac resistance of the matching network elements.

The above discussion has dealt with a very specific but often encountered impedance matching network. Reference 131 provides formulas for dissipative loss in more general cases involving broadband impedance matching circuits or filters which consist of a number of cascaded resonant sections (e.g., low pass prototype circuits with Chebyshev or maximally flat characteristics). For a specific impedance matching network, with many reactive lossy elements, a computer analysis may be practical to estimate the matching circuit efficiency as a function of frequency given the Q values for the individual elements. This is especially desirable in broadband circuits since the Qs themselves are functions of frequency. In a paper on broadband matching to microwave power GaAs MESFETs, Honjo et al. [80] analyze the loss in several impedance matching networks which are used to match from 50Ω to as low as 1Ω over fractional bandwidths (BW) up to 60%. The circuits analyzed are lumped-element low-pass-type Chebyshev networks with pure resistive loads. All have the same transformation ratio but a different number of reactive elements, N. An interesting result is that the losses are nearly independent of N for N ranging from 2 (BW = 20%) to 6 (BW = 60%).

As a final comment on circuit losses, it is worthwhile to consider the dependence of the element Q values on frequency. For inductors the Q is given by $\omega L/R_{ac}$. R_{ac} is proportional to the square root of frequency. Consequently the inductor Q increases as the square root of frequency. For microwave capacitors made of high quality dielectric materials, the dominant losses occur in the metal plates through which the displacement current must pass. The equivalent resistive element is in series with the capacitance so that the Q is given by $(\omega C R_{ac})^{-1}$. Since R_{ac} increases as the square root of frequency as before, the Q of the capacitor decreases as $\omega^{-3/2}$. At high frequencies low circuit losses in lumped capacitors become more difficult to achieve. In GaAs monolithic circuits, overlay MIM (metal-insulator-metal) or interdigital capacitors fabricated on the GaAs chip are employed. Preliminary reports from several laboratories on the losses associated with monolithic overlay capacitors generally indicate lower Qs ($\leqslant 50$) than would be expected from metal losses only. For these cases, losses in the dielectric (or in the GaAs substrate) may be significant also [32]. Finally, for microstrip distributed circuits where metallization losses are dominant, the unloaded Q factor for a transmission line is expressed as $Q_u = \beta/2\alpha \cong \omega Z_o/v_{ph}R$ where R is the resistance per unit length, Z_o the characteristic impedance and v_{ph} the guide phase velocity. For a given microstrip line Z_o and v_{ph} are constant with frequency while R increases as the square root of frequency. Consequently, as in the case of the inductor, the Q of the microstrip transmission line increases as the square root of frequency.

Frequency Selectivity, Bandwidth Considerations.

In the design of impedance matching networks for FETs the specific choice of circuit configuration may be dictated by bandwidth requirements. General theoretical bandwidth limitations and circuit design techniques, applicable to FET linear amplifiers were discussed in Section 1. It is also useful to consider some specific bandwidth limiting factors from a more practical circuit design point of view. This section briefly discusses some of these factors which affect the frequency selectivity of typical FET circuits including the physical realizability of the circuit elements.

For a given design situation, the choice of circuit configuration must be made on the basis of good engineering judgment, taking into consideration those factors which are most relevant to the design objectives. Although the statement is obvious, its practice is generally not.

For example, to characterize a new FET it may be necessary to evaluate directly (i.e., not through a set of S-parameters) the maximum available gain (MAG) of the device. In this case relatively narrow-band impedance matching circuits are sufficient, provided they are flexible enough to present a wide range of impedances to the input and output terminals of the device and can be conveniently adjusted at several different frequencies. A fixed microstrip circuit alone is not adequate unless it is already optimized for the given FET, in which case it is likely the measurement would not be required. Frequently a microstrip-to-coax transition or a microstrip to-coax-to-waveguide transition and a coax or slide-screw tuner respectively, are used to adjust the impedance levels presented to the FET. Even in this case, the microstrip and the bond-wire inductance are usually approximately correct for the given FET type and the tuners used for fine adjustment only. In addition, metallized ceramic tuning chips placed on the microstrip circuit at appropriate locations (found by experiment) can also be used to optimize the impedance for achieving MAG. Since tuning is done several wave-lengths away from the device, especially in the case of the waveguide slide-screw tuners, the reactance slope presented to the device terminals is high (i.e., reactance is a rapidly changing function of frequency) and the resultant frequency sensitivity of the amplifier response is also high (3 dB fractional BW $< 10\%$). Nevertheless, the circuit configuration is well adapted to the spot measurement of MAG and evaluation of the FET.

In contrast, the design of a compact, broadband, and reproducible FET amplifier requires impedance matching in close proximity of the FET device and a circuit configuration which is easily reproduced and realizes with sufficient accuracy the required circuit element values. Microstrip distributed or lumped circuit elements, or both, can be used, the choice often being made on the basis of physical realizability of the component values. Initial broadband impedance matching designs are often developed on the basis of lumped circuit elements simply because most of the available tabulated data [130, 131, 40] assumes lumped element networks. Conversion of the reactive lossless lumped elements to the approximate equivalent distributed elements can then be made using essentially a frequency change of variable [179]. Lumped inductors are replaced by shorted lengths of line ($<\lambda/4$) and capacitors are replaced by open lengths of line ($<\lambda/4$). (These techniques are discussed thoroughly in the literature [131] on filter and impedance matching design and are not repeated here. A good summary of basic microwave filter design can be found in Reference 70, for example.) A distributed circuit will depart significantly from its approximate lumped equivalent if the frequency is high enough, and generally will have a frequency repetitive pass-band characteristic due to the cyclic nature of distributed impedance functions. However, if the range of realizable distributed circuit element values and corresponding tolerances are sufficient, the first distributed pass-band can be made as large as the fundamental lumped element pass-band, and vice versa. From a more practical viewpoint a broadband FET amplifier network can be optimized using CAD techniques on a distributed circuit basis as well as on a lumped element basis [21, 105, 245].

In FET power amplifiers chip level power combining necessitates the use of impedance matching networks with large transformation ratios, n. Generally these networks become more frequency selective with increasing n. A specific example can illustrate this point. For the simple case of conjugate impedance matching to the input equivalent circuit of an FET using a lumped element T-matching network, as shown in Figure 29a, the relative frequency selectivity of the network can be inferred from an evaluation of the external quality factor, Q_{ext}. Q_{ext} is directly proportional to the reactance slope [146] as indicated in Figure 29a. As shown in the figure the circuit is viewed from the generator end and the calculations made with respect to the terminal pair which connects to the generator and source impedance. Again it is assumed that the capacitive reactance of the FET is series resonated at mid-band by the bond wire inductance and that the symmetric lumped element T-matching network is used in Figure 29a so as to be able to compare it directly with its distributed equivalent shown in Figure 29b. For these calculations the element values of the matching networks are chosen so as to achieve a perfect match at the center frequency, $f_o = 2\pi\omega_o$, $Z(fo) = R_o$, the source impedance. Q_{ext} is evaluated to be

$$Q_{ext} = \frac{|X_{FET}|}{R_{FET}} + \sqrt{n}\left[1 - \frac{1}{n}\right] \quad \text{(lumped elements)} \tag{40}$$

where n is given by $n = R_o/R_{FET} \geqslant 1$. A similar calculation made for the distributed case shown in Figure 29b where the T-network is replaced by a $\lambda/4$ section of transmission line, yields the result

$$Q_{ext} = \frac{|X_{FET}|}{R_{FET}} + \frac{\pi}{4}\sqrt{n}\left[1 - \frac{1}{n}\right] \quad \text{(distributed)} \tag{41}$$

From Equations 40 and 41 it is seen that both the lumped element matching network and the distributed network have comparable Q_{ext}'s. As the transformation ratio, n, increases, Q_{ext} does also and the networks become more frequency selective. Note that the expressions for the external quality factor consist of the Q of the FET input equivalent circuit, $Q_{FET} = |X_{FET}|/R_{FET} = (\omega C_{FET} R_{FET})^{-1}$ plus the contribution to the Q from the matching network. Obviously, Q_{ext} would have additional terms and the network have greater fre-

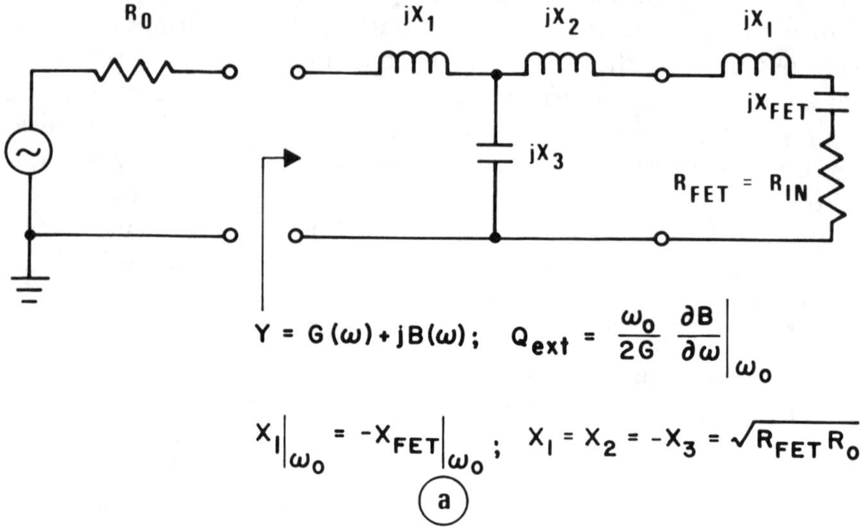

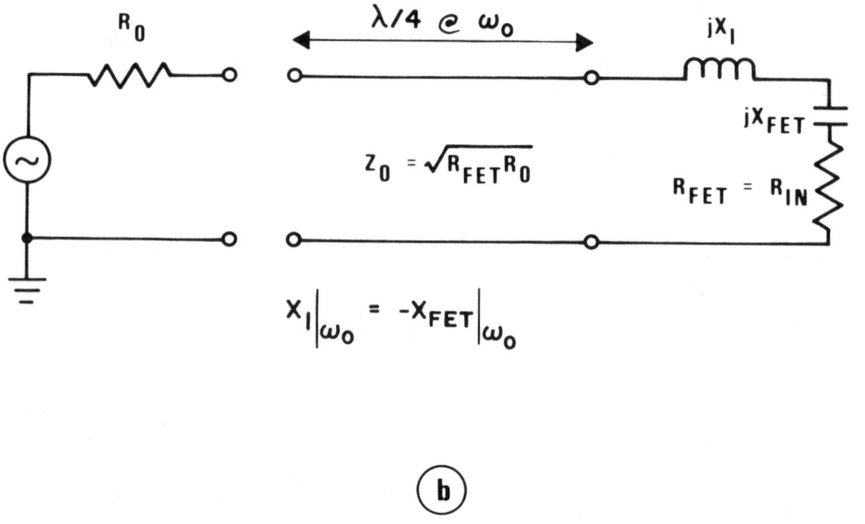

Figure 29/*Comparison of Q_{ext} for two impedance matching networks.*

a) Using lumped circuit elements;

b) Using lumped and distributed circuit elements

quency sensitivity if some or all of the impedance matching were done remotely through additional lengths of transmission line. For large gatewidth power FETs, Q_{FET} also tends to increase because of additional reactive parasitics. In terms of broadbanding capability the motivation for in-package tuning becomes apparent.

Although many other "paper" design and circuit configurations are possible for impedance matching to large gate-width FETs, the primary limitations for broadbanding the actual physical circuit arise from the difficulty of realizing reproducibly the necessary circuit elements. If further increase in power output is required and the low impedance limit of chip level combining is reached, circuit level power combining is employed. Commonly used microstrip two-way power combiners are in-phase (Wilkinson power splitter [244]) and quadrature (Lange 3 dB coupler [116]) combiners. These can be cascaded into a corporate structure to achieve 2^k power division, but losses limit these approaches to $k \leqslant 3$ for frequencies in the range of 5-20 GHz. Both types of combiners can be realized to have usable bandwidths approaching 40% at X-band frequencies. Recent developments in direct 1:m (m = 12) power division have also been obtained [185] applicable for frequencies at least up to 10 GHz. Circuit level power combining is a subject in itself, however, and will not be covered here. A good review of these techniques can be found in Reference 181.

Physical realizability of distributed or lumped circuit elements depends in part on the particular fabrication technology being used for the amplifier construction and its compatibility with the frequency of operation. Typical broadbanding limitations which are associated with the physical realization of circuits at microwave frequencies include the following:

- Achievable tolerance of the physical circuit element values limit the use of complex matching networks involving a large number of reactive circuit elements. For lumped elements, typical commercially available chip capacitors may have tolerances no better than ~ 10%. Similarly, bond wire inductance is difficult to control to better than ~ 10%. Furthermore, a lower limit exists on the values achievable for lumped inductances and capacitances because of parasitics. Inductances of 0.2 nH and capacitances of ~ 0.1 pF are approximate lower limits on the lumped element values which can be achieved reproducibly using conventional MIC hybrid technology such as implemented at X and Ku-band frequencies. As discussed earlier in this section, limits on the range of realizable characteristic impedances of distributed circuits also prevail, as shown in Table 6. For thin substrates like those used in GaAs monolithic circuits, high characteristic impedances are difficult to achieve and are limited to less than $\sim 100\Omega$. These design constraints restrict to about four or five the practical number of reactive circuit elements that can be used for a specific impedance matching network. More elements generally will not improve bandwidth performance.

- Packaged devices have reactances which are fixed and often cannot be incorporated into optimized impedance matching networks. The impedance matching must be done external to the package terminals, and the packaged device Q is higher than the intrinsic Q of the FET equivalent circuit.
- Parasitic reactances such as stray capacitances and inductances store electric and magnetic energy. As a consequence circuit Qs are higher and fractional bandwidths reduced. Miniaturization of lumped circuit components and the use of well controlled and repeatable distributed circuit elements reduce reactive parasitics and permit predictable broadband circuit design at higher frequencies. A distinct advantage and prime motivation for monolithic integration of microwave circuits is its potential for circuit reproducibility and the control of parasitics.

Technologies which improve the physical realizability and reproducibility of distributed and lumped circuit elements will permit the realization of broadband microwave circuits at higher frequencies and at lower cost. Although monolithic integration of microwave circuits on GaAs has its own design constraints [171], this technology is just beginning to emerge and holds great promise for future microwave circuit development. If fabrication processes are sufficiently improved then this technology has the highest potential for reproducibly fabricating precision microwave circuits with superior performance at lower cost.

5.3 Examples of FET Amplifier Technology

We present next some representative examples of FET amplifier technology taken out of the technical literature. Because of the profuseness of available literature on this subject only a small sample can be included here. For further reading, the reader is encouraged to refer to the numerous fine technical articles included in the bibliography. The examples begin by illustrating conventional MIC technology using discrete FETs and progress towards higher levels of integration, and end with some recent developments in GaAs monolithic FET circuits.

5.3.1 Amplifiers using Chip or Packaged FETs

In 1974 Liechti [123] gave a systematic design approach to the development of an X-band amplifier which has high gain, low noise, and moderate bandwidth. The amplifier described achieves 20 ± 1.3 dB gain with a 5.5 dB typical noise figure (6.9 dB maximum) over the frequency range of 8-12 GHz. The minimum output power for 1 dB gain compression is 13 dBm. Three identical gain stages using 1.0 μm gate-length, 500 μm gate-width FETs are cascaded in a single ended configuration. The input coupling network for each stage provides a noise match (rather than a conjugate impedance match) over the 8-12 GHz band using a Chebyshev band-pass impedance matching network to match the generator impedance for minimum noise. The output coupling network provides a conjugate impedance match at the upper edge of the band and uses a frequency selective dissipative network to compensate for the increased FET gain at lower frequencies. The dissipative circuit consists of a short circuited stub in series with an 85Ω resistor. This combination is placed in shunt at the output circuit. The stub is $\lambda/4$ long at the upper band edge and therefore the resistor dissipates no power. At lower frequencies the dissipation increases at the same rate as the FET gain ($\sim$ 4.7 dB under conditions of input noise match) thus yielding a flat gain response across the 8-12 GHz frequency range. This output coupling network maintains a reasonable output VSWR ($<$ 2.0:1) over the frequency band of interest and ensures a well-behaved and predictable multistage amplifier when identical gain stages are cascaded. The circuits are realized on 0.63 mm sapphire substrates which are soldered onto a ground plate and are separated by a ridge on which the transistor chip is mounted. Although the amplifier performance results are somewhat outdated by recent progress in FET device technology, this early article covers the basic design considerations and approaches relevant to FET amplifier development.

FET amplifiers have important applications in the telecommunications area; they are replacing older tube triode versions in typical microwave telephone equipment. Schroeder and Gewartowski [188] describe a 2 W, 4 GHz GaAs FET transmitter amplifier suitable for radio relay systems (TD-2 and TD-3). The amplifier produces a minimum of 21 dB of gain using three stages, with a single packaged GaAs FET in each stage. The three FETs used have gate-widths of 1, 3, and 6 mm for the first, second, and third stages, respectively. The 6 mm gate-width device has a nominal output power of 3 W with 7 dB gain at 4 GHz. For the specific application, a linearity requirement consistent with digital transmission systems is imposed on

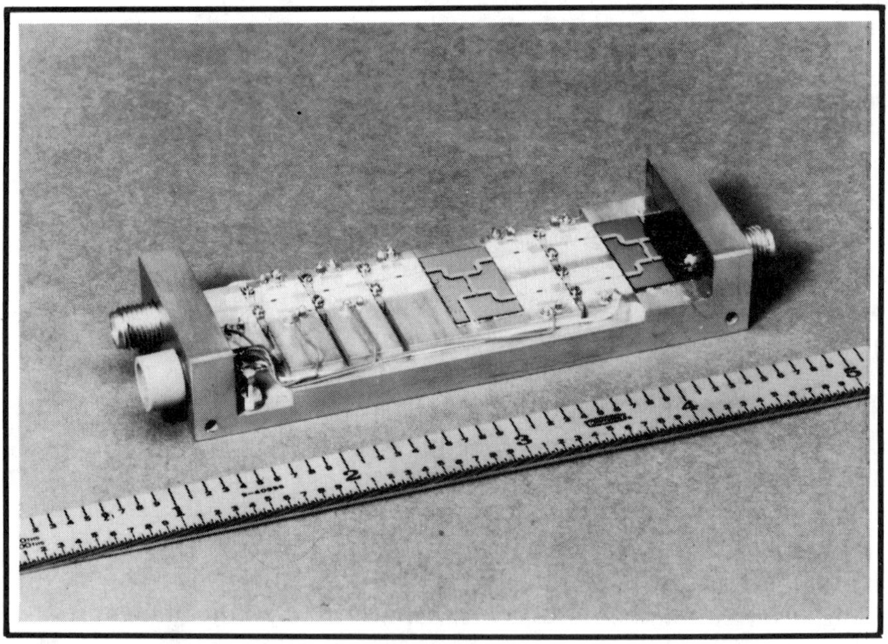

Figure 30/A 2-Watt, 30 dB gain, FET amplifier module suitable for X-band phased array applications.

the design. Measurements of the contours of constant gain and C/I in the load impedance plane made on the third stage transistor reveal the compromise that must be made between maximum output power and low C/I (the ratio of the carrier to third-order intermodulation products). The optimized load impedance chosen allows the 6 mm gatewidth transistor to achieve a minimum C/I ratio of 20 dB and a minimum gain of 6.5 dB over the 3.7 to 4.2 GHz frequency range. The impedance matching networks are fabricated in microstrip. Several different types of matching networks were used with satisfactory results: cascaded quarter-wavelength transformers, shunt open-circuited stubs and series interdigital capacitors. All of these networks are easily realized in microstrip. Copper carrier plates provide the mounting surface for the hermetically sealed packaged FETs as well as the alumina substrates which are soldered to the carrier plates and contain the microstrip input and output maching networks.

FET microwave amplifiers are especially suited for airborne or space applications because of their small size and weight and proven reliability. An important application is in the area of airborne phased array radars where a large number of identical power amplifier modules are required. Tserng [222, 224] describes a 2 W, 30 dB gain amplifier module operating over the 9-10 GHz frequency range with a power added efficiency of 25%. Figure 5.9 shows the four stage amplifier with the cover removed. The module housing accommodates a three stage, single-ended driver amplifier plus a balanced output stage. The three stage driver amplifier uses two 300 μm gate-width FETs in the first and second stages and a 1.2 mm gatewidth FET in the third stage. The amplifier provides 24 dB gain at a 0.5 W power output level. The fourth balanced stage uses two 2.4 mm gate-width FETs which are circuit power combined by means of a pair of 3 dB interdigital (Lange) quadrature couplers [116]. This amplifier stage yields 2 W output power with 6 dB gain. All of the FETs are in chip form and impedance matching is accomplished using bond-wire inductance and distributed microstrip circuits fabricated on 0.25 mm thick alumina. For lower loss, the quadrature couplers are fabricated on 0.38 mm thick fused quartz substrates. The interdigital structure of these couplers uses 0.08 mm wide strips and 0.015 mm wide gaps. Approximate loss for such a single 3 dB microstrip coupler is about

TWO STAGE, BALANCED GaAs MONOLITHIC POWER AMPLIFIER

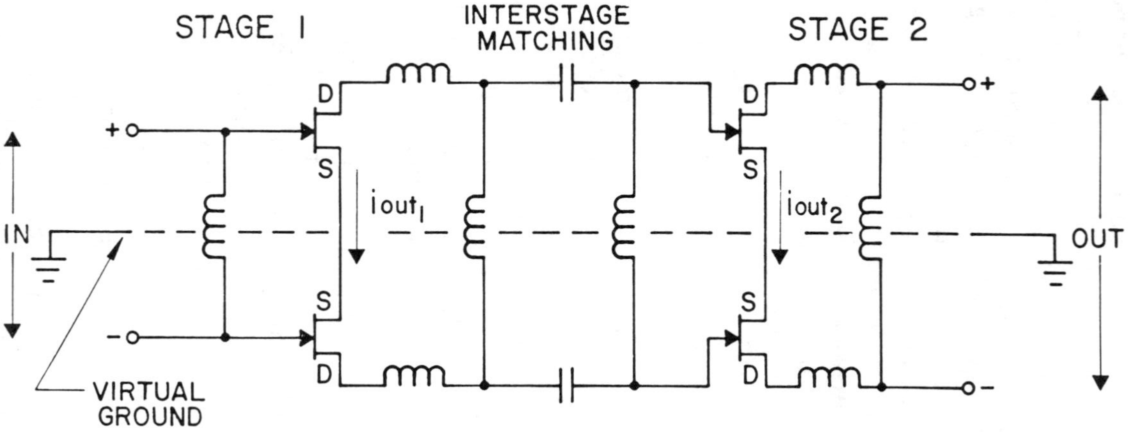

L_{Series} (nH)			0.8 (each)					0.3 (each)	
L_{Shunt} (nH)	0.9			1.0		0.6			1.5
C (pF)					1.0 (each)				
Total Gatewidth FET (mm)		0.6 (each)					1.2 (each)		

Figure 31**/A monolithic two-stage push-pull amplifier for X-band operation* ***a) *Schematic Diagram.*

0.15 dB to 0.2 dB at X-band (8-12 GHz) frequencies using chrome-gold (200 Å, 2.5 μm) metallization. All of the FETs in the amplifier operate at the same drain and gate voltages (~ 8.5 V and – 1.5 V respectively). This facilitates pulsed operation wherein the gate voltage is driven from pinch-off (~ – 6 V) to the nominal operating voltage, with the drain voltage left constant. For dc blocking between stages, 47 pF beam lead capacitors are used. Another amplifier using the same configuration but larger gatewidth FETs (300 μm, 600 μm, 2.4 mm for the driver amplifier stages and a pair of 4.8 mm FETs in the balanced stage) has achieved 5 W of output power with 30 dB gain at 9.5 GHz.

FET amplifier technology continues to progress towards higher power, higher frequency, and greater bandwidths. The following two examples are representative of recent developments in these specific performance objectives. Sone et al. [203, 204], have shown the feasibility of medium gain, high power amplification at Ku-band (12-18 GHz) frequencies. They developed a three stage single-ended 14-14.5 GHz 1W amplifier with 8 dB gain having a power-added efficiency of 14.6%. The small signal gain is 13 dB. Two 1.5 mm gate-width FETs (NE 8692) are used for the first two stages and a 3.0 mm gate-width FET (NE 8694) is used for the last stage. Impedance matching circuits are fabricated on 0.64 mm thick alumina substrates. For the design of the input matching circuit small signal S-parameters were used. An open-circuited stub followed by a quarter-wave impedance transformer matches the large inductive reactance and low resistance of the input to the 1.5 mm gatewidth FET(Z_{in} = 3.6 + j18 Ω). The output impedance is determined under large signal conditions and the corresponding distributed matching circuit is designed on the basis of the large signal measurements. Linearity characteristics of the amplifier are presented in terms of the two-tone third-order intermodulation tests and the one-tone AM to PM conversion measurements. At 400 mW output (G = 11 dB) the third order products are 17 dB below the carrier while at 100 mW (G = 13 dB) the C/I ratio is 33 dB. AM to PM conversion is below 2 degrees/dB for output powers up to 1W.

Very broadband FET amplifiers covering UHF through Ku-band frequencies have also been developed. Such amplifiers are important for electronic countermeasure as well as instrumentation applications.

Niclas et al. [155, 156] describe a 350 MHz to 14 GHz 3 dB gain single-stage amplifier which makes use of both negative and positive feedback. A series connection of an inductor, L_{FB}, and a resistor, R_{FB} are used as the frequency selective feedback circuit. This feedback circuit connects to the output node (which ties to the FET drain terminal via a series inductor L_D) and the gate terminal. The inductor, L_D, is chosen to compensate for the capacitive portion of the transistor's output impedance at the upper band edge. L_{FB} reduces the effectiveness of the negative feedback with increasing frequency. Below 4 GHz, L_D and L_{FB} have little effect on the frequency response and the amplifier stage behaves as a conventional feedback amplifier. As the frequency increases above 4 GHz the phase of the feedback voltage changes such that positive feedback is realized at the upper end of the band (above 7.2 GHz). One of the advantages of this approach to achieving very broadband amplifiers over usual techniques where reactive mismatch is used at the lower frequencies to maintain a constant gain, is that low input and output VSWR's can be maintained throughout the full amplifier bandwidth. For the amplifier described, a maximum input voltage reflection coefficient of $|S_{11}| = 0.45$ and a maximum output voltage reflection coefficient of $|S_{22}| = 0.54$ is achieved over the entire band of 350 MHz to 14 GHz. It is interesting to note that the measured small signal maximum available gain of the FET used in the amplifier is about 23 dB at 1 GHz and 7.6 dB at 14 GHz. For amplifier fabrication, fused quartz, 0.38 mm thick, is used as the substrate material for the input and output impedance matching circuits. The circuit pattern is gold while the feedback resistor, $R_{FB} = 160\ \Omega$, is realized by a thin film of tantalum nitride evaporated on the quartz substrate. The FET is an 800 μm gate-width, and 1 μm gate-length device optimized for broadband operation (with special emphasis paid to reduction of parasitics).

An important application for FET amplifiers in Europe and Japan will be their use in home TV satellite broadcast systems which are now under development. The main FET amplifier application will be low noise microwave receivers for the individual home earth stations operating in the upper X-band frequency range. However, because of the potential reliability and performance of power FETs it is also feasible that FET power amplifiers will replace the TWT transmitter amplifiers in the broadcasting satellite. Such a power FET amplifier for satellite TV transmitter application is considered in a paper by Nakatani et al. [147]. The laboratory model is a five stage, 24 dB gain, 1W amplifier operating at 12 GHz. For reliability and reproducibility, packaged FETs are incorporated in unit amplifiers which are constructed on a single gold-plated brass carrier plate. The carrier plate supports the packaged FET and the input and output microstrip impedance matching circuits which are fabricated on 0.64 mm thick alumina substrates. The amplifier assembly is composed of five such unit amplifiers which are mounted and inter-connected in a single amplifier housing. The first stage or unit amplifier, is a balanced module using a pair of small-signal, 400 μm gate-width FETs combined in quadrature by a 3 dB interdigital coupler. The first stage provides low noise preamplification. The second stage is a single-ended driver composed of a medium power ($\sim$ 200 mW) 800 μm gate-width FET. The third and fourth unit amplifiers use the 800 μm gate-width FETs in a balanced configuration for medium power amplification (1 dB gain compression point is 400 mW, 6.0 dB gain at 18% power-added efficiency). The last unit amplifier is a single-ended stage employing a 2.4 mm gate-width FET which provides 1 W at 3 dB gain. For the total five-stage amplifier, a linear gain of 27 dB and a 1 dB gain compression point of 0.8 W is achieved. The saturated output power is 1 W with an associated gain of 24 dB and a 1 dB bandwidth of 350 MHz. Power added efficiency is 7.0%. Linearity characteristics include an AM to PM conversion of 2.5°/dB and a C/I ratio of 16 dB for third order intermodulation distortion measured at the 0.8 W power output level. At 100 mW output the C/I ratio is 44 dB. Using the same amplifier housing but replacing the last two unit amplifiers with improved high power single-ended packaged flip-chip MESFETs (2.4 mm and 4.8 mm gate-width FETs respectively) a saturated output power of 2 W with an associated gain of 23 dB and a power added efficiency of 16% is obtained for the five stage transmitter amplifier.

5.3.2 Amplifiers Using Internally Matched Packaged FETs

Honjo et al. [80] present design considerations for the development of C-band (4-8 GHz) single stage amplifiers using high power FETs incorporating in-package impedance matching networks. Such a package amplifier using an 11.2 mm gate-width FET has a 2.5 W power output at 1-dB gain compression and a small signal gain of 5.5 ± 1.5 dB from 4.2 to 7.2 GHz without the use of external impedance matching. The package dimensions are about 9.0 x 8.8 mm. Inside the package are found the FET chip, a high dielectric ceramic substrate for the input impedance matching circuit, and an alumina substrate for the output impedance matching circuit. Both substrates and the chip are mounted to the carrier plate of the package. Input matching networks of the lumped-element two-section low-pass type are employed. The input circuit design is based on the small signal S-parameters. Bond wire inductance and metal-insulator-metal (MIM) capacitors formed on a 0.1 mm thick barium titanate substrate (ϵr = 39.5) realize the two section input matching network. All of the MIM capacitors are fabricated on a single substrate. Distances between the capacitors are determined according to the bonding wire length required to realize the necessary inductances. Capacitor values of up to 8.6 pF and inductor values of up to 0.48 nH were realized. For the output matching network large signal characterization using load-pull measurements and empirical optimization were used for the design. The output circuit employs a one-section semi-distributed network designed in microstrip and fabricated on a 1 mm thick alumina substrate. Bond wire inductance and shunt open microstrip stubs are used. The internally matched 11.2 mm gate-width FET develops 4.4 W saturated power output at 6.25 GHz with 20% power added efficiency and 4 dB gain. Using basically the same approach as above, recent results from the same laboratory (Higashisaka, et al. [74]) have demonstrated a 6 GHz, 25 W packaged FET which used 4 chips in parallel for a total gatewidth of 60 mm. The internally matched four-chip device gave output powers of 23 W (4 dB gain) and 25 W (3 dB gain) with a power-added efficiency of 23% at 6 GHz.

Yamamura et al. [247] report the development of a 4-8 GHz single-stage GaAs FET low-power balanced amplifier constructed on a 4 x 5 mm sapphire substrate which is mounted onto a metal carrier plate. This unit amplifier employs a pair of 300 μm gate-width FETs in a balanced configuration. The advantage of the balanced approach is that good input and output VSWR can be maintained for the unit amplifiers. This enables one to cascade the amplifiers for higher gain, with minimum overall amplifier gain ripple. The 3 dB coupler is of the interdigital quadrature type and is fabricated on the sapphire substrate. For miniaturization the matching networks are realized with lumped elements and are of the L-C-L, T-network design. The capacitors are formed on the sapphire substrate and are of the MIM type where the insulator is a 0.5 μm silicon dioxide film deposited by chemical vapor deposition of SiH_4 and O_2. The Q values of the capacitors were measured to be above 80 up to 8 GHz and the capacitance values used in the T-networks are about 0.5 pF and 0.3 pF. For the inductances 30 μm track-width two-turn and three-turn spiral inductors etched on the sapphire substrate are employed. Inductances values of up to 3 nH are used in the T-networks and a three-turn 8 nH spiral inductor is used in the bias circuit. Three such unit amplifiers were cascaded to form a 21.5 ± 1 dB gain amplifier from 4.5 to 8 GHz. The authors of this paper argue that, from the standpoint of cost versus performance, this approach to miniaturization and reducing cost of GaAs FET microwave amplifiers is superior to that of monolithic integrated circuits using GaAs substrates.

5.3.3 Monolithic GaAs Microwave Amplifiers

Although GaAs has been recognized for over a decade [170, 133] as a suitable material for the integration of active and passive microwave devices, only recently, with the development of the FET, a renewed interest in monolithic GaAs FET analog circuits has emerged in several major laboratories all over the world [1, 172, 164, 165, 166, 199, 200]. The ability to fabricate microwave circuits on single chips has potentially significant cost and performance advantages for many applications where a large number of identical circuits are required including, for example, phased array radars.

In contrast to the more conventional FET microwave amplifier technology described previously, GaAs monolithic FET integrated circuits are at an early stage of development. As a consequence, the examples given below are early laboratory results which demonstrate the feasibility of this approach to the development of both small signal and power FET amplifiers.

Pengelly and Turner [165] describe a small signal, signal-stage 7.0-11.7 GHz, 4.5 ± 0.9 dB gain, monolithic amplifier fabricated on a single GaAs chip whose dimensions are 1.2 x 1.8 x ~ 0.18 mm. A lumped element input matching network that includes loop inductors and interdigital capacitors is fabricated on high-resistivity gallium arsenide. The transistor is fabricated on the same chip but is defined within an etched mesa region containing the epitaxially grown active channel and buffer layers. Shunt circuit components and the source terminal of the common source FET are connected to ground metallizations on the edges of the chip. Mesh gold wire is then used to connect this ground metallization to the grounded package header. All chip metallization is defined photolithographically including the 1 μm-long transistor gate, and the pattern is formed using metal lift-off. Because of the metal lift-off technique, the metallization thickness is limited to about 1.15 μm (note that at 10 GHz the skin depth in gold is on the order of 0.8 μm). Track widths for the interdigital capacitors and the loop inductors are 30 μm. The largest monolithic inductor value is 0.7 nH, while the largest interdigital capacitor value is 0.25 pF. The output impedance matching is accomplished using a single bond wire which connects to the external alumina circuit. This allows some tuning flexibility and adjustment of the center-frequency. The 1 dB gain compression point occurs at 13 mW. In contrast to a similar amplifier using semi-lumped elements and more conventional MIC technology, the monolithic amplifier has about 0.6 dB less gain, which is attributed to the losses in the monolithically fabricated input matching network. More recent results from the same laboratory have shown the feasibility of single stage X and Ku-band broadband, low noise monolithic amplifiers [166]. A typical J-band (10-20 GHz) single-stage amplifier consists of a single FET and input and output impedance matching networks fabricated on a chip whose plan dimensions are 1.2 x 1.8 mm. The impedance matching networks are designed based on S-parameter measurements made on discrete FETs of the same type as those used in the monolithic amplifiers and on extensive characterization of lumped passive elements fabricated on alumina and GaAs [167]. Loop inductors and interdigital capacitors fabricated on the GaAs chip are used for the passive lumped elements. Chips are mounted onto nickel/gold-plated copper package headers for easy installation into microstrip circuits and for cascading. A single monolithic amplifier on a chip provides about 6 dB gain from 8-18 GHz. Two X-band monolithic amplifiers cascaded in a single package and biased for best noise figure achieve about 10 dB gain from 8-12.4 GHz with 5 dB noise figure. Seven such amplifier chips cascaded in a single-ended configuration achieve 28 ± 1 dB gain from 8-12 GHz. Small lengths of microstripline are placed between stages and a small amount of conventional tuning is used to minimize gain ripple in the seven-stage amplifier.

The design of GaAs monolithic power amplifiers requires some special design considerations [171, 201]. Because large gate-widths are employed, heat sinking of the FETs becomes critical and thinner substrates must be used. Good RF grounding is also a critical requirement since impedance levels are low and ground currents high. At the time of this writing output powers of up to 2 W for single-stage and 1.5 W for two-stage monolithic amplifiers operating in X-band have been obtained.

Sokolov, Williams, and Shaw [199] describe a single-stage and a two-stage monolithic power amplifier fabricated on a 1 x 2 x 0.1 mm and 2 x 2 x 0.1 mm GaAs chip respectively. Both circuits operate as push-pull amplifiers using a pair of transistors for each stage. The two-stage chip uses a pair of 600 μm gate-width FETs and a pair of 1.2 mm gate-width FETs for the first and second stages respectively. To test the amplifier external baluns are needed at input and output; 180° hybrid rings fabricated on alumina are used for this purpose. These circuits transform a signal on a single 50 Ω unbalanced microstrip line to two anti-phase signals on a pair of 50 Ω coplanar transmission lines, and vice versa.

For the two-stage amplifier a small signal gain of 13 dB and a 1 dB gain compression point of 1.1 W at 9.5 GHz is achieved. The saturated power is 1.25 W with 10 dB gain. In a more recent article [200] Sokolov and Williams describe in detail the push-pull power amplifier and also present preliminary results on a monolithic paraphrase GaAs amplifier which is eventually to replace the input 180° hybrid ring.

The schematic diagram of the two-stage, four-transistor push-pull design is shown in Figure 31a. Figure 31b shows the corresponding chip layout. The series inductors and capacitors are monolithically integrated on the chip while the shunt inductors are realized with 25 μm diameter bonding wire. This procedure allows some flexibility in initial studies. For each of the multi-cell FETs a plated "air-bridge" or source overlay is used which interconnects the individual source

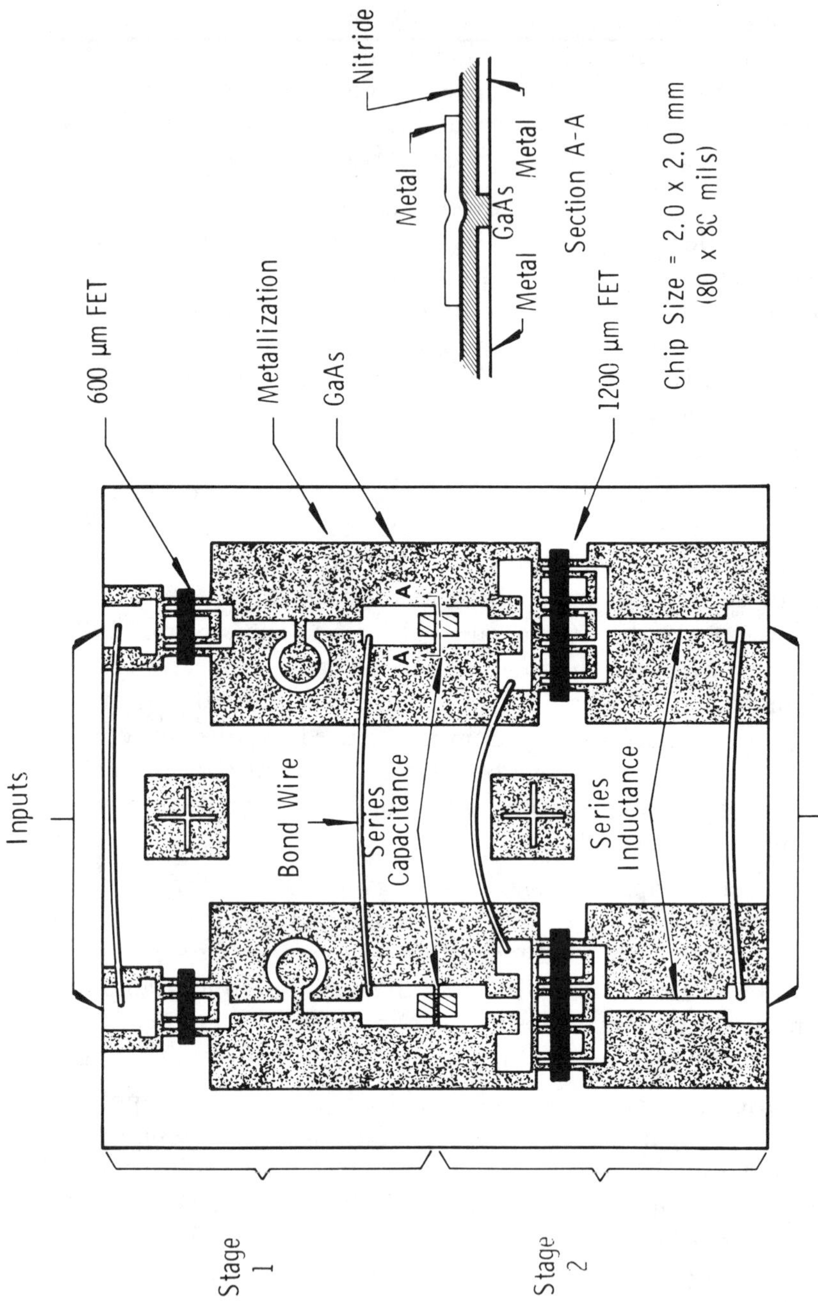

***Figure 31**/A monolithic two-stage push-pull amplifier for X-band operation* *b) Chip Layout.*

pad metallization islands and connects them to the top grounding metallization. The input and output ports of the chip are designed to interface with a pair of 50 Ω anti-phase transmission lines or, more simply, a 100 Ω balanced line.

Two advantages of the monolithic push-pull configuration which are especially significant for microwave power amplifier applications are utilized in this chip design. First, since the push-pull arrangement is essentially a series connection, the 2.4 mm of total gatewidth in the output stage is basically a series connection to two 1.2 mm devices. This relieves some of the impedance matching problem by halving the required transformation ratio. Second, as depicted in Figure 31a, current at the fundamental frequency flows between the source of each transistor and the source of its push-pull counterpart. Coplanar grounding metallization connects all sources so that these RF currents are primarily confined to the chip's surface. If each transistor pair is closely matched, then only a small fraction of the total RF source current needs to flow to true chassis ground. Consequently, connection of the chip's grounding metallization to the chassis ground is not as crucial as in the case of a conventional single ended amplifier.

Since the amplifier operates in the push-pull mode a virtual ground exists symmetrically between the two halves of the circuit (see Figure 31a). Consequently the design of the matching network can proceed by considering a two-stage amplifier with a 600 μm FET for the first stage and a 1.2 mm FET for the second stage. Any shunt inductance is then doubled in value for the actual push-pull circuit. Two series capacitors are used for dc blocking between the first and second stages. The capacitor is an overlay type using silicon nitride as the dielectric ($\epsilon_r \approx 7$) and is nominally 1 pF. Its reactance is taken into account in the inter-stage matching network. The silicon nitride thickness is nominally 4000 Å. Quality factors for the overlay capacitors were measured to be about 30 at 6 GHz, while the loop inductor Qs are estimated to be about 20 based on top conductor losses in microstrip fabricated on 0.1 mm GaAs. Since relatively small fractional bandwidths were considered (~ 10%), circuit design was based on simple Smith Chart techniques and on small and large signal impedance levels of comparable discrete FETs.

Figure 32 shows the monolithic two-stage chip without the shunt bond wire inductances. At 9.0 GHz, the best amplifier chip yielded a saturated output of 1.5 W with 13 dB gain and 19% power-added efficiency. The small signal gain was 17 dB with a 1 dB gain compression point of 1.3 W. It is interesting to note that comparable gate-width FET amplifiers constructed using conventional MIC technology achieve slightly higher power (~ 0-0.5 dB) with typical power added efficiencies of 25%.

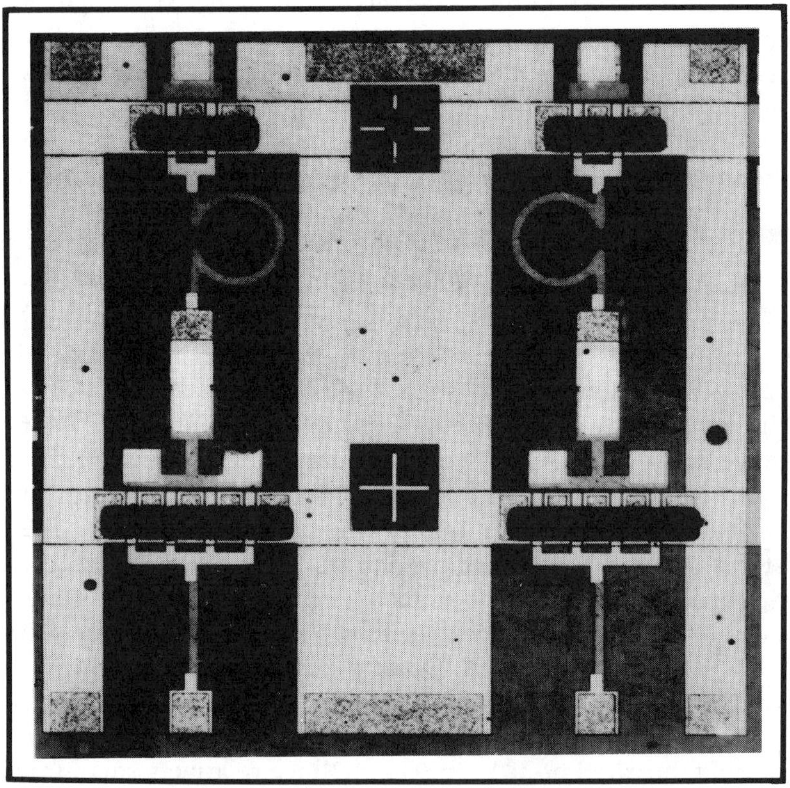

Figure 32*/The two-stage push-pull amplifier chip without the shunt bond wire inductances.*

Pucel et al. [172, 173] describe a single-stage monolithic power amplifier which employs four FETs on a single chip having a total gate width of 6.4 mm. The monolithic amplifier chip has dimensions 4.8 x 6:1 x 0.1 mm and develops 2 W with 4 dB gain at 10 GHz, with a power added efficiency of 15%. Linear gain is 5 dB. The design approach is somewhat different from the monolithic examples given above. First, no bond wires, except for dc bias connections, are used. Second, via holes etched in the GaAs substrate and subsequently plated through form the grounding connections to the metal carrier plate. Since the amplifier operates in a conventional common source single-ended configuration, RF grounding is critical, and the via-holes etched in close proximity to the sources of the FETs serve this function. Third, distributed circuit elements are used in much the same way as conventional MIC amplifiers employ them, except that all of the elements are fabricated monolithically on the 0.1 mm thick GaAs. All metallization including the device and the surrounding circuitry is gold. Plating is used to thicken the evaporated metal layer to minimize ohmic losses in the microstrip lines and the source overlay or "air-bridge" which connects all of the FET source pad islands.

The development of the four transistor chip evolves from the design of single and dual FET monolithic amplifiers employing total gate-widths of 1.6 and 3.2 mm respectively. A single-stage, single-transistor monolithic amplifier chip is fabricated on a chip whose dimensions are 5 x 6 x 0.1 mm. The chip is rather large due to the necessity of large grounding metallization pads which connect to metal pressure clamps at the edges of the chip. Incorporation of the via hole technology helped to reduce the chip size considerably in subsequent designs. An output power of 0.5 W with 4 dB gain and a 1 dB bandwidth of 1.8 GHz is obtained for the single-stage device. Linear gain is 8 dB. The circuit design employs sections of series transmission line and open-circuited stubs designed in microstrip on GaAs. Impedance matching at input and output is to 50Ω. No provision for tuning is used and none was required, inasmuch as the center frequency was within 5% of the value predicted by computer analysis. The two FET monolithic amplifiers are fabricated on a 2.5 x 3.8 x 0.1 mm chip and consists of two 1.6 mm gate-width FETs each matched to 100Ω and combined in parallel to impedance match to 50Ω input and output circuits. Via holes are etched in three pad areas adjacent to the devices which connect to the sources of the FETs. An output power of 1 W with 4 dB gain and 18% power added efficiency is achieved. Finally, the four-transistor 2 W amplifier consists of essentially a parallel combination of a pair of two-FET circuits, with intervening impedance-leveling circuitry fabricated on a 4.8 x 6.1 x 0.1 mm GaAs chip.

REFERENCES

[1] Abbot, D.A. et al., *Microwave System News,* pp. 73-96, August 1979.

[2] Agarwal, K.K. and Y.L. Kuo, *Electron Lett.,* 14:282, 1978.

[3] Agarwal, K.K. and Y.L. Kuo, *1978 IEEE MTT-S International Microwave Symposium Digest,* 393, 1978.

[4] Aihara, S., and H. Katoh, *NEC. Res. and Dev.* (Japan), 44:16, 1977.

[5] Aihara, S., et al., *1979 IEEE MTT-S International Microwave Symposium Digest,* 484.

[6] Akello, R.J., B. Easter, and I.M. Stephenson, *Electron. Lett.,* 13:160, 1977.

[7] Anastassiou, A. and M.J.O. Strutt, *IEEE Trans. Microwave Theory and Tech.,* MTT-21:419, 1973.

[8] Arai, Y. et al., *IEEE Trans. Microwave Theory and Tech.,* MTT-24:381, 1976.

[9] Arai, Y., S. Mural, and H. Komizo, *Proceedings of the 8th European Microwave Conference,* Sept. 1978.

[10] Arai, Y., S. Mural, and T. Sakane, *Injitsu Scientific & Technical Journal,* Sept. 1979, p. 63-82.

[11] Baechtold, W., *Electron. Lett.,* 7:275, 1971.

[12] Baechtold, W., and W. Jutzi, *Proceedings of the 1981 European Microwave Conference,* 1971.

[13] Baechtold, W., *IEEE J. Solid-State Circuits,* Sc 8:54, 1973.

[14] Barlow, H.M., and J. Brown in *Radio Surface Waves,* Oxford: The Clarendon Press, 1962.

[15] Basawapatna, G.R., *Proceedings of the 8th European Microwave Conference,* Sept. 1978.

[16] Baudet, P., M. Parisot, and R. Vellex, *Agard Conference Proceedings No. 197* (on New Devices, Techniques and Systems in Radar), 1977.

[17] Bechtel, N.G., W. Hooper, and G.D. Vendelin, *IEEE International Conference on Communications,* Vol. I, June 1973.

[18] Beneking, H., J. Jahncke, and J. Naumann, *Proceedings of the 3rd Biennial Cornell Electrical Engineering Conference on High Frequency Generation and Amplification Devices and Application,* August 1971.

[19] Besser, L., and S. Swenson, *Microwave Syst. News,* 7:97, 1977.
[20] Binet, M., B. Kramer and M. Parisot, *Microwave Syst. News,* 6:82, 1976.
[21] Besser, L., *Microwave Syst. News,* 9:28, 1979.
[22] Blanco, C., *Elect. Lett.,* 15:31, 1979.
[23] Bode, H.W., *Network Analysis and Feedback Amplifier Design.* New York: Van Nostrand, 1945.
[24] Borrego, J.M., R.J. Gutmann, M. Chudzicki and S. Moghe, *Electron. Lett.,* 13, 1977.
[25] Brehm, G.E. and G.D. Vendelin, *Microwaves,* 13:38, 1974.
[26] Brunet-Brunol, D, *Rev. Phys. Appl.,* 13:180.
[27] Bura, P., D. Cowan and K. Betharon, 1976 IEEE International Solid-State Circuits Conference, *(Digest of Technical Papers).*
[28] Camisa, R.L., J. Goel, and I. Drukier, *Electron. Lett.,* 11:572, 1975.
[29] Camisa, R.L., and E. Mykietyn, *RCA Tech. Not.,* TN1176:1, 1977.
[30] Camisa, R.L., et al., *1978 IEEE MTT-s International Microwave Symposium Digest,* 147-9, June 1978.
[31] Carlin, H.J., and J.J. Kamiak, *IEEE Trans. Microwave Theory and Tech.,* MTT-27:93, 1979.
[32] Caulton, M., S.P. Knight, and D.A. Daly, *IEEE Trans. on Electron Devices,* ED-15:459, 1968.
[33] Ch'en, D.R., et al., 1976 IEEE International Solid-State Circuits Conference, *(Digest of Technical Papers).*
[34] Chen, W.K., *Theory and Design of Broadband Matching Networks.* Elmsford: Pergamon Press, 1976.
[35] Chiang-Liang, H., et al., 1979 IEEE International Solid-State Circuits Conference, *(Digest of Technical Papers).*
[36] Cohen, M., E. Geller, and J. Schellenberg, *1979 IEEE MTT-S International Microwave Symposium Digest,* 292.
[37] Cooper, J.F., and M.S. Gupta, *IEEE J. Solid-State Circuits,* SC-12:325, 1977.
[38] Courtney, W.E., *IEEE Trans. on Microwave Theory and Techniques,* MTT-25:697, 1977.
[39] Cowan, D.A., P. Mercer, and A.B. Bell, *IEEE Trans. Microwave Theory and Tech.,* MTT-25:995, 1977.

[40] Cristal, E.G., *IEEE Trans. Microwave Theory Tech.*, MTT-13: 693, 1965.
[41] Crowne, K.M., *Electron. Lett.*, 15:197, 1979.
[42] Cusack, J.M., S.M. Perlow, and B.S. Perlman, *IEEE Trans. Microwave Theory Tech.*, MTT-22:1146, 1974.
[43] Czech, J., M. Schunzel, and W. Wurscher, *Nachrichtentech. Z. (NTZ)* (Germany), 30:407, 1977.
[44] Dance, M., *Electron, Ind.*, 4:12, 1978.
[45] Das, M.B., and P. Esgreda, *IEEE Transaction on Electron Devices,* ED 24:757, 1977.
[46] Debrecht, R.E., *1973 G-MTT International Symposium Digest of Technical Papers,* 309, June 1973.
[47] de Koning, J.G., W.M. Kelly, and H. Tokuda, *Microwaves,* 16:60, 1977.
[48] Denlinger, E.J., *IEEE Trans. Microwave Theory and Techniques,* MTT-17:235, 1969.
[49] De Los Reyes, E., E. Camargo, and R. Soares, *Electron. Lett.,* 14:378, 1978.
[50] Dornan, B., et al., ISSCC, 1980, p. 162-163.
Transponder ISSCC", 1980, p. 162-163.
[51] Derewonko, H., M. Laviron, and J. Lepage, *Electron. Lett.,* 15:8, 1979.
[52] Drago, F., and J. Goel, IEEE, Eascon '78, Sept. 1978.
[53] Drukier, L., *Microwave Journal,* 23:59, March 1980.
[54] Eisenberg, J., *Microwave Syst. News,* 7:45, 1977.
[55] Eisenberg, J.A., 1975 IEEE International Symposium on Circuits and Systems, 408-10
[56] Emery, F.E., *Microwave J.,* 20:79, 1977.
[57] Engelmann, R.W.H., and C.A. Liechti, *IEEE Trans. Electron Devices,* ED-24:1288, 1977.
[58] Estabrook, P., et al., *1978 IEEE MTT-S International Microwave Symposium Digest,* 129-31.
[59] Everitt, W.L., and G.E. Anner, *Communication Engineering.* New York: McGraw-Hill Book Company, 1956.
[60] Fano, R.M., *J. Franklin Inst.,* 249:57, 1960.
[61] Filensky, W., H.J. Klein, and M. Beneking, *IEEE J. Solid-State Circuits,* SC-12:276, 1977.
[62] Fukuta, S., Y. Arai, and I. Haga, *NEC. Res. and Dev.* (Japan), 48:67, 1978.

[63] Fukuta, M., et al., *IEEE Transactions on Electron Devices,* ED-23:388, 1976.
[64] Furuya, T., et al., *Proceedings of the 8th European Microwave Conference,* 645-9, Sept. 1978.
[65] Hewitt, S.J., and R.S. Pengelly, *Radio and Electron. Eng.,* 46:463, 1976.
[66] Goel, J., and R. Camisa, *Electron. Lett.,* 12:493, 1976.
[67] Goell, J.E., *Bell Syst. Tech. J.,* 53:177, 1974.
[68] Gupta, R.K., P.A. Goud, and G.G. Englefield, *Electronics Letters,* 15:741, 1979.
[69] HP Application Note # 95 containing the following articles: "Microwave Transistor Characterization including S-parameters," J. Lange; "Scattering parameters speed design of high frequency transistor circuits," F. Weinert, *Electronics,* Sept. 5, 1966; "S Parameters Techniques for Faster More Accurate Network Design", *Hewlett Packard Journal,* Feb. 1967, Vol. 18; "Combine S-Parameters with Time Sharing", L. Besser, *Electronic Design,* 16, Aug. 1, 1968; "Quick Amplifier Design with Scattering Parameters", L.H. Froehner, *Electronics,* Oct. 16, 1967; "Two Port Power Flow Analysis using Generalized Scattering Parameters", G.E. Bodway, *Microwave Journal,* Vol. 10, No. 6, 1967; "Circuit Design and Characterization of Transistors by Means of Three-Part Scattering Parameters", G.E. Bodway, *Microwave Journal,* Vol. 11, No. 5, 1968.
[70] Harry Diamond Laboratories, *Microwave Journal,* 22:87, 1979.
[71] Hasegawa, H., and H. Okizaki, *Electron. Lett.,* 13:663, 1977.
[72] Heiter, G.L., *IEEE Transactions on Microwave Theory and Techniques,* MTT21:797, 1973.
[73] Hennies, S.R., 1978 IEEE International Solid-State Circuits Conference, *(Digest of Technical Papers),* 124-5.
[74] Higashisaka, A., et al., *1980 IEEE MTT-S International Microwave Symposium Digest,* 9-11.
[75] Higgins, J.A., *1978 IEEE MTT-S International Microwave Symposium Digest,* 138-41.
[76] Higgins, J.A., A. Gupta, G. Robinson, and D.R. Choen, 1979 IEEE International Solid-State Circuits Conference, *(Digest*

of Technical Papers), 120-1, Feb. 1979.
[77] Higgins, J.A., and R.L. Kuvas, *IEEE Transactions on Microwave Theory and Techniques,* MTT 28:9, 1980.
[78] Ho, P., C. Pham, and R. Mencik, *1979 IEEE MTT-S International Microwave Symposium Digest,* 128-30.
[79] Ho, P., *1978 IEEE MTT-S International Microwave Symposium Digest,* 142-144.
[80] Honjo, K., Y. Takayama, and A. Higashisaka, *IEEE Trans. Microwave Theory and Tech.,* MTT-27:3, 1979.
[81] Honjo, K., et al., *1979 IEEE MTT-S International Microwave Symposium,* 289-291.
[82] Hornbuckle, D.P., and L.J. Kuhlman, Jr., *IEEE Trans. Microwave Theory and Tech.,* MTT-24:338-42, 1976.
[83] Hornbuckle, D.P., *1978 IEEE MTT-S International Microwave Symposium Digest,* 288-289, June 1978.
[84] Hosseini, N.M., and H.V. Shurmer, *Radio and Electron. Eng.,* 48:85, 1978.
[85] Hoult, D.I., and R.E. Richards, *Electron. Lett.,* 11:596, 1975.
[86] Hower, P.L., and N.G. Bechtel, *IEEE Trans. Electron Devices,* ED-20:213, 1973.
[87] Huang, H.C., et al., *Electron. Lett.,* 11:508, 1975.
[88] Hundley, R.E., P. Estabrook, and E.J. Crescenzi, Jr., *Microwave J.,* 21:47, 1978.
[89] Ida, M., T. Sugeta, and K. Shimada, *Rev. Electr. Commun. Lab.* (Japan), 23:1193, 1975.
[90] Ida, M., T. Sugeta, and K. Shimada, *Electr. Commun. Lab. Tech. J.* (Japan), 24:335, 1975.
[91] Jahncke, J., *Nachrichtentech. Z. (NTZ)* (Germany), 26:193, 1973.
[92] Jutzi, W., *Arch. Elek. Ubertrag,* (Germany), 25:159, 1971.
[93] Kadowaki, Y., M. Nakatari, and T. Ishii, *1979 IEEE MTT-S International Microwave Symposium Digest,* 486.
[94] Keister, F.Z., *IEEE Trans. on Microwave Theory and Techniques,* MTT-16:469, 1968.
[95] Kitamura, M., S. Fukuda, and I. Haga, *Trans. Inst. Electron. and Commun. Eng., Jpn., Sect. E* (Japan), E60:501, 1977.
[96] Knerr, R.H., and C.B. Swan, *1977 International Microwave Symposium Digest,* 195.

[97] Knerr, R.H., and C.B. Swan, *Bell Syst. Tech. J.,* 57:479, 1978.

[98] Kodera, H., Y. Kaneko, and H. Sato, *1977 International Microwave Symposium Digest,* 277-80.

[99] Komizo, H., Y. Arai, and T. Saito, *1977 International Microwave Symposium Digest,* 243. K.L. Kotzebue and E.R. Ehlers, *Electronic Letters,* 15:237, 1979.

[100] Kotzebue, K.L., and E.R. Ehlers, *Electronics Letters,* 15:237, 1979.

[101] Kotzebue, K.L., *Electronics Letters,* 12:74, 1976.

[102] Kotzebue, K.L., *IEEE Transactions on Microwave Theory and Techniques,* MTT-24:975, 1976.

[103] Kramer, B., M. Parisot, and A. Collet, *Radio and Electron Eng.,* 48:23, 1978.

[104] Krumm, C.F., H.T. Suyematsu, and B.L. Walsh, *1978 IEEE MTT-S International Microwave Symposium Digest,* 383.

[105] Ku, W.H., et al., *Proc. 1975 International Microwave Symposium,* 62.

[106] Ku, W.H., and P.L. Clouser, *Proceedings of the 1971 European Microwave Conference,* 1971.

[107] Ku, W.H., W.C. Petersen, and A.F. Podell, *Proc. Ninth Asilomar Conference,* 211, 1975.

[108] Ku, W.H., and W.C. Petersen, *1977 International Symposium on Circuits and Systems Proceedings,* 704.

[109] Ku, W.H., and W.C. Petersen, *IEEE Trans. on Circuits and Systems,* CAS-22:523, 1975.

[110] Ku, W.H., *1977 International Microwave Symposium Digest,* 312.

[111] Ku, W., H. Willing, 1980 IEEE MTT-S International Microwave Symposium, 390.

[112] Kuono, T., Y. Aral, and H. Komizo, *Trans. Inst. Electron. and Commun. Eng., Jpn. Sect. E,* (Japan), E60:500, 1977.

[113] Kurokawa, K., *IEEE Trans. Microwave Theory and Techniques,* MTT-13:194, 1965.

[114] Kurokawa, K., *Bell System Technical Journal,* 44:1675, 1965.

[115] Lane, D., *Microwave J.,* 21:126, 1978.

[116] Lange, J., *IEEE Trans. on Microwave Theory and Techniques,* MTT-17:1150, 1969.

[117] Lee. Y.S., W.J. Getsinger, and L.R. Sparrow, *IEEE Trans.*

Microwave Theory and Techniques, MTT-27:655, 1979.

[118] Leighton, W.H., R.J. Chaffin, and J.G. Webb, *IEEE Trans. Microwave Theory and Tech.,* MTT-21:809, 1973.

[119] Leost, J.C., and B. Loriou, *Onde Electr.,* (France), 54:514, 1974.

[120] Levy, R., *Proc. IEEE,* 111:1098, 1964.

[121] Lewin, L., *Proc. IEEE* (London), 107:163, 1960.

[122] Liechti, C.A., and R.L. Tillman, *Digest of Technical Papers of the 1973 IEEE International Solid State Circuits Conference,* 74.

[123] Liechti, C.A., and R.L. Tillman, *IEEE Trans. Microwave Theory and Tech.,* MTT-22:510, 1974.

[124] Liechti, C.A., *IEEE Trans. Microwave Theory and Tech.,* MTT-23:461, 1975.

[125] Liechti, C.A., R.B. Larrick, and D.J. Mellor, *5th European Microwave Conference,* 306, 1975.

[126] Lindauer, J.T., and N.K. Osbrink, *Microwave Syst. News,* 6:63, 1976.

[127] Maeda, M., and Y. Minai, *IEEE Microwave Symposium Digest of Technical Papers,* 351, 1974.

[128] Maeda, M., and Y. Minai, *IEEE Trans. Microwave Theory and Tech.,* MTT-22:1226, 1974.

[129] Markle, R.E., *1977 International Conference on Communications Digest,* 3:40.1.

[130] Matthaei, G.L., *Proc. IEEE,* 52:939, 1964.

[131] Matthaei, L. Young, and E.M.T. Jones, *Microwave Filters, Impedance-Matching Networks, and Coupling Structures.* Dedham: Artech House, 1980.

[132] Medley, M.W., Jr., *Microwaves,* 16:54, 1977.

[133] Mehal, E.W., and R.W. Wacker, *IEEE Trans. on Microwave Theory and Techniques,* MTT-16:451, 1968.

[134] Mellor, D.J., and J.G. Linvill, *IEEE Trans. Microwave Theory and Techniques,* MTT 23, 1975.

[135] Mellor, D.J., *Proc. 9th Asilomar Conference,* 206, 1975.

[136] Minasian, R.A., *IEEE Trans. on Microwave Theory and Techniques,* MTT 28:1, 1980.

[140] Mitama, M., and H. Katoh, *Electron. Lett.,* 14:319, 1978.
[141] Mitama, M., and H. Katoh, *NEC Res. and Dev.* (Japan), 50: 69, 1978.
[142] Mitsui, Y., et al., *1980 IEEE MTT-S International Microwave Symposium Digest,* 6.
[143] Mitsui, T., et al., *Electronics Letters,* 16:205, 1980.
[144] Mokari-Bolhassan, and W.H. Ku, *IEEE Trans. Microwave Theory and Tech.,* MTT-25:837, 1977.
[145] Monroe, J.W., J.G. deKoning, W.M. Kelley, and H. Tokuda, *Microwaves,* 16:60, 1977.
[146] Montgomery, C.G., R.H. Dicke, and E.M. Purcell, *Principles of Microwave Circuits.* New York: McGraw-Hill Book Company, 1948.
[147] Nakatani, M., Y. Kadowaki, and T. Ishii, *IEEE Trans. on Microwave Theory and Techniques,* MTT-27:1066, 1979.
[148] Neider, R.E., and H.A. Willing, *IEEE Trans. Microwave Theory and Tech.,* MTT-24:342, 1976.
[149] Neidert, R.E., H.A. Willing, and B.E. Spielman, *Naval Research Laboratory Interim Report #8388,* NRL, Washington, D.C., March 1980.
[150] Nevin, L., and R. Wong, *Proceedings of the 8th European Microwave Conference,* 140.
[151] Niclas, K.B., *Electron. Lett.,* 14:29, 1978.
[152] Niclas, K.B., R.B. Gold, W.T. Wilser, and W.R. Hitchens, 1978 IEEE International Solid-State Circuits Conference *(Digest of Technical Papers),* 128.
[153] Niclas, K.B., R.B. Gold, W.T. Wilser and W.R. Hitchens, *IEEE J. Solid-State Circuits,* SC-13:520, 1978.
[154] Niclas, K.B., 1979 IEEE International Solid-State Circuits Conference, *(Digest of Technical Papers),* 122, 1979.
[155] Niclas, K.B., W.T. Wilser, R.B. Gold, W.R. Hitchens, *1980* IEEE International Solid State Circuits Conference, p. 169-165.
[156] Niclas, K.B., et al., *IEEE Trans. on Microwave Theory and Techniques,* MTT-28:285, 1980.
[157] Niclas, K.B., et al., *IEEE Trans. on Microwave Theory and Techniques,* MTT-28:172, 1980.
[158] Nowak, M.M., P.A. Terzian, and R.D. Fairman, *Electron. Lett.,* 13:159, 1977.

[159] Ohkawa, S., K. Suyama and H. Ishikawa, *Fujitsu Sci. and Tech. J.,* 11:151, 1975.

[160] Ohta, K., et al., *Microwave Journal,* Nov. 79, pp. 66-67.

[161] Owens, R.P., J.E. Aitken, and T.C. Edwards, *IEEE Trans. Microwave Theory and Techniques,* MTT-24:499, 1976.

[162] Kyu, Park Han, *J. Korea Inst. Electron. Eng.,* 13:126, 1976.

[163] Pearson, R.E., Agard Conference Proceedings No. 197 on New Devices, Techniques and Systems In Radar, 2/1-9, 1977.

[164] Pengelly, R.S., *Electron. Lett.,* 11:58, 1975.

[165] Pengelly, R.S., and J.A. Turner, *Electronics Letters,* 12:251, 1976.

[166] Pengelly, R.S., *MicroWaves,* pp. 57-65, July 1979.

[167] Pengelly, R.S., and D.C. Richard, Seventh European Microwaves Conference Proceedings, Copenhagen, Denmark, pp. 460-464, 1977.

[168] Pierro, J., *IEEE Trans. Microwave Theory and Tech.,* MTT-24:972, 1976.

[169] Pucel, R.A., *Elect. Lett.,* 14:204, 1978.

[170] Pucel, R.A., D.J. Masse, and C.P. Hartwig, *IEEE Trans. Microwave Theory and Techniques,* MTT-16:342, 1968.

[171] Pucel, R.A., Seventh Biennial Conference on Active Microwave Semiconductor Devices and Circuits, 1979. Cornell University, Ithaca, New York.

[172] Pucel, R.A., et al., International Electron Devices Meeting *(1979 Technical Digest),* pp. 266-268, December 1979.

[173] Pucel, R.A., *Microwave Journal,* pp. 51-56, March 1980.

[174] Pucel, R., P. Ng, and J. Uorhaus, *1979 IEEE MTT-S International Symposium Digest,* 387-389.

[175] Ragan, G.L., *Microwave Transmission Circuits.* New York: McGraw-Hill Book Company, 1948.

[176] Rauscher, C., and R.S. Tucker, *Electron. Lett.,* 12:701, 1977.

[177] Rauscher, C., and H.A. Willing, *1979 IEEE MTT-S Int. Microwave Symp. Digest,* pp. 401-404.

[178] Rauscher, C., and H.A. Willing, *Proc. 9th European Microwave Conference,* pp. 287-290, 1979.

[179] Richards, P.I., *Proc. IRE,* 36:217, 1948.

[180] Rossel, P., and J.J. Cabot, *Electron. Lett.,* 11:150, 1975.

[181] Russel, K.J., *IEEE Trans. on Microwave Theory and Techniques,* MTT-27:472, 1979.

[182] Saal, R., and E. Ulbrich, *IRE Trans. Circuit Theory*, CT-5: 284, 1958.

[183] Saito, Y., et al., *1978 IEEE MTT-S International Microwave Symposium Digest*, 27-29.

[184] Sakane, T., Y. Arai, H. Komitto, 1980 IEEE International Solid State Circuits Conference, p. 160-161.

[185] Schellenberg, J.M., and M. Cohn, 1978 IEEE Int. Solid State Circuits Conf., *Digest of Technical Papers*, February 1978.

[186] Schilling, W., and S.S. Lehrfeld, *Microwaves*, December 1968, January 1969, March 1969.

[187] Schneider, M.V., *Bell System Technical Journal*, 5:1421, 1969.

[188] Schroeder, W.E., and J.W. Gewartowski, *1978 IEEE MTT-S International Microwave Symposium Digest*, p. 279-281.

[189] Sechi, F.N., and R.W. Paglione, 1976 IEEE International Solid State Circuits Conference, *(Digest of Technical Papers)*, 162-3.

[190] Sechi, F.N., Ho C. Huang, and V. Riginos, 1977 International Solid-State Circuits Conference, *(Digest of Technical Papers)*, 164-5, Feb. 16-18, 1977.

[191] Sechi, F., H. Huang, and B. Perlman, *Proceedings of the 8th European Microwave Conference*, 473-7.

[192] Sechi, F.N., and R.W. Paglione, *IEEE Journ. Solid State Circuits*, SC-12:285, 1977.

[193] Siddiqui, J., 1979 IEEE International Solid-State Circuits Conference, *(Digest of Technical Papers)*, 116-7.

[194] Slaymaker, N.A., and J.A. Turner, *Colloquium Digest on Solid-State Microwave Amplifiers*, 15/2 PP, March 16, 1973.

[195] Slaymaker, N.A., and J.A. Turner, 1973 European Microwave Conference, Vol. 1, 1973.

[196] Slaymaker, N.A., R.A. Soares, and J.A. Turner, *IEEE Trans. Microwave Theory and Tech.*, MTT-24:329, 1976.

[197] Soares, R.A., and J.A. Turner, *Electron. Lett.*, 11:474, 1975.

[198] Sokolov, V., and H.Q. Tserng, 1976 National Telecommunications Conference, Nov. 29-Dec. 1, 1976.

[199] Sokolov, V., R.E. Williams, and D.W. Shaw, 1979 IEEE International Solid-State Circuits Conference, *(Digest of Technical Papers)*, 118-19.

[200] Sokolov, V., R. Williams, *IEEE Transactions on Electron. Devices,* Ed-27:1164, 1980.
[201] Sokolov, V., and R.E. Williams, Seventh Biennial Conference on Active Microwave Semiconductor Devices and Circuits (Proceedings), Cornell University, Ithaca, New York, 1979, pp. 223-235.
[202] Sone, J., and Y. Takayama, and Y. Aono, *Electron. Lett.,* 15:212, 1979.
[203] Sone, J., and Y. Takayama, *Electronics Letters,* 15:562, 1979.
[204] Sone, J., and Y. Takayama, *1980 IEEE MTT-S International Microwave Symposium Digest,* 180-182.
[205] Steyskal, H., Sr., *Proceedings on the 8th European Microwave Conference,* 130-4.
[206] Stogner, D., *Teleph. Eng. and Manag.,* 81:60, 1977.
[207] Strid, E.W., and T.C. Duder, Eleventh Annual Asilomar Conference on Circuits Systems and Computers, 146-51, Nov. 7-9, 1977.
[208] Strid, E.W., and T.C. Duder, *1978 IEEE MTT-S International Microwave Symposium Digest,* 135-7.
[209] Surana, D.C., and J.G. Gardner, *Electronics Letters,* 6:403, 1970.
[210] Suzuki, H., et al., *Proceedings of the 9th Conference on Solid State Devices,* 1977.
[211] Takayama, T., 1976 IEEE MTT-S International Microwave Symposium, 218-20.
[212] Takayama, Y., et al., 1977 IEEE International Solid-State Circuits Conference, pp. 166-167.
[213] Takayama, Y., and K. Honjo, 1980 IEEE MTT-S International Microwave Symposium, 496-498.
[214] Takayama, Y., T. Ogawa, and Y. Aono, *Electronics Letters,* 15:326, 1979.
[215] Terman, F.E., *Radio Engineers' Handbook.* New York: McGraw-Hill Book Company, 1943.
[216] Tohyama, H., and H. Mizuno, *Trans. Inst. Electron. and Commun. Eng., Jpn. Sect. E,* (Japan), E59:35, 1976.
[217] Tohyama, H., *1977 International Microwave Symposium Digest,* 246-8.
[218] Tsai, W.C., IEEE Eleventh Annual Asilomar Conference on Circuits Systems and Computers, 152-6, 1977.

[219] Tsai, W.C., *1978 IEEE MTT-S International Microwave Symposium Digest,* 285-287.
[220] Tserng, H.Q., et al., *IEEE Trans. Microwave Theory and Tech.,* MTT-24:936, 1976.
[221] Tserng, H.Q., and H.M. Macksey, 1980 IEEE International Solid State Circuits Conference, p. 166-167.
[222] Tserng, H.Q., et al., 1978 IEEE International Solid-State Circuits Conference, *(Digest of Technical Papers),* 122-3.
[223] Tserng, H.Q., and H.M. Macksey, *1978 IEEE MTT-S International Microwave Symposium Digest,* 282-4.
[224] Tserng, H.Q., *Microwave Journal,* pp. 94-100, June 1979.
[225] Tucker, R.S., *IEEE Trans. Microwave Theory and Techniques,* MTT 21:322, 1973.
[226] Tucker, R.S., and C. Rauscher, *Elect. Lett.,* 13:508, 1977.
[227] Ulrich, E., *Microwaves,* 17:66, 1978.
[228] van Heuven, J.H.C., *IEEE Trans. Microwave Theory and Techniques,* MTT-22:841, 1974.
[229] Van Tuyl, R.L., *1978 IEEE International Solid-State Circuits Conference (Digest of Technical Papers),* 72-3.
[230] Vendelin, G.D., and M. Omori, *Electron. Lett.,* 11:60, 1975.
[231] Vendelin, G.D., and M. Omori, *Microwaves,* 14:58, 1975.
[232] Vendelin, G.D., *IEEE Trans. Microwave Theory Tech.,* MTT-24:383, 1976.
[233] Vendelin, G.D., IEEE Eleventh Annual Asilomar Conference on Circuits Systems and Computers, 139-141, 1978.
[234] Vendelin, G.D., *Microwaves,* 17:40, 1978.
[235] Vincent, G.A., *Frequency Technology,* 1969, p. 15-21.
[236] Wade, P.C., 1978 IEEE International Solid-State Circuits Conference *(Digest of Technical Papers),* 120-1.
[237] Walker, M.G., F.T. Mauch, and T.C. Williams, *Microwaves,* 14: 1975.
[238] Walker, M.G., F.A. Marki, and H.M. Abramowitz, *Microwave Syst. News,* 6:39, 1976.
[239] Walker, M.G., and E.J. Crescenzi, Jr., 1976 IEEE MTT-S International Microwave Symposium, 107-9.
[240] Wemple, S.H., M.L. Steinberger, and W.D. Schlosser, *Electron. Lett.,* 16:459, 1980.

[241] Williams, R.E., and D.W. Shaw, *IEEE Trans. Electron Devices,* ED-25:600, 1978.
[242] Williams, R.E., and D.W. Shaw, *Electron. Lett.,* 13:400, 1977.
[243] Willing, H.A., C. Rauscher, and P. deSantis, *IEEE Trans. Microwave Theory Tech.,* MTT-26:1013, 1978.
[244] Wilkinson, E., *IRE Trans. on Microwave Theory and Techniques,* MTT8:116, 1960.
[245] Woo, A.N., and D.R. Ch'en, *Microwave J.,* 17:26, 1974.
[246] Youla, D.C., *IEEE Trans. Circuit Theory,* CT-11:30, 1969.
[247] Yamamura, S., et al., 1979 IEEE MTT-S International Microwave Symposium, 335-337.
[248] Yokouchi, H., et al., *1978 IEEE MTT-S International Microwave Symposium Digest,* 276-278.

Chapter 8

Digital Integrated Circuits

Introduction

GaAs FETs are revolutionizing both the RF monolithic integrated circuit and the digital or logic integrated circuit (IC) areas. The RF integrated circuit technology briefly described in Chapter 7 is still in its infancy. Based on the conventional depletion mode FETs as active devices, this technology uses the established technology of Microwave Integrated Circuits (MICs) with dielectric substrates as the circuit and integration base. GaAs digital integrated circuits, referred to as GaAs ICs or just ICs, have been developed around depletion mode FETs, enhancement mode FETs (both the p-n junction gate type and the Schottky gate type), Metal-Oxide-Semiconductor FETs or Metal-Insulator-Semiconductor FETs with Schottky diodes as switching elements. This digital IC technology has silicon VLSI (Very Large Scale Integration) technology as its base and is paving the way for higher frequency LSI circuits using GaAs as an extension of the present silicon VLSI technology.

Of the various approaches used for GaAs ICs, the MOSFET technology is probably the least mature technology from the practical IC point of view. This chapter, therefore, omits MOSFETs and only covers those approaches which are deemed more practical at this time.

The first part of the chapter written by Wilson and Phillips describes the IC technology known as E/D FET logic. E/D FET logic utilizes both the conventional depletion mode FETs (DFETs) and the enhancement mode FETs (EFETs). The enhancement mode FETs are structurally the same as the depletion mode FETs discussed in Chapters III and IV, except for the n-type active layer, which, in the case of EFET, is very thin and is therefore completely depleted at zero gate voltage (Vp = 0). This technology, which has shown some success, relies heavily on advanced GaAs processing techniques and more stringent starting material parameter control for yield improvement.

The discussion of DCFL (Direct Coupled FET Logic, also known as EFET Logic) by Lehovec and Zuleeg, examines EFET devices with both the p-n junction gates and Schottky gates. This is the most widely accepted approach for the GaAs ICs. The authors summarize the various design considerations and the state-of-the-art of direct coupled FET logic technology.

The third section, by Eden and Welch, covers the IC approach which uses Schottky diodes in conjunction with the conventional depletion mode FETs; hence the name Schottky Diode FET Logic (SDFL). The authors describe the concept of SDFL, the technology involved, and its performance status. The fabrication technology for SDFL entails the same underlying technology of the other approaches.

If one were to identify the one technology which has contributed the most to the development of GaAs ICs, it would probably be the ion-implantation technology. The other important areas for ICs are the semi-insulating GaAs substrate characteristics and small geometry lithography.

The fast growth of GaAs digital ICs has led to a jargon of overlapping acronyms. To facilitate comprehension of the literature in this area, the following definitions are provided.

Depletion Mode FET (DFET, D-FET, D-MESFET, MESFET, GaAs MESFET): Conventional Metal-Semiconductor Field Effect Transistors using a Schottky gate on relatively thick active layers.

Enhancement Mode FET (EFET, E-FET, ENFET, 'Normally Off'

GaAs MESFET): Field Effect Transistors with relatively thin active layers (such that the active layer is fully depleted at zero gate voltage, leading to Vp = 0) with either p-n junction gate or Schottky gate.

Junction FET (JFET, J-FET): FETs with p-n junctions as gates.

Metal-Oxide-Semiconductor or Metal-Insulator-Semiconductor or Insulated Gate FET (MOSFET, MISFET, IGFET): FETs with an oxide or other insulator under the gates.

Direct Coupled FET Logic (DCFL, EFET Logic, ENFET Logic): A digital integrated circuit approach in which EFETs which are 'normally off' are used. The direct coupling of two such inverters is possible because their output is small and positive.

Schottky Diode FET Logic (SDFL): A digital IC approach in which Schottky diodes are used as switching elements and DFETs are used as inverters and amplifiers.

Buffered FET Logic (BFL, DFET Logic): A digital IC approach using DFETs only and no Schottky diodes.

Enhancement/Depletion MESFET Logic (E/D FET Logic): A digital IC approach using both enhancement and depletion mode FETs.

Enhancement/Depletion GaAs FET Logic

D. Wilson
Harris, Sunnyvale, California

D. H. Phillips
The Aerospace Corporation, Los Angeles, California

1. INTRODUCTION

With the availability of better quality GaAs substrates, ion implantation and planar GaAs processing technology, it has become possible to fabricate enhancement and depletion devices on the same substrate [1, 2] by incorporating a highly uniform etch (± 20Å) of the GaAs surface into the fabrication process. Further, it is relatively easy to fabricate and interconnect all-depletion-mode circuitry adjacent to enhancement/depletion mode circuitry and therein realize the advantages that each type has to offer. In doing this, a high speed IC technology has been realized.

Enhancement/depletion (E/D FET) ICs with normally-off logic have the advantages of lowest gate complexity (number of transistors per gate), lowest power, and highest density relative to alternative GaAs logic gate implementations. The yet unanswered question is what particular approach is best for the E/D FET type implementation in which the combination of high speed, low power, and high density is needed for LSI and VLSI applications. A comparison of DFET and E/D FET approaches are given in this chapter along with some measured performance results. An outline of the important process parameters for fabricating high performance E/D FET ICs are given, along with a discussion of the circuit design techniques which are appropriate for DFET and E/D FET devices on the same wafer.

At present, GaAs MESFET devices are easier to fabricate and have produced higher operational speeds than MOSFETS and JFETS. Lack of a suitable gate oxide for GaAs [3] has hampered MOSFET device development and implementation of the shallow ($\leqslant$ 1000 Å) p-type dopant suitable for $\leqslant$ 1 micron gate length devices has impeded JFET high speed capability. Also, fabrication complexity in producing H.J.FETS has hampered this circuit implementation. Unquestionably, the GaAs MESFET structure is currently the best choice when fabricating DFET devices (all depletion mode) from the standpoint of ease of processing, high speed, and reliability.

2. BASIC PRINCIPLES AND PROCESS

A high degree of control of V_p (pinch off voltage) and I_{DS} (drain source current) is required to produce GaAs ICs with reasonable yields and circuit performance. Figure 1 shows the relationship between V_p, circuit complexity and the basic process tools required to realize a particular technology. The region under the sloped line indicates the possible complexity. Table 1 gives additional information for the three GaAs FET technologies (assuming Dual 1 micron gates with 50 micron gate widths).

TABLE I

Technology Comparison

Technology	V_p	Pwr/Gate	Area/gate	V_{bias}	t_{pd}	EGC
DFET	−2.5V	20 mw	10 mil^2	+4.5V and −3.5V	50 ps	100
SDFL	−1.0V	2 mw	7 mil^2	+2.0V and −1.5V	80 ps	1000
E/DFET	0.0V	0.2 mw	4 mil^2	+1.0V	100 ps	10,000

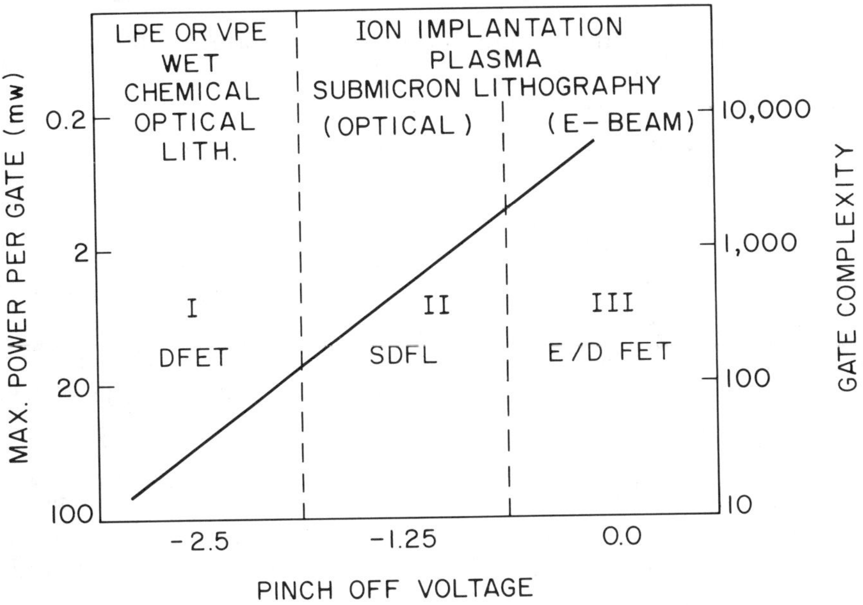

Figure 1/GaAs technology progression.

The early DFET, typified by Column I parameters, utilized liquid or vapor phase GaAs epitaxial material. Control of the epi layer thickness was difficult for channel thicknesses, a, ⩽ 2000 Å, which necessitated a pinchoff voltage of ≈ − 2.5 volts, as described by the relations:

$V_p = U_o + V_D$ where: V_p = pinchoff voltage, (1)
V_B = barrier voltage

$U_o = \frac{Nqa^2}{2\epsilon\epsilon_o}$ where: q = electronic charge (2)
N = doping density,
$\epsilon\epsilon_o$ = permittivity of GaAs

This technology resulted in the ≈ 20 mw/gate power dissipation limitation and SSI complexity limitation [4]. Wet chemical etching was used to define the necessary "mesas" which defined the active areas.

Using ion implantation techniques, control of the uniformity of impurity concentration, N, and channel thickness, a, were improved, allowing pinchoff voltages of − 1.0 volt to be achieved in ICs along with increased complexity [5]. Plasma etching techniques were found to be necessary to precisely define the small geometries required for the column II technology.

For the Column III (Figure 1) technology, e-beam lithography has been added to the list of required process tools. While not absolutely required for a normally-off ($V_p \approx 0.0$ volts), the reasons for its efficacy will be shown below. However, the dramatic reduction in power/gate and increased circuit complexity possible with E/D FET technology are readily apparent. Note also that gate switching speeds decrease as the pinchoff voltage increases. (Values shown are for $L_g \approx 1\mu$). We will show in another section that control of the channel thickness to $\Delta a \pm 20$ Å and $\Delta N/N$ to ± 2% is required to realize good E/D FET yields across a wafer. Measurements taken at the Lockheed Microelectronics Center and other published data [6] show that this degree of control, though difficult at present, can be obtained.

Techniques for obtaining a passivating dielectric on the surface of GaAs which are satisfactory for high density ICs have been developed. Most approaches use either Si_3N_4 or SiO_2 or combinations of the two. Si_3N_4 can be used to prevent the GaAs surface from decomposing at elevated temperatures, in addition to serving as an insulator. SiO_2 is also a satisfactory insulator for GaAs. In addition, thicker gate metallizations can be patterned using the SiO_2 in conjunction with high resolution photoresists in "enhanced" liftoff techniques.

As an added benefit, matching the SiO_2 thickness to the gate and first layer interconnect metallization thickness results in a very planar structure which is desirable when second layer metal interconnect structures are to be fabricated. The second-layer metallization, which crosses over the first layer metals, must be both physically continuous and electrically isolated from the first layer metal. For this purpose, a second SiO_2 layer can be deposited which must be pinhole free to prevent electrical shorts. This second dielectric layer can then be used to form metal-over-metal capacitors useful to the circuit designer. Ion milling techniques may be used to define the second layer metal pattern. Figure 2 shows a completed GaAs IC cross-section from

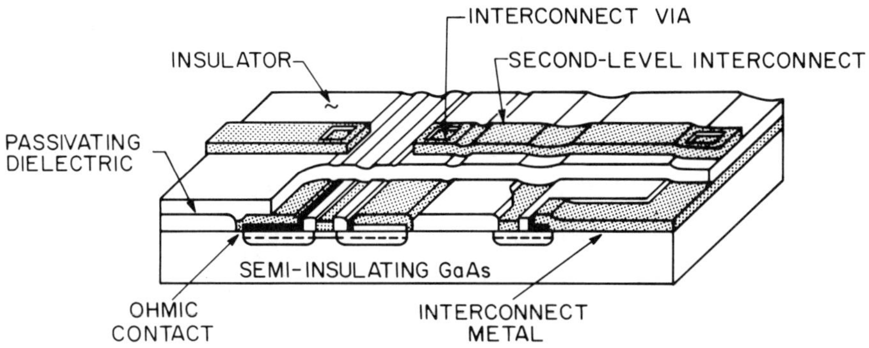

Figure 2/GaAs IC cross-section.

which all of the basic circuit elements required for SSI to VLSI complexities are formed.

3. DEVICE DESIGN AND FABRICATION

Having summarized the basic process requirements and some of the fabrication methods currently available, we will now consider in more detail the design and fabrication of an E/D FET device. In order to fully appreciate the interrelated parameters which must be controlled when fabricating these high performance devices, one must first consider the device model (shown in Figure 3), along with the basic equations for frequency response (f_t) and (f_{max}), noise figure (NF) [7] and drain-source current (I_{DS}):

$$f_t = \frac{gm}{2\pi C_{gs}} \tag{3}$$

$$f_{max} = \frac{f_t}{2}\sqrt{\frac{g_c^{\,1}}{g_{ds}}} \quad \text{where } g_c^{\,1} = \frac{1}{R_c + R_s} \tag{4}$$

$$NF = 1 + fK\left(\frac{N}{a}\right)^{1/6} (L_g)^{5/6} (R_s + R_g + R_c)^{1/2} \qquad (5)$$

$$I_{ds} = \frac{Nq\mu aWUo}{Lg}\left\{\frac{V_{ds}}{Uo} - \frac{2}{3}\left[\frac{V_{ds} + V_B - V_{gs}}{Uo}\right]^{3/2} + \frac{2}{3}\left[\frac{V_B - V_{gs}}{Uo}\right]^{3/2}\right\} \qquad (6)$$

(where μ is the carrier mobility and V_{ds} and V_{gs} are the drain-source and gate-source voltages, respectively).

It is immediately apparent that C_{gs}, R_s, R_g, R_c, and L_g should be minimized in order to obtain better f_t, NF, and f_{max} performance. From Equation (7) it can be seen that reducing L_g, the gate length, is of primary importance, as well as reducing the W/a ratio [8].

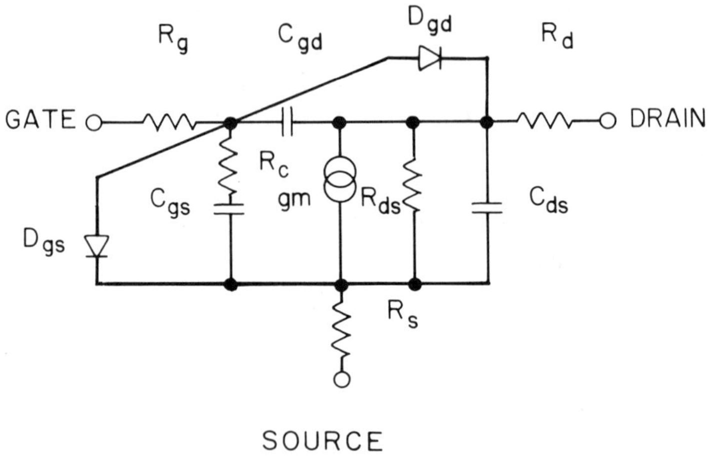

Figure 3/GaAs FET device model.

$$C_{gs} = \frac{3\epsilon\epsilon o W L_g}{a} \left[\frac{1 + \sqrt{\dfrac{(V_B - V_{gs})}{Uo}}}{\left(1 + 2\sqrt{\dfrac{(V_B - V_{gs})}{Uo}}\right)^2} \right] \quad (7)$$

However, using Equation (6), reducing W also reduces I_{ds}. A trade-off is necessary. This will be considered in more detail below.

Two more equations are introduced to present the full spectrum of trade-offs involved in the device design:

$$P_{dm} \geqslant \frac{1}{2} C_L (\Delta V)^2 \qquad \text{(dynamic switching energy)} \quad (8)$$

$$\Delta t = \frac{C_L \Delta V}{I_{ds}} \qquad \text{(slew rate limited rise or fall time)} \quad (9)$$

From Equation (8), we see that both the switching load capacitance (C_L) and the voltage swing (ΔV) should be minimized for power considerations. Also, reducing C_L and ΔV will minimize Δt (for highest speed performance).

For the device to have a high frequency response, a low noise figure, and, simultaneously, dissipate a minimum amount of energy, the design must achieve minimum values of C_L and ΔV, in addition to minimizing C_{gs}, R_s, R_g, R_c, and L_g. By fabricating the device on a semi-insulating substrate, the stray fringing capacitance of interconnects is low and — with proper design — the total load capacitance C_L can be reduced to approximately the C_{gs} of the next device to be driven. To determine the logic voltage swing (ΔV), the logic low voltage must turn the device off, so $V_{lo} \leqslant V_p$. For GaAs FET devices, the gate should not be driven higher than V_B.

$$V_{hi} \leqslant V_B \quad (10)$$

Therefore, a reasonable voltage swing would be

$$\Delta V_m = V_{hi} - V_{lo} = V_B - V_p = -U_o \quad (11)$$

U_o is the channel pinch-off voltage. Lowering the dynamic switching energy can therefore be accomplished by reducing U_o. This compromises the switching speed Δt (Equation 6) via I_{DS}, so tradeoffs are necessary.

From a processing standpoint, reducing the parameters R_c, R_s, C_{gs}, R_g, L_g, and U_o depend upon limits which change with the process state-of-the-art. Current photo-lithographic limits place L_g at about 1 micron (about 0.5 microns for e-beam). R_g is determined by the resistivity and thickness of the gate metallization. U_o is limited by the ability of the process engineer to control the variation in the pinchoff voltage ΔV_p (for IC yield considerations).
The contact resistance, R'_c, becomes a primary limitation as surface geometries shrink to accommodate LSI → VLSI gate densities. In Equation 12, where r_c is the specific contact resistance and ρ is the GaAs resistivity, it can be seen that reducing the ohmic contact area (w•d) of a device increases the contact resistance [9].

$$R'_c = \left[\frac{1}{W}\sqrt{\frac{r_c\rho}{a}}\right]\left[\coth\left(\sqrt{\frac{\rho}{ar_c}}\cdot d\right)\right] \tag{12}$$

Therefore, care must be taken in both processing and device design in order to minimize R'_c. The series gate resistance, R_g, is given by

$$R_g = \rho_m \frac{W}{L_g \cdot t_m} \tag{13}$$

where ρ_m is the metallization resistivity and t_m is the thickness. R_g must also be considered for either large gate widths or for long interconnect lines. In order to minimize L_g, either W must be reduced or t_m increased (process limitation) to keep R_g within acceptable bounds.

By combining Equations (8) and (9), we obtain the power product term which must be minimized for LSI and VLSI circuits:

$$P\Delta\tau \approx \frac{1}{2}\frac{C_{gs}^{\,2}\,U_o^{\,3}}{I_{max}} \tag{14}$$

where $I_{max} = \dfrac{2\epsilon\epsilon o\mu\, WU_o^{\,2}}{3aL_g}$. For a one-sided junction,

$$C_{gs} = L_g W \sqrt{\frac{q\epsilon\epsilon o N}{2(V_B - V_{gs})}} . \tag{15}$$

Using Equation (2), we obtain

$$P\Delta\tau \approx \frac{3}{8}\left(\frac{L_g{}^3}{\mu}\right)\left(\frac{WNqa\ Uo}{V_B - V_{gs}}\right) . \tag{16}$$

The need to reduce L_g is readily apparent as well as minimizing W and maximizing the mobility, μ.

An example of the device design parameters needed for a high performance E/D FET is given below. Consider the cross sections of a dual gate (Figure 4) and a single gate (Figure 5) MESFET structure constructed with a dual implant into semi-insulating GaAs. An important feature is that the GaAs FET gate overlaps the deep implant by an amount approximately equal to the mask alignment tolerance of ± 0.25 micron. Also, the SiO_2 dielectric thickness equals the gate metallization thickness (~0.5 microns). With 1 micron gate metal length photomask dimensions, it is simple to define ≈ 0.5 micron spacings in the deep implant directly under the gate, resulting in significantly reduced drain-source resistance (when the gate is fully forward biased).

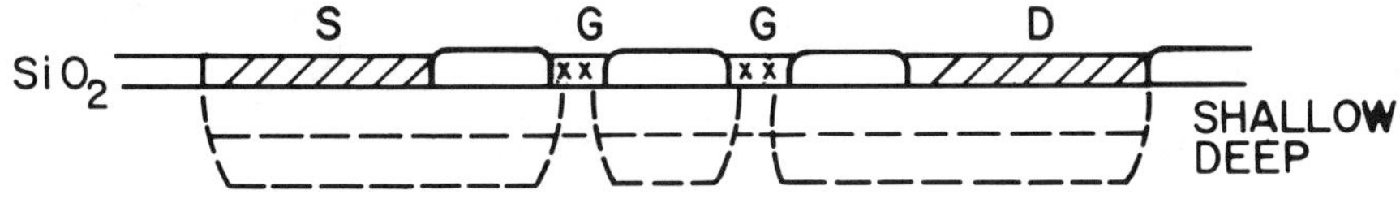

Figure 4/Dual gate GaAs FET cross-section.

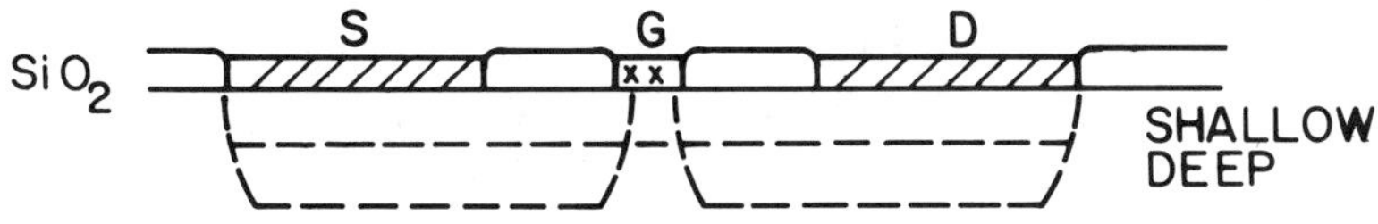

Figure 5/Single gate GaAs FET cross-section.

Table II gives the relevant calculated resistances, saturation currents and saturation voltages for the two structures in N–ON (−0.8V = V_p) and N–OFF (+0.1V = V_p).

TABLE II

V_p	Structure	R'_c	R_{sg}	R_c	R_{gd}	R_{ds}	I_{DSS} @ V_{gs} = +0.7V	$V_{ds\ Sat}$
− 0.8V	Single gate	6.9Ω	30Ω	20Ω	30Ω	87Ω	2.0 ma	0.5 V
−0.8V	Dual gate	6.9Ω	30Ω	70Ω	30Ω	137Ω	1.5 ma	0.7 V
+0.1V	Single gate	6.9Ω	30Ω	25Ω	30Ω	92Ω	1.0 ma	0.5 V
+0.1V	Dual gate	6.9Ω	30Ω	70Ω	30Ω	137Ω	0.8 ma	0.7 V

Assumptions:
1. 1000 Ω/□ shallow implant, 300 Ω/□ deep implant
2. $r_c = 10^{-6}\ \Omega\ cm^2$
3. L_g = 1 micron (metal); 0.5 μ (electrical)
4. W = 25 micron

Based on Table II for E/D FET devices (single and dual gate), one can conclude:

(i) $V_{ds\ sat}$ can be set at values of ⩽ 0.7 volts;

(ii) Dual input NAND is feasible with E/D FET;

(iii) With dual implant it is possible to minimize the series resistance *independent* of the pinchoff voltage.

The inference to be drawn from the evidence is that in order to realize low power, high speed ICs, $V_{ds\ sat}$ and C_{gs} must be minimized, and $I_{ds\ sat}$ must be maximized. Increasing the electron mobility and decreasing the gate length is necessary to achieve all of these goals. In the previous example, we showed that low contact resistance and a low sheet resistance deep implant are required to achieve low $V_{ds\ sat}$ values. In addition, by reducing the "effective" gate length below the actual metal gate, both R_{ds} and $V_{ds\ sat}$ are reduced, while $I_{ds\ sat}$ is increased. The design of E/D FET circuits is feasible using these device and processing techniques.

4. CIRCUIT DESIGN

Some of the basic circuit building blocks available to the E/D FET designer are shown in Figure 6.

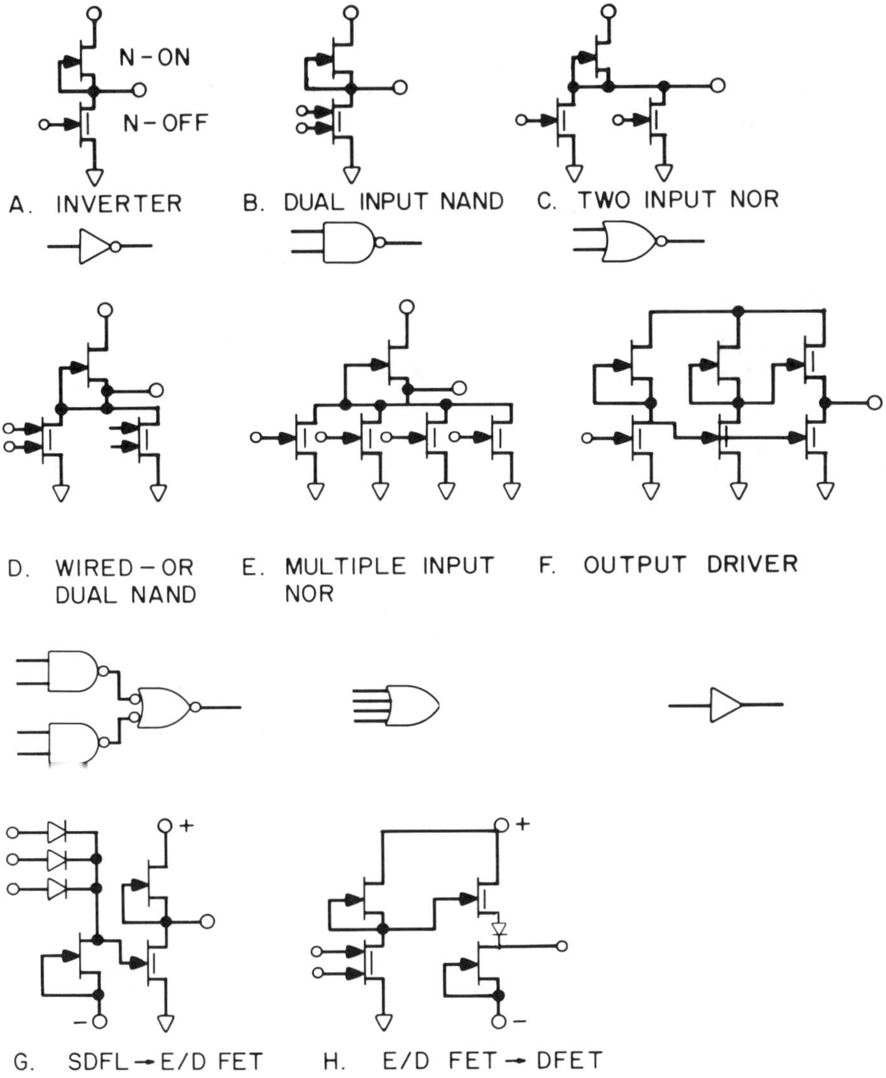

Figure 6/E/D FET circuit elements:

The first step in the circuit designer's job is to select an approach compatible with his speed, power and die size requirements. If these requirements identify the E/D FET approach, the second concern is capacitance loading, which determines the required drive currents (hence, driver width) from one gate to the next. Then one must derive the proper ratios of load to driver widths which will provide appropriate values of V_{lo} and V_{hi} while attaining gate speed, power and size goals. Assuming a DFET active load, the sequence for deriving this ratio would be:

(a) Obtain the actual $I_{ds}-V_{ds}$ data for the individual depletion and enhancement devices;

(b) Obtain $V_{ds\ sat}$ and R_{ds} values for both types of devices;

(c) Determine the ratio of W_{DRIVER}/W_{LOAD}, (W_D/W_L), that provides acceptable V_{LOW} from measured data; or

(d) Calculate voltage transfer curves (E_{out} vs. E_{in} to determine the W_D/W_L ratio that provides acceptable V_{low}.

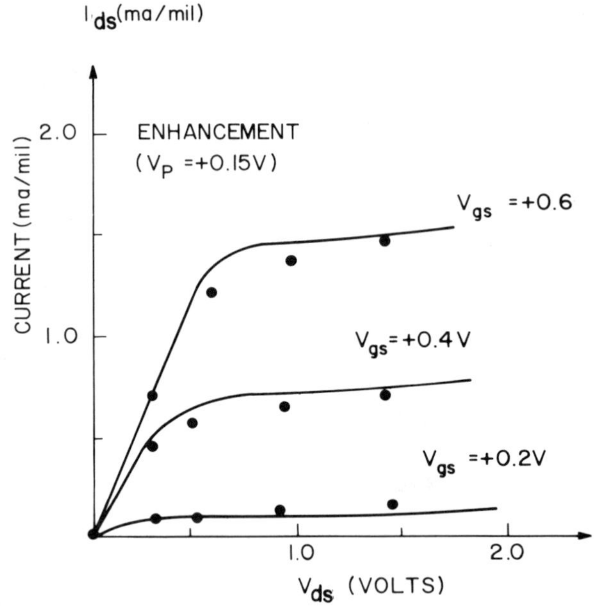

Figure 7/Enhancement mode characteristics.

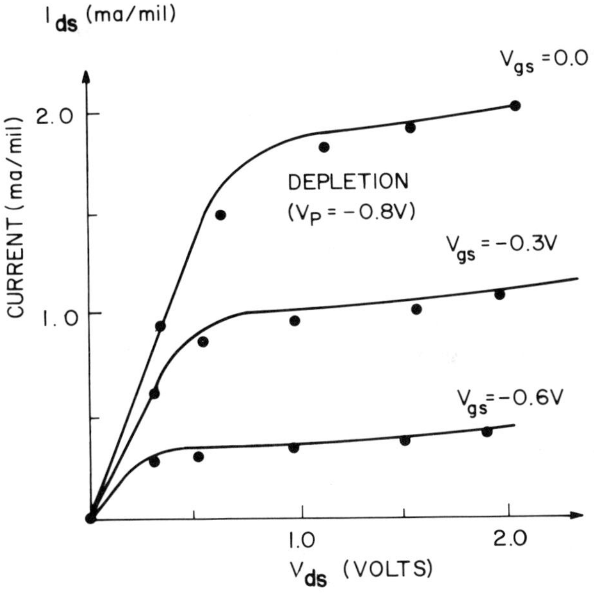

Figure 8/Depletion mode characteristics.

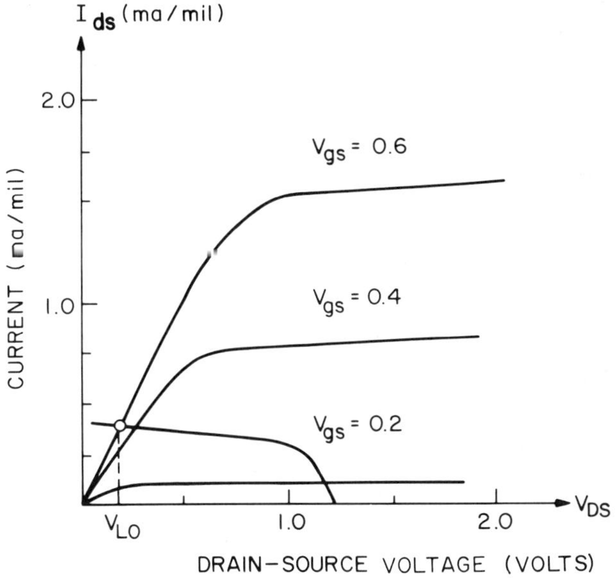

Figure 9/Characteristics of an enhancement driver with a depletion mode.

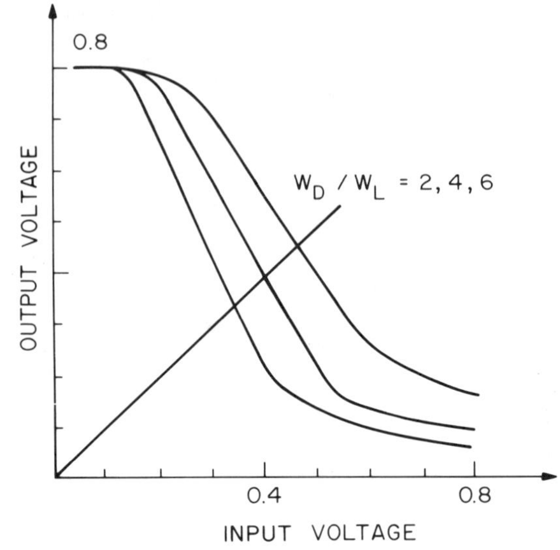

***Figure 10**/Transfer curves for single input gate E/D FET inverter.*

Figures 7 and 8 show measured and calculated characteristics for the enhancement and depletion GaAs FETs. Figure 9 depicts the enhancement driver connected to a depletion load (E/D FET inverter). Figure 10 gives the voltage transfer curves calculated for the E/D FET inverter. Figure 11 shows the voltage transfer curves for the dual gate inverter configuration.

It is evident from Figures 7 through 10 that W_D/W_L ratios of 4 and 6 are sufficient for V_{lo} = +0.10V for the single and V_{lo} = +0.20V for the dual input E/D FET inverter gates, respectively.

Approximate values for Δt (Equation (9)), P (Equation (8)) and $P\Delta\tau$ (Equation(17)) are obtained using the following assumptions: (1) stray capacitance is small compared to the input gate capacitances; (2) the gate capacitances are constant; and, (3) the currents supplied by the load and driver are constant and equal to $I_{ds\ sat}$. (More rigorous and accurate calculations require circuit simulation methods.) Figure 12 shows a plot of the calculated (and measured) transfer curves from which we obtain $I_{ds\ sat}$ values for the GaAs FET driver and load. The measured values of C_{gs} are 0.8 to 1 ff/micron^2 for high performance devices.

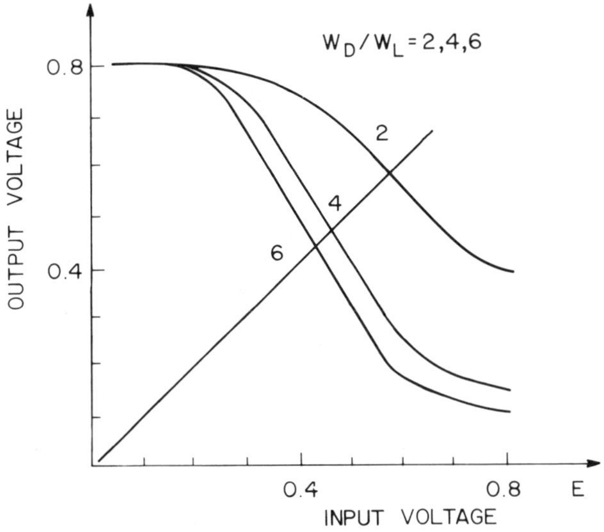

Figure 11/Transfer curves for dual gate inverter.

Using the inverter shown in Figure 13, with $W_D/W_L = 3$ and FO = 2 (allowing 15 ff capacitance for the pull-up device and 10 percent stray capacitance), we can calculate the output voltage rise time (input driver off):

$$\Delta t_{rise} = \frac{C_L(V_{DD} - V_{LO})}{I_{DS_{load}}}$$

$$= \frac{C_L(V_{DD} - V_{LO})}{0.6\ \text{ma}\left\{\dfrac{V_{DD} - V_{LO}}{1.46} - \dfrac{2}{3}\left[\dfrac{V_{DD} - V_{LO} - 0.8}{1.46}\right]^{3/2} + \dfrac{2}{3}\left[\dfrac{0.8}{1.46}\right]^{3/2}\right\}} \tag{17}$$

where the NqμaWUo/L values have been obtained from experimental GaAs FET I_{ds} vs. V_{gs} curves given in Figure 12 for N–ON and N–OFF devices fabricated on the same wafer. Similarly,

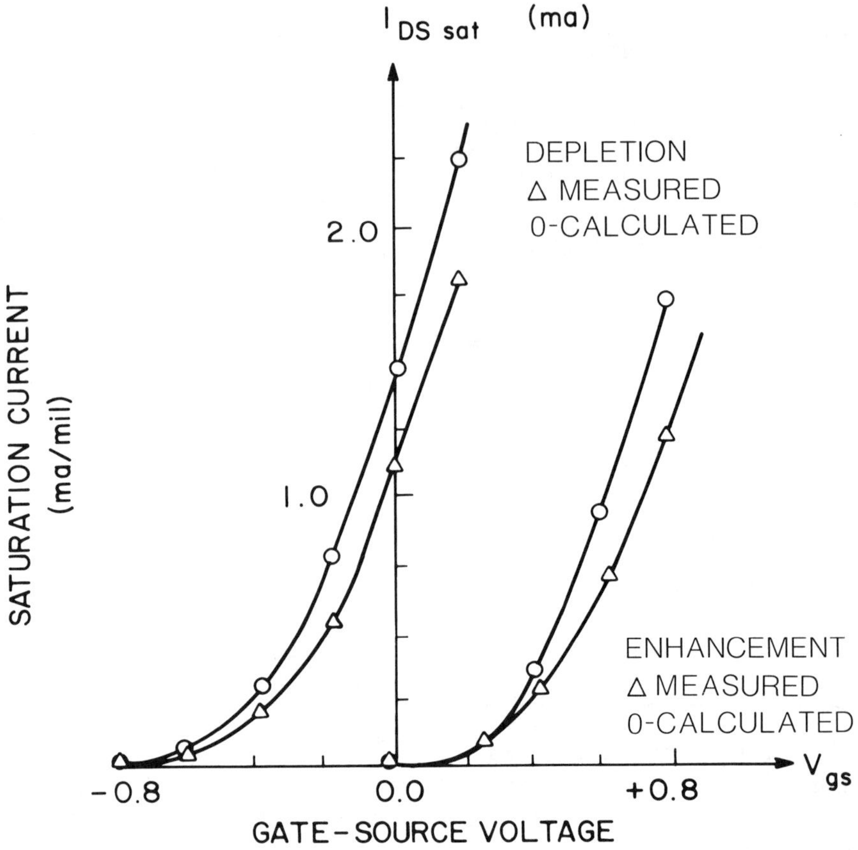

Figure 12/I_{ds}(sat) vs V_{gs} transfer characteristics.

$$\Delta t_{fall} = \frac{C_L (V_{DD} - V_{lo})}{\underset{\text{Driver}}{I_{ds}} - \underset{\text{Load}}{I_{ds}}}$$

$$= \frac{C_L (V_{DD} - V_{lo})}{1.44\ \text{ma} \left\{ V_{lo} - \frac{2}{3}\left[\frac{V_{lo} - 1.4}{1.0}\right]^{3/2} + \frac{2}{3}\left[\frac{1.4}{1.0}\right]^{3/2} \right\} - 0.60\ \text{ma} \left\{ \frac{V_{DD} - V_{lo}}{1.46} - \frac{2}{3}\left[\frac{V_{DD} - V_{lo} - 0.8}{1.46}\right]^{3/2} + \frac{2}{3}\left[\frac{0.8}{1.46}\right]^{3/2} \right\}} \tag{18}$$

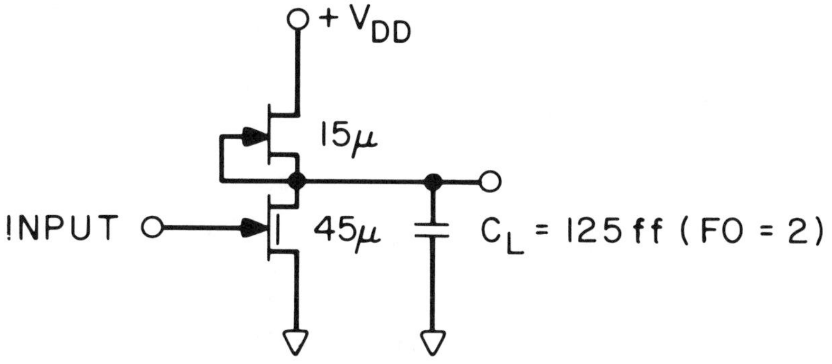

Figure 13/E/D FET inverter.

Values for Δt_{rise} and Δt_{fall} curve are calculated and plotted in Figure 14. It is readily apparent that the rise and fall characteristics are more controllable and provide better performance for the ratio W_D/W_L = 4 than for W_D/W_L = 3. A computer simulation can also be generated which plots the output voltage waveforms for various W_D/W_L ratios as a function of time, in which C_L, ΔV_{in}, and V_{DD} are treated as variables in order to optimize circuit performance.

Note that propagation delays in the region of 250 ps are feasible, where gate lengths are 1 micron, FO = 2, with power dissipations of ≈ 200 microwatt per gate. An input voltage of V_{MAX} = 0.6 volts was used in this calculation which, if extended to $V_{MAX} \leqslant$ +0.75 volts, would result in an approximate doubling in speed performance.

Finally, it should be noted that experimental $I_{DS\,sat}$ values were used in the calculations and that further significant reductions in the series gate, source, contact and drain-source resistances are possible. Of all the competing GaAs technologies, the E/D FET technology will benefit the most from continued processing improvements.

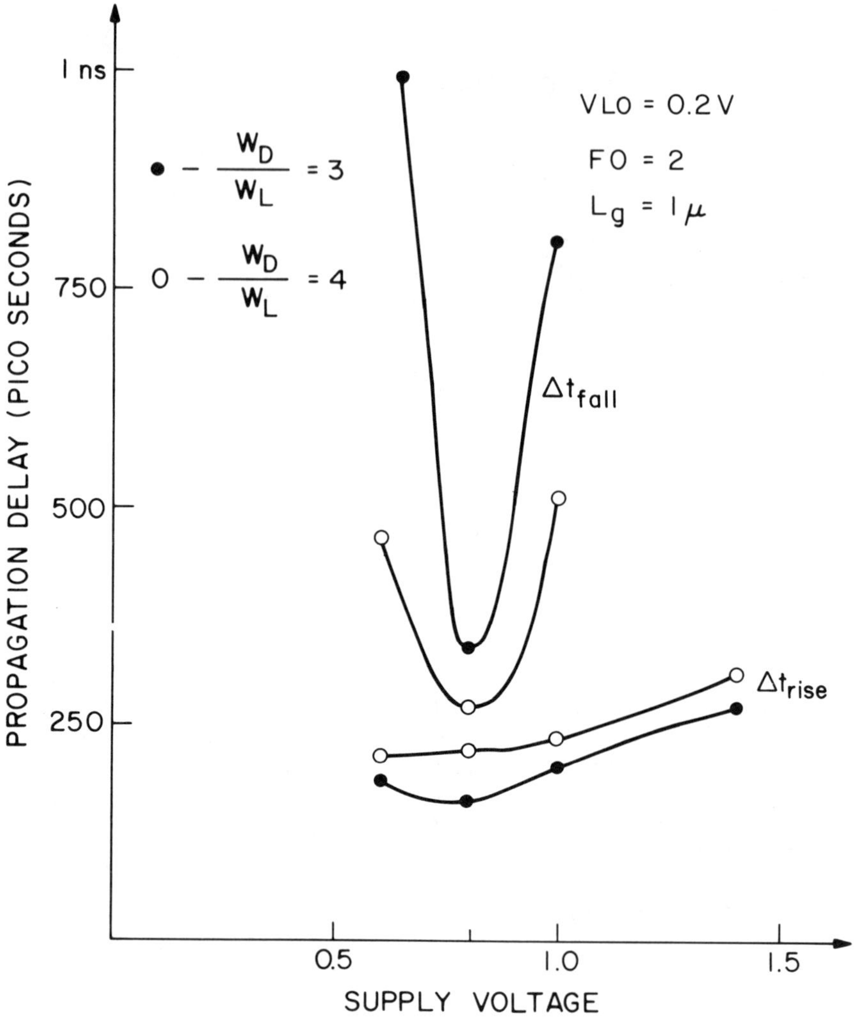

Figure 14/Propagation delay vs applied supply voltage for the E/D FET inverter.

5. CONCLUSIONS

The GaAs E/D FET approach for high speed, low power LSI or VLSI circuit applications is very promising and is currently being developed in several major laboratories. While somewhat more difficult to process than the DFET or SDFL approaches because of the tighter control of pinchoff voltage required, this control (±30 mv to 50 mv) has been achieved on selected wafers fabricated with the planar ion implantation process. DFET and E/D FET ICs of up to 30 EGC complexity have been simultaneously fabricated on the same wafer, tested, and are operational. Interfacing DFET and E/D FET gates on the same wafer is a straightforward procedure and offers a degree of flexibility appreciated by circuit designers.

Mention should be made of the utility of the E/D FET approach in a wide range of applications in the low power, high frequency regions inaccessible to silicon-based technologies, i.e., the ≈500 MHz to 3 GHz frequency range where LSI to VLSI technologies require the designer to limit gate dynamic energies to the range of 1 picojoule to 0.03 picojoule, to prevent excessive power dissipation. ED/FET devices fit this requirement very nicely and, in addition, can be operated from a single-voltage battery source.

The primary limiting factors in producing GaAs IC circuits with good performance and reasonable yields are the quality of the GaAs starting materials and the uniformity of the implant/anneal process. The E/D FET technology is the most vulnerable of the three or four alternative approaches due to relatively low signal levels, but, conversely, it stands to gain the most from the GaAs material and processing improvements that are constantly occurring.

The relatively low signal voltage swings (≈ 0.5 volt) of E/D FET ICs are a great advantage for LSI to VLSI complexities because of stringent power dissipation constraints. The larger forward gate voltages possible with the E-JFET and MOSFET approaches will adversely increase power dissipation/gate with little or no gain in speed. Figure 15 shows an E-JFET cross-section [10]. The fabrication procedure is similar to that for the E/D FET, with the major exception of P+ diffusion under the gate. Given the ≈ 1 micron dimension limits imposed by present lithography systems, and the lateral spreading diffusion of the operation, present gate lengths for E–JFETS are limited to 3 microns. Availability of an ion implanted P+ process should alleviate the problem; however, a ≈ 1 micron photolithographic limitation, coupled with alignment errors of ± 0.25 microns for P+ implant-to-gate-metal, will result in a minimum electrical gate length $L_g \gtrsim 1.5$ microns for the E-JFET.

This will result in a gate capacitance for the "1 micron" JFET which is larger than a E/D FET of similar lithographic geometry. The additional forward-voltage bias ($V_B \approx 1.0$ volts) possible with the E-JFET approach will result in an increase in power dissipation without an increase in speed performance comparable to the E/D FET.

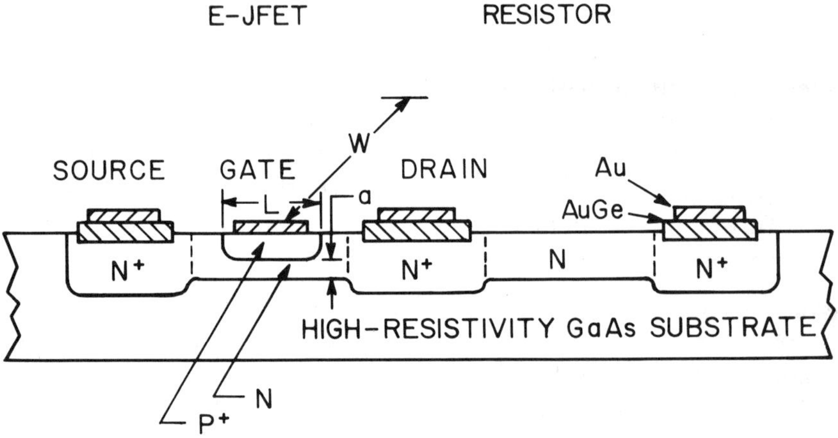

Figure 15/Cross-section of planar GaAs E-JFET inverter made by selective ion implantation.

Another significant technical advantage that the E/D FET may enjoy compared to the E-JFET is better control of the pinchoff voltage due to processing variations. On n-type channels, pinch-off voltage of $V_p = -(0.8 \pm 0.07)$ volts across the entire wafer surface (as implanted and annealed) has been found. Plasma etching techniques are capable of producing etch uniformities across a wafer surface of $< \pm 5\%$. This means that, for a 300 Å etch which transforms the DFET ($V_p = -0.8$ volts) into the E/D FET,

$$\Delta V_p \cong \sqrt{(\Delta V_{implant})^2 + (\Delta V_{etch})^2} \tag{19}$$

or

$$\Delta V_p \approx \pm 0.076 \text{ volts} \quad \text{(for E/D FET)} \tag{20}$$

where $\Delta V_{implant} = \pm 0.070$ volt. (1 σ)

The E-JFET threshold variation will be greater due to the need to control two separate implants and activations. Again, using the $\Delta V_{implant(n)}$ ± 0.070 volts for the n-type and assuming the availability of a well controlled, shallow, p-type implant capable of producing $\Delta V_{implant(p)}$ ± 0.070 volt pinchoff variation, then, using Equation (26) we find, in the ideal case,

$$\Delta V_{p(p/n)} \cong \pm 0.099 \text{ volts} \quad (1\ \sigma) \tag{21}$$

Naturally, if the p-type implant is not well controlled, the pinch-off voltage variation will be larger than the value in Equation (21).

In summary, for LSI and VLSI applications, the E/D FET approach has advantages over the DFET and JFET approaches, provided the N–OFF etch uniformity is held to ± 5 percent or less. With the dual implantation/planar process, very low $V_{ds\ sat}$ voltages are feasible ($\approx$ 0.5 volt) (when coupled with good metallization techniques) for normally-off GaAs FET operation and the dual gate NAND structure is equally feasible when coupled with the dual implant process and $\leqslant$ 1 micron gate lengths. We have demonstrated methods for deriving the optimum driver/load gate width ratio, shown to be $\approx$ 4 for V_p(load) = – 0.8 volts and V_p(driver) = + 0.15 volts (fan out = 2). The need to provide low contact resistance ($10^{-6}\ \Omega cm^2$) and low source to gate resistance has also been shown.

The E/D FET approach offers many possible gate configurations for the circuit designer and techniques to extend the fan-in and fan-out capability without seriously affecting the gate propagation delay. We have shown the need for high speed, low power, and LSI GaAs ICs [11, 12, 13]. Finally, Figure 16 shows a power vs. delay plot from which power products below 10 fj and propagation delays of 40 ps may be predicted for the new generation ($L_g \approx$ 0.5 micron) devices which use deep UV or e-beam lithographic techniques.

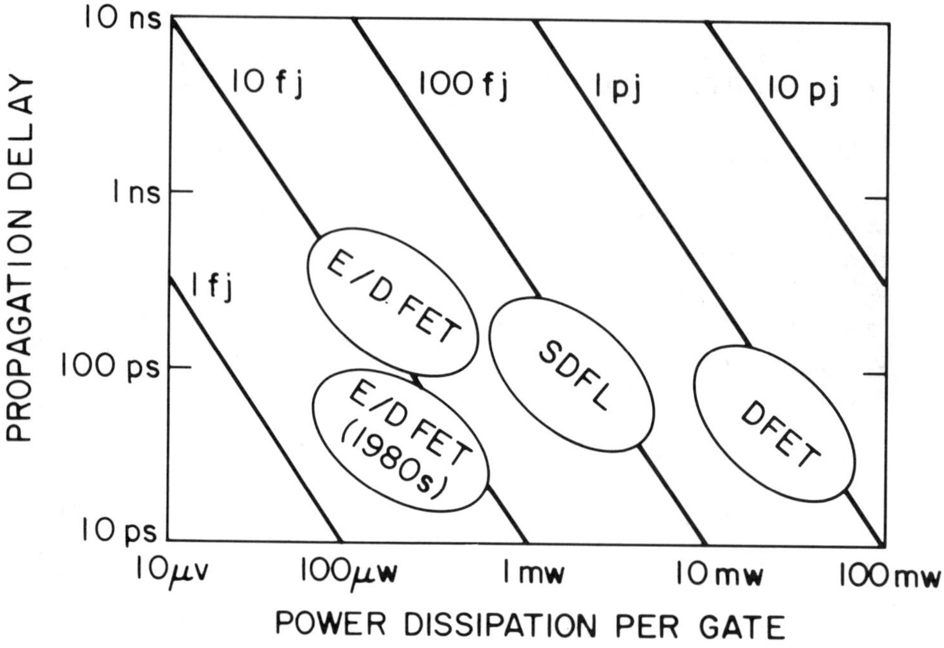

Figure 16/GaAs FET IC performance.

REFERENCES

[1] Wilson, D.O., W.R. Scoble, R.P. Mandal and H.L. Petersen, "A Comparison of High-Speed Enhancement and Depletion Mode GaAs MESFETS," International Electron Devices Meeting, Washington, DC, Dec. 5, 1978, pp. 601-603.

[2] Mandal, R.P., and W.R. Scoble in *Gallium Arsenide and Related Compounds (St. Louis) 1978* (Inst. Phys. Conf. Ser. no. 45) pp. 462-471. London: Adam Helger/The Institute of Physics, 1979.

[3] Phillips, D.H., W.W. Grannemann, L.E. Coerver, and G.J. Kuhlmann, "Fabrication of GaAs MIS Capacitors Using a Thermal Oxidation Dielectric-Growth Process," presented at the 141st National Meeting of the Electrochemical Society, Houston, Texas, May 7-12, 1972; *Journal of the Electrochemical Society,* 120:1087, 1973.

[4] Leichti, C., R. Van Tuyl, R.E. Lee, and E. Gowen, *IEEE Journal Solid State Circuits,* SC-12:485, 1977.

[5] Eden, R.C., "GaAs Integrated Circuits MSI Status and VLSI Prospects," 1978 Int. Electron Devices Meeting, *Tech. Digest,* Dec. 1978, pp. 6-11.

[6] Eden, R.C., et al., *IEEE Journal Solid State Circuits,* SC-14: 221, 1979.

[7] Hewitt, B., et al., *Electron. Lett.,* 11:309, 1976.

[8] Hartgring, C.D., W.G. Oldham, and Tsu-Yin Chiu, *Solid State Electronics,* 23:121, 1980.

[9] Macksey, H.M., and R.L. Adams, "Fabrication Processes for GaAs Power FETs," *Proc. Fifth Biennial Cornell Elect. Eng. Conf.,* Cornell University, Ithica, NY, 1975.

[10] Zuleeg, R., J. Notthoff, K. Lehovec, *IEEE Trans. on Electron. Devices,* ED-25:628, 1978.

[11] Phillips, D.H., *Military Electronics/Countermeasures,* 5:24, 1979.

[12] Phillips, D.H., "Outlook for the Development of Gallium Arsenide Circuits," Paper No. 1, IEEE Gallium Arsenide Integrated Circuit Symposium, Lake Tahoe, Nevada, September 27, 1979.

[13] Phillips, D.H., and H.L. Petersen, "High-Speed Gallium Arsenide Integrated Circuit Development for Satellite Systems," published by New York University and AIAA, *Progress in Astronautics and Aeronautics,* 67:391, 1979.

Direct-Coupled FET Logic (DCFL)

K. Lehovec
University of Southern California, Los Angeles, California

R. Zuleeg
McDonnel Douglas Astronautics Co., Huntington Beach, California.

1. INTRODUCTION

The inverter is the basic unit of logic circuits which perform the functions AND, OR, NAND, and NOR, and provide memory. The prototype inverter circuit consists of a transistor switch in series with a load component (Figure 1). The negative terminal of the supply voltage V_{DD}, i.e., the transistor source, is grounded for convenience. The transistor is conducting when the "high" input voltage $V_I = V_{IH}$, corresponding to logic "1", applied at the transistor gate, exceeds the threshold voltage V_T. The output voltage is then low, i.e., $V_O = V_{OL}$ is close to zero since the transistor resistance is small compared to the load resistance. The transistor is non-conducting when the "low" input voltage $V_I = V_{IL}$, corresponding to logic "0", is less than the threshold voltage V_T. The output voltage $V_O = V_{OH}$ is then close to the supply voltage V_{DD} provided that no current is drawn from the output node.

Since the low output voltage V_{OL} of an inverter is inherently positive, direct coupling of the low output of a first inverter to the input of a second inverter, i.e., $V_{OL}^{(1)} = V_{IL}^{(2)} < V_T$ requires a "normally-off" enhancement mode FET of positive threshold voltage. In this chapter we shall be concerned with this Direct Coupled FET Logic (DCFL) [1].

The logic swing

$$L_S = V_{OH} - V_{OL} = V_{IH} - V_{IL} \tag{1}$$

is limited by the forward gate bias V_{GM} at which the transistor gate junction becomes highly conducting. This occurs at about 0.7 V for metal Schottky barrier gate field effect transistors [2] and at about 1.2 V for junction gate field effect transistors (JFET) [3].

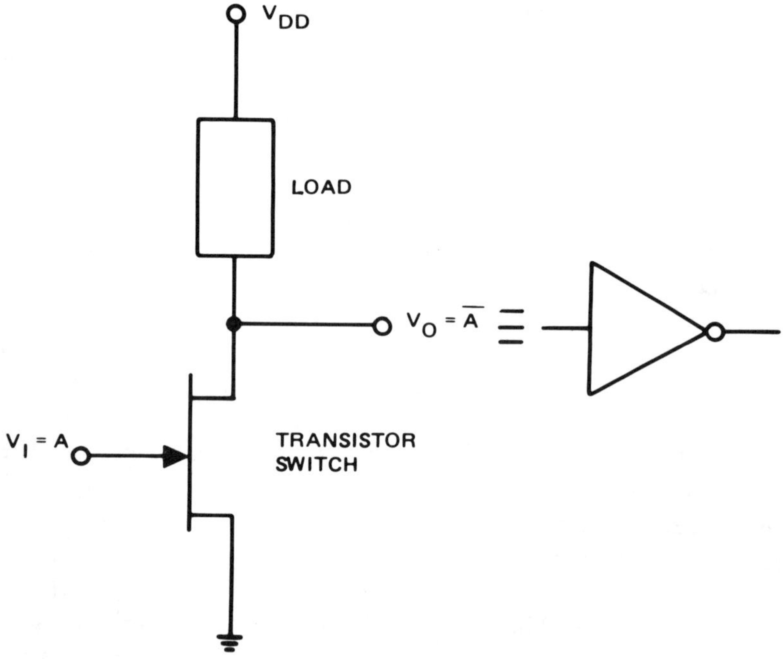

Figure 1/Basic inverter switch.

The NAND circuit with fan-in 2 shown in Figure 2 uses two series connected switching transistors. The two transistors can be combined into a single dual-gate transistor [4, 5, 6]. The NOR circuit of Figure 3 comprises several parallel switching transistors. Figure 4 shows a flip-flop composed of two inverter circuits to provide a memory cell [1]. Figure 5 shows a master-slave flip-flop connected as a "divide-by-two" circuit composed of inverter subcircuits.

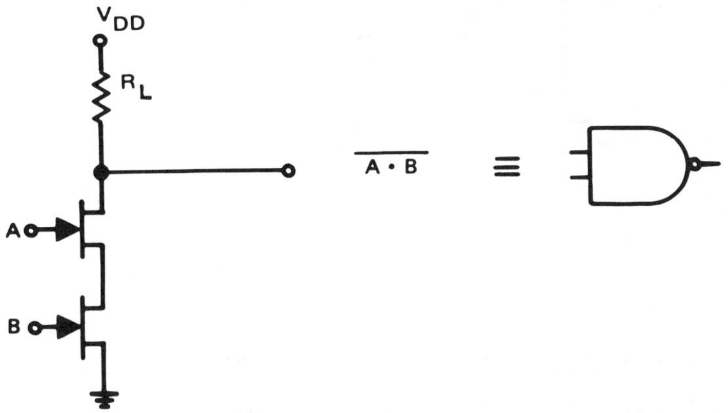

Figure 2/NAND gate with fan-in 2 using two transistors in series which can be combined into a single dual-gate transistor.

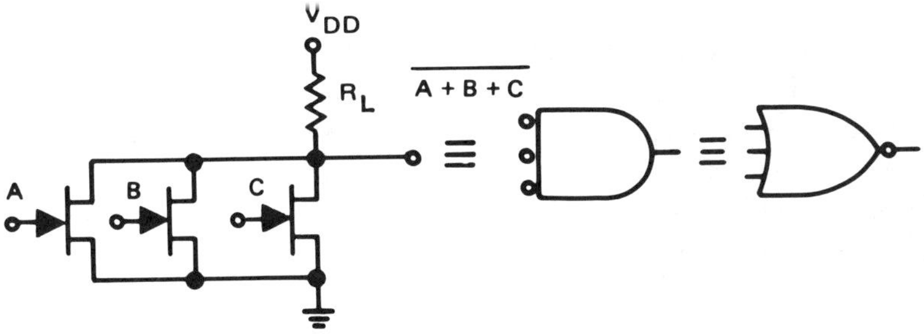

Figure 3/NOR circuit of fan-in 3.

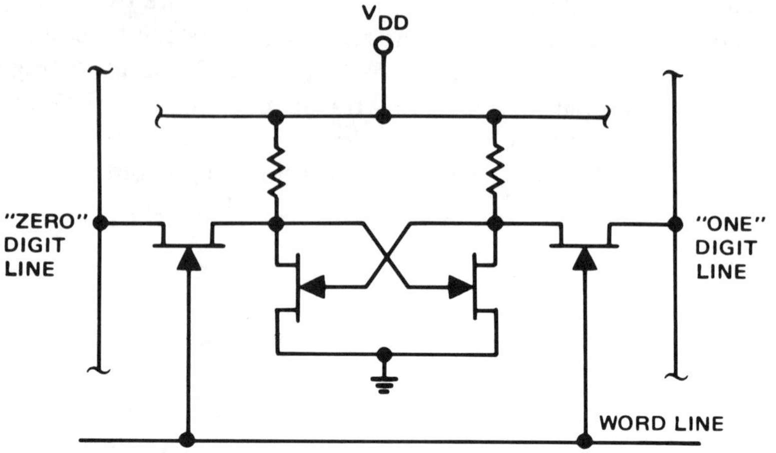

Figure 4/Memory cell for DCFL (according to Jutzi [1]).

2. THE GaAs FIELD EFFECT TRANSISTOR

Before analyzing the performance of the inverter circuit, we shall briefly review pertinent properties of the n-channel GaAs FET.

2.1 Threshold Voltage, Pinch-off Voltage and Built-in Voltage

The GaAs field effect transistor uses the depletion layer under a gate electrode to modify the width of an undepleted n-channel which affects the electron flow between a source and a drain contact to the channel. The steady state transistor current I_T depends on the voltages V_G and V_D applied at the gate and drain, respectively, with respect to the grounded source. The current vanishes when the depletion layer under the gate near the source extends across the entire n-layer, thereby pinching off the electron flow at the source. The potential across the fully depleted n-layer is the pinch-off voltage

$$V_P = (q/\epsilon) \int_0^\infty N(y)y\,dy \qquad (2)$$

where ϵ is the permittivity of GaAs and N(y) is the distribution of the dopant concentration in the direction normal to the gate elec-

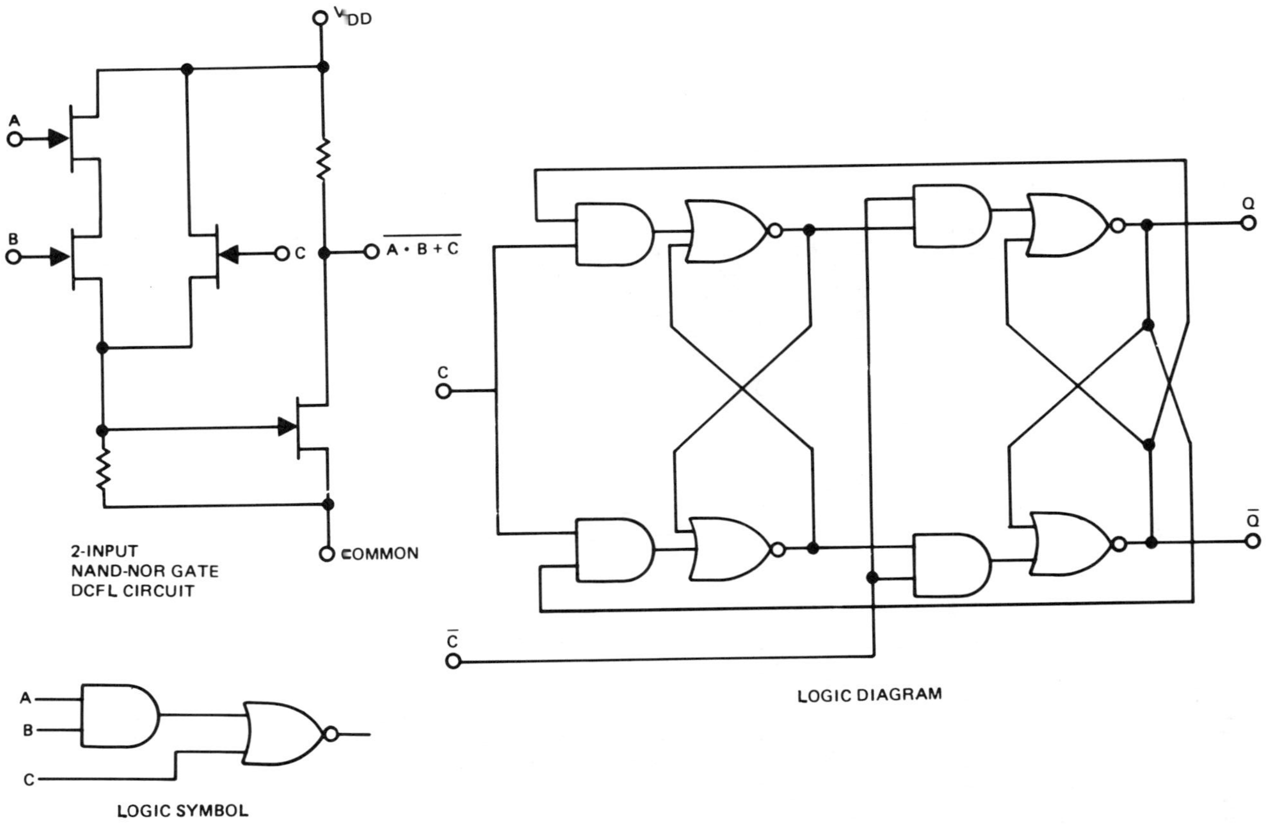

***Figure 5**/DCFL master-slave flip-flop connected as frequency divider. The basic unit is the input buffered NAND-NOR gate shown at the upper left.*

trode. The threshold voltage V_T is the gate voltage at which the depletion layer just pinches off the channel at the source. The threshold voltage differs from the pinch-off voltage because of the built-in voltage V_B between gate and n-channel.

$$V_T + V_B = -V_P, \text{ i.e., } V_T = |V_B| - V_P \tag{3}$$

The negative sign in front of V_P results from our definition of V_P as the voltage across the depletion layer with respect to gate, while V_G, V_T, and V_B are defined with respect to channel. There are confusing notations in the literature for the externally applied gate-to-source voltage which just pinches off the conducting channel at the source (our threshold voltage V_T) [1, 7, 8, 0], and for the internal potential developed across the fully depleted n-layer (our pinch-off voltage V_P). Some authors [4, 5, 10] refer to our threshold voltage V_T as the pinch-off voltage V_P. This confusing notation originates with Shockley's paper [3], which neglects the built-in potential, in which case the pinch-off voltage equals the threshold voltage: $V_T \cong -V_P$. At that time, the neglect was justified, since the pinch-off potentials of the early transistors were much larger than the built-in potential. However, this is no longer the case. The absolute value of the built-in potential is in fact larger than the pinch-off potential in the normally-off, enhancement mode FETs used in DCFL.

The built-in voltage arises from the contact potential difference between gate and channel. For the p^+-gate of a JFET, the built-in voltage to a channel of the typical dopant concentration 10^{17} cm^{-3} is about −1.4 V, while that for the metal Schottky barrier gate of a Schottky gate FET is −0.8 V. The Schottky barrier built-in voltage is nearly independent of the type of metal electrode (e.g., Al, Ti-Pt-Au, Cr-Pt-Au), since the Fermi level at the GaAs surface is pinned by surface states to a position at about 2/3 the forbidden band below the bottom of the conduction band [11]. The pinch-off voltage for a uniformly doped n-layer of thickness a is, according to Equation (2),

$$V_P = Nqa^2/2\epsilon \tag{4}$$

However, ion-implantation is now used almost exclusively for doping, resulting in the approximately Gaussian dopant distribution [12, 13]

$$N = N_o \exp[-(y - R_p + \ell)^2/2\sigma_p^2] \tag{5}$$

for which

$$V_P = \frac{N_o q}{2\epsilon} (R_p - \ell)\sigma_p \sqrt{\frac{\pi}{2}} \left[1 + \text{erf} \frac{R_p - \ell}{\sigma_p\sqrt{2}}\right.$$

$$\left. + \sqrt{\frac{2}{\pi}} \frac{\sigma}{R_p - \ell} e^{-(R_p-\ell)^2/2\sigma_p^2}\right] \tag{6}$$

Here ℓ (assumed to be less than the projected range R_p) is the depth of the metallurgical junction below the wafer surface for the p^+-gate of a JFET, while $\ell = 0$ for the Schottky gate FET. Depending on implant dosage and depth, n-channel transistors can be fabricated with either positive or negative threshold voltages. Transistors with positive threshold voltage are normally-off, i.e., non-conducting at zero gate input voltage, and require a positive gate voltage $V_G > V_T > 0$ to become conducting. Thus, normally-off transistors are operated in the enhancement mode, i.e., they are made conducting by a positive gate voltage which biases the gate in the forward direction. Transistors with negative threshold voltage are normally-on, i.e., they are conducting at zero gate input voltage, and are made non-conducting by a negative gate voltage $V_G < V_T < 0$, which biases the gate in the blocking direction. Thus, normally-on transistors are operated in the depletion mode for turning the transistor off.

2.2 Steady State Current-Voltage Characteristics

The steady state I_T-V_D-V_G characteristics for a uniformly doped channel have been derived by Shockley [3], assuming that electron drift velocity is proportional to the electric field. According to this theory, the current saturates for

$$V_D \geqslant V_{SAT} = V_G - V_T \tag{7}$$

when the channel is pinched off at the drain. For ion-implanted channels, the MOSFET-like relation

$$I_T = 2K[(V_G - V_T)V_D - V_D^2/2] \tag{8}$$

with

$$K = W\mu\epsilon/2a_{eff}L \tag{9}$$

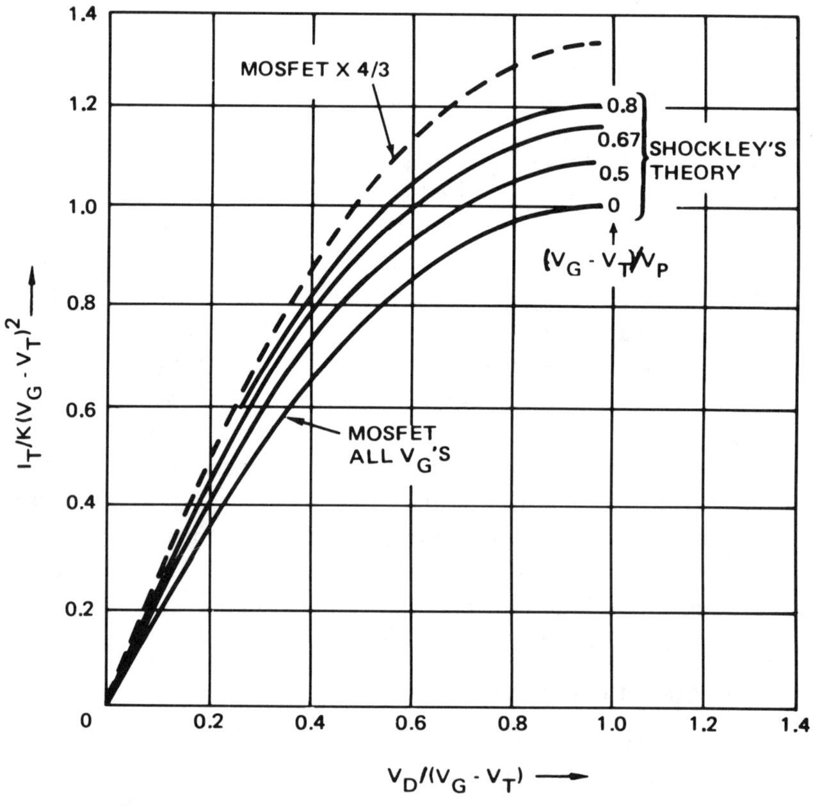

Figure 6/Comparison of Shockley's transistor characteristics [3] with that of a MOSFET, equation 8. The dotted line is an approximation proposed by reference 14.

and Equation (3),

$$a_{eff} = R_p + 2\sigma_p \sqrt{\frac{2}{\pi}} \tag{10}$$

appears more appropriate [8], since the width of the depletion layer varies little along most of the channel and with applied gate voltage, its boundary being pinned more or less to the peak of the dopant concentration. The current also saturates for $V_D \geqslant V_{SAT} = V_G - V_T$, at

$$I_{SAT} = K(V_G - V_T)^2 \tag{11}$$

Figure 6 compares the transistor characteristics of Shockley's theory [3] with that of a MOSFET, Equation (8).

The saturation of current for $V_D > V_{SAT}$ is not complete, however, since an effective channel length which decreases with increasing $V_D - V_{SAT}$ replaces the actual channel length L in Equation (9). For $V_G \leqslant V_{GM}$ the gate current may cause a significant voltage drop across a parasitic source resistance R_S located between source contact and the channel opening. This causes decreasing increments in current for equal increments in gate voltage, as expected from Equation (11) when replacing V_G by $V_G - R_S \cdot I_{SAT}(V_G)$.

2.3 Hot Electron Effects

At fields where $E \geqslant 5 \times 10^3$ V/cm, which occurs over a large portion of the channel for $V_{SAT} > 0.5$ V at L = 1 μm, the conduction electrons become "hot" so that they do not maintain thermal equilibrium with the crystal lattice; the drift velocity is then no longer proportional to the field, but instead passes through a maximum of $v_s = 2 \times 10^7$ cm/s with increasing field, and decreases to a field-independent saturation value of about 1×10^7 cm/s [15]. This decrease is caused by hot electrons populating an upper conduction band of large effective mass.

When drift velocity saturation is taken into account, the channel cannot be fully pinched off at the drain if a finite current is to flow, and the drain voltage V_{SAT} at which the current saturates is thus less than $V_G - V_T$.

The expression for the saturation current (11) can be modified for drift velocity saturation by replacing the average drift velocity of thermal electrons, $\mu V_{SAT}/L = \mu(V_G - V_T/L)$, by v_s, which results in

$$I_{SAT} = (W\epsilon v_s/2a)(V_G - V_T) \qquad (12)$$

Comparison with Equation (11) shows that the saturation current is reduced by drift velocity saturation, and that the saturation current increases linearly in case of a drift velocity saturation with gate voltage in excess of threshold voltage, and increases quadratically in the absence of drift velocity saturation. Thus, drift velocity saturation can be noted by equal increments of saturation current for equal increments of gate voltage, as contrasted to increasing increments in saturation current for equal gate voltage increments in the absence of drift velocity saturation. Shockley's theory [3] has been extended to incorporate drift velocity saturation [16] by using the Troffimenkoff expression [17]

$$v = \mu \partial V/\partial x \cdot (1 + \mu \partial V/\partial x/v_s)^{-1} \qquad (13)$$

with $v = \mu \delta V/\delta x$ and $v = v_s$ as the limiting cases for fields small and large respectively, compared to $v_s/\mu \cong 5 \times 10^3$ V/cm. The current of the Schockley theory is reduced by the factor $1+\mu V_D/Lv_s = 1+zV_D/V_P$ where $z = \mu V_P/Lv_s$ characterizes the importance of drift velocity saturation, $z \ll 1$, providing the Shockley case. However, the Trofimenkoff expression gives a much more gradual transition from ohmic behavior to drift velocity saturation than has been observed in GaAs [15]. A more abrupt transition is provided by [18]

$$v = 2\mu \partial V_c/\partial x[1 + \sqrt{1 + (2\mu \partial V_c/\partial x/v_s)^2}]^{-1} \qquad (14)$$

Another widely used but less convenient expression for the $I_T - V_D$ characteristics with drift velocity saturation is given in reference 19. A stationary Gunn domain, which may form in the high field region near channel pinch-off at the drain side of the gate [20-23] could reduce the saturation current. However, this effect should not be significant in VLSI logic circuits anticipated to operate at low supply voltages.

2.4 Transient Behavior

A change of the gate voltage from below to above the threshold voltage or vice versa causes a change in the number of free electrons in the channel. This requires a period during which drain and source currents differ. The duration of this period following an abrupt change of the gate voltage is the intrinsic response time of the transistor, which is only about 10 ps for a transistor of 1 μm gate length. It is, therefore, permissable to use the steady state $I_T - V_G - V_D$ relationship with the instantaneous values of $V_G(t)$ and $V_D(t)$ in the calculation of the fall time of the inverter, and $I_T = 0$ for the rise time. The delay caused by the change of electron number in the channel requires that source and drain currents differ (which is ignored using the steady state current). This can be accounted for by adding an effective "channel capacitance" C_c to the node capacitance. This capacitance is chosen so that the associated charge change during rise and fall time, i.e., $C_c(V_{OH} - V_{OL})$, equals:

(i) in the case of fall time, the entire mobile channel charge, $Q_c(V_D = V_{OL})$, and

(ii) in the case of rise time, half of the mobile channel charge. The latter results from the fact that during the rise time half the channel electrons exit through the drain, and the other half exit through the source, while during the fall time all electrons enter from the source. Using the "MOSFET approximation" where all channel electrons and donors are assumed to be located in a lamella at distance a_{eff} from the gate junction, and considering that the transistor operates in the linear regime at $V_D = V_{OL} \ll V_{SAT}$, we have

$$|Q_c| = C_{GS}(V_{GH} - V_T) \tag{15}$$

with

$$C_{GS} = \epsilon WL/a_{eff} \tag{16}$$

so that, for DCFL where $V_{GH} = V_{OH}$,

$$C_c(\text{fall}) = C_{GS}(V_{OH} - V_T)/(V_{OH} - V_{OL}) \tag{17}$$

while C_c(rise) = C_c(fall)/2. A typical value for the gate capacitance for L = 1 μm channel length transistors is C_G/W = 1 fF per micron channel width, since $a_{eff} = 10^{-5}$ cm.

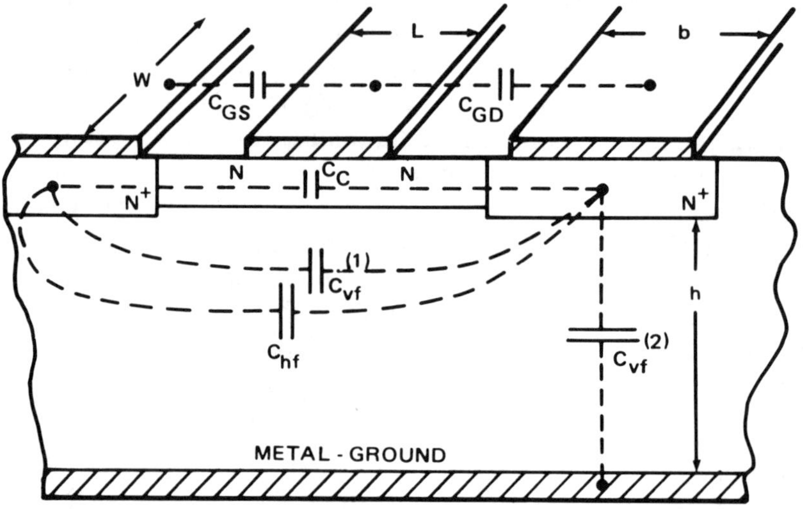

Figure 7/Transistor capacitance.

Parasitic drain-to-source capacitances (Figure 7) are associated with

(i) vertical field lines fringing through the semi-insulating substrate, and

(ii) field lines passing from source to substrate electrode and then from the substrate electrode to the drain [24]. Approximate expressions for (i) are [25]

$$C_{vF}^{(1)} = (W\epsilon/\pi)\ln[4(1 + 2b/L)] \tag{18}$$

and for (ii)

$$C_{vF}^{(2)} = (\pi W\epsilon/2)/\ln(h/b) \tag{19}$$

where W x b is the area of the contact lines, $b > L$ has been assumed, and $h \gg b$ is the substrate thickness. These capacitances amount typically to a small fraction of one fF per μm gate width W for a channel length of 1 μm. The subscript vF stands for vertical fringe capacitance. In addition, there is a lateral fringe capacitance between the terminations of the source and drain contacts [24] which, unlike the vertical fringe capacitances, is not reduced by reducing the width W of the transistor.

3. INVERTER PERFORMANCE

3.1 The dc Transfer Function

The dc transfer function relates output voltage to input voltage in the steady state. When the input voltage of the inverter is below the threshold voltage V_T, which is about +0.2 V in Figure 8, the output voltage is "high". The high output level V_{OH} in Figure 8 is less than V_{DD} because of the loading of the output node by the forward biased gate input junction of the next logic stage indicated in the inset. When the input voltage exceeds threshold, the transistor becomes conducting, and the inverter output voltage decreases by the voltage drop across the load resistor. The voltage gain

$$G_M = -\partial V_O/\partial V_I > 1 \tag{20}$$

(slope of dotted line in Figure 8) increases the ramp speed of an output pulse relative to that of an input pulse, since $\partial V_O/\partial t = \partial V_O/\partial V_I \cdot \partial V_I/\partial t$. This effect can be used for pulse sharpening [27, 28]

Noise margins (NM) can be defined as the voltage difference, measured on the input voltage axis (abscissa) of the transfer characteristic between the operating point and the nearest unity gain point, i.e., the point where the slope $\partial V_O/\partial V_I = -1$. Almost identical values for the noise margins are obtained by substituting for the unity gain points either those points where the output voltages reach 10% and 90% of the logic swing, or else the intersection points of the tangent in the inflection point with the "0" and "1" levels, as has been done in Figure 8. The noise margin for logic "0" input (lower noise margin) in Figure 8 is $NM^0 = 0.06$ V, and that for logic "1" input (upper noise margin) is $NM^1 = 0.27$ V.

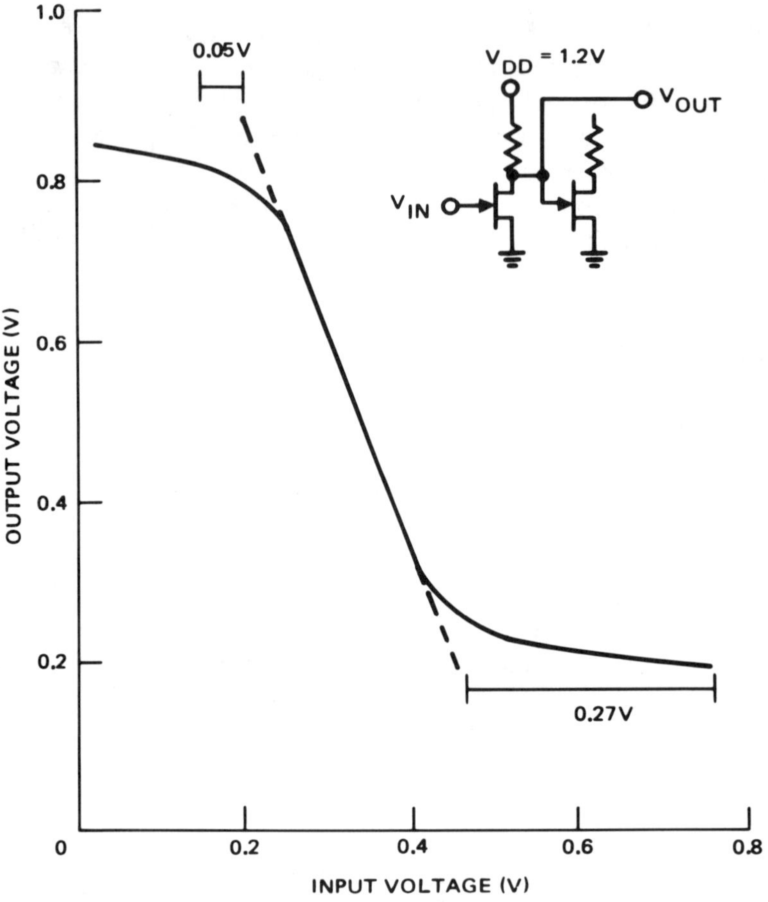

Figure 8/dc transfer function of a DCFL ring oscillator section; noise margins indicated (according to Ishikawa et al. [26]).

Since internally generated noise increases with the logic swing, "noise immunity" – the ratio of noise margins to logic swing – is sometimes used when comparing different circuits. However, the importance of wide noise margins arises neither from overcoming noise sources inherent to circuit components, nor from spurious signals or cross-talk. Wide noise margins are required to obtain high production yields with state-of-the-art processes which do not as yet maintain close control over threshold voltage. Thus noise margin requirements will be reduced as production technology improves.

3.2 Node Capacitance

Transfer functions are constructed for steady stage operation, i.e., for equal transistor and load currents, and do not apply to fast transients. Switching the inverter output from a low to high level or vice versa involves charging or discharging capacitances associated with the output node. These capacitances include (Figure 9a): transistor gate-to-source, and drain-to-gate capacitances C_{DS} and C_{DG}, channel capacitance C_c, load device capacitance C_L, input capacitance C_I of the subsequent logic gate or gates, and the parasitic ("stray") capacitance C_{St}. The charge stored at the output node is

$$Q_N = (C_{DS} + C_I + C_{St})V_O + C_{DG}(V_O - V_I) - C_L(V_{DD} - V_O) \tag{21}$$

Conservation of charge requires that

$$\partial Q_N/\partial t = C_N\, \partial V_o/\partial t = I_L - I_T \tag{22}$$

where

$$C_N = C_{DS} + C^*_{DG} + C_L + C_{St} + C_I \tag{23}$$

is the lumped node capacitance (Figure 9b). The effective drain-to-gate capacitance

$$C^*_{DG} = C_{DG}(1 - \partial V_I/\partial V_O) \tag{24}$$

does not differ significantly from the geometrical capacitance C_{DG}, since $-\partial V_I/\partial V_O = G^{-1}_{max} < 1$. We are here concerned with the effective drain to gate capacitance as seen from the transistor drain. In the well-known Miller effect, which views the gate-to-drain capacitance from the gate, C^*_{DG} is increased by the large factor $(1 + G_{max})$. The Miller effect applies to the contribution to C_I from the gate input of the next stage. Changes of the capacitances C_{GS} and C_I with voltage V_O have been neglected here as second order effects, but should be taken into account in more accurate, computer based calculations [9].

The purely capacitive input impedance of the stage or stages to which the inverter output is connected neglects the conductive current through the forward biased Schottky or p-n junction gate barrier. This conduction current can be taken into account by shunting the capacitance C_I in Figure 9a, and the capacitance C_N in Figure 9b with a forward biased diode. This will be necessary in DCFL if V_{DD} is larger than V_{GM} which is above 0.7 V for Schottky barrier gates, and 1.2 V for p^+-n junction gates.

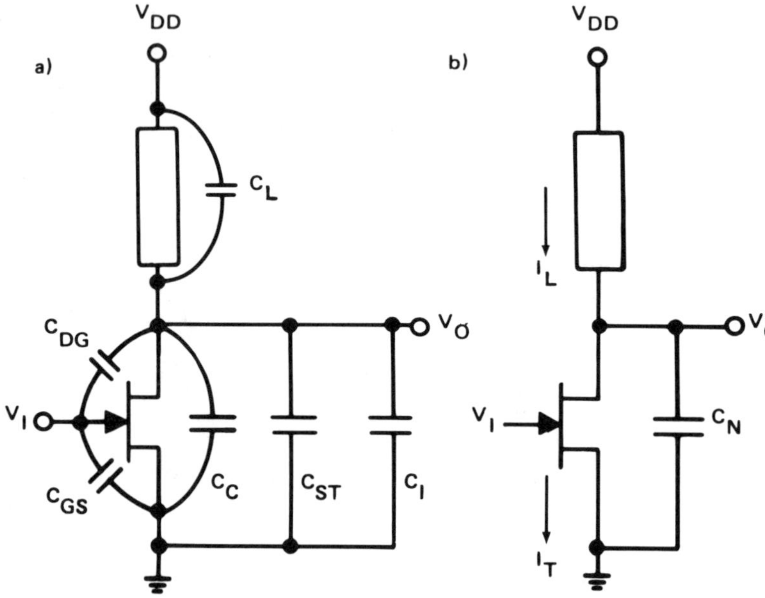

Figure 9/(a) Basic inverter circuit of Figure 1 showing various capacitances; (b) The capacitances have been lumped into the node capacitance for transient analysis.

Some of the contributions to the node capacitance are in proportion to the width W of the transistor contacts, while others (e.g., the lateral fringing capacitance between the terminations of source and drain contacts) are not. The magnitude of the fixed capacitance is revealed by a plot of the product of channel width W and delay time t_D vs. W for the same circuit with transistors of different W [24] (Figure 10). The abscissa intercept W_o of the straight line fit through experimental points shows that we can obtain the sum of the fixed and width-dependent contributions to the capacitance by ignoring the fixed capacitance and increasing the width of the W-dependent portion by using the intercepts $W_o = 6.4$ and 7.8 microns, respectively, for the two circuit layouts analyzed.

3.3 Rise, Fall, and Delay Times

Figure 11 depicts the response of an inverter to a voltage pulse. The time interval during which the output falls from 90% to 10% of the difference between logic "1" and Logic "0" is the fall time t_F. The time interval during which the output rises from 10% to 90% of this

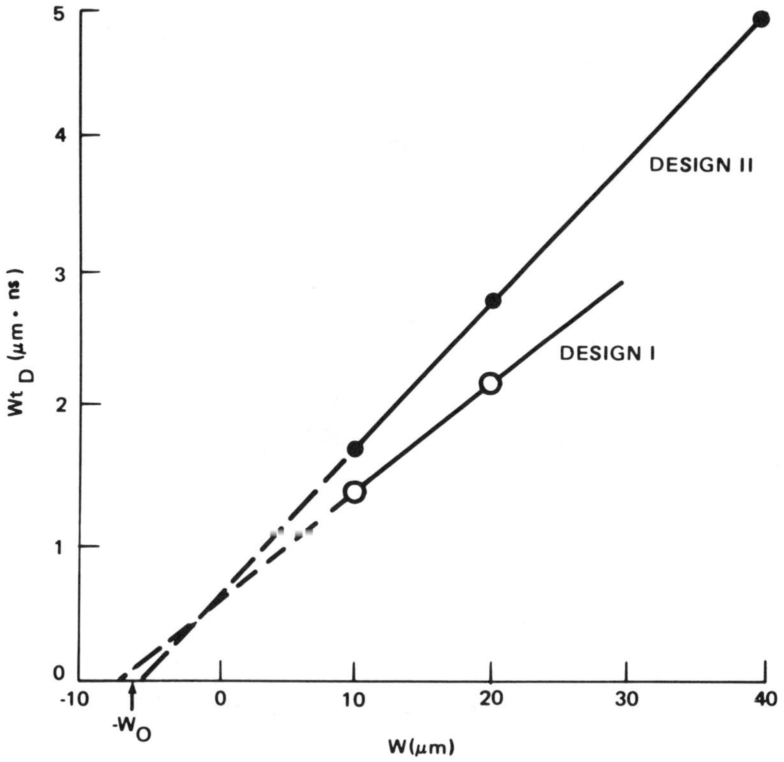

Figure 10/Separation of the capacitance portion which depends on the transistor width W from a fixed residual capacitance equivalent to a transistor width W_o = 7 μm. Experimental points are from reference 24.

difference is the rise time t_R. The time interval between the input and output levels being halfway between logic "0" and logic "1" is the delay time t_D. The delay time is of paramount interest in systems involving a large number of logic gates in series. The total delay between input and output of the system is then a large multiple of the delay time per gate.

A direct measurement of t_D, t_R, and t_F is described in Reference 4. However, the delay time per gate is usually determined indirectly by measuring the oscillation frequency f_o of an odd number N_R of inverters connected into a ring counter. The delay time t_D is calculated from the relation

$$t_D = (2N_R f_o)^{-1} \tag{25}$$

A chain of inverters generates an almost symmetric output pulse which is the result of an equilibrium established between pulse sharpening due to voltage amplification, implicit in the transfer function, and pulse spreading due to the finite rise and fall times of the inverters.

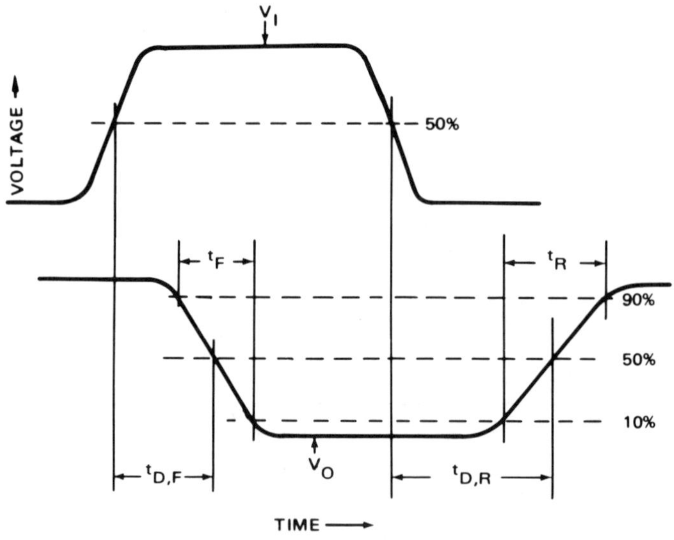

Figure 11/Definition of rise, fall and delay times.

The output voltage transient $V_O(t)$, after a change in V_I, is obtained by integrating Equation (22):

$$t = \int_{V_O(t=0)}^{V_O(t)} C_N (I_L - I_T)^{-1} \partial V_O \qquad (26)$$

where I_L and I_T are the instantaneous steady state currents which are functions of $V_O(t)$, and, in the case of I_T, also of $V_I(t)$. The rising transient of the inverter output is governed by the charging of the node capacitance through the load ($I_L \gg I_T \cong 0$), while the falling output is governed by the discharge of the node capacitance through the transistor ($I_T > I_L$). Rise time t_R and fall time t_F are obtained by inserting appropriate values for $V_O(t{=}0)$ and $V_O(t)$, namely:

$$V_O(t{=}0) = V_{OL} + 0.1(V_{OH} - V_{OL}) \text{ and } V_O(t_R) = V_{OH} - 0.1(V_{OH} - V_{OL})$$

for the rise time, and

$$V_O(t{=}0) = V_{OH} - 0.1(V_{OH} - V_{OL}) \text{ and } V_O(t_F) = V_{OL} + 0.1(V_{OH} - V_{OL})$$

for the fall time. The integration limits for the rise delay time are $V_O(t{=}0) = V_{OL}$ and $V_O(t_{R,D}) = V_{OL} + 0.5(V_{OH} - V_{OL})$. Thus Equation (26) transforms into

$$t_F = 0.8(V_{OH} - V_{OL})[C_N/(I_T - I_L)]$$

$$t_R = 0.8(V_{OH} - V_{OL})[C_N/I_L] \qquad (27)$$

$$t_{R,D} = 0.5(V_{OH} - V_{OL})[C_N/I_L]$$

where the brackets indicate the average value over the range of integration. Table 1 lists rise and rise-delay times for two load components, one a fixed resistor and the other providing a constant current of such magnitude that the dc power dissipation is the same in both cases. The node capacitance has been assumed to be a constant in both cases also. If this capacitance were the same, the constant current load would provide faster rise and rise-delay times.

The rise and rise-delay times of the fixed resistor load can be shortened by overdrive, i.e., by choosing $V_{DD} > V_{GM}$. The V_{OH} level is then determined by the intersect with the forward biased gate diode characteristic of the next stage, rather than by that with the switching transistor characteristic (Figure 12). Thus the RF-type charging curve is clipped at the V_{GM} level, causing shorter rise and rise-delay times, as illustrated in Figure 13. Figure 14 shows computed values and Figure 15 gives actual measurements. Integration of $(I_T - I_{L,SAT})^{-1}$ using the Shockley $I_T - V_D - V_G$ characteristics [3] for the transistor and a constant load current over the integration interval $V_{OL} + 0.1(V_{SAT} - V_{OL})$ to V_{SAT} provides

$$\langle [I_T - I_{L,SAT}]^{-1} \rangle = \eta (I_{SAT} - I_{L,SAT})^{-1} \tag{28}$$

with the form factor $\eta = 1.56 \pm 0.04$ over the remarkably wide range $0.1 \leqslant I_{L,SAT}/I_{SAT} \leqslant 0.5$ and $0.2 \leqslant d_S/a \leqslant 0.6$ where d_S is the depletion layer width at the source, and a is the n-layer height.

3.4 Power Dissipation and Maximum Clock Frequency

The power dissipation of an inverter circuit driven by a clock of frequency f comprises a static component P_{st} which is independent of f, and a dynamic component P_d which is proportional to f [30]. The static power dissipation

$$P_{st} = I_L(V_L = V_{DD} - V_{OL})V_{DD}/2 \tag{29}$$

results from resistive heating of the transistor and load component during the periods when the transistor is conducting, assumed to be half the time. Additional power

$$P^*_{st} = I_L(V_L = V_{DD} - V_{GM})V_{DD}/2 \tag{30}$$

is dissipated by current flow through the gates of circuits connected to the output node in the case where $V_{DD} > V_{GM}$. The dynamic power dissipation

$$P_d = fC_N V_{DD}^2 \tag{31}$$

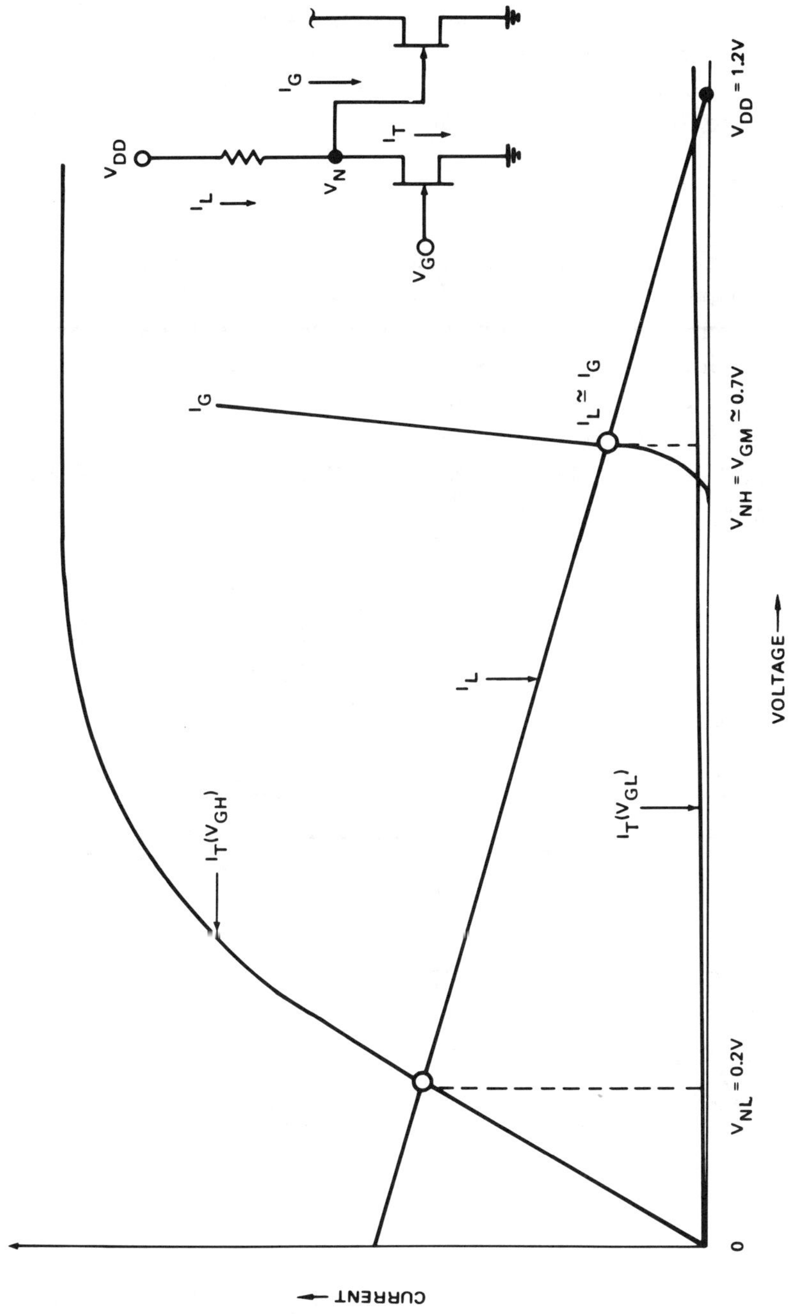

Figure 12/*Load line diagram for DCFL GaAs FET inverter with resistive load operated in overdrive. V_{NH} is determined by the gate voltage V_{GM} at which substantial gate current flows.*

results from the periodic charge and discharge of the node capacitance $V_{OL} \cong 0$ and $V_{OH} \cong V_{DD} < V_{GM}$. The delay time limits the maximal clock frequency to about $f_{max} \cong (t_D)^{-1}$. Thus, the product of delay time and dynamic power dissipation

$$t_d P_d = C_N V_{DD}^2 \tag{32}$$

is solely determined by node capacitance and supply voltage. The delay time-dynamic power product, also called speed power product, or dynamic switching energy, is the minimum energy a gate can dissipate during a logic transition, assuming two transitions per cycle, and thus provides a significant figure of merit for comparing various circuits for VLSI applications [31].

A less general but more detailed analysis will now be presented for a fixed resistor load and absence of current loading of the output node.

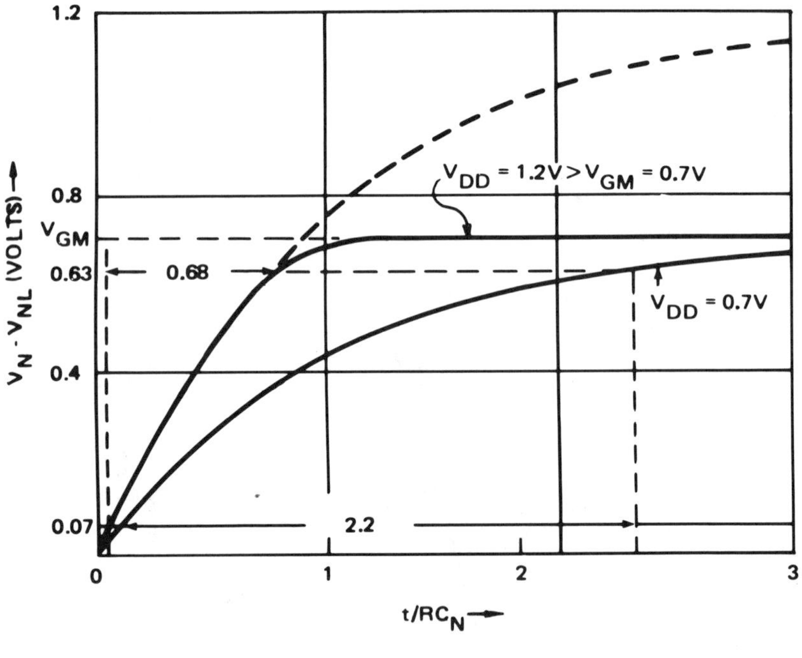

Figure 13/Decrease of rise time by raising the supply voltage V_{DD} above the gate voltage V_{GM} at which substantial gate current flows (overdrive).

To obtain maximal clock frequency, the load resistance should be chosen so that $f_{max} = (2t_R)^{-1} = (4.4RC_N)^{-1}$, while $P_{st} = V_{DD}^2/2R$ and $P_d(f = f_{max}) = 0.8V_{DD}^2/(4.4R)$, the factor 0.8 resulted from the limited voltage swing to 90% of V_{DD}. The total power dissipation is somewhat larger than P_d but less than $P_d + P_{st}$, so that $t_R P = C_N V_{DD}^2$. Both low power dissipation and maximal clock frequency benefit from lowering the node capacitance. Power dissipation is greatly reduced by lowering the supply voltage level. V_{DD} has no explicit effect on maximal clock frequency in the above analysis. However, the choice of R to provide $t_R = t_F$ depends implicitly on V_{DD}.

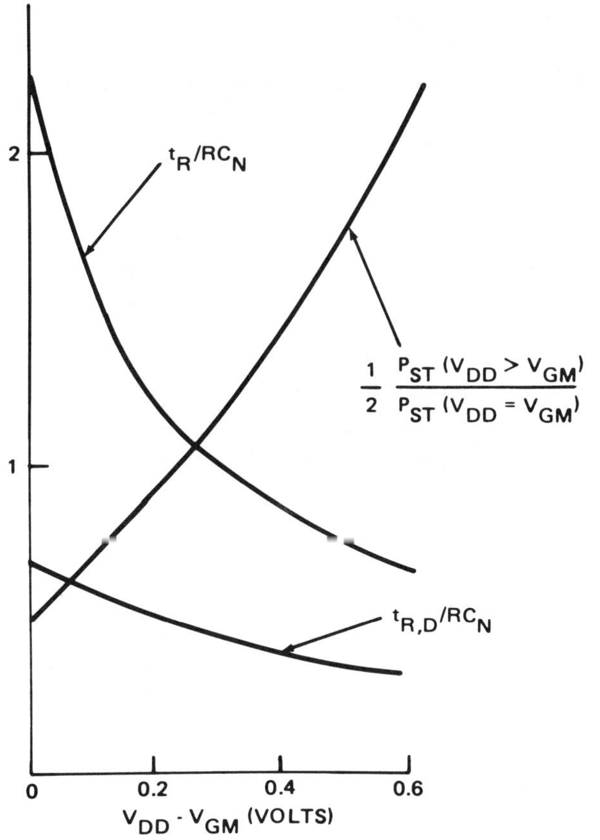

Figure 14/Rise, rise delay, and static power dissipation vs. supply voltage overdrive computed for V_{GM} = 0.7 V and V_{NL} = 0.

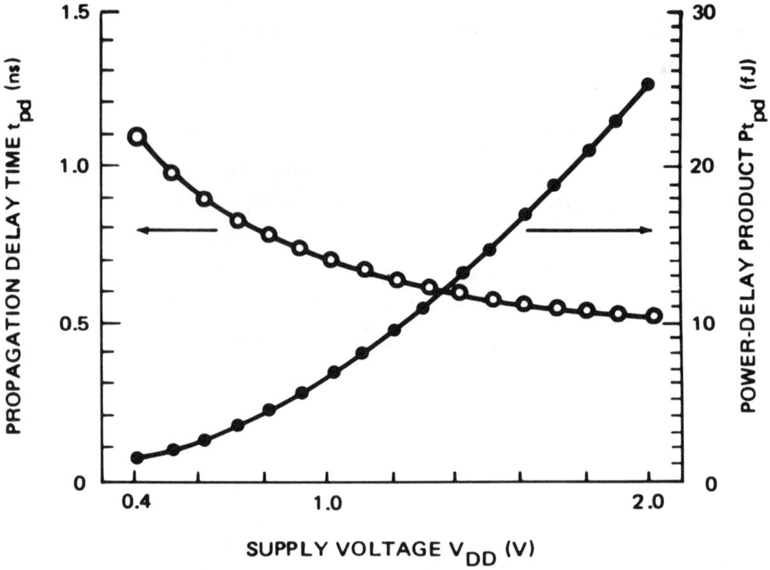

Figure 15/Influence of supply voltage on propagation delay time and power-delay product measured by reference 29.

3.5 Active Load Components

Speed can be enhanced without increasing static power dissipation, i.e., the power delay time product can be reduced, if load elements have non-linear I-V characteristics called "active loads." This has been demonstrated in Table 1 for the load component providing a constant current.

TABLE I

Rise and rise-delay times after an abrupt gate voltage drop for a fixed resistor load R and a constant current load $I_{L,SAT} = (V_{DD} - V_{OL})/R$ chosen so as to provide the same dc power dissipation

Load	$r = t_R/RC_N$	$T_{D,R}/RC_N$
Fixed resistor	2.2	0.7
Constant current	0.9	0.5

Pull-up Transistor Load. Such a load component can be constructed from a normally-on, depletion mode transistor of negative threshold voltage by connecting its gate to its source [4, 10]. This load component, called "pull-up transistor", supplies the constant current $I_{SAT} = K_L V_{T,L}^2$ for voltages $V_{DD} - V_O > V_{L,SAT} = -V_T$. The capacitance contributed by this load component to the node capacitance is its gate-to-source capacitance C_{GS}, given by Equation (16), with appropriate values of W_L and L_L.

Two Terminal Current Limiter [32]. Another load component which provides a constant current (Figure 16) consists of an ordinary n-channel terminated by two n^+ regions with low resistance contacts operated at voltages which cause drift velocity saturation for the channel electrons. Traveling domain formation is prevented by choosing the product of dopant concentration and channel length to be less than 10^{12} cm^{-2}. The current at L = 1 μm should saturate at voltages in excess of about 0.5 V. Figure 16 shows current saturation at 1.5 V due to an unusually low mobility of about 1,000 cm^2/V s (assuming 50% conversion into donors of 2 x 10^{11} cm^{-2}). This current limiter has a very small capacitance, and is simpler to build than the pull-up transistor load. Overshoot effects of drift velocity over the steady state value of 2 x 10^7 cm/s are to be expected only at times of a few ps where electrons are incapable of being scattered into the energy levels of the upper low mobility band during their short flight time.

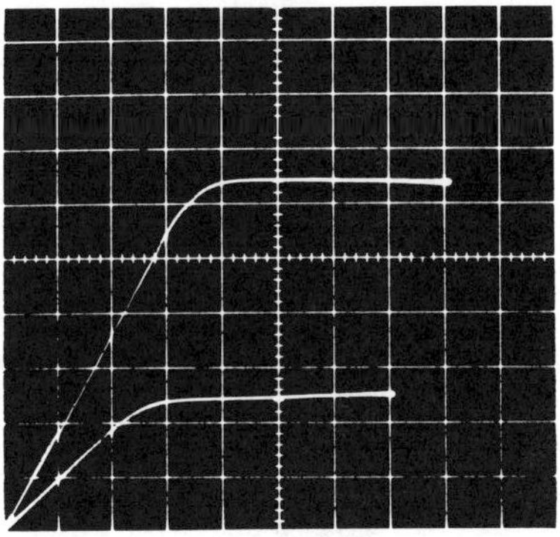

SCALE: VERT. 200 μA/DIV.
HORIZ. 0.5 v/DIV.

Figure 16*/I-V characteristics of a two terminal GaAs current limiter (L = 1 μm, W = 2.5 and 5μ, respectively) according to Zuleeg [32].*

Tunnel Diode Load [32]. Figure 17 shows transistor characteristics for various gate voltages and a tunnel diode load line. For a supply voltage V_{DD} about equal to the voltage drop across the tunnel diode pertaining to the valley current, i.e., $V_{DD} \cong 0.5$ V, a low steady state current at V_{OL}, and, thus, low dc power dissipation is combined with a large pull-up current, which approaches the tunnel diode peak current during a major portion of the rise transient. The aging encountered with early tunnel diodes is associated with forward current flow over the junction barrier rather than tunneling through it [34], and should thus not occur in circuits which limit the diode voltage to a range not much higher than the valley voltage.

The tunnel diode introduces a large load capacitance which contributes significantly to the node capacitance C_N. Figure 18 shows computed dependencies of rise and fall times on the ratio of tunnel diode peak current to transistor saturation current, on load capacitance C_N^* (excluding the tunnel diode capacitance) and on the "speed" of the tunnel diode, i.e., the ratio of its peak current to its capacitance. Speeds greater than 10 mA/pF have been achieved [34]. The comparatively large capacitance of the tunnel diode makes the inverter circuit with tunnel diode rather insensitive to increases of the remaining node capacitance by the input capacitances of multiple fan-out. Thus, while the tunnel diode load circuit may be somewhat slower than conventional load circuits for fan-out of one, it should be faster for large fan-outs without buffering. The peak tunnel diode current must be kept below the transistor current at V_{GH}, or else the circuit locks itself into one of two steady state operating points at the same input voltage. The excellent noise margins of the tunnel diode load, $NM^1 = V_G^{(5)} - V_G^{(2)}$ and $NM^0 = V_G^{(4)} - V_G^{(1)}$ are determined by the transistor characteristics which separate monostable from bistable operations (dotted lines in Figure 17). The tunnel diode load circuit operated in the "lock-in" mode (voltage $V_G^{(3)}$ in Figure 17) is of interest as a memory element.

3.6 Buffers

Buffers protect the output node between switching transistor and load element from being loaded by the input impedance of connected stages. This loading is particularly severe for large fan-out. If no buffer is used and there is a multiple fan-out, the node capacitance

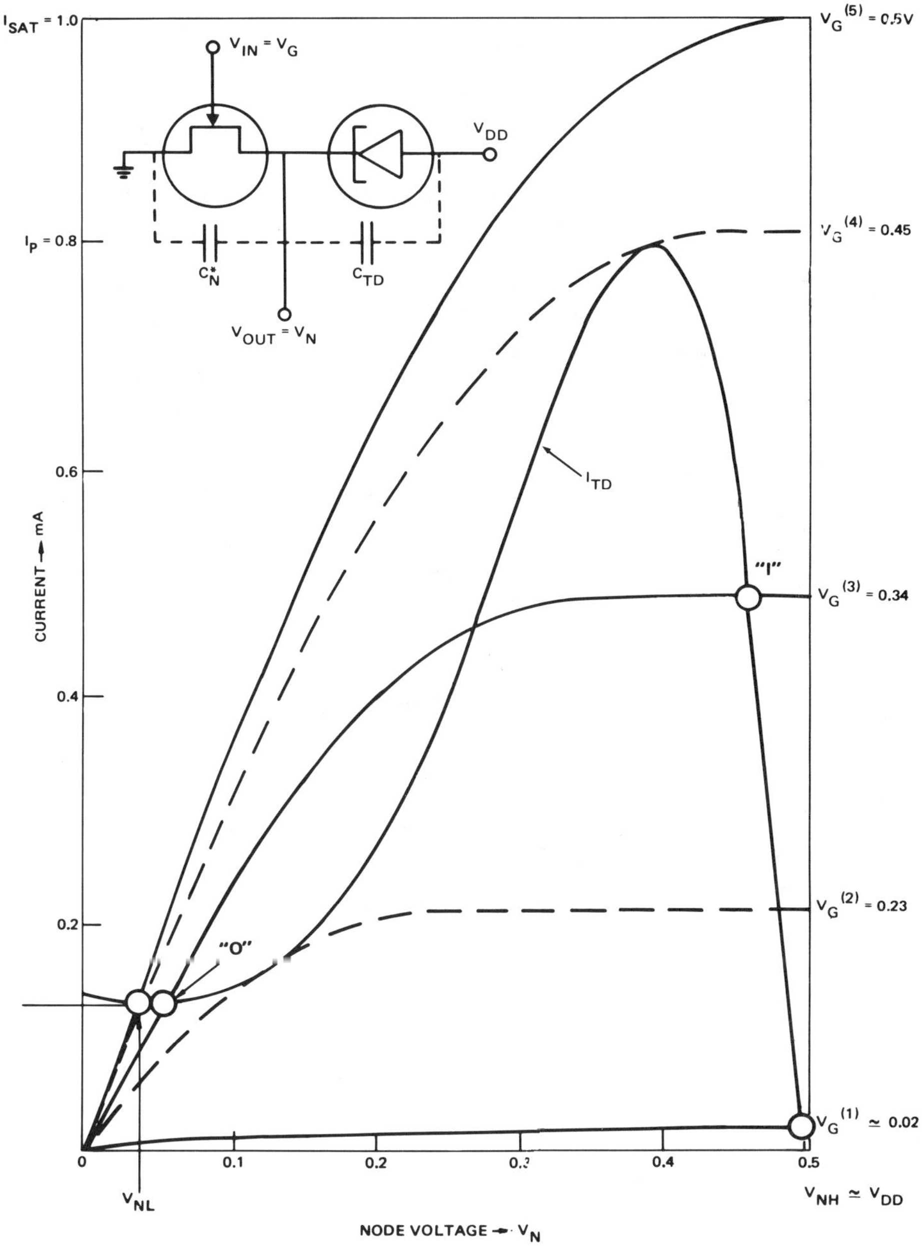

***Figure 17**/I-V characteristics of an enhancement mode FET at five gate voltages with tunnel diode load line. The inset shows the basic TDFL circuit which can be operated as inverter ($V_G^{(1)}$) and $V_G^{(5)}$), or as a memory cell ($V_G^{(3)}$) according to Lehovec [33].*

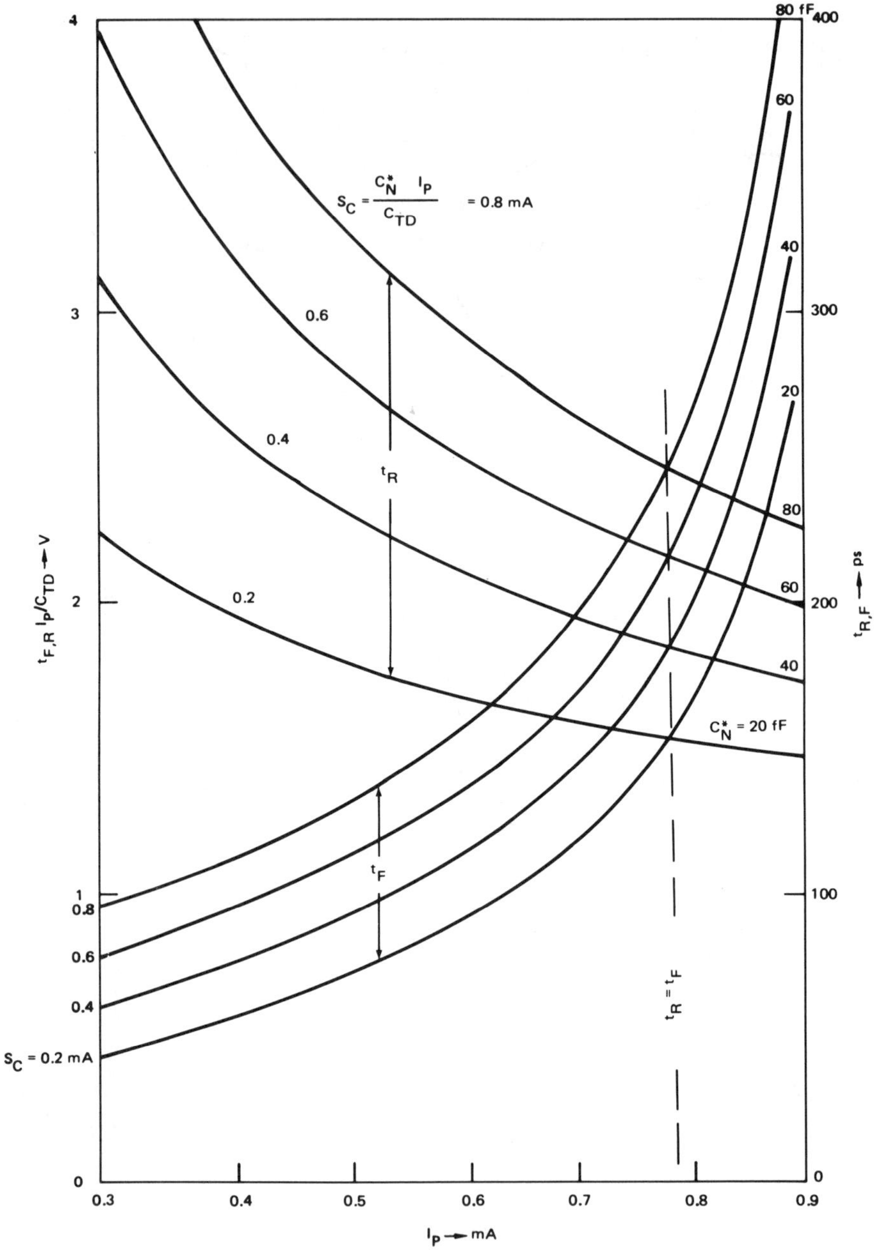

Figure 18/*Rise and fall times for the TDFL inverter of Figure 17 vs. tunnel diode peak current I_p computed for $I_{SAT} = 1$ mA, $V_T = 0$, and $V_{GH} \cong V_{DD} = 0.5$ V. Parameter is the speed-capacitance product $S_C = C_N^* I_p/C_{TD}$, where $C_N^* = C_N - C_{TD}$. Right hand scales assume $I_p/C_{TD} = 10$ mA/pF.*

increases linearly with the fan-out number, as do the rise, fall and delay times [4, 7, 9, 35], according to Equation (27).

The buffer comprises a FET in source follower connection with a load component. The voltage at the node between source and load component varies in unison with the gate voltage because of the large amplification of the transistor in case of a resistor source follower load. In case of a diode load as used in the level shift circuit of Buffered FET Logic (BFL) [4], the voltage at the output node varies also in unison with the gate voltage because a constant current source is inserted in the source follower circuit. In both cases, the gate-to-source charge does not change with gate voltage, which eliminates the loading of the inverter output node by the gate-to-source capacitance of the buffer transistor. The buffering is done at the input side in the circuit in Figure 19a. The circuit in Figure 19b [8, 37] uses the output voltage of a first inverter circuit, comprising Q_1 and R, as gate voltage for the n-channel load transistor Q_2 of a second inverter circuit consisting of Q_2 and Q_3. The load transistor

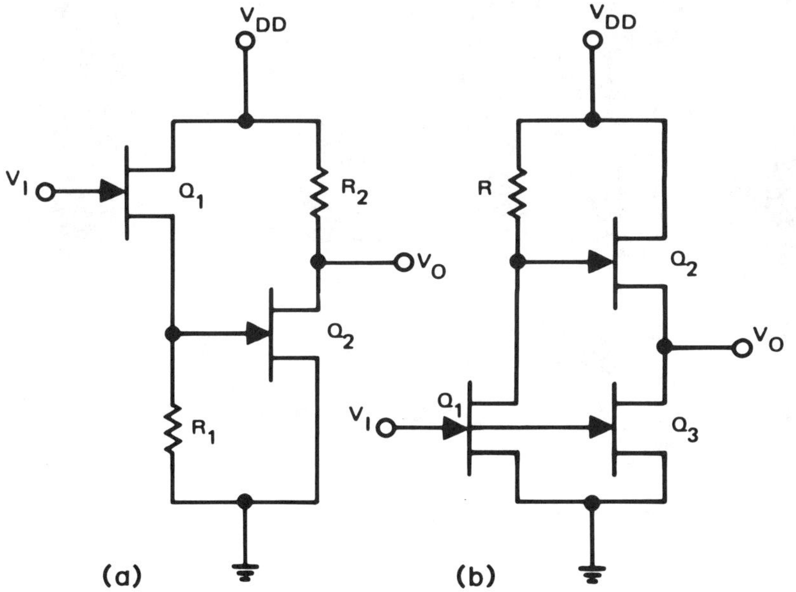

Figure 19/Input buffered (left) [36, 44] and quasi-complementary (right) inverter circuits.

Q_2 becomes non-conducting when its switching transistor Q_3 is conducting, and vice versa, because the gate voltage of Q_2 is the inverted gate voltage of Q_1 and Q_3. In this manner, a quasi-complementary inverter circuit with zero dc power dissipation in the second inverter circuit is obtained, provided that no dc current is drawn from its output node. In addition, buffering of the drain node of the first inverter circuit with respect to any capacitive output load connected to the second inverter circuit output has been achieved. Figure 20 shows a 9-stage ring oscillator constructed from direct coupled quasi-complementary inverters [37] used as a test vehicle for inverter performance evaluation.

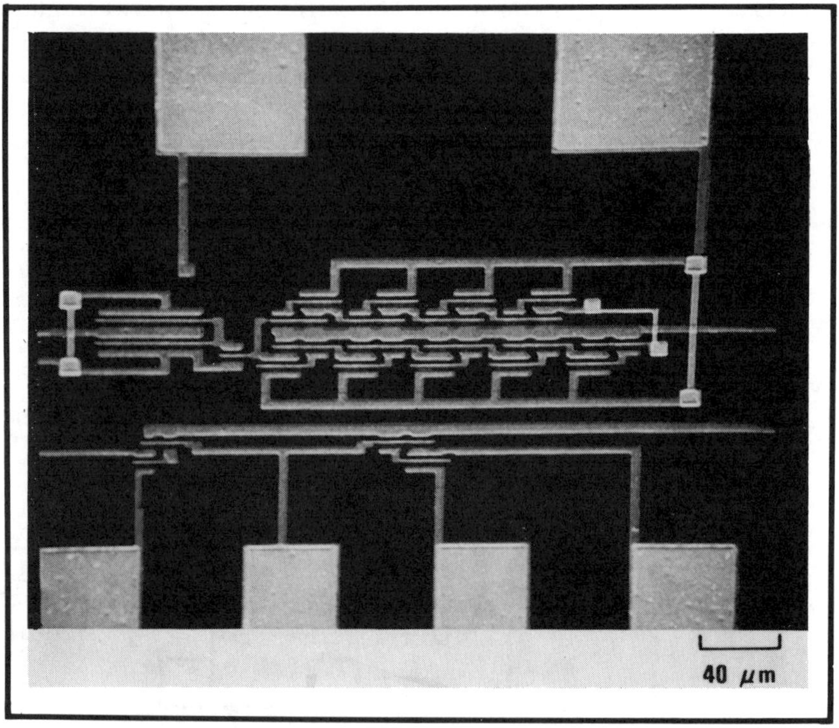

***Figure 20**/Scanning electron micrograph of a DCFL nine stage ring oscillator using direct coupled quasi-complementary inverters of gate dimensions (in μm): 1 x 6 (Q_1), 1 x 10 (Q_2), 1 x 5 (Q_3), and R = 14 Kohm according to reference 37.*

4. STATE-OF-THE-ART OF DCFL

4.1 General Considerations

The n-channel doping of GaAs components is usually achieved by ion implant, either directly into the high resistivity (10^8 – 10^9 ohm-cm) Cr-doped substrate [8, 10], or into a quasi-intrinsic buffer layer grown epitaxially on the substrate [24, 38]. Peak dopant for the switching transistor channel, about 2 x 10^{17} cm^{-3}, is close to degeneracy. The channel height is typically about a tenth of a micron for the normally-off enhancement mode devices, and several tenths for the normally-on depletion mode devices. Gate lengths have been reduced to 1 micron by photolithographic processing, with typical source-to-drain contact separations of from 3 to 5 microns. Efforts are now underway to achieve submicron gate lengths by electron beam processing [39, 40, 41]. Switching transistor widths range typically from 10 to 20 microns; below this, current decreases more strongly than capacitance, thus impairing speed [24, 39, 42]. As a rough rule-of-thumb, transistor capacitance can be taken as 0.7 fF/micron times gate width in microns plus a fixed value, which is about 7 fF for BFL [24]. Typical saturation currents range from 50 to 150 μA/micron · V^2 times gate width in microns times the square of gate voltage in excess of threshold voltage [37]. These values pertain to a gate length of 1 micron. Figure 21 shows a cross-section of a DCFL JFET inverter with resistive load.

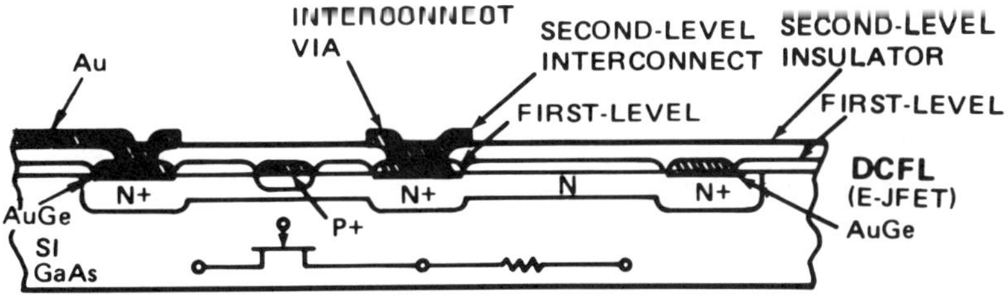

Figure 21/Cross-section through a planar JFET inverter according to reference 8.

The main obstacle to process control is the achievement of reproducible conversion of ion-implanted impurities into donors. The GaAs lattice provides two non-equivalent substitutional positions for impurities, and therefore a larger variety of lattice defects and defect-impurity associations than does the silicon lattice. The implanted impurity (Se, S) can lodge in several configurations after anneal, not all of them behaving as donors. The lack of dopant control affects the pinch-off potential and consequently the threshold voltage. This control problem is particularly serious for the small positive threshold voltages of the switching transistors in DCFL and for pull-up transistors of small negative threshold voltage. Novel production techniques such as laser anneal, or crystallization of amorphous doped layers at low temperatures [43] offer some hope for improved conversion control for ion-implanted impurities. In the meanwhile, the threshold voltage can be adjusted to a desired value by GaAs removal by anodization [29], which stops when the depletion layer encompasses the entire n-doped layer. In JFET DCFL the dopant level after n-implant and anneal is determined from measurements of the sheet conductance and Hall effect in test slices; the p^+-implant is then adjusted to provide the desired threshold voltage [37]. Also, circuits have been designed which allow for a larger variation of threshold voltage, albeit at the expense of circuit complexity [44] (Section 4.4) and power consumption.

Another production problem arises from out-diffusion of Cr during the anneal after ion-implant[45]. This out-diffusion appears to be related to the capping of the GaAs required to prevent As loss. The exact cause of, and the mechanism for out-diffusion are still under investigation. The Cr-depleted surface layer may become n-type due to the presence of parasitic donors, resulting in unwarranted surface conduction between circuit components. This problem can be eliminated by improving the quality of bulk crystals, or else by depositing a high quality epitaxial buffer layer on lower quality bulk material [38] before ion implant.

DCFL circuits can be subdivided according to transistor type (JFET or Schottky gate FET), load element type (resistor, pull-up transistor, two terminal current limiter or tunnel diode), and use of auxiliary circuitry.

4.2 JFET vs. MESFET

The JFET has a higher built-in voltage and therefore allows a larger forward bias before significant gate current flows. Thus JFET circuits provide a larger logic swing than MESFET circuits. The fall time is proportional to the ratio of node charge, $C_N L_S$, to transistor saturation current $I_{SAT} = K(V_{GM} - V_T)^2$, so that $t_F = \eta C_N V_{GM} / K(V_{GM} - V_T)^2$ decreases with increasing V_{GM}. This favors JFET. On the other hand, the rise time usually limits speed, and clipping by the smaller V_{GM} of the MESFET when operating in the overdrive mode reduces the rise time more strongly for MESFET than for JFET at equal $V_{DD} \gg V_{GM}$. Table 2 compiles delay times, power dissipations, and delay time-power dissipation products achieved with DCFL ring oscillators of gate length about 1 micron. All circuits have been fabricated photolithographically, except in the case marked E-beam. So far, the most effort has gone into MESFET logic; only Reference 37 used JFETs in conjunction with quasi-complementary buffering. The advantages of this type of buffering in a ring oscillator of fan-out 1 is not obvious to the writer, particularly since the oscillator was operated in overdrive, causing current drain from the output node of the buffer. This unusual circuitry, and the extremely short gate widths used, prevent a simple comparison of the JFET to the MESFET based on data from Table 2. All circuits use resistor loads. However, these resistors may not always have been chosen so that $t_R = t_F$, which would result in the shortest delay times. Nevertheless, the following conclusion can be drawn from Table 2:

(i) Increasing the supply voltage reduces the delay time somewhat, but greatly increases power dissipation and the delay time-power product.

(ii) Increasing transistor width reduces delay time, but increases power consumption and delay time power product.

The decrease of rise and rise delay times by overdrive $V_{DD} > V_{GM}$ was discussed in Section 3.3. The initial linear range of an RC-type charging curve,

$$t_R = 0.8RC_N(V_{GM} - V_{NL})/(V_{DD} - V_{NL}) \tag{33}$$

decreases about inversely with $V_{DD} > V_{GM}$. However, the additional dc power dissipation at V_{NH} is $V_{DD}(V_{DD} - V_{GM})/R$, and it thus increases quadratically with V_{DD}. The dynamic power dissipation $(2t_R)^{-1} C_N V_{DD}^2$ increases even more strongly as the cube of V_{DD}.

A significant advantage of the JFET over the MESFET arises from the larger threshold voltage which can be employed in JFET circuits without unduly restricting the logic swing. The larger threshold voltage improves the lower noise margin which roughly equals the threshold voltage. V_T values of about 0.2 to 0.3 V [8] which are quite tolerable for a JFET inverter leaving gate voltage excursion beyond threshold of still about one volt, would reduce the logic swing to about half this value in a MESFET inverter. Thus DCFL MESFET circuits use threshold voltages of about zero to 0.1 V [26, 29]. Lack of control over this threshold voltage may result in depletion mode MESFETs, rather than in the intended enhancement mode devices. The current flowing through a depletion mode MESFET of slightly negative threshold at small positive gate voltage is still so small that no significant power dissipation results at $V_{GM} < V_{DD}$. However, V_{OH} would decrease significantly unless it were pinned to V_{GM} by overdrive at $V_{DD} > V_{GM}$. The sensitivity of DCFL inverters to variations of the threshold voltage can be reduced by auxiliary circuitry [37] discussed in Section 4.4.

The advantage of JFET over MESFET resulting from larger logic swing will disappear once power supply levels will be decreased to less than 0.7 V. This will be required for high speed ultra-large scale integrated circuits (Section 5.4), but will necessitate a better control on threshold voltage than is presently achievable. In the meantime, production problems may influence the choice between MESFET and JFET. Aluminum gates, which are used increasingly in lieu of Ti-Pt-Au, or similar alloys, can also be used as low resistance contacts to highly degenerate source and drain n^+ layers, so that MESFET technology eliminates JFETs. Tunnel diode loads do require a p^+ layer so that JFETs with tunnel diode load should not be more complicated to produce than MESFETs with tunnel diode load.

4.3 Choice of Load Component

The *pull-up transistor* load requires less area than a fixed resistor load. However, pull-up transistor loads with gate connected to source are not suited to DCFL since two different threshold voltages would have to be controlled. Moreover, in order to achieve the constant current load line over much of the logic swing, the negative threshold voltage of the pull-up transistor would have to be only a few tenths of a volt, resulting in considerable variation of current and thus power dissipation if threshold voltage should vary by ±0.1 V.

A pull-up transistor load with positive threshold voltage and gate transistor load with positive threshold voltage and gate connected to drain provides a parabolic, concave upwards load line which is undesirable because of lower speed and smaller lower noise margin than a fixed resistor load.

The commonly used load for DCFL is the *fixed resistor* [1, 8, 26, 29, 37, 39, 40, 44, 46]. At an implant level of 10^{12} cm^{-2}, using a mobility of about 4000 cm^2/V s, the sheet conductance is about 6 x 10^{-4} ohms per square. A square area provides a load resistance of 1.6 kΩ and $t_R = 2.2RC_N = 100$ ps for $C_N = 30$ fF. The fixed resistor load is easily produced and has a low capacitance.

The *two terminal current limiter load* [32] is as easily produced as the fixed resistor load, requires less area, and provides a faster rise time. The current limiter has a rather large off-set voltage, about 0.5 V at a length of 1 micron, so that the advantages of a constant current load line would not be realized over most of the logic swing of a MESFET unless it is used in overdrive, $V_{DD} > V_{GM}$. The current limiter is therefore very suitable for DCFL circuits with JFETs, as long as supply voltages remain above 1 V. Because of its lower capacitance, the current limiter should be superior to the constant current pull-up transistor in SDFL [10] and BFL [4, 24] circuits which have larger logic swings than DCFL circuits.

The projected advantages of the *tunnel diode load* [32] have been discussed in Section 3.5. However, before these advantages can be realized, considerable progress in threshold voltage control has to be made, and methods for building tunnel diodes in microcircuits will have to be worked out. Control of tunnel diode peak current promises to be an even more formidable problem than that of the transistor threshold voltage, so that tunnel diode FET logic (TDFL) is at least several years away.

4.4 Quasi-Normally Off MESFET Logic

This logic has been developed to allow for a wider spread in transistor threshold voltages without affecting inverter circuit performance. The circuit (Figure 22) will operate over the wide range of threshold voltages −0.4 to +0.1 V. A representative standard deviation for the output voltage $V_O = V_I$ at which switching occurs is only 68 mV at a threshold voltage of 224 mV. This remarkable result is achieved by a level shift circuit with the diode. However, the circuit is operated in

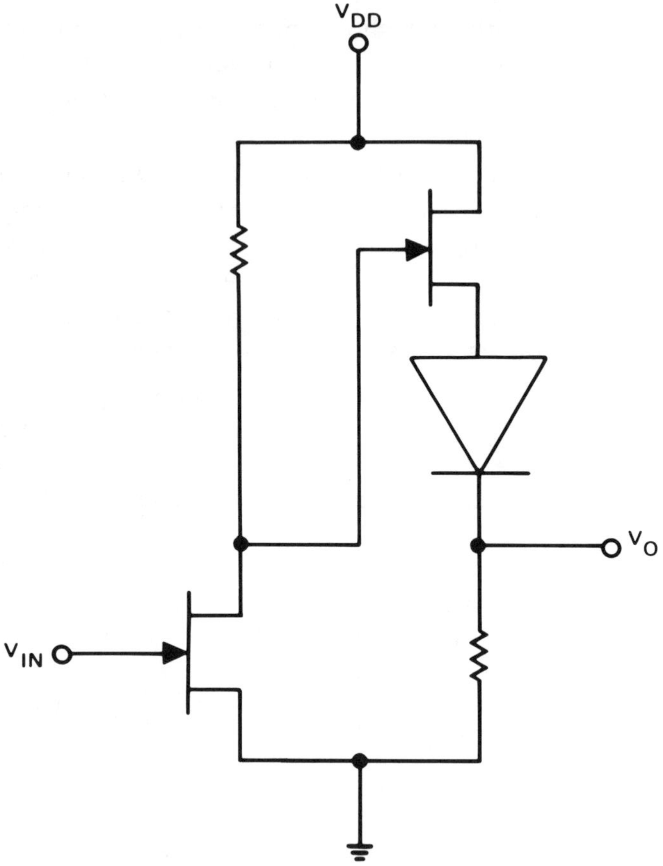

Figure 22/Quasi-normally-off inverter circuit according to reference 44.

strong overdrive, V_{DD} = 2.5 V, and therefore consumes considerable power (Table 2). Its performance is slightly worse than that of SDFL [10]. Unlike the level shift circuits of BFL and SDFL, the quasi-normally-off MESFET logic does not require a second, negative supply voltage.

5. COMPARISON OF DCFL WITH SDFL AND BFL

5.1 General Considerations

In this section we discuss the advantages and disadvantages of DCFL vs. SDFL and BFL, and make projections for the future. Circuits will be evaluated according to electric performance (speed, power dissipation), capability for very large scale integration (area per gate), and yield (production control).

Table 2 lists area per gate and maximal power per gate for various degrees of integration, assuming a chip area of 0.25 cm^2, and a power dissipation of 5 W per chip. At the present integration level of low to medium scale integration, neither power dissipation per gate, nor area per gate are particularly restrictive. Thus emphasis at these integration levels is placed on production, yield, and speed.

5.2 Production Yield

Ignoring catastrophic failures, such as shorts, the excessive variation of operating points V_{OL} and V_{OH} primarily causes rejection. This variation results from lack of threshold voltage control.

The tolerance of a circuit for variations of the input voltages is expressed by the noise margins. BFL [4, 34] and SDFL [10] have wide noise margins as a result of a large logic swing which is not limited by V_{GM}. These circuits use a second, negative supply voltage V_{ss} in order to shift the positive voltage V_N of the node between switching transistor and load component by a large amount S to a negative value $V_G = V_N - S < V_T < 0$, which is then applied to the input gate of the switching transistor of the next stage. BFL level shifting is performed at the output side of the inverter, while SDFL level shifting occurs at the input side where the current is lower, thereby achieving some power economy. Figure 23 shows I_{SAT} $V_G = V_O$) for a BFL inverter where $V_O = V_N - S$. The value of S is chosen so that the saturation current for the low output level applied at a transistor gate is either zero, because it falls below the negative threshold voltage, or else, as in Figure 23, $V_G - V_T > 0$, but

$$I_{SAT}(V_G = V_{OL}) = K(V_{NL} - S - V_T)^2 \tag{34}$$

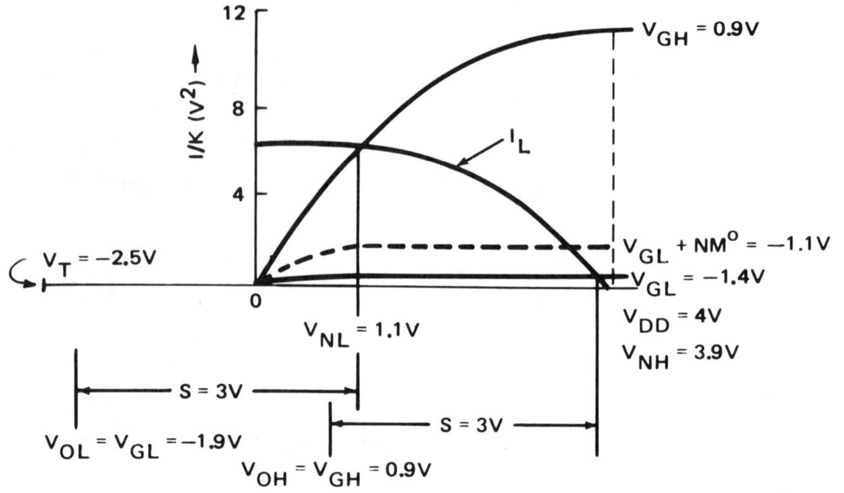

Figure 23/Switching transistor characteristics and pull-up transistor load line I_L for a BFL inverter. The level shift S from V_N to $V_O = V_G$ is marked at the abscissa. The dotted characteristic pertains to the lower noise margin.

is still sufficiently small to provide a large lower noise margin. Unlike BFL and SDFL, the lower noise margin of DCFL depends critically on the difference $V_T - V_{OL}$. With V_T a small positive value, and since $V_{OL} = 0$, the lower noise margin is necessarily small. Among various DCFL circuits, those with JFETs are capable of larger lower noise margins than those with MESFETs, and those with current limiter loads are superior to those with fixed resistor loads. The dependence of the lower noise margin on V_T can be reduced by the auxiliary level shift circuit of the quasi-normally-off logic [44] (Section 4.4), albeit at the expense of power dissipation and area. Tunnel diode load circuits have extremely large noise margins (Section 3.5), but production yield will be adversely affected by the additional problem of tunnel diode peak current control.

5.3 Switching Speed

Speed of the inverter proper, i.e., of the switching transistor with load component, depends on the I-V characteristics and on the node capacitance. Additional speed limitations due to the level shift circuitry of BFL and SDFL will be neglected in the following considerations.

The contribution to the node capacitance from the input capacitance of the subsequent stage or stages is eliminated in BFL [24] by the use of the buffer transistor in source follower configuration. The same can be done in DCFL by placing a source follower buffer transistor between stages (Figure 19a) [36]. The contribution to the node capacitance by the load component, fixed resistor, or two terminal current limiter in DCFL is less than that of the pull-up transistor load in BFL or SBFL. However, two terminal current limiter loads could be used also in these circuits. Thus, we shall assume equal node capacitances in all three circuit types in what follows.

The maximal achievable clock frequency is obtained by choosing the load line so that rise and fall times given by Equation (27) become equal. For a constant load current as obtained by the pull-up transistor load, or by the two terminal current limiter, the maximal clock frequency

$$f_{max} = I_{SAT}/[1.6(V_{OH} - V_{OL})C_N(1 + \eta)] \tag{35}$$

occurs at $I_{L,SAT} = I_{SAT}/(1 + \eta)$ where η is the form factor defined in Equation (28). In absence of drift velocity, saturation I_{SAT} increases quadratically with $V_{GH} - V_T$ so that maximal clock frequency increases linearly with the logic swing. Using $K = 100\ \mu A/\mu m\ V^2$, $W = 10\ \mu m$, $C_N = 30$ fF, and $\eta = 1.5$, we obtain for DCFL with $V_{OH} = V_{GM} = 1.2$ (JFET) and 0.7 V (MESFET) and $V_{OL} \simeq V_T \simeq 0.1$ V: $f_{max} = 9$ (JFET) and 5 GHz (MESFET). However, these estimates neglect drift velocity saturation which would become more significant at the larger voltage swing of the JFET. Drift velocity saturation most certainly occurs at the much larger logic swings of SDFL and BFL. Once drift velocity saturation becomes dominant, I_{SAT} increases linearly with $V_{GH} - V_T$, (Equation (12)), and the maximal clock frequency tends to level out at

$$f_{max} = \frac{KLv_s}{2(1 + \eta)C_N\mu} \cdot \frac{V_{GH} - V_T}{V_{NH} - V_{NL}} \tag{36}$$

Using $KL/\mu = W\epsilon/2a_{eff} = 5 \times 10^{-11}$ F/cm, corresponding to W = 10 μm and $a_{eff} = 0.1\ \mu$m, $C_N = 30$ fF, $v_s = 2 \times 10^7$ cm/sec, and $\eta = 1.5$, we obtain $f_{max} = 7$ GHz assuming that $(V_{GH} - V_T)/(V_{NH} - V_{NI}) = 1$. Clock frequencies of 4.5 GHz have been reached using BFL [24] for a complementary clocked binary divider using NAND/NOR gates for which $f_{max} = (2\tau_D)^{-1}$. A complementary clocked binary divider using NOR gates, for which $f_{max} = (4\tau_D)^{-1}$, has been built with DCFL [46] and provided $f_{max} = 0.61$ GHz, however, at a power dissipation about 100 times less than that of the divider with BFL [24]. Power dissipation and f_{max} of a single-clocked D flip-flop frequency divider using SDFL [42] were intermediate between those of BFL and DCFL circuits. The η-factors for DCFL, SDFL, and BFL should not differ much, since most of the transistor characteristics during the fall time lie in the non-saturation regime for all three circuits. This is obvious for DCFL where $V_{NH} - V_{SAT} = V_T \ll V_{SAT} = V_{NH} - V_T$ and is found also to be true for BFL by inspection of Figure 23. Figure 24 shows that delay times achieved with BFL [24, 34], SDFL [10] and DCFL [26] of L = 1 μm transistors differ by less than a factor of 3. Part of the longer delay times of the DCFL circuits shown results from the fixed resistor load and from the absence of buffer transistors. Also, the resistor value may not have been optimized for maximal speed, but was chosen rather for low power dissipation at reasonable speed. Since drift velocity saturation is already reached at the V_{DD} levels of DCFL, no large speed advantage can be expected for BFL or SDFL over DCFL. This is confirmed by Figure 24 [32].

Before leaving the subject of switching speed, we will discuss briefly the choice of W and L. The following comments apply to all three circuit types. Decrease of the switching transistor width W below about 15 microns affects the switching speed adversely, since transistor current decreases in proportion to W, while node capacitance contains a fixed width-independent offset value arising, among other things, from lateral fringing (Figure 10).

Transistor current in the absence of drift velocity saturation increases inversely as the decrease in channel length, while gate and channel capacitances decrease in proportion to channel length. Both relations are beneficial for speed increase by reduced channel length. However, once drift velocity saturation occurs, the transistor saturation current becomes independent of gate length, and some contributions to node capacitance such as the vertical fringing capacitance $C_{vf}^{(1)}$ and

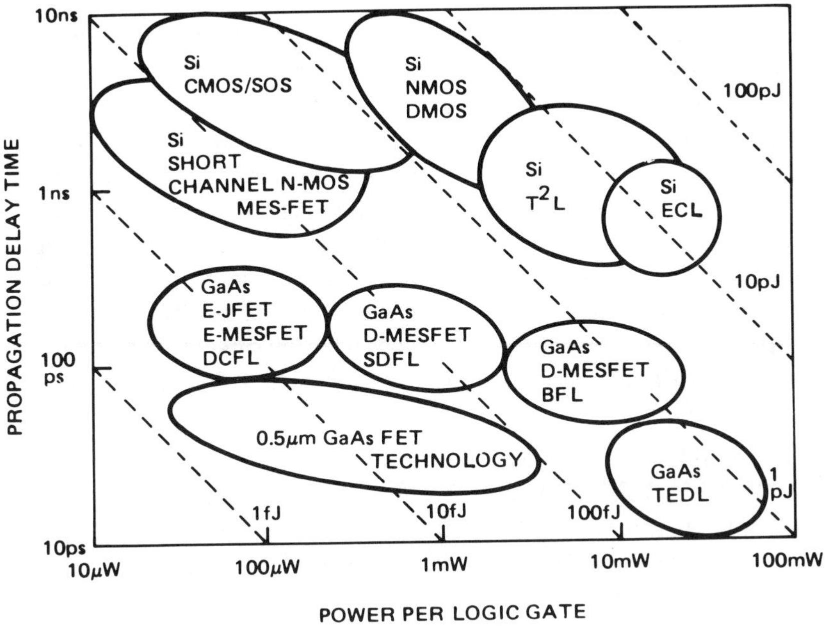

Figure 24/Propagation delay time vs. power dissipation per logic gate for GaAs and Si circuits according to reference 32.

the lateral fringing capacitance, C_{lf}, increase with decreasing gate length. Nevertheless, we still expect a speed increase about inversely proportional to gate length, at least down to L = 0.3 μm.

5.4 VLSI Prospects [32, 42]

At the VLSI, and particularly, at the ULSI levels, small area per logic gate and low power dissipation per gate become stringent requirements (Table 3). While the chip size may well be increased beyond 0.25 cm^2, only a fraction of the chip area (estimated to be 30% [42]) can be used for actual logic gates so that the area per gate listed in Table 3 appears realistic. The areas for present BFL and SDFL circuits are 7000 and 600 μm^2 per gate, respectively [42]. This would exclude such BFL circuits from VLSI and such SDFL circuits from ULSI. The SDFL is particularly suitable for large fan-in OR and NOR circuits, since it replaces the FI switching transistors of BFL or DCFL by one switching transistor, one constant current transistor, and FI Schottky diodes; the diode area is much smaller than a transistor area. Future efforts may be able to reduce the area requirements of BFL and SDFL to enable larger scale integration. However, it is questionable whether such efforts would be worthwhile, in view of the power requirements of these designs (to be discussed below). DCFL, particularly with two terminal current limiter or tunnel diode load, has no difficulty meeting the ULSI area per gate restriction. The discussion so far applies to 1 micron gate length photolithographically manufactured circuits. Electron beam fabrication of submicron gate length devices can increase packing densities by an order of magnitude. However, economic considerations favor the parallel produced photolithographic circuits over the sequentially produced E-beam circuits.

At f_{max} = 5 GHz and C_N = 30 fF, satisfying the power per gate restriction for VLSI in Table 3 would require logic swings of less than 1.8 V, according to Equation (31), without allowing for additional power losses by overdrive and by level shifting. For ULSI, logic swing will have to be restricted to less than 0.6 V, based on similar considerations. Actual power dissipations per gate in divide-by-two circuits were 40 mW/gate for BFL [24], 2.5 mW/gate for SDFL [42], and 0.21 mW/gate for DCFL [46]. Thus, only the DCFL divider satisfies the power restrictions for VLSI of Table 3. Table 2 shows that high power consumption eliminates the quasi-normally-off DCFL [44] for VLSI, and overdrive for ULSI. Some of the data from References 39 and 40 listed in Table 2 satisfy both the low power requirements for ULSI and also the delay time less than 200 ps required for 5 GHz clock rates. However, to apply existing DCFL for VLSI and ULSI at reasonable yields requires improved threshold voltage control.

Since VLSI circuits will not require simultaneous operation of all gates at the maximum clock rate, it may be argued that the power restriction can be relaxed. Nevertheless, the trend, both with respect to area per gate, power dissipation per gate, and V_{DD} level, as well as circuit simplicity and power supply considerations, clearly favors DCFL as one proceeds from LSI to VLSI and ULSI, and efforts on DCFL for VLSI and ULSI appear to be most fruitful.

Two terminal current limiters [32] will be less useful as the supply voltage shrinks, since the offset voltage of 0.5 V of these constant current load devices with 1 micron gate length would be a large fraction of the logic swing. These load devices could remain viable, however, with submicron technology to 1/2 micron, at which point a lower limit for the offset voltage of about 0.25 V would be reached due to optical phonon energy and upper band separation.

Tunnel diode loads [33] would become attractive at supply voltage levels of about 0.5 V, provided that production methods and adequate control can be worked out. Tunnel diode loads would be particularly well suited for large memory circuits.

6. COMPARISON OF GaAs AND Si BASED LOGIC FOR VLSI HIGH SPEED APPLICATIONS

The main advantages of GaAs based circuits over bulk Si based circuits are the five to six times larger electron mobility in GaAs devices at room temperature, and the existence of a reasonably high resistivity, so-called semi-insulating, substrate of GaAs. The semi-insulating substrate provides isolation between circuit components with lower parasitic capacitance than that of the p-n junction isolation used in bulk Si circuits. However, silicon-on-sapphire technology also provides low isolation capacitance, so that the principal advantage of GaAs over Si for high speed logic (Figure 23) resides in the larger electron mobility. The resulting speed advantage of GaAs of a factor of five or six (other things being equal) is lost at drift velocity saturation, since the saturation drift velocity of electrons in GaAs is only about twice that in Si. However, there are several reasons why the low field electron mobility, rather than the high field drift velocity, is the decisive factor for high speed VLSI circuits [9, 42], namely:

(i) The node voltage excursion includes the linear range of transistor operation $V_N = V_D \ll V_{SAT}$ in which drift velocity saturation is absent.

(ii) The supply voltage levels required for the low power dissipation at VLSI should exclude significant drift velocity saturation in Si, where it occurs at higher fields ($\cong 2 \times 10^4$ V/cm, i.e., $V_D \cong 2V$ at $L = 1\ \mu m$) than in GaAs ($\cong 5 \times 10^3$ V/cm).

(iii) The transition from absence of drift velocity saturation to occurrence of drift velocity saturation with increased field is rather abrupt in GaAs (Figure 16), while in Si it is more gradual [42]. Thus the adverse effect of drift velocity saturation on speed, compared to the hypothetical case of its absence at a given operating voltage, is more severe in Si.

Equations (33) and (11) show that the maximum clock frequency in the absence of drift velocity saturation is in proportion to K, and thus to the electron mobility. The gate-to-channel capacitance of GaAs field effect transistors and of silicon MOSFETs of equal gate dimensions is comparable. Furthermore, both insulator thickness (oxide in case of MOSFET, and depletion layer width a_{eff} in case of GaAs FETs) and its dielectric constant cancel in the ratio K/C_G which enters the maximal clock frequency for $C_N \simeq C_G$. Speed of Si circuits can be improved by operating at larger voltage swings than those in the case of GaAs before drift velocity saturation sets in. However, this implies larger V_{DD}, and thus larger power dissipation. The low hole mobility in GaAs excludes fast p-channel devices, and thus complementary circuitry. Therefore, there is a steady state power dissipation in GaAs logic circuits at V_{NH} which is absent in CMOS logic. However, when operating at high clock frequencies, the dynamic power dissipation of CMOS becomes much higher than that of GaAs circuits of equal speed, since the Si CMOS must be operated at about five times higher voltage level (in absence of drift velocity saturation) to achieve equal speed, and thus it has 25 times more dynamic power dissipation. Thus GaAs DCFL VLSI circuits will not only replace Si VLSI logic in the clock frequency range beyond 2 GHz, which is inaccessible for Si [47], but GaAs VLSI logic also has power advantages in the high megahertz clock range which is still accessible to silicon-on-sapphire circuits.

7. CONCLUDING REMARKS

(1) The capacitance of silicon-on-sapphire MOSFET transistors of several hundred angstrom oxide thickness is about the same as that of GaAs MESFETs or JFETs of equal gate dimension. The superior speed of GaAs circuits arises, thus, solely from an electron mobility about five times larger at low fields. Clock frequencies beyond 2 GHz appear to be out of reach for SOS circuits with 1 micron channel length devices; in this clock frequency range, GaAs should reign supreme.

(2) GaAs circuits will be superior to SOS circuits even at clock frequencies below 2 GHz in VLSI and VHSI because GaAs offers lower power dissipation at equal speeds. $V_G - V_T$ for silicon devices has to be about twice as high as that for GaAs transistors of equal geometry

to produce the same saturation current, and about five times as high to produce the same channel conductance in the linear regime. Thus, GaAs circuits can achieve the same speed as Si circuits at a lower logic swing, and hence have a lower supply voltage and less power dissipation.

(3) Considering now the various GaAs circuits, the immediate need is for improved production control, particularly with respect to the effective dopant concentration, and thus threshold voltage. As the control improves, supply voltage levels will be reduced, and VLSI and ULSI will become feasible. At supply voltage levels below one volt, DCFL is favored because of its inherent simplicity. At this low voltage level, hot electron effects will become unimportant for photolithographically manufactured circuits.

(4) In the meanwhile, production control will dictate circuit selection. Here SDFL and BFL have the edge over DCFL, since they are less affected by small changes in threshold voltage and have larger noise margins. They are adequate for LSI.

(5) At supply voltages above one volt the saturated drift velocity constant current limiter [32] should replace the pull-up transistor and fixed resistor loads presently used, since the current limiter has a lower capacitance than the pull-up transistor, provides higher speed than the fixed resistor, and offers excellent production control.

(6) For VLSI and ULSI DCFL circuits operating at about 0.5 V supply voltage, which may become available in the late eighties, tunnel diodes [33] appear highly attractive in their use as load elements and in memory elements, provided that adequate production control can be achieved.

TABLE II
DCFL Ring Oscillator Performance

Author	L (μm)	W (μm)	V_{DD} (V)	τ_D (ps)	P (μW)	$\tau_D \cdot P$ (fJ)	
Bert et al. [29]			0.4	1100	1.8	2	
			0.8	800	6.2	5	
			1.2	650	15.4	10	
			2.0	520	52	27	
Ishikawa et al. [26]	1		0.8	350	74	26	
			1.2	300	160	50	
			1.4	280	260	74	
Fukuta et al. [46]	1.2	20		300	100	30	
Mizutani et al. [39]	0.8	40	2.0	77	977	69	
(E-beam)	0.8	40	0.4	163	20	3.3	
	0.8	20	0.4	200	8	1.6	
Lundgren et al. [40]	1	25		186	21	4	
Bert et al. [44] (Quasi-normally-off)	1	35	2.5	≅100	≅2200	≅220	
Troeger et al. [37]	1	5, 6, 10	1.4	400	30	12	Quasi-
			2.5	200	500	100	Complementary
McDonnell Douglas (unpublished)	1	10	2.5	142	100	14.2	Follower-
			3.0	113	280	32	Inverter

TABLE III
Area and power dissipation per gate for logic circuits of various integration scales based on a chip area of 0.25 cm² and a power dissipation per chip of 5 W.

Integration Scale	Number of Gates per Chip	Area per Gate	Maximum Average Power Dissipation per Gate (mW)
small (SSI)	10	2.5 (mm^2)	500
medium (MSI)	10^2	0.25 (mm^2)	50
large (LSI)	10^3	0.025 (mm^2)	5
very large (VLSI)	10^4	2500 (μm^2)	0.5
ultra large (ULSI)	10^5	250 (μm^2)	0.05
giant (GSI)	10^6	25 (μm^2)	0.005

REFERENCES

[1] Jutzi, W., *Archiv El. Übertragung AEU,* 25:595, 1971.

[2] Mead, C.A., *Proc. IEEE,* 54:307, 1966.

[3] Shockley, W., *Proc. Inst. Radio Engrs.,* 40:1374, 1952.

[4] Van Tuyl, R.L., and C.A. Liechti, *IEEE J. Solid-State Circuits,* SC-9:269, 1974.

[5] Wilson, D.O., R.F. Mandal, W.R. Scoble, and H.L. Petersen, Intnl. Electron Device Conf., 600, 1978.

[6] Furutsuka, T., M. Ogawa, and N. Kawamura, *IEEE Trans. Electron Devices,* 25:580, 1978.

[7] Muta, H., et al., *IEEE Trans. Electron Devices,* 23:1023, 1976.

[8] Zuleeg, R., J.K. Notthoff, and K. Lehovec, *IEEE Trans. Electron Devices,* 25:628, 1978.

[9] Solomon, P.H., Intnl. Electron Device Conf. Digest, 201, 1978.

[10] Eden, R.C., B.M. Welch, and R. Zucca, *IEEE J. Solid-State Circuits,* SC-13:419, 1978.

[11] Mead, C.A., *Solid-State Electronics,* 9:1023, 1966.

[12] Lehovec, K., and R. Zuleeg, *Gallium Arsenide and Related Compounds (Edinburgh) 1976* (Inst. Phys. Conf. Ser. no. 33a) p. 263. London: Adam Helger/The Institute of Physics, 1977.

[13] Taylor, G.W., H.M. Darley, R.C. Frye, and P.K. Chatterjee, *IEEE Trans. Electron Devices,* ED-26:921, 1978.

[14] Jutzi, W., and M. Reiser, *IEEE Trans. Electron Devices,* ED-19:314, 1972.

[15] Ruch, J.G., and G.S. Kino, *Phys. Rev.,* 174, 3:921, 1968.

[16] Lehovec, K., and R. Zuleeg, *Solid-State Electronics,* 13:1415, 1970.

[17] Trofimenkoff, F.N., *Proc. IEEE,* 53:1765, 1965.

[18] Himsworth, B., *Solid-State Electronics,* 15:1353, 1972.

[19] Turner, J., and B. Wilson, Proc. 1968 Symp. GaAs, 195.

[20] Engelmann, R.W., and C.L. Liechti, *IEEE Trans. Electron Devices,* ED-24, 1288, 1977.

[21] Yamaguchi, K., S. Asai, and H. Kodera, *IEEE Trans. Electron Devices,* ED-23:1283, 1976.

[22] Engelmann, R.W.H., and C.A. Liechti, *1976 IEDM Digest Tech. Papers,* 351.

[23] Shur, M.S., and L.F. Eastman, *IEEE Trans. Electron Devices,* ED-25:606, 1978.

[24] Van Tuyl, R.L., C.H. Liechti, R.E. Lee, and E. Gowen, *IEEE J. Solid State Circuits,* SC-12:485, 1977.
[25] Wolfe, P.N., *Proc. IRE,* 50:2131, 1964.
[26] Ishikawa, H., H. Kusakawa, K. Suyama, and M. Fukuta, *IEEE Intnl. SSCC Digest,* 200, 1977.
[27] Kohn, E., *Solid-State Electronics,* 20:29, 1977.
[28] Filensky, W., H.J. Klein, and H. Beneking, *IEEE J. Solid-State Circuits,* SC-12:276, 1977.
[29] Bert, G., G. Nuzillat, C. Arnodo, *Electronic Letters,* 13, 21:644, 1977.
[30] Müller, R., J. Pfleiderer, and K.U. Stein, *IEEE J. Solid-State Circuits,* SC-11:657, 1976.
[31] Josephs, H.C., *Microelectr. Reliability,* 4:345, 1965.
[32] Zuleeg, R., 11th Intnl. Conference on Solid State Devices, Tokyo, 1979.
[33] Lehovec, K., *J. Solid State Circuits,* SC-14:797, 1979.
[34] Gold, R.D., and L.R. Weisberg, *Solid-State Electron.,* 7:811, 1964.
[35] Zuleeg, R., J.K. Notthoff, and A.F. Behle, IEEE/MTT-S Intnl. Microwave Symposium, 1979.
[36] Notthoff, J.K. and C.H. Vogelsang, First Annual Gallium Arsenide Symposium, No. 10, Lake Tahoe, Nevada, 1979.
[37] Troeger, G.L., A.F. Behle, P.E. Friebertshauser, K.L. Hu and S.H. Watanabe, Intnl. Electron Devices Meeting, Washington, D.C., *Technical Digest,* 497, 1979.
[38] Cos, H.M., J.V. DiLorenzo, and L.A. D'Asaro, First Annual Gallium Arsenide Integrated Circuit Symposium, No. 10, Lake Tahoe, Nevada, 1979.
[39] Mizutani, T., et al., 11th Intnl. Conference on Solid State Devices, Tokyo, *Digest of Technical Papers,* B-1-4, 1979.
[40] Lundgren, R.E., C.F. Krumm, and R.F. Lohr, Jr., First Annual GaAs Integrated Circuit Symposium, No. 8, Lake Tahoe, Nevada, 1979.

[41] Greiling, P.T., et al., *IEEE Trans. Electron Devices,* ED-25: 1340, 1978.
[42] Eden, R.C., B.M. Welch, R. Zucca, and S.I. Long, *IEEE Trans. Electron Devices,* ED-24:299, 1979.
[43] Eckhardt, G., C.L. Anderson, M.N. Colborn, L.D. Hess, and R.A. Jullens, *Extended Abstracts,* 79-2:1277, Am. Electrochemical Soc. Meeting, Los Angeles, CA, 1979.
[44] Bert, G., T. Pham Ngu, G. Nuzillat, and M. Gloanec, First Annual Gallium Arsenide Integrated Circuit Symposium, No. 7, Lake Tahoe, Nevada, 1979.
[45] Huber, A.M., G. Morillot, N.T. Linh, P.N. Favennec, B. Deveaud, and B. Toulouse, *Appl. Phys. Lett.,* 34(12):858, 1979.
[46] Bosch, B.G., *Proc. IEEE,* 67:340, 1979.

Planar Localized Implantation Process for Fabricating GaAs SDFL, LSI and VLSI Circuits

R. C. Eden
GigaBit Logic
Culver City, California

B. M. Welch
Rockwell International Microelectronics
Research and Development Center
Thousand Oaks, California

1. INTRODUCTION

The objective of the effort described in this chapter is to extend the $\tau_d \sim 100$ ps performance levels demonstrated for GaAs FET SSI and MSI out to much higher, LSI to VLSI, complexity circuits. The basic approach is to retain the demonstrated performance and yield advantages of the depletion mode FET active device approach, but to use these devices only where necessary (for logic inversion and gain), using instead ultra high speed Schottky barrier switching diodes for most logic functions. This Schottky diode FET logic (SDFL) gate configuration is very simple, uses very little chip area (as low as 600 μm^2/NOR gate), and requires little power ($P_D \sim 100$ μW to 2 mW/gate), all factors of prime importance for VLSI. At the same time, the propagation delays demonstrated for the SDFL gates ($\tau_d = 62$ ps in ring oscillators or $\tau_d = 110$ ps in D flip-flop dividers) are comparable to anything demonstrated in 1 μm gate length depletion mode FET circuits, and the dynamic switching energies ($P_D \tau_d$ = 68 fJ at $\tau_d = 62$ ps in NOR gates of $P_D \tau_d = 16$ fJ at $\tau_d = 136$ ps in inverters) are outstanding.

The requirement of the SDFL circuit approach for extremely small, high performance switching diodes makes it impossible to fabricate optimized circuits with only a single active layer, the usual approach for GaAs FET ICs. This requirement for multiple active layers optimized for different types of devices (e.g., low power depletion mode FETs and small high conductance switching diodes) has been met by the development of a sophisticated planar, localized-implantation fabrication process. This process has been used since 1977 [1] for the fabrication of scores of types of GaAs integrated circuits from SSI to LSI; some of these are listed in Table I. The most complex circuits successfully demonstrated to date are 1008-gate 8 x 8 bit parallel multiplier chips, 2.475 mm x 2.7 mm in size [17, 18, 19]. Extensions of this technology to the capability for 10,000 gate VLSI circuits is planned.

The key for successfully achieving these high levels of circuit complexity is combining a circuit approach of sufficient simplicity and density, and of low enough gate power to make VLSI possible, with a process capable of achieving the very high yields necessary to realize it. This requires from the process both an extremely high yield of circuit features (e.g., interconnect overcrossings or 1st level to 2nd level VIAs), since there may be several hundred thousand of these in a VLSI circuit, and extremely tight control on device parameters (e.g., FET pinchoff voltage or diode forward voltage drop) in order that the statistical variations of device parameters not limit yields by exceeding allowable margins in the circuit design [17, 19]. (This can be aided, of course, by a circuit approach with inherently large margins.) With our planar implanted process, standard deviations of GaAs FET pinch-off-voltage as low as σ_{V_p} = 41.5 mV over an entire wafer, or less than σ_{Vp} = 30 mV over a chip, have been obtained, with diode forward voltage drop standard deviations of σ_{VF} = 10.3 mV (1.3%). In comparison to the typical 0.5 V to 0.75 V noise margin of SDFL circuits, these uniformities are very good. (The attainment of high yields in VLSI parts with tens of thousands of devices requires a peak-to-peak parameter tolerance "window," even with optimum mean, of nearly 10 times the standard deviation for the parameter.) In addition, because of the substantial sizes of even high-density VLSI chips, the density of "fatal" defects, both process-induced and starting material related, must be kept very low.

In this chapter we will briefly describe the SDFL circuit approach used for our GaAs digital IC work, then discuss in detail our planar, localized implantation process with which the circuits are fabricated, and finally review the performance results obtained with these GaAs ICs.

2. SCHOTTKY DIODE-FET LOGIC CIRCUIT APPROACH

The achievement of LSI or VLSI requires high gate densities, attained in part in this work by high resolution (1 μm) lithographic techniques. Another important factor limiting the number of logic gates attainable per unit chip area is the circuit design. Whereas in all previously published GaAs FET digital IC work the GaAs FETs themselves are used as the non-linear logic elements [20], with this planar, multiple localized implantation approach, optimized GaAs Schottky barrier switching diodes can be fabricated. Inasmuch as GaAs Schottky barrier diodes are among the fastest-switching semiconductor devices existent and their switching energies ($\sim 10^{-15}$ joules), required chip areas (1 μm x 2 μm active area) and capacitances ($\sim$ 2 fF) are very low, their use as logic elements in GaAs digital ICs seems quite desirable. We have developed a logic circuit approach using such high speed GaAs Schottky barrier switching diodes [2, 21] for most logic functions with, in its simplest form, GaAs Schottky gate FETs used primarily for inversion and gain, as shown in Figure 1(a).

Figure 1(a) shows the circuit diagram for a simple 4-input Schottky diode-FET logic NOR gate. The logical OR function (prior to inversion) is accomplished by the very small high conductance GaAs Schottky barrier diodes. In addition to their logic function, these diodes provide $\sim$ 0.75 V of level shifting between the positive logic levels from previous stages and the negative gate voltages required by the normally-on, depletion mode Schottky gate GaAs FETs. For very low pinchoff voltage GaAs FETs ($V_p \approx -0.5$ V), the switching diodes (D_A - D_D) alone provide adequate level shifting for very low voltage and very low power ICs. By adding the level shift diode, D_s, in Figure 1, GaAs FETs with slightly higher pinchoffs ($V_p \sim -0.7$ to -1.4 V) may be used with higher supply voltages and logic swings; the data presented here are for gates with this shift diode.

The pulldown (PD) to $-V_{SS}$ sinks the bias current for the diodes and

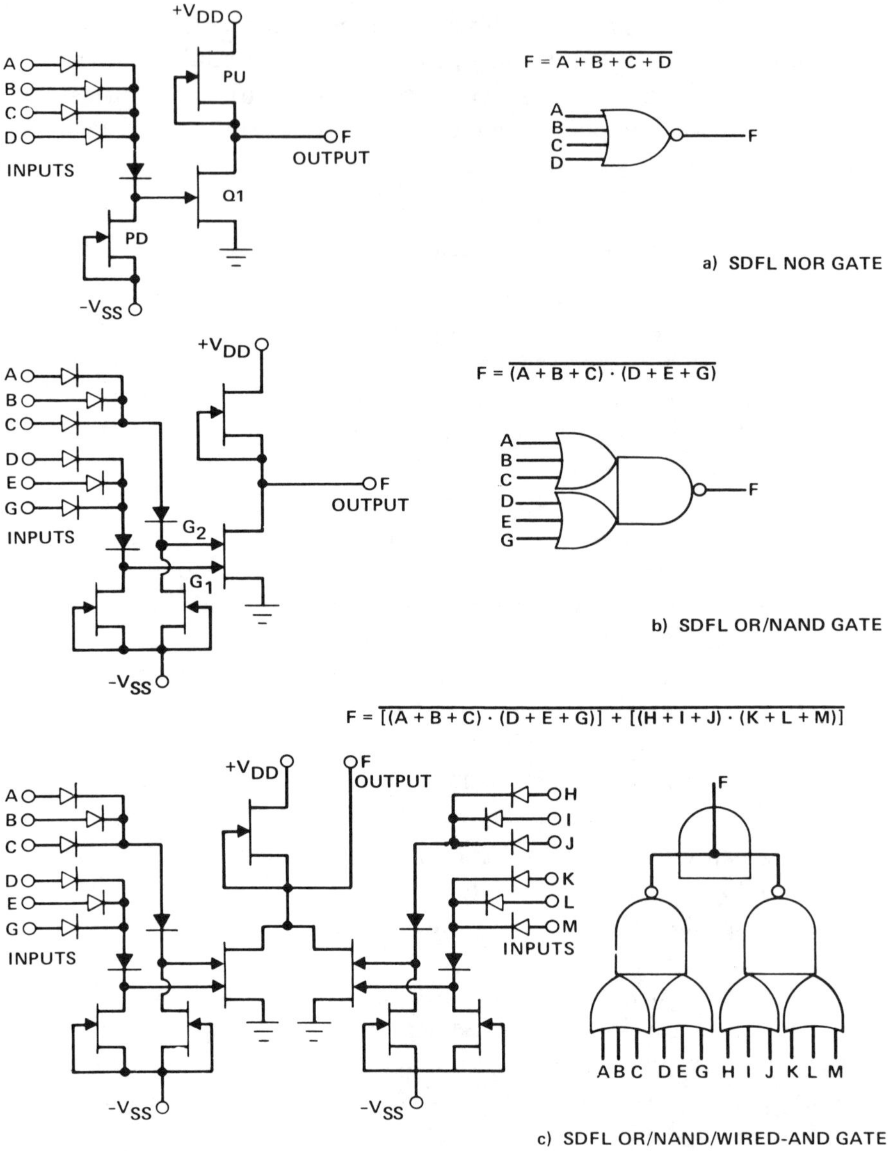

Figure 1*/Schottky diode-FET logic (SDFL) gate configurations for 1-, 2-, and 3-logic level gates. All FETs are depletion-mode, typically* $-1.5\ V < V_p < -0.5\ V$*; unshaded diodes are very small high speed switching Schottky diodes, while shaded diodes are larger area, higher capacitance voltage shifting diodes.*

provides most of the turn-off current for the gate of Q1. PD is kept small (typically $\leqslant$ 20% of the width of Q1). However, since in this current-sourcing logic the dc fanout is limited to the current ratio between PU and PD (typically 3 or 4), larger fanouts are achieved, where necessary, by adding a source-follower or other buffer configuration to the gate. Because of the very small size of the logic diodes, and because of the fact that the diodes, which are two-terminal devices, require many fewer overcrossings and vias than circuits using 3-terminal devices like FETs, multi-input SDFL NOR gates require only a fraction of the chip area needed by FET-logic approaches. This results in lower parasitic drain and stray capacitances for multi-input SDFL NOR gates, for improved dynamic switching performance in comparison to multi-input gates in FET logic.

In the interests of efficiency and speed it is often desirable to implement logic functions with multi-level gates. In FET logic, the use of combinations of series (NAND) or parallel (NOR) FET configurations to achieve two-level logic functions is common. For example, in earlier GaAs work, Van Tuyl et al. [3] demonstrated depletion-mode GaAs FET buffered-FET logic (BFL) NAND/NOR (or NAND/WIRED-AND) gates, with up to two NAND input terms (series or dual-gate FETs) and up to two of these NAND functions drain dotted together, which have propagation delays as low as τ_d = 110 ps. Four of these 2-level gates were utilized to implement a fast ($f_c \sim 1/(2\tau_d)$) complementary-clocked ÷ 2 frequency divider stage which gave toggle frequencies up to 4.5 GHz, for $\tau_d \sim$ 111 ps effective gate delays, at a $P_D \sim$ 40 mW/gate power level [3]. With single-level gates, the fastest NOR-gate implemented complimentary-clocked divider circuit uses 8 gates and has a maximum speed of $f_c = 1/(4\tau_d)$. In the current approach, we have fabricated more easily used, single-clocked T-connected D flip-flops using 6 NOR gates, achieving maximum toggle frequencies of f_c = 1.9 GHz at P_D = 2.5 mW/gate, which corresponds to τ_d = 110 ps ($f_c = 1/(4.85\ \tau_d)$) in this configuration [7]. Hence, while the SDFL NOR gates gave the same propagation delays as the BFL gates (and at much lower power levels), the architecture of the NAND/NOR dividers allowed much higher toggle frequencies to be reached; this has been (inaccurately) cited as a significant limitation of the SDFL circuit approach.

In fact, such multi-level logic gate configurations may indeed be realized in SDFL [5] with up to 3-level gates with many (10-20) inputs; it is not restricted to 4-input 2-level gates as in BFL. Thus, the Schottky diode-FET logic subsumes normal FET logic, to which is added the additional level of logic provided by the ultra high speed switching diodes.

In our earlier work [2, 4, 6], the FET logic function utilized was principally the inverter, so that, with the diode-OR, a NOR gate function was realized (Figure 1(a)). On the other hand, by using such diode-OR clusters on each gate of a dual gate FET (or series-FET connection), an OR/NAND 2-level gate is achieved (Figure 1(b)) which is the complement of the 2-level NAND/NOR gate realized in BFL, except that the number of first level terms in the SDFL version is no longer restricted to two. Further, drain dotting can be used between two of these SDFL OR/NAND gates to achieve a 3-level OR/NAND/WIRED-AND gate configuration (Figure 1(c)), achieving the (positive) logic function

$$F = \overline{[(A + B + C) \cdot (D + E + G)] + [(H + I + J) \cdot (K + L + M)]} \qquad (1)$$

or equivalently,

$$F = \overline{[(A + B + C) \cdot (D + E + G)]} \cdot \overline{[(H + I + J) \cdot (K + L + M)]} \qquad (2)$$

While the application of the two-level SDFL OR/NAND gates of Figure 1(b) is obvious because of its complementary relationship to the widely-used NAND/NOR configuration, the utility of the three-level SDFL OR/NAND/WIRED-AND gates (Figure 1(c)) is less clear and best illustrated with an example. One of the key goals of our GaAs IC effort was to demonstrate an ~ 1000 (NOR) gate 8 x 8 bit parallel multiplier. Both 3 x 3 bit and 5 x 5 bit multipliers, implemented with 75 and 260 SDFL NOR gates, respectively, were fabricated to test the 12-gate, 3 τ_d to carry full adder cell configuration selected for this design before the 8 x 8 bit parallel multiplier was built. The speed of the multiplier could be improved by reducing the number of logic gate propagation delays, τ_d, required to accomplish the full-adder function. Even with NOR gates, a 2 τ_d delay for both sum and carry can be achieved, but at the (prohibitive) expense of an

18-gate cell complexity. By using 2-level and 3-level SDFL gates, a full adder circuit (Figure 3 of Reference 5) can achieve a 1 τ_d to carry (C or $\overline{C}$) and 2 τ_d to sum (S or $\overline{S}$) delay with only 6 gates (4 OR/NAND and 2 OR/NAND/WIRED-AND). This would reduce the delay for an 8 x 8 bit parallel multiply from the 35 τ_d of the NOR implementation to 21 τ_d, a 40% reduction. The power should be greatly reduced, since only half as many gates are required, although the yield would not be expected to be improved, since similar numbers of logic FETs are required. The essential speed advantage of the 3-level OR/NAND/WIRED-AND gate in generating $\overline{C}_{OUT}$ from C or C_{OUT} from $\overline{C}$ in only 1 τ_d can be utilized in fast ripple-adders without generating all four (S, $\overline{S}$, C_{OUT}, $\overline{C}_{OUT}$) terms by changing from propagating C_{OUT} to $\overline{C_{OUT}}$ in alternate stages. This approach (illustrated in Figure 4 of Reference 5) has been demonstrated in a fast ripple carry 4-bit adder chip (Table I). Figure 2 shows a microscope photograph of a section of this adder circuit showing 1-, 2-, and 3-level SDFL gates.

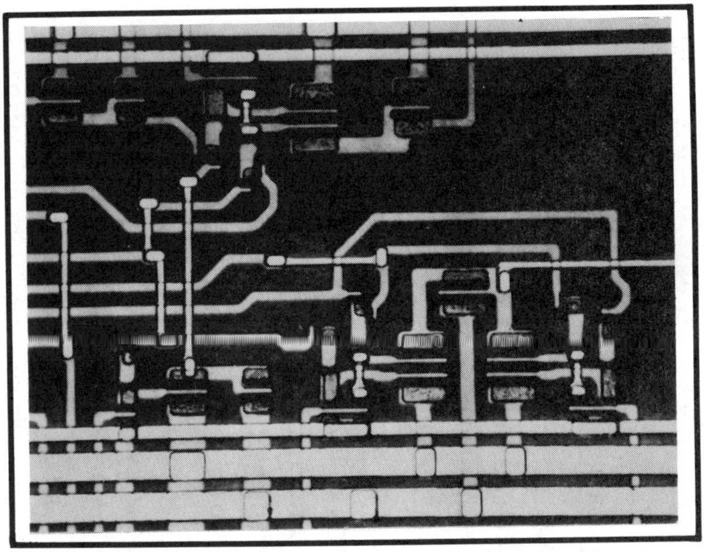

Figure 2*/Microscope photograph of portion of planar SDFL circuit (Figure 4 of Reference 5) showing 1-, 2- and 3-level SDFL gates. At the lower left is an inverter (1-input NOR gate); at the top center is a 2-level OR/NAND gate, while at the lower right is a 3-level OR/NAND/WIRED-AND gate.*

TABLE I

Review of Rockwell Planar GaAs IC Demonstration Circuits

FULLY WORKING, SUCCESSFUL TESTED CIRCUITS	
8 x 8 PARALLEL MULTIPLIER	1000 GATES
5 x 5 PARALLEL MULTIPLIER	260 GATES
8-STAGE SHIFT REGISTER/217 BIT P/N CODE GENERATOR	96 GATES
3 x 3 PARALLEL MULTIPLIER	75 GATES
4-BIT FAST RIPPLE CARRY ADDER USING 2- AND 3-LEVEL GATES	43 (63 EQUIV)
÷ 10/11 OR ÷ 20/22 VAR. MOD. DIVIDER FOR FREQ. SYNTHESIZER	37 GATES
8:1 DATA MULTIPLEXER WITH ADDRESS GENERATOR	64 GATES
1:8 DATA DEMULTIPLEXER WITH ADDRESS GENERATOR	60 GATES
1 GHz ANALOG SAMPLE/HOLD OR TRACK/HOLD CIRCUIT	25 COMPONENTS
DIGITAL FM DEMODULATOR CIRCUIT	65 COMPONENTS
÷ 8 FREQUENCY PRESCALER (D-FLIP FLOP RIPPLE DIVIDER)	25 GATES
÷ 8 SYNCHRONOUS COUNTER IMPLEMENTED WITH D-FFs	33 GATES
NOR GATE RING OSCILLATORS IN W = 3, 5, 10 AND 20 μm WIDTHS	9 GATES
OR/NAND GATE, 3 PROPAGATION MODE RING OSCILLATORS, 5 & 10 μm	7 GATES

3. PLANAR GaAs IC PROCESSING

The realization of high speed, low power GaAs LSI/VLSI requires a fabrication process technology capable of extremely high yields and excellent device uniformity and control [7]. While the superior electronic properties of GaAs relative to Si have provided the motivation for the development of GaAs ICs [7, 20], few efforts have really attempted to address LSI, or in fact could reasonably be expected to satisfy LSI requirements, due to the limiting nature of their fabrication approach.

Very early (before 1970) GaAs IC efforts were doomed to complete failure due to the lack of any process-useful native oxide of GaAs (such as that of Si), the lack of a controllable n-type diffusion technology and the inferior quality of the GaAs material available at the time. More recent work (evolving during the 1970s) has been more promising, particularly due to the development of a useful GaAs ion implantation technology [8, 9], leading to the initial development of GaAs ICs [1, 3]. However, the majority of these efforts have been restricted to SSI/MSI applications because of limited fabrication approaches based on extending existing GaAs microwave FET processes to IC applications.

Rockwell's goal, from the inception of the program, has been to develop a GaAs LSI technology. This goal provided strong motivation for developing a planar fabrication process using the low power Schottky diode FET logic (SDFL) circuit approach [2]. In the following, we will discuss a planar GaAs integrated circuit fabrication technology capable of achieving LSI/VLSI complexity circuits. The specific fabrication approaches and techniques were chosen to satisfy LSI requirements for high yield, high density, and low power [10, 11]. Extension of the current GaAs technology to the LSI/VLSI domain will depend critically on device yield which will be dictated by the GaAs material properties and by the fabrication processes used. The purpose of this section is to describe a planar GaAs IC process technology which combines advanced planar device and multri-level interconnect structure with state-of-the-art LSI processing techniques, including multiple localized ion implantations, reduction photolithography, plasma etching, reactive ion etching, and ion milling.

3.1 Fabrication Approach, Process Techniques and Device Structure

The development of an ion implantation technology in GaAs has made it possible to conceive new fabrication approaches not viable within the limitations of epitaxial layers. GaAs implantation advancements coupled with the observation that the rapid development of Si LSI was greatly facilitated by planar fabrication techniques using ion implantation and dielectric passivation, suggested that similar developments in GaAs would enhance the development of LSI/VLSI. The schematic of Figure 3 represents a cutaway view of the Rockwell planar IC approach. The planar circuits are fabricated by using multiple localized ion implants directly into semi-insulating GaAs substrates. Hence, individual devices can be optimized by using different implants with the unimplanted GaAs substrate providing isolation between adjacent devices. This fabrication method ideally suits the Schottky diode-FET logic (SDFL) circuit approach since SDFL requires the GaAs FETs and Schottky diodes to have different implanted active layers. The sophistication of this process technology is illustrated (Figure 3) by the use of a 1 μm wide n^+ implant region placed between dual 1 μm GaAs FET gates in order to lower the channel resistance between the dual gates. Another important aspect of this approach is the use of dielectrics. The dielectric regions shown in Figure 3 are utilized for post implantation annealing, protecting the GaAs surfaces during processing, and passivating the surface. Also, this fabrication approach is compatible with any number of implantation steps. Therefore, this process technology has, in principle, the capabilities for producing mixed device types, such as enhancement mode and depletion mode FETs on the same GaAs IC chip.

Fabrication of the planar GaAs IC shown in Figure 3 utilizes many state-of-the-art processing techniques. Since GaAs ICs are designed with 1 μm features, a dry LSI/VLSI compatible process has been developed. The only wet chemical process steps are associated with the photolithography and photoresist lift-off technique.

Table II summarizes the key LSI processing techniques used in the fabrication of these GaAs ICs. The material processes are characterized by multiple, localized implanted, non-epitaxial methods providing planar device structures which are automatically isolated by non-implanted areas due to the use of high quality semi-insulating GaAs substrates. The use of any number of localized Se, S, or Si implants or combinations of implants provides the process flexibility for optimizing various LSI circuits. All of the dielectrics and metals use conventional deposition techniques, including RF diode and magnetron sputtering, E-beam evaporation and plasma enhanced CVD techniques. The appropriate deposition technique is chosen in order to optimize any particular process step.

Resist delineation is accomplished by using reduction projection photolithography for defect free, micron resolved photoresist patterns. Pattern replication is accomplished without any chemical etching. A unique high yield replication process has been developed by Rockwell specifically for planar GaAs ICs. This fabrication process incorporates plasma etching in conjunction with an intermediate lift-off technique providing extremely high yield self aligned metal/dielectric patterns. Second level metal interconnects are replicated using a dry etched ion milling technique. The resulting planar structures are used to great advantage in complex multi-level GaAs ICs.

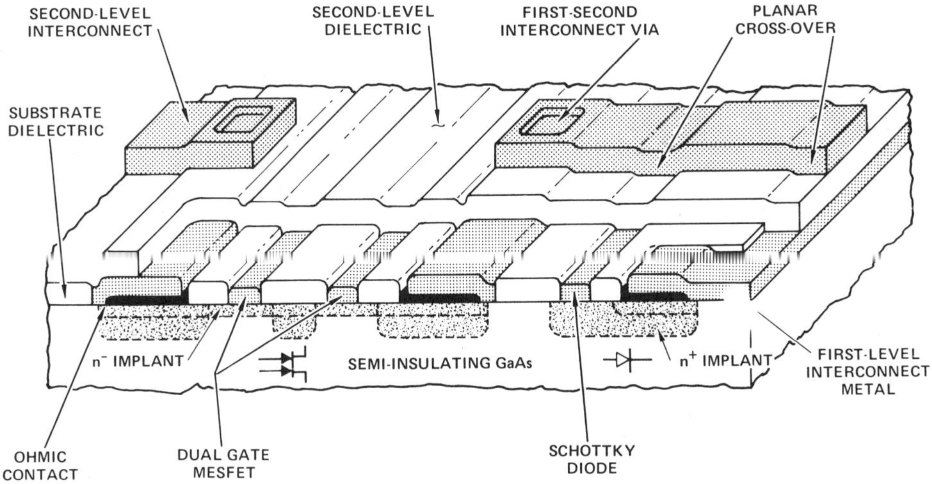

Figure 3/Cutaway view of a planar GaAs IC showing a dual gate FET, a diode, and interconnects.

TABLE II

Fabrication Techniques for Planar GaAs ICs

PROCESS	MATERIAL	TECHNIQUE
ACTIVE LAYERS	Se, S, Si DOPED	LOCALIZED IMPLANTS
ISOLATION	SEMI-INSULATING GaAs SUBSTRATE	PLANAR STRUCTURE
DIELECTRICS: 1st LEVEL 2nd LEVEL	SiO_2, Si_3N_4 SILICON NITRIDE	RF DIODE SPUTTERING PLASMA ENHANCED CVD
METALLIZATIONS: OHMIC CONTACTS 1st LEVEL 2nd LEVEL	 AuGe/Pt Ti/Pt/Au Ti/Au	 E-BEAM EVAPORATION MAGNETRON SPUTTERING
RESIST DELINEATION	POSITIVE RESIST	4x PROJECTION PHOTOLITHOGRAPHY
PATTERN REPLICATION	DIELECTRICS METALLIZATIONS	REACTIVE ION ETCHING PLASMA ETCHING DIELECTRIC ASSISTED LIFT-OFF ION MILLING

Figure 4 shows the Rockwell planar, dielectric passivated, double implanted GaAs FET structure compared to a typical mesa, single implanted GaAs FET structure [10, 11]. The planar GaAs FET is fabricated with two implants: one shallow, lightly doped n^- implant forming the channel region, and the other deeper n^+ implant forming the source and drain regions. This design minimizes source and drain series resistance by effectively narrowing the high resistance channel gap between gate and source/drain, and facilitates the fabrication of ohmic contacts. In this manner, excellent device characteristics are obtained without resorting to recessed gate techniques (shown in the GaAs FET in the lower portion of Figure 4). The simplest alternative to the Rockwell approach is to utilize a single, ultra thin, lightly doped layer. However, such a GaAs FET structure is subject to poor operating characteristics due to the high series resistances resulting from the high sheet resistance ($\sim 2000\Omega/\square$) of the channel regions between the gate and the source/drain contacts. The most widely used compromise, which minimizes the effects of high series resistance, consists of using a deeper implant and recessing the Schottky gate into the GaAs surface by chemical etching or anodization. This technique can provide good FET characteristics, but the uniformity, control, and yield of the resulting devices for LSI/VLSI applications are in serious question. For example, the difficulties encountered in obtaining adequate device uniformity using implanted layers with recessed gate structures for enhancement mode FET devices has led workers to explore innovative, less demanding (in terms of device uniformity) circuit concepts such as quasi-normally-off GaAs FET

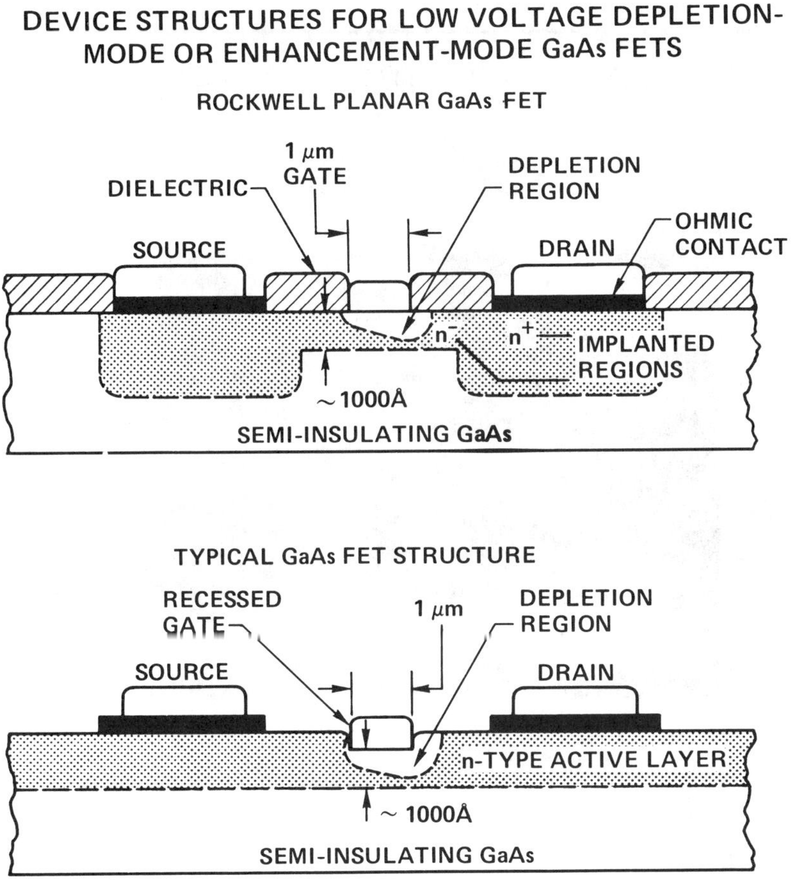

Figure 4/Comparison of a planar GaAs FET fabricated by localized ion implantation with a recessed gate device fabricated on a uniform layer.

logic [12]. It would appear that enhancement or depletion mode FET processing technologies utilizing chemical etched recessed gate structures may not be LSI/VLSI yield compatible.

An important feature of the planar Rockwell approach is the role of the dielectric as an integral element of the device structure. The dielectric is used for high temperature post implantation annealing and for protecting the GaAs surface during subsequent processing steps.

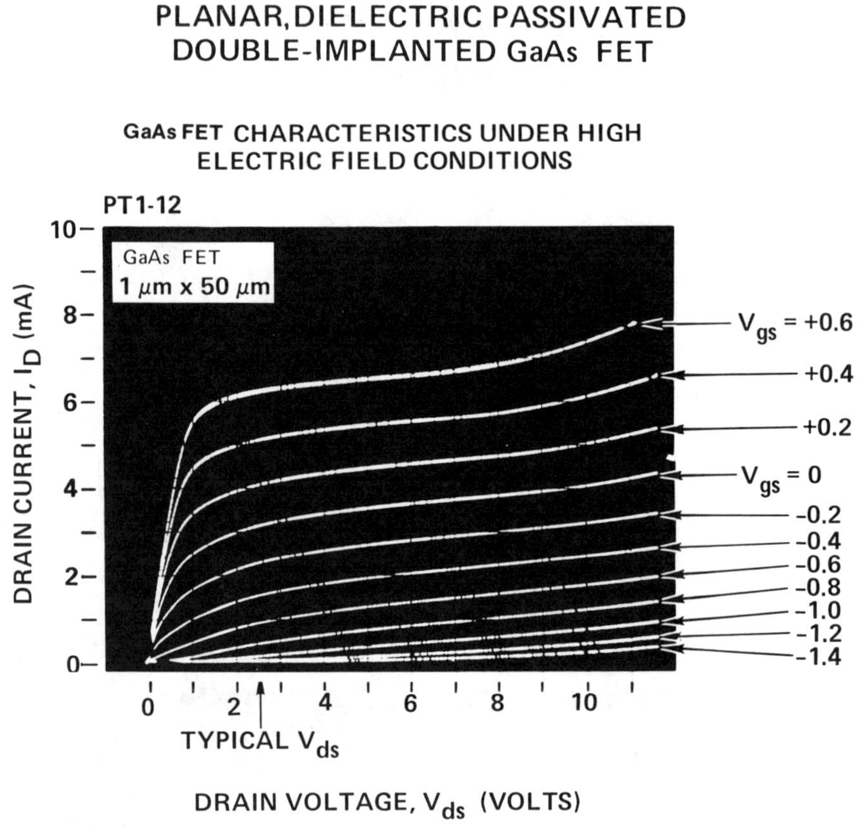

Figure 5/I-V characteristics of a planar GaAs FET showing the excellent high electric field behavior of the device. Note that the operating V_{ds} typically does not exceed 2.5 V in digital IC applications.

This encapsulation layer, which remains on the critical GaAs channel surface region between the gate and source-drain contacts of the FET, contributes positively to device characteristics (as evidenced by very moderate looping and light sensitivity and by the absence of current "lag" effects). These problems, sometimes observed on non-passivated devices, seem to be associated with exposing bare GaAs surfaces to numerous process steps. The dielectric structure used in these planar GaAs FETs helps allow these devices to withstand high electric fields without premature breakdown. This is demonstrated in Figure 5 which shows the I-V characteristics of a planar GaAs FET operating under high electric field conditions. The device shown is biased from +0.6 V (enhancement) to − 1.4 V (depletion) with a 12 V drain voltage (as opposed to the typical 2.5 V_{ds} operation in ICs). Similar devices have operated at up to 20 V without failure. These data show that GaAs FETs fabricated with this planar process are very rugged and do not fail easily under adverse conditions (e.g., supply voltage transients). This reliable performance (no premature breakdown) is attributed to the heavier doped implanted region under the ohmic contacts, the use of ion implanted layers (no epitaxial/substrate interfaces) and the channel surface protection provided by the dielectric between the gate and the source-drain electrodes. These features contribute to improved device reliability and should enhance the yield of GaAs LSI.

3.2 Multiple Localized Ion Implantation Processes

Ion implantation is now an established material technology for GaAs ICs and for many discrete GaAs devices as well. Many workers, however, still appear to be using mesa, single implanted fabrication processes. Since mesa structures are only easily implemented with a single implantation step, this approach inherently restricts device and circuit optimization. The power of ion implantation was fully exploited in this work by developing a multiple localized implant technology ideally suited to SDFL logic, which requires, by design, optimization of individual device (GaAs FET and Schottky diode) active layers.

The implant process steps used in the work are shown in Figure 6. Initially, the GaAs is coated with a thin layer of Si_3N_4 which remains on the wafer throughout all of the subsequent processing steps. The first process steps are the two localized implantations carried out

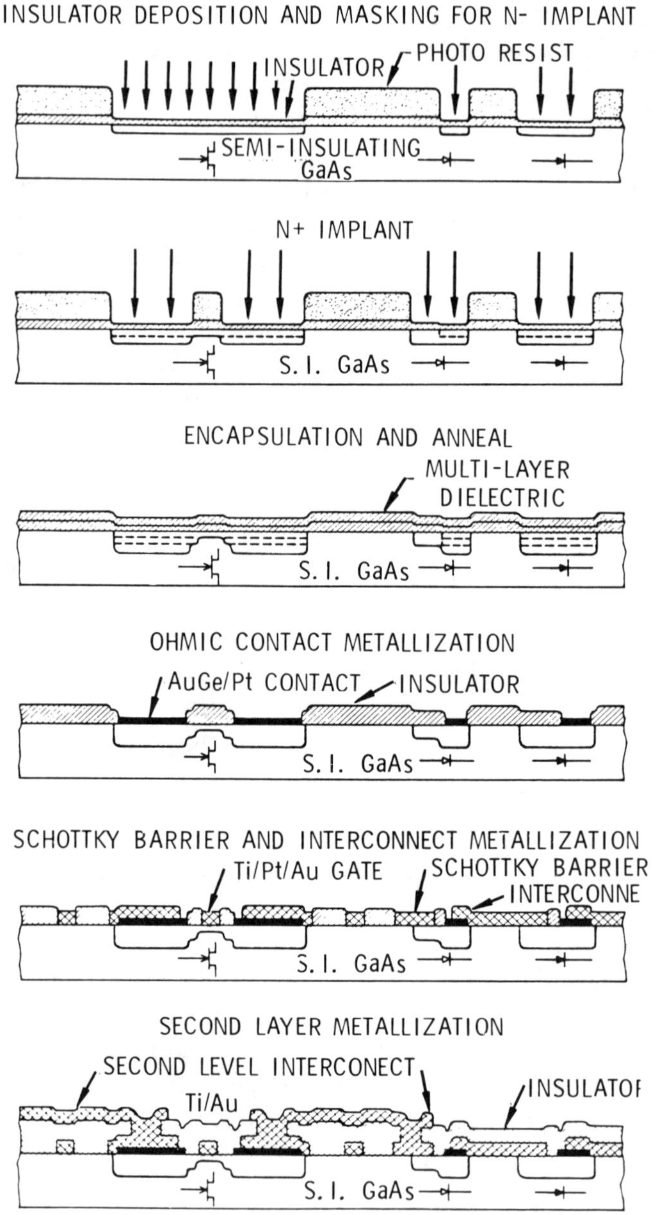

Figure 6/Planar SDFL GaAs IC fabrication steps.

through the thin Si_3N_4 layer using thick photoresist as the ion beam mask. After each implant, a shallow step is plasma etched into the Si_3N_4 for registration of the implanted regions. Following the implants, additional dielectric (SiO_2) is added prior to the post implantation annealing step. The same dielectric used in annealing also provides us with an intermediate layer lift-off medium used in the processing of first level metal interconnects (to be discussed below).

Normally, a sequence of two implants is used in this planar IC process; however, the process is capable of any number of implantation steps necessary to optimize individual devices. An example would be an additional high dose (n^{++}) implantation step to heavily enhance the doping under ohmic contacts. In light of the yield requirements of LSI/VLSI, a determination of the relevant trade-offs between improved performance resulting from an additional implant step and reduced yield must be made.

Typical implantation parameters and the resulting electron concentration doping profiles for both IC implants conducted through the Si_3N_4 cap are shown in Figure 7. The shallow 400 keV Se implantation profile peaks near the GaAs surface, due to the energy absorbed in penetrating the thin Si_3N_4 layer. Except for the shift in the peak of the implanted profile, little influence has been observed as a result of implanting through thin dielectric layers. In spite of theoretical predictions of enhanced doping from "knock-ons" from the Si_3N_4 cap during ion implantation, this has not been identified in our implanted profiles nor has any shift in the peak of the Se profile been observed as a result of small variations in dielectric thickness (the Si_3N_4 thickness is held to ±20Å). The through-the-cap implantation technique used in the fabrication of these GaAs ICs has proven to be a very workable process.

As shown in Figures 6 and 7, the S (n^+) implant provides a conductive layer much deeper (~ 4000Å) than the shallow (~ 1500Å) low threshold (V_p ~ - 1V) FET channel (n^-) implant. The S implanted layer is ideally suited for the high speed switching diode required in SDFL. Both n^- and n^+ implants are used for level shifting diodes and for enhancing the doping under ohmic contact regions. The S profile is deeper and the resulting electron carrier concentration much lower (~ 30% activation) than the predicted, LSS-calculated profile for this energy and dose. These differences can be attributed to a cap dependent enhanced diffusion mechanism. Low S activation and the relatively flat implant profile are due to sulfur atoms outdiffusing to the GaAs/Si_3N_4 interface [13].

Recently, continued development and refinement of planar localized implant processes for LSI performance and yield optimization has led to investigating the use of Si as an alternate ion implant species for both Se and S. Silicon appears to be the only possible ion species that could in fact serve as a "universal" n-type implant ion in GaAs.

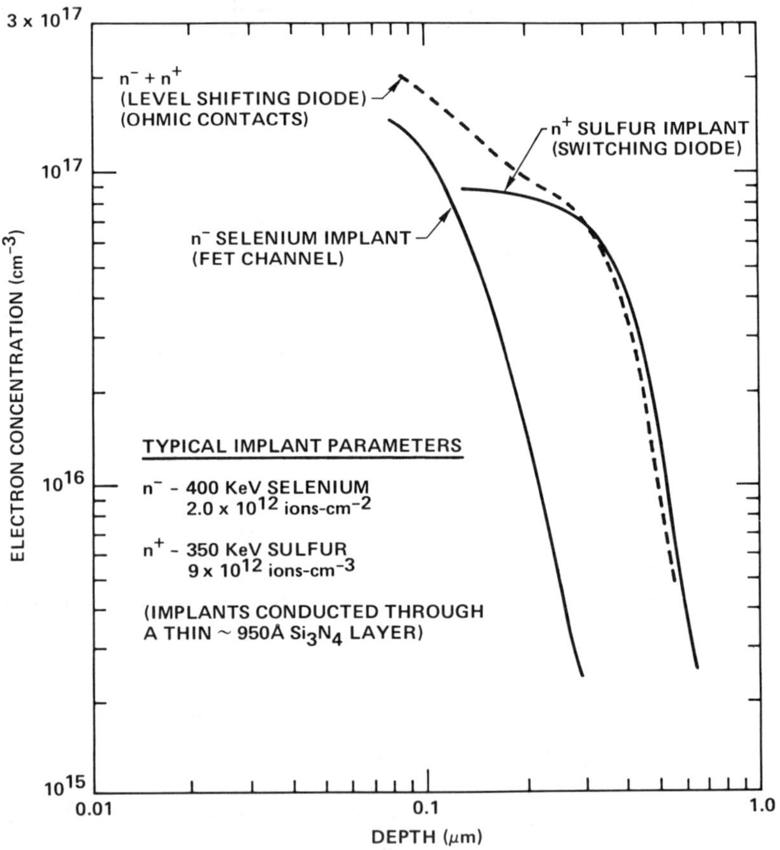

Figure 7*/Active layer profiles (for the n^- FET channel Selenium implant and the n^+ high-speed switching diode sulfur implant).*

The insert table and profiles shown in Figure 6 illustrate how Si implanted profiles can satisfy both the GaAs FET and Schottky switching diode requirements. As Figure 8 demonstrates, the appropriate choice of Si implant energy and dose result in profiles similar to the Se and S profiles normally used. However, merely replacing the current Se and S profiles with Si implanted profiles is not the issue. In the case of GaAs FET channel profiles, and to a lesser extent the Schottky diode profiles, reproducibility is the key issue for further investigation. It is possible that the lower mass of Si may result in less implantation damage than that caused by Se for room temperature implants (room temperature implants are desired in the process because of the use of photoresist as the implant mask); there is also the probability of fewer Si and N "knock-ons" as a result of implanting through the Si_3N_4 cap, also due to the lower mass. Thus Si may provide even better reproducibility than Se and S n-type profiles. Si profile reproducibility and uniformity are being investigated in our current IC work. Implanted planar GaAs ICs with direct comparison of Si vs. Se fabricated GaAs FETs, and Si vs. S fabricated Schottky diodes are already under evaluation.

GaAs IC device reproducibility and uniformity depends both on the quality of the semi-insulating GaAs material used and on the degree of control available from the fabrication processes. Both factors must be considered to be of equal importance. From our analysis of reproducibility data over the past three years, it has become clear that the variations of doping profiles arise, to a large extent, from subtle differences between semi-insulating GaAs substrates. Profile reproducibility is excellent when implant profiles are compared on wafers all from a single ingot. Figure 9 shows a group of low dose Se implanted profiles measured on 16 different slices from the same GaAs ingot. The standard deviation of the voltages required to deplete the layers to 10^{16} cm^{-3} is only 0.11 V with the standard deviation of profile depth (measured at 10^{16} cm^{-3}) of only 64Å. The variation of GaAs FET pinchoff voltages from wafer to wafer would be expected to be of the same magnitude as the depletion voltages measured on these C-V profiles.

Since a single IC qualified GaAs ingot results in good doping profile reproducibility for that ingot, in practice we routinely perform small implantation dose adjustments in order to accommodate slight varia-

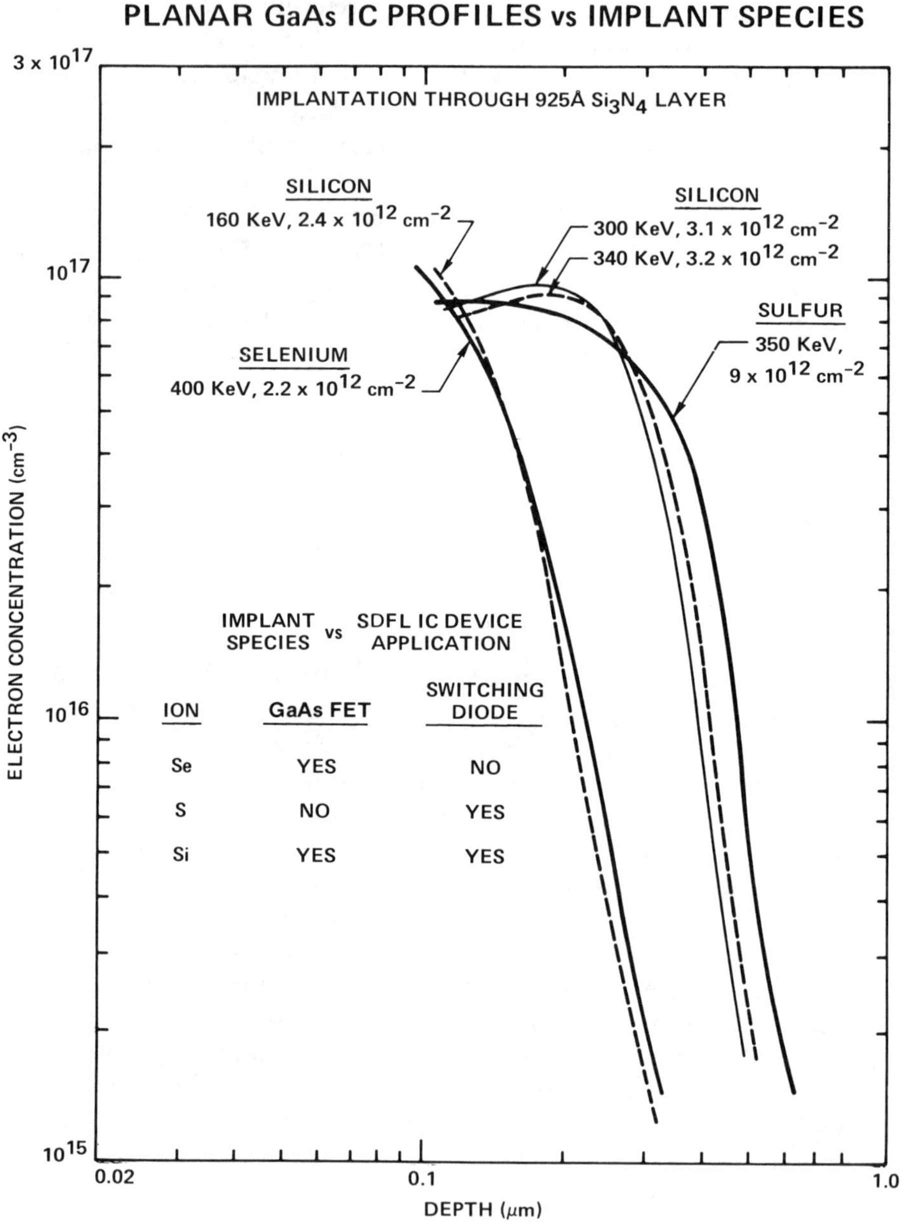

Figure 8/Carrier concentration profiles for Se, Si, and S implants.

tions between different GaAs ingots. Test profiles measured as part of the IC material qualification procedures provide the information used to tailor the implantation dose for individual ingots. In this manner, good reproducibility is achieved between various IC process lots using wafers from different GaAs ingots. Although this is a manageable approach, future large scale manufacturing efforts will require more consistency in the GaAs material (e.g., LEC wafers).

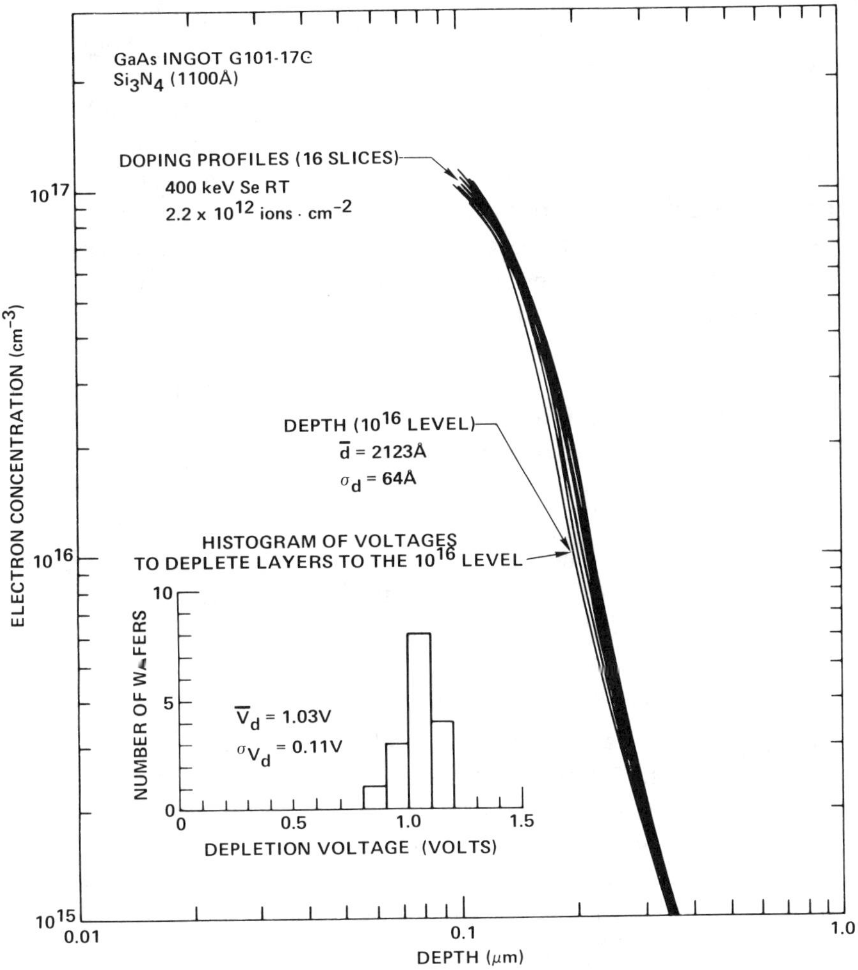

Figure 9/Doping profiles obtained by Se implantation through a thin cap into semi-insulating GaAs substrates, all from the same ingot.

The most critical requirement for LSI/VLSI is the uniformity of device characteristics on a single wafer. As indicated from previous discussions, the most difficult layer to control, and the key to device performance, is the shallow, lightly-doped, high-resistance n^- GaAs FET channel layer. This active layer directly controls the pinchoff voltage, V_p, and is the key parameter in the optimization of power-delay product of the logic gates.

Excellent active layer uniformity and control of the planar process is evidenced by the histograms of typical V_p distributions of planar, double implanted, dielectric-passivated GaAs FETs, shown in Figure 10. The long range V_p distribution of 71 FETs from a full GaAs IC wafer are characterized by a V_p of -1.24 V with a standard deviation of 94 mV. The area encompassed by these distributed V_p measurements would accommodate over 10^5 GaAs SDFL gates. Short range statistics of 81 FETs distributed over an equivalent 50 gate area show an outstanding 37 mV standard deviation. It should be emphasized that these are not our best data. Standard deviations as low as 41.5 mV have been measured over an entire wafer, and values as low as 23 mV have been observed in short range statistics [14]. This excellent device uniformity reflects the present state of GaAs material, implantation, and fabrication processes.

3.3 Circuit Lithography

The first level circuit lithography (process steps 6-8 shown in Figure 6) encompasses the most crucial steps in the fabrication of planar GaAs ICs. Since GaAs ICs are designed with 1 μm features to achieve speed goals, a dry pattern replication process capable of extremely high LSI/VLSI compatible yields has been developed.

The following summarizes the first level circuit lithography steps used to define ohmic contacts, Schottky barriers, gates, and first-level interconnects. At this point in the process, the entire surface of the GaAs wafer is covered with dielectric. Subsequent ohmic contact and Schottky metallization layers are deposited within windows in this dielectric. This process represents a significant departure from microwave GaAs FET mesa fabrication and more closely resembles Si planar IC processing. The contacts for the FETs, active loads, and Schottky diodes use a AuGe/Pt or AuGe/Ni metallization, whereas the Schottky barriers, gates, and first level interconnects use a Ti/Au

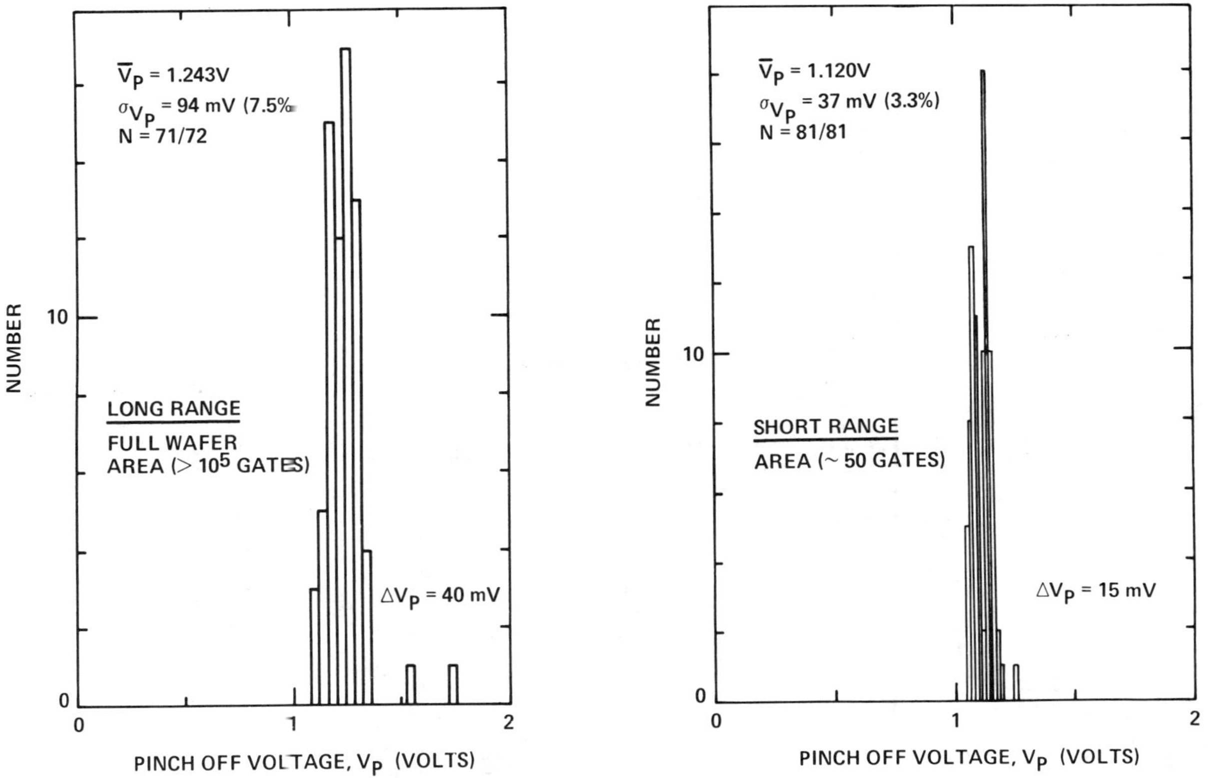

Figure 10/*Histogram of pinchoff voltages of test FETs comparing the uniformity for GaAs FETs uniformly distributed over the entire wafer area (long range) with that for 81 GaAs FETs packed in the area of a small chip (~ 300 μm x 300 μm – short range).*

or Ti/Pt/Au system. (Reliability studies of Ti/Au vs. Ti/Pt/Au and AuGe/Pt vs. AuGe/Ni are in progress.) The circuits are fabricated by defining the desired patterns in photoresist and plasma etching the underlying Si_3N_4/SiO_2 dielectric to the GaAs surface. The appropriate metal contact or interconnect layer is then evaporated and the circuits are defined using a photoresist lift-off technique. This technique results in the first level circuit metals being self-aligned precisely within the dielectric window.

Common to all of the GaAs IC process steps is the delineation of the fine line resist patterns required for the fabrication of these circuits. Since 1977 fabrication of our GaAs ICs has been accomplished by using a Canon FPA 141 4X projection mask aligner. With the recent emphasis on VLSI, the Si industry is also gradually adopting this state-of-the-art photolithography technique (in the form of 10X direct step on wafer (DSW) systems). Reduction projection photolithography has several advantages: there is no mask wear as in contact printing; any mask step-and-repeat error or small defect on the mask is minimized by a factor of 4 at the wafer level; and alignments are precise, typically within 0.5 μm. An example of the 1 μm resolution capability of this photolithography technique is shown in the scanning electron micrograph in Figure 11. In principle, as shown in this edge view of dual 1 μm gate openings, relatively vertical photoresist side walls and excellent resolution can result from this photolithographic technique.

Although Figure 11 demonstrates that excellent photoresist patterns can be obtained, practical process considerations dictate that variations in photoresist resolution and edge acuity across wafers must be anticipated. This is highly dependent on wafer flatness, photoresist thickness variation, local circuit topography, and other photoresist process variations.

Typically, replication of 1 μm geometries on GaAs has been accomplished by using direct photoresist lift-off methods. When direct photoresist lifting is used for defining metal patterns, excellent vertical edge acuity of the photoresist profile is absolutely necessary. However, in practice, such good edge acuity is difficult to attain with reproducibility through reduction projection photolithography – or any other photolithography method – at the 1 μm resolution level.

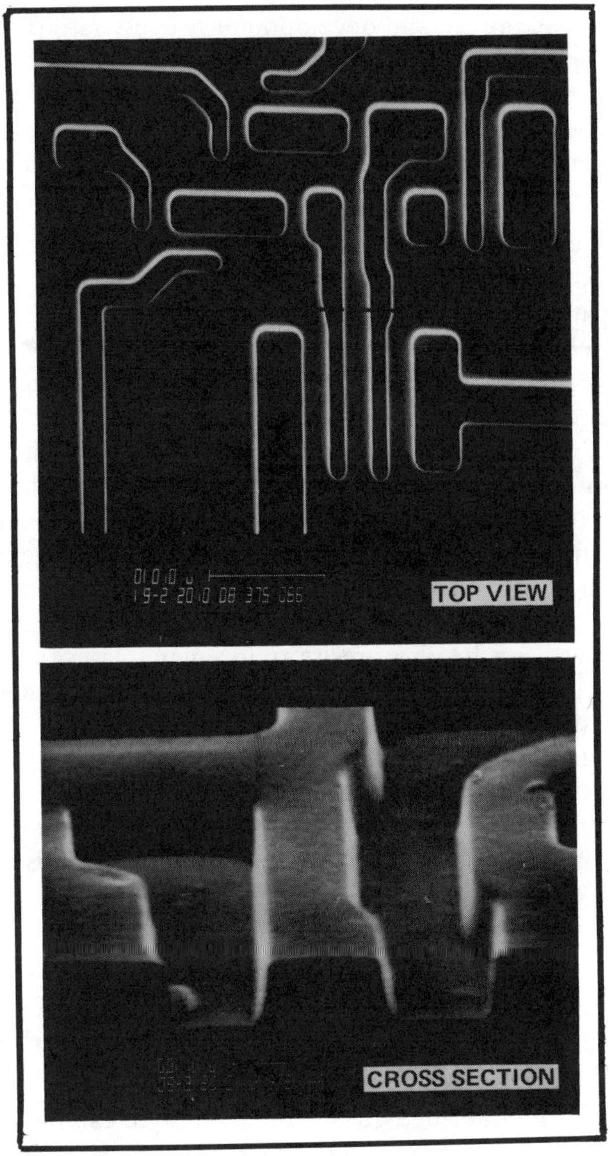

Figure 11/SEM photograph highlighting dual 1 μm gate photoresist patterns defined by projection photolithography.

This is due to exposure and development process variations and to the small depth of focus (±1.5 μm) required in this photolithography technique which is inconsistent with the ability to maintain wafer flatness to better than ±1.5 μm. Therefore, in order to ensure high process yields, enhanced lift-off techniques have been developed [11].

The process described here makes use of the dielectric covering the wafer as an intermediate lift-off medium. A schematic of the resulting intermediate layer lift-off technique is shown in Figure 12 (left side) compared with a conventional direct lift-off process (right side). The intermediate lift-off technique using dielectric is easy to implement and very reproducible. It eliminates two fundamental limitations of direct lift-off, namely, the need to use thin metal films (< 5000Å), and the need to use strong resist solvents and harsh mechanical agitation which can lead to marginal process yields. The resulting metal profile using direct lift-off technique is usually very coarse, as illustrated in the right side of Figure 12. The explanation for this coarse, low yield structure is as follows: during the metal evaporation process, metal deposited on the GaAs substrate also fills up the side wall of the photoresist and in many cases (without special photoresist undercut profile techniques) becomes attached to the metal deposited on the photoresist top surface (seen in Figure 12, right side). Hence, there is the need for thin metal films and strong agitation during the lift-off process to tear this metal sidewall attachment. In contrast, the intermediate lift-off process does not suffer from these limitations. Since the underlying dielectric is plasma etched, the deposited metal (metal thickness $\leqslant$ dielectric thickness) lies below the photoresist layer within windows in the dielectric, and is actually slightly shadowed by the photoresist during evaporation, resulting in deposited metal films which are not connected to the metal deposited on the top of the photoresist. The unwanted metal is easily lifted away by simply dissolving the photoresist in acetone.

The merits of the intermediate lift-off technique can be judged in the drawings and scanning electron micrographs (SEM) of Figure 13. In the top portion a smooth Schottky metal structure is precisely aligned within the dielectric opening (illustrating the application of the high yield plasma etching/lift-off technique). The smoothness of this structure can be compared with an example of the coarse profile resulting from direct lift-off shown in the lower portion of Figure 13.

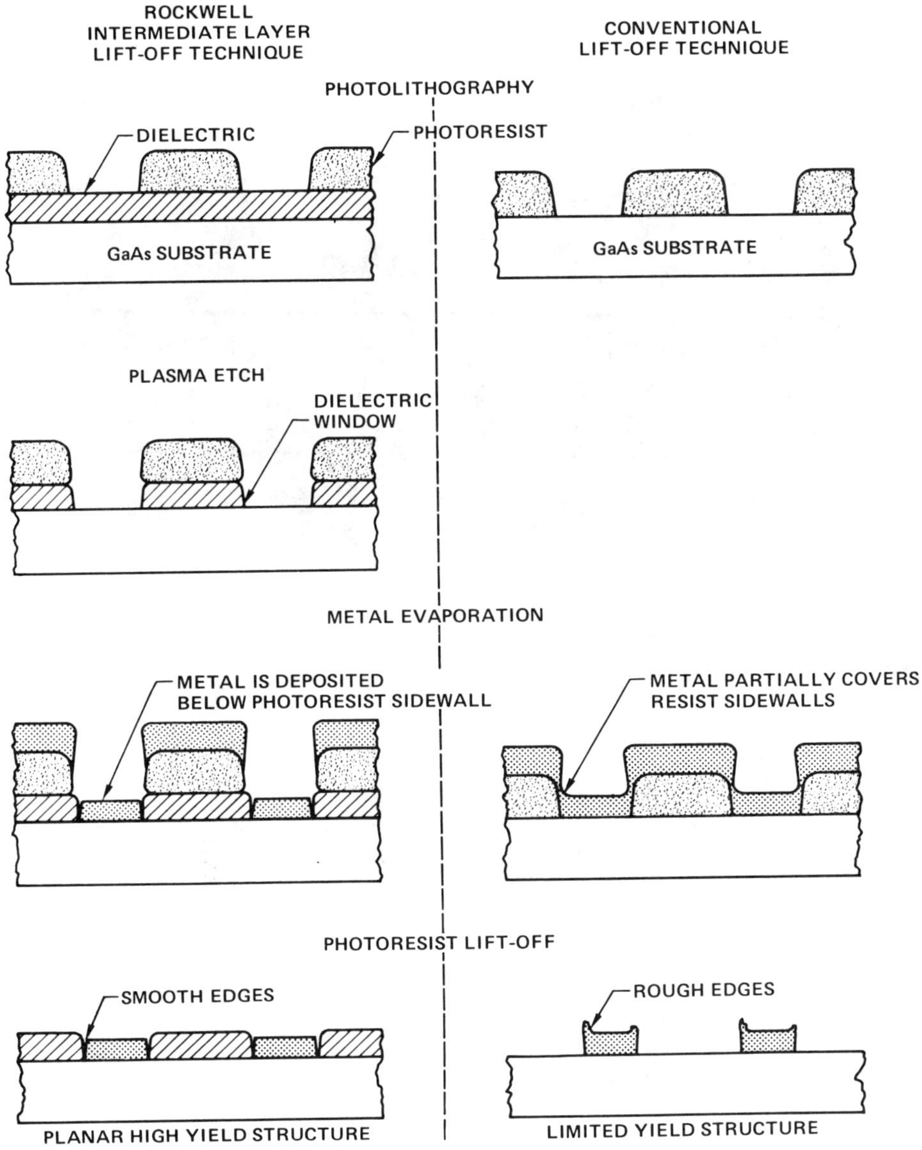

***Figure 12**/Comparison of the results of the direct photoresist liftoff and the intermediate dielectric layer liftoff techniques.*

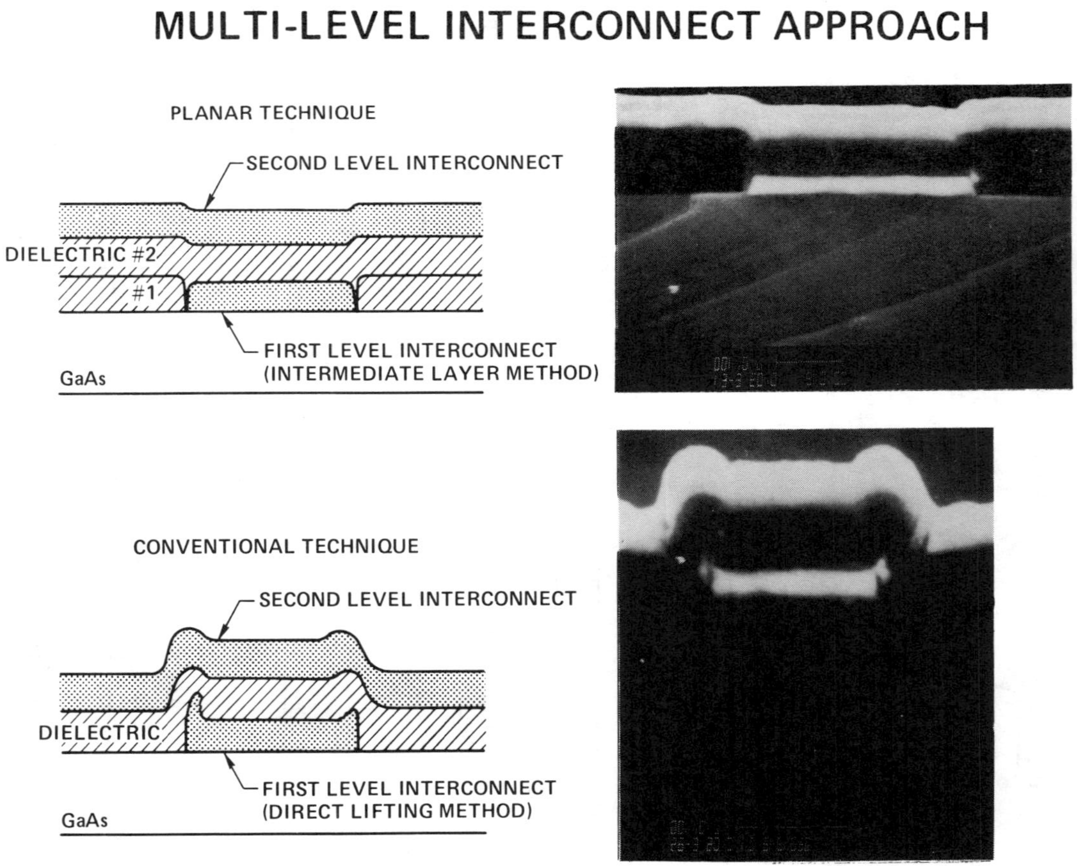

Figure 13/*Multi-level interconnect techniques comparing the planar Rockwell process with a conventional direct liftoff process.*

The lower SEM photograph illustrates the yield limiting coarse metal edge often resulting from using direct lift-off techniques and illustrates the potential step coverage and crossover liabilities.

The dielectric intermediate layer lift-off technique has demonstrated high yields with the added advantage that metal contacts and interconnects are automatically and precisely registered within dielectric windows. The sophistication of this powerful lithographic technique for fabricating planar, high yield, dense 1 μm structures precisely aligned within dielectric windows is highlighted in Figure 14. The scanning electron micrograph of a portion of a planar GaAs IC shows an SDFL gate containing a dual 1 μm gate GaAs FET. Recent work has demonstrated that this same process, when used in conjunction with E-beam lithography, is fully compatible with fabricating 0.5 μm long gates.

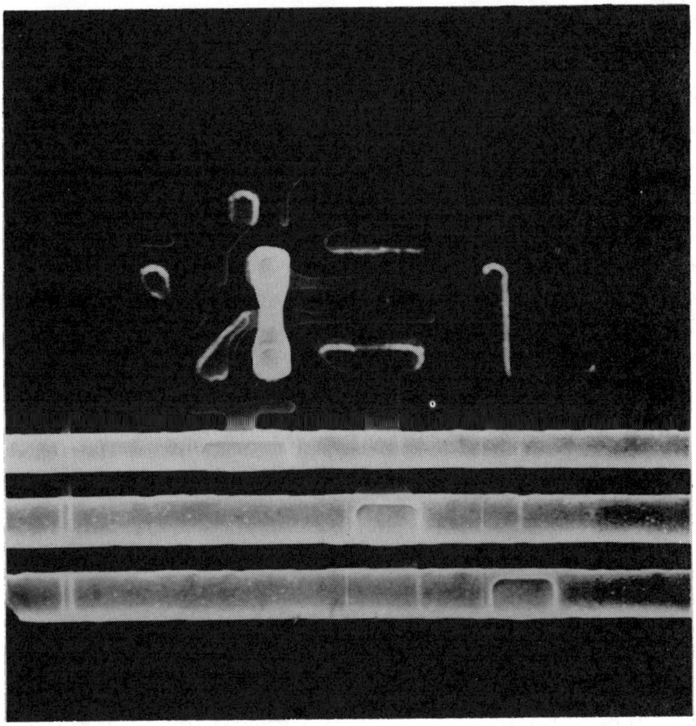

***Figure 14**/SEM photograph of a portion of an IC (a two-level logic gate as in Figure 1 (b)) which has a dual gate FET.*

Another important benefit from this process approach stems from the resulting planar structure. As shown in Figure 13, fabricating the first-level metal within windows in the first-level dielectric and maintaining the first-level metallization thickness close to the dielectric thickness results in a planar structure which greatly facilitates the fabrication of complex multilayer interconnects. The upper portion of Figure 13 shows the smooth crossovers that result from this planar multi-layer fabrication approach. The planar crossovers eliminate any potential problems such as shorts between first and second level interconnects, and high resistances or open interconnects resulting from poor step coverage. The lower portion of Figure 13 illustrates how these interconnect difficulties could easily exist when poorly defined first level metal is used in conjunction with conventional multi-level crossover techniques.

3.4 Planar Multi-Level Interconnects

Recent advances in GaAs IC fabrication technology have made possible high speed, low power GaAs digital ICs with complexities currently at the MSI level. Extension of the current technology to the LSI/VLSI domain requires a refined multilevel interconnect process capable of $> 10^5$ crossovers with virtually 100% yield. A totally dry etched, low pinchoff voltage compatible, multi-level interconnect process is under development at Rockwell, specifically addressing GaAs LSI/VLSI [15]. Important features of this process include a unique planar crossover structure, the use of high quality plasma Si_3N_4 dielectric, and reactive ion etching and ion milling for pattern replication.

Figure 13 is a schematic comparing the Rockwell planar multi-level interconnect approach with a conventional IC approach. One key feature of the Rockwell approach is the smooth planar crossover structure, which eliminates potential crossover problems such as shorts between first and second level metal interconnects, and/or opens resulting from poor step coverage. Another advantage of this approach is that during the 2nd level interconnect processes the critical GaAs FET channel surface region between the gate and source-drain contacts is protected by the first level dielectric. This surface protection contributes to the high yield of low pinchoff voltage devices ($V_p \approx -1.0$ V) utilized in low power digital ICs. As shown in Figure 15, the process can be further refined by filling the 1st to 2nd level via windows with metal. This via filling can be easily accomplished using a process similar to the first layer metal process already described. Such refinement is capable of providing a fully planar via interconnect structure.

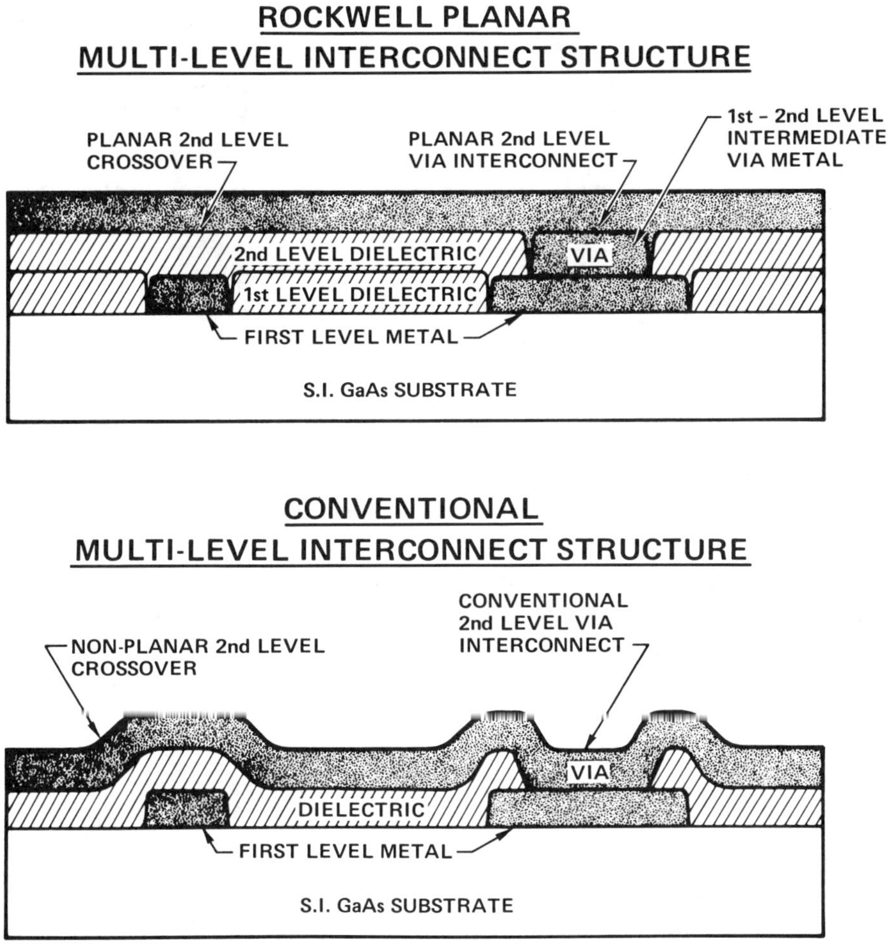

Figure 15/Super planar vs. conventional multi-level interconnect approach showing the planar crossovers as well as the planar via hole structure.

The 2nd level dielectric layer shown in Figure 15, which serves as insulation between the first and second level metal, must be a low stress, pinhole free, high dielectric strength film and exhibit good adherence to metals, GaAs, and other dielectrics. The deposition process must not cause any thermal or electrical damage to the devices, and the dielectric must be easily etched for via openings. All of these requirements are satisfied by plasma enhanced CVD deposited silicon nitride films. The resulting films are pinhole free over 10^{-3} cm^2 test areas, with an average dielectric constant of ~ 8, for a crossover capacitance per unit area of .15 ff/μm^2 for 5,000Å thick films. Dielectric strengths of 5 x 10^5 Volts/cm^2 have been measured for these films, both on larger area capacitors and on test patterns containing several hundred crossovers.

Via openings (shown in Figure 13) required for connecting the first level interconnects to the second level interconnects are opened by reactive ion etching (RIE). The first level metal (Au) serves as an automatic etch stop; it also provides excellent adherence to the second level metal since a small amount of Au is sputtered during the RIE process. The reactive ion etching of plasma nitride can be tailored to produce isotropic or anisotropic etched profiles depending on the plasma etching technique used. Currently both methods are used depending on the desired via process. In conventional processing, sloped holes (isotropic) are necessary since the current path between the second level metal and first level metal must pass over the sidewalls of the via. The yield and reliability of the via interconnect depends on the continuity of the sidewall coverage. Planetary e-beam evaporation or magnetron sputtering is used to ensure good metal sidewall coverage. The disadvantage of using isotropic via sidewalls is that the dimensions of via holes are actually larger than needed. For ultimate densities and highest yields, anisotropic etching of via windows with via filling, as shown in Figure 15, has been developed.

Figure 16 shows an anisotropically etched via hole with nearly vertical dielectric sidewall and intermediate via hole metal. The via hole metal is defined by depositing a metal film of slightly less thickness than the second level dielectric, followed by a lift-off process. This process scheme results in a planar structure, eliminating any need for sidewall coverage, hence increasing the yield and reliability of second level metal to first level metal interconnects.

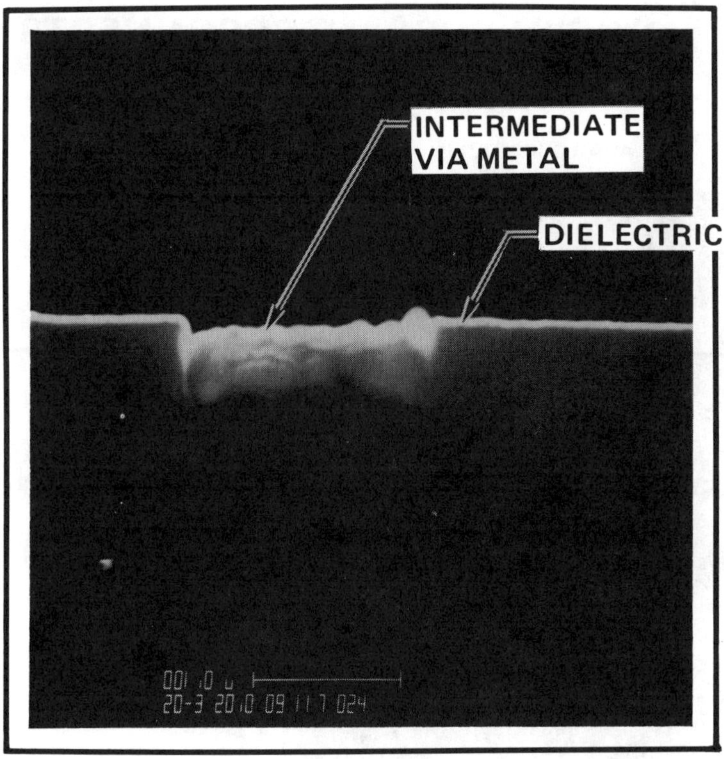

***Figure 16**/Scanning electron micrograph of the cross section of an anisotropically etched, metal filled via window.*

After vias have been fabricated, the wafers are coated with a 6000Å thick Ti/Au second level metal. Ion milling is used to replicate the fine line (1.5 μm min dimensions) multi-level interconnects. Samples are ion milled at an angle of incidence of 70° in order to minimize redeposited material, often observed along the edges of the photoresist patterns when normal (90°) angle of incidence is used.

An example of ion milled planar second level interconnects is shown in Figure 17. With this dry etching technique, precise linewidth control with near vertical walls can be maintained for Ti/Au films up to 1 μm thick. As was shown in Figure 15, due to the planar structure utilized, there is no appreciable step where the second level interconnect crosses over the first level interconnect. Extremely high process

MULTI-LEVEL INTERCONNECTS

GaAs IC
PLANAR CROSSOVER

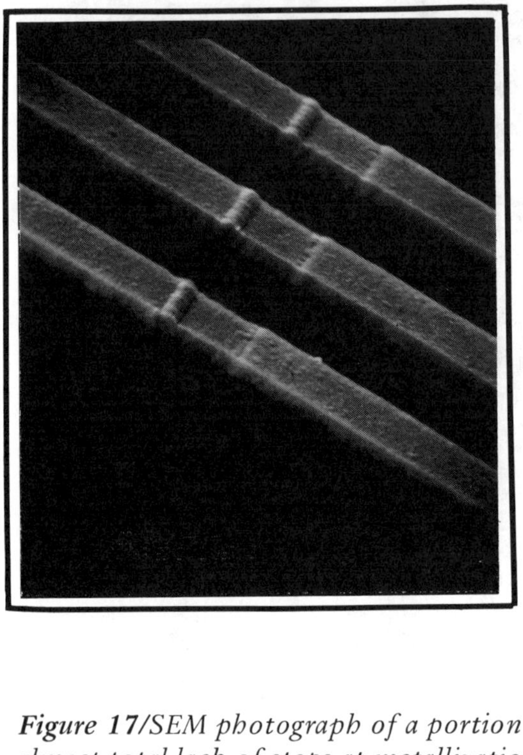

PROCESS

METAL REPLICATION
ION MILLED

SECOND LEVEL METAL
5500Å TiAu

SECOND LEVEL DIELECTRIC
5000Å PLASMA NITRIDE

Figure 17*/SEM photograph of a portion of a planar IC showing the almost total lack of steps at metallization crossovers.*

yields have been encountered using these multi-level interconnect process techniques.

Figure 18 shows a very compact GaAs IC that utilizes, for the sake of density, parallel first-to-second level crossovers. The ability to accommodate this type of circuit layout without compromising process yield or circuit performance demonstrates the power of this planar crossover approach. Furthermore, this planar multi-level interconnect structure provides, in principle, the capability for fabricating more than two levels of interconnects. This is desired in VLSI for simplified circuit layouts and reduced chip areas.

The SEM photograph in Figure 18 of a portion of a planar IC highlights the planar, featureless GaAs surface and the resolution and alignment capabilities of the overall process. This planar IC is composed of both 5 μm and 10 μm wide GaAs FETs which have 1 μm long gates precisely aligned within a gap between n^+ implanted regions and source and drain contacts. In a number of locations, two extremely small (1 μm x 2 μm) logic diodes are used for 2-input logic "OR" function. Diodes and GaAs FETs requiring different implantations are separated by as small as 3 μm isolation gaps provided by the semi-insulating substrate.

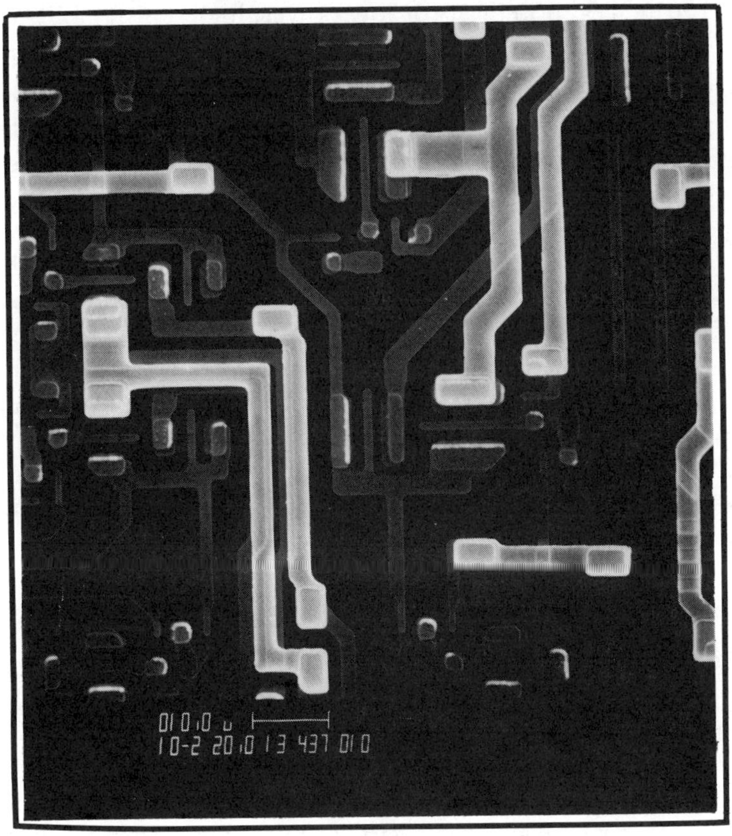

Figure 18/A SEM photograph of a portion of a planar IC showing excellent lithography and high density. Note the second level interconnects running parallel above first level interconnects.

3.5 Process Status and Conclusion

The integrated circuit performance advantages which can be obtained by virtue of gallium arsenide's high electron mobility, aided by the availability of semi-insulating GaAs substrates, have motivated the research in GaAs IC processing technology presented here. Results of this research have provided significant advances in GaAs processing techniques, allowing the full development of a planar "silicon-like" LSI fabrication technology. This processing technology has already been utilized to fabricate GaAs ICs with hundreds of gates (e.g., Figure 23), exhibiting excellent performance with good yield, and demonstrating real potential for low power, high density LSI/VLSI.

In summary, we have presented a GaAs fabrication process which is well on the way to meeting the goals of future GaAs VLSI. The excellent prospects of GaAs VLSI are enhanced by the fact that the basic process steps used in fabricating GaAs ICs are generally fewer than those used in Si ICs. This is possible mainly because of the availability of semi-insulating substrates which provide minimal parasitic capacitances and direct device isolation without any of the additional complicated processes found in Si NMOS, CMOS, or CMOS/SOS. These factors should have a significant favorable impact on the ultimate yield attainable for GaAs VLSI applications. Despite differences in the materials, the fabrication techniques required for ultra-high density and extremely fine-line Si VLSI and GaAs VLSI are quite similar. Continued progress in the GaAs material technologies of substrate growth and ion implantation can be anticipated. Therefore, from a fabrication point of view, the prospects for GaAs VLSI appear as realistic as for Si VLSI.

4. EXPERIMENTAL PERFORMANCE RESULTS

As indicated in Table 1, we have fabricated many different circuits, from SSI to LSI, to demonstrate the capability of this planar SDFL GaAs IC technology. The simplest demonstration circuits, very useful for establishing base-line gate propagation and dynamic switching energy values, are ring oscillators. Ring oscillators are formed of an odd number, N_R, of inverting gates (e.g., NAND, NOR or inverter gates) [4]. Hence, with the proper logic state applied to inputs not in the ring signal path, no stable state exists for the ring and the outputs must oscillate at a frequency determined by the gate delay, τ_d ($f_{osc} = 1/(2N_R\tau_d)$). Table III summarizes the results of ring oscillator propagation delay measurements on our planar SDFL NOR and OR/NAND logic gates with nominal $L_g = 1\ \mu m$ GaAs FET gate lengths. The highest speed obtained in low power ($P_D = 1.1$ mW) gates was $\tau_d = 62$ ps at $P_D\tau_d = 68$ fJ, while the lowest speed-power products were measured in very small geometry, W = 3 μm inverter gates with $P_D\tau_d = 16$ fJ at $\tau_d = 136$ ps. Of particular interest in Table III is the comparison between the simple SDFL NOR gate (Figure 1(a)) and the 2-level SDFL OR/NAND gate (Figure 1(b)) results for $L_g = 1\ \mu m$ ring oscillators of W = 10 μm and W = 5 μm sizes measured on the same chip of the same IC wafer. These OR/NAND ring oscillators [5] were fabricated with FI = 4 gates (like Figure 1(b) but with diode inputs B and G missing) with the output connected to both G_1 and G_2 inputs of the following stage (i.e., F connected to C and D in Figure 1(b)). Hence, if a positive control line signal is applied to the common "A" inputs in Figure 1(b), oscillation through G_1 (the FET gate closest to the source) is obtained, while if the "A" inputs are low and the common "E" inputs are high, oscillation through the G_2 gates is obtained. If both the "A" input and "E" input control lines are low, then simultaneous propagation through G_1 and G_2 is required (the worst case, since it simulates a 2 μm gate length FET). As indicated in Table III, the speed penalty for going to the 2-level SDFL OR/NAND gates is quite modest, particularly for the W = 10 μm gate sizes. Hence, these configurations should be extremely attractive for circuit applications favoring the use of multi-level gate implementations. For example, extrapolating the τ_d = 105 to 130 ps ring oscillator speeds to a fast complimentary-clocked divider ($f_c \sim 1/2\ \tau_d$) implemented with OR/NAND gates would give a maximum clocking frequency on the order of $f_c \sim 4$ GHz. As mentioned in Section 2, the use of 3-level SDFL gates has been demonstrated for 1 τ_d carry delay ripple adder circuits and would be attractive for the parallel multiplier as well. Of course, for many applications it would be desirable to eliminate the fan-in restrictions at the 2nd and 3rd levels of logic (typically FI = 2 for high circuit speeds) inherent in FET logic implementations, but with the unlimited diode fan-in available at the first logic level in SDFL many very useful functions can be achieved.

TABLE III

Comparison of GaAs IC Ring Oscillator Results for $L_g = 1\ \mu m$ SDFL OR/NAND and SDFL NOR Gates

	GATE POWER	MODE	f_{OSC} (MHz)	τ_d (ps)	$P_D\tau_d$ (fJ)
W = 10 μm OR/NAND	$P_D = 1.7$ mW	G_1	640	111.6	201.1
GATE 7-STAGE RO	$V_{DD} = 2.34V$ @ 625 μA	$G_1 + G_2$	550	129.9	216.9
FI = 4, FO = 2	$V_{SS} = -0.86V$ @ -250 μA	G_2	680	105.0	179.6
W = 10 μm NOR GATE	$P_D = 865$ μW				
9-STAGE RING OSC	$V_{DD} = 1.675V$ @ 460 μA		740	75	65
FI = 2, FO = 1	$V_{SS} = -0.95V$ @ -100 μA				
W = 5 μm OR/NAND	$P_D = 475$ μW	G_1	405	176.4	83.9
GATE 7-STAGE RO	$V_{DD} \simeq 1.9V$ @ 200 μA	$G_1 + G_2$	348	205.2	93.7
FI = 4, FO = 2	$V_{SS} \simeq -1V$ @ ~ -95 μA	G_2	410	174.2	94.4
W = 5 μm NOR GATE	$P_D = 372$ μW				
9-STAGE RING OSC	$V_{DD} = 1.57V$ @ 218 μA		502	110.7	41.1
FI = 2, FO = 1	$V_{SS} = -0.61V$ @ -48 μA				
(ALL OF THE ABOVE WERE MEASURED ON THE SAME CHIP OF THE SAME WAFER FOR COMPARISON)					
HIGHEST SPEED:					
W = 10 μm NOR GATE	$P_D = 1.10$ mW		896	62	68
9-STAGE FI = 2, FO = 1	$V_{DD} = 1.7V$ @ 546 μA				
LOWEST $P_D\tau_d$:					
W = 3 μm INVERTER	$P_D = 110$ μW		817	136	16
9-STAGE FI = 1, FO = 1					

While ring oscillators are useful devices for measurement of the basic speed-power properties of logic gates, a more realistic index of performance is the operation of gates in real sequential or combinatorial logic circuits. These circuits generally require use of gates with multiple inputs and fanouts of two or more. For this study, circuits using D flip-flops were chosen as representative examples of sequential logic circuits; a parallel multiplier, implemented with half and full adders, was selected as a combinatorial logic circuit example. Data multiplexer and demultiplexer circuits with on-chip synchronous address generators, representing a mixture of sequential combinatorial functioning, were also fabricated [6, 7].

Circuits in the MSI to LSI range of complexity have been designed and fabricated using the planar SDFL approach, including the achievement of fully-working 1008-gate LSI latched parallel multiplier circuits in 1980. All high speed measurements required for testing the circuits have been made at wafer probe level, although many packaged devices have also been tested, frequently with slightly better performance results. Output buffers of various types (source follower, open drain, etc.) are located on-chip to drive the output package capacitances or transmission line interconnections, as required, to prevent loading of the internal circuitry. Binary ripple counters or frequency dividers have been made by using the D flip-flop (D-FF) implemented with 6 NOR gates, as shown in Figure 19, as a basic building block. By connecting the D (data) input to the $\overline{Q}$ output, the clock input will produce an output transition for every full clock cycle (÷2; this is called a toggle or T-connected D-FF). Logic simulation of this D-FF configuration indicates that correct operation should be obtained up to a clock frequency of $1/4.85\ \tau_d$, where τ_d is the propagation delay for the logic gates in the circuit.

A simple, multi-stage binary frequency divider configuration is the ripple divider in which the Q output of the first T-connected D-FF is connected to the clock input of the second, etc. We have fabricated three stage ripple dividers (÷8) containing 25 SDFL NOR gates and have operated these up to a clock frequency of 1.9 GHz. Figure 20(a) shows "low frequency" (f_c = 100 MHz) ÷2, ÷4 and ÷8 outputs, while Figure 20(b) shows the ÷8 output at 237 MHz from this divider when the clock input is at f_c = 1.9 GHz. This corresponds to an equivalent propagation delay of 110 ps, in good agreement with ring oscillator data for 10 μm SDFL NOR gates from the same wafer. The observed dynamic switching energy varies over the 0.25 to 0.45 pJ/gate range, depending on the bias conditions and wafer pinchoff voltage. Power dissipations are as low as 45 mW for the three stage dividers.

In addition, larger MSI circuits, including an 8-input data multiplexer containing 64 gates, have been evaluated (see the logic diagram shown in Figure 3 of Reference 15). This circuit would be useful, for example, for parallel-to-serial conversion or a high speed data transmission link. A 1 input to 8 output data demultiplexer containing 60 gates was also fabricated and evaluated. Both circuits utilize a presettable 3-stage synchronous counter [15] to generate the address for

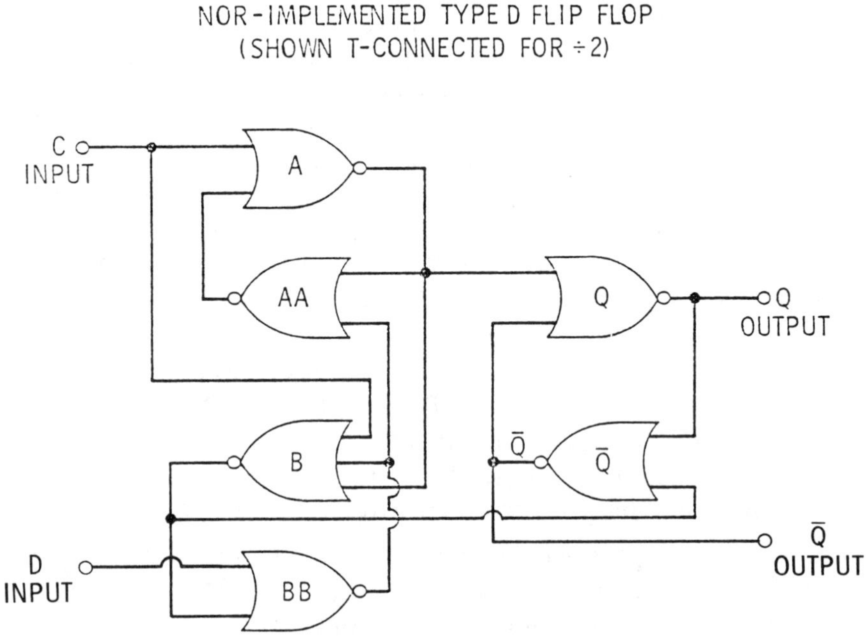

***Figure 19**/Logic diagram for the NOR-implemented type D flip-flop used in various shift register and divider circuits in this work. In ripple divider applications the D-FF is T-connected as a ÷2 by connecting the $\overline{Q}$ output to the D (data) input.*

the multiplexer or demultiplexer gate arrays. (A ripple counter could not be used here since the outputs from the 3 stages are not "valid" at the same time; in the synchronous counter where all of the D-FFs are clocked simultaneously, all of the Q and $\overline{Q}$ outputs form a valid address for the multiplexer.) A SEM photograph of a multiplexer chip is shown in Figure 10.4 of Reference 7; the chip size, excluding probe pads, is only 0.77 mm x 0.54 mm. Operation of both the multiplexer and demultiplexer chips has been demonstrated at a clock frequency of 1.1 GHz. Power dissipation of the multiplexer circuits varied from 75 mW to 375 mW for wafers with pinchoff voltages of $V_P = -0.5$ and -1.45 V, respectively.

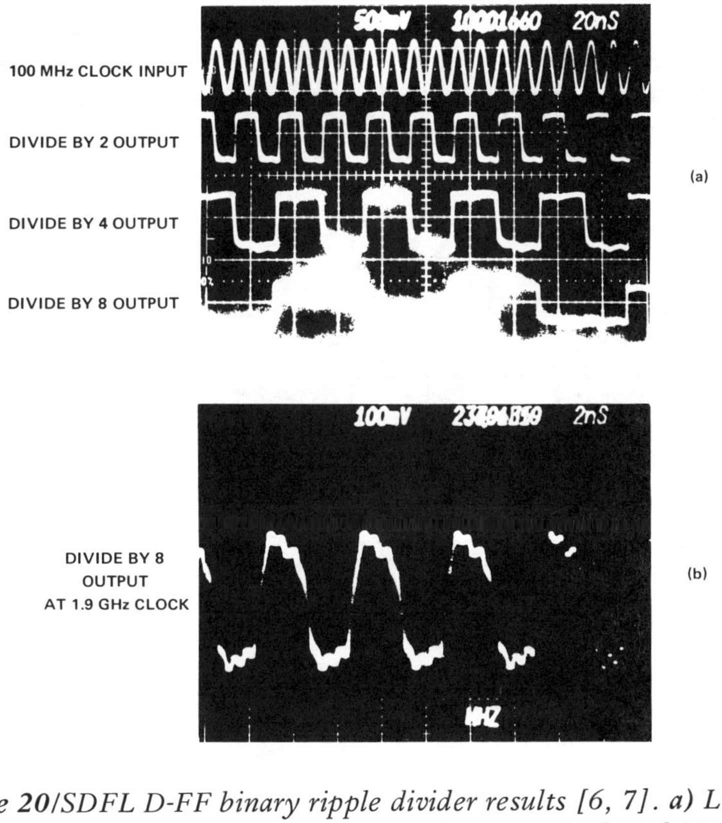

Figure 20/SDFL D-FF binary ripple divider results [6, 7]. a) Low frequency operation showing 100 MHz sinewave clock and ÷2, ÷4 and ÷8 circuit outputs; b) ÷8 output of this ripple divider operating at a clock frequency of 1.9 GHz.

Since multiplication frequently represents a bottleneck in signal processing and computer systems, a particularly attractive candidate for a representative combinatorial logic part is a parallel multiplier. Typical sizes for a high-speed parallel multiplier would be 8 x 8 bits to 16 x 16 bits, with larger products formed of combinations of these. A straight parallel multiplier, N x N bits, without carry look-ahead or more complex (e.g., Wallace tree) approaches, requires N(N- 2) full adders and N half adders in its implementation. And it requires a total of (N- 1) sum delays plus (N- 1) carry delays to obtain the product. A goal of our ARPA GaAs IC program is to construct an ~ 1000 gate 8 x 8 parallel multiplier chip (760 gates multiplier plus I-O latching and buffering). As a first step toward this goal, a 75-gate 3 x 3 bit parallel multiplier was fabricated, followed by a 260-gate 5 x 5 bit version. Based on this design/test data, the full 1008-gate latched parallel multiplier was designed, fabricated and successfully tested in 1980.

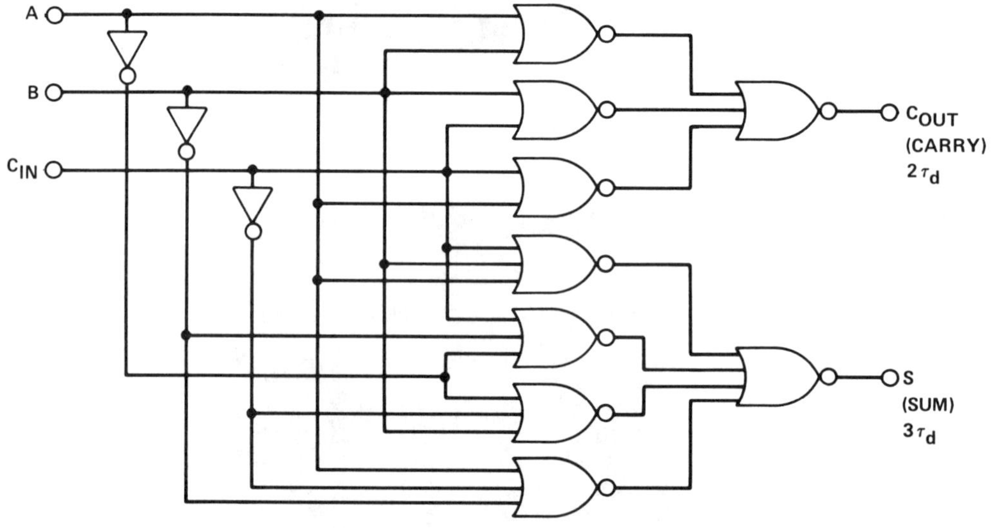

Figure 21*/Full adder cell used for array multipliers, implemented with 12 SDFL NOR gates. Carry delay is 2 τ_d and sum delay is 3 τ_d where τ_d is the basic NOR gate propagation delay.*

Since the bulk of a large multiplier is full adder cells, the speed of these cells dictates the multiply time. Figure 21 shows the logic diagram for a NOR-implemented full adder cell. By using a full minterm expansion for both sum and carry, the propagation delays are kept to 2 τ_d for the carry and 3 τ_d for the sum (where τ_d is the NOR gate propagation delay). A total of 12 gates (9 NORs and 3 inverters) is required per full adder cell. Figure 22 shows the logic diagram for the 5 x 5 parallel multiplier chip (FA stands for a full-adder cell and HA for a 5-gate half-adder cell). For this N = 5 case, 5 half adders and 15 full adders are used. A microscope photograph of the 5 x 5 multiplier GaAs IC chip is shown in Figure 23.

The wafer probe test procedure for these multiplier chips allows for both static and dynamic performance measurements. Initial probe measurements for gross functionality involve the application of a specific input code with a square-wave signal on one of the inputs. This should produce a square-wave output, either in-phase or inverted, on all of the product bits (e.g., for the 3 x 3, 111 x 10S = $S\overline{SSS}SS$ where S is the square wave). If proper gross functionality is observed, complete low-speed functional testing is carried out for all possible combinations of operand inputs, using a MACSYM II computer system to generate inputs and measure and verify outputs. Wafers of both 75-gate 3 x 3 and 260-gate 5 x 5 multipliers have shown good yields of fully functional circuits and fully functioning 8 x 8 bit multipliers have been obtained, both on horizontal Bridgeman-grown GaAs, and on 3 in. diameter wafers of liquid encapsulated Czochralski (LEC)-grown substrate material. The problem of measuring multiply speed at wafer probe on such high-speed circuits was dealt with by adding a few extra gates to the multiplier chip to make an on-chip feedback path. On the 3 x 3 bit multiplier, for example, under control of an external "RO disable" input, $\overline{P}_5$, the invert of the most significant product bit (Figure 22), may be fed back to the least significant bit of the multiplicand, A_o. (As shown in Figure 22, in the 5 x 5 bit version, P_9 is fed back to B_0.) With B = 111 and A = $10A_0$, the correct P_5 output equals A_0. Hence, in the feedback mode with A = $10\overline{P}_5$, the 3 x 3 multiplier behaves like a ring oscillator (with 9 τ_d in the ring for this configuration so that $f_{osc} = 1/18\tau_d$; in the 5 x 5 it is 22 τ_d for $f_{osc} = 1/44\tau_d$, while in the 8 x 8 the feedback path is 40 τ_d for $f_{osc} = 1/80\tau_d$). Initial measurements on the 3 x 3 multipliers gave f_o = 323 MHz, corresponding to τ_d = 172 ps at an average gate

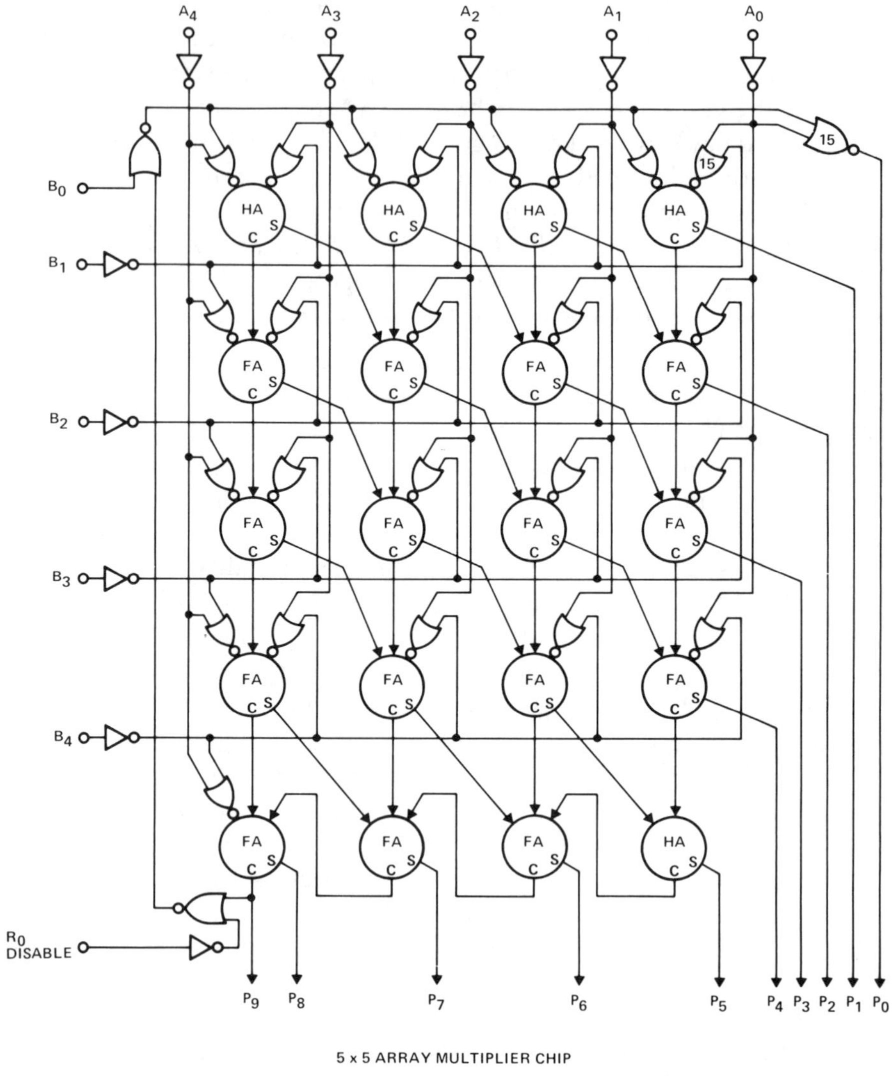

Figure 22/*Logic diagram of the 5 x 5 bit parallel multiplier chip. Note the "ring oscillator" mode feedback path from* P_9 *to* B_0 *for testing propagation delay through the multiplier.*

dissipation of 750 μW/gate (V_{DD} = 2.284 V, V_{SS} = −0.79 V, P = 56 mw for 75 gates), or $P_D \tau_d$ = 128 fJ. Lower pinchoff wafers gave reduced powers, P_D = 420 μW/gate, at slightly lower speeds, τ_d = 225 ps (f_o = 246 MHz, $V_{DD} \sim 1.8$ V, $V_{SS} \sim -0.9$ V and $P_D \sim 31.5$ mW), for $P_D \tau_d$ = 95 fJ. The 260 gate 5 x 5 bit multipliers show good chip yields, up to 20%, with similar gate delay and power/gate results.

The 1008-gate 8 x 8 bit multiplier differs from the earlier 3 x 3 bit and 5 x 5 bit chips in that it consists not only of the 8 x 8 bit parallel multiplier array (an expanded version of Figure 22 containing 760 SDFL NOR gates), but has two 8-bit input operand data latches and a 16-bit product output data latch as well [18, 19]. This architecture facilitates pipelined use of the multiplier in slower-speed Si ECL

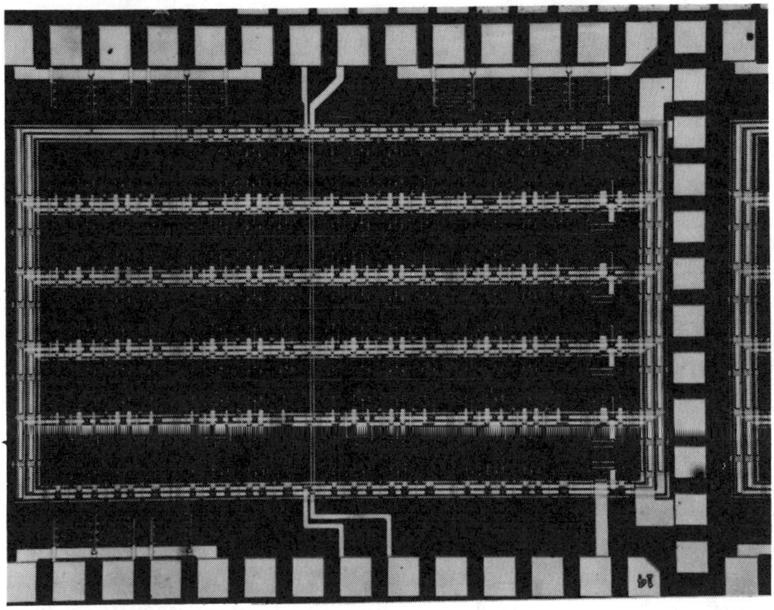

Figure 23/Microscope photograph of 5 x 5 bit parallel multiplier chip implemented with 260 SDFL NOR gates. Area (including pads) is 1.125 mm x 1.35 mm (0.044" x 0.053") and the chip power is as low as 45 mW (depending on wafer pinchoff voltages). Good wafer yields of completely operational 5 x 5 multipliers have been obtained.

processing systems, since the GaAs multiplier can carry out one multiply while the system is passing on the product of the previous multiply and fetching the operands for the next multiply. Both the input and output data latches may be disabled (made "transparent") with latch mode control lines. While the 8 x 8 bit parallel multiplier array uses the same full adder cell as the 3 x 3 and 5 x 5 bit chips (Figure 21), the layout was improved somewhat for better performance. The NOR gate propagation delay measured on-chip in the "ring oscillator" mode was as low as τ_d = 150 ps at a total chip dissipation of 2 watts. Since the worst-case multiply time in the 8 x 8 bit parallel multiplier array is 35 τ_d, this corresponds to a multiply time of 5.25 ns, or a rate of 190 million multiplies/second. This is about an order of magnitude faster than commercial silicon multipliers, at a fraction of the power. As discussed previously, using multi-level SDFL gates would reduce the full adder time from 3 τ_d sum, 2 τ_d carry to 2 τ_d sum, 1 τ_d carry so that the 8 x 8 parallel multiply time (7 sums and 7 carries) is reduced by 40% (from 35 τ_d to 21 τ_d). This alone would increase, for the same τ_d = 150 ps value, the multiply rate from 190 million multiplies/second to 320 million multiplies/second. It is felt that with modest further improvements in gate delay and improved multiplier architecture, 8 x 8 bit multiply times of 2 ns should be attainable (0.5 G multiplies/second).

In addition to the parallel multiplier combinatorial logic LSI demonstration circuits, we have fabricated large sequential logic circuits as well. As noted in Table I, we have fabricated and successfully tested a 96-gate 8-bit shift register/217 bit P/N code generator, implemented with eight D-FFs of the type shown in Figure 19. In the shift register, the Q-output of a stage is connected to the D-input of the succeeding stage and all flip-flops are clocked together. This chip is set up for various modes of shift register operation (serial load, parallel preset, recirculate, etc.) In addition, an exclusive-OR gate with inputs taken from the 5th and 8th stages, and with its output connected to the serial input at the 1st stage, makes the circuit into a pseudo-noise (P/N) code generator. Depending on the initial state set in with the parallel preset inputs, P/N codes of 217 bit, 31 bit, and 7 bit lengths may be obtained. Proper operation in the shorter code lengths can be verified by directly examining the P/N code generator output on an oscilloscope, but this is rather marginally practical for the 217 bit case. It has proven easier for these high speed P/N code generator circuits to verify operation by examining the output on a spectrum analyzer (a Tektronix 7L13, using 0 – 1860 MHz) for the proper overall amplitude envelope (as calculated from the Fourier transform of the P/N code) and the proper spacing of the frequency lines in the output spectrum (e.g., $f_c/217$, $f_c/31$, or $f_c/7$).

More recently, we have fabricated much larger, 550 gate, 64-stage shift registers of this D-FF implemented type. A microscope photograph of this chip is shown in Figure 24. The chip is left-right divided into two 32-bit serial shift registers, which can, by mode control inputs, also be selected to operate as a single 64-bit shift register. Each side also has taps at the bit 17 and bit 32 positions for inputs to an exclusive-OR (EOR) gate which feeds back to the stage 1 input on that half. Hence, the chip can be operated as a dual P/N code generator with a pattern length of 4,292,868,097 bits. One reason for selecting this configuration is testability, since it would be almost impossible to verify the correctness of this rather long code by means other than direct comparison with another generator. (On the chip an EOR gate comparator is provided to compare the left and right P/N code generator outputs. If both generators are preset simultaneously, then the comparator output will flag any code errors in either generator, or any jumps in phase within the code.) The spectrum analyzer technique cannot be used here, of course, since even at 1 GHz the P/N code does not repeat until 4.3 seconds, so that the spectral lines in this UHF/microwave pseudo-noise output are so closely spaced (0.233 Hz) they would not be observable even with a good quality *audio* spectrum analyzer. As an additional convenience, a sync output is provided from each generator to mark its return to the initial preset code. Note that the 1.35 mm x 2.7 mm overall chip size for this 550 gate circuit is about half of the 2.475 mm x 2.7 mm (0.097″ x 0.106″) size of the 8 x 8 bit latched parallel multiplier chips.

5. SUMMARY

In summary, we have described a super planar fabrication process capable of optimizing two or more types of devices as required for sophisticated LSI circuit approaches. This process has demonstrated the capability for achieving tight parameter control on low power

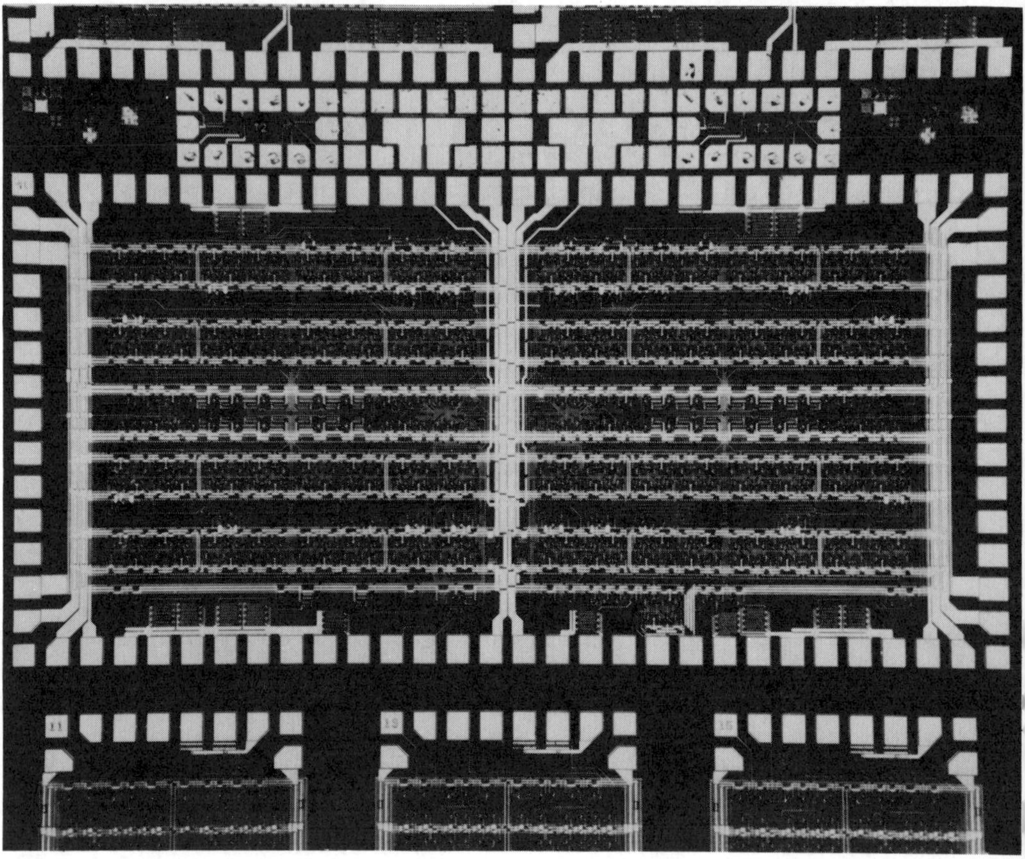

Figure 24/*Microscope photograph of a 64-bit shift register chip implemented with 550 SDFL NOR gates. Chip area is 1.35 mm x 2.7 mm (0.053" x 0.106"). This chip is half of the size of the 8 x 8 bit latched I/O parallel multiplier chip.*

GaAs FETs and high speed switching diodes and for achieving good yields on lower LSI complexity (hundreds of gates) circuits. The Schottky diode-FET logic circuit approach used in this work has demonstrated high speeds (τ_d as low as 62 ps), low power ($P_D \sim$ 100 μW to 2 mW/gate), high densities (NOR gate areas as low as 600 μm^2), and outstanding $P_D \tau_d$ products (as low as 16 fJ in ring oscillators, or 95 fJ even in large MSI parts). The simplicity, high density, and large device parameter margins of SDFL, along with its speed and power characteristics, make it ideal for ultra high speed VLSI. We have demonstrated scores of different SDFL circuits, from SSI to lower LSI (Table I, Figures 23, 24). We have obtained 20% chip yields on 260-gate 5 x 5 parallel multiplier chips and have demonstrated fully working 1008-gate LSI 8 x 8 bit I/O latched parallel multipliers, both on the smaller boat-grown GaAs substrate material and on 3 inch wafers of the new LEC-grown GaAs substrate. Based on the results to date, extension of this planar GaAs IC technology to the capability for 10,000-gate VLSI circuits seems a reasonable goal, and efforts to achieve this goal are planned.

ACKNOWLEDGMENT*

The authors acknowledge the invaluable contributions made by numerous workers involved in the GaAs integrated circuit programs at Rockwell. In particular we wish to thank P. Asbeck, C. Kirkpatrick, and Y.D. Shen for GaAs material studies, ion implantation and processing, and S. Long, G. Kaelin, E. Shen, R. Zucca, F. Lee and C.P. Lee for circuit design, testing, and device analysis. F. Eisen has made strong contributions in GaAs ion implantation technology and, along with F. Blum, has provided guidance to the IC programs. Finally, the GaAs IC program at Rockwell is indebted to the encouragement, support, and technical guidance provided by Richard Reynolds and Sven Roosild through the Defense Advanced Research Project Agency.

*This work was supported in part by the Advanced Research Projects Agency of the Department of Defense and was monitored by the Air Force Office of Scientific Research under Contract F49620-77-C-0087.

REFERENCES

[1] Welch, B.M., and R.C. Eden, Int. Electron Devices Meeting, Tech. Digest, pp. 205-208, Dec. 1977.

[2] Eden, R.C., B.M. Welch and R. Zucca, 1978 Int. Solid State Circuits Conf., Digest of Tech. Papers, pp. 68-69, Feb. 1978.

[3] VanTuyl, R.F., C.A. Liechti, R.E. Lee and E. Gowen, *IEEE J. Solid State Circuits,* SC-12:485, 1977.

[4] Eden, R.C., B.M. Welch and R. Zucca, *IEEE J. Solid State Circuits,* SC-13:419, 1978.

[5] Eden, R.C., et al., *ISSCC Digest of Technical Papers,* pp. 122-123 and 265-266, Feb. 1980.

[6] Eden, R.C., in 1978 Int. Electron Devices Mtg., Tech. Digest, pp. 6-11, Dec. 1978.

[7] Eden, R.C., et al., *IEEE J. Solid State Circuits,* SC-14:221; also *IEEE Trans. Electron Devices,* ED-26:299, 1979.

[8] Welch, B.M., F.E. Eisen and J.A. Higgins, *J. Appl. Physics,* 45:3685, 1974.

[9] Eisen, F.E., et al. in *Applications of Ion Beams to Materials 1975,* ed. G. Carter, J.S. Colligon and W.A. Grant, p. 64. London: The Institute of Physics, 1976.

[10] Welch, B.M. et al., 1979 Int. Electron Devices Mtg., Tech. Digest, pp. 493-496, Dec. 1979.

[11] Welch, B.M., et al., *IEEE Trans. Electron Devices,* ED-27: 1116, 1980.

[12] Nuzillat, G., et al., *IEEE Trans. Electron Devices,* ED-27: 1074, 1980.

[13] Eisen, F.E. and B.M. Welch, Proc. 5th International Conference on Ion Implantation in Semiconductors and Other Materials, Boulder, CO, August 1976.

[14] Zucca, R., et al., *IEEE Trans. Electron Devices,* ED-27:1109, 1980.

[15] Shen, Y.D., B.M. Welch and R. Zucca, ECS meeting, St. Louis, Missouri, May 11-16, 1980. To be published in the *Journal of ECS.*

[16] Long, S.I., et al., *IEEE Trans. Microwave Theory and Techniques,* MTT-28:466, 1980.

[17] Long, S.I., et al., *J. Vac. Sci. Technol.,* 19:531, 1980.

[18] Long, S.I., et al., *Proceedings of the IEEE,* 70:35, 1982.

[19] Eden, R.C., and B.M. Welch in *VLSI Electronics: Microstructure Science,* Vol. 3, ed. N. Einspruch. New York: Academic Press, 1982.

[20] Eden, R.C., *Proceedings of the IEEE,* 70:5, 1982.

[21] Eden, R.C., U.S. Patent 4,300,064, "Schottky Diode FET Logic Integrated Circuit," issued Nov. 10, 1981.

Chapter 9

A Look into the Future

Introduction

As a fitting conclusion to this book, the editors thought it appropriate to depart from the concepts expressed in the previous chapters. These concepts are, in essence, a description of the evolutionary development of GaAs FETs from silicon based semiconductor devices and technology. This departure or "a look into the future" by J.R. Barker is a discussion of some "revolutionary" concepts which may play an important role in devices and circuits five to ten years hence. Barker reviews where we are now and points out where we may be going. The regime of very small and very fast devices are addressed pointing out such phenomena as ballistic and transient electron motion, electron velocity overshoot effects, ultra-small devices, and the impact of micro-refrigeration.

New materials are discussed in terms of what is made possible by the new growth techniques which allow precise control both of doping fluctuations and compositional variations. An example of this is the GaAs-GaAlAs superlattice configuration, which has electronic properties controlled by quantum size effects. J.R. Barker looks even further into the future by speculating that as circuits approach an ultra-large scale in size, radically different device and system concepts may develop. For example, Barker points out that "synergism" exists in many natural biological systems; as technology approaches the complexity levels of biological systems, "synergetic electronics" may become a reality.

The editors agree that "synergetic electronics" may not appear relevant to the *Principles and Technology of GaAs FETs*, but as Barker points out, using a quote from Antonio Machado (1875-1939), "Traveller, there is no path, paths are made by walking".

A Look into the Future

J. R. Barker

Physics Department, Warwick University, Coventry, England.

"Caminate, no hay camino
Se hace camino al andar"

Traveller, there is no path
Paths are made by walking.

Antonio Machado 1875 - 1939

1. INTRODUCTION

Thomas Kuhn, in his book *The Structure of Scientific Revolutions* [1], argues that the normal role of science (and technology) involves puzzle solving. Problems are posed within a well-defined conceptual framework and the solutions are worked out and catalogued within the corps of mainstream scientific and technological knowledge. Occasionally, anomalies or problems are uncovered which do not succumb to established tradition. New concepts are evolved in a high-risk transition period in which competition occurs between the new and the old, and indeed between the new alternatives. This, so-called revolutionary period ends with the establishment of a new consensus or paradigm.

The "new" science is then explored as before by "normal" scientific methods. Not everyone would agree with these views [2] ; and there is a nasty catch: how does one know when revolution is imminent, what form will it take and is it really necessary? Nevertheless, Kuhn's analysis provides a useful perspective for looking at what has been achieved and what might emerge in the future development of electronics.

At some risk of over-simplification, and with the almost certainty that long-range projections are doomed to failure [3] , let us try and map out what the future might hold, both from the "normal" science point of view: continuous evolution (or extrapolation); and from the "revolutionary" stance: discontinuous evolution – the emergence of new branches in electronics.

2. WHERE WE ARE NOW

2.1 Out of the Past

The history of electronics provides one of the most fascinating opportunities for seeing the complex interaction between science, engineering and the perceived needs of society [4] . Certainly, many of Kuhn's concepts of "normal" and "revolutionary" phases can be clearly observed. In many cases new scientific developments have rescued applied science from crises brought about by the need for new or more sophisticated technology. For example, once the generation and propagation of radio waves became possible, the requirement for radio communication systems focused attention on devices/systems capable of detection, rectification and amplification of signals. The cat's whisker, a metal-semiconductor point contact rectifier was one such device, a fore-runner of true solid state devices, but quickly superseded by the invention of vacuum tube electronics. The general requirements for communications, control systems, radar and information processing have dominated lines of development in electronics. Crises in these areas have often stimulated revolutionary developments. The scientific invention of the transistor, from attempts to understand and control the metal-semiconductor interface, is an example of a revolutionary change which came at exactly the right time to rescue the infant computer science from disasters brought about by the unreliability of vacuum tube components in complex processing systems. This example illustrates a key factor in applied science: whenever a new device or circuit concept has appeared, its ultimate importance has been decided by the ability to economically control the growth and dimensional/compositional constraints of the material supporting its physical operation, rather than by whatever elegant or novel physical phenomena which it exploits. In simplistic terms, the trade-off between perceived applications, exploitable physical phenomena and the economic control of the relevant materials/fabrication process has largely determined which line of development is fruitful. One might argue that these three areas provide the crucial regimes for spotting future trends. However, there is a fourth area [5-9], which may also be significant: it concerns the prejudices underlying our presently conceived schemes for organizing and operating complex systems [5].

At the heart of present day technology is the transistor. It has two crucial properties: the ability to store charge and the ability to switch current. These meet two requirements for digital information processing: a localised memory unit, and a localised logic unit. Combinations of transistors interconnected by line elements are thus physical representations of abstract logic systems. Until recently these logic systems have been based on Boolean, largely sequential (or von Neumann) architectures, temporally ordered by synchronous clocking schemes. The lack of discrete transistor standardization in the late 1950's led to a commercial crisis which was solved by the development of integrated circuits: a first step in recognizing the importance of organization in complex systems.

But apart from the standardised, unified fabrication and the complexity level, modern integrated circuits are not dissimilar to systems of the era of discrete electronics. The localised circuit elements are separately determined and then isolated within the chip by appropriate design rules. Completion of the interconnect net then determines the operational function of the system. The novel features are: the integration of a well-controlled economic fabrication process, and the use of scaling rules to scale the device/interconnect geometry, doping levels, control voltages, etc., so as to maintain transistor/system operation but at larger packing densities for better speed, power dissipation, and complexity.

Integration has followed a classic learning curve, typical of Kuhn's "normal" science periods, to achieve the present 1-2 micron design rules, with semiconductor circuits containing as many as 250,000 transistors per chip. Silicon has dominated this field, not because of its superior electronic properties, but because it is easily refined and economically processed and has a natural high quality insulator, silicon dioxide, ideally suited for passivation and incorporation into active device structures.

The emphasis has been on improving fabrication resolution, processing, and tolerances, and clever circuit design. This continuous development has been sustained by a relatively sound and broad base of scientific knowledge in the fields of bulk solid-state physics, materials science, planar fabrication, and classical design methodology. Fundamental physical limitations have not yet proved important.

2.2 New Crises?

As sub-micron geometries are approached, there are significant signs of crisis in a number of inter-related areas which confront the progression to very large integration densities [5-9, 10-12], and high speed systems.

First, major changes in fabrication technology are required to supersede the old combination of photolithography, high-temperature diffusion and wet etching, if finer patterning is to be possible. It is not known whether the newer techniques can be made economical. Second, the demand for very high speed processing systems in communications, control, and computer science has focused attention on alternative materials and devices to those based on silicon. This move is already well established in microwave optical and optoelectronic applications, where flexible control of the physical properties of the III-V compounds and their alloys (band gaps, refractive indices, mobilities, radiative recombination efficiencies, etc.) have evident advantages over silicon [13]. The development of magnetic bubble, Josephson superconducting media, and gallium arsenide integrated circuit technologies is perhaps more significant because they are beginning to be competitive in areas traditionally dominated by silicon [14]. A major development however, concerns the III-V based materials which [15] can now be configured into a variety of heterostructures with a wide range of exploitable physical phenomena. This step marks the first serious move to artificial material having "atomic" architectures specially configured for their electronic properties. Similarly, recent advances in polymer physics and chemistry, molecular engineering, and genetic engineering has opened prospects for radically new ultra-small circuit elements: molecular gates and conducting chains [16, 17].

A third problem concerns the recent departure from the traditional sequential "random" logic system architectures [18]. This is occurring because of constraints which are imposed by the planar physical technology of the circuit representations of high density memory/logic and by the power dissipation, delay and synchronization properties of the interconnect paths. More concurrency and regularity in the circuit layout is also a necessity to handle the design of VLSI circuits [18]. Very little theoretical basis exists for exploiting highly concurrent architectures.

Fourth, serious soft and hard errors due to natural and artifical radiation hazards, as well as intrinsic defect and fluctuation phenomena, pose problems for very high density circuits [19]. Some redundancy schemes are already being deployed in memory chips, but it may well be that a radical change from localised, single-bit information processing will be necessary [9]. After all, the natural processing systems in, say, the eye or brain show an astonishing stability and capacity to avoid effects of local change; indeed, natural systems appear to be organized in quite different ways from man-made systems [20, 21].

Finally, although our list could be made much longer, as miniaturization proceeds below 1 micron scales, the characteristic bulk semiconductor/insulator/metal properties give way gradually to processes and phenomena which are appropriate to very non-equilibrium finite-sized heterostructures intermediate to the macroscopic solid state and the domain of atomic/molecular physics [5-9]. The problem is illustrated in Figure 1, where we sketch a few physical scales (for silicon) plotted against lithography limits of current fabrication techniques. It should be obvious that devices configured to scales of the order of the bulk mean free path, classical screening length, or de Broglie wavelength will have different properties, or may not work at all, compared to their bulk counterparts. Lack of basic scientific knowledge at this intermediate scale causes problems for the evaluation of individual device limitations, but more importantly for the limitations in a complete integrated system of devices [5-9].

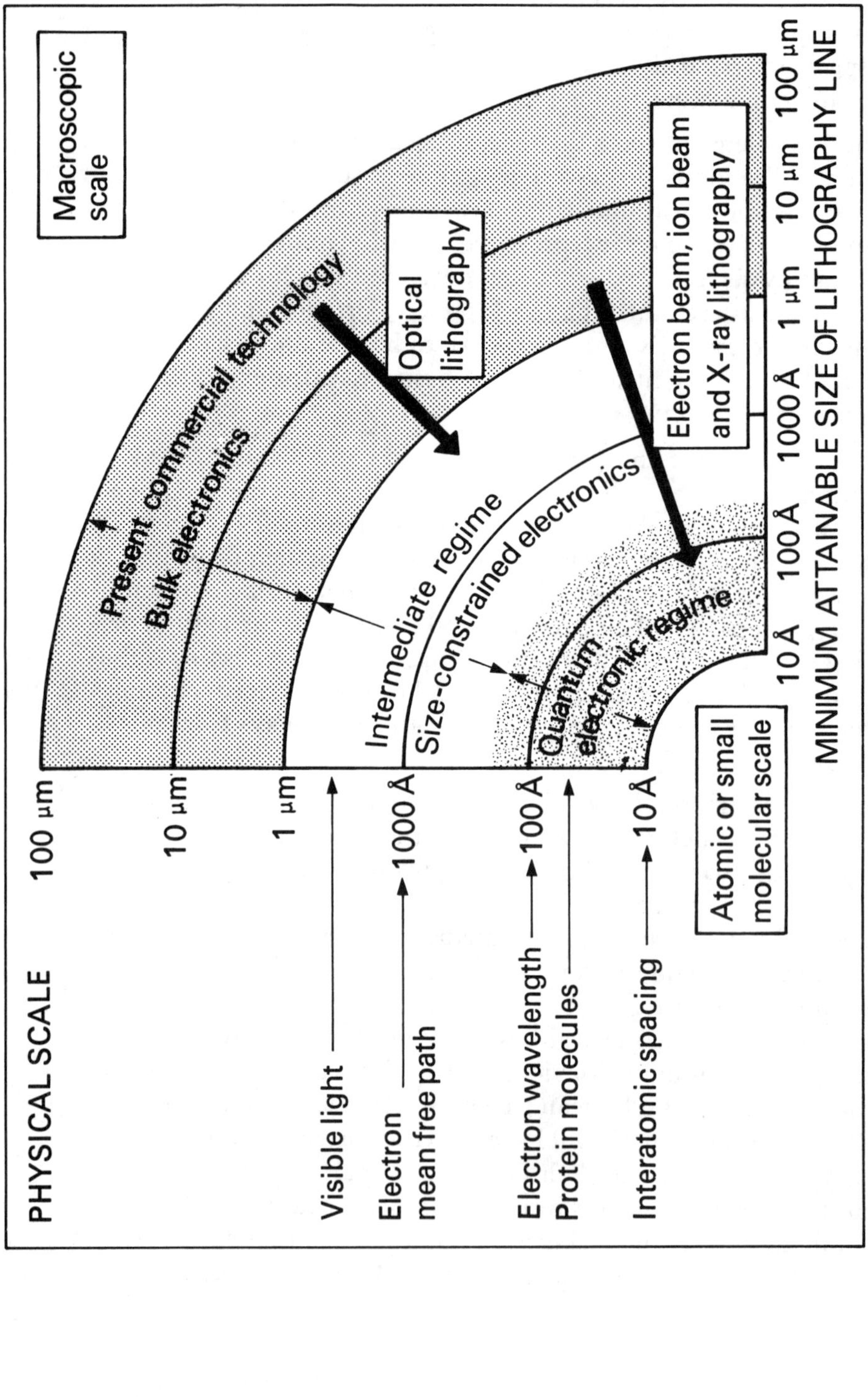

Figure 1/Physical scales and lithography limits for silicon.

From a practical viewpoint, the development of electronics is at an important cross-roads. Certainly, some continuous lines of development can be forseen, but considerable uncertainty surrounds what materials, fabrication, electronic processes, systems, etc., are to be significant at the end of the decade. Indeed new market opportunities, aside from the enormous scope of microprocessor systems such as bio-electronics, are beginning to appear but cannot yet be evaluated. Some radical developments do appear possible, and we shall discuss these in later sections.

3. VERY SMALL AND VERY FAST

The mainstream approach to high-speed signal processing and supercomputer systems continues to emphasize very fast devices with low net power dissipation in the devices and interconnects. Speed can be achieved by deploying high-mobility materials, with or without low temperature operation; using very short channel devices; using superconducting logic; and, finally, by using parallel, array, or highly concurrent processing which gives fast net system speed. In all these cases, the fullest advantage accrues from making very small devices with short interconnect paths to minimize propagation delay and power dissipation.

There are very definite physical limits to the ultimate small scales at which any particular device structure can continue to function [5, 6, 9]. However, the *absolute* limits [9] appear to be in the range of 50 - 200 Angstroms, unless molecular devices [17] are deployed. This is quite close to the present finest resolution achievable by electron-beam, X-ray, or ion-beam lithography. The real problems concern what device structures are to be used to achieve high speed/low power and whether or not the interconnect delay can be reduced to similar levels. The answers will depend to a considerable extent on the intended application and what materials/fabrication technology is used. For example, in terms of computer processing power (device speed x number of devices per chip), there are indications that GaAs FET circuits will not rival silicon MOSFET circuits until the end of the decade, despite their intrinsically higher speed.

The major difficulty confronting fast sub-micron device design is that device performance cannot be extrapolated from equivalent large (bulk) scale devices. The problem arises primarily from the fact that no scaling theory exists which can simultaneously scale all the physical factors affecting device operation and performance. For example, the impracticalities of temperature scaling mean that control voltages cannot be scaled with certainty below kT/e (a much higher figure is actually necessary for reliability [5]), so that electric field strengths must eventually rise (for T = 300 K, V = kT/e, and channel length L = 0.25 μm, we find a *minimum* field strength between 1-10 kV cm^{-1}). The inhomogeneous self-consistent field in a minimum sized silicon MOSFET structure has been predicted to have a maximum value of around 100 kV cm^{-1} [8] . High field effects such as hot electron phenomena are therefore to be expected, with consequent reliability problems and changes to the intrinsic transport processes [8] .

A second problem is that the device transit time τ_d approximately scales with channel dimension. Assuming, for instance, an average carrier with velocity of 10^7 cm s^{-1} crossing a 0.1 μm channel gives a transit time of 1.0 ps; a time comparable, and in many cases, less than typical bulk mean free times τ between collisions, and comparable to the collision duration τ_c of certain scattering processes. Conditions for which $\tau \leqslant \tau_d$, or equivalently the bulk mean free path exceeds the channel or other device dimensions cannot be described by steady-state flows, and in extreme situations cannot be described by classical Boltzmann transport theory. This regime occurs between 0.1 - 1 μm channel lengths in silicon and GaAs, and is favoured by very high mobility material and small effective masses. It has been called the ballistic or transient regime.

3.1 Ballistic and Transient Effects

The transient, or middle scale regime, occurs when injected carriers do not have time to reach a dynamic steady-state with the scattering system before exciting the device. Steady drift is not achieved and the transport is governed by the transient current response. Inertial effects predominate; the local electron velocity is not slaved by the bulk velocity-field relation to the instantaneous local electric field. This regime is not necessarily a true ballistic regime [24] which would imply that carriers are freely accelerated without any scattering across the full device length. Scattering still occurs, since there is always a distribution of scattering times, and indeed the transport process is highly sensitized to space-charge (space-charge limited transport), velocity distribution in the contact regions, back-scattering near the contacts, and surface/interface scattering on the encroaching boundaries (surface to volume ratios increase with down-scaling). The presence of high fields can distort the collision processes [22, 8] via intracollisional field effects, since there exists a wide distribution of collision durations available to accelerated electrons. This particular effect derives from energy/momentum exchange between field and carrier during the collision process and its proper treatment requires quantum transport theory although the effects are largely perturbative at middle scales.

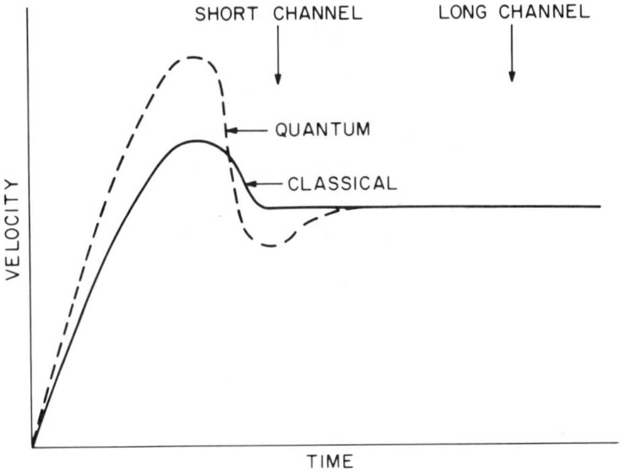

Figure 2/Velocity overshoot in response to a stepped high field.

3.2 Overshoot Effects

Figure 2 gives a qualitative sketch of the average velocity response for carriers subjected to a sharp step function field applied across a short channel device [23], illustrating the classical and quantum transport velocity overshoot effect (the latter shows faster overshoot, due to intra-collisional field effects and retardation of the scattering processes: non-local response, see [23]). At a short distance into the channel, the majority of carriers are freely accelerating; further in, the faster momentum relaxation processes pull the carriers into an intermediate quasi-steady state which is finally released to the true steady state by the slower energy relaxation processes. The precise energy dependence of these processes is critical to the details, and it is now believed that diffusion and carrier temperature show a similar overshoot structure. Figure 3 is an example of how the mean velocity (channel length divided by the transit time) varies with field for two short channel devices, assuming uniform fields. Figure 4 shows that overshoot effect, again qualitatively, in its dependence on pulse shape. The maximum overshoot occurs for sharp pulses: however, in real circuits the pulse rise/fall time is determined by speed restrictions in other devices and by RC time constants, so that pulse softening occurs which can substantially diminish overshoot structure [25]. Any attempt to exploit high velocities/short transit times from overshoot must take these circuit factors into account.

All these spatio-temporal relaxation effects are conditioned by the *self consistent fields* within the device volume, which – due to fringing effects, the device geometry, contact/gate boundary conditions, diffusion space-charge, and transient dielectric relaxation – are also inhomogeneous in space and time. No device models have yet been investigated which take all these effects into account. Current models range from simple space-charge limited "ballistic" transport to high field quantum transport in homogeneous fields.

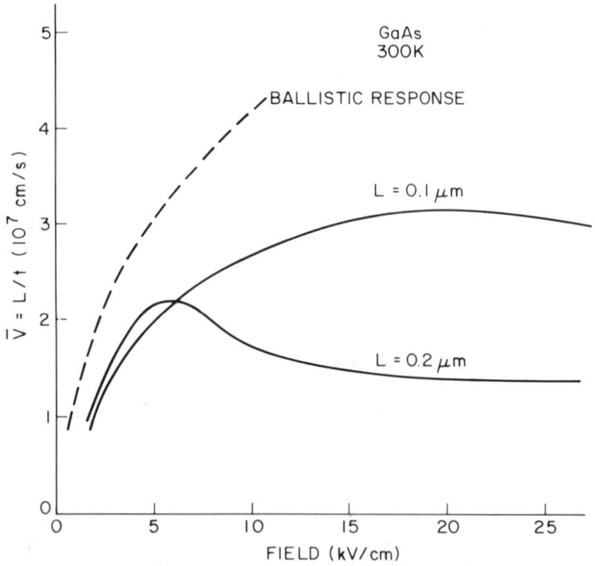

Figure 3*/Mean velocity versus field for short channel devices.*

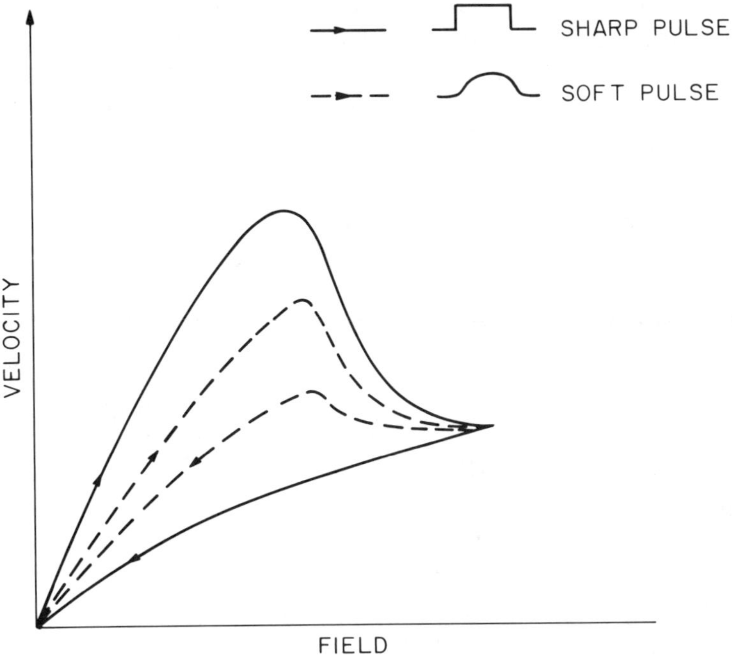

Figure 4/Effects of pulse shape on velocity overshoot.

One should note that the transient regime may well involve a substantial fraction of lower energy carriers with de Broglie wavelengths comparable to the channel dimension and hence mean a free path. If the device boundaries are sharply profiled, some wave interference may occur, as reflected waves attempt to form size-localised states. Of course, the electric field will mix these states to produce a net current, but some increase in net transit time could occur due to the resonances. This type of effect is already under consideration in simple models for MOMOM structures.

A number of novel sub-micron device structures have begun to appear, including the permeable base transistor, MOM . . . structures, hot electron canal transistors, high mobility transistors, "ballistic" transistors, and many others, which seek to exploit the transient regime for speed, high-power, or high-frequency operation. So far these structures are not well understood theoretically, particularly so in time circuit configurations. Reliability of operation, manufacture, integration, etc., has not yet been demonstrated and, since little power dissipation can actually occur within the active volume of a "ballistic" device, it is not known where, or with what effect, the heat dissipation is actually distributed.

Parasitic effects, short channel failure modes, and radiation damage effects for sub-micron devices have been discussed recently [15] in the literature, but many new problems are likely to emerge.

3.3 Ultra-small Devices

At the smallest scales so far conceived, with channel lengths less than 250 Å (silicon) or less than 1000 Å (GaAs), the concept of mean free time becomes meaningless. At this quantum scale, neither the effects of the electric fields nor the influence of the device environment (surfaces, interfaces, contacts, boundaries, and other devices) can be treated perturbatively [8]. Coherent and tunneling quantum transport processes become possible. Devices on this scale are unavoidably "box-like". Size quantization, which will be highly sensitive to the local electric fields, could lead to energy spacings in the 1 meV to 1 eV range. Band-bending is also important, particularly at the very much higher carrier densities ($\leqslant 10^{19}$ cm^{-3}) which would be required to keep a reasonable number of carriers in the device. Current injection/extraction will be problematic because of the discrete nature of the carriers and due to tunneling/interference phenomena. Transport in short, narrow interconnects scaled to widths of 100-200 Angstroms may be more appropriately described by de Broglie waveguides or transmission lines, rather than controlled by classical point particle trajectors. A very wide-range of non-classical transport processes is available on these scales depending on the device configuration [6, 8]. Projected limitations to device operation on these scales has been reviewed in the literature [5, 9].

Some general comments on future quantum scale devices are worthwhile. Contact regions will be particularly problematic. For example, in a hypothetical 250 Å channel silicon FET [8], ohmic contacts cannot be used because doping and size-quantization effects are likely to raise the channel region a few lengths of an electron volt in carrier energy and so deplete the channel *regardless* of gate voltage. Blocking contacts are thus a necessity to contain the carriers. Injection is then by tunneling with a negligible admixture of thermonic emission.

The choice of materials is important when considering the isolation (or partitioning) of ultra-small device regions. For example, there is an important trade-off between speed and size in small devices. High speed is generally a property of light mass media such as GaAs: however, in FET's it is crucial to minimize tunneling currents across the gate insulator barrier or, more significantly, the junction depletion region, and this is controlled by the effective mass. If x is the width and U the energy of the barrier, then, for negligible tunneling, we require $x \gg X = ½(h^2/2m^* U)^{½}$. For $U = 1$ eV, $X \sim 100$ Å for $m^* = 0.01\ m_e$. Thus, devices which utilize low mass materials for speed may have speed advantages on large scales, but on limiting scales will occupy more area than equivalent high mass devices.

The smallest scale at which a localized circuit element would be capable of bistable operation for use in a memory or logic function has been studied by several groups [5]. Physically, the generic bistable device has two energy minima separated by a potential barrier. The barrier must be high enough to prevent thermal fluctuations or quantum tunneling from mixing the two equilibrium states. Conversely, the barrier must be sufficiently low that state transfer can be achieved by an input signal. The subsequent trade-off between chance and necessity indicates a minimum linear dimension of about 10 Å. For a quantum scale device with a switching time of $\tau_s = 1$ ps and barrier height $U = 100$ kT, Landauer's [26] formula, $\tau_s/\tau_o = \cosh(U/2kT)$, yields a storage lifetime $\tau_o = 30$ years, evidently an acceptable value even for high complexity systems.

3.4 Micro-Refrigeration

An alternative approach to high speed which might avoid some of the problems of extensive down-scaling is to re-examine a current prejudice: is room temperature device electronics really a necessity? It is already well established that low temperatures provide improved performances for semiconductor devices and the power dissipation qualities of integrated circuits. There also exists a wide range of potentially explorable quantum electronics effects at low temperatures; for example, the size quantization, quasi-two dimensional levels in layered structures and MOS devices, and *a fortiori* in Josephson logic structures. If controllable, localised cooling on-board integrated circuits could be achieved, there would exist a much wider range of options for viable high-speed, high-density electronics.

This question has been addressed recently by Little and Richardson [27] in an examination of the possibility of scaling refrigerator systems to microstructure dimensions. Existing, small, closed-cycle gas refrigerators have a relative capacity to power dissipation ratio of at least 1000:1.

Therefore, the use of microfabrication techniques to construct a thermally matched refrigeration system has been suggested. Little has proposed using photolithographic techniques to etch gas lines, heat exchange, and valves into a silicon wafer sealed off by an anodically bonded cover plate. The idea is to construct a miniature refrigeration system based on the Joule-Thompson open cycle. Richardson has gone further [28] in proposing a miniaturized nuclear-demagnetization rig for operation in the millidegree Kelvin range. Any exploitation of these ideas – in particular, combining the fabrication technology for the gas line construction with the integrated circuit construction – provides difficult, but not necessarily intractable, problems for future research.

4. MATERIALS

The search for new device materials is likely to accelerate in the next stages of microelectronics. It has already been projected [10, 11, 14, 15], that future super-computers will use magnetic-bubble and Josephson junction devices, particularly for memory and, in the latter case, for fast logic, in addition to scaled silicon MOS devices. Long term projections have suggested that the fastest memory access times will be achieved using devices based on insulating substrates, such as gallium arsenide. Indeed, gallium arsenides' potential for VLSI gigabit circuitry with gate switching times of less than 100 ps and dynamic switching energies of less than 0.1 pJ makes an eventual break with silicon an attractive possibility. Of course, GaAs and related materials have traditionally been much more important than silicon in microwave and optical device applications; only recently has GaAs looked competitive for memory/logic applications. In a sense, the use of alternative semiconductors to silicon is part of the *continuous* evolution of microelectronics. The devices, circuit concepts, and semiconductor properties are similar to those of the silicon LSI era, and the switch to GaAs stems directly from the low power, high speed advantages of GaAs FET circuits which derive from the high electron mobility and the semi-insulating substrate. It is likely that LSI and VLSI applications will benefit from the recently developed fully planar GaAs process using multiple localised implantations directly into a semi-insulating substrate so as to avoid circuit restrictions, yield, and producibility difficulties inherent in mass GaAs IC fabrication, so even the planar technology is retained. As already noted in optical and microwave applications, the III-V compound and solid solution families have considerable advantages in that sharp transistions in band gaps, refractive indices, etc., can be controlled by compositional changes. The trend to non-silicon semiconductors is thus a move to materials with better bulk properties.

But, looking further ahead, there are more fundamental changes at work which could have major consequences for revolutionary developments in electronics.

These are most obvious in GaAs related materials. The ability of certain families, e.g., GaAs and $Al_xGa_{1-x}As$, to form lattice matched systems of different composition has allowed substantial changes in the electrical and optical properties in very small regions without serious distortion of the total crystal structure. The development of *molecular beam epitaxy* has enhanced these capabilities for thin-layered structures by providing extremely fine dimensional control of both compositional and doping variations. Indeed, essentially artificial high quality crystal structures can be fabricated in this fashion with novel periodic structures, e.g., compositional or modulation-doped one-dimensional superlattice configurations (Figure 5) [16, 29-32].

The fine dimensional control of MBE provides a base for a new type of materials (and device) technology: *quantum well architectures*, in which materials can be fabricated with electronic properties controlled by quantum size effects. For example, it has been shown (reviewed in [16]) that single or multiple undoped GaAs/n-AlGaAs heterojunctions exhibit an accumulation of electroncs originating from the doped layers, but residing in the undoped GaAs layers near the heterojunction interface: a spatial separation of donors and carriers. The resulting two-dimensional electron gas can have very much higher mobilities than equivalently doped bulk GaAs. More recently, it has been suggested that real-space transfer [43] of electrons out of the quantum wells into the neighbouring layers could be controlled by exciting the well electrons into high energy states using carrier heating by an electric field aligned along the layer, followed by scattering/diffusion into the low mobility layer [33] (Figure 6). Such a non-equilibrium process is predicted to generate negative differential mobility (Figure 7). Negative conductivity and indeed mini-band filling have also been suggested as controllable features of electron motion perpendicular to the 1-dimensional superlattice layers due to mini-gap formation.

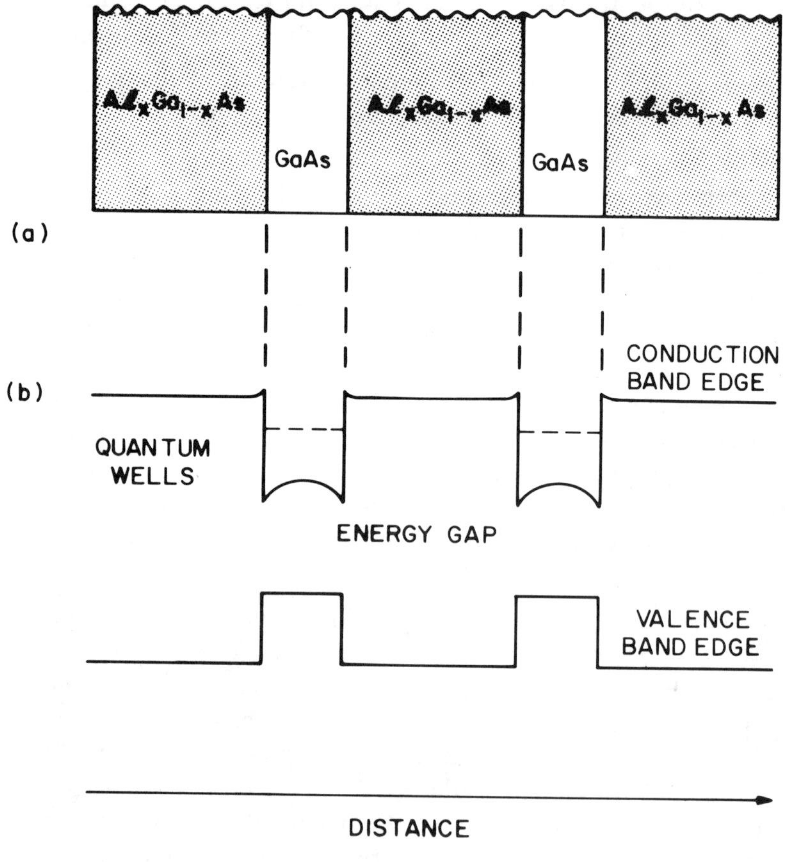

Figure 5/(a) *Layered structure;* ***(b)*** *Equivalent energy scheme.*

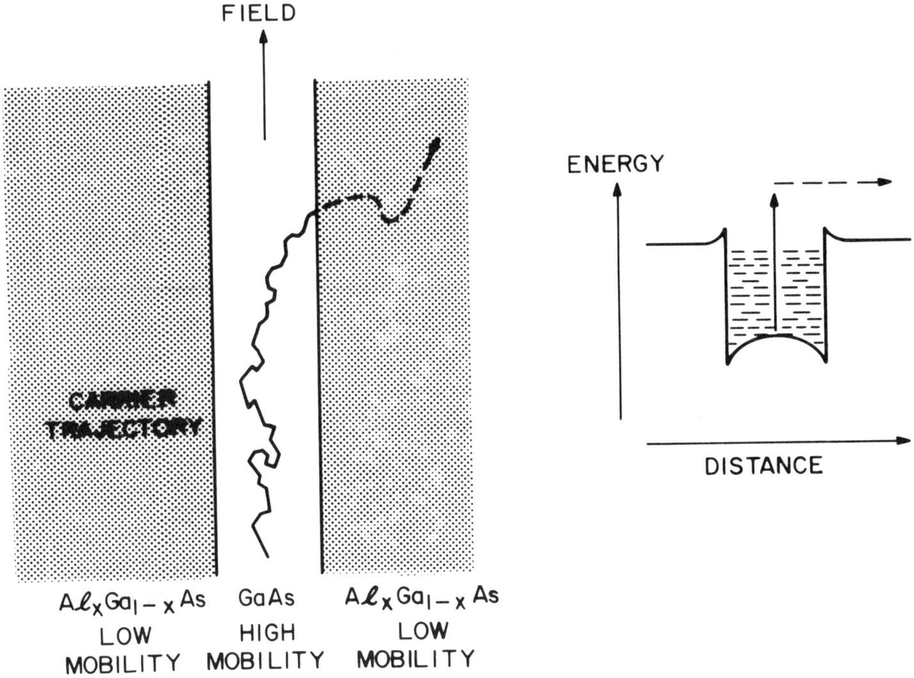

Figure 6/Real space transfer.

The high mobility, modulation doped well structures are already being deployed for improved GaAs microwave devices. Simple integratable current rectifiers have also been devised from GaAs-Al_xGa_{1-x} GaAs heterostructures [34] having a very thin $Al_xGa_{1-x}As$ layer with a graded band gap.

In the long term, these developments suggest that *layered* heterostructured artificial crystals may have considerable advantages for novel electronic applications, a situation which may culminate in the development of three-dimensional heterostructures if lateral patterning/lithography can be achieved on successive layers or if implantation can yield vertical inserts (such as p-n islands) into the layers.

global level, devices can only interact via the built-in interconnects. If the partition holds, the integrated circuit is equivalent to a locally communicating discrete system in which the operational function of the system is determined completely by the topology of the interconnection net and the model device elements.

Modern circuit design rules represent one method for attempting to preserve partitioning under scaling. Exceptions to partition have long been recognized, e.g., parasitic Miller and node capacitances, and may often be minimized or renormalized into the effective interconnect net. However, in VLSI, isolation is more difficult to achieve at any level, a problem exacerbated by high integration densities and the trend to minimized interconnect path lengths and regular architectures. Partition evidently fails first at microscopic and local levels, primarily because scaling enhances the surface-to-volume ratio of the active device regions.

Some examples serve to illustrate known and projected failures of the partition principle for sub-micron configured systems. At the local level, the inequality $\tau_d > \tau > \tau_c$ fails for middle scale devices and intradevice processes always in the transient regime, for which normally second order effects such as velocity, diffusion, and temperature overshoot become important, a situation which is exacerbated by geometry induced inhomogeneous high field regions. Static bulk derived velocity-field (uniform) characteristics are thus poor indicators for small device performance. On sufficiently small scales a nearly ballistic regime may be encountered. The reduction of scattering events and the possibility of strong band bending in inversion layers enhances the possibility of vertical size quantization.

Transverse size quantization is also encouraged if some carriers have sufficiently long de Broglie wavelengths and mean free times to build up interference of coherent incident and reflected wave trains within the device boundaries. Surface roughness and associated scattering degrade this effect. A strong driving field obviously mixes any nearly size localised states (as does scattering) so that current flow will still occur but damped oscillatory current voltage characteristics and transit time delays due to interference effects may be expected on sufficiently small scales. This regime is already accessible in certain MOM, FET structures.

At the neighbourhood level, partition failure is expected at the middle scale due to the interaction of carriers with regions outside the device volume. Examples include shift of the overshoot regime into the source constant regions due to hot electron motion against the driving field; hot electron injection into interface tunnelling states and into the insulator regions with subsequent charge trapping; and remote phonon scattering [36] at interfaces. The latter occurs in MOS devices due to the penetration of fringing fields from oxide region surface polar phonons and plasmons into the inversion layer; the effect modifies the dynamical carrier mass, induces remote energy/momentum loss and influences local dielectric properties. All these phenomena are strongly geometry dependent and conditioned by doping profiles and field tailoring. Some mechanisms will prohibit device operation. We have already mentioned an interesting neighbourhood effect concerning the magnitude of the overshoot velocity in GaAs devices which is often quoted as giving additional speed advantages over silicon technology. Most overshoot estimates are based on the response to a rectangular pulsed field. However, actual driving pulses originate from other devices and have finite rise times proportional to the minimum delay times in the system. Overshoot is substantially diminished by employing realistic pulse shapes. The effect is architective dependent via the RC and driver delays.

Global failure of partition is most serious [8, 9] on quantum scales when many forms of device-device coupling (e.g. tunnelling, superlattice effects, capacitative coupling, charge spill-over, etc.) may occur. However, interconnection cross-talk and pattern sensitive cross-talk in the read/write operation of VLSI memories is already known.

In general, the failure of partition at any level can at least degrade device/system performance and at most can change the functional organization of the system or lead to device/system failure.

5.2 Inter-device Coupling

Simple capacitative coupling between circuit elements is already known in high density systems, and indeed electromagnetic coupling between neighbouring interconnects is to be expected in high speed systems. However, more subtle inter-device interactions are possible which could degrade device and system performance if not controlled. For example, in any device which has a channel length greater than the mean free path for intra-device energy relaxation, the predominant power dissipation will occur over an extended region in the vicinity of the device. We distinguish power dissipation into adjacent isolation regions (substrate, gate insulator phonon modes, etc.) from power dissipation along the direct communication paths (contacts, gate electrodes, diffusion, and metallised pathways). For very high speed systems, this could lead to thermal hot spots between devices which could introduce thermal cross-talk. If the hot spot homogenisation times are slow compared to the signal pulse separation, switching and injection thresholds could be seriously affected. More seriously, interface phonon selection rules could slow down phonon thermalisation sufficiently that a non-equilibrium phonon bottleneck might occur in insulator/semiconductor interfaces leading to lattice and subsequent interface instabilities. Long-range coherent polarization modes involving large energy storage and subsequent structural deformation have long been studied in polar molecular media having dielectric structural properties similar to quantum scale integrated circuits. Charge coupling between adjacent interconnect channels can occur by an indirect effect in which an electron in one channel excites a phonon in the isolation region (via the general, very strong, electron-phonon coupling in insulators) which in turn is absorbed by an electron in the second channel. Evidently, this effect can only be controlled if the relevant interactions and coherence lengths are properly investigated. In regular structures, a long range interaction could be built up by coherent parasitic coupling.

Some interdevice tunnelling is to be expected in high-density systems. Phenomena such as hot electron real-space transfer, charge spill-over in homogeneous doped and modulation doped GaAs-GaAlAs MBE layers are also known, and are likely to have counterparts in quantum scale integrated circuits. Tunnelling barriers in narrowly spaced systems are likely to be strongly controlled by the voltage pulses present in the system: indeed a new type of trade-off between charge confinement and unwanted size-quantization effects may lead to additional leakage currents between devices /interconnects.

Recent exploratory studies of a range of interdevice couplings [6-9] suggest that the coupling may be classified as: (i) stabilizing: the charge transfer fluctuation tends to relax back into the original device; (ii) resonant: charge transfer into the interdevice region with possible trapping; (iii) unstable: permanent transfer into adjacent device regions, e.g., real space transfer; (iv) cooperative: coherent build up unstable fluctuations to set up collective modes, e.g. charge density waves, particularly in regular structures; (v) super-lattice effects on intra-device properties due to wavefunction overlap between devices in locally regular structures. Various indirect correlation effects mediated by phonons, polarization waves, etc., which may be dissipative or non-dissipative are also anticipated. These general circuit interactions and feedback processes cannot be handled by conventional parasitic modelling and are dependent on the actual distributed current, charge, voltages, and local temperatures in the system, i.e., on the information state of the system. The effects are consequently non-linear pattern sensitive, and inhomogeneous. Some effects could be modelled by parasitic, non-linear positive, and negative feedback elements.

These problems will require considerable study if quantum scale integrated circuits are to be forthcoming.

5.3 Complexity

The design bottleneck is already recognized in LSI, with the gate density of logic chips considerably lagging behind the more simply structured memory chips. If the true potential of VLSI is to be realized, there must clearly be a breakaway from "random" logic architectures towards a more regular systematic and simpler design philosophy. High circuit density is inevitable, anyway. For example, future ultra-fast computers based on switching times on the order of picoseconds require CPU's with minimal dimensions of a few tens of centimeters to achieve signal coherance. There are also applications in the fields of array processing, pattern recognition robotics, and artificial intelligence which lead to requirements for some form of concurrent, as opposed to sequential, architectures. From a system viewpoint, scaling is not always beneficial to VLSI "unstructured" architectures [37]. In particular, line resistance, and hence line delay and power dissipation, rise with decreasing scale. Indeed, the main consumption of time, space, and energy is in steering the data flow. It is therefore crucial to minimize the interconnect area rather than the number of gates, as in conventional optimization schemes. Highly concurrent architectures, such as parallel, pipeline, or tree processors, are therefore more favorable to VLSI [10, 15, 37]. It has also been argued [37] that problems of temporal management of complex architectures militate against synchronous clocking schemes, and that self-timed concurrent subunits should be used. At a more detailed level, there is also controversy [37] over what logic schemes should be used, e.g., the replacement of level restoring logic of the more simple steering logic structures. One interesting possibility is to use conservative logic based on Fredkin gates, which carries the advantages of reversibility (logical) and minimum system switching energy. From a more practical viewpoint, the defect hazard (intrinsic, fabrication, or radiation induced) could be serious for sequential structures, and either crude redundancy methods or concurrent architectures which could "switch out" defective areas are likely to assume prominence in the future.

All these considerations point towards the emergence of much more dense and much more regular system topologies. Physically, this may possibly present new limitations to VLSI. As discussed earlier, architectures with ultra-small line widths and device separations are prone to additional parasitic linkages between circuit elements which depend partially on the currents and voltages present in the system, i.e., on the information state of the system. Such parasitic linkages are pattern sensitive to such architecture and information flow. In irregular systems the parasitic connection matrix will not necessarily degrade the function of the system, although it provides an additional noise source. However, in regular systems or with de-stabilizing feedback between devices/interconnects in irregular systems, the competition between parasitic and hardwired connection nets may severely de-stabilize the system to the extent that the intended processing function is changed. Very little study exists of the stability properties of highly concurrent non-linear systems, but analogies with natural systems suggest [38, 39] cooperative functional transitions could occur (non-equilibrium "phase" transitions, characterized by spatially or temporally inhomogeneous new modes appearing in the system [39]). The intrinsic non-linearity of the parametric connection net may also induce "chaotic" behaviour [39] as noticed in the vacuum tube parasitics of "universal" circuits [40]. These are not simply abstract non-linear system problems because the parasitic net is strongly conditioned by its physical origins and depends on the detailed nature of the chip heterostructure and control voltages/currents. With regular structures, even if the parasitic inter-element coupling can be confined to nearest neighbour interactions, the coherent build-up of long-range pair-wise coupling is possible, as already noted for functional superlattice phenomena [41] and well-known in cooperative phenomena in the many-body systems of pure physics and chemistry [39, 42].

These systems problems could be regarded either as severe drawbacks or as possible new opportunities to rethink information processing structures.

5.4 2-D and 3-D Superlattices

As we have seen, the trend in circuit design is towards dense, considerably regular (and in many cases lattice structured) architecture, particularly in memory and concurrent processing schemes. If each element is small and in close proximity to its neighbours, it becomes logical to enquire whether super-lattice effects can occur as already known in layered structures [30]. If super-lattice effects can occur, how will this affect individual device operation, and indeed how will this affect the total system behaviour?

The concept of two-dimensional lattices [41, 44] is a natural development of the pioneering work of Esalsi and co-workers [30] which led to one-dimensional super-lattices in layered systems. The basic one-dimensional super-lattice arises in the following way. Suppose there exist two isolated potential wells of finite depth, with a width much greater than the lattice constant but smaller than the electron mean free path and bulk de Broglie wavelength: quantum ballistic "devices". If the wells are widely separated, the electrons will occupy well-defined discrete quantum states defined by the effective mass Hamiltonian for the well structure. Bringing the wells closer together will result in significant wave-function tunneling through the intervening barrier so that electron waves from each well will overlap. The degeneracy is thus lifted, and the quantum levels in each well will split. Adding further wells to form a periodic structure gives rise to a special version of the Kronig-Penney one-dimensional solid.

The original band structure will now exhibit mini-gap formation and sub-band (mini-band) formation controlled by the barrier height, well separation and the original electron effective mass.

The basic idea of a system super-lattice is that the electrons in each device occupy a two-dimensional potential well. If devices become packed regularly with spacings of the same order as device width and characteristic wavelengths, some tunneling will be possible, as well as two-dimensional super-lattice effects with mini-gap and mini-band formation associated with each device. We noted earlier that if U is the energy, x is the width of the tunneling barrier, and k is the electron wave-vector, tunneling is negligible if and only if $2|k|x \gg 1$. Or since $k = (2m^* E/h^2)^{1/2}$, $x \gg X = \frac{1}{2}(h^2/2m^* U)$. For a large barrier height of one electron volt, $X \sim 10$ Å for free electrons and $X \sim 100$ Å for a low effective mass $m^* \sim 0.01\, m_e$. This most unfavourable case illustrates that super-lattice affects can visually be expected near the limiting resolution of advanced lithography, and that light mass materials exacerbate the effect.

One should note that effective tunneling barriers are highly sensitive to control signals (voltages, charge density, currents) and coded electrical information passing through the system. So that, at the other extreme, we have the possibility of a dynamically disordered super-lattice — a major reliability problem for system design, but potentially useful for new system concepts or for studying Anderson localization.

Deliberately fabricated two-dimensional super-lattices have been studied recently [41, 44]. One type envisages a multiple gate metal-insulator-semiconductor device in which the buried gate is partitioned into a regular lattice having a spacing, d, comparable to the average electron de Broglie wavelength, and smaller than the screening length in an inversion layer at the insulator-semiconductor interface. A global metal gate (Figure 8) provides control over the buried electrodes, without requiring critical alignment of successive levels. For a p-substrate, and positive bias V on the upper gate, there will be an induced positive bias on the super-lattice gates and the surface potential for electrons under the gates will decrease. Thermal carrier generation in the depletion layer will invert the surface under the electrodes and the surface potential in the mid-region will be both raised and flattened. At large bias, tunneling should be possible between the valence and conduction bands, with resulting wave-function mixing and mini-gap formation (controlled via d, V, and m^*).

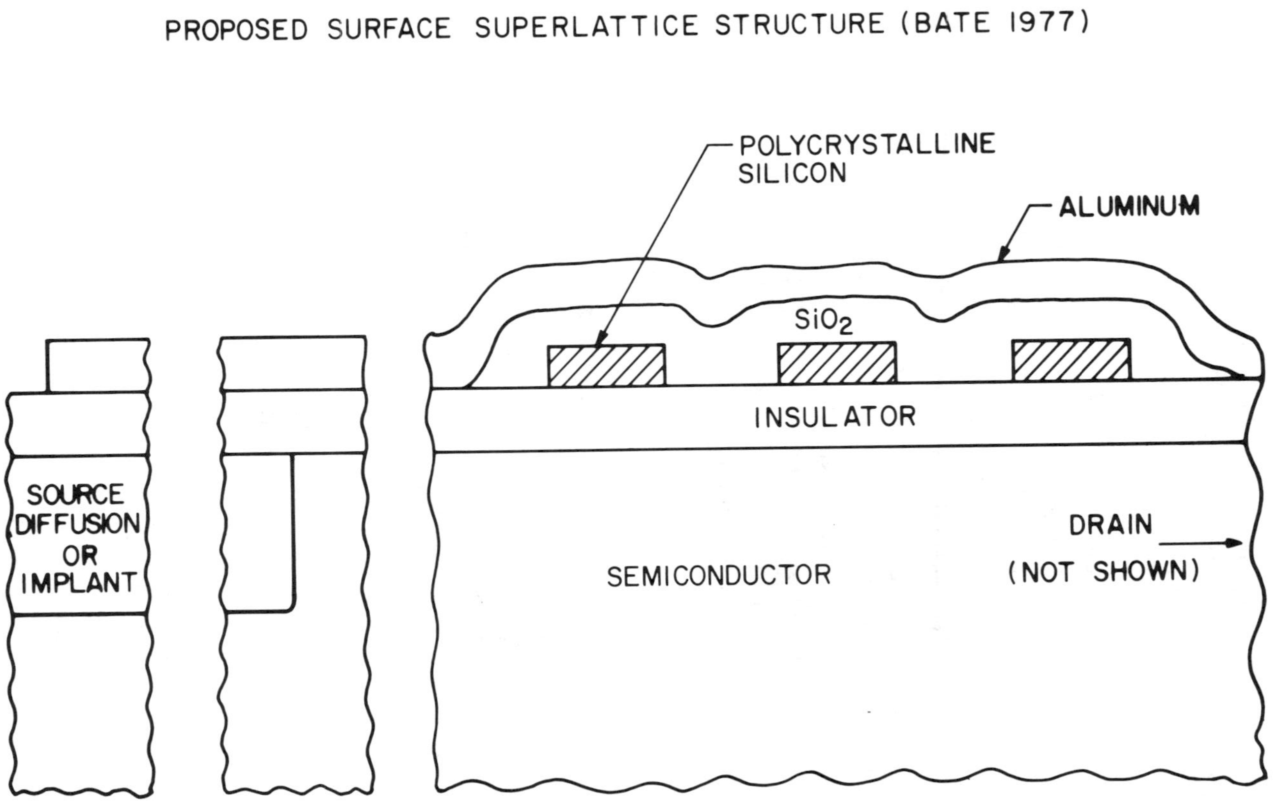

Figure 8/A possible 2-D superlattice (after Bate 1977).

Detailed studies [41] suggest that silicon would require 100 - 200 Å spacings for a sizeable mini gap (0.1 eV) at an energy of say 0.03 eV. GaAs, InAs, InP, and InSb show considerable promise for super-lattice effects at d $\sim$ 1000 Å.

A second method would be to implant narrow, density packed n (or p) type doping islands into a p (or n) type substrate.

Three dimensional super-lattices could be configured by deploying patterned islands into MBE layered one-dimensional super-lattices.

An interesting effect is predicted by considering a generalization [41] of Kroerner's [45] arguments on one-dimensional super-lattices. The basic idea centers on the electron contribution to the dielectric function. The latter exhibits a sharp maximum (Kohn anomaly) when the wave number of a periodic signal (in space) equals twice the Fermi wave number. The actual free electron contribution depends upon the screening/descreening character of the electrons, i.e., the sign can change depending upon on which side of the mini gap the electrons are resident. Under appropriate conditions, the super-lattice electrodes could represent such a periodic signal. The subsequent Kohn anomaly would lead to a dramatic reduction in the mini-gap.

On the other hand, it has been shown [41] that, if population inversion can be achieved in the vicinity of a virtual mini gap, a zero can arise in the dielectric function such that a net spatially periodic effective potential appears on the electrodes at zero applied bias. This effect generates a super-lattice instability in which the appearance of a charge-density wave brings the super-lattice into existence spontaneously without any bias field. This is an example of a self-organizing or synergetic effect [42]. One should note that population inversion can be achieved optically, or more significantly in multivalley structures by hot electron effects in the channel. The most significant feature of this gedanken experiment is that it points the way to global switching of a circuit architecture by means of a simple encoding of the information flow.

5.5 Synergetic Electronics

We have seen from our previous discussion that a wide range of non-linear parametric inter-device couplings exist on sufficiently small scales. Advanced fabrication and patterning techniques deployed in quantum well heterostructures offer the opportunity to deliberately fabricate partially isolated small device structures in which the inter-device coupling is strongly modulated by certain of the control signals in the system. The new science of synergetics [39, 42] has established that in many natural biological, chemical, physical, highly concurrent, complex systems, the competition between stabilizing and de-stabilizing modes can stabilize the net system into successive, highly ordered, very nonequilibrium, dissipative structures which are controlled by a few order parameters. (The laser provides a seminal case but other examples are Bernoulli phenomena in liquids, pattern formation in non-linear chemical reaction-diffusion systems, morphogenesis in cellular biology, pattern formation in the information processing units of the visual cortex, etc. [39, 42]). Is it possible, with VLSI systems, to tailor the requisite inter-device coupling by appropriate fabrication techniques, combined with selective control fields and currents, in order to devise cooperative (synergetic) device systems with regimes of strong structural stability and possible net switchable architectures? Such systems would represent truly holistic integrated systems [8, 9].

In a recent mathematical study [46], we have investigated a model structure comprising a large, regular array of identical quantum scale devices coupled by nearest neighbour interactions. A small amount of tunneling or capacitive coupling between devices is capable of generating a variety of change concentration waves in the total system, which are sensitive to the "pattern" of control signals across the groups. The basic discrete equations can be approximated for large systems by a continuum model in which the charge density variables within each group satisfy non-linear reaction-diffusion equations of the type known to have solutions corresponding to bifurcation into various self-organized steady spatial patterns, as well as non-linear wave propagation. The range of self-organizing patterns is considerably enhanced by using weakly coupled bistable devices, or by adding feedback linkages into the hard wired net. This suggests that, with appropriate minimal control patterns, and by using the charge density variations to gate the hard-wired diffusion paths, a global re-structuring of the logic steering architecture could be achieved. In addition, it appears possible to use pattern formation to both store/retrieve and process distributed information (rather than serial bit data) in a form which is stable when confronted with quite substantial local distortion of some device properties. The robustness of synergetic systems against noise and local disturbances (e.g. damage) is well-known, and, of course, the use of feedback to relax element tolerances has already been discussed for VLSI circuits [5, 9].

Synergetic electronics is, then, one new, radical possibility for future microelectronics, and we have already seen that the 2-D superlattice is an archetypal example. When considering new devices, we might single cut the possibility of a passive holistic device whose function and performance might be controlled by renormalization of the intradevice properties by altering the global features of the system, e.g., super-lattice voltage levels in the neighbouring device array. On the other hand, the deliberate use of feedback – either hard-wired or deriving from inter-device coupling – may be crucial to the operation of highly concurrent architectures so as to non-linearly mode-lock the effective system architecture and provide reliable performance. In the distant future, there is the exciting possibility of completely revising the logic schemes for representing and processing information. At the simplest level, for instance, this could involve a truly programmable logic array in which the logic gate pattern was set up by a small set of analogue control signals. At the most sophisticated level, the potential for an enormous information density capable of being packed into a highly concurrent system could be used to process patterns in analogue fashion and offers a variety of applications for artificial intelligence applications. Possible applications to bioelectronics have been described elsewhere [47].

6. CONCLUSIONS

We have seen that physical factors exist which may set ultimate limits to the conventional development of micro-electronics. It is already apparent that the highly successful silicon MOS technology will reach the end of its evolutionary travel at about the 0.1 μm scale. The move to new materials, especially heterostructures, promises many new devices, and the revolution in system design toward highly concurrent architectures suggests new problems of system reliability, but also radical new system concepts. In the long term, as technology approaches the complexity levels of biological systems, it becomes pertinent to ask what changes in microelectronics concepts will be required if we are ever to match the power, efficiency and reliability of information processing systems such as the eye or brain.

The technological problems are challenging but offer exciting prospects for the future of electronics.

REFERENCES

[1] Kuhn, T.S., *The Structure of Scientific Revolutions,* second edition. Chicago: Chicago Univ. Press, 1970.

[2] See for example, M. Bunge's critique in *Foundations of Physics.* New York: Springer-Verlag, 1967.

[3] Certain "doomsday" predictions for micro-electronics limits, e.g. 10 μm limiting feature sizes (ref. [5]) have already been passed. Even in A. Toffler's book, *Future Shock,* a futurist's warnings of rapid technological change failed completely to assess the explosive technological growth brought about by the microprocessor – for the simple reason that it had not been invented at the time.

[4] Braun, E., and S. MacDonald, *Revolution in Miniature.* Cambridge: Cambridge Univ. Press, 1978.

[5] Wallmark, J.T., *Sol. St. Devices,* Inst. Physics, London, conf. 25, 113, 1975.

[6] Barker, J.R., in *Physics of Non-linear Transport in Semiconductors,* ed. D.K. Ferry, J.R. Barker, C. Jacoboni. New York: Plenum Press, 1980.

[7] Barker, J.R., and D.K. Ferry, *Proc. Int. Conf. Cybernetics and Society,* IEEE Cat. 79, Ch. 14 23-1 SMC, 762, 1979.

[8] Barker, J.R., and D.K. Ferry, *Sol. State Electron.,* 23:519, 531, 1980.

[9] Barker, J.R., "New Trends in Integrated Circuits," *Proc. of Int. Conf.,* Paris, 1981, in press.

[10] Ballantyne, J.M., ed. *Proc. N.S.F. Workshop on Opportunities for Microstructure Science, Engineering and Technology.* Ithaca: Cornell Univ. Press, 1979.

[11] Solid State Sciences Committee, National Research Council, *Microstructure, Science, Engineering and Technology,* Nat. Acad. Sciences, Washington, DC, 1979.

[12] Robinson, A.L., *Science,* 208:1246, 1980. See also the special issue on microscience, *Physics Today,* November 1979.

[13] Woodall, J.M., *Science,* 208:908, 1980.

[14] Hnatek, E.R., *Computer Design,* 19:147, 1980.

[15] Barbe, D.F., ed., *Very-large Scale Integration.* New York: Springer-Verlag, 1980.

[16] Parish, M.B., *Science,* 208:916, 1980.

[17] Carter, F., *Proc. Molecular Electronics Workshop,* NRL 1981, in press.

[18] Mead, C.A., and L. Conway, *Introduction to VLSI Systems.* New York: Addison-Wesley, 1980.

[19] May, T.H., and M.H. Woods, *Proc. 16th IEEE Rel. Phys. Symp.,* vol. 33, 1978; J.F. Ziegler and W.A. Lamford, 1980 *Int. Sol. St. Circuits Conf. 70,* Digest of Tech. Papers, 1980.

[20] Segel, L.A., *Mathematical Models in Molecular and Cellular Biology.* Cambridge: Cambridge Univ. Press, 1980.

[21] Winfree, A.T., *The Geometry of Biological Time.* New York: Springer-Verlag, 1980.

[22] Barker, J.R., in *Physics of Non-linear Transport in Semiconductors,* D.K. Ferry, J.R. Barker, C. Jacoboni, eds. New York: Plenum Press, Ch. 5, 1980.

[23] Ferry, D.K., and J.R. Barker, *Sol. St. Electronics,* 23:545, 1980.

[24] Barker, J.R., D.K. Ferry, H.L. Grubin, *IEEE Electron Device Letters,* EDL-1:209, 1980.

[25] Grubin, H.L., D.K. Ferry, G. Iafrate and J.R. Barker, in press.

[26] Landauer, R., *IBM U. R&D,* 5:183, 1961.

[27] Little, W.A., in *Future Trends in Superconductive Electronics,* B.S. Denver, ed. *A.I.P. Conf. Proc.,* vol. 44, 1978, p. 421; see also Ref. [11], p. 31.

[28] Richardson, R.C., *Physics Today,* 32:29, 1979.

[29] Gossard, A.C., P.M. Petroff, W. Wiegmann, R. Dingle, *Appl. Phys. Letters,* vol. 29, p. 323, 1976.

[30] Esaki, L., and R. Tsu, *IBM Journ. Research & Dev.,* 14:61, 1970.

[31] Dingle, R., et al., *Appl. Phys. Lett.,* 33:665, 1978.

[32] Chang, L.L., et al., *J. Vac. Sci. Technol.,* 16:1504, 1979.

[33] Glisson, T.H., et al., *J. Appl. Phys.,* 51:5445, 1980.

[34] Allyn, C.L., A.C. Gossard, W. Wiegmann, *Appl. Phys. Lett.,* 36:373, 1980.

[35] Smith, H., et al., *Physics Today,* 31:21, 1979; M.W. Geis, D.C. Flanders, H.I. Smith, *Appl. Phys. Lett.,* vol. 35, p. 71, 1979.

[36] Ferry, D.K., K. Hess, and P. Vogl, in *Microstructure Science and Technology/VLSI,* N. Einspruch, ed. New York: Academic Press, to be published.

[37] Mead, C.A., and L. Conway, *Introduction to VLSI Systems.* Reading: Addison-Wesley, 1980.

[38] Barker, J.R., *Advances in Physics,* to be published.

[39] Haken, H., *Synergetics,* 2nd ed., New York: Springer-Verlag, 1979.

[40] Andronov, A.A., A.A. Vitt, S.E. Khaitin, *Theory of Oscillators.* Elmsford: Pergamon Press, 1966.

[41] Barker, J.R., R. Bate, R. Dingle, D.K. Ferry, to be published.

[42] Haken, H., ed. *Dynamics of Synergetic Systems.* New York: Springer-Verlag, 1980.

[43] Hess, K., H. Morkoc, H. Shichijo and B.G. Streatman, *Appl. Phys. Lett.*, 35:469, 1979.

[44] Bate, R.G., *Bull. Am. Phys. Soc.,* 22, 1977.

[45] Kroemer, H., *Phys. Rev.,* B15:880, 1977.

[46] Barker, J.R., to be published.

[47] Barker, J.R., *Spectrum,* U.K. Central Office of Information, Vol. 170, p. 2, 1980.

INDEX